Carbon Dioxide Capture for Storage in Deep Geologic Formations – Results from the CO_2 Capture Project

Geologic Storage of Carbon Dioxide with Monitoring and Verification

Volume 2

Elsevier Internet Homepage – http://www.elsevier.com
Consult the Elsevier homepage for full catalogue information on all books, major reference works, journals, electronic products and services.

Elsevier Titles of Related Interest

AN END TO GLOBAL WARMING
L.O. Williams
ISBN: 0-08-044045-2, 2002

FUNDAMENTALS AND TECHNOLOGY OF COMBUSTION
F. El-Mahallawy, S. El-Din Habik
ISBN: 0-08-044106-8, 2002

GREENHOUSE GAS CONTROL TECHNOLOGIES: 6TH INTERNATIONAL CONFERENCE
John Gale, Yoichi Kaya
ISBN: 0-08-044276-5, 2003

MITIGATING CLIMATE CHANGE: FLEXIBILITY MECHANISMS
T. Jackson
ISBN: 0-08-044092-4, 2001

Related Journals:
Elsevier publishes a wide-ranging portfolio of high quality research journals, encompassing the energy policy, environmental, and renewable energy fields. A sample journal issue is available online by visiting the Elsevier web site (details at the top of this page). Leading titles include:

Energy Policy
Renewable Energy
Energy Conversion and Management
Biomass & Bioenergy
Environmental Science & Policy
Global and Planetary Change
Atmospheric Environment
Chemosphere – Global Change Science
Fuel, Combustion & Flame
Fuel Processing Technology

All journals are available online via ScienceDirect: www.sciencedirect.com

To Contact the Publisher
Elsevier welcomes enquiries concerning publishing proposals: books, journal special issues, conference proceedings, etc. All formats and media can be considered. Should you have a publishing proposal you wish to discuss, please contact, without obligation, the publisher responsible for Elsevier's Energy program:

Henri van Dorssen
Publisher
Elsevier Ltd
The Boulevard, Langford Lane
Kidlington, Oxford
OX5 1GB, UK

Phone: +44 1865 84 3682
Fax: +44 1865 84 3931
E.mail: h.dorssen@elsevier.com

General enquiries, including placing orders, should be directed to Elsevier's Regional Sales Offices – please access the Elsevier homepage for full contact details (homepage details at the top of this page).

Carbon Dioxide Capture for Storage in Deep Geologic Formations – Results from the CO_2 Capture Project

Geologic Storage of Carbon Dioxide with Monitoring and Verification

Edited by

Sally M. Benson
Lawrence Berkeley Laboratory
Berkeley, CA, USA

and Associate Editors

Curt Oldenburg[1], Mike Hoversten[1] and Scott Imbus[2]
[1]Lawrence Berkeley National Laboratory
Berkeley, CA, USA
[2]Chevron Texaco Energy Technology Company
Bellaive, TX, USA

Volume 2

ELSEVIER
2005

Amsterdam – Boston – Heidelberg – London – New York – Oxford
Paris – San Diego – San Francisco – Singapore – Sydney – Tokyo

ELSEVIER B.V.
Radarweg 29
P.O. Box 211, 1000 AE Amsterdam
The Netherlands

ELSEVIER Inc.
525 B Street, Suite 1900
San Diego, CA 92101-4495
USA

ELSEVIER Ltd
The Boulevard, Langford Lane
Kidlington, Oxford OX5 1GB
UK

ELSEVIER Ltd
84 Theobalds Road
London WC1X 8RR
UK

First edition 2005

Library of Congress Cataloging in Publication Data
A catalog record is available from the Library of Congress.

British Library Cataloguing in Publication Data
A catalogue record is available from the British Library.

ISBN: 0-08-044570-5 (2 volume set)

Volume 1: Chapters 8, 9, 13, 14, 16, 17, 18, 24 and 32 were written with support of the U.S. Department of Energy under Contract No. DE-FC26-01NT41145. The Government reserves for itself and others acting on its behalf a royalty-free, non-exclusive, irrevocable, worldwide license for Governmental purposes to publish, distribute, translate, duplicate, exhibit and perform these copyrighted papers. EU co-funded work appears in chapters 19, 20, 21, 22, 23, 33, 34, 35, 36 and 37. Norwegian Research Council (Klimatek) co-funded work appears in chapters 1, 5, 7, 10, 12, 15 and 32.

Volume 2: The Storage Preface, Storage Integrity Preface, Monitoring and Verification Preface, Risk Assessment Preface and Chapters 1, 4, 6, 8, 13, 17, 18, 19, 20, 21, 22, 23, 24, 25, 26, 27, 28, 29, 30, 31, 32, 33 were written with support of the U.S. Department of Energy under Contract No. DE-FC26-01NT41145. The Government reserves for itself and others acting on its behalf a royalty-free, non-exclusive, irrevocable, worldwide license for Governmental purposes to publish, distribute, translate, duplicate, exhibit and perform these copyrighted papers. Norwegian Research Council (Klimatek) co-funded work appears in chapters 9, 15 and 16.

♾ The paper used in this publication meets the requirements of ANSI/NISO Z39.48-1992 (Permanence of Paper).
Printed in The Netherlands.

Preface

Gardiner Hill, BP, plc, Sunbury-on-Thames, UK

Chairman, CO_2 Capture Project Executive Board

We are seeking solutions to one of the great international challenges – reducing carbon emissions and their impact on climate change. Over the past decade, the prospect of climate change resulting from anthropogenic CO_2 has become a matter of deep and growing public concern. Many believe that the precautionary principle is the appropriate response at this time and there is increasing consensus that the action to mitigate this human induced climate change will require not just reducing anthropogenic CO_2 emissions, but more importantly stabilizing the overall concentration of CO_2 in the earth's atmosphere. There are many technology options that can help, but it appears that almost all will add cost to the price we pay for energy.

Given the scale of the climate change challenge and the need to continue to provide affordable energy in many different cultural, social and operational settings, a portfolio of approaches will be required. The best solution will not be the same in each case. It seems that the full portfolio of energy technologies will be required. Yet, one option that has broad potential application is the technology of CO_2 capture and geological storage. Capture technology is already in use, but only at small scale. While this technology is proven, it needs considerable development to enable scale-up for industrial application and to reduce the cost of what is a very expensive technology today. Geological storage, on the other hand, builds on the oil and gas industries' considerable experience of injecting gas for enhanced oil recovery (EOR), gas storage operations and reservoir management, which are all today successfully managed at scale. Capturing and storing CO_2 from the combustion of coal, oil and natural gas could deliver material reductions in greenhouse gas emissions and provide a bridge to a lower carbon energy future.

That is why the participants of the CO_2 Capture Project (CCP) decided to work together and collaborate with governments, industry, academic institutions and environmental interest groups, to develop technologies that will greatly reduce the cost of CO_2 capture and to demonstrate that underground, geological storage is safe and secure. The goal is to reduce the environmental impact of fossil fuel based energy production and use – over the same period of time when global energy demand is forecast to continue to grow strongly – in the most cost effective manner.

Three governments and eight companies have jointly funded, and actively participated in, the CCP. The best minds and research laboratories have been brought together to identify and pursue the most promising of the CO_2 capture technologies that could be commercially ready in the 2012 time frame. A wide range of academic and commercial institutions, all subject to open and comprehensive peer review, have provided breakthrough thinking, concepts and technology. The views of external bodies, such as environmental groups have been incorporated. Through international public–private collaboration, we believe the CO_2 Capture Project has made a real difference by stimulating rapid technology development and creating the new state of the art.

The CCP book contains technical papers and findings from all contractors involved in the first phase of the project. This work is the combined effort of over 70 technology providers, 21 academic institutions, six NGOs and each of the eight participating companies. In addition, the work benefited from the input and guidance from our four participating government organisations. The book is compiled in two volumes: Volume 1 covers capture technology development, our work in the area of capture and storage policy, the Technology Advisory Board project review and the common economic model that was developed to enable us to compare performance on a common basis and present the economic results. Volume 2 covers the geological storage program which we called SMV – Storage, Monitoring and Verification. These two volumes should serve as a valuable reference document for a wide spectrum of industry, academia and interested stakeholders on technology development for CO_2 capture and geological storage.

The CCP has achieved its Phase I goals for lower cost CO_2 capture technology and furthered the safe, reliable option of using geological storage. The results speak for themselves; delivering upwards of a 50% reduction in the cost of CO_2 capture in a 3 year time frame, is a considerable accomplishment. The results also offer promise that further significant improvements are likely in the performance and costs of this technology. The geological storage program has pioneered the risk-based approach for geological site selection, operation and abandonment. The program has made a major contribution overall to the confidence of CO_2 geological storage integrity and has developed some exciting new monitoring tools. There is now a much deeper understanding of the important role carbon capture and geological sequestration can play in a carbon-constrained future, particularly in a future that involves stabilization of the concentrations of CO_2 in the earth's atmosphere.

The industrial participants in the CCP would like to thank all of the people who have worked with us over the past 4 years and who have supported the delivery of our encouraging results. The list is long and includes people from academia, technology providers, the NGO community, industry and governments. The degree of cooperation, and hard work by those involved has been gratifying and has helped enormously in finding our way through the many challenges that lay in our path. The CCP project has succeeded because of extreme hard work from the whole extended multi-disciplinary team.

I would like to especially acknowledge the US DOE's National Energy Technology Laboratory, The European Union's DGTREN and DGRES programs, and the Norwegian Research Council's Klimatek program, without whose support the CO_2 Capture Project would not have been possible. Finally, I would like to formally thank the companies who were the project industrial participants – BP, ChevronTexaco, EnCana, ENI, Hydro, Shell, Statoil and Suncor – for their proactive engagement and strong leadership of the program. All the participants were engaged, active, and willing partners working towards the project goals.

The two volumes that you hold in your hand are the result of many thousand hours of effort. It is the Executive Board's hope that the technologies described here will form the basis of a vibrant and important industry for the benefit of mankind.

Acknowledgements

Helen Kerr, BP, p.l.c.

Program Manager, CO_2 Capture Project

The CO_2 Capture Project results reported here were delivered with the help of an exceptional technical team, who all deserve a special mention, but in particular I would like to acknowledge with thanks the CCP technical team leaders past and present: Henrik Andersen (Hydro), Mike Slater (BP), John Boden (BP), Odd Furuseth (Statoil), Henriette Undrum (Statoil), Robert Moore (BP), Torgeir Melien (Hydro), Ivanno Miracca (Eni), Mario Molinari (Eni), Craig Lewis (ChevronTexaco), Scott Imbus (ChevronTexaco), Arthur Lee (ChevronTexaco) and Iain Wright (BP).

The contracting and procurement support staff who handled over 100 contracts were magnificent: Robert Sloat, John Woods, John Hargrove, Sheetal Handa (BP) & Ole Morten Opheim (Statoil), Donna Douglas (Accenture, BP), Svein Berg (Statoil) and Stuart Green (Atkins, Faithful & Gould).

The Technology Advisory Board provided timely sage advice and the benefits of their collective experience to help the project succeed. Special thanks to Chairman Vello Kuuskraa (Advanced Resources International, ARI) for your outstanding commitment and personal support.

The project could not have happened without the support from our partners in government who co-funded the program. A special thanks to the project managers, Philip Goldberg and David Hyman (US DOE, NETL), Dennis O'Brien and Vassilios Kougionas (EU DGTREN & EU DGRES) and Hans-Roar Sorheim (NRC Klimatek).

These volumes were edited by two exceptional people, David Thomas (ARI) and Sally Benson (Lawrence Berkeley National Laboratory). Thank you for your hard work on behalf of the CCP.

CONTENTS

VOLUME 1

SECTION 2: PRE-COMBUSTION DE-CARBONIZATION TECHNOLOGY

SECTION 3A: OXYFUEL COMBUSTION TECHNOLOGY

SECTION 3B: CHEMICAL LOOPING COMBUSTION (CLC) OXYFUEL TECHNOLOGY

FUTURE RESEARCH NEEDS

VOLUME 2

SECTION 1: GHG, CLIMATE CHANGE AND GEOLOGICAL CO_2 STORAGE

SECTION 2: STORAGE INTEGRITY

SECTION 3: STORAGE OPTIMIZATION

SECTION 4: MONITORING AND VERIFICATION

SECTION 5: RISK ASSESSMENT

SECTION 1:
GHG, CLIMATE CHANGE AND GEOLOGICAL CO_2 STORAGE

CO_2 STORAGE PREFACE

Sally M. Benson

Lawrence Berkeley National Laboratory, Berkeley, CA, USA

From 2000 to 2003, the CO_2 Capture Project (CCP) sponsored a number of interrelated research projects on the topics of geologic storage of CO_2 in deep geologic formations. Topics ranged from literature surveys of existing technologies employed in the oil and natural gas industry that are relevant to geologic storage of CO_2 to original research on storage integrity, storage optimization, monitoring methods and risk assessment. Over 30 individual projects were carried out by a team of scientists and engineers from across the world. The team consisted of academic researchers, scientists from government-sponsored laboratories and private sector scientists and engineers. The original projects funded as part of this project were selected on a competitive basis. Once the team was established, collaborative efforts grew, allowing groups to leverage unique skills and knowledge of others on the team. The chapters presented here reflect that evolution.

Geologic storage of CO_2, as a method to eliminate atmospheric emissions of CO_2 from combustion of fossil fuels, has evolved remarkably quickly—from merely a concept to one of the most important methods for dealing with the climate change problem. In the mid 1990s, Statoil pioneered the first geologic storage project at Sleipner Vest in the North Sea. Nearly 1 Mt CO_2/year is removed from natural gas and injected into a salt-water filled sandstone formation deep under the North Sea. By the year 2010, we expect that over 10 Mt CO_2 will be stored annually in a number of projects across the world. In part, this remarkable progress can be attributed to leveraging the tremendous experience and technological acumen of the oil and gas industry. Simulation methods, monitoring technologies, well drilling techniques and injection technologies can all be applied to geologic storage of CO_2. However, the rapid progress cannot be explained on this basis alone. There are a number of issues unique to geologic storage that require additional knowledge, new technological approaches and new perspectives. Most notably, the need to provide assurance that CO_2 can and is stored safely underground for hundreds to thousands of years or longer creates significant challenges that must be addressed. Progress here can only be made by investment in research and development unique to geologic storage. This book documents both the foundation on which geologic storage is built and the progress that the CCP research team has made in addressing challenges that are unique to geologic storage. As such, we provide a snapshot in time of this rapidly evolving field.

Perhaps most importantly, rapid progress is being made because the time has come when the private sector, academic organizations and government-sponsored research laboratories have come together to solve one of the most challenging problems of our lifetime—dealing with global climate change. Geologic storage of CO_2 is just one of the methods needed. More efficient use of energy and renewable sources of power such as solar, wind and geothermal are all needed. But geologic storage of CO_2 is a powerful addition to the arsenal of tools that can be used to combat global climate change. By providing deep reductions in fossil fuel emissions from point sources such as power plants, refineries and gas processing plants, while continuing to use the fossil fuel resources available today, we can build a bridge to a healthy, safe and secure energy future.

ACKNOWLEDGEMENTS

The editor would like to acknowledge the extraordinary efforts of Jennifer Rosado of Lawrence Berkeley National Laboratory as the Production Manager for Volume II. She worked cheerfully and with great professionalism to bring the volume together from materials from around the world. In addition, the editor would like to recognize her associate editors Curt Oldenburg, Scott Imbus and Mike Hoversten and all those who contributed to the volume as authors or reviewers. Finally, it is important to recognize the important role that Craig Lewis played in getting this project started. His leadership shaped the content and team spirit that this volume represents.

Carbon Dioxide Capture for Storage in Deep Geologic Formations, Volume 2
D.C. Thomas and S.M. Benson (Eds.)

Chapter 1

OVERVIEW OF GEOLOGIC STORAGE OF CO_2

Sally M. Benson

Lawrence Berkeley National Laboratory, Berkeley, CA, USA

ABSTRACT

This paper presents an overview of geologic storage of CO_2. Topics addressed include the nature and extent of formations that could be used for geologic storage, the physical and chemical processes responsible for geologic storage, risks of geologic storage, and demonstration projects underway today. In addition, this chapter introduces the topics that are covered in this book.

INTRODUCTION

Over the past several hundred years, atmospheric CO_2 concentrations have steadily increased and have now risen to over 370 ppm from the pre-industrial level of 280 ppm. Increases in CO_2 concentrations are mainly attributed to burning of coal, oil, and natural gas for electrical generation, transportation, industrial and domestic uses. Today, globally, over 20 billion tons of CO_2 are emitted into the atmosphere. There is a growing consensus that increases in CO_2 concentrations will disrupt the earth's climate, cause sea level to rise enough to flood many low-lying coastal regions and damage sensitive ecosystems. Experts believe that to avoid significant disruption of the climate system and ecosystems, CO_2 concentrations must be stabilized within the next several decades. At today's emission rates, atmospheric CO_2 concentrations will continue to grow rapidly and, within 50 years, may exceed the levels needed to protect sensitive ecosystems and avoid flooding in low-lying coastal areas. This situation is even more urgent when we consider that over the next 50 years CO_2 emissions are expected to double as the developing world's economies grow and the standard of living increases. To address this challenge, we need a multi-pronged approach for decreasing CO_2 emissions—more efficient production and use of energy, solar power, wind energy, biomass, switching to fuel sources with lower or negligible CO_2 emissions, and CO_2 capture and storage (CCS), the subject of this book.

CCS is a four-step process where: first, a pure or nearly pure stream of CO_2 is separated and captured from flue gas or other process stream; next it is compressed to about 100 atm; it is then transported to the injection site; and finally, it is injected deep underground into a geological formation such as an oil and gas reservoir where it can be safely stored for thousands of years or longer (see Figure 1). Volume I of this two-part book provides detailed discussion of recent innovations in capture and compression technology. This volume (Volume II) focuses on transportation and storage-related issues.

That CO_2 could be separated from flue gases and stored from the atmosphere emerged in the open literature in the late 1970s [1,2]. However, it was not until the early 1990s that R&D in CO_2 storage began in earnest. Since that time, however, progress has been accelerating through a combination of industrial, academic, and public-sector efforts. A number of factors contribute to the rapid progress in this area, specifically:

- industrial experience in the oil, gas, and gas-storage industry can provide the expertise and technology needed for CO_2 transportation, injection, performance assessment, and monitoring;
- several collateral economic benefits are possible, including CO_2-enhanced oil and gas recovery and enhanced coalbed methane recovery;
- suitable geologic formations, including oil, gas, saline, and coal formations are located near many CO_2 sources; and

- geologic analogs such as natural CO_2 reservoirs demonstrate that geologic structures can store CO_2 over very long times.

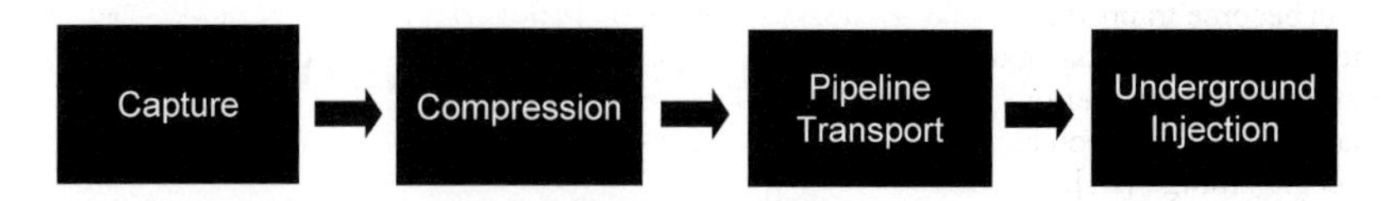

Figure 1: Schematic showing the major steps in the CO_2 capture and storage process.

Over the past decade, CCS has emerged as one of the most promising options for deep reductions in CO_2 emissions, so much so that, in fact, today 1 million tons of CO_2 is being stored annually at the Sleipner Project beneath the North Sea. Several more commercial projects are underway or in the advanced stage of planning: the In Salah project in Algeria, the Gorgon Project in Australia, and the Snohvit Project in the continental shelf offshore of Norway. In addition to these, more are under development.

STORAGE FORMATIONS AND PROCESSES

Sedimentary basins, created by the gradual deposition and compaction of sediments eroded from mountain ranges, are the mostly likely location for storing CO_2. Deposits, as thick as many thousands of meters, have accumulated in sedimentary basins around the world. Typically, sedimentary basins consist of alternating layers of coarse (sandstone) and fine-textured sediments (clay, shale, or evaporites). The sandstone layers, which provide the storage reservoir, have high permeability, allowing the CO_2 to be injected into the storage reservoir. The shale or evaporite layers have very low permeability and act as seals to prevent CO_2 from returning to the surface. Naturally occurring CO_2 reservoirs exist in North America, Australia, China, and Europe, demonstrating that CO_2 can be stored underground for millions of years or longer. In addition, many oil and gas reservoirs also contain large quantities of CO_2 confirming that oil and gas reservoirs can also store CO_2 over geologic time scales.

The conceptual framework and opportunity for storage of CO_2 in saline formations and depleted oil and gas formations were presented in early papers by Koide et al. [3–5], Winter and Bergman [6], Van der Meer [7, 8], Gunter et al. [9], Hendriks and Blok [10,11], Holloway and Savage [12], Holt et al. [13], Bachu et al. [14], Bergman and Winter [15], Omerod [16], and Weir et al. [17]. In 1996, Gunter et al. [18] described a process by which coalbed methane production could be enhanced while simultaneously storing CO_2. Studies by Byrer and Guthrie [19,20,34] and Stevens et al. [21,22] suggest that worldwide CO_2 coalbed methane recovery may also significantly add to the capacity for geologic storage of CO_2. Today, four principle types of geologic formations are widespread and are considered to have significant potential for storing large amounts of CO_2:

- active and depleted oil reservoirs;
- active and depleted gas reservoirs;
- saline formations; and
- deep coal seams and coalbed methane formations.

Other geologic formations such as marine and arctic hydrates, CO_2 reservoirs, mined cavities in salt domes and oil shale may increase storage capacity or provide niche opportunities but are likely to be developed only after the storage formations listed above are utilized.

CO_2 can be stored in these geologic formations by four principal processes [23,24].

- CO_2 can be trapped as a gas or supercritical fluid under a low-permeability cap rock, similar to the way the natural gas is trapped in gas reservoirs or the gas is stored in aquifer gas storage. Immediately after CO_2 is injected, this is likely to be the most important storage mechanism.

- CO_2 can dissolve into the fluid phase. This mechanism is referred to as solubility trapping. The relative importance of solubility trapping depends on a large number of factors, such as the sweep efficiency of CO_2 injection, formation of fingers, and the effects of formation heterogeneity.
- CO_2 can become trapped as a residual, non-wetting phase in the pore spaces of the rock. This mechanism is referred to as residual gas trapping. Once the saturation of CO_2 drops below the residual "gas" saturation, it is no longer mobile and consequently will remain trapped. The importance of this trapping mechanism has only been recognized recently and is expected to contribute significantly to the security of geologic storage [24].
- CO_2 can react, either directly or indirectly, with the minerals and organic matter in the geologic formations to become part of the solid mineral matrix. Formation of carbonate minerals such as calcite, siderite, or alumino-carbonates such as dawsonite and adsorption onto coal are examples of mineral trapping. Mineral trapping will create stable forms of carbon that are unlikely to return to the biosphere and will increase storage security by eliminating the risk of unexpected leakage of CO_2 to the surface.

Over time, the contribution of each of these processes to provide secure long-term storage will change as illustrated in Figure 2. Initially, physical trapping will be the dominant mechanism for keeping CO_2 in the storage formation. As CO_2 migrates away from the injection well it will displace some fraction of the in situ fluids. Simultaneously, CO_2 will dissolve in the pore fluids that are left behind. Over time, as the CO_2 plume grows, larger amount of CO_2 can dissolve, thus increasing the extent of solubility trapping. Over very long periods, small-scale convection cells created by density differences between the CO_2 saturated brine and the in situ fluids will dissolve even more CO_2 [25]. The extent and evolution of CO_2 trapped as a residual phase will depend on the petrophysical properties of the storage formation. Recent studies have shown that the residual saturation may be as high as 20–30% of the pore space. In this case, the CO_2 plume tends to be compact and remains trapped near the injection well. If the residual saturation is much lower, in the range of 5–10%, residual gas trapping will increase over time as the plume migrates over a greater volume. Mineral tapping is expected to be slow but, over long time scales, may trap a significant fraction of the CO_2—the extent of which will depend on the mineralogy of the formation. Storage formations composed of a large fraction of feldspar minerals will have a higher degree of mineral trapping.

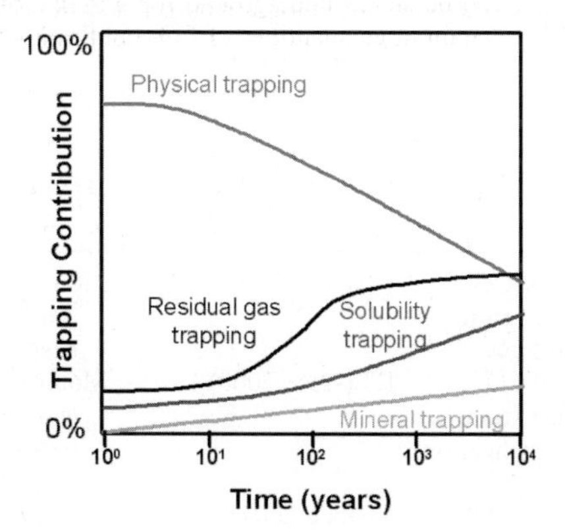

Figure 2: Schematic illustrating that residual gas trapping, solubility trapping and mineral trapping increase storage security over time.

STORAGE CAPACITY

Several worldwide and national assessments demonstrate the significant potential for geologic storage of CO_2 in saline formations, coal formations, and depleted oil and gas reservoirs [16]. Subsequent studies have focused on assessing important aspects of regional geologic formations that may be suitable for storage [18, 26–33]. Global storage capacity estimates are summarized in Table 1. While the range of estimates is large, there is a consensus that the largest potential capacity is in deep saline formations in large sedimentary basins. It is estimated that saline formations have the capacity to accommodate hundreds of years at the current CO_2 emission rates. However, these capacity estimates have not yet been validated by regional or site-specific field experiments.

TABLE 1
SUMMARY OF WORLDWIDE STORAGE CAPACITY ESTIMATES

Formation type	Capacity estimate ($Gt\ CO_2$)	Source
Depleted oil and gas reservoirs	~450	Stevens et al., 2001: GHGT 6, pp. 278–283
Coalbed methane reservoirs	60–150	Stevens et al., 1999: GHGT 5, pp. 175–180
Salt-water filled formations	300–10,000	IEA Greenhouse Gas R&D Programme, 1994 [16]

EXISTING AND PLANNED CO_2 STORAGE PROJECTS

Today there are four active geologic storage projects and at least two more are planned (see Table 2). These demonstrate the range of current experience with CCS. In all but two of these projects, the source of the CO_2 is natural gas. CO_2 is separated from the natural gas because some natural gas reservoirs contain too much CO_2 to sell on the open market unless the CO_2 is removed first. In addition to these projects, which were developed for the specific purpose of CCS, about 20 million tons per year of CO_2 is injected annually to recover oil from over 50 oil fields, primarily from carbonate formations in West Texas.

All the CO_2 storage projects listed in Table 2 are being used to one degree or another as demonstration projects. International teams of scientists, funded by private and government sources, are deploying monitoring technologies, computer simulation models, and risk assessment methods to assess the safety of these projects, improve our understanding of geologic storage, and develop advanced technologies for

TABLE 2
SUMMARY OF CURRENT AND PLANNED CCS PROJECTS

Project (operator)	Application	Mass of CO_2 (million tons/year)	Demonstration activities	Storage formation
Sleipner, North Sea (Statoil)	Storage of CO_2 stripped from natural gas	1 (since 1996)	Monitoring, modeling, best practices	Offshore salt-water sand formation
Weyburn, Canada (Encana)	EOR and CO_2 storage from coal gasification	1.7 (since 2000)	Monitoring, risk assessment, performance assessment	On-shore oil reservoir in carbonate rock
In Salah, Algeria (BP)	Storage of CO_2 stripped from natural gas	1 (since 2004)	Monitoring, risk assessment	On-shore gas reservoir in sandstone
Gorgon, Australia (ChevronTexaco)	Storage of CO_2 stripped from natural gas	4 (planned for 2006)	To be determined	Island salt-water sandstone formation
Snohvit, Offshore Norway (Statoil)	Storage of CO_2 stripped from natural gas	0.7 (planned for 2006)	To be determined	Offshore salt-water sandstone formation
San Juan Basin, New Mexico (Burlington)	Enhanced coalbed methane production		Performance assessment, risk assessment	On-shore coalbed

monitoring CO_2 storage projects. None of these existing projects is as large as would be required to capture and store the 8 million tons per year of CO_2 from a typical 1000 MW coal-fired power plant. However, the scale-up of individual projects ranging from 1–4 million tons per year to 8 million tons per year should be achievable and these projects provide substantial experience on which future projects can build.

RESEARCH AND DEVELOPMENT NEEDED TO ADVANCE GEOLOGIC STORAGE

While rapid progress has been made in the development of geologic storage of CO_2 since its inception in the 1990s, additional knowledge is needed in a number of areas to support widespread implementation of this technology. This book addresses many of these topics, which can be broadly grouped under the following topics.

Storage security and integrity. Additional knowledge is needed about the processes that contribute to long-term storage of CO_2. These include physical trapping beneath low-permeability cap rocks, trapping as an immobile residual phase in the pore spaces of the storage reservoir, and geochemical trapping in fluids or rocks. Information about and strategies to preserve the long-term integrity of well construction materials are needed to assure that the wells penetrating the storage reservoir do not fail and provide a short circuit for CO_2 back to the atmosphere. Geomechanical stresses on the cap rock that could compromise the integrity of the rock by reactivating faults or fractures need to be better understood. The influence of other gases such as H_2S, SO_x, and NO_x, which may be stored along with CO_2 need to be understood. This book addresses all these issues, both by evaluating existing analogues for CO_2 storage, such as naturally occurring CO_2 reservoirs, CO_2-enhanced oil recovery and natural gas storage, as well as presenting the results of original research on this topic.

Storage optimization. Geologic storage of CO_2 can be optimized economically by combining it with enhanced oil and gas recovery. Revenues from enhanced oil and gas recovery can be used to offset the cost of storage, and capital investments can be used to help build the infrastructure for CO_2 storage. Optimization can also be achieved by assuring efficient use of the underground storage space and applying best practices learned from related activities such as natural gas storage. This book addresses both these issues, by both evaluating existing analogues for CO_2 storage, such as naturally occurring CO_2 reservoirs, CO_2-enhanced oil recovery and natural gas storage and presenting the results of original research on this topic.

Monitoring and verification. Monitoring has been identified as one of the highest priority needs to provide safe and secure storage of CO_2. Monitoring CO_2 migration in the subsurface plays several diverse and critical roles in the development and acceptance of geologic storage. First, it is essential for accounting purposes. That is, it will be necessary to verify the net quantity of CO_2 that has been stored in the subsurface. Second, it is necessary for monitoring sweep efficiency and determining whether the available storage capacity is being used effectively. Third, it is needed for optimizing EOR and enhanced coalbed methane recovery. Finally, it is necessary to ensure the safety of storage projects by demonstrating that CO_2 is retained in the formation into which it was injected. This book provides information on monitoring technologies that can serve all these purposes, by both drawing from relevant experience across a number of monitoring applications and presenting the results of original research on this topic. Specific topics include: surface monitoring of rates and compositions of injected and produced gases and liquids; atmospheric CO_2 concentration and flux monitoring; ecosystem monitoring; surface (including 3D seismic methods), surface-to-borehole, single-well, and cross-borehole time-lapse seismic methods; electrical methods such as electrical resistance tomography and cross-well electromagnetic methods; reservoir pressure and temperature measurements; and natural and introduced chemical tracers that will provide additional information needed to quantify hydrodynamic, solubility, and mineral trapping rates and processes.

Risk assessment and mitigation. Assessing risks and developing a risk mitigation strategy are an essential part of the process for selecting and obtaining permits for a geologic storage project. The nature of the risks must be understood fully. Scenarios for both secure and leaking CO_2 storage projects must be developed. Reliable and accepted methods for quantitative probabilistic risk assessment are needed. In addition, methods for mitigating risks, including monitoring and remediation must be developed. Over the past several years, significant progress has been made in this area, particularly with regard to the application of the features, events, and processes (FEP) methodology for risk characterization and assessment. This book describes this methodology and provides examples of its application to a number of storage projects. Significant progress has also been made in understanding the consequences of leaking geologic storage projects on ecosystems and humans. Models have been developed to quantify how CO_2 behaves when released into the near surface environment and escapes back to the atmosphere. The potential impact to underground microbial communities has also been assessed. A compilation of potential remediation options, based on analogous experience in natural gas storage and disposal of liquid wastes, has also been developed. Taken together, these studies provide the foundation for risk assessment and mitigation for CO_2 storage projects in deep geologic formations.

CONCLUSIONS

Geologic storage of CO_2 in underground formations has quickly advanced from a mere concept to a reality. Significant progress has been made in the critical areas of storage security and integrity, storage optimization, monitoring and verification, and risk assessment and mitigation. More remains to be accomplished before widespread application of this technology takes place, but the results of research conducted in this project and others continue to demonstrate that this technology can make large contributions to reducing CO_2 concentration in the atmosphere. This book highlights accomplishments in the areas listed above, and in each case, identifies additional research and development needed to further advance this technology.

ACKNOWLEDGEMENTS

The author gratefully acknowledges the support by the CO_2 Capture Project and the US Department of Energy. In addition, the author acknowledges the tremendous intellectual contributions of many scientists and engineers around the world for advancing the state of knowledge about geologic storage of CO_2, and so quickly turning this technology from a mere concept to reality.

REFERENCES

1. C. Marchetti, Geoengineering and the CO_2 problem, *Clim. Change* **1** (1977) 59–68.
2. C.F. Baes Jr., S.E. Beall, D.W. Lee, The collection, disposal and storage of carbon dioxide, in: W. Bach, J. Pankrath, J. Williams (Eds.), Interactions of Energy and Climate, D. Reidel Publishing Co, Dordrecht, Holland, 1980, pp. 495–519.
3. H. Koide, Y. Tazaki, Y. Noguchi, S. Nakayama, M. Iijima, K. Ito, Y. Shindo, Subterannean containment and long-term storage of carbon dioxide in unused aquifers and in depleted natural gas reservoirs, *Energy Convers. Manage.* **33** (5–8) (1992) 619–626. R. Korbol, A. Kaddour, Sleipner vest CO2 disposal—injection of removed CO2 into the Utsira Formation, *Energy Convers. Manage.* **36** (6–9) (1995) 509–512.
4. H. Koide, Y. Tazaki, Y. Noguchi, M. Iijima, K. Ito, Y. Shindo, Underground storage of carbon dioxide in depleted natural gas reservoirs and in useless aquifers, *Eng. Geol.* **34** (3–4) (1993) 175–179.
5. H. Koide, Geological Storage and Microbiological Recycling of CO_2 in Aquifers, *Proceedings of the Fourth International Conference on Greenhouse Gas Control Technologies*, 30 August–2 September, 1998, Interlaken, Switzerland, 1998.
6. E.M. Winter, P.D. Bergman, Availability of depleted oil and gas reservoirs for disposal of carbon dioxide in the United-States, *Energy Convers. Manage.* **34** (9–11) (1993) 1177–1187.
7. L.G.H. Van der Meer, Investigations regarding the storage of carbon dioxide in aquifers in The Netherlands, *Energy Convers. Manage.* **33** (5–8) (1992) 611–618.
8. L.G.H. Van der Meer, The conditions limiting CO_2 storage in aquifers, *Energy Convers. Manage.* **34** (9–11) (1993) 959–966.
9. W.D. Gunter, E.H. Perkins, T.J. McCann, Aquifer disposal of CO2-rich gasses: reaction design for added capacity, *Energy Convers. Manage.* **34** (1993) 941–948.
10. C.A. Hendriks, K. Blok, Underground storage of carbon dioxide, *Energy Convers. Manage.* **34** (9–11) (1993) 949–957.
11. C.A. Hendricks, K. Blok, Underground storage of carbon dioxide, *Energy Convers. Manage.* **36** (1995) 539–542.

12. S. Holloway, D. Savage, The potential for aquifer disposal of carbon dioxide in the UK, *Energy Convers. Manage.* **34** (1993) 925–932.
13. T. Holt, J.I. Jensen, E. Lindeberg, Underground storage of CO2 in aquifers and oil reservoirs, *Energy Convers. Manage.* **36** (N6–9) (1995) 535–538.
14. S. Bachu, W.D. Gunter, E.H. Perkins, Aquifer disposal of CO_2: hydrodynamic and mineral trapping, *Energy Convers. Manage.* **35** (1994) 269–279.
15. P.D. Bergman, E.M. Winter, Disposal of carbon dioxide in aquifers in the U.S, *Energy Convers. Manage.* **36** (1995) 523–526.
16. W. Omerod W. IEA greenhouse gas R&D programme, carbon dioxide disposal from power stations, 1994, IEA/GHG/SR3.
17. G.J. Weir, S.P. White, W.M. Kissling, Reservoir storage and containment of greenhouse gases, *Energy Convers. Manage.* **36** (6–9) (1995) 531–534.
18. W.D. Gunter, S. Bachu, D.H.-S. Law, V. Marwaha, D.L. Drysdale, D.E. MacDonald, T.J. McCann, Technical and economic feasibility of CO2 disposal in aquifers within the Alberta sedimentary basin, Canada, *Energy Convers. Manage.* **37** (1996) 1135–1142.
19. C.W. Byrer, H.D. Guthrie, Assessment of world coal resources for carbon dioxide (CO2) storage potential—while enhancing potential for coalbed methane, US Department of Energy, Greenhouse Gas Mitigation, Technologies for Activities Implemented Jointly, *Proceedings of Technologies for Activities Implemented Jointly*, 26–29th May, Vancouver, Canada, 1997, 573–576.
20. C.W. Byrer, H.D. Guthrie, Carbon dioxide potential in coalbeds: a near-term consideration for the fossil energy industry, US Department of Energy, *The Proceedings of the 23rd International Technical Conference on Coal Utilization and Fuel Systems*, 9–13 March, 1998, Clearwater, FL, 1998, 593–600.
21. S.H. Stevens, D. Spector, Enhanced coalbed methane recovery: worldwide applications and CO2 storage potential, Report prepared for IEA Greenhouse Gas R&D Programme, IEA/CON/97/27, 1998.
22. S.H. Stevens, J.A. Kuuskraa, D. Spector, CO_2 storage in deep coal seams: pilot results and worldwide potential, *Fourth International Conference on Greenhouse Gas Control Technologies*, 30 August–2 September, 1998, Interlaken, Switzerland, 1998.
23. B. Hitchon (Eds.), Aquifer Disposal of Carbon Dioxide, Hydrodynamic and Mineral Trapping—Proof of Concept, Geoscience Publishing Ltd, Sherwood Park, Alberta, Canada, 1996.
24. C. Doughty, S.M. Benson, K. Pruess, in: Capacity investigaton of brine-bearing sands for geologic sequestration of CO_2, J. Gale and Y. Kaya (Eds.), Greenhouse Gas Control Technologies, 2003, 1645–1648.
25. J. Ennis-King, L. Patterson, in: Rate of dissolution due to convection mixing in the underground storage of carbon dioxide, J. Gale and Y. Kaya (Eds.), Greenhouse Gas Control Technologies, 2003, 507–510.
26. S. Tanaka, H. Koide, A. Sasagawa, Possibility of underground CO_2 storage in Japan, *Energy Convers. Manage.* **36** (6–9) (1995) 527–530.
27. N. Gupta, B. Sass, J. Sminchak, T. Naymik, Hydrodynamics of CO2 disposal in a deep saline formation in the Midwestern United States, *Proceedings of the Fourth International Conference on Greenhouse Gas Control Technologies*, 30 August–2 September, 1998, Interlaken, Switzerland, 1998.
28. B. Sass, et al., IEA Book, 1998.
29. L. Liu, G.H. Huang, A. Chakma, Environmental impacts and risks of CO2 injection for enhanced oil recovery in Western Canada, *Proceedings of the Fourth International Conference on Greenhouse Gas Control Technologies*, 30 August–2 September, 1998, Interlaken, Switzerland, 1998.
30. S. Bachu, W.D. Gunter, Storage capacity of CO_2 in geological media in sedimentary basins with application to the Alberta Basin, *Proceedings of the Fourth International Conference on Greenhouse Gas Control Technologies*, 30 August–2 September, 1998, Interlaken, Switzerland, 1998.
31. T.A.M. McKean, R.M. Wall, A.A. Espie, Conceptual evaluation of using CO_2 extracted from flue gas for enhanced oil recovery, Schrader Bluff Field, North Slope, Alaska, *Proceedings of the Fourth International Conference on Greenhouse Gas Control Technologies*, 30 August–2 September, 1998, Interlaken, Switzerland, 1998.
32. R.P. Hattenbach, M. Wilson, K.R. Brown, Capture of carbon dioxide from coal combustion and its utilization for enhanced oil recovery, *Proceedings of the Fourth International Conference on Greenhouse Gas Control Technologies*, 30 August–2 September, 1998, Interlaken, Switzerland, 1998.

33. S.M. Benson, Comparison of three options for geologic sequestration of CO2—a case study for California, Proceedings, *Fifth International Conference on Greenhouse Gas Control Technologies*, CSIRO Publishing, Australia, 2001, 299–304.
34. C.W. Byrer, H.D. Guthrie, Coal deposits: potential resource for storing carbon dioxide emissions from power plants, US Department of Energy, *Fourth International Conference on Greenhouse Gas Control Technologies (GHGT4)*, 30 August–2 September, 1998, Interlaken, Switzerland, 1998.

Carbon Dioxide Capture for Storage in Deep Geologic Formations, Volume 2
D.C. Thomas and S.M. Benson (Eds.)

Chapter 2

TECHNICAL HIGHLIGHTS OF THE CCP RESEARCH PROGRAM ON GEOLOGICAL STORAGE OF CO_2

S. Imbus

ChevronTexaco Energy Technology Company, Bellaire, TX, USA

ABSTRACT

This chapter provides an overview of the contents of this volume and the technical contributions of the CCP research team. Key results from 32 individual research projects are described. Contributions are discussed under four headings: storage integrity; storage optimization; monitoring; and risk assessment.

INTRODUCTION

The geological storage research program focused on four areas: storage integrity, storage optimization, monitoring and risk assessment. The following sections summarize progress in each of these areas.

STORAGE INTEGRITY

The storage "integrity" theme studies are directed towards better understanding elements of natural (reservoir and cap rocks, overburden, faults, etc.) and engineered (well materials) features that permit safe and effective geological storage of CO_2. Implications of an important industrial analog, natural gas storage, are also examined.

Assessment of Storage Integrity from Natural Geological Systems

Features of natural systems that are "effective" in accumulating and retaining large quantities of CO_2 are examined by Stevens (Chapter 3). Three large US CO_2 reservoirs were evaluated: (1) McElmo Dome, CO (30 Tcf at 2300 m; carbonate reservoir with thick halite cap rock), (2) Jackson Dome, MS (2 Tcf at 4700 m; sandstone with some carbonate reservoir with carbonate cap rock) and (3) St Johns, AZ (14 Tcf at 500 m; sandstone reservoir and anhydrite cap rock). Production and operations data were gathered for each of the sites. Key findings of the study are given below.

- CO_2 storage is a natural process that occurs where reliable reservoir seals such as thick evaporites or shales are present.
- Production of CO_2 from natural reservoirs provides insights for CO_2 storage.
- Efficient CO_2 storage operations will require specialized practices and technologies.

Recommendations include further analog studies focusing on classification of site suitability for storage, site characterization, modeling of injection process and monitoring.

The study by Shipton et al. (Chapter 4) on leaking natural CO_2 reservoirs systems focuses on CO_2-charged geysers from Western Colorado Plateau (East Central Utah). A three-dimensional (3D) model for CO_2 sources, travel paths and fate in the subsurface was developed by integrating multiple geologic data sets. Compositional and isotopic data suggest that CO_2 originates from clay–carbonate reactions at 100–200 °C (1.0–1.5 km in Upper Paleozoic or Triassic strata), migrates to a sequence of shallow, stacked reservoirs (300–500 m) with shale cap rocks and escapes to the surface through fractures associated with faults.

Features and distribution of travertine deposits in the area indicate that the system has been leaking since before historical times. Anthropogenic activity, such as drilling through faults, has created additional leakage pathways to the surface and appears to have altered the location and episodicity of CO_2-charged eruptions in the area. No untoward ecological or human health effects attributed to CO_2 release to the surface have been recorded. The study demonstrates the utility of constructing 3D geological and fluid history models to assess the suitability of geologic systems for CO_2 storage.

The Next Generation Capture and Storage (NGCAS) project comprised a multi-scale, integrated assessment of the Forties Field (UK North Sea) for CO_2 storage (Cawley et al., Chapter 5). The study workflow moved from 2D basin scale hydrogeology models to 3D fluid flow simulation around the field to reservoir simulation of CO_2 – water simulations interactions (e.g., diffusion). Risk evaluation applied a series of sensitivity tests that took data uncertainties into account. It was found that the potential for CO_2 escape via geological pathways by various mechanisms (diffusion and advective flow through cap rock, dissolution and transport if CO_2 into the underlying aquifer) is low due to the quality and thickness of the cap rock and overburden and the very slow, compaction driven natural fluid velocity in the reservoir and surrounding are. Although the geologic features of the Forties Field combine to comprise an excellent venue for CO_2 storage, the risks associated with well leakage and seepage need to be examined in detail.

Streit et al. (Chapter 6) reviews the methods used to predict and monitor geomechanical effects of CO_2 injection. Increases in formation fluid pressure due to CO_2 injection decrease the effective rock stress thereby increasing the likelihood of fault reactivation or rock failure. Assessment of the geomechanical stability of reservoir rocks and top seals and faults requires predictions of in situ stresses, fault geometries and rock frictional strength. Commercial tools exist to predict the maximum sustainable fluid pressure for rocks and faults (e.g. FAST™ or TrapTester™). Fault stability is also predicted by mapping fault geometry and constructing fault-failure plots. In assessing the suitability of a CO_2 storage site in a depleted oil or gas field, it is necessary to analyze for the effects of both depletion (from production) and recharging (from injection). Stress–seismic velocity relationships are used to detect poroelastic changes in rocks due to fluid injection. Recent development of new waveforms and data processing techniques may improve the accuracy of seismic techniques used for detecting stress changes. Installation of downhole seismic monitoring instruments may provide rapid, early detection of faulting or fracturing induced by effective stress changes.

Changes in geophysical attributes and mineral stability with CO_2 injection are the subject of the experimental study by Schütt et al. (Chapter 7). Using a triaxial cell and autoclaves to reproduce reservoir PT conditions, supercritical CO_2 was injected into rock samples to assess how suitable existing geophysical models are for predicting rock mechanical changes and whether or not mineral dissolution occurs. Seismic data show that both the bulk and shear modulus depend on the CO_2 saturation and differential pressure. The Gassmann model underestimates the fluid substitution effect that reflects the modulus dispersion between the static (Gassmann) and ultrasonic (laboratory) regimes. The dependence of shear modulus on fluid composition is not predicted by the Gassmann model. The higher pressure sensitivity of the shear modulus, compared to the bulk modulus, may permit discrimination of pressure and saturation effects through simultaneous use of compressional and shear waves. Seismic wave attenuation may be used to infer saturation. The experiments corroborate numerical models that predict fluid-front instabilities. Improvements in the standard models using these data may enhance seismic monitoring techniques. The geochemical results suggest that major elements essential for rock stability and minor elements of importance to water quality are mobilized by CO_2 injection.

Johnson et al. (Chapter 8) used reactive transport geochemical and distinct element geomechanical models to infer long-term effects of CO_2 injection on cap rocks. It was shown that CO_2 influx-triggered mineral dissolution and precipitation reactions in typical shale cap rocks reduce microfracture apertures whereas pressure and effective stress evolution initially increase and then slowly decrease them. For a given shale composition, the extent of geochemical alteration (to reduce permeability) appears nearly independent of key reservoir properties (permeability and lateral continuity) and CO_2 influx parameters (rate, focality and duration). In contrast, the extent of geomechanical degradation (to increase permeability) is highly dependent on the reservoir and influx parameters as they control the magnitude of pressure perturbation. One implication of this study is that natural CO_2 accumulations, which have not been subjected to large stress changes, may not be good analogues for man-made CO_2 storage reservoirs. Stress changes that could threaten the security of a CO_2 storage project can be avoided by appropriate reservoir selection (e.g. large, unconfined) and adhering to safe operation parameters (e.g. injection rate).

Storage Integrity of Engineered Systems
In the survey of the natural gas storage industry operational experience in North America and Europe, Perry (Chapter 9) draws important parallels to a future CO_2 storage industry. Through operation of ~600 natural gas storage facilities in North America and Europe over the past 90 years, only nine gas migration incidents are recorded (all in the US). These include three cap rock failures, five wellbore failures and one case of poor reservoir selection. The review of natural gas storage technologies with possible implications for CO_2 storage includes the following.

- *Field integrity determination.* This involves selecting a structure that has a competent seal and structurally adequate closure. Broadly structured sites are favored because those with tight structuring have often developed faults and fractures. Pump testing of structures to ensure cap rock integrity is often performed. A modified pump test may be feasible for CO_2 cap rock testing.
- *Monitoring and leak detection.* This involves monitoring via observation wells for occurrence of gas above and lateral to the structure. Similar approaches may be used for CO_2 storage although gas migration may not be as readily detected.
- *Response to leakage.* Leak mitigation measures for natural gas leaks include shallow gas recycle, aquifer pressure control and cap rock sealing. For CO_2 storage, the former approaches are relevant but the latter approach needs further development.

Based on experience from the natural gas storage industry, the CO_2 sequestration industry should investigate the "science of observation wells," integration of existing technologies for locating and sealing leak, and methods to test field integrity. Claims that gas will never be found outside of the containment area should not be made.

The well integrity issue, particularly as it relates to cement stability, is addressed by Scherer et al. (Chapter 10). Well leakage scenarios are defined, and modeling of the potential number of wells expected to be exposed to a plume of injected CO_2 are tested. Degradation of annular and casing plug cements through poor construction, age or acid attack provides multiple pathways for CO_2 leakage with potential impacts on shallower resources and surface ecosystems. In the high well density Viking Formation (Alberta Basin), a modest CO_2 injection plume of 5 km radius is expected to contact an average of 240 wells. Rates and mechanisms of cement attack by carbonated water are tested experimentally and by simulation of subsurface conditions. Experiments with cements (class H with 0, 6 and 12% bentonite) included exposure of slices of sandstone and limestone cores with cement cores to static carbonated water (3% NaCl) at a range of temperature and pH. Post-exposure, the samples were analyzed for compositional (chemical, mineralization), structural (porosity, cracks) and hardness changes. Cured cement cylinders were exposed to carbonated water to test changes in permeability. Cement pastes were tested to determine the rate of leaching and permeability. The experiments demonstrated that carbonated water attack on cements is rapid. The reaction rim showed increases in porosity, and extensive removal of Ca and changes in Fe redox state (II to III) were noted. Preliminary simulations of plume delivery rate and attendant changes in water composition and pH indicated that the rate of acid attack on cements is most intense with the arrival of the plume but eventually stabilizes to a lower rate. Acid attack on cement is most severe when fresh carbonated water is continually delivered to the exposed cement. The study highlights the need to develop well leakage, cement stability and fluid flow scenarios prior to CO_2 injection in high well density areas.

The storage integrity studies contribute useful protocols for site assessment and considerations for operating and monitoring planning. The natural and industrial analogs are reassuring in terms of safety and provide practical operations and intervention information. The issue of well integrity is increasingly recognized as critical, probably more so than geological systems' integrity.

STORAGE OPTIMIZATION

The storage optimization studies are aimed at realizing operational efficiencies or cost savings that might make CO_2 storage a technical and economic success.

Industry CO_2 Injection Experience

Grigg (Chapter 11) surveyed the performance of Permian Basin (West Texas, Southeast New Mexico) CO_2 EOR operations over the past 30 years to assess what can be learned from the projects and where further research is needed. Data from operator surveys and the literature were tabulated by reservoir/seal type, performance issues such as injectivity, oil response and gas breakthrough and containment. There is significant industry experience in the safe separation, compression, transportation and injection of CO_2. In general, for well-characterized reservoirs in which previous operation problems were noted (e.g. during water flood), CO_2 behavior is consistent with reservoir simulations. In the short term (compared to geological time), behavior is consistent with predictions from reservoir simulations. In the short time that CO_2 has been injected into reservoirs, seals are maintaining their integrity and CO_2 is retained in the injection formation. The Permian Basin CO_2 EOR survey is a valuable "lessons-learned" exercise for CO_2 storage efforts given the extensive and unique collective experience of such operations.

Injection of acid gas (CO_2 and H_2S) from natural gas processing has been practiced without incident for 15 years at over 40 sites in the Alberta Basin of western Canada. Bachu and Haug (Chapter 12) describe the wide range of acid gas compositions injected, reservoir characteristics and operating conditions involved. Site selection criteria, including proximity to source, confinement of gas, effect of the gas on reservoir rock, protection of energy, mineral and groundwater resources, equity interests, wellbore integrity and public safety are outlined. Well completions, testing, operations and abandonment regulatory requirements have been established to ensure safe storage. The acid gas injection experience is encouraging for the prospects of safe and secure CO_2 storage as the presence of H_2S in the former poses a much greater hazard. The remaining issues include long-term containment and the applicability to larger scale operations.

Simulation of CO_2 Injection Performance in Coal Beds

Coal bed storage of CO_2 offers an economic offset from expected enhanced coal bed methane (ECBM) recovery. In the case study of a Colorado Plateau gas field (Tiffany) currently under N_2 flood, Wo and Liang (Chapter 14) outline considerations for the success of CO_2 ECBM in terms of reservoir performance and the potential for seepage. History matching of the N_2 injection shows that coal/CO_2 contact is limited. N_2 injection has caused coal fracturing and the development of preferred gas pathways from the injection to production wells. Methane seepage has already been noted in the San Juan Basin area. A representative seepage model for the Fruitland coal simulated conditions under which outcrop seepage of CO_2 and methane can be expected. Placement of injection wells within 2 miles of an outcrop could result in seepage of injected and mobilized gases.

Cost Reduction

Heggum et al. (Chapter 15) focused on designing safe and cost-effective systems and operational parameters for the compression and transportation of CO_2 under various conditions (e.g. offshore versus onshore, distance, presence of cooling water, CO_2 impurities). The principal goal of the study is to assess the utility of inexpensive carbon steel in settings, such as offshore Norway (hydrated, cool) as opposed to the better-known US situation (dehydrated, onshore). Based on water solubility in supercritical CO_2 experiments it is suggested that the proposed dehydration specification for LNG (50 ppm) might easily be relaxed to 600 ppm (the existing US Kinder-Morgan specification). Thermodynamic calculations of free water precipitation from supercritical CO_2 indicate that the specification might be further relaxed to 1300 ppm.

The Seiersten and Kongshaug (Chapter 16) study provided experimental results for CO_2 corrosivity to carbon steel. Experimental data obtained at higher pressure (up to 50 bar) showed that corrosion rates in CO_2 systems containing water and those containing water and MEG inhibitor are considerably lower than that predicted by existing corrosion models, particularly at low temperatures typical for subsea pipelines in northern waters. The study provides the basis for operational constraints for CO_2 transport in inexpensive carbon steel pipelines which may improve the economics of CO_2 storage offshore.

The study by Sass et al. (Chapter 17) on CO_2 impurities' tradeoffs serves as a link between storage studies and those examining transportation and capture. The substantial cost-saving potential in CO_2 capture of delivering CO_2 contaminated with impurities such as SO_x, NO_x and others (e.g. N_2, O_2, hydrocarbons, Hg) is balanced with potential operational complications and damage to surface facilities such as compressors, pipelines and injection systems. Absorption and regeneration characteristics of amines and other solvents used for CO_2 capture are adversely affected by acid gas impurities. Compression of gas mixtures may be complicated by the presence of higher boiling constituents, which may limit the ability to achieve adequate interstage cooling and damage compressors and related processing equipment. Materials used in separation, compression and transmission are subject to corrosion by carbonic, sulfuric, nitric and nitrous acids. Although corrosion mechanisms and their effects are fairly well understood, further work needs to be done on phase behavior of gas mixtures and their effects on compression and piping. Once the likely gas composition ranges from the capture process are defined, experiments and thermodynamic modeling can proceed to better predict possible adverse effects of impure gas streams and approaches devised to prevent them.

Bryant and Lake (Chapter 18) examined the possible subsurface implications of injecting CO_2 with impurities (e.g. SO_x, NO_x) into a saline formation (dissolution/precipitation affecting injectivity) and for CO_2 EOR. It was found that injecting CO_2 with impurities is unlikely to degrade injectivity even in the worst case scenario. Increased acidity from the nitric or sulfuric acid might even improve injectivity (temporarily). Impurities in CO_2 EOR injection are unlikely to affect performance as there is a tradeoff between lowering MMP and increasing the mobility ratio. The study suggests that CO_2 impurities (particularly, soluble species such as SO_x and NO_x) are not of particular concern in aquifer injectivity or EOR performance. Other gases such as N_2, however, would present operational difficulties and degrade performance.

The optimization studies provide direct industry analogs for safe and effective CO_2 injection. Simulations of CO_2 behavior in the subsurface document the rates and mechanisms of CO_2 immobilization. Reducing costs of CO_2 transportation and storage may become critical determinants in implementing CO_2 storage projects.

MONITORING

The monitoring studies were intended to examine the efficacy of a wide range of techniques, commercially available and under development, applied remotely, near the surface or in the subsurface.

Arts and Winthaegen (Chapter 19) conducted a broad survey of geophysical and geochemical monitoring techniques for the purpose of recommending "optimal" techniques for various CO_2 storage venues. Monitoring well technologies include pressure and temperature sensors, electrical resistivity, TDT, microseismic, VSP, crosswell seismic and fluid sampling. Surface geophysical methods include 4D seismic, subbottom profiling and sonar (marine), gravity, electromagnetics (EM), gravity, InSAR and tiltmeters. Geochemical monitoring includes groundwater sampling, tracer surveys, atmospheric detection and geobotanical hyperspectral remote sensing. The applicability of the various monitoring techniques was matched to specific features, events and processes (FEPs) such as those related to seal, casing/cement or well failure. The study provides a useful assessment of available technologies to monitor CO_2 leakage in a variety of settings and potential failure modes.

Atmospheric

In addition to surveying the state of the art in atmospheric monitoring systems, Shuler and Tang (Chapter 20) evaluate in detail the capability of various ground-based instruments to detect CO_2 leakage. The target detectable leak rate of 1% over 100 years (0.01%/year) was used as a base case. The detectability of leaks of this magnitude depends on the amount of leakage with time (flux), size of the affected area, mode of leakage (diffuse or point source) and atmospheric conditions. Currently available instruments can detect if the atmospheric CO_2 concentrations increase 10 ppm over background. Nomograms are used to predict the "excess" CO_2 present in the atmosphere for a given situation. Open path instruments (laser spectrometers) may be a cost-effective means of detecting small CO_2 leakage over a field-sized area (a few km^2). A spreadsheet application produced for the study permits matching of analytical instruments suitable for detecting CO_2 under various leakage scenarios.

The "eddy covariance" (micrometeorological perturbation) method, a technology used to establish baseline CO_2 flux from plant photosynthesis and respiration cycles, was evaluated for its applicability to CO_2 leak detection at the field scale by Miles et al. (Chapter 21). The technology is based on laser spectrometers mounted on towers (~10 m) that could be set up in an array at the field scale. This technology has been widely applied and is considered reliable and robust. Its applicability and expense should be compared with similar ground-based detection, given field size and the type and magnitude of CO_2 leakage.

The "hyperspectral geobotanical" remote sensing study by Pickles and Cover (Chapter 22) uses aerial data acquisition and processing to indirectly detect CO_2 leakage through CO_2 effects on plants and soils. Case studies include a satellite survey of the Mammoth Lakes, CA area where substantial volcanogenic leakage is known to have caused tree kills, and an aerial survey of Rangely Field, CO where low CO_2 leakage due to EOR operations is postulated. Hyperspectral images of Mammoth Lake and Rangely correlated well with ground-based CO_2 measurements and observations of vegetation effects. The Rangely Field surveys included pre- and post-rain images that showed marked differences in the (sparse) vegetation patterns but no obvious indications of CO_2 leakage. Detection of CO_2 leakage at Rangely Field will require further development and be mindful of the results of an independent Colorado School of Mines soil gas survey that showed little to no CO_2 leakage from the EOR operation (however, a possibly significant methane flux was detected). Additional processing and interpretation might reveal soil changes due to long-term CO_2 leakage and the location of hidden faults.

Geophysical and Geochemical Techniques
The "novel geophysical" monitoring study conducted by Hoversten and Gasperikova (Chapter 23) evaluates the resolution and applicability of seismic and non-seismic geophysical techniques to detecting CO_2 leakage. The Schrader Bluff, Alaska and S. Liberty, Texas reservoirs were used to model the spatial resolution of various geophysical CO_2 detection techniques. The significant changes in water with increasing CO_2 saturation might be detectable using seismic amplitude and AVO analysis. Ground-based gravity modeling shows that resolution is insufficient but might be improved with permanent sensor emplacement coupled with surface deformation measurements. Borehole gravity instrumentation emplaced up to 1200 ft above the reservoir might be sufficient to directly map the areas of net density changes caused by injecting CO_2 into water. The electrical resistivity changes attending CO_2 dissolution in water are easily detectable using EM techniques. This technique is currently available, inexpensive compared to seismic and most applicable to CO_2/brine systems. The streaming potential (SP) method has been successfully modeled in 2D for the Liberty Field and experimental results show promise. Unlike the other techniques, however, further developments in instrumentation and interpretation are needed. The novel geophysical techniques show considerable technical promise for CO_2 performance and leakage modeling whether by adding value to time-lapse seismic data or by development of inexpensive non-seismic techniques.

The utility and cost of using noble gas additives to monitor CO_2 movement and leakage in subsurface were studied by Nimz and Hudson (Chapter 24). The West Texas Mabee Field was used as a model for the study. Among the factors considered in selecting noble gases are cost, availability, subsurface transport characteristics and "distinctiveness" relative to the atmosphere and noble gases native to the reservoir. The Xe "system" (10 isotopes) was considered to meet these criteria. Given the volume of CO_2 injected into the reservoir and the detectability limits of the Xe isotopes, it is calculated that it would cost ~$0.18/tonne CO_2 stored to adopt this monitoring system for the Mabee field. Further work on the subsurface partitioning behavior of noble gases relative to CO_2 is a prerequisite of effectively applying this technique.

The monitoring studies have surveyed diverse techniques in various stages of development. Near-term application of ground-based techniques is feasible. Further development of other technologies is warranted as these techniques may not be universally applicable and considerable cost savings might result.

RISK ASSESSMENT

The risk assessment studies have evolved from earlier lessons-learned analyses of natural and industrial analogs to scenario development and modeling of specific elements of systems to whole system comprehensive methodologies.

HSE Analogs, Regulatory and Intervention/Remediation

Benson (Chapter 25) produced a comprehensive compendium of information relevant to CO_2 storage (directly or by analog) via experiences with deep well injection of industrial wastes, natural gas storage, geologic repositories for nuclear waste and other information. Human health and ecosystem responses to various levels of CO_2, which are the most immediate concerns associated with CO_2 capture, transportation, injection and leakage, are also addressed. The lessons learned are as follows.

1. There is an abundant base of experience to draw on that is relevant and suggests that CO_2 can be stored safely if storage sites are selected carefully and monitored (natural gas storage, deep injection of liquid and hazardous waste, enhanced oil recovery).
2. The human health effects of exposure to elevated concentrations of CO_2 are well understood and occupational safety regulations are in place for safe use (confined spaces, transportation, food additive, pipeline transportation). Ecosystem impacts of elevated CO_2 concentrations in soils are not as well understood and may need additional study.
3. The hazard presented by CO_2 depends more on the nature of the release rather than on the size of the release (volcanic eruptions, ecosystem fluxes, fire suppression, limnic releases).
4. Experience from industrial analogs predicts that the biggest risks from CO_2 storage will be from leakage from poor quality or aging injection wells, leakage up abandoned wells, leakage through poorly characterized cap rocks and result from inconsistent or inadequate monitoring that could have been used for early intervention.
5. Regulatory paradigms and approaches vary and none address all the issues that are important for CO_2 storage (leakage between geologic units, performance versus practice-based requirements, state versus federal regulatory oversight, short versus long-term monitoring).

Recommendations for risk management approaches include development of a single set of consistent regulations, identification and investigation of the effectiveness of multi-barrier concepts, development of well completion, abandonment procedures and methods and of a risk management strategy that couples monitoring requirements with performance confirmation. Risk mitigation and remediation methods should also be developed. The lessons-learned study was an early SMV contribution that guided selection of subsequent risk assessment projects.

Hepple (Chapter 26) surveys data on human health impacts and ecosystem effects from exposure to elevated CO_2 concentrations. CO_2 is ubiquitous in the environment and an essential part of all living things. Humans can tolerate up to 1% concentration without suffering adverse effects. Exposure to concentrations of 3% begins to have acute, but reversible, physiological effects. Concentrations of greater than 10% can lead to death. Regulatory guidelines have been established for the safe use of CO_2. Ecosystem impacts due to high soil gas CO_2 concentrations are not well understood and more information on potential impacts may be needed.

The evolution and status of US federal and state laws and regulations to protect underground sources of drinking water from industrial and municipal wastes and their likely applicability to CO_2 storage are discussed by Apps (Chapter 27). Application of Type I well standards may be used for CO_2 injection. Under the Type I classification, CO_2 injection could be classified as non-hazardous, unless impurities such as H_2S and Hg are present. The buoyant character of supercritical CO_2 in the subsurface, however, would present a containment risk that is not addressed by current Type I well regulations governing disposal of hazardous waste. A new category of injection well, designed specifically for CO_2 storage projects may be desirable to ensure safe and effective storage, while facilitating the application of this technology.

Early detection and remediation of CO_2 leakage from storage sites is an understudied topic that Benson and Hepple (Chapter 28) address. The objective of this scoping study was to identify (1) monitoring approaches for early detection of CO_2 leakage, (2) remediation options that could be used to eliminate or manage risks after leakage has been detected and (3) additional information and R&D necessary to develop new remediation approaches. Scenarios for CO_2 leakage from storage sites include damaged injection wells, over-pressured reservoirs and accumulation in groundwater. The consequences of leakage include groundwater and surface water contamination by acidification and toxic element mobilization, vadose zone accumulation and surface releases. Remediation options applicable to leaking CO_2 storage projects are available from natural gas storage, oil and gas production, groundwater remediation and soil gas/vadose zone clean-up experience. HSE concerns become relevant not only for large leaks but also for chronic small leaks that may cause CO_2 to accumulate in structures. The study establishes a framework from which CO_2 leakage scenarios can be developed for specific storage sites and outlines technologies needed to manage such leaks and lessen their consequences. A site-specific plan that includes such contingencies will be essential for acceptance of CO_2 storage by NGOs, regulators and the public.

A coupled modeling framework has been developed by Oldenburg and Unger (Chapter 29) to simulate CO_2 leakage and seepage in the subsurface and atmosphere for risk characterization. The coupled model framework is built on the integral finite difference multi-phase and multi-component reservoir simulator (TOUGH2), and models CO_2 and air in both subsurface and atmospheric surface layer regions simultaneously. The model is demonstrated for a coupled subsurface–surface layer system and shows that seeping CO_2 can reflux into the subsurface as a dissolved component in infiltrating rainwater. Whereas CO_2 concentrations in the subsurface might be high, surface layer winds act to reduce CO_2 concentrations via dilution to low levels for the fluxes investigated (e.g. the Rio Vista, CA area which is characterized by strong persistent winds). High CO_2 levels persisting in the vadose zone, however, are a threat to ecosystems and for humans occupying poorly ventilated, low lying structures. The coupled subsurface–surface leakage and seepage modeling framework is likely to attract the attention of stakeholders in proposed CO_2 storage projects as the behavior of CO_2 at the surface is of the most immediate concern.

Onstott (Chapter 30) assessed potential impacts of CO_2 injection on subsurface organisms. The deep biosphere extends to $\sim$3.5 km with decreasing number of organisms with depth. These organisms are primarily methanogens, sulfur and iron reducers, and fermentative anaerobes. Genetic testing (16S rDNA) suggests that only about one-third of these subsurface organisms have been identified. By defining microbial assemblages and determining “microbial power” (free energy of redox reactions and availability of nutrients), a forward model is used to predict the impact of CO_2 injection on microbes in different environments (reservoir lithologies, ground water types) over three reservoir temperatures and constrained pCO_2 and pH/Eh. Fe (III) reducers and fermentative anaerobes are not favored by the presence of CO_2 but there is an increase in methanogenesis and acetogenesis. In general, the impact on microorganism’s growth in carbonate systems is expected to be most significant.

Risk Assessment Methodologies

A methodology for risk assessment was developed by Stenhouse et al. (Chapter 31) for the IEA Weyburn Monitoring and Storage Project to determine the long-term fate of CO_2 injected into the reservoir. An interdisciplinary effort involving geology, hydrogeology, geochemistry, geomechanics, reservoir modeling and wellbore technology was made to assess the potential for CO_2 migration, via natural and artificial pathways, from the reservoir to the environment. The core of the long-term assessment is the systems analysis approach which includes definitions, development of internal, external and inter-relating FEPs, construction of scenarios and description of how FEP–FEP interactions will be accommodated in the consequence analysis modeling undertaken for each scenario. The results of this in-progress study are expected to quantify the length of CO_2 containment in the system and potential consequences of containment loss.

Risk assessment has become a critical issue for advancing CO_2 transportation and storage. Although the behavior and HSE impacts of CO_2 leakage are understood, the challenge is to predict the likelihood and impact of such leakage at specific sites. Further research and development on intervention and remediation technologies are needed to assure avoidance of leaks and effectively treat those that might do occur.

Wo et al. (Chapter 32) have developed a mathematical model for probabilistic risk assessment for the Tiffany Field, CO which is presently under N_2 flood for ECBM production. The risk assessment methodology includes four major elements (hazard identification, event and failure quantification, predictive modeling, risk characterization) and the mathematical model includes six functional constituents (initiators, processes, failure modes, consequences, indicators and inference queries). To demonstrate the applicability of the methodology and model, a prototype application, capable of performing scenario and Monte Carlo simulations, was developed in Microsoft Access™. The geomechanical study revealed processes that lead to risks of developing leakage paths at each step of CO_2 storage in coal beds. It was found that risk of leakage is higher for old wells that were converted to injectors and that the most likely mechanism of leakage path formation is slip on preexisting discontinuities cross-cutting coal seams. Predictive quantitative modeling demonstrates that elevated pressure resulting from N_2 injection caused the coal fractures on the preferred permeability trends to expand and extend from injectors to producers. This could increase the risk of early gas (N_2 plus CO_2) breakthrough and under certain conditions the risk of CO_2 seepage from the outcrop is increased if CO_2 injection is placed within 2 miles of an outcrop. The importance of evaluating the effects of processes employed prior to CO_2 injection (e.g. coal bed depressurization and dewatering, N_2 injection) on CO_2 movement is highlighted. Further testing of the methodology on additional, candidate CO_2 storage venues and benchmarking with other risk assessment models will strengthen the application and make it more universally accepted by regulators and the public.

Wildenborg et al. (Chapter 33) have developed a comprehensive methodology for long-term safety assessment of underground CO_2 storage that is available for application. The three basic components of the methodology are: (1) scenario analysis, which includes a comprehensive inventory of risk factors or FEPs that are selected as appropriate to a given venue, (2) model development, which enables a quantitative safety assessment and (3) consequence analysis. A performance assessment (PA) model based on the large number of simulations with physical models comprised of multiple compartments has been developed. The PA model is capable of a statistical analysis that predicts CO_2 concentrations and fluxes in the biosphere, and therefore established whether or not they are likely to exceed acceptable levels. The methodology has been applied to a reference scenario (combined on- and offshore case, The Netherlands). The scenario was run without mitigation efforts and therefore represents a worst case scenario. The results showed that seepage of CO_2 to the biosphere would not occur in the 10,000 year timeframe simulated for all 1000 parameter realizations considered. Further development of the surface (hydrosphere and atmosphere) components and benchmarking with other risk assessment models will improve its reliability and acceptance by regulators, NGOs and the public.

CONCLUSION

The CCP geological storage program addressed many of the technical gaps evident at its inception. Future work should aim to integrate concepts, models and simulation into a comprehensive methodology for storage site assessment, process optimization, near- and long-term monitoring and verification strategies and credible risk assessment protocols.

SECTION 2:
STORAGE INTEGRITY

STORAGE INTEGRITY PREFACE

Curtis M. Oldenburg

Lawrence Berkeley National Laboratory, Berkeley, CA, USA

One of the fundamental concepts behind geologic capture and storage of carbon dioxide (CO_2) is that injected CO_2 will be isolated from the atmosphere and separated from the near-surface environment by virtue of its emplacement within deep, low-permeability geological formations. However, while this concept is intuitively appealing, the capacity of geologic formations to provide a seal against upward migration over large lateral areas and for long periods of time is by no means guaranteed in every geological environment—and seal effectiveness depends on a wide range of processes and features. For example, in areas of active volcanism or tectonics, igneous intrusion, faulting and associated crystal deformation generally cause an increase in large-scale vertical permeability through fracturing of otherwise low-permeability rock, an environment to be avoided in selecting geologic CO_2 storage sites. In general, heterogeneity of subsurface formations suggests that the sealing capacity of low-permeability formations will vary spatially, thereby complicating the overall integrity of sealing formations over large areas. Finally, man-made features such as wells and the associated geomechanical effects of historical hydrocarbon extraction may also affect seal integrity.

On the other hand, fluid migration in the deep crust is closely associated with mineralization reactions that can cause effective plugging (e.g. vein-filling) or enhancement (e.g. dissolution) of fracture permeability. These effects create the potential for dynamic changes in seal integrity as geochemical reactions occur during fluid migration. Clearly, a diverse set of coupled processes acting over a wide range of timescales is relevant to questions about the integrity of geologic formations relative to CO_2 containment.

Given the wide range of processes and features that can affect CO_2 storage integrity, the CCP initiated and supported a multifaceted research program to build the understanding of the potential flow and reactive transport of injected CO_2 in the deep subsurface. Fortunately for the nascent field of geologic CO_2 storage, scientists can transfer a considerable amount of knowledge and research experience in industrial hydrocarbon extraction and storage operations to the geologic storage field to answer questions about CO_2 storage integrity.

The summary chapters collected here report on a diverse set of research efforts, focused on the common goal of assessing the integrity of deep geologic storage sites for sequestering CO_2. These chapters range from review of industrial experiences with underground gas storage operations, to reports of cutting-edge field, laboratory and simulation studies. In the first two chapters, the focus is on natural CO_2 reservoirs, systems that can provide information on what features control whether a system will leak and what features will create a sealed system. The third chapter presents results of experience in characterizing and evaluating the North Sea Forties oil field for containing CO_2 that could be injected for CO_2 enhanced oil recovery (EOR). The next three chapters report on laboratory, theory, modeling and field experience of geochemical and geomechanical issues at the heart of storage integrity. The seventh chapter presents the results of a survey of gas-storage operators, reporting on the experience of industry in underground natural gas storage with respect to methane leakage. Finally, the last chapter presents the importance of man-made wells and boreholes for seal integrity.

From this wide array of research on storage integrity, the impression arises that while questions remain, there already exists a foundation of understanding upon which reasonable projections of storage integrity can be made. From these research efforts, recommendations for future research areas are laid out by the authors in each chapter. Ongoing and future research into these and other areas will lead to further understanding of long-term and large-scale storage integrity, an understanding that is critical to moving forward with geologic CO_2 storage.

Carbon Dioxide Capture for Storage in Deep Geologic Formations, Volume 2
D.C. Thomas and S.M. Benson (Eds.)

Chapter 3

NATURAL CO_2 FIELDS AS ANALOGS FOR GEOLOGIC CO_2 STORAGE

Scott H. Stevens

Advanced Resources International, Inc., Arlington, VA, USA

ABSTRACT

Our study evaluated three underground gas fields in the USA that have been effective CO_2 traps for millions of years: the Jackson, McElmo, and St. Johns Domes. Together, these fields stored 2.4 billion t of CO_2, equivalent to more than 1 year of USA power plant emissions. Because CO_2 in these fields has been commercially extracted for industrial uses, the fields offer data on natural CO_2 reservoirs, cap rocks, and production operations. M0cElmo Dome, the largest and most important field, originally stored 1.6 billion t of supercritical CO_2 within a carboniferous carbonate reservoir at a depth of 2300 m. Carbon isotope data indicate the CO_2 originated from a nearby igneous intrusion dated to 70 Ma. Its cap rock is a 400-m thick sequence of salt (halite), which is finely layered and unperturbed by faults which cut the underlying reservoir; there is no evidence of CO_2 leaking into the overlying strata. McElmo Dome has two decades of safe operational history. It currently produces 15 million t/year (800 MMcfd) of 99%-pure CO_2, which is transported 900 km via pipeline to depleted oil fields for re-injection and enhanced recovery. However, the three fields in our study represent a small sampling of geologic situations, insufficient for defining universal criteria for cap rock integrity. Building scientific and public acceptance for geologic CO_2 storage may be facilitated if proposed projects each had a local or regional natural analog.

INTRODUCTION

Geologic storage has been proposed as a promising option for reducing net emissions of CO_2. But is geologic storage a safe and long-term disposal option? Since the early 1980s, in the USA and several other countries, CO_2 has been injected on a large scale into depleted oil fields for enhanced oil recovery (EOR). The safety record of this activity has been excellent and industry's two decades of experience with EOR represents an invaluable tool for assessing the near-term performance of geologic storage projects [1].

However, the long-term safety and performance of geologic storage is still unknown. This effectiveness still must be quantified to demonstrate storage feasibility as well as to win public acceptance [2]. One approach is to numerically simulate the flow and storage of CO_2 in candidate storage sites. This approach requires an extremely large data set on reservoir properties, as well as upgrading simulation codes to better model long-term geochemical reactions, but is only feasible at well-documented depleted oil and gas fields [3,4].

A parallel empirical approach, taken by this and several other studies, is to examine sites where large volumes of nearly pure CO_2 have naturally accumulated and have been stored in geologic formations over a demonstrably long time period (millions of years). These naturally occurring CO_2 deposits provide unique natural analogs for evaluating the long-term safety and efficacy of storing anthropogenic CO_2 in geologic formations. CO_2 has been trapped for millions of years in reservoirs with effective cap rocks, such as thick salt or shale deposits. In other settings, CO_2 springs and fluxes developed where cap rocks were breached or faulted. Understanding why certain natural geologic settings are effective CO_2 traps while others are not can help guide the screening and designing of engineered sites for CO_2 storage. Production operations at CO_2 fields also provide proven and low-cost technologies applicable to engineered geologic storage sites. These natural analogs offer unique natural laboratories for studying the long-term storage of

CO_2 underground and can help in the screening of candidate sites for geologic storage. Natural analogs in the Colorado Plateau (USA), Europe, and Australia are currently undergoing study in separate research projects [5–7].

Our study focused on three commercial CO_2 fields in the USA. The petroleum industry has been exploiting natural CO_2 fields for over two decades, yet little technical information has been published about this activity [8]. Our study had three objectives that are consistent with CCP's goals of understanding geologic storage and developing long-term, cost-effective verification and monitoring technologies:

- *Establish CO_2 storage as a natural process.* Studying natural analogs documents that CO_2 storage in geologic formations is indeed a natural process in many geologic settings.
- *Document long-term impacts of CO_2 storage.* More convincingly than any model or laboratory experiment, natural analogs can demonstrate empirically the long-term chemical and physical interactions of CO_2 with reservoir rocks and fluids. Dating the emplacement of non-leaking CO_2 deposits can uniquely establish the integrity of geologic storage over very long time periods (thousands to millions of years).
- *Assess surface and subsurface CO_2 handling technologies.* Many of the production, monitoring, and safety techniques and facilities developed by the commercial CO_2 production industry can be adapted for long-term geologic storage of CO_2. These technologies and their costs have never been comprehensively documented.

METHODOLOGY

To conduct this study, we assembled geologic and engineering data from each of the three fields into a geographic information system database for mapping and analysis. We also conducted gas sampling and analyzed molecular composition, as well as stable and noble gas isotopes. Figure 1 shows the location of the three fields, while Table 1 provides a summary of each site's characteristics. This section discusses the key aspects of our study, including: geologic setting; CO_2 origin, timing, and storage; cap rock integrity; production operations; and implications for geologic storage for each of the three natural analog fields.

Figure 1: Location map of the three CO_2 fields.

RESULTS AND DISCUSSION

Geologic Setting

St. Johns Dome

The St. Johns Dome is a large (1800 km^2), asymmetrical, faulted anticline located on the southern part of the Colorado Plateau in east-central Arizona and west-central New Mexico [9]. CO_2 is trapped within

TABLE 1
SUMMARY OF NATURAL CO_2 FIELD STUDY SITE CHARACTERISTICS

Field	State	Operator	Original CO_2 in place		2003 CO_2 production		Reservoir lithology	Depth (m)	Cap rock	Years stored
			10^6 t	Tcf	10^6 t/year	MMcfd				
St. Johns	AZ	Ridgeway	730	14	0.02	1	Sandstone	500	Anhydrite	0–6 Ma?
Jackson Dome	MS	Denbury Resources	100	2	3.5	185	Sandstone Some carb	4700	Carbonate	70 Ma?
McElmo Dome	CO	Kinder Morgan	1600	30	15	800	Carbonate	2300	400 m salt	70 Ma?
All 3			2430	46	18.5	986				

sandstones of the Permian Supai Formation. Overlying and intercalated evaporite deposits (anhydrite, gypsum) and shales act as cap rocks and local seals (Figure 2). The Supai Formation thins over the dome, concordant with the structure, demonstrating that the dome began as an older (280 Ma) basement structure which later intensified during the Laramide Orogeny (Cretaceous, 65 Ma). The St. Johns field is 30 km northeast of the Springerville Volcanic Field (SVF), a large Plio-Pleistocene (0.3–5 Ma) igneous feature, but we have no data directly linking the two.

Jackson Dome
The Jackson Dome is an igneous intrusion of Late Cretaceous age (70 Ma) located in central Mississippi. Numerous CO_2 deposits occur on the eastern flank of this structure. The largest is the Pisgah Dome CO_2 field, a symmetrical, faulted anticline located in the onshore Gulf of Mexico province [10]. CO_2 is trapped within sandstone and carbonate reservoirs of the Jurassic Buckner, Smackover, and Norphlet Formations by structural closure and permeability barriers.

McElmo Dome
The McElmo Dome is a large (800 km^2) anticline located on the Colorado Plateau in southwestern Colorado. CO_2 is trapped within the Carboniferous (Mississippian) Leadville Limestone [11]. McElmo Dome is only a few km north of the Sleeping Ute Mountain laccolith, a large dacitic igneous intrusion dated at 70 Ma, which may be the source of the CO_2 deposit, as discussed below.

CO_2 Storage
As part of this project we sampled CO_2 production wells at each of the three study fields and analyzed the gases for major chemical composition and stable carbon isotopes. Although additional noble gas analyses are underway, early results are presented here.

St. Johns Dome
The Supai Formation contains an estimated 730 million t (13.9 Tcf) of CO_2 which, due to its relatively shallow depth (300–750 m), is stored in a free gas state. The field's reservoir architecture is complex, with multiple, vertically dispersed reservoirs consisting of sandstone, siltstone, and vuggy dolomite (porosity 11–20%, permeability 0.5 to >100 mD) that are separated by thin, impermeable anhydrite seals ($k < 0.010.25$mD). CO_2 concentrations vary from 83 to 99%, averaging 92%. Other constituents include nitrogen (N_2: 6.6%), argon (Ar: 0.2%), and commercially significant quantities of helium (He: 0.6%).

Jackson Dome
The Pisgah anticline originally contained an estimated 100 million t (2 Tcf) of CO_2, making it the smallest of the three study sites. With a reservoir depth of about 4700 m, the CO_2 is stored in the supercritical state. Its reservoir architecture is complex, with fluvial and eolian sandstones with 8–15% porosity and up to 1 D

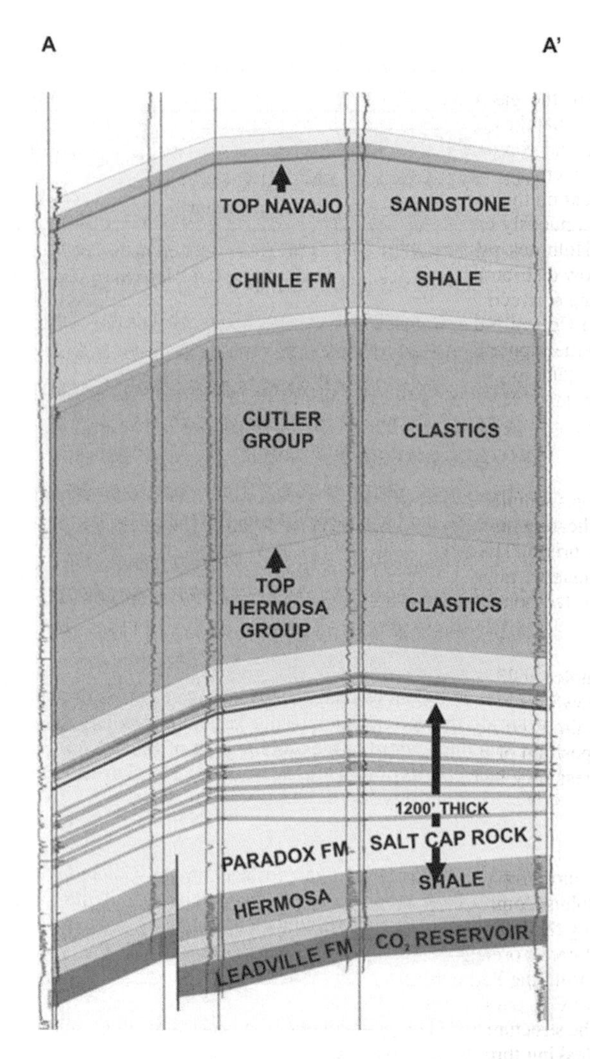

Figure 2: Structural cross-section of the CO_2 reservoir and cap rock at McElmo Dome.

permeability. The CO_2 concentration averages 99%, with minor methane (CH_4), N_2 and significant hydrogen sulfide (H_2S) of up to 1%.

McElmo Dome

The Leadville Formation originally contained 1.6 billion t (30 Tcf) of CO_2 stored in a supercritical state at a depth of 2300 m. Reservoir architecture is complex, with interbedded dolomite (porous, permeable) and limestone (impermeable) capped by an erosional unconformity. The reservoir porosity averages 11% and permeability 20 mD. The CO_2 concentration of this deposit ranges from 96 to 98%, along with minor N_2 (1.6–2.2%), CH_4 (0.2–0.9%), and H_2S (0–15 ppm). Traps are provided by structural closure, permeability barriers in the Leadville, the water/CO_2 contact, and a 400-m thick salt cap rock; faults in the Leadville die out in the lower portion of this salt cap rock.

CO_2 Origin

St. Johns Dome

$\delta^{13}CO_2$ values within the gas were uniform at the three wells we sampled across the St. Johns field (−3.8‰), suggesting that the CO_2 was generated from a single source or well-mixed multiple sources and that internal barriers and compartmentalization are minimal. Major gas composition within the reservoir exhibits significant reverse gravity segregation, with heavier CO_2 concentrated at the crest and lighter He and N_2 more prevalent on the northern flank. There are two possible explanations. The more likely is that CO_2 and He are continuously emanating from beneath the halite–anhydrite boundary at the southeastern edge of the nearby Holbrook salt basin, entering the northwest portion of St. Johns Dome. This origin is supported by heat flow distribution, which is low over the salt basin and high (>100 mW/m^2) over the St. Johns field, suggesting convective flow. Another possible explanation for the geochemical trends is that the lighter (and smaller) He and N_2 components have preferentially escaped from the crest of the structure, leaving behind extremely pure (99%) CO_2. This is less likely given that all wells sampled have similar overburden thickness (500 m), so the flank wells should be just as likely to leak (or not leak) He as the crest. The SVF is another potential CO_2 source, but we have not yet found evidence of a direct connection with St. Johns Dome, as there appears to be with the Holbrook salt basin.

Jackson Dome

$\delta^{13}CO_2$ values from gas sampled in 10 wells range from −3.55 to −2.57‰. The $^3He/^4He$ ratio ranged from 4.27 to 5.01 Ra, indicating strong mantle signature. The $^4He/^{40}Ar$ ratios range from 1.26 to 2.52, also indicative of mantle origin. These noble gas isotope data demonstrate that the CO_2 at Pisgah Dome was outgassed from the mantle, rather than derived from thermal decomposition of carbonate [12]. The most likely source was the Jackson Dome intrusion.

McElmo Dome

$\delta^{13}CO_2$ from gas sampled at 28 wells within the field are quite uniform (−4.3 to −4.5‰), demonstrating no significant internal flow barriers or compartments. However, a subtle gradation is apparent, emanating away from the Ute Mountain laccolith. The CO_2 likely formed by direct outgassing from Ute Mountain rather than thermal decomposition of the Leadville Limestone (which has $\delta^{13}CO_2$ value of −0.64‰). Our noble gas analysis in progress may help to resolve this uncertainty.

CO_2 Timing

St. Johns Dome

Noble gas analysis currently underway is the best hope for resolving the origin and timing of CO_2 emplacement at St. Johns Dome. Our geologic mapping suggests that the earliest likely storage of CO_2 was immediately following the Laramide Orogeny (65 Ma) that generated the current structural closure. Given the thinning of the Supai Formation over the dome, it is even possible that a modest structural closure existed as early as during the Permian (280 Ma). On the other hand, there is no data establishing the most recent possible time of CO_2 emplacement. CO_2 and other gases could even be continuing to fill the St. Johns Dome, overspilling the structure and charging the overlying Glorieta Sandstone and San Andres Limestone, without necessarily leaking through the cap rock.

Jackson Dome

Our noble gas data, previously cited, suggest that timing was coeval with the Jackson Dome intrusion, which is dated to about 70 Ma.

McElmo Dome

Under either CO_2 origin scenario (outgassing or decomposition), the nearby Ute Mountain laccolith (70 Ma) is the most likely source for CO_2 emplacement at McElmo Dome.

Cap Rock Integrity

St. Johns Dome

The multiple but thin, mainly anhydrite cap rocks at St. Johns Dome—while a reasonably good seal for preserving commercial quantities of CO_2 and He—may not form as thorough a long-term seal as the thicker halite cap rocks at McElmo and Jackson Domes. Four wells at the field encountered voids (karsts?) or lost circulation while drilling through the San Andres Formation and had to be abandoned. CO_2 is widely present (in non-commercial quantities) in the overlying Permian Glorieta Sandstone and San Andres Limestone, entering these formations either by gradual seepage through the cap rock matrix porosity, or by overspill and lateral migration, or migration along fault planes. On the other hand, the presence of He, a light and small molecule particularly prone to leakage, in high concentrations (up to 1.1%) indicates that the cap rock seals over the north flank of the field. But the low He concentration (0.1%) at the crest of the structure, the reverse expected under gravity segregation, suggests that it and N_2 may have preferentially leaked through the cap rock, leaving behind nearly pure (99%) CO_2. Detailed sampling and analysis of noble gases in the reservoir, along with soil gas analysis above the field, is needed to fully evaluate cap rock integrity.

Jackson Dome

Sudden geopressuring with depth—50% higher than the hydrostatic gradient—strongly suggests that the Bruckner Carbonate is an excellent cap rock seal to underlying CO_2 reservoirs. However, this cap rock has not been cored, usually is not fully logged, and thus remains poorly characterized.

McElmo Dome

The 400-m thick halite unit above the Leadville CO_2 reservoir apparently has acted as an excellent cap rock for millions of years. There is no significant evidence of CO_2 locally above the Leadville or in the ground water. Faults that cut the Leadville die out in the lower portion of this salt layer, as indicated by thinly layered shales within the salt unit that are unaffected by faulting (Figure 2). However, detailed sampling and analysis of noble gases in the reservoir, along with soil gas analysis above the field, is needed to fully evaluate cap rock integrity.

CO_2 Production Operations

This section is based on internal company documents and operating procedures discussed in Ref. [13] or not previously documented.

St. Johns Dome

Ridgeway Petroleum Corp., the field operator, has drilled 21 CO_2 production wells since discovering the field in 1994. At present, due to limited local market demand and lack of a CO_2 pipeline, only one well is on line producing approximately 50 t/day (1 MMcfd). The production wells were drilled with air or with a fresh water and starch-based mud to avoid formation damage and were completed in one or more of the three CO_2-bearing zones (Ft. Apache, Amos Wash, and Upper Abo/Granite Wash) at an average depth of 850 m. The wells were completed using 11.4-cm diameter casing, consisting of amine carbon gauze fiberglass or conventional carbon steel lined with high-density polyethylene (HDPE) to minimize corrosion. The St. Johns wells were considerably less expensive to drill and complete ($300,000) than the deeper wells at McElmo and Jackson Domes, making them good analogs for shallow geologic storage projects. Since discovery in 1994, CO_2 exploration and production activities at the field have been accident free. Given the extremely low population density at the field (<0.1 residents/km^2), impacts on the natural and human environment have been negligible.

Jackson Dome

Field operator Denbury Resources Inc. currently produces about 3.5 million t/year (185 MMcfd) of CO_2. Production wells, on 2.6 km^2 spacing, are equipped with up to 10,000-psi working pressure wellheads, which can be operated remotely by a control station (Figure 3). Stainless steel is used for the production casing and downhole fittings, the high-pressure wellheads, and surface flow lines to the central facility. Production costs (including depreciation and amortization) average about $0.007 m^{-3} ($0.20 Mcf^{-1}). Produced CO_2 is dehydrated (<0.27 kg H_2O/tCO_2), compressed, and transported to EOR fields via a 293-km, 50-cm diameter carbon steel pipeline with no internal protective coating. It is maintained at a supercritical state at all times to preclude hydrate formation.

McElmo Dome

Shell, the original field operator, and current operator Kinder Morgan have drilled a total of 60 CO_2 production wells since 1976. Currently the field produces 15 million t/year (800 MMcfd) of CO_2 from 41 wells. Early wells were completed using perforated carbon steel casing across the Leadville production zone, with high-chromium steel (13% Cr) production tubing to convey CO_2 to the surface. Recent wells

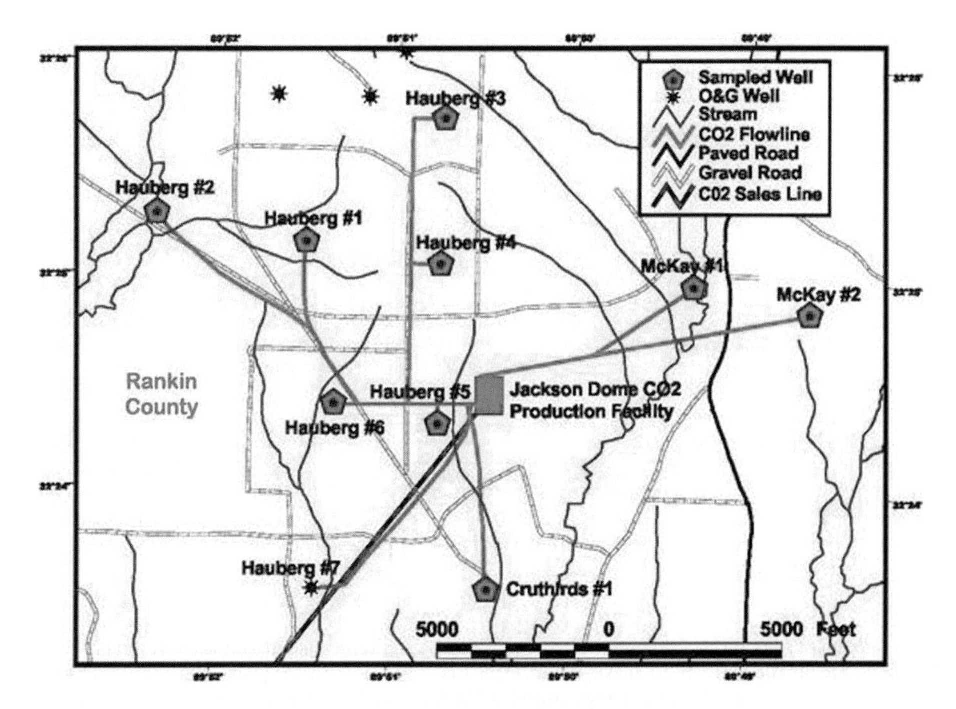

Figure 3: CO_2 gathering system at the Pisgah field, Jackson Dome.

employ tubingless completions using 17.8-cm chromium steel casing to increase per-well production (up to 1600 t/day) while lowering cost. A field-wide supervisory control and data acquisition (SCADA) system provides communications and control of the facilities from the Cortez field office, including capability to remotely open/close wellhead and cluster shutdown valves, compressors, and the central facilities. Processing facilities reduce water content of produced gas to <0.09 kg H_2O/tCO_2 (10 lb/MMscf). The dry CO_2 is compressed to 14.5 MPa (2100 psi) in two-stage, electrically driven compressors. The dry, supercritical CO_2 is cooled to 16–38 °C and then transported via the 800-km, 76-cm carbon steel Cortez pipeline to EOR projects in the Permian basin. Since commercial production began in 1983, over 235 million t (4.2 Tcf) of CO_2 has been produced with no safety or environmental incidents (Figure 4).

Implications for Geologic Storage

St. Johns Dome

St. Johns Dome is a large but operationally immature CO_2 field. It is not yet as well defined as the Jackson or McElmo Dome CO_2 fields. In addition, the thin anhydrite cap rocks, the large bounding fault that reaches to surface, and the presence of CO_2 in groundwater across this fault (but less so apparently directly above the field itself) all suggest that cap rock integrity may be somewhat less secure than at the two other sites (or, alternatively, that too much CO_2 was generated causing overspill of the structure into the adjoining fault block). On the other hand, St. Johns is the only field studied that traps significant He, a highly fugitive molecule. The field is a good analog for storage sites with cap rock uncertainty, as well as shallow settings where gas is stored in the free state.

Jackson Dome

Jackson (Pisgah) Dome is significant in that it securely contains CO_2 at extreme pressure, 50% above the hydrostatic gradient. Yet, its cap rock is neither salt nor shale, but rather carbonate. Jackson Dome is a good

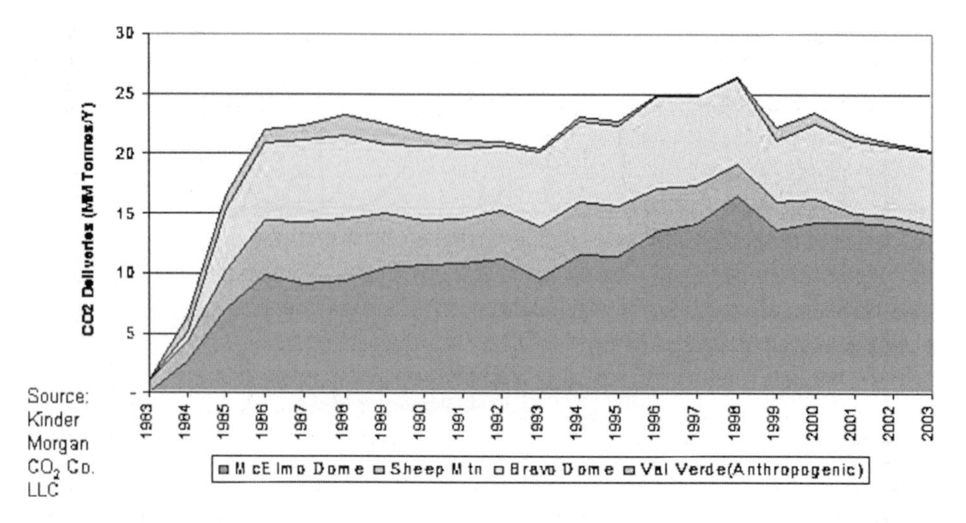

Figure 4: Production from McElmo Dome and other CO_2 fields.

analog for Gulf of Mexico area storage sites. The challenging and somewhat dangerous operating conditions (deep, overpressured, toxic H_2S) make its accident-free operation an excellent example of industry's capability to handle relatively tame storage projects (which are likely to be developed in low-pressure, well-characterized reservoirs).

McElmo Dome

McElmo Dome is the largest and operationally most mature commercial CO_2 field and possibly the best analog for future geologic storage sites, particularly in the Colorado Plateau. CO_2 has been stored at this field for approximately 70 million years, implying that geologic storage can be sufficiently long term in favorable settings. The porous, permeable dolomitized carbonate reservoir is continuous and CO_2 floats atop a regional aquifer; in this regard it is an excellent model for storage reservoirs. The thick salt cap rock at McElmo Dome appears to resist and accommodate faulting. The field's 20-year safe and environmentally sound operating record provides a good foundation for permitting storage sites. The field's tubular, cement, and monitoring technologies are appropriate for the several-decade field life, but upgrades would be needed for long-term ($>$10,000 year) CO_2 storage.

CONCLUSIONS

1. CO_2 accumulation is a natural process in many geologic settings. Prior to being developed, the three study sites stored 2.4 billion t (46 Tcf) of CO_2, equivalent to over 1 year of power plant emissions in the USA. They are comparable in size to the largest proposed individual storage sites. This evidence provides justification for industrial-scale geologic storage as an environmentally compatible GHG mitigation option.
2. Reliable reservoir seals require evaporites, shales, or low-permeability carbonates as the cap rock. Complementing parallel modeling and laboratory studies, study of natural analogs demonstrates empirically that, in favorable settings, CO_2 has been stored essentially "forever" (on human timescales; possibly 70 million years at McElmo and Jackson Domes) with no major adverse impacts on reservoir and cap rock. Thick salt cap rocks (such as the 400 m of halite at McElmo Dome) appear nearly impermeable and self-sealing to faults over geologic time in tectonically stable locations. Anhydrite (St. Johns Dome) and carbonate (Jackson Dome) also can be highly effective cap rocks. Remarkably, Jackson Dome's cap rock has contained excess pressures 50% above hydrostatic levels, probably for millions of years. We recognize that every geologic setting is unique and it is not

realistic to formulate universal criteria for cap rock integrity based on our limited study. Nevertheless, this information can provide guidelines useful for screening candidate CO_2 storage sites, particularly in similar geologic settings. To build confidence, early storage site selection would benefit from CO_2 field analog characteristics, such as the presence of thick and secure evaporite, shale, or carbonate cap rocks.

3. Natural CO_2 production practices provide valuable "lessons learned" for CO_2 storage. During the past two decades, the commercial CO_2 production industry has developed safe and cost-effective CO_2 production, processing, monitoring, and safety techniques and equipment that can be adapted for long-term storage of CO_2. The study fields are collectively producing 18.5 million t/year (986 MMcfd) of CO_2 for commercial use, mainly EOR. Corrosion control is achievable with chromium steel, carbon steel with amine carbon gauze coatings, batch corrosion inhibitors (e.g. $NaHCO_3$), or cathodic protection of flowlines; corrosion surveillance using boroscope, ultrasound, weight-loss coupons, and other methods. Wireline-set plugs downhole can automatically shut in the well in case of accidental damage to the wellhead. However, certain components (e.g. well casing, cements, etc.) would need to be upgraded to withstand the much longer time scale required for geologic storage projects. For example, CO_2-resistant cements may be adequate for short-term applications (decades), but require advancements to withstand the longer lifespans of geologic storage.
4. Efficient operation of CO_2 storage will require its own set of practices and technologies. Despite the encouraging evidence and lessons learned of long-term secure CO_2 storage at the three study fields, future geologic storage sites will differ in several important respects. For example, a depleted oil and gas field will have significant remaining hydrocarbons, whereas the studied natural analogs are essentially pure CO_2 with minimal contaminants. Also, the natural accumulation took many thousands of years to fill, yet storage sites may inject CO_2 for a few decades or less.

RECOMMENDATIONS

The three natural CO_2 fields assessed in this study have yielded considerable information relevant to long-term anthropogenic CO_2 storage, including storage capacity, storage period, cap rock type and other factors (Table 2). However, there are areas where natural analogs fail to provide needed data for evaluating geologic storage, such as the impact of rapid fill rates or long-term well cementing technology. Furthermore, the three natural analogs alone cannot prove the case for safe, long-term storage in every geologic province. Additional work identified by this study that could help advance CCP's goals in this area include:

- Develop a worldwide database of natural CO_2 deposits to help identify geologic provinces that are particularly suitable for long-term storage, as they already have demonstrated natural CO_2 trapping. The database also would provide a set of storage analogs that could be used to evaluate the potential effectiveness of projects in similar formations and structures.
- Profile other natural analogs in high-priority storage basins located near major anthropogenic CO_2 sources (such as Appalachia, Alaska, the Middle East, Russia, China, Southeast Asia, etc.). Even if they are smaller deposits or have lower CO_2 concentrations, they are more likely to closely resemble local storage projects and thus could help strengthen scientific and public confidence.
- The CO_2 fill rate of natural analogs was probably very slow (thousands, perhaps millions of years) compared to the decades likely for engineered storage sites. A well-characterized depleted natural CO_2 field (e.g. McElmo Dome) should be simulated to model the efficiency and safety of rapid re-fill rates, including hysteresis effects and tensional stress changes on the cap rock.
- Natural CO_2 field cap rocks are not normally cored, thus there is little direct data on their composition, texture, fracturing, and chemistry that make them such excellent seals. The cap rock of a well-characterized depleted CO_2 field should be cored for detailed analysis.
- Soil gas analysis has not been performed at the study sites, yet this information could help confirm or disprove cap rock integrity.
- Develop new CO_2-resistant cements designed to withstand exposure for $> 10,000$ years, rather than the current time scale of decades.

TABLE 2
SUMMARY OF KEY SIMILARITIES AND DIFFERENCES BETWEEN NATURAL CO_2 FIELD ANALOGS AND FUTURE GEOLOGIC SEQUESTRATION SITES

Factor	Natural analog	Sequestration site	Assessment	Work needs
Storage capacity	0.1–2.4 Gt	Comparable	Good analog	None
Storage period	Millions of years?	$>10,000$ years	Good analog	Noble gas analysis
Cap rock	Salt best; shale or anhydrite good	Comparable	Good analog	Coring and characterization of cap rock at natural analog sites
Fill rate	Slow	Fast	Poor analog	Model re-filling a depleted CO_2 field
Operation objective	Withdrawal	Injection	Poor/Fair	Conduct test injection at analog
Cement life	Decades	$>10,000$ years	Poor	CO_2-resistant cements

NOMENCLATURE

‰	parts per thousand
CCP	CO_2 Capture Project
cm	centimeter
D	Darcy
Fm	Formation
GHG	greenhouse gas
GIS	geographic information system
kg	kilogram
km	kilometer
lbs	pounds
m	meter
mD	millidarcy
Ma	million years ago
mW/km^2	milliwatt per square kilometer
MMcfd	million cubic feet per day
MMscf	million standard cubic feet
MPa	megapascal
ppm	parts per million
psi	pounds per square inch
SVF	Springerville Volcanic Field
t	metric ton
Tcf	trillion cubic feet

ACKNOWLEDGEMENTS

The author thanks the Carbon Capture Project and U.S. Department of Energy for funding this study and Denbury Resources Inc., Kinder Morgan CO_2 Co. LLC, and Ridgeway Petroleum Corp. for contributing data. The author also recognizes the contributions of Chris Ballentine, Keith Bowman, Chuck Fox, Stuart Gilfillan, Stephen Melzer, Curt Oldenburg, Martin Schoell, Scott Tye, and anonymous reviewers.

REFERENCES

1. S.H. Stevens, V.K. Kuuskraa, J. Gale, Storage of CO_2 in depleted oil and gas fields: global capacity and barriers to overcome. *Fifth International Conference on Greenhouse Gas Control Technologies, Cairns, Australia, August 13–16, 2000.*
2. S. Holloway, Safety of the underground storage of carbon dioxide, *Energy Convers. Manage.* **38** (Suppl.) (1997) S241–S245.
3. J.W. Johnson, J.J. Nitao, J.P. Morris, Reactive transport modeling of long-term cap rock integrity during CO_2 injection for EOR or saline-aquifer storage. *Second Annual Conference on Carbon Storage*, U.S. Department of Energy, Alexandria, Virginia, USA, May 5–8, 2003.
4. I. Czernichowski, B. Sanjuan, C. Rochelle, K. Bateman, J. Pearce, P. Blackwell, Analysis of the geochemical aspects of the underground disposal of CO_2, in: J.A. Apps, C.F. Tsang (Eds.), *Deep Injection Disposal of Hazardous and Industrial Waste*, Academic Press, New York, 1996, pp. 565–583.
5. S.P. White, R.G. Allis, D. Bergfeld, J.N. Moore, T.C. Chidsey, C. Morgan, K. McClure, S. Rauzi, Evaluating the seal integrity of natural CO_2 reservoirs of the Colorado Plateau. *Third Annual Conference on Carbon Storage*, U.S. Department of Energy, Alexandria, Virginia, USA, May 3–6, 2004.
6. J.M. Pearce, J. Baker, J. Beaubien, S. Brune, I. Czernichowski-Lauriol, E. Faber, G. Hatziyannis, A. Hildenbrand, B. Krooss, S. Lombardi, A. Nador, H. Pauwels, B. Schroot, Natural CO_2 accumulations in Europe: understanding long-term geological processes in CO_2 storage. *Sixth International Conference on Greenhouse Gas Control Technologies*, Kyoto, Japan, September 30–October 2, 2002.
7. M. Watson, Natural accumulations of carbon dioxide in reservoir rock as analogues for carbon dioxide storage, Masters thesis, University of Adelaide, Australia, 2002.
8. R.F. Weeter, L.N. Halstead, Production of CO_2 from a reservoir—a new concept, *J. Pet. Technol.* (1982) 2144–2148.
9. S.L. Rauzi, Carbon dioxide in the St. Johns–Springerville area, Apache County, Arizona. *Arizona Geological Survey*, Open-File Report 99-2, 1999.
10. J.R. Studlick, R.D. Shew, G.L. Basye, J.R. Ray, A giant carbon dioxide accumulation in the Norphlet Formation, Pisgah Anticline, Mississippi, in: Barwiss, McPherson, Studlick (Eds.), *Sandstone Petroleum Reservoirs*, Springer, New York, 1990.
11. D. Rice, Carbon dioxide in Mississippian rocks of the Paradox and adjacent areas, Colorado, Utah, New Mexico, and Arizona. U.S. Geological Survey, Bulletin 2000-H.
12. Z. Zhou, C.J. Ballentine, M. Schoell, S. Stevens, Noble gas tracing of subsurface CO_2 origin and the role of groundwater as a CO_2 sink (abstract). American Geophysical Union, Fall Meeting, San Francisco, California, USA, December 8–12, 2003..
13. S.H. Stevens, C. Fox, T. White, S. Melzer, C. Byrer, Production operations at natural CO_2 fields: technologies for geologic storage. *Sixth International Conference on Greenhouse Gas Control Technologies*, Kyoto, Japan, September 30–October 2, 2002.

Carbon Dioxide Capture for Storage in Deep Geologic Formations, Volume 2
D.C. Thomas and S.M. Benson (Eds.)

Chapter 4

NATURAL LEAKING CO_2-CHARGED SYSTEMS AS ANALOGS FOR FAILED GEOLOGIC STORAGE RESERVOIRS

Zoe K. Shipton[1], James P. Evans[2], Ben Dockrill[3], Jason Heath[2], Anthony Williams[2], David Kirchner[4] and Peter T. Kolesar[2]

[1]Centre for Geosciences, Division of Earth Science, University of Glasgow, Glasgow G12 8QQ, UK
[2]Department of Geology, Utah State University, Logan, UT 84322-4505, USA
[3]Department of Geology, Trinity College, Dublin 4, Ireland
[4]Department of Earth and Atmospheric Sciences, Saint Louis University, St. Louis, MO 63013, USA

ABSTRACT

Analysis of leaky CO_2 reservoirs in the northern Paradox Basin, Utah has allowed us to develop a model for the shallow subsurface CO_2 flow system. The results provide information on how CO_2 migrates and reacts with groundwater and reservoir rocks in the subsurface, and what the effects on surface environments are when CO_2 leaks to the surface. A series of shallow fluvial and eolian sandstone groundwater reservoirs are charged with CO_2 derived mostly from clay–carbonate reactions in Paleozoic source rocks within the basin (depths greater than 1.5 km). The CO_2-charged groundwater builds up in a north-plunging anticlinal trap with fault sealing on its southern margin. Top seal is provided by shale-rich formations, but fractures related to the fault damage zone provide conduits through the top seal. This geometry has resulted in a series of stacked reservoirs, and ultimately in escape of the natural CO_2 into the atmosphere. The CO_2 escapes through a series of springs and geysers along the faults, and through wellbores that have penetrated the reservoir. At the surface, rapid degassing of CO_2-charged groundwater results in the formation of travertine mounds around active springs. The presence of deeply incised ancient mounds attests to the long lifespan of this leaky system. There is no evidence of adverse effects of this leakage on wildlife or humans, and the springs provide (somewhat saline) water for plants in this high desert environment. Studies on the effect of long-term leakage both in the subsurface and at the point of leakage to the surface provide data on factors that affect the safety and feasibility of future CO_2 injection projects and should guide the design and implementation of geologic storage projects.

INTRODUCTION

For geological CO_2 storage to be effective, we need to be able to monitor the flow of CO_2 in the subsurface, and to ensure that little or no CO_2 leaks to the Earth's atmosphere over periods of thousands of years [1]. We consider a geological CO_2 storage "system" to consist of four main components: (1) a relatively porous and permeable reservoir lithology acting as a storage "tank", (2) a low-permeability and capillary-entry-pressure sealing lithology that is a barrier to flow out of the reservoir (cap rock and/or fault seal), (3) in the case that the seal fails, the likely migration pathways through the overburden and possible secondary reservoirs where gas may be trapped, and (4) the vadose zone and Earth's surface. Potential negative consequences of CO_2 leakage and seepage from the storage reservoir may potentially be felt if it infiltrates aquifers, and if it interacts with plants, animals, and humans. For accurate risk assessment we need to understand each step of migration from "tank" to surface, to quantify the rates and volumes of gas released to the atmosphere in the case of a leak, to determine the environmental impact of escaped gas on the surface biota, and to design mitigation strategies for the effects of any leakage. Analyses of natural leaky CO_2-rich systems are ideal for determining how CO_2 migrates and reacts with groundwater and reservoir rocks in the subsurface, and what the effects are when it leaks to the surface. These studies provide data on the factors that affect the feasibility and safety of future CO_2 injection projects and should guide their design and implementation.

The Paradox Basin, in the Colorado Plateau region of the United States, contains a number of natural CO_2 reservoirs, which provide analogs for understanding the integrity of stored gas systems [2]. Many of these fields have stored CO_2 for long periods of time, but others leak gas into the atmosphere, primarily along faults. In this paper we review studies of the hydrology, stratigraphy, structural geology, and geochemistry of a naturally degassing CO_2 reservoir in Utah. The CO_2 discharges along the Little Grand Wash and Salt Wash faults, creating a series of CO_2-charged springs and geysers, travertine deposits (both active and ancient), and carbonate-filled veins. A number of abandoned hydrocarbon boreholes also act as active conduits for CO_2 to the surface. This multidisciplinary study summarized here examines the controls and processes active in such leaky systems, and the effect of leakage in surface environments.

GEOLOGICAL SETTING

The Little Grand Wash and Salt Wash normal faults are situated in the northern Paradox Basin (Figure 1). This basin is defined by the extent of organic-rich Pennsylvanian and Permian limestones, shales and evaporites, which cover a large area of southern Utah and western Colorado. A basin-wide system of salt anticlines and faults initiated during Pennsylvanian/Permian uplift of the Uncompaghre plateau to the northeast, and were reactivated during several episodes of deformation ranging from the Triassic to Quaternary [8,9]. Many of the CO_2 reservoirs have accumulated within these salt anticlines, including the leaky reservoir in this study.

The Paradox Basin is filled with a series of clastic and carbonate sedimentary rocks analogous to those in North Sea oil and gas fields, and a number of good reservoir and seal systems exist in the basin. The regionally important Permian White Rim Sandstone reservoir is capped by the shale-rich Triassic Moenkopi and Chinle Formations. The overlying fluvial and eolian redbed reservoir units of the Lower Jurassic Navajo, Kayenta and Wingate Sandstones are capped by the marine limestones of the Carmel Formation, or the shale-rich Dewey Bridge Member of the Entrada Formation. The Middle Jurassic Entrada and Curtis Formations are the youngest good reservoir units in the basin, and are capped by interbedded fluvial and eolian siltstones and sandstones and gypsum seams of the Middle Jurassic Summerville Formation. The remaining overlying sequence does not contain any large reservoir units, but does contain several sand-rich units. The Upper Jurassic Morrison Formation consists of stacked fluvial channels of the Salt Wash Sandstone member, overlain by the bentonite-rich lacustrine shales of the Brushy Basin member. The Lower Cretaceous Cedar Mountain Formation black lacustrine shale is overlain by the Upper Cretaceous Dakota Sandstone conglomeratic channel sandstones. The youngest formation exposed in the field area is the Upper Cretaceous Mancos Formation, a dark organic-rich marine shale.

The east–west trending Little Grand and Salt Wash faults cut an open, north plunging anticline (Figures 1 and 2). The 61 km long, 70–80° south-dipping, Little Grand Wash fault is a complex fault zone comprised of several anastomosing normal faults defining structural terraces with varying dips (Figure 2a). Total vertical separation at the center of the fault is 180–210 m, most of which is accommodated by the southern fault strand. The Salt Wash faults (sometimes termed the Tenmile Graben) are a set of 290° striking dip-slip normal faults that form a graben over 15 km long (Figure 2b). Well data from abandoned oil wells and water wells have been used to constrain the subsurface geometry of the north-plunging anticline and faults.

PRESENT-DAY LEAKAGE

Active Springs and Wellbore Leakage

CO_2-charged groundwater effuses from a number of natural springs and leaky wellbores along the faults. Almost all of these effusions occur to the north (footwall) of both faults (Figure 2). The wellbores are mostly abandoned oil exploration drill holes and a few water wells. The most dramatic of these leaks is the Crystal Geyser on the eastern bank of the Green River in the footwall of the Little Grand Wash fault zone (Figure 3a). This cold-water geyser has erupted at 4–12 h intervals since the Glen Ruby #1-X well was drilled to the base of the Triassic section (TD 801 m) in 1935. The well was spudded into a 21.5 m thick

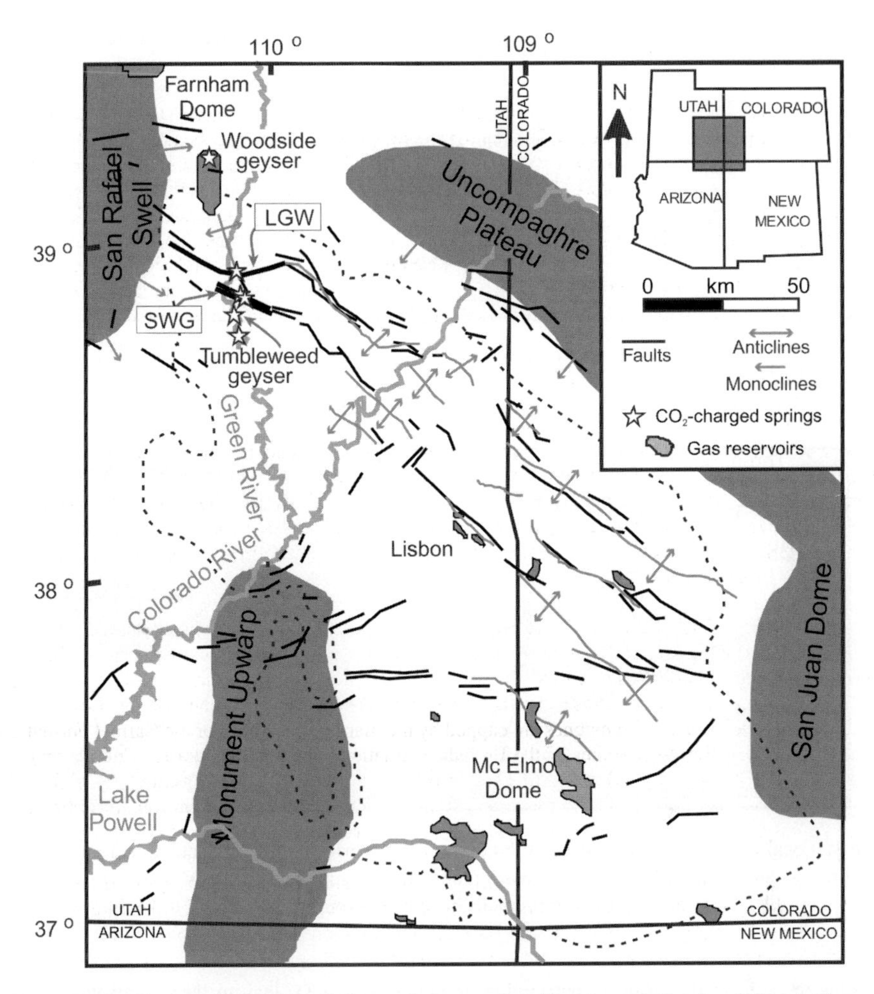

Figure 1: Regional geologic setting of the study area, after Nuccio and Condon [3] and Cappa and Rice [4]. Dotted line marks the extent of the Paradox Basin. LGW, Little Grand Wash fault; SWG, Salt Wash Graben.

travertine mound [10], so the spring system must therefore have been active for a considerable length of time prior to the well being drilled. This is corroborated by reports of "satin spar" at this location in 1869 by the Powell expedition along the Green River [11]. Three other springs within 10 m of the wellhead effuse periodically throughout each geyser eruption. These pools could represent the location of pre-well CO_2-charged springs or could be due to escape of the CO_2-charged waters from the well bore at shallow levels.

Smaller intermittent CO_2 fluxes occur in the Green River, where a line of small bubbles can be observed along the trace of the fault. Approximately 1 km east of the Crystal Geyser and ~ 100 m north of the fault zone, dry gas seeps audibly from the ground. Although there is no surface water at this location, the soil is commonly wet even in the dry season. These observations suggest that a diffuse flux of CO_2 may exist in the vicinity of point-source leaks (springs and wellbore seeps). In the absence of detailed flux monitoring it is

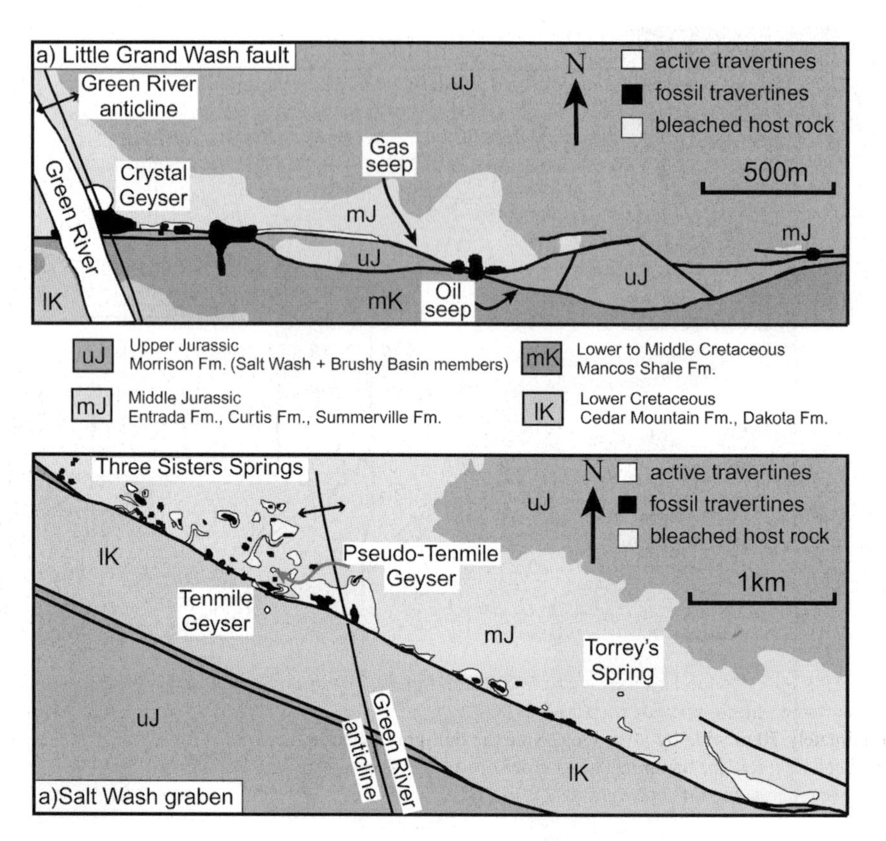

Figure 2: Local geological map of the distribution of active springs, travertine and reduction zones along (a) the Little Grand Wash and (b) the Salt Wash faults (after Dockrill et al. [5]; Williams [6] and Doelling [7]). Note that geologic formations have been grouped to simplify the map.

impossible to estimate the relative contribution of each type of CO_2 leak to the overall flux out of the reservoir (i.e. point sources vs. diffuse seeps under the Green River and possible degassing through soil).

Five CO_2 springs or small geysers occur along the northern Salt Wash fault (Figure 2). The westernmost Three Sisters springs flow continuously, but there is relatively little carbonate deposition at the site (Figure 3b). These springs lie in a 3–4 ha topographic low with saltpan crusts. Water can be found within 10 cm of the surface throughout the region, and we suggest that the surface seeps are a smaller manifestation of a broader gas leakage system. The Tenmile Geyser erupts infrequently with 1–1.5 m high eruptions, is located 200 m south of the northern fault and is the only visible point-source of CO_2 effusion that occurs within the graben (Figure 3c). It is centered on an abandoned well, which may penetrate the fault into the footwall reservoir (unfortunately no drilling records are available for this well). Pseudo-Tenmile geyser, a mineral-charged spring that vents a constant stream of CO_2 bubbles, sits on a low mound 100 m north of the fault. Torrey's Spring is in the footwall of the northern Salt Wash fault (Figure 3d) and is associated with an abandoned drill hole. This spring flows and bubbles continuously and has developed a small carbonate mound ~15 m in diameter.

Several other CO_2-charged springs occur in the northern Paradox Basin, all of which are associated with wellbore leakage from abandoned water wells (Figure 1). The once spectacular Woodside Geyser,

Figure 3: Sites of active leakage along the Little Grand Wash and Salt Wash faults. (a) View of Crystal Geyser facing north, trace of fault marked with a dashed line. The orange active travertine deposit is approximately 70 m wide by 80 m long. Note the dull gray inactive travertine exposed in riverbed and on west bank of river. (b) Sampling water from one of the Three Sisters springs. (c) Tenmile Geyser with remains of well casing. (d) Torrey's spring. Note the more restricted size of these mounds, and the lack of well-developed terraces.

approximately 40 km north of the study area, now only erupts sporadically to a height of a few meters from an abandoned oil well. The Tumbleweed and Chaffin Ranch geysers to the south of the faults in this study erupt occasionally from water wells. These other springs fall along the line of the regional north-plunging anticline axis, as do the geysers and springs along the faults, suggesting that the flow of CO_2 or CO_2-charged groundwater is focused along the anticline axis.

Travertine deposits are developed to various degrees around all the active springs. The most well-developed mound is at Crystal Geyser, which consists of down-stepping lobes, which radiate outward from the central wellhead, covered in rimstone terraces (Figure 4a). The other natural springs have smaller, less well-developed travertine mounds (Figure 3b–d). The wellbore leakage sites are surrounded by cemented Quaternary material and thin, friable, poorly developed travertine drapes. We suggest that the degree of travertine development reflects the length of time the spring has been active.

Water Composition

Water samples were collected from seven locations according to the detailed field sampling and measurement methods that Heath et al. [12] developed for sampling high-CO_2 groundwaters. All water samples had in situ temperatures less than 18 °C, confirming that CO_2 degassing is the only driving mechanism for the geysers (rather than high heat flow). The low effusion temperature of the spring waters suggests a shallow source, assuming the waters did not cool during ascent. The δD and $\delta^{18}O$ for the sampled groundwater do not show an $\delta^{18}O$-isotopic shift away from the local meteoric water line, implying that they

Figure 4: (a) The foreground of this photo shows the surface texture on the active Crystal Geyser terrace, while a typical geyser eruption is occurring in the background. Height of water column in this photo ~ 10 m. (b) Eroded ancient porous terrace travertine cut by a white banded vein with radiating crystals pointing inward from vein walls. (c) Highly altered Summerville Formation below an inactive travertine mound with a dense box-work of white-banded veins.

are meteoric and have not exceeded temperatures of > 100 °C. Given local geothermal gradients [3], the water for the springs along these faults is therefore likely to have come from the Wingate and Navajo Sandstone at around 300–500 m depth [13,14].

The waters are saline and slightly acid, with 13,848–21,228 mg total dissolved solids (TDS) per liter and pH values from 6.07 to 6.55. The $\delta^{13}C$ values of total dissolved carbon from three springs or geysers range from 0.0 to 1.2‰. The waters are supersaturated with respect to calcite, aragonite, dolomite, and hematite, and are undersaturated with respect to anhydrite, gypsum, halite, and quartz, consistent with the carbonate minerals found at the locations of the emanating waters. The carbonate precipitation may be a result of degassing effects that bring the waters to supersaturation with respect to the carbonate phases when the waters reach the surface. All of the waters are closely grouped in the sodium chloride chemical facies (Figure 5) suggesting a similar chemical evolution history of all the waters in the study area. Springs on the southern side of the Salt Wash graben, however, have lower bicarbonate contents, lower salinities and are more alkaline. We suggest that these springs tap a local flow regime and not the regional CO_2-charged groundwater system. Shipton et al. [14] showed that the salinity of the Crystal Geyser water decreases during and after an eruption suggesting that as gas and water are discharged through the water column, fresher water drains into the wellbore.

Gas Composition

Gas samples were collected from seven sites according to the detailed field sampling and measurement methods of Heath et al. [12,13] using both diffusion samplers and glass bottle samplers. The sample sites include three abandoned drill holes and four natural bubbling springs. The gases emanating from all the springs are 95.66–99.41% CO_2 by volume with minor amounts of Ar, O_2, and N_2. A small amount of atmospheric gases are probably entrained during geyser eruptions and the vigorous bubbling of the emanating waters. The $\delta^{13}C$ values of the CO_2 gas phase range from − 6.42 to − 6.76‰ (SD 0.13‰). This indicates that the CO_2 gases may all come from the same source and that the travel path may not greatly alter the carbon isotopic values, even though the gases are emanating from three distinct areas nearly 10 km apart. Thus, the same type of gas may be ubiquitous in the northern part of the Paradox Basin.

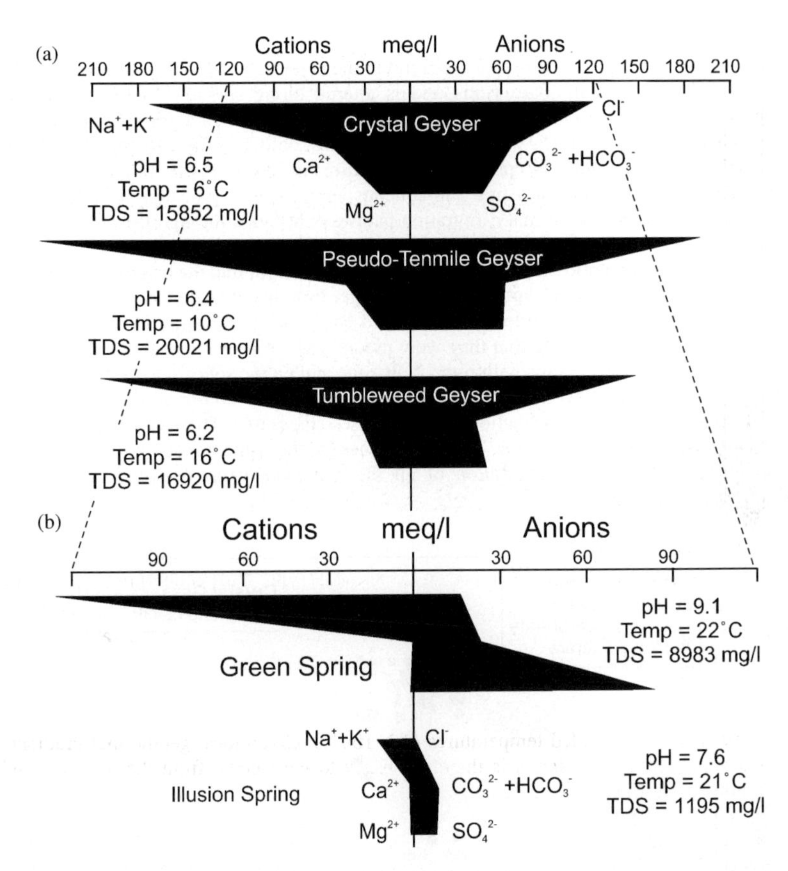

Figure 5: Stiff diagrams showing chemical compositions of the waters from various springs in milliequivalents per liter (after Heath et al. [13]). (a) Springs associated with CO_2 leakage and which erupt in geyser-like eruptions. For all of these springs pHs are slightly acidic, all have low temperatures and fairly high total dissolved solids. All the waters are chemically similar, with Na^+ and Cl^- as the major ions and fall in the same chemical facies. All waters have high levels of bicarbonate indicating high CO_2. (b) Springs on the south strand of the Salt Wash fault, note the change in scale. The pHs of the south strand springs are higher and their total dissolved solids are much lower.

INACTIVE SPRINGS

A series of partial to complete remnants of ancient travertine mounds runs parallel to the Little Grand Wash and Salt Wash fault traces (Figure 2). All the ancient travertine along both faults is situated either on, or to the north of, the fault zones. The ancient travertines are up to 4 m thick and occur up to 30 m above the present level of the actively forming deposits, forming a resistant cap on top of a series of buttes, reflecting the progressive down-cutting of the Green River.

Cross-sectional exposures through well-developed ancient travertine mounds demonstrate the processes that are likely to be active below the surface of the modern mounds [5]. The base of all the ancient travertine mounds consists of carbonate-cemented sediment suggesting that colluvium surrounding the leak site was cemented by the erupting spring waters. The main body of the travertine mounds consists of interbedded layers of sub-horizontal layered carbonate and cemented colluvium. The carbonate layers have a distinctive terraced texture that is similar to the rimstone textures of the active travertines (Figure 4b), so we suggest this unit represents the aggrading surface of a fossil travertine deposit. The host rock underlying the ancient travertine deposits is cut by carbonate veins and is altered along sporadic beds. In sand-dominated lithologies, the veins are 4–15 cm thick and usually have a reduction halo 1–5 cm wide. In mud-dominated lithologies a box-work of thin veins (5–20 mm thick) can almost obliterate the host-rock fabric (Figure 4c). Thick white-banded veins (5–80 cm) cut through the entire thickness of the mounds. These veins are interpreted to precipitate predominantly in a sub-aqueous environment from rapidly degassing CO_2-rich waters and therefore represent the primary migration pathways of fluids through the deposit.

Carbon and oxygen isotopic compositions of travertines provide insight into the origin of the water and CO_2 and the precipitation conditions of the carbonate over a longer timespan than do the analyses for the modern water and gas. The $\delta^{13}C$ and $\delta^{18}O$ values of the layered carbonate facies from both active and inactive travertine overlap (Figure 6), implying that they were precipitated from parental fluids that have remained isotopically consistent over time. Veins within the fault zone and on the south side of the faults have much lighter $\delta^{13}C$ values than the travertine mounds indicating that the non-travertine-related veins were generated from different sources and fractionation processes (Figure 6). The CO_2-charged water appears therefore to have not crossed the fault zone. The $\delta^{18}O$ values for the white-banded veins and non-travertine veins are similar indicating that precipitation of all the carbonate veins occurred at comparable low temperatures from meteoric waters.

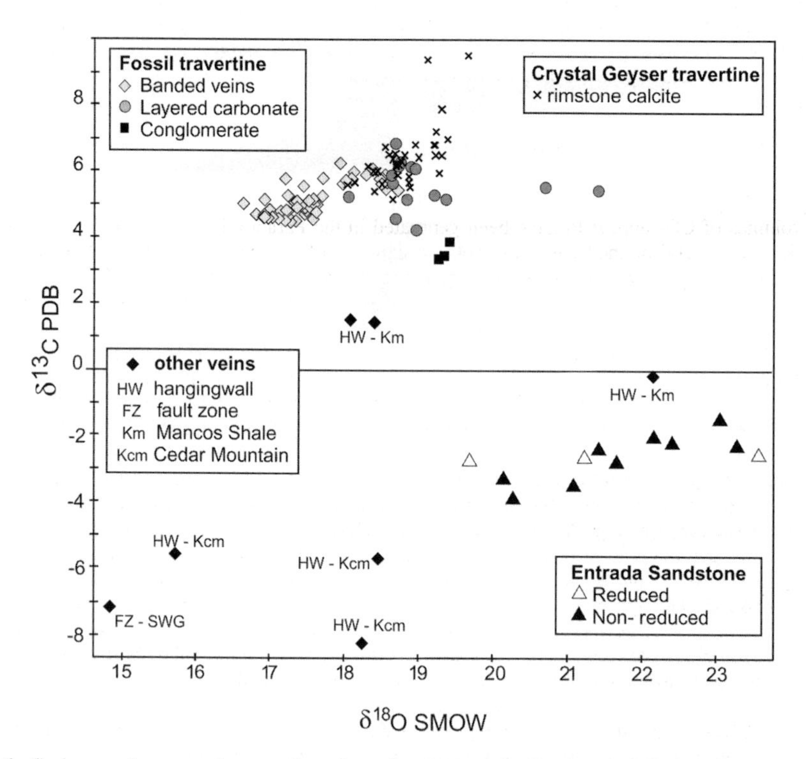

Figure 6: Carbon and oxygen isotope data from fossil travertine and active Crystal Geyser travertine (after Dockrill et al. [5]). Samples for veins from the Salt Wash fault zone (FZ-SWG), and north of the Little Grand Wash fault (HW-Km) and Salt Wash Graben (HW-Kcm) are also plotted. The stable isotope signatures of these veins are very different from the travertine-related veins, and are interpreted to be precipitated from different source waters. Data for red unreduced Entrada Sandstone and reduced Entrada Sandstone are also shown.

IRON OXIDE REDUCTION AND HYDROCARBON STAINING

Areas of the Entrada and Curtis Formation red beds near the Little Grand Wash and Salt Wash faults have been altered to pale yellow or white. This is due to extra-formational reducing fluids stripping out Fe^{2+}- and Mn^{2+}-bearing minerals causing an apparent bleaching of the unit (see Refs. [8,15]). These reducing fluids could include hydrocarbons, organic acids, methane, or hydrogen sulphide [15,16]. The iron oxide reduction is focused in structural highs (i.e. where the anticlinal crest is cut by both faults) and at leak points (i.e. below travertine deposits) along both faults. The reduced zones are interpreted to represent the migration pathways of buoyant fluids through this faulted system. The reduced sandstones are isotopically consistent with the non-reduced red sandstones and are isotopically distinct from any precipitates related to the overlying travertine mounds (Figure 6). This indicates that the presently erupting carbonate-rich fluids are not responsible for the reduction zones and that one or more earlier reducing fluids must have migrated through this faulted system to cause the reduction zones. It is interesting to note that similar altered rocks are seen on a large-scale across the Paradox Basin [15,17].

A fresh oil seep is located to the east of the active seeps along the Little Grand Wash fault and the Salt Wash Member sandstones close to this oil seep contain patches of bitumen staining. This seep has been active since at least the 1940s [18] and the freshness of the oil indicates that there is active flow of petroleum to the surface. Carbon isotopes of the oil from the seep with respect to saturated and aromatic hydrocarbons are -28.47 and -29.26‰, respectively [19]. These values are much more depleted than the $\delta^{13}C$ of the CO_2 gas. Without a detailed paradiagenetic study it is not clear whether the flow of hydrocarbons is related to the flow of CO_2, but the close spatial association of the CO_2 leaks, the oil seep and the reduced sandstone suggests that similar pathways are used by hydrocarbons, the present day CO_2 flow, and possibly a separate phase of reducing fluids.

DISCUSSION

Source of the CO_2

Unusual volumes of CO_2 appear to have been generated in the Paradox basin and the likely sources are discussed in Refs. [13,14] on the basis of the isotopic signature of the gas and carbonates. Measured helium R/R_a values of ~ 0.3 [14] are well out of the range for mantle helium signatures of 7–21 and are similar to crystal values [20]. It should be noted, however, that transport properties of He and CO_2 are distinct and that He and CO_2 might be expected to fractionate during migration. Although hydrocarbon source rocks occur in the Paradox Formation [3], the thermal degradation of organic matter during diagenesis and catagenesis results in values of $\delta^{13}C_{CO_2}(g)$ from -8 to -12‰ (Ref. [21]), lower than those measured from the springs (-6.42 to -6.76‰). Production of CO_2 from the degradation of organic matter through sulfate reducing or methanogenic bacteria produces more depleted values of $\delta^{13}C_{HCO_3}$ than are seen in our analyses [22].

Clay–carbonate diagenetic reactions at temperatures of about 100–200 °C during deep burial of impure carbonate sedimentary rocks can generate large amounts of CO_2 gas [23,24]. By assuming isotopic equilibrium between the source carbonates and the gases, Heath et al. [13] showed that the clay–carbonate reactions involving rocks with $\delta^{13}C_{CaCO_3}$ values of $+1$ to -3‰ (close to the average $\delta^{13}C$ marine carbonates) could have produced the CO_2. Occurrences of metamorphic CO_2 have been identified elsewhere in the Paradox Basin (e.g. Figure 1). The average $\delta^{13}C_{CO_2}$ isotopic value of -6.60‰ from the effusing gases is ~ 2‰ more negative than would usually be expected from the thermal decomposition of marine carbonates. To establish if the gases do derive from metamorphic sources, travel pathways from the source to the area must be identified such as faults and the structural grain of the basin.

Most of these CO_2 sources come from relatively deep in the basin (depths of 1–1.5 km in upper Paleozoic or Triassic rocks) and it is likely that faults provide pathways for flow of CO_2 through normally sealing lithologies such as the Paradox Salt. This scenario requires the generation of a gas phase that can migrate away from the gas source to accumulate in shallow aquifers. One scenario involves the generation of a free-phase of CO_2 when gas-charged groundwater rises above ~ 2 km depth. The free-phase CO_2 can migrate by diffusive and advective flow much faster as a separate supercritical or gas phase, depending on depth. We suggest that the rapid uplift and erosion of the Colorado Plateau has brought CO_2 source waters to shallow depths which has facilitated generation, migration, and accumulation of CO_2 in shallow reservoirs.

Shallow Flow Pathways

Our observations of fault structure and aqueous and carbonate geochemistry enable us to construct a conceptual model of the regional groundwater flow in the upper 1.5 km of the basin (Figure 7).

Potentiometric surface data from groundwater wells show that regional groundwater flows from the northwest to the southeast [25]. Water temperature and stable isotope data for springs along both faults show

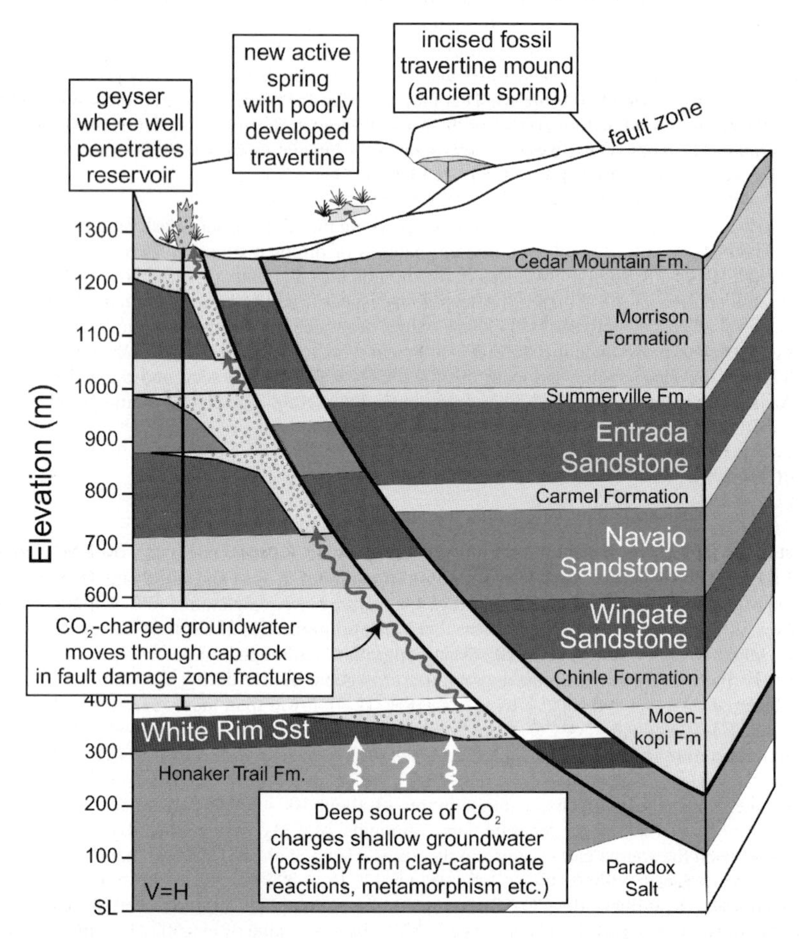

Figure 7: Schematic north–south cross-section through the Little Grand Wash fault (after Williams [6]), used to illustrate the conceptual reservoir model developed in this study. The subsurface geology, specifically unit thicknesses, is constrained by wells. CO_2-charged groundwater (small circles) is pooled in a north plunging anticlinal trap against the south-dipping fault. Although the fault zone geometry is much more complex than is indicated by this cross-section, water-chemistry data and stable isotopes from veins on the south side of the fault show that little or no cross-fault CO_2 migration is occurring. From HCO_3 concentration in different wells, the CO_2 gas may have infiltrated into many of the sandstone formations such as the Entrada, Navajo, Kayenta, and Wingate (schematic CO_2-filled reservoirs are not shown to scale). Fractures related to the faulting allow infiltration of the CO_2-charged groundwater through the otherwise sealing cap rock (arrows). Springs and geysers mark points where CO_2-charged groundwater escapes along natural fractures, or through wellbores that penetrate the reservoirs.

that the CO_2 is charging a reservoir approximately 300–500 m below the surface. Conversely, relatively short flow paths in a local flow system are indicated by the geochemistry of springs on the south strand of the Salt Wash fault. All of the modern and ancient CO_2 leakage points lie on the structural high where the north-plunging anticline is cut by the faults. Therefore, the faults are acting as flow barriers to southeast directed CO_2-charged groundwater flow, and CO_2-charged groundwater is accumulating against the faults within the folded reservoir. Within the framework of this model for the geometry of the shallow CO_2 storage system, the observations and data collected at the leaky faults can give us insight into each of the four components of the CO_2 system.

CO_2 reservoir
In our model, shallow groundwater reservoirs are charged from below by CO_2 generated at depth. These shallow reservoirs are the high-porosity eolian or fluvial Jurassic sandstones. The extent of reaction between the reservoir rocks and the CO_2-charged fluid is unclear, but the continued effusion of the springs and geysers at the same locality through time shows that the high porosity must be maintained, and that the porosity is not being clogged by the products of diagenetic reactions.

Cap rock and fault seal
The topseal for upward movement of fluid from the shallow aquifer reservoirs is provided by shale-rich Jurassic units. A lateral seal is provided by the fault rocks. The difference between the stable isotope signatures of veins in the hanging wall and footwall of the faults shows that the faults have acted as effective barriers for cross-fault flow at depth. The nature of the fault rocks at depth is partly dependent on the type of rocks that are juxtaposed across the faults and the amount of displacement that those fault rocks have undergone. The faults in this area offset a series of clean sandstones and shale-rich rocks and therefore could be expected to produce a clay-rich fault gouge which would be expected to act as a barrier to cross-fault flow as discussed by Yielding et al. [26]. It must be emphasized that predicting fault seal from throw distributions is prone to error, and a small variation in fault-zone thickness or properties can create an apparent "hole" where fluid can leak through the fault.

In contrast to the fault rocks, the shale-rich units that provide the topseal are leaking. Lithologically similar cap rocks have retained their integrity in CO_2 reservoirs elsewhere in the Paradox Basin (e.g. McElmo Dome, Lisbon Dome, Figure 1); therefore an explanation must be sought for why the cap rocks have failed at this location. Prior to drilling of the well, the leakage was focused in the immediate footwall to the faults. We suggest that fractures that formed in the cap rock as part of the damage zone to the faults are providing a conduit for leakage. It is also possible that an increase in CO_2 volume at shallow depths leads to hydrofracturing, therefore enhancing fracture permeability. The fractures through the cap rock must have stayed open for substantial amounts of time (i.e. they are not self-sealing). The strength and mechanical behavior of the cap rock units and the hydrodynamic behavior of CO_2-rich fluid at shallow depths are poorly understood. Without such data, a reliable prediction could not presently be made of the integrity of cap rocks in similar structural settings.

Migration pathways
In our conceptual shallow reservoir model, the CO_2-rich waters are sourced from the Wingate and Navajo Formations. Chemical analyses of groundwater from oil and gas exploratory and development wells, water wells and springs within ~ 100 km of the field area indicate that high dissolved CO_2 concentrations are common in many formations from the Devonian Elbert Formation to the Jurassic Entrada Sandstone as well as the Navajo, Kayenta and Wingate Sandstones [27]. This distribution of CO_2 content suggests that there is a sequence of stacked CO_2-charged aquifers above the primary CO_2 source. This is a critical issue for CO_2 storage since leakage and migration of CO_2 into formations overlying the intended storage formation may provide secondary trapping sites, reducing the overall flux to the surface.

The drilling of the oil and gas wells has provided pathways for rapid transport of gas-charged groundwater through the cap rock. The only leakage south of the sealing fault occurs through a well that may have penetrated the fault and into the footwall reservoir (Tenmile Geyser). Conversely, the two wells that penetrate the Little Grand Wash fault in the Triassic section (sealing lithologies) do not leak. Most of the wellbore leakage is from abandoned oil and gas exploration wells, and no record exists of the type of cement or casing that was used in these wells. Injection wells drilled for future storage projects are likely to be specially engineered to avoid leakage in this manner. However, any oil- and gas-rich area, which may present an attractive target for CO_2 disposal is likely to contain a large number of abandoned wells.

Effect of Leakage on the Surface

Leakage of CO_2 to the surface has occurred in this system for at least some portion of the Holocene, and thus any effects on the local biological system should be evident. Our initial observations show that there is little or no impact of the CO_2 emissions on the local ecosystems, although more work needs to be done to quantify these observations. The region lies in a high, cold desert, so the natural populations of organisms are limited. We observed no changes in plant mortality around any of the leakage sites. Indeed, slightly enhanced growth of salt tolerant plants occurs at several sites due to the increase of water at the surface (Figures 3 and 4). The water is very high in TDS, S^{2-}, and Cl^-, thereby limiting the type of plant that can tolerate the areas near the springs. Although we might expect to see local effects from the higher salinity groundwater that effuses from the Crystal Geyser, Mayo et al. [28] showed that it does not have a significant effect on the downstream salinity of the Green River. The CO_2 effusion has resulted in no reported casualties (from analysis of historical records and oral histories acquired by historian D. Martindale, Utah State University, personal communication), even though the area is visited by locals and tourists.

Much of the leaking CO_2 is vented to the atmosphere, but some is trapped by the formation of the carbonate travertine mounds. From the groundwater composition, we can estimate the amount of calcite that would precipitate out given the amount of CO_2 that remains in solution (i.e. run a reaction from the supersaturated initial condition to equilibrium). The Crystal Geyser averages 50–100 m^3 of water per eruption, and we estimate that 0.90 g of calcite are precipitated per liter of H_2O, and 3.60 g CO_2/L H_2O are released to the atmosphere. If we assume that the reaction is run to completion, we find that about 10% of the carbon flux is trapped in the travertine. This estimate is a maximum, however, as it assumes that the reaction that precipitates the calcite achieves equilibrium, does not consider the changing water chemistry as it flows over the ground surface, and does not account for the free gas phase present in the system. Thus, our analysis shows that for this natural leaking system, very little of the escaped carbon is presently trapped at the surface. The trapping efficiency could be increased by adding reactive cations (Ca^{2+}, Mg^{2+}, Sr^{2+}) and raising pH, but for a large-scale leak, such an effort may not be effective. It appears that in the Little Grand Wash–Salt Wash fault system, the rate of CO_2 transport to the surface, in both the natural and industrially developed parts of the system, is faster than the rate of mineral precipitation. Thus, in the present study, surface mineralization due to leaking CO_2 does not seal the system.

CONCLUSIONS

We have integrated a variety of geologic data sets and methodologies to examine the sources, travel paths, and fate of CO_2 from a subsurface reservoir to the Earth's surface. The geological and structural analysis shows that the three-dimensional structure of the system consists of an open, north-plunging anticline cut by northwest-trending normal faults. These faults cut a Mesozoic section of clastic rocks that range from high-porosity and permeability eolian and fluvial sandstones, which are the dominant aquifers of the area, and low-permeability shales that appear to form effective top seals to a series of stacked CO_2-charged reservoirs. Although the faults provide a barrier to cross fault flow, the footwall reservoir has leaked for >150 years through the fault-related fractures in the damage zone. Typically in these types of rocks, analyses of fault seal capacity would predict that these faults would be barriers for cross-fault flow. In contrast, fractures in the damage zone associated with the faults appear to provide a conduit for CO_2 leakage through the cap rock units. The sealing characteristics of faults are therefore a key to understanding the storage capacity in these settings. More recent leakage is focused around abandoned oil wells and water wells.

Long-term leakage appears to have had an insignificant effect on surface biota, and no adverse effect on the salinity of the Green River. Despite the fact that the Crystal geyser is visited by tourists and locals, there are no reported casualties from the high CO_2 concentrations. However, water in the aquifers above the CO_2 source tends to have high values of TDS and chemistries that classify them as "contaminated" water.

RECOMMENDATIONS

1. *Leaky faults and fracture systems.* Experience with ongoing geologic CO_2 storage projects has highlighted that each storage site has a very specific set of circumstances requiring detailed structural characterization. Faults and fracture systems pose a leakage risk to any proposed geologic storage site. The risk of encountering sub-seismic scale faults and fractures means that detailed structural characterization and an understanding of cap rock integrity is an essential component of any future CO_2 disposal project.
2. *Wellbore leakage.* The leakage around the wellbores has been continuous since 1935, though we suspect that natural leakage has occurred for much longer. Much anecdotal information regarding damage done to the Ruby well and crude attempts at plugging the well indicate that free CO_2 gas is a robust component in a water gas system to depths of less than 500 m. Although injection and monitoring designed for a storage system would be specially engineered, it is clear that older wells pose a potential for long-term leakage that must be examined.
3. *Surface trapping.* We show that groundwater flow can result in transport of CO_2 for some distance before precipitation results. In the present study, a relatively large amount of CO_2 is vented to the atmosphere relative to the estimated amount precipitated in a mineral phase at the surface. The kinetics of carbonate reactions must be well understood before a "self-sealing" scenario can be proposed for shallow leakage.
4. *Cap rock integrity.* Careful analysis of seal integrity is critical in designing a geologic storage program because fractures and faults can provide pathways for gas migration. The few outcrops of the shale-rich "seal" in the present study all contain abundant veins and fractures, indicating that much more work is needed to understand this part of the system. There are few geomechanical and geochemical data sets that constrain the ability of shale and siltstone cap rocks to prevent the transport of CO_2.
5. *Future studies of CO_2 leakage.* In addition to the dramatic localized fluxes of CO_2 at the geysers and springs near the Green River, there may also be diffuse CO_2 fluxes over a broader area surrounding the fault zone. This could occur if either the cap rock integrity has been compromised by distributed fracturing, or if the CO_2 is spreading out within the vadose zone from localized fractures. The only way to address these questions is to fully characterize the flux of CO_2 from all springs, dry seeps and elevated "background" soil CO_2 with a campaign of monitoring fluxes and concentrations. This could be done locally above the leaky reservoirs identified in this study, and across the Colorado Plateau above other known CO_2 reservoirs. In addition to increasing our understanding of CO_2 flow in the shallow subsurface and vadose zone environments, a monitoring program in this area would be useful for testing instruments and methods, and for assessing hazards associated with elevated CO_2 concentrations.

ACKNOWLEDGEMENTS

This study funded by the CO_2 Capture Project, with extra data from Exxon. B. Dockrill was additionally supported by grants from Trinity College, Dublin. Water and gas geochemistry samples were run in the labs at Brigham Young University with the assistance of Stephen Nelson. Comments from two anonymous reviewers helped to improve the paper.

REFERENCES

1. C.A. Rochelle, J.M. Pearce, S. Holloway, in: R. Metcalfe, C.A. Rochelle (Eds.), *Geol. Soc. Spec. Pub.* 157, 1999, pp. 117–129.
2. R. Allis, T. Chidsey, W. Gwynn, C. Morgan, S. White, M. Adams, J. Moore, U.S. Department of Energy, NETL compact disk DOE/NETL-2001/1144, 2000.
3. V.F. Nuccio, S.M. Condon, *U.S. Geol. Surv. Bull.*, 2000-O, 1996.
4. J.A. Cappa, D.D. Rice, *U.S. Geol. Surv. Bull.*, 2000-H, 1995.
5. B. Dockrill, D. Kirschner, Z.K. Shipton, submitted to *Chem. Geol.* (2004).
6. A.P. Williams, MS Thesis, Utah State University, 2004.
7. H. Doelling, *Utah Geol. Surv. Map*, M-180, 2001.
8. M.A. Chan, W.T. Parry, J.R. Bowman, *Am. Assoc. Petrol. Geol. Bull.* **84** (2000) 1281–1310.
9. N.C. Davatzes, A. Aydin, *J. Struct. Geol.* **25** (2003) 1795–1813.
10. J.L. Baer, J.K. Rigby, *Utah Geol.* **5** (2) (1978) 125–130.

11. J.W. Powell, The Canyons of the Colorado (1895); (now published as The exploration of the Colorado River and its canyons), Penguin Books, 1997.
12. J.E. Heath, T.E. Lachmar, P.T. Kolesar, S.T. Nelson, in preparation for *Appl. Geochem.* (2004).
13. J.E. Heath, T.E. Lachmar, J.P. Evans, P.T. Kolesar, S.T. Nelson, *Chem. Geol.* (2004).
14. Z.K. Shipton, J.P. Evans, D. Kirschner, P.T. Kolesar, A.P. Williams, J.E. Heath, Analysis of CO_2 leakage through "low-permeability" faults from natural reservoirs in the Colorado Plateau, Southern Utah, in: S.J. Baines, R.H. Worden (Eds.), *Geological Storage of Carbon Dioxide*, Geological Society, London (2004) Special Publications, **233**, 43–58.
15. I.R. Garden, S.C. Guscott, S.D. Burley, K.A. Foxford, J.J. Walsh, J. Marshall, *Geofluids* **1** (2001) 195–213.
16. R.D. Elmore, R. McCollum, M.H. Engel, *Bull. Assoc. Petrol. Geochem. Exp.* **5** (1989) 1–17.
17. J.E. Huntoon, P.L. Hansley, N.D. Naeser, *Bull. Am. Assoc. Petrol. Geol.* **83** (1999) 467–495.
18. E.T. McKnight, *U.S. Geol. Surv. Bull.* 908, 1940.
19. P.G. Lillis, A. Warden, J.D. King, *U.S. Geol. Surv. Digital Data Series DDS-69-B*, 2003.
20. B.M. Kennedy, Y.K. Kharaka, W.C. Evans, A. Ellwood, D. J. DePaolo, J. Thordsen, G. Ambats, R.H. Mariner, *Science* **278** (1997) 1278–1281.
21. J.M. Hunt, *Petroleum Geochemistry and Geology*, W.H. Freeman and Company, New York, 1996.
22. W.W. Carothers, Y.H. Kharaka, *Geochim. Cosmochim. Acta* **44** (1980) 323–332.
23. A.L. Mayo, A.L. Muller, *J. Hydrol.* **194** (1996) 286–304.
24. I.E. Hutcheon, H.J. Abercrombie, *Geology* **18** (1990) 541–544.
25. J.W. Hood, D.J. Patterson, *Utah Dept. Nat. Res. Tech. Pub.* **78** (1984).
26. G. Yielding, B. Freeman, D.T. Needham, *Am. Assoc. Petrol. Geol. Bull.* **81** (1997) 897–917.
27. J.W. Gwynn, *Utah Geol. Surv. Circular*, C-87, 1995.
28. A.L. Mayo, D.B. Shrum, T.C. Chidsey, Jr., in: T.C. Chidsey Jr., (Ed.), *Geology of East-central Utah: Utah Geol. Assoc. Pub.*, 19, 1991, pp. 335–342.

Carbon Dioxide Capture for Storage in Deep Geologic Formations, Volume 2
D.C. Thomas and S.M. Benson (Eds.)

Chapter 5

THE NGCAS PROJECT—ASSESSING THE POTENTIAL FOR EOR AND CO_2 STORAGE AT THE FORTIES OILFIELD, OFFSHORE UK

S. J. Cawley[1], M. R. Saunders[2], Y. Le Gallo[3], B. Carpentier[3], S. Holloway[4], G.A. Kirby[4], T. Bennison[5], L. Wickens[5], R. Wikramaratna[5], T. Bidstrup[6], S.L.B. Arkley[7], M.A.E. Browne[7] and J.M. Ketzer[8]

[1]BP Exploration, Aberdeen, UK
[2]BP Exploration, Sunbury-on-Thames, UK
[3]Institut Francais du Petrole (IFP), Rueil Malmaison, France
[4]British Geological Survey, Nottingham, UK
[5]ECL Technology Ltd, Winfrith, UK
[6]Geological Survey of Denmark and Greenland (GEUS), Copenhagen, Denmark
[7]British Geological Survey, Edinburgh, UK
[8]Universidade Federal do Rio Grande do Sul, Rio Grande do Sul, Brazil

ABSTRACT

The Next Generation Capture and Storage Project studied the potential to store underground 2 million tonnes of CO_2, approximately half the annual CO_2 emissions from the Grangemouth refinery and petrochemicals complex near Edinburgh, Scotland. The study concluded that the best potential storage site for these emissions was the Forties oilfield in the UK sector of the North Sea. Numerical simulation indicated that enhanced oil recovery using a WAG process and CO_2 as the injection gas would yield significant incremental oil. A Features–Events–Processes (FEP) identification process was used to narrow down the risks to storage at the Forties field. Numerical modelling was then used to assess the risks of CO_2 escape. It was concluded that the geological risks of CO_2 escape were negligible, but it was not possible to analyse the chances of CO_2 escape via pre-existing wells. The wells are perceived as the main uncertainty in the analysis and it is recommended that a comprehensive risk assessment methodology for wells is developed.

INTRODUCTION

The Next Generation Capture and Storage Project (NGCAS) was conceived in 2000 as a case study to test the potential for geological storage of large volumes of anthropogenic CO_2 captured from industrial plants.

By the time the project was conceived, there was already an indication that an attractive option for CO_2 storage on this scale might be to use CO_2 for enhanced oil recovery (EOR) in Europe's major oil province, the North Sea. The additional oil recovered would partly offset the costs of storage and there would be a more rational use of resources as a greater proportion of the oil resources in place would be produced. The Forties field stood out amongst BP's North Sea assets in terms of its storage volume: at least 75 million tonnes (Mt) of CO_2 could be stored underground as a result of EOR [1], with further potential if storage was continued for its own sake after EOR. Looking to the future, if a case could be made to build a CO_2 pipeline to the North Sea oilfields, there would be every opportunity for further CO_2-EOR projects because many North Sea oilfields appear technically suitable [1,2]. This would accrue further benefits to the EU in terms of import reduction and security of supply. This option clearly merited further investigation, but it was decided that the choice of a storage site should not be completely pre-judged as it was uncertain whether Forties really was the most cost-effective storage option, given the long transport distances from even the most northerly major industrial point sources of CO_2 in the UK and the high costs of offshore operations. Therefore onshore and nearshore areas which might have CO_2 storage potential were also considered.

Given the above it was considered that the source(s) of CO_2 selected for the study should be chosen from amongst the largest industrial point sources in the northern UK because these are relatively close to the oilfields, which lie in the Northern and Central North Sea. An obvious possibility was BP's Grangemouth refinery and petrochemical complex, some 30 km west of Edinburgh in Scotland, where a study was taking place to determine CO_2 capture costs. The Grangemouth site emits about 4 Mt of CO_2 annually. The Longannet and Cockenzie coal-fired power plants, which emitted 8.76 and 2.47 Mt CO_2, respectively, in 2000, are nearby. It was clear that these could be supplementary sources if required: they also added a "Cleaner Coal Technology" dimension to the project. Thus the Grangemouth site was selected as the nominal source for the project. It was arbitrarily assumed to have an additional 25 years of production, and thus it would emit roughly 100 Mt of CO_2 in the future. If half of this were to be made available to be stored rather than emitted to the atmosphere, the operation of the capture plant would emit an additional 600,000 tonnes/year of CO_2 [3]. The net CO_2 avoided would be 1.4 Mt/year and the net emission reduction would be 35%. It has been estimated that this might cost about \$50–60 per tonne CO_2, representing a total cost of about \$100–120 million/year [3]. However, the cost is sensitive to the price of natural gas, which is used as fuel in the capture plant.

STUDY METHODOLOGY

Strategy for Finding a Storage Site

The selected storage site needed to be able to receive 2 Mt CO_2 per year (half the CO_2 emissions from the Grangemouth site) and required a total capacity exceeding 50 Mt CO_2. The paramount requirement was that storage should be safe and secure. Thereafter, the minimum cost solution would be sought. Socio-economic factors, such as public acceptance and planning issues, would not be considered in the analysis. It was decided to investigate the storage possibilities in the onshore area around Grangemouth first, on the grounds that there would be low transport costs. If this proved fruitless, the search would move to the nearby offshore area to the east of the Firth of Forth known as the Forth Approaches, and then to Forties (Figure 1), which was the nearest oilfield known to have sufficient CO_2 storage capacity and where the greater transport costs might in part be offset by the potential revenue from EOR.

Potential for CO_2 storage onshore, near the Grangemouth site

The area around the Grangemouth site, known geologically as the Midland Valley of Scotland, is about 90 km wide and 300 km long and its long axis trends ENE (Figure 1). It is the most densely populated part of Scotland, containing the cities of Glasgow and Edinburgh and four-fifths of the Scottish population; approximately 3.9 million people.

Geologically, the Midland Valley of Scotland is a complex graben. It is bounded to the south by the Southern Uplands (which comprise mainly highly faulted and folded Ordovician to Silurian "greywacke" sandstones) and the north by the Scottish Highlands (mostly Dalradian and older metamorphosed sedimentary and igneous rocks). The geological contacts between the three terrains are defined by the major fault complexes of the Southern Upland and Highland Boundary faults. None of the indurated rocks of the Southern Uplands or the Scottish Highlands are suitable for CO_2 storage.

Within the Midland Valley of Scotland itself are several partly superimposed sedimentary basins of Upper Palaeozoic (i.e. Devonian, Carboniferous and Permian) age. Four major synclines, each containing Namurian and Westphalian Coal Measures, occur along its length. The easternmost of these is entirely offshore, in the Firth of Forth. Thus at first glance there seemed to be two possibilities for storing CO_2 in the Midland Valley: as free CO_2 in the pore spaces of sandstone reservoir rocks and adsorbed onto coal.

In general, the Devonian and Carboniferous sandstones of the Midland Valley of Scotland have low to fair porosity (up to 20%), very low primary permeability but sometimes quite significant secondary (fracture) permeability. They are probably unsuitable as CO_2 storage reservoirs because their permeability and porosity are thought to be too low at depth [4]. The Permian sandstones have good reservoir characteristics but are unsuitable because they are not sealed.

Coal seams are plentiful in the Midland Valley in both the Namurian Limestone Coal Formation and the Westphalian Coal Measures. Additionally, individual economically important coal seams occur

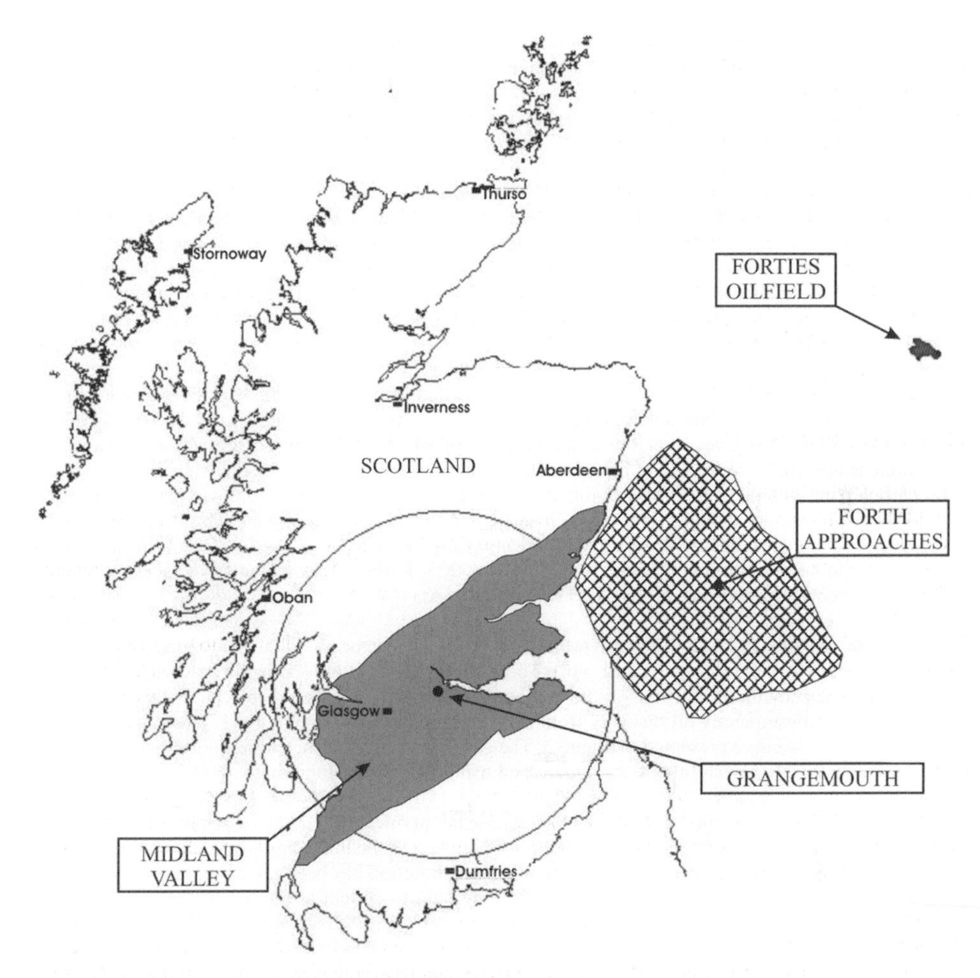

Figure 1: Location of the Grangemouth plant, the Midland Valley of Scotland, the studied area of the Forth Approaches and the Forties oilfield.

sporadically in other Carboniferous formations. However, the coal seams themselves may not be sufficiently permeable for large scale CO_2 storage. Moreover, the scale of operations that would be required to store 2 Mt of CO_2 per year indicates that storage in coal seams would be impractical at present [4].

The Midland Valley contains more than 100 boreholes that are in excess of 700 m deep and hundreds of abandoned coal mines. Both boreholes and mines are concentrated in the coalfields. There is clear potential for leakage through these, and also via pathways to the surface created by mining subsidence and natural migration pathways such as faults. Furthermore, groundwater abstraction and coal mining represent alternative uses of the subsurface that might take precedence over CO_2 storage.

Having considered and rejected the onshore Midland Valley as a potential storage site, attention was directed to the area immediately offshore, in the Forth Approaches (Figure 1). The Carboniferous and Devonian geology of this area is thought to be similar to that of the Midland Valley itself, but there was a possibility that the highly porous and permeable Permian sandstones seen onshore might be capped by Zechstein evaporites that would provide a good seal to retain any injected CO_2. However, it became clear that there was significant uncertainty as to whether the Permian sandstone reservoir was present and whether the evaporites would form a good cap rock, because in places the evaporites had clearly been subject to dissolution. Consequently, the Forth Approaches area was also rejected as a potential storage site and attention was focused on the Forties field.

Simulation of CO_2 Injection and EOR at the Forties Oilfield

The first step in the investigations at the Forties field was to use a numerical simulation model to investigate the optimisation of incremental oil recovery and storage of CO_2 at the field. The study was conducted using the VIP compositional simulation model to represent a sector of the Forties Charlie Sand. The model was initially waterflooded and then subjected to WAG with CO_2 as the injection gas. A range of simulations was performed to investigate different WAG strategies, timing of initiation of postflood gas injection, well placement, and well completions.

An existing VIP sector model of the Charlie Sands, provided by BP, was used as a starting point for the modelling. The model represents a volume of 1500 m by 500 m by 38 m, divided into 60 by 5 blocks areally and with 76 layers. This gives a total of 22,800 blocks, of which 16,618 blocks are active. Average grid block dimensions are 25 m by 100 m by 0.5 m. The total pore volume in the model is 26.3 MMrb, with a hydrocarbon pore volume of 21.3 MMrb; this corresponds to a stock tank oil initially in place (STOIIP) of 17.8 MMstb. The model includes two wells, located at either end of the model. The fluid behaviour is represented by an EoS model with seven components (CO_2, $N_2 + C_1$, C_2–C_3, C_4–C_5, C_6–C_{13}, C_{14}–C_{19} and $C_{20}+$) and has a saturation pressure of 1165 psia at 205 °F (96 °C). Representative values of the fluid densities at reservoir conditions are included in Appendix A.

The oil–water relative permeabilities that were used in the BP sector model are presented graphically in Figure 2a (linear plot) and b (logarithmic plot). These show that a waterflood is likely to yield an efficient piston-like displacement. A portion of the remaining oil could then be displaced by the continuing waterflood also. Consequently oil recovery from the waterflood should be very good. This can also be seen in the fractional flow curve presented in Figure 3. The gas–oil relative permeabilities are shown in Figure 4. Three-phase relative permeabilities were determined using Stone's calculation method 2.

A 3D view of the whole model showing the horizontal permeability (K_x) is presented in Figure 5. Although the detail of the permeability distribution is not clear in the black and white illustration, it can be seen that the model is very heterogeneous. Note that the z-direction has been greatly exaggerated. A slight incline can be seen in the 3D view, so mobile gas may tend to collect at the top of the model near the producer.

The model initially contains undersaturated oil at an average reservoir pressure of 3220 psia with no free gas. The simulation starts on 1st January 1976 with no production until June 1976. Waterflood is then performed until January 2005. The production rate is specified as 1900 rb per day, which represents production of approximately 3% of the initial hydrocarbon pore volume per annum. Voidage replacement is used to maintain the reservoir pressure at 3000 psia, and the fluid remains above its saturation pressure. Recovery of stock tank oil to January 2005 represents 67.1% of STOIIP. A summary of the fluids in place on 1st January 2005 is presented in Table 1. The high oil recovery is consistent with the oil–water relative permeabilities discussed earlier. Following the waterflood, a WAG process is simulated, with the simulations continuing until January 2050.

Simulation cases

In addition to the base case waterflood, which was run until 2050, a number of variant cases were run to investigate a range of different issues. The main variants are described briefly in the following sections.

Base case WAG. Gas injection was commenced in 2005, with 10 WAG cycles (each consisting of 2 years gas injection followed by 2 years water injection, with equal volumes of gas and water injected). The final WAG cycle finished in 2044 and was followed by a waterflood until 2050. It is noted that in this study we did not investigate the effect of shorter WAG cycles; there may be potential for improving the recovery by optimising the cycle length.

(a)

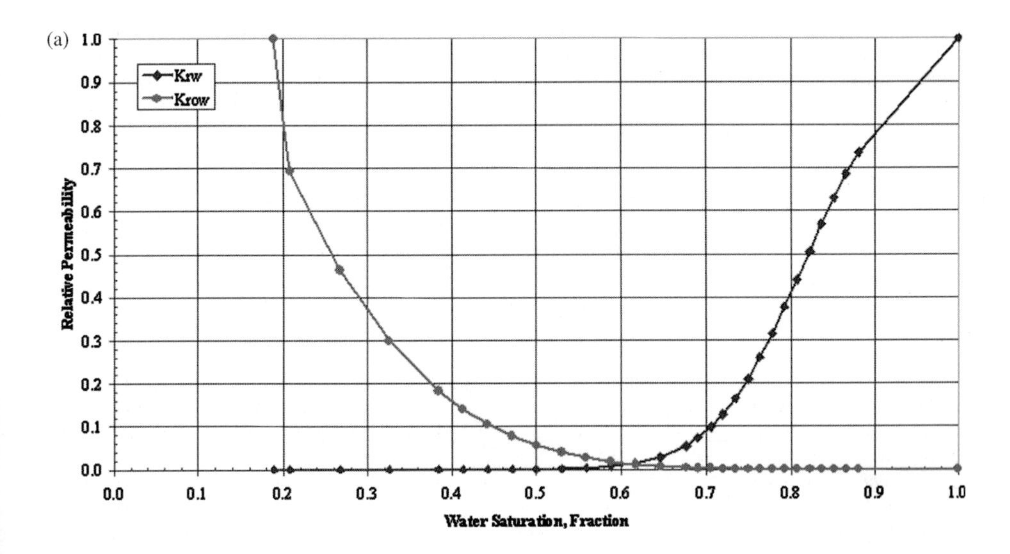

(b)

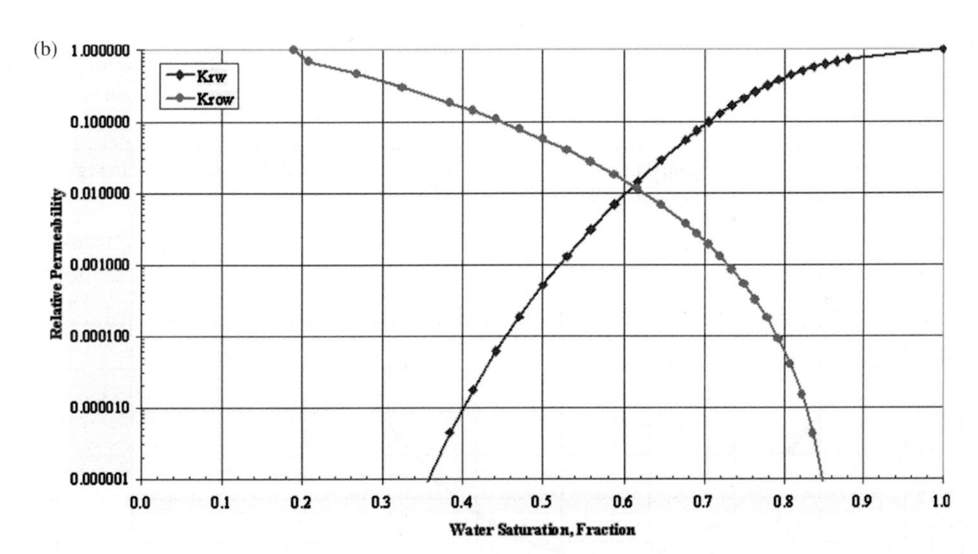

Figure 2: (a) Oil–water relative permeability (linear plot). (b) Oil–water relative permeability (logarithmic plot).

Postflood with gas. This case was similar to the base case WAG until 2025 (5 WAG cycles). The fifth WAG cycle was followed by postflood gas injection which continued till the end of the run.

3-Cycle WAG. In this case, the length of the gas injection periods was increased by a factor of three (giving a gas to water volume ratio of three to one for each WAG cycle) and the number of cycles reduced to three. The final WAG cycle concluded in 2028. Two alternatives were considered beyond this. In the first case,

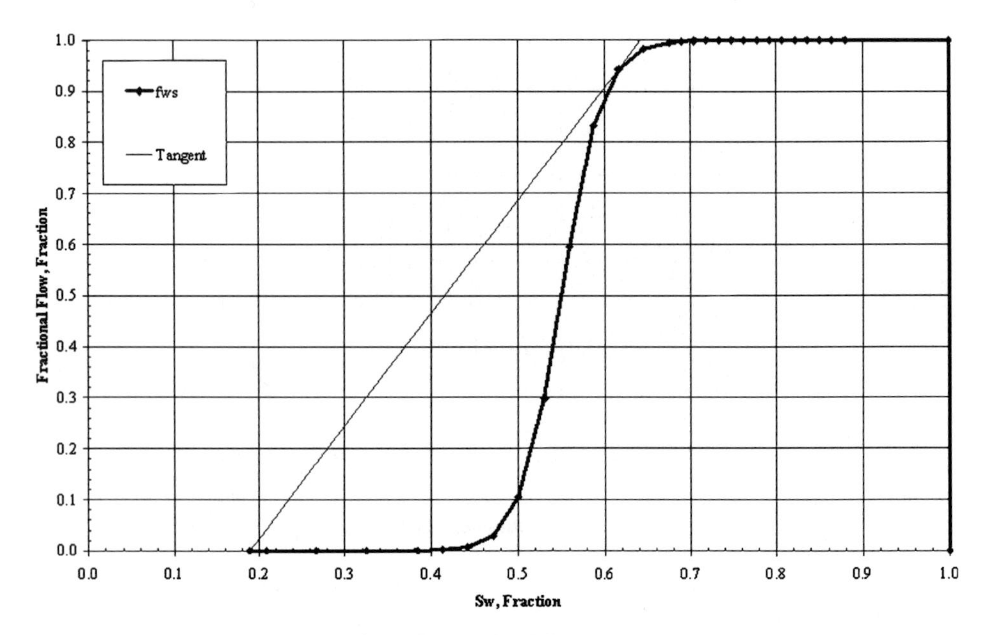

Figure 3: Fractional flow curve.

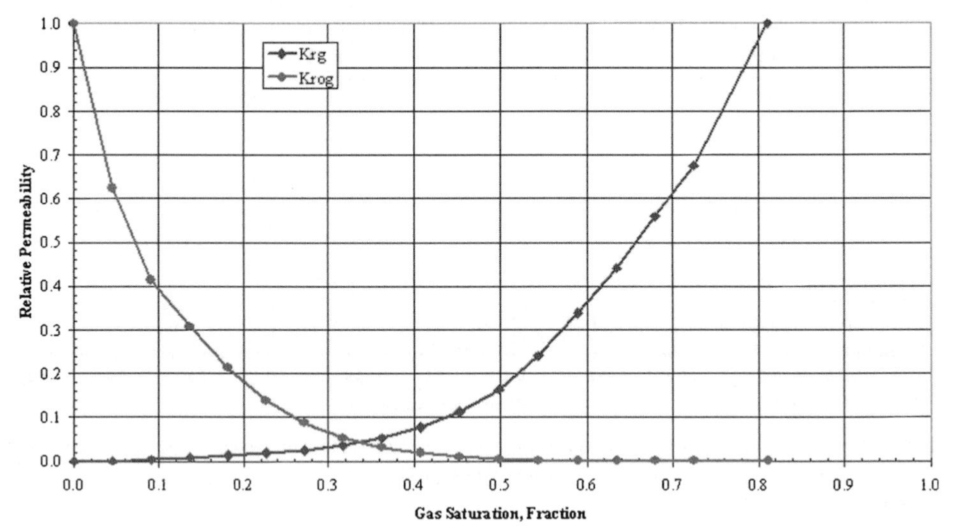

Figure 4: Gas–oil relative permeabilities.

the final WAG cycle was followed immediately by postflood gas injection, while in the second it was followed by three units of water injection and then the postflood gas injection.

Other sensitivities. Further sensitivity cases were undertaken to investigate the sensitivity to flow rate, injection well location and the position of completions as well as to look at the redistribution of fluids beyond 2050 once all the wells were shut-in.

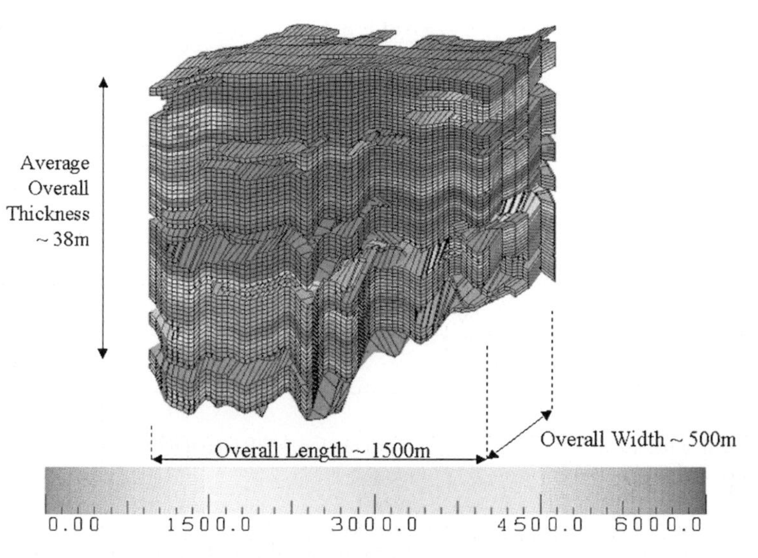

Figure 5: Model permeability (K_x) showing heterogeneity.

TABLE 1
FLUIDS IN PLACE ON 1ST JANUARY 2005

Fluid	Amount in place
Oil	5.9 MMstb
Oil recovery	67.1% STOIIP
Average So	26.7%
Water	18.7 MMstb
Average Sw	73.3%

Results of simulations

The 10-cycle WAG case showed an increase in oil recovery compared to the base case waterflood, of 6.5%, with almost 25% of the hydrocarbon pore volume containing CO_2 at the end of the run (see Table 2). The effect of the postflood gas injection was to increase the CO_2 storage while also resulting in a further small increase in oil recovery.

The 3-cycle WAG resulted in an acceleration of oil production, while further increasing the amount of CO_2 stored in the reservoir. Inclusion of the extra water slug at the end of the final WAG cycle had a detrimental effect on the oil recovery.

The maximum volume of CO_2 stored in the reservoir was around 50% of the hydrocarbon pore volume; this required approximately 1.2 hydrocarbon pore volumes of gas injection, so that a significant proportion of the injected gas was produced, indicating a need for gas recycling.

The 3-cycle WAG (postflood gas) case was run beyond 2050 with all wells shut-in. This showed a marked redistribution of fluids over a period of more than 100 years after shut-in. Gas moved upwards until trapped under shales or at the top of the reservoir, while oil continued to accumulate through the effects of gravity drainage.

TABLE 2
IOR AND AMOUNT OF CO_2 SEQUESTERED IN 2050

Case	Oil recovery (% STOIIP)	IOR (rel waterflood, % STOIIP)	CO_2 stored (% HCPV)
Waterflood	73.0	0.0	<0.1
10-Cycle WAG	79.5	6.5	22.7
10-Cycle WAG (postflood gas)	80.9	7.9	43.7
10-Cycle WAG (postflood gas) and relocate injector	82.8	9.8	48.9
3-Cycle WAG (postflood gas)	79.8	6.8	44.4
3-Cycle WAG (waterslug, then postflood gas)	79.2	6.2	40.1

It was observed that the use of WAG accelerated oil production, while maximizing the storage of CO_2 led in turn to an increase in oil recovery.

Modelling of Regional Fluid Flow to Underpin Risk Assessment

Following the favourable conclusions of the numerical simulation of WAG using CO_2 as the injection gas, it was recognized that risk assessment would be necessary, both to evaluate the likely safety and security of storage of CO_2 within the field and to secure any putative carbon credits for the storage of CO_2.

Assuming that there are no faults or other fractures in the cap rock, CO_2 stored in hydrocarbon fields can escape from the trap via the following routes: (1) abandoned or producing wells, (2) diffusion through the cap rock, and/or (3) dissolution and transport of CO_2-charged waters along the aquifer by groundwater flow (see below). In order to underpin the assessment of the risk of CO_2 escaping via route 3, a study of the regional (i.e. basin-scale) fluid flow patterns was undertaken, to constrain the likely rates of fluid flow along aquifers and across aquitards surrounding the Forties field. This would simulate the likelihood of CO_2 migrating out of the reservoir and eventually reaching the seabed.

For a realistic fluid flow model to be constructed, a regional cross-section from basin margin to basin centre, that passed through or close by the Forties oilfield, was required. The section needed to detail the geometry of key geological horizons, emphasizing sandy stratigraphic units and the connectivity of these with the Forties reservoir rocks. A seismic profile from a speculative survey shot by WesternGeco, passing within 20 km of the Forties field was chosen to provide this information. It is 250 km long and data quality is good. The profile is oriented approximately EW, from a position 42 km off the coast of NE Scotland near Fraserburgh, eastwards into Norwegian waters (Figure 6). It passes through or close by many key hydrocarbon wells and hence geological control on the interpretation of reflectors is good.

Seismic interpretation

Interpretation was concentrated from the western limit of the seismic profile to, and slightly beyond, the North Sea basin centre; the Forties field lies on the western basin margin, and any escape of CO_2 would be either driven upwards across the stratigraphy by buoyancy, or westwards by regional fluid flow due to compaction, using porous and permeable layers within the stratigraphic succession as conduits.

The interpreted seismic profile is presented in Figure 7 and reveals the complex nature of the basin fill. The Forties field lies to the south of the profile but the Forties-Montrose Ridge, on which the Forties field lies, can be clearly seen. The stratigraphic level of the Forties reservoir and hence of the potential CO_2 injection is indicated. The base of the Cretaceous Chalk is marked, and below this the succession is faulted and very variable in thickness. The lowest horizon is the Variscan unconformity, regarded for modelling purposes as impermeable and the base of the model. The Chalk extends across the faulted succession and, although folded, appears essentially unfaulted. Above the top of the Chalk, the Cenozoic to Recent succession extends to the seabed and can be divided into prograding basin margin successions that progressively move further

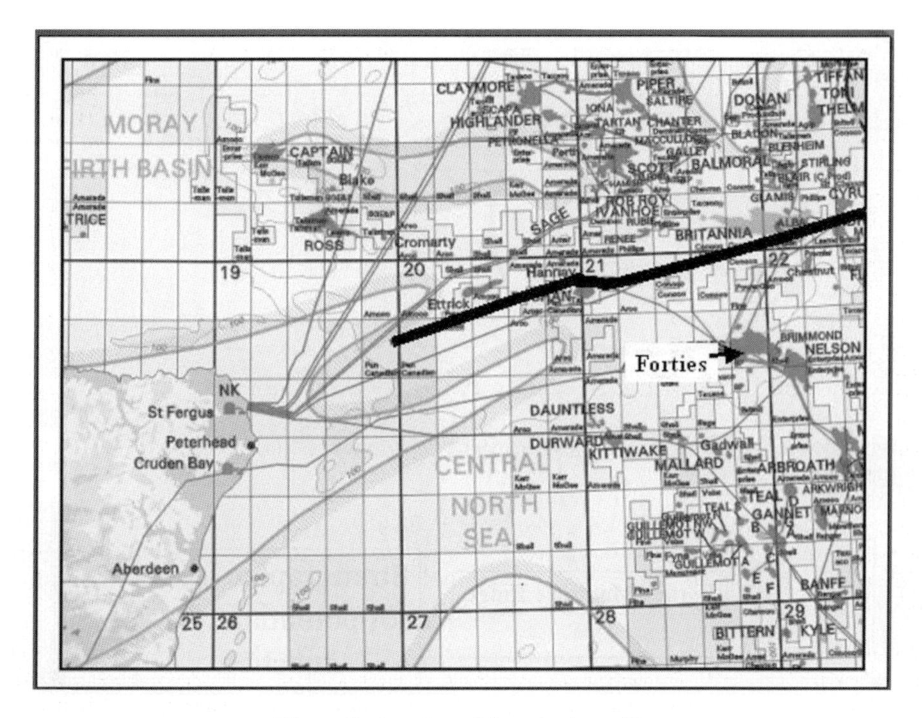

Figure 6: Location of the seismic profile.

towards the basin centre, and basin fill successions that infill and lap out against the prograding successions. Wells permit the identification of particularly sandy intervals, and these are shaded. It is immediately evident that there is a thick, and dominantly shaley and therefore assumed sealing, succession above the Forties reservoir rocks. Detailed analysis of well logs through this interval suggests that in the Forties field location, and also farther towards the basin centre, there are a few slightly sandier horizons within this shaley unit. However, they do not appear to be laterally continuous and are thought therefore to provide little significance in the way of migration paths for any CO_2 that might escape vertically from the Forties field.

Of particular importance in the Cenozoic succession is the thick sequence of stacked sandy basin floor fans immediately above the Chalk in western and central areas and the overlying prograding sandy interval. A more detailed image of this section of the seismic profile, circled in Figure 7, is presented in Figure 8.

The topmost of the stacked basin floor fans is the Forties Sand, which forms the lower of the two main reservoirs in the Forties field. This appears to pinch out to the west under a thick and dominantly shaley and therefore "sealing" prograding succession. It is thought that any CO_2 migrating laterally out of the Forties Sandstone would not have an obvious path to the surface via this route. The upper reservoir in the Forties field, the Charlie Sand, appears to form the distal "toe sets" of the prograding Dornoch Delta succession immediately to the west. Wells penetrating this interval show that this prograding unit is largely sandy in its upper part, which is known as the Dornoch Sandstone. Seismic evidence therefore suggests that the Charlie and Dornoch Sands may be contiguous and therefore provide a potential migration path for any CO_2 that may move laterally out of the Forties field.

Basin modelling
The main aims of the regional 2D basin modelling were to give an indication of the basin-scale natural fluid flow to be expected within the various rock layers surrounding the Forties field reservoir, and evaluate

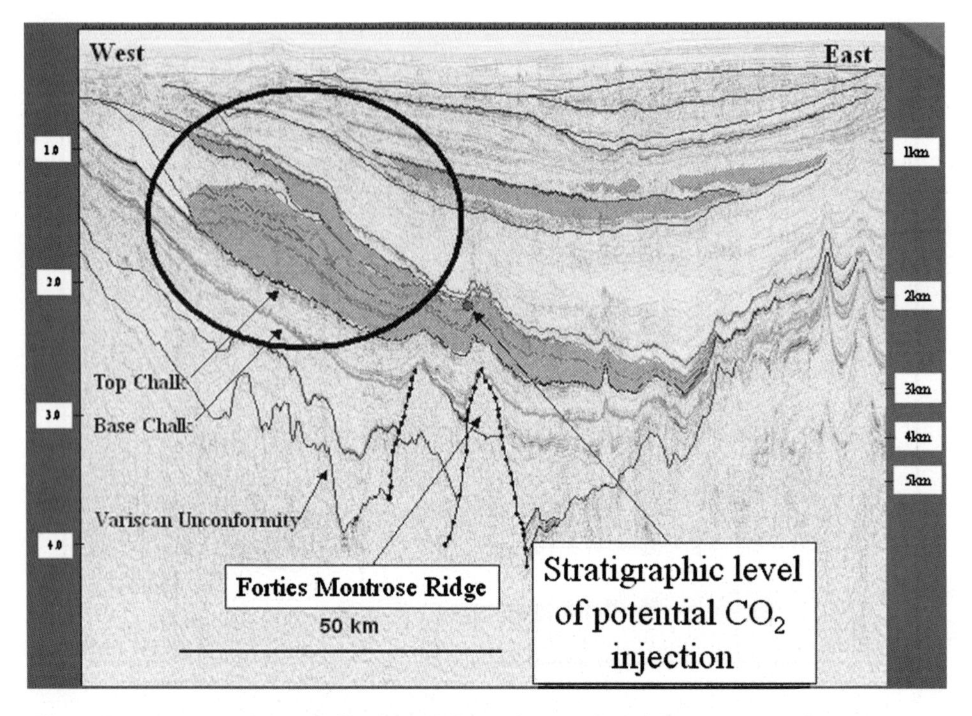

Figure 7: Interpreted seismic profile. Potentially sandy units shaded. Data courtesy of WesternGeco.

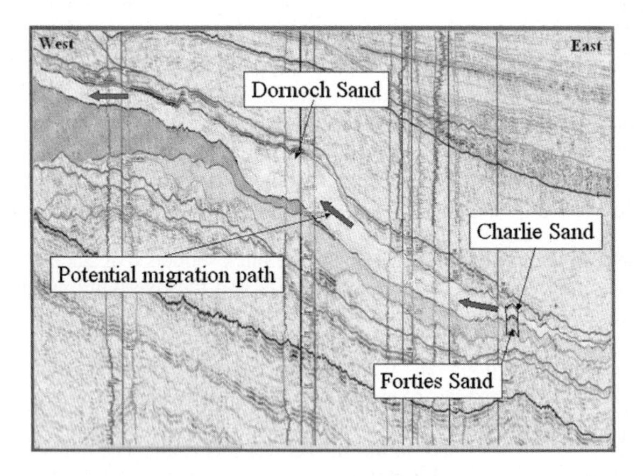

Figure 8: Detail of the lateral extent of the Forties reservoir sands.

pressure boundary conditions for input into a more detailed 3D model of the field itself. The modelled profile does not cross the Forties field and therefore was not necessarily expected to match the pressure data measured at Forties exactly. PETROMOD basin modelling software was used. The methodology used was to set up a "base case" model using the geological framework obtained from the interpreted seismic profile and the boundary conditions described below. The model was then run to calculate the pore fluid pressure distribution within the various rock layers making up the profile. Subsequently, scenarios were run in which the geology or other parameters of the model were varied, to assess how these might affect the pore fluid pressure distribution.

Model construction. The interpreted seismic section was depth-converted using check-shot surveys from seven of the in-line wells. The section was then extended to the west, using seabed outcrop data so that the termination of the stratigraphic layers at the seabed could be defined, which is important for the flow calculations. The depth-converted horizons were input into the basin modelling program, together with information on the lithological composition of the seismic intervals, and the likely age of the horizons. This resulted in a "base case" 2D model consisting of 62 "events" (model layers) and 108 grid points along the 2D section. The geometry and lithology of the base case model are shown in Figure 9. Each layer has a standard pre-defined lithology, porosity, and permeability and each event was assigned a duration, to allow different scenarios to be modelled.

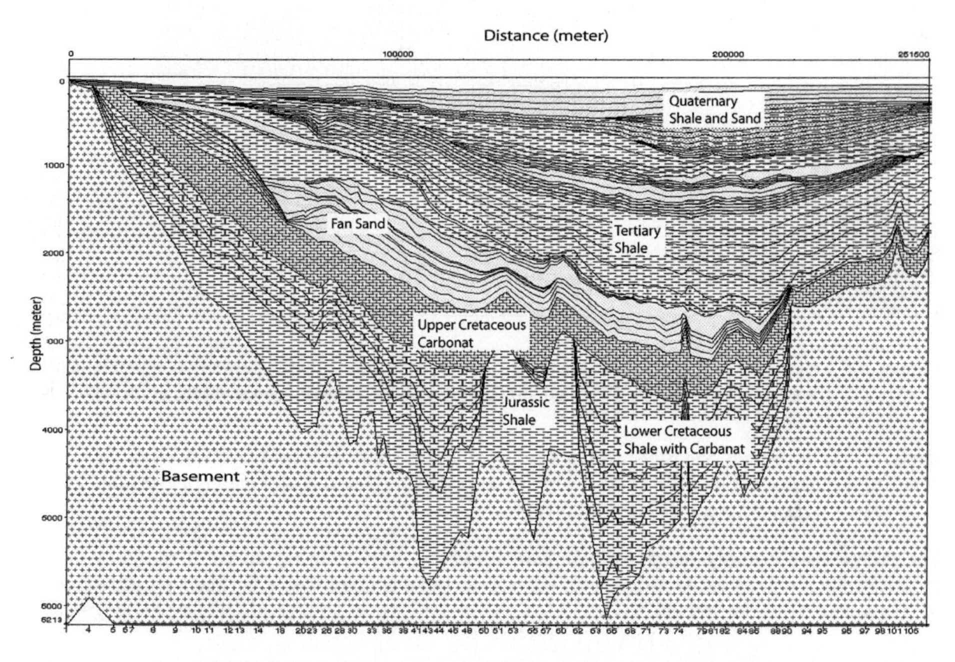

Figure 9: Geometry and lithology of the "base case" model.

Boundary conditions. The flow boundary conditions assumed for the 2D model were open, except for a no-flow boundary imposed at the impermeable basement rocks at the base and the western side of the model. The section was chosen such that the depocentre for the Cenozoic shale generating most of the overpressure in the Cenozoic section is located some 50 km from the eastern boundary (Figure 10). Fluids were expected to move away from this area. Towards the western end of the profile fluids were expected to move to the seabed, particularly along carrier beds. At the eastern end of the profile fluids were expected to move to the east through the open boundary.

Base case. The base case model (Figure 9) was based on the following simplified geological assumptions: The base of the Jurassic is taken as a no-flow boundary. The lowest layer in the model is the Jurassic section, which is represented as shale with source-rock properties. The overlying Lower Cretaceous section also contains shale, but it also contains some carbonate, especially in the South Viking Graben, therefore the lithology chosen for this event was shale with a little carbonate. This is followed by a section of Upper Cretaceous Chalk. On top of the Chalk there is a series of fan sands of Palaeocene to Eocene age. These

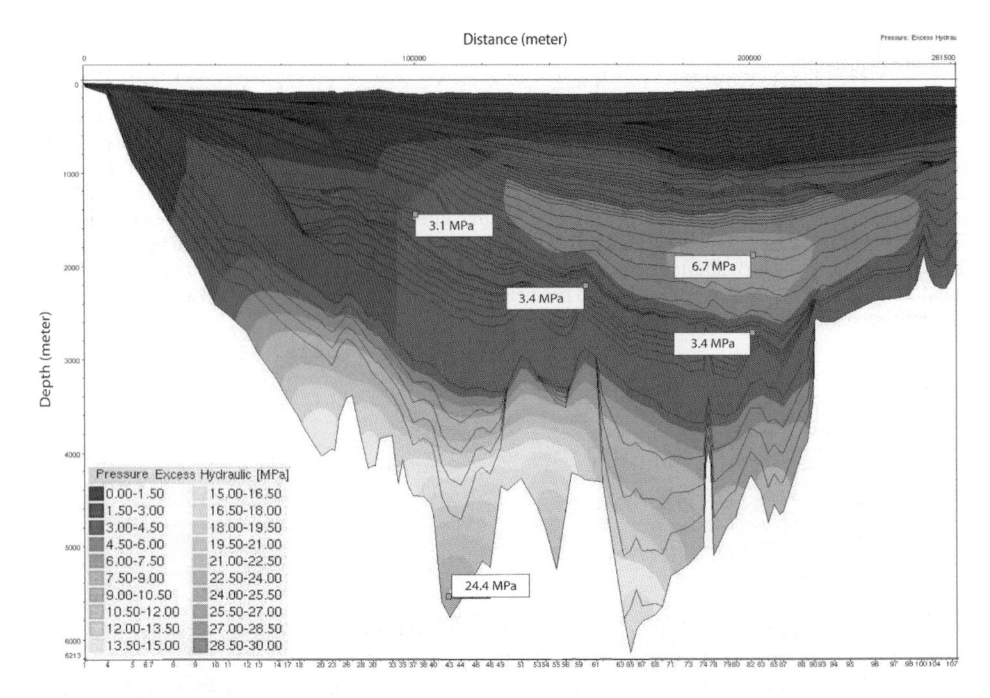

Figure 10: Calculated overpressure distribution for the "base case" model shows two separate centres of overpressure—one in the Cenozoic and one in the Jurassic/Lower Cretaceous.

sands include the Forties Sandstone. A thick series of shales with a few sand layers overlies these sands—none of the sand layers has direct contact with the seabed in the base model (Figure 9). The lithologies used in the base case model were standard lithotypes.

The overpressure distribution for the base case model can be seen in Figure 10. It shows two different pressure centres; one in the Jurassic/Lower Cretaceous and one in the Cenozoic, and a pressure gradient in the fan sand of 0.4 MPa/100 km.

Model scenarios. Different scenarios were simulated, first by adding shale layers in the fan sand, then lowering the sedimentation rate in the last part of the Quaternary from approximately 50 m/100,000 years to 25 m/100,000 years and finally removing the active petroleum system in the Jurassic. This lowers the overpressure in the Jurassic almost 20 MPa, but the overpressure in the fan sand drops less than 1 MPa and the pressure gradient is only changed 0.2 MPa/100 km.

Scenarios with permeability in the fan sand, lowered by up to two orders of magnitude, show that the overpressure in the Forties position varies from 3.4 MPa in the "base case" model to 2.3 MPa in the case with the lowest permeability. The pressure gradient changes by only 0.1 MPa/100 km.

In a scenario with no active hydrocarbon system and with Cenozoic shales with one order of magnitude higher permeability, the overpressure in the Cenozoic shales drops approximately 5 MPa. However, the pressure in the fan sand only drops 1.7 MPa and the pressure gradient in the sands is unchanged. This scenario was further modified with more sand, a better hydraulic connection to the seabed and a leaking fault approximately 70 km east of the Forties position. The overpressure drops to almost hydrostatic (0.4 MPa) at the location similar to the Forties and the pressure gradient becomes almost zero and it actually reverses close to the location of the fault. The calculated pressure distribution for this case can be seen in Figure 11.

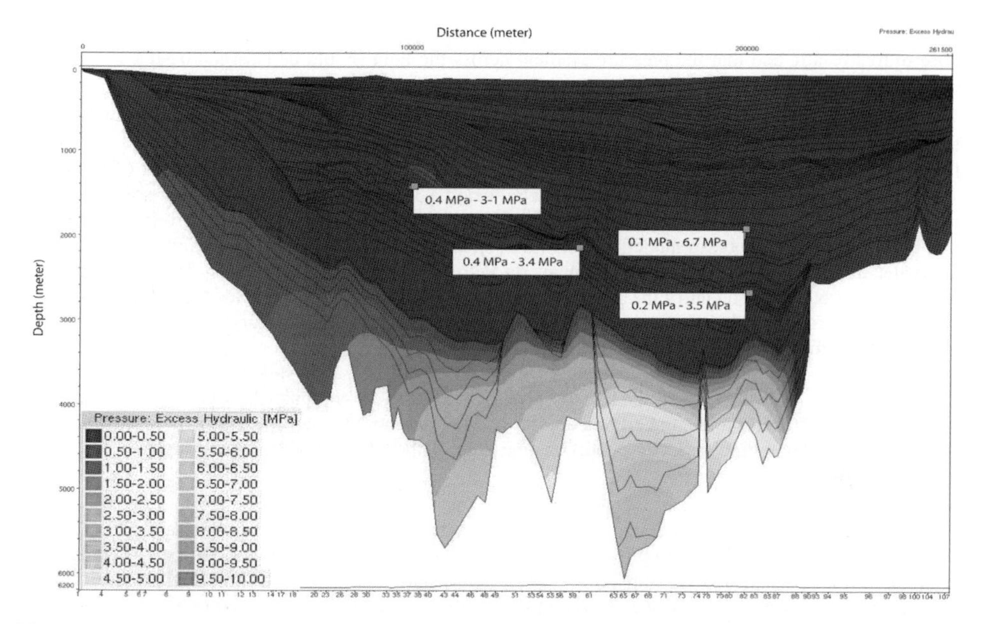

Figure 11: Calculated distribution of overpressure for case with no active hydrocarbon system, Cenozoic shales with one order of magnitude higher permeability, good hydraulic connection to the seabed and a leaking fault approximately 70 km east of the Forties position. Numbers to the left, this case; to the right, "base case" model.

Results. The 2D modelling reveals two almost separate pressure systems: (1) a Jurassic–Lower Cretaceous system which is related to grabens with tight shale and active hydrocarbon systems (2) a Palaeocene/Neogene system of which the Forties Sand is a part. Pressure changes in the Jurassic of 20 MPa only generate a pressure change in the fan sand of 0.3 MPa, demonstrating the small degree of communication between the two systems.

Whilst one of the main factors influencing the Jurassic–Lower Cretaceous system is hydrocarbon generation, a variety of factors influence the Palaeogene–Neogene system. The most important factor is the composition of the Eocene–Miocene shales, but factors such as Quaternary loading and composition, geometry of the fan sand and distance to leaks also have influence on the pressure system.

The results from the 2D modelling show that the potential range of pressures in the Forties Sands along strike from the Forties Field may vary from almost hydrostatic (zero overpressure) to an overpressure of 3.5 MPa. The pressure gradient is very stable, being less than 1 MPa/100 km in all scenarios, with an average for all scenarios close to 0.5 MPa/100 km. The regional pressure gradient from West Central Graben to Inner Moray Firth is 10 MPa over 320 km which, with average Forties Sandstone parameters from the Montrose Field (permeability 80 mD, porosity 23%), gives a flow velocity of 33 cm/year (330 m per thousand years) [5]. Using the same parameters, the modelled pressure gradient in this study of less than 1 MPa/100 km results in even smaller flow velocities.

The scenario that assumes a connection between the Charlie Sand and the Dornoch Sand, and thus a better hydraulic connection to the seabed, gives almost hydrostatic pressure in the Forties Sand on the Forties-Montrose ridge, which is in good agreement with published data [6] and pressure data from the Forties Field.

The regional 2D modelling shows that the pressure boundary conditions for the 3D modelling should be close to hydrostatic and that the pressure gradient across the field should lie between 0 MPa/100 km and 1 MPa/100 km.

Modelling Potential Escape Routes for CO_2 from the Forties Field

The next step in the workflow necessary to underpin the assessment of the potential risks of CO_2 leakage from the Forties field was to model the field in detail using a 3D model. We defined a multi-scale approach comprising two stages: (1) 3D simulation of the fluid flow in the Forties field and its surrounding drainage area using the TEMIS3D basin modelling software [7]; and (2) simulation of the interaction between CO_2 and water (diffusion) using the Institut Francais du Petrole (IFP) SIMUSCOPP fluid flow simulator. The basin model was used to quantify the groundwater flow pattern (direction and velocity) within the Forties main drainage area. This flow pattern was then used to determine pressure boundary conditions for reservoir simulations performed by the IFP SIMUSCOPP code.

The scope of the fluid flow modelling with SIMUSCOPP was to compute the CO_2 escape routes and quantify the CO_2 transfer to the underlying aquifer and the overburden once the CO_2 is in place within the Forties Field. At prevailing reservoir conditions the bulk of the CO_2 would be supercritical.

Geological setting

The Forties field consists of an anticline with sandy reservoirs, known as the Forties Sand and Charlie Sand, located in the Central North Sea (Figure 11). The original oil–water contact was at 2217 m [8]. Temperature in the reservoir is 90 °C and pressures are close to hydrostatic. There are no large faults present in the reservoir or the overburden sequence and the minor faults encountered are believed to have no significant influence on reservoir continuity or production [8].

Basin model (TEMIS3D). Primarily designed to simulate compaction, source-rock maturation and hydrocarbon migration, TEMIS3D can also be used, as here, as a single-phase water flow simulator able to quantify the development of overpressures and the direction and magnitude of water flow. Fluid flow is defined by Darcy's law, using permeability as a function of porosity through the Kozeny–Carman law. The TEMIS3D block construction followed the classical numerical modelling steps [7].

Database. Provided by BP, the database included 10 isobath maps, high-resolution petrophysical data from seismic inversion of a 3D reservoir block, well logs and core description.

Gridding. Due to CPU limitation, a model grid of approximately 400×400 m^2 in horizontal plan was chosen in order to compromise between cell size and total number of cells (Figure 12).

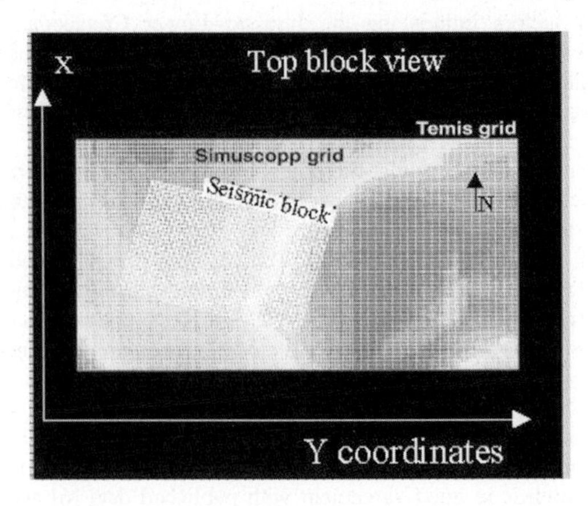

Figure 12: Optimised TEMIS3D horizontal grid pattern.

Layers. The 3D TEMIS initial block comprised nine seismically derived layers. In order to take advantage of the more detailed information that was available and facilitate importation of the high-resolution lithological data, 31 additional interpolated layers were created, particularly in the intervals associated with the Forties reservoir (14 layers) and the cap rock (five layers).

Lithology. For each layer, lithology maps were created. These maps were elaborated from the studied wells. Additional high-resolution lithological information, obtained from seismic inversion, was incorporated into the TEMIS3D block using a geo-modeller (Figures 13 and 14). Porosity vs. depth curves originated from default TEMIS3D lithology types or were extracted from literature [9] and references therein.

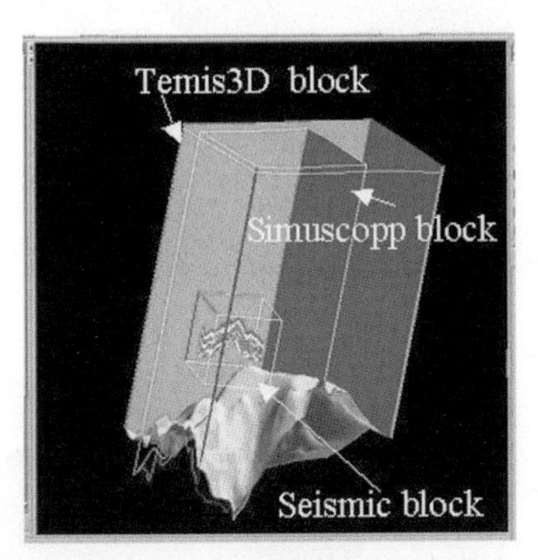

Figure 13: Model block location.

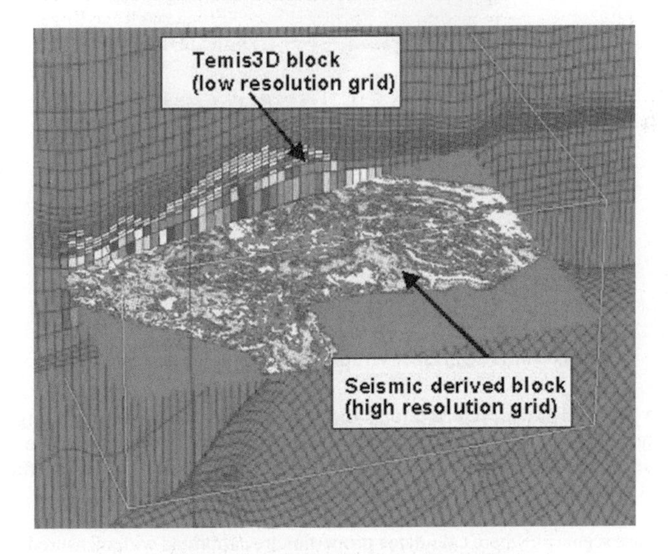

Figure 14: Distribution of grid blocks.

Boundary conditions. Data from the Forties reservoir indicates pressure close to hydrostatic [10]. This agrees with the results of the 2D numerical modelling described above. However, regional studies suggest that there is a small pressure gradient between the depocentre of the basin and the region to the NW (Figures 15 and 16).

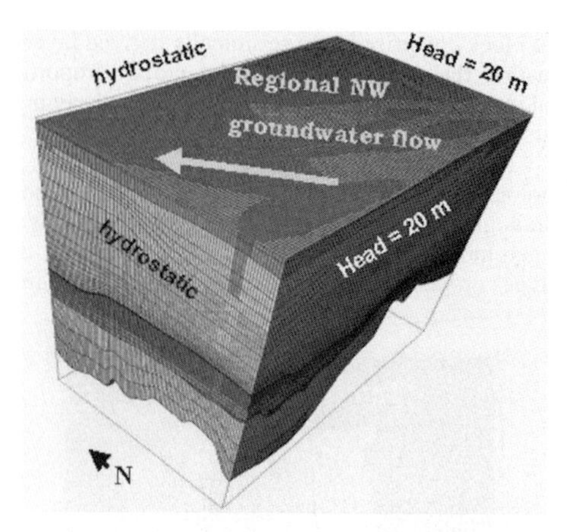

Figure 15: TEMIS3D boundary conditions (base model).

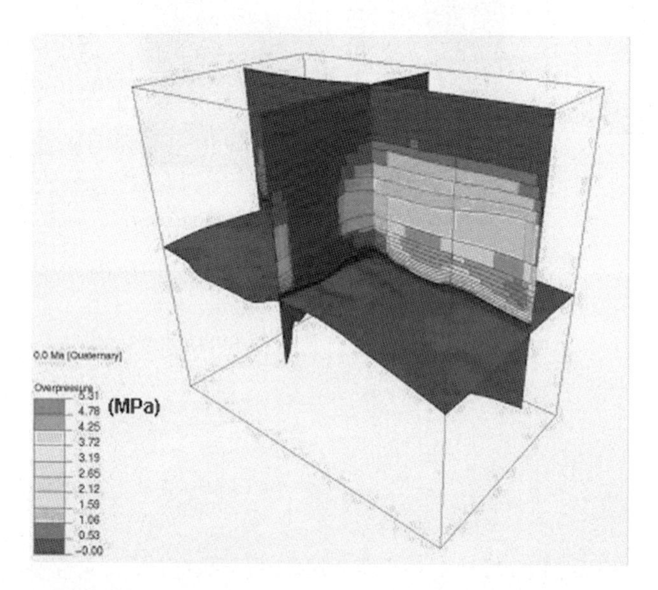

Figure 16: TEMIS3D overpressure distribution, present day.

Sensitivity tests. A first simulation, referred to as the "base case", was performed using the above (most realistic) input data. Additionally, as part of sensitivity tests, 12 simulations were made to encompass the uncertainty of the input parameters (mainly permeability—both of the reservoir and overburden intervals—and the water head boundary condition).

Results. The base case scenario, where calculated properties are calibrated with measured data, suggests that in the overburden interval, overpressures are up to 5.3 MPa, and mainly located in the centre and southern and eastern parts of the model, approximately between 1800 and 2300 m depth (Figure 16). In the centre of the model overpressure drops to <1 MPa. In the Forties reservoir there is an overall decrease in pressure from the south and east to north and west. This pressure distribution pattern exerts a strong influence on the present-day water circulation pattern. The base case scenario indicates that at present, the water in the

Forties reservoir flows dominantly horizontally, from southeast to northwest, at a velocity lower that 500 m/Ma (Figures 17 and 18). Water flow in the upper overburden interval (0 to approximately 1400 m) typically ranges from below 60 m/Ma in the mudstones up to perhaps 150 m/Ma in the siltstones, with flow vectors pointing predominantly upwards (Figure 17). In the lower overburden interval (>1400 m) water flow typically ranges from below 10 m/Ma in the mudstones up to perhaps 30 m/Ma, with flow vectors pointing predominantly downwards. Water flow in the interval below the reservoir (Cretaceous and Danian) is less than10 m/Ma, in part because of the low permeability values encountered (<0.01 mD). The sensitivity tests showed no significant changes in the orientation of the regional water flow, and pressure in the reservoir interval was always near hydrostatic (i.e. overpressure <0.01 MPa).

Figure 17: TEMIS3D fluid flow distribution at present day (base case).

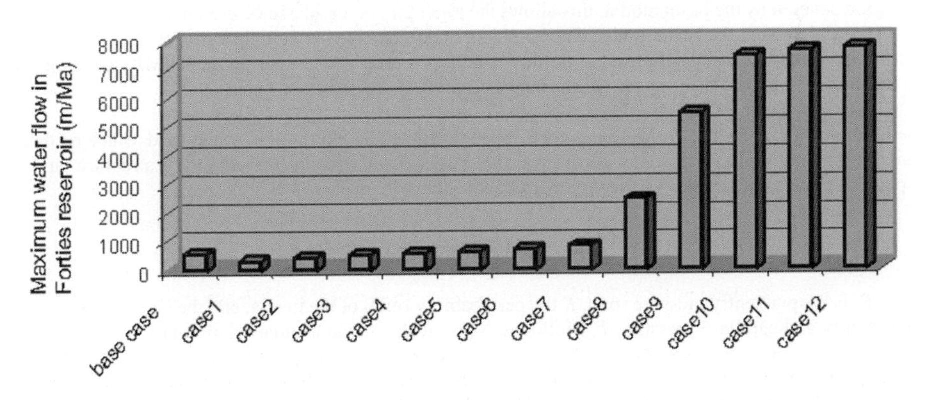

Figure 18: Sensitivity tests—maximum water flow at present day.

The modelling results suggest the following.

1. The present-day water flow is dominantly horizontal, from southeast to northwest, with velocity likely to be lower than 500 m/Ma.
2. The sensitivity tests show that with the worst case scenario water flow reaches 8000 m/Ma (Figure 18). This means that water flow is still too slow to remove significant amounts of CO_2 from the reservoir

by dissolution in the aquifer (maximum horizontal displacement = 8 m in the time framework of the storage-1000 years).
3. The near hydrostatic pressures lead to the conclusion that the boundary conditions for reservoir simulation (SIMUSCOPP IFP software) can be set as hydrostatic.

Fluid Flow Simulator (SIMUSCOPP)

SIMUSCOPP is a 3D, 3-phase compositional fluid flow simulator, which uses Darcy's law and mass conservation to compute pressure and saturation variation over the whole model. The conservation equations are solved using the classical finite volume method. SIMUSCOPP assumes a block-centred grid but easily handles local grid refinement and dual media to better characterize fluid flow. It is designed to handle complex, laterally variable aquifers. Fluid flow properties can be described through user-defined data or by using a governing equation such as the Peng–Robinson equation of state to compute phase density variation or the Lorehnz–Bray–Clark correlation to compute phase viscosity variation with pressure. In its current version, SIMUSCOPP is isothermal. However, SIMUSCOPP handles CO_2 dissolution in water through (tabulated user-defined) equilibrium constants and also CO_2 diffusion in the water phase. SIMUSCOPP has been successfully applied to model CO_2 behaviour [11,12]. In its current version, SIMUSCOPP does not handle any fluid-rock chemical reaction. SIMUSCOPP simultaneously solves for pressure and saturation within the whole field either using a fully implicit numerical scheme or a numerical scheme implicit for pressure and explicit for saturation. To compute its initial conditions, SIMUSCOPP assumes a capillary-gravity initial equilibrium. Then, using mass conservation and Darcy's law, it computes the evolution of pressure and saturation with time.

Fluid flow data

In order to understand CO_2 behaviour in Forties after the CO_2 injection period, CO_2 diffusion and dissolution in water must be modelled. Furthermore, the overburden and underlying strata must be characterized in terms of their lateral extension and petrophysical properties such as permeability and porosity and, most importantly, capillary pressure behaviour and relative permeability.

The fluid flow model (SIMUSCOPP) takes significantly longer to run for an equivalent size of problem than the basin model (TEMIS3D). Therefore the fluid flow calculations had to be carried out on a smaller grid. To avoid any upscaling issues between the basin and the local scale, the fluid flow model only covers a part of the region covered by the basin model; this allows the same grid block size to be used with the same porosity and permeability as in the basin model. The boundary conditions for the fluid flow model are assumed to be hydrostatic on all lateral boundaries and no-flow conditions otherwise, based on the TEMIS3D results and the location of the fluid flow model within the basin model.

The multiphase data for the Forties Sand are derived from the reservoir model of the field. Since no data is available for multiphase flow properties in the shale (overburden), the capillary pressure and its displacement pressure are derived form a permeability based correlation [13]

$$P_e = 7.37K^{-0.43}$$

where P_e is the pore entry pressure (psi), K the permeability (mD) of the media, and the CO_2–water relative permeability and capillary pressure (P_c) follow the classical Van Genuchten relation [14]

$$k_{rg} = \sqrt{S_g^*}\left\{1 - (1 - [S_g^*]^{1/\lambda})^{\lambda}\right\}^2 \quad \text{and} \quad P_c = -P_0([S_g^*]^{1/\lambda} - 1)^{1-\lambda}$$

where λ is the Land exponent of the shale and S_g the dimensionless gas saturation.

The petrophysical model was reduced to only two rock-types: sand, i.e. the Forties Sand within the reservoir and shale anywhere else.

The CO_2 thermodynamic properties are derived from literature [15] whilst the CO_2 equilibrium solubility is computed from Duan equation of state [16].

After CO_2 injection, the Forties reservoir is assumed to be at its original pressure (22.7 MPa) and the CO_2/water contact is located at the initial water–oil contact (2217 m). The CO_2 saturation is assumed uniform and constant at 50%.

Fluid flow results

The goal of the fluid flow simulation was to compute a "reasonable" worst case scenario and determine the CO_2 escape rate out of Forties over a 1000-year period. To achieve this, some the model parameters were varied within reasonable limits to maximize the CO_2 leakage rate, e.g. by assuming the same effective diffusion within the sand and the shale, i.e. assuming a uniform and constant tortuosity and bulk diffusion coefficient.

CO_2 does not break the capillary barrier of the overburden (2125 m) at any time during the 1000-year period, since no gaseous nor supercritical CO_2 is seen above the Forties reservoir (Figure 19). However, CO_2 diffusion within the water phase transports CO_2 upward but only less than 50 m into the first layer of the overburden (Figure 20). Due to CO_2 dissolution in water and high permeability within strata underlying the Forties reservoir, CO_2-saturated water migrates downward, driven by the density contrast with undersaturated aquifer water. In this base case scenario, only 3.6% of the original mass of CO_2 migrated out of the reservoir over the 1000-year period.

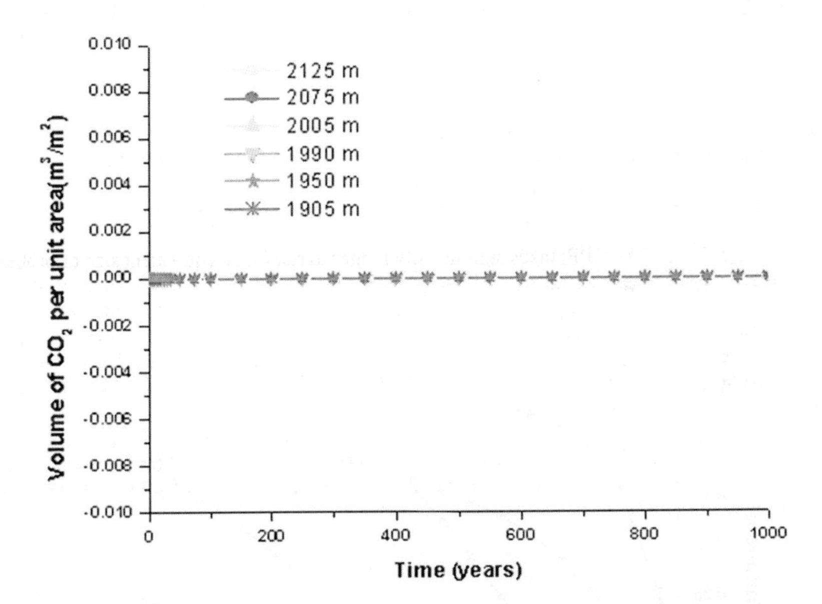

Figure 19: Volume of CO_2 versus time at different depths within the overburden of Forties.

In the worst case scenario approach, assuming a complete capillary barrier failure of the cap rock (i.e. zero pore entry pressure in the shale) the CO_2 migrates upward both in gaseous form (Figure 21) but more importantly through diffusion within the water phase (Figure 22). The upward migration of CO_2 is quite significant since the dense supercritical CO_2 rises almost 175 m within the overburden during the 1000-year period (Figure 21). The influence of water diffusion is still quite significant since dissolved CO_2 rises almost 350 m in the same period (Figure 22). In this worst case scenario, nearly 37% of the original mass of CO_2 migrated out of the reservoir over the 1000-year period.

Despite the quite extreme assumption of complete failure of the cap rock capillary barrier, CO_2 migration from Forties is limited to an area up to 350 m above the reservoir. Under a more realistic set of assumptions,

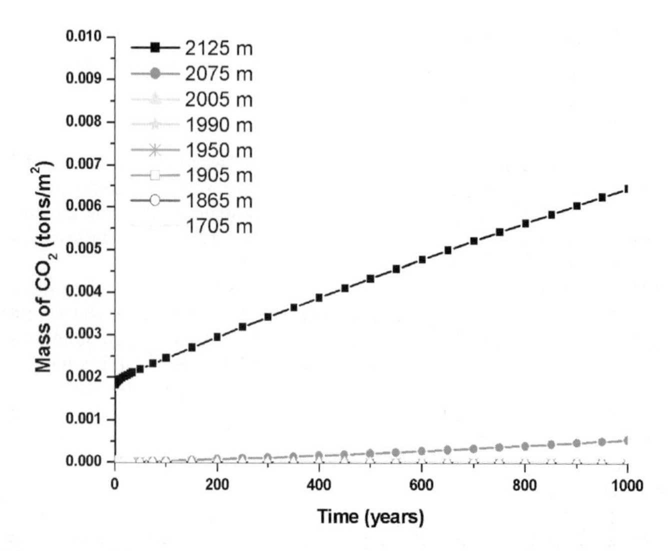

Figure 20: Mass of CO_2 versus time at different depths within the overburden of Forties.

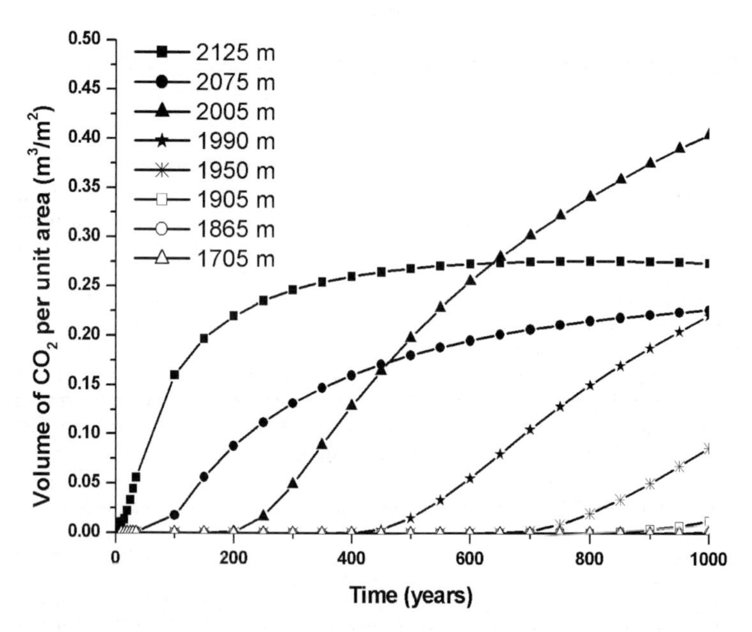

Figure 21: Volume of CO_2 versus time at different depths within the overburden of Forties cap rock failure case.

minimal CO_2 migration above the Forties cap rock is predicted. Due to the absence of major faults, hydrostatic conditions (mainly due to its offshore location), and the thickness and very low permeability of its overburden, Forties is an appropriate structure for CO_2 storage as long as significant cap-rock characterization (capillary and permeability properties) is available to validate the model hypothesis.

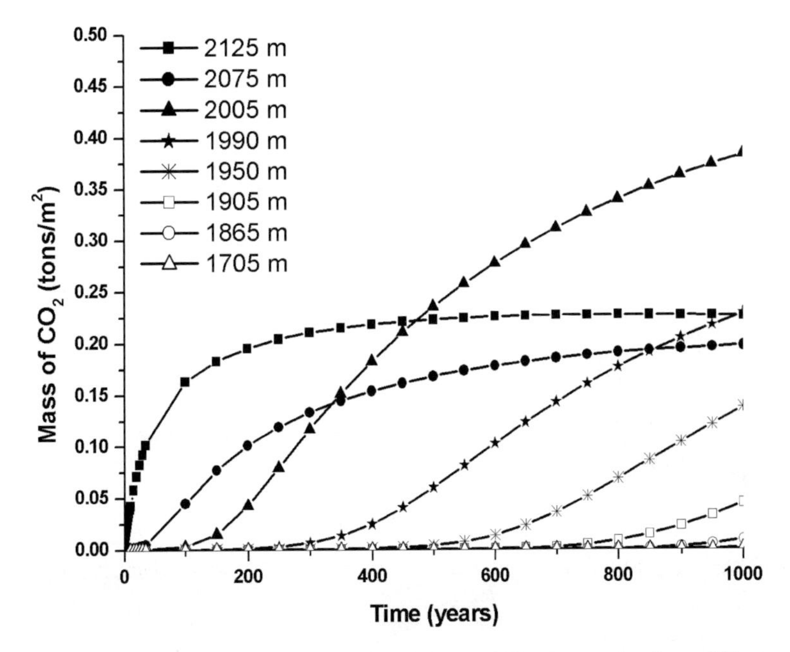

Figure 22: Mass of CO_2 versus time at different depths within the overburden of Forties cap rock failure case.

Risk Assessment

This section describes an assessment of the risks associated with long-term geological storage of carbon dioxide in a depleted oil reservoir. Illustrative calculations were undertaken using data representative of the Forties reservoir, in which it is assumed that a WAG process would be applied for CO_2 injection; the approach that was adopted and the results of the calculations are outlined in the Appendices. A similar approach could be taken in applying the methodology to a different field or to a different injection strategy; however, the particular calculations used to bind the risks from any specific pathway may need to be modified for such a case. There may be some pathways where a very simplistic calculation can provide an adequate bound on the flux in one-field situation, whereas a much more detailed and complex calculation may be required in applying the same approach in a different field. Equally, there may be particular pathways where the outcome of the risk assessment for another field may be very different, perhaps leading to a different conclusion about the suitability of the field for CO_2 storage.

The main steps in the risk assessment process can be identified as follows.

(i) Identify potential pathways for release (FEP analysis—identification of features, events and processes (FEP) that may impact on the release rates and/or the risk).
(ii) Use analytical models and/or numerical simulation to establish bounds on the release rates and/or the risk for different pathways and potential release scenarios. As a general principle, the approach is to use the simplest model that permits an adequate bound to be established for the magnitude of the release and/or the risk.

FEP analysis

In undertaking an FEP analysis of the problem, the objective is to identify the potential escape routes without making judgments about the relative significance of the different routes. It is important to be as comprehensive as possible during this initial stage of the process. The assessment of the significance of potential escape routes forms a separate stage of the risk assessment process; it will be seen in practice that many of the potential escape routes identified during the FEP analysis can in fact be demonstrated to be insignificant at the assessment stage.

Figure 23 captures in schematic form the key FEP that need to be assessed in order that CO_2 release routes and potential release rates can be determined. The diagram can be divided into three main areas: CO_2 storage, CO_2 escape and CO_2 migration to surface. Each of these is briefly summarized below.

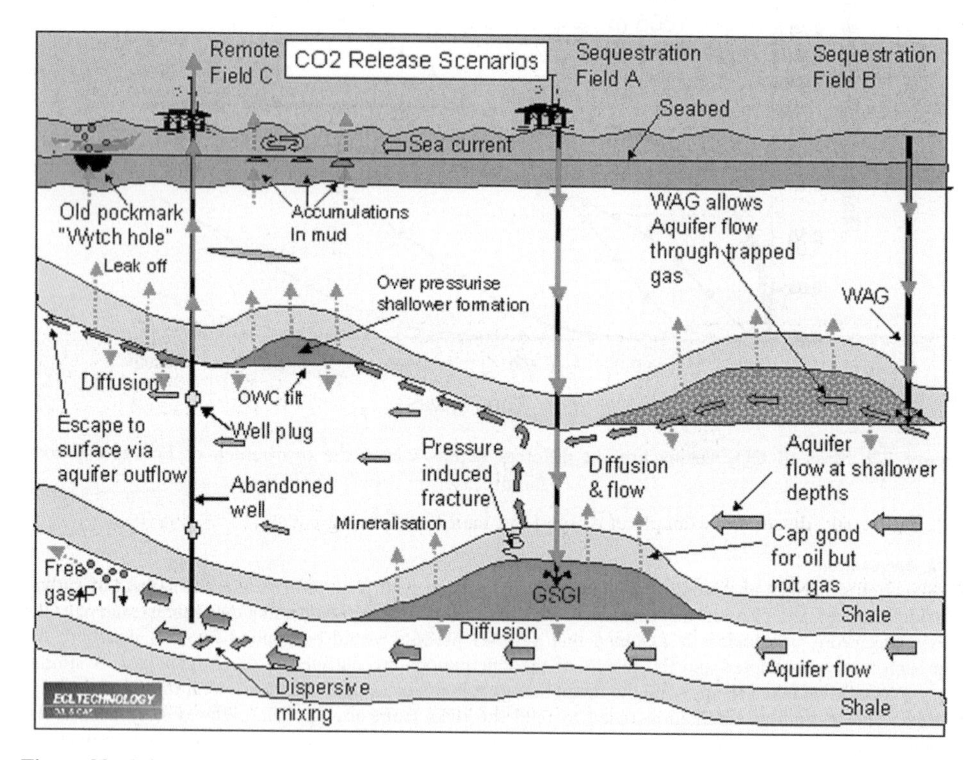

Figure 23: Schematic of the key features, events and processes that need to be considered when assessing the potential release paths and release rates of CO_2 sequestered into a subsea oil reservoir.

Storage of CO_2 in an oil reservoir with the subsidiary aim of enhancing oil recovery could typically be by either a gravity stable gas injection (GSGI), vertical sweep process or a water alternating gas (WAG) horizontal sweep process. The key difference between these approaches is that WAG alternates CO_2 slugs with water slugs to help control gas mobility whereas GSGI injects only CO_2. The WAG process has generally been favoured for EOR. After closure of the storage facility, the objective is to retain the CO_2 over an extended period of time (many hundreds or thousands of years).

The geological trap into which the CO_2 is stored has kept oil and any associated gas cap in place for, in many cases, millions of years. The key issue is therefore whether CO_2 behaves differently. Figure 23 illustrates routes for potential CO_2 escape from the trap.

Once CO_2 has escaped from the trap, the migration routes to surface depend on the regional geology, the extent of CO_2 transport by aquifer flow (either dissolved or as bubbles), and the availability and condition of man-made pathways such as wells.

Assessment of risks for key pathways in Forties

In assessing the risks associated with complex pathways to the surface, it can be useful to construct a fault tree to represent both the sequence of events that might lead to a release and also the interactions that might occur between those events. The features and processes identified in the FEP analysis provide a means to help compute potential release rates for scenarios identified in the fault tree. Figure 24 is a schematic example of a high-level fault tree. The circles at the bottom represent lower level events in the fault tree that are not shown explicitly in this diagram.

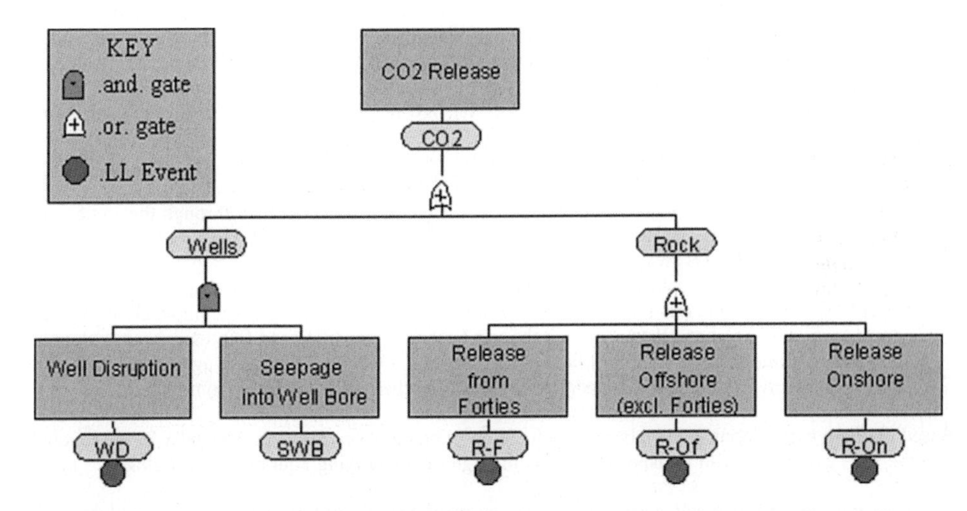

Figure 24: Illustrative example of a high-level fault tree showing potential CO_2 release paths.

Figure 25 shows some of the key parameters relating to the location, depth and dimensions of the Forties reservoir that are of relevance to the risk assessment. Appendix A details the values of key parameters (relating to Forties) that have been used in the risk calculations.

In considering the potential for escape of CO_2 that has been stored in Forties, the pathways can most conveniently be considered in three groups: pathways through the underlying aquifer; pathways through

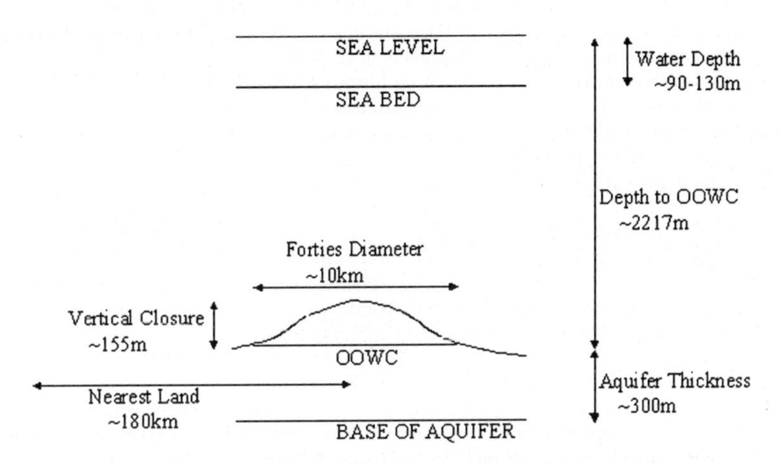

Figure 25: Key parameters, relating to the location, depth and dimensions of the Forties reservoir.

the cap rock and overburden; and well pathways. Appendices B, C and D, respectively, detail the calculations that were undertaken to assess the risks associated with each of these three groups of pathways.

Pathways through the underlying aquifer. Pathways that have been considered include convective and diffusive transports of dissolved CO_2, and transport of supercritical liquid-phase CO_2, either with or through the aquifer water.

(i) Transport of dissolved CO_2 through the underlying aquifer represents one potential pathway. Calculations described in Appendix B show that the advective flux of dissolved CO_2 in the aquifer water is insignificant compared to the volumes that are stored, and that the distance over which the dissolved gas might be transported in 1000 years is insignificant compared to the size of the Forties reservoir. Bounds calculated on the diffusive flux are negligible by comparison with the bound calculated for the advective flux.
(ii) The very low groundwater flow velocities in the Forties aquifer indicate that transport of liquid-phase CO_2 entrained in the aquifer flow makes no significant contribution.
(iii) There are two other potential causes of supercritical liquid-phase CO_2 flow through the aquifer that need to be considered. The first is that high injection pressures at the wells may lead to a downward flow of liquid-phase CO_2 away from the injection points and out of the Forties trap; we note that this is a transient effect that applies primarily during the injection period and perhaps for a short time thereafter. The second is that the total volume of CO_2 injected is sufficient to completely fill the trap down to the spill point and that as a result the trap becomes over-filled and CO_2 escapes. It is assumed that the CO_2 injection strategy will be designed in such a way as to mitigate against these possible effects and therefore there will be negligible impact on the risk assessment from them.

Based on these calculations and analysis, which are described in more detail in Appendix B, we conclude that the risks associated with transport pathways through the underlying aquifer are negligible.

Pathways through cap rock and overburden. There are a number of issues relating to pathways through the cap rock and the overburden.

(i) One such issue is the increase in overpressure due to replacing the oil originally present in the reservoir with CO_2. It is shown in Appendix C that the levels of overpressure in the reservoir are unlikely to be sufficient to allow liquid-phase CO_2 to escape into the cap rock.
(ii) There may be local increases in pressure around the injection wells during the period of injection. It is assumed that the injection strategy will be designed to ensure that these short-term levels of overpressure are such that they do not cause any problems during the injection period. Pressures in the immediate vicinity of the injection wells will tend to fall once injection ceases.
(iii) Analytical calculations (see Appendix C) show that the vertical diffusive flux of dissolved CO_2 through the cap rock and into the overburden is negligible.
(iv) Arguments outlined in Appendix C, based on historical observations, demonstrate that the risk of damage to the seal as a result of earthquakes or other seismic activity is negligible.
(v) Chemical reactions involving CO_2 are considered to have negligible risk associated with them in the short and medium terms. However, in the longer term (timescales of hundreds of years) we do not as yet have any experience of the effect of injected CO_2 on the seal in oil reservoirs.

Based on these calculations and analysis, which are described in more detail in Appendix C, we conclude that the risks associated with transport pathways through the cap rock and through the overburden are negligible. Some further work may be required to consider the long-term effects of CO_2 on the seal in the reservoir; this may need to include both field and laboratory studies to improve our understanding of the processes involved, followed by the development of appropriate models.

Well pathways. Pathways involving wells represent the biggest remaining area of uncertainty in the risk analysis. The Forties cap rock has been penetrated by several hundred wells, of which perhaps half have been abandoned to date. Appendix D outlines some calculations relating to levels of overpressure that might be anticipated in well bores, and make comparisons with the fracture pressures for the formation; similar comparisons might be made for cement plugs that are placed in the well bore on abandonment.

A more comprehensive assessment of the risks associated with well pathways requires a detailed audit of all the wells, which needs to focus in particular on the abandonment strategy that has been adopted in each of the wells that has been abandoned to date and also any changes to the abandonment strategy to be applied in the future, especially in the light of the potential for CO_2/water/rock, CO_2/water/cement and CO_2/water/steel reactions. Issues that need to be considered in relation to well pathways include

(i) circumstances under which CO_2 might enter an abandoned well bore in the reservoir;
(ii) how easily the CO_2 might move up the well bore, and how far it might travel;
(iii) location and circumstances under which CO_2 might escape from the well bore into the overburden, the sea or back to the platform.

Conclusions from Forties risk assessment
In this study we have identified potential pathways for escape of CO_2 from the Forties reservoir, and made an assessment of risks associated with those pathways. The risk assessments that have been made are based on a combination of analytical models and numerical simulation, and the results of these assessments are specific to the particular reservoir and the particular assumptions that have been made. It should be noted that a similar approach might be used to assess the risks associated with CO_2 injection in a different reservoir; however, the results of the risk assessment and the relative importance of the different risk factors depend on the particular circumstances that applied. The main conclusions from the risk assessment of CO_2 storage in the Forties reservoir are as follows.

(i) There are remaining uncertainties about well integrity and potential pathways to seabed through abandoned well bores. These need to be addressed through an audit of the well abandonment strategies that have been adopted to date and a review of well abandonment strategies to be applied in the future.
(ii) The risks associated with the escape of CO_2 through the cap rock and into the overburden (relating in particular to levels of overpressure and sealing of cap rock) have been shown to be negligible. A particular requirement for further work would be to address the long-term integrity of the seal in the presence of CO_2. This is an area where there is little historical experience to date.
(iii) Transport pathways through the underlying aquifer have been shown to have no significant areas of concern in the longer term. There are some possible short-term issues relating to the levels of overpressure around the injection wells and the detailed injection strategy that would need to be addressed as part of the design of the particular gas injection strategy that is adopted.

CONCLUSIONS FROM THE STUDY AS A WHOLE

The workflow necessary to select and characterize a site for storage of captured CO_2 emissions from a major industrial site has been illustrated by a case study based on the emissions from the Grangemouth refinery and petrochemicals complex.

Having selected the Forties oilfield as the most suitable storage site, a multi-scale, integrated approach was used to evaluate possible long-term leakage of geologically stored CO_2 in this mature oilfield. This approach was based on the use of commercial software. The workflow moved from the regional (basin) scale to the site-specific (field) scale, allowing a reliable reconstruction on the fluid flow pattern around the gas storage target.

The approach comprised several stages:

- simulation of the fluid flow at basin-scale using a 2D model
- simulation of the fluid flow in the aquifers around the field by 3D modelling
- evaluation of CO_2 and water interactions (diffusion) using a reservoir simulator
- risk evaluation using sensitivity tests taking into account the uncertainties of the data.

Using this novel approach the most significant risks of CO_2 escape from the Forties field can be bounded numerically using a combination of numerical simulation and scoping calculations. The potential for escape of CO_2 via geological pathways (diffusion and advective flow through the cap rock, dissolution of CO_2 into the aquifer below the oilfield and transport of CO_2-charged waters along the aquifer) is regarded as low. This is mainly due to:

- the quality and thickness of the cap rock and the overburden
- the very slow natural fluid flow velocity in the Forties reservoir and surrounding strata, controlled here by the sediment compaction rate.

Given that the risk of CO_2 escape by geological pathways appears to be very low, the potential for escape of CO_2 from the Forties field via active or abandoned wells, which could not be assessed meaningfully within the scope of the project, is perceived to be the most important unknown in the risk analysis.

Provided the risk from wells can be demonstrated to be acceptable, the Forties field appears to be an excellent potential location for CO_2 storage.

RECOMMENDATIONS

It is recommended that a comprehensive methodology for assessing the risks of leakage of stored CO_2 via wells is developed.

ACKNOWLEDGEMENTS

The authors thank WesternGeco for permission to publish Figures 7 and 8. S. Holloway, G.A. Kirby, M.A.E. Browne and S.L.B. Arkley publish with the permission of the Executive Director, British Geological Survey. The project was funded by the CO_2 Capture Project and the EU Energy, Environment and Sustainable Development Programme.

APPENDIX A: KEY DATA USED IN RISK CALCULATIONS

This appendix specifies the values of key parameters pertaining to Forties. The parameter values listed here have been used in the risk calculations that are detailed in Appendices B, C and D.

Geometry
Depth to original oil–water contact = 2217 m.

Thickness of Forties aquifer ~300 m.

Height from original oil/water contact to highest point in Forties reservoir = 155 m.

Closed area of Forties structure ~90 km^2.

Volumetrics
Forties STOIIP = 6.5×10^8 m^3.

Volume of Charlie Sand = 1.8×10^8 m^3.

Reservoir Conditions
Representative Forties temperature ~205 °F.

Representative Forties pressure ~220 bar.

Fluid Densities
Density of brine at reservoir conditions = 1030 kg/m^3.

Density of oil at reservoir conditions = 750 kg/m^3.

Density of CO_2 at reservoir conditions = 540 kg/m^3.

Volume conversion for CO_2 surface to reservoir conditions $\sim(1/300)$ rm^3/sm^3.

Representative density of CO_2 at 1000 m depth, temperature 320 K = 450 kg/m^3 (note that the CO_2 density can be sensitive to the temperature value chosen).

Aquifer Flow

Estimates of Forties fluid flow patterns and rates are described in the section "Modelling of Regional Fluid Flow to Underpin the Risk Assessment". The maximum regional flow velocities in the aquifer underlying the Forties field were estimated from the TEMIS3D results. Based on these results, the values used in the risk analysis were as follows:

Maximum Darcy velocity for Forties aquifer flow $\sim$6000 m/million years (from TEMIS3D simulation) $\sim$0.006 m/year.

CO_2 Solubility and Diffusion Parameters

CO_2 solubility in water (at 200 bar, 212 °F) $\sim$150 scf/rbbl $\sim$26 sm^3/rm^3.

Diffusion in water phase (using Tyn and Calus correlation) $\sim$0.0001 m^2/d $\sim 10^{-9}$ m^2/s.

Frequency of Earthquakes

Historical data concerning the magnitude and frequency of UK earthquakes[1] enable a first pass assessment of earthquake likelihood and magnitude in the Forties area to be made.

The magnitude and frequency of Forties earthquakes can be estimated (see Table A1) using the relative size of the Forties area and the area used to compile the UK data (the Forties area is assumed to correspond to a zone of radius 100 km, beyond which seismic events will have little effect). We note that over a 22-year period the Forties area has had no seismic events above magnitude 4.0.

TABLE A1
FREQUENCY OF EARTHQUAKES OF DIFFERENT MAGNITUDES

Magnitude of earthquake	Frequency—UK area (year)	Frequency—Forties area (year)
>3.7	1	25
>4.7	10	250
>5.6	100	2500

Effects of Earthquakes of Different Magnitude

Table A2 summarizes the perceived effects at the Earth's surface from earthquakes of different magnitudes.

TABLE A2
EFFECT OF EARTHQUAKES OF DIFFERENT MAGNITUDES

Magnitude of earthquake	Effect
7	Moderate damage to buildings (chimneys fall, cracks in walls)
6	People run out in alarm, slight damage to buildings (plaster cracks)
5	Felt by most indoors, small objects fall over
4	Felt by many indoors, windows and doors rattle
3	Felt by few

[1] R.M.W. Musson, A catalogue of British earthquakes, BGS Technical Report No. WL/94/04, 1994.

Other Parameters, Assumptions
Acceleration due to gravity = 10 m^2/s.

Conversion factor 1 N/m^2 = 1 Pa = 10^{-5} bar.

Assume no overpressure in aquifer.

Assume base of concrete plug set at 1000 m depth (in well overpressure calculations).

APPENDIX B: RISK CALCULATIONS—PATHWAYS THROUGH UNDERLYING AQUIFER

This appendix details the calculations of risk for each of the pathways considered through the underlying aquifer. Note that the parameter values listed in Appendix A have been used where appropriate in these calculations without further detailed justification; reference should be made as required to Appendix A for details of the parameter values that have been used.

Dissolution in Aquifer Water—Transport by Convection
Diameter of Forties (assuming circular, based on closed area) = 10,700 m.

Upper bound on flux of water through Forties aquifer (upper bound Darcy velocity times cross-sectional area for flow) = 0.006 m/year × 300 m × 10,700 m = 19,260 m^3/year.

Suppose that all water flowing in aquifer beneath Forties is saturated with CO_2 (in practice this is an upper bound on the concentration) then maximum advective flux of CO_2 in the aquifer away from Forties = 19,260 m^3/year × 26 sm^3/rm^3 = 5 × 10^5 sm^3/year. At reservoir conditions, this corresponds to 5 × 10^5 sm^3/year × (1/300) rm^3/sm^3 = 1670 rm^3/year.

Suppose that just 15% of Forties oil volume replaced with CO_2, then reservoir volume of CO_2 stored = 10^8 rm^3. Lower bound on time to remove this amount of CO_2, by advection alone = 10^8 rm^3/(1670 rm^3/year) = 60,000 years. Proportion of stored CO_2 removed in 1000 years by advection of dissolved CO_2 in aquifer water is at most 2%.

Suppose that just 15% of Charlie Sand volume replaced with CO_2, then reservoir volume of CO_2 stored = 2.5 × 10^7 rm^3. Lower bound on time to remove this amount of CO_2, by advection alone = 2.5 × 10^7 rm^3/(1670 rm^3/year) = 15,000 years. Proportion of stored CO_2 removed in 1000 years by advection of dissolved CO_2 in aquifer water is at most 6.7%.

Note that both these calculations represent significant over-estimates of the amounts of CO_2 removed since the calculations used the maximum possible concentration of dissolved CO_2 whereas the average concentration in the aquifer water will be significantly lower.

The pore water velocity in the underlying aquifer can be calculated by dividing the Darcy velocity by the porosity of the aquifer. A different way of looking at the advective transport of dissolved CO_2 is to estimate the maximum distance that the water will move in a given time. In 1000 years, this will be at most 600 m, even if the porosity were as low as 0.01.

The above calculations have demonstrated that the advective flux of dissolved carbon dioxide in the aquifer water is insignificant compared with the volumes that are stored, and that the distance over which the dissolved gas might be transported in 1000 years is insignificant compared with the size of the Forties reservoir.

Dissolution in Aquifer Water—Transport by Diffusion
Model
A model of radial diffusion in a hollow cylinder was used (see Figure B1). The hollow represents that part of the Forties aquifer underlying the Forties structure, and we assume that the water in this

region is fully saturated with CO_2 (leading to a boundary condition on the inner surface of the cylinder).

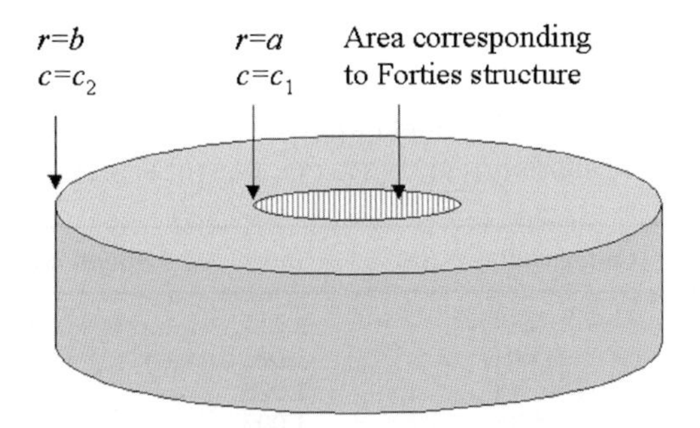

Figure B1: Model of radial diffusion in a hollow cylinder.

The governing equation is the radial diffusion equation

$$\frac{\mathrm{d}}{\mathrm{d}r}\left(r\frac{\mathrm{d}c}{\mathrm{d}r}\right) = 0 \qquad a < r < b$$

with boundary conditions $c = c_1$ at $r = a$and $c = c_2$ at $r = b$.

This has the general solution:

$$c = \frac{c_1 \log\left(\frac{b}{r}\right) + c_2 \log\left(\frac{r}{a}\right)}{\log\left(\frac{b}{a}\right)}$$

The diffusive flux through unit length of the cylinder is

$$Q = \frac{2\pi D(c_2 - c_1)}{\log\left(\frac{b}{a}\right)}$$

and through a cylinder of height h is

$$Qh = \frac{2\pi hD(c_2 - c_1)}{\log\left(\frac{b}{a}\right)}$$

where D represents the diffusion coefficient in the medium comprising the hollow cylinder.

Result of calculation
Radius of Forties (assuming circular, based on closed area) = 5350 m (a).

Concentration at inner radius = 26 sm^3/rm^3 = (26/300) rm^3/rm^3 = 0.087.

Take zero concentration at outer radius (b).

Upper bound on diffusive flux, reservoir conditions (in m^3/s) is:

$$\frac{2\pi(300\ \mathrm{m})(10^{-9}\ \mathrm{m}^2/\mathrm{s})(0.087)}{\log(b/a)}$$

Upper bound on diffusive flux (in rm^3/year) is:

$$\frac{2\pi(300\ \text{m})(0.0315\ \text{m}^2/\text{year})(0.087)}{\log(b/a)}$$

TABLE B1
BOUNDS ON DIFFUSIVE FLUX OF CO_2 IN THE FORTIES AQUIFER

Forties radius *a* (m)	**Outer radius *b* (m)**	**(*b*/*a*)**	**Bound on diffusive flux (rm^3/year)**
5350	6500	1.215	26.5
5350	5850	1.093	57.8
5350	5500	1.028	186.8
5350	5400	1.009	555.3

Table B1 shows a range of values for the bound on the diffusive flux, using the Forties radius as the inner radius and a range of different outer radii. These bounds can be compared with the bound calculated for the advective flux of 1670 rm^3/year. Even at a distance of just 50 m beyond the Forties footprint area, the diffusive flux is therefore small by comparison with the bound already calculated for the advective flux. The bound on the diffusive flux reduces as the distance from the Forties footprint increases, and becomes negligible at distances greater than about 500 m beyond the Forties footprint area.

Transport of liquid-phase CO_2 entrained in the aquifer flow
The pore water velocity in the underlying aquifer can be calculated by dividing the Darcy velocity by the porosity of the aquifer. In 1000 years, the lateral distance moved by the aquifer water will be at most 600 m, even if the porosity were as low as 0.01. This provides an upper bound on the distance moved by the supercritical liquid-phase CO_2. If liquid-phase CO_2 were entrained in the aquifer flow and convected along with the aquifer water, then the distance moved by the CO_2 in a given time would at most be equal to the distance moved by the aquifer water. It should be noted that in the situation being considered here, any buoyancy effects will tend to retain the liquid-phase CO_2 within the Forties reservoir, preventing it from being carried downwards and out of the reservoir by the very slow groundwater flow. There are other circumstances (e.g. in a steeply dipping formation) where buoyancy effects might enhance the flow rate of the liquid-phase CO_2, and in such a case the buoyancy effect would need to be taken into account.

Effect of injection pressure on transport of liquid-phase CO_2
At the injection wells, the pressure will clearly be higher than in the surrounding region. There is a possibility that the increased pressure in the region of the injection wells may result in downward flows of supercritical liquid-phase CO_2 away from the injection point, with a possibility of eventually escaping from the Forties reservoir as a liquid-phase flow in the underlying aquifer.

This is clearly a short-term issue, which will need to be addressed as a component part of the planning of the injection phase of the project. If the location of the injection wells and/or the rates of injection are such that CO_2 is able to escape below the original oil–water contact and out of the reservoir, then this will not be an acceptable injection strategy. It is assumed that the injection strategy will be planned in such a way that this is not an issue during the injection period.

Note that once injection has ceased, pressures will decline over time and buoyancy forces will then tend to transport the CO_2 back up towards the top of the reservoir. It is sufficient therefore to consider the issue during the injection period alone. If it is not an issue during the injection period, this should ensure that there are no issues at later times.

APPENDIX C: RISK CALCULATIONS—PATHWAYS THROUGH CAP ROCK AND OVERBURDEN

This appendix details the calculations of risk for each of the pathways considered through the cap rock and overburden. Note that well pathways are considered separately, in Appendix D.

Note that the parameter values listed in Appendix A have been used where appropriate in these calculations without further detailed justification; reference should be made as required to Appendix A for details of the parameter values that have been used.

Effects of Increased Overpressure Due to CO_2 in the Reservoir

When Forties was initially filled with oil, there was a certain level of overpressure below the cap rock, arising as a result of the density difference between oil and water. If the oil is subsequently replaced by CO_2, then there will potentially be an increase in the level of overpressure, due to the fact that at reservoir conditions the density of the supercritical liquid-phase CO_2 is less than that of the oil.

Calculation

An estimate of the overpressure at the highest point of the reservoir due to a column of fluid density ρ_f (assuming no overpressure in the aquifer) is given by:

$$P_0 = (\rho_b gh - \rho_f gh)$$

For initial Forties conditions (oil-filled) this gives a maximum overpressure of 4.3 bar. Following CO_2 storage, the overpressure will be less than that due to replacing all the oil with CO_2, which gives an upper bound of 7.6 bar. The increase in overpressure due to CO_2 storage will therefore be significantly less than 3.3 bar.

Significance of overpressure

The initial pressure in Forties is approximately 220 bar. The pressure maintenance scheme that has been implemented during production of the Forties reservoir was designed to keep reservoir pressures above 170 bar. Hence we might expect to see differential pressures of tens of bars without significant leakage of fluids into Forties from overlying shale. When considering the potential escape of CO_2 from the Forties reservoir through the cap rock and into the overburden, the differential pressure has the opposite sign; however, it is instructive to make a comparison of the magnitude of the differential pressure that might be generated in each case. An increase in overpressure of 3 bar is very small in magnitude compared to the differential pressures (specifically, underpressures) that have been generated historically within Forties, and it is therefore considered unlikely to have a significant impact on the seal integrity.

Effects of Increased Overpressure Due to CO_2 Injection

At the injection wells, the pressure will clearly be higher than in the surrounding region. There is a possibility that the increased pressure in the region of the injection wells may result locally in overpressures, at the top of the reservoir, that are higher than the maximum steady state overpressures that have been calculated above. This is clearly a short-term issue, which will need to be addressed as a component part of the planning of the injection phase of the project. It is considered unlikely to be a serious issue since the injection does not commence until after the reservoir pressures have declined significantly. Note that this represents a short-term transient effect only. Once injection has ceased, pressures will decline over time. It is sufficient therefore to consider the issue during the injection period alone.

Potential for Damage to Seal Due to Earthquake

Appendix A includes a discussion of frequency and magnitude of earthquakes in the UK. Damage to the seal due to earthquakes is considered unlikely. This can be demonstrated using a historical argument, based on the relatively low likelihood and magnitude of earthquakes, and the fact that the seal is good—evidenced by the fact that on discovery the Forties trap was full to the spill point.

Potential for Damage to Seal Due to Chemical Reactions

In the short and medium term this is considered unlikely to be an issue, based on current field experience. It is known that the seal has long-term resistance to reactions with hydrocarbon gas, evidenced by the fact that the trap must have existed over extremely long timescales for the hydrocarbon to have accumulated. There is evidence from CO_2 injection for EOR that putting CO_2 into oil reservoirs does not cause seal damage in the short to medium term (i.e. tens of years). In the longer term there is as yet no field experience on which to base any assessment (either for or against). We may need to undertake further work to demonstrate that this is not an issue on the 100–1000 year timescale.

Potential for CO_2 Escaping through Cap Rock

Estimates of Forties fluid flow patterns and rates are described in the section "Modelling of Regional Fluid Flow to Underpin the Risk Assessment". The SIMUSCOPP simulator was used to assess the potential for CO_2 escaping through the cap rock and into the overburden. SIMUSCOPP is a 3-phase, 3D, compositional porous medium flow simulator, which can model the effects of dissolution and diffusion. The model, which covered the Forties footprint area, was based on a submodel taken from the basin-scale TEMIS3D model. The rock properties and boundary conditions for the model were extracted from the appropriate region of the TEMIS3D model. The SIMUSCOPP model was initialised with brine and CO_2 only; the initial CO_2 saturation was 0.5 in all grid blocks within the Forties reservoir that lay above the Forties initial oil–water contact and zero elsewhere. This initial condition was set up to be a simplified representation of the conditions at the end of the storage phase. The SIMUSCOPP model was used to provide an estimate of the leakage of CO_2 through the cap rock, including the effects of dissolution and diffusion of dissolved CO_2, for a number of different scenarios. The conclusion from these calculations is that escape through the cap rock is most unlikely to represent a significant risk of release of CO_2 into the overburden and ultimately to the seabed.

Base case

In the base case run, the shale properties were set up to be representative of Forties; the capillary entry pressure for CO_2 to enter the shale was taken to be 4 bar (equivalent to the original overpressure in Forties when it was oil-filled). In this case, there is negligible escape of CO_2 into the cap rock and overburden (of the 3.9×10^8 tonnes of CO_2 in the SIMUSCOPP model, less than 0.2% enters into the overlying layers on a 1000 year timescale).

Sensitivity cases

Two sensitivity cases were carried out. The first of these was identical to the base case except that the capillary entry pressure for the shale layers overlying the reservoir was set to zero. The second sensitivity case also had zero capillary entry pressure and in addition the vertical permeability was increased by a factor of 10 in the shale layers. The modifications made in the sensitivity cases were designed to make it significantly easier for the CO_2 to escape from the reservoir and are not considered to be realistic. As expected, both sensitivity cases showed some CO_2 escaping through the cap rock, but in neither case did the CO_2 get anywhere near the surface. Even in the worse case, on a timescale of 1000 years the maximum vertical distance moved by any of the CO_2 was less than half-way to the seabed.

Diffusion of Dissolved CO_2 through Overburden

It is possible to perform an analytical calculation to estimate a bound on the vertical diffusive flux of dissolved CO_2 through the cap rock and into the overburden.

Analytical model

A model of 1D (vertical) diffusion was used. The base of the model represents the base of the Forties cap rock, and we assume that the water in this region is saturated with CO_2 (leading to a boundary condition at $z = 0$). The diffusive flux at a height h above the top of Forties is bounded by the flux calculated from the solution of the linear steady-state diffusion equation

$$\frac{d}{dz}\left(D\frac{dc}{dz}\right) = 0 \qquad 0 < z < h$$

with boundary conditions $c = c_0$ at $z = 0$ and $c = 0$ at $z = h$, where D represents the diffusion coefficient in the medium comprising the cylinder.

Hence

$$D\frac{dc}{dz} = A$$

and

$$c = \left(\frac{A}{D}\right)x + B$$

Applying the boundary conditions:

$$c = c_0 - \frac{Ax}{Dh}$$

The quantity $(-A)$ represents the vertical diffusive flux per unit area of the reservoir.

Results of calculation
Concentration at base of cap rock = 26 sm^3/rm^3 = (26/300) rm^3/rm^3 = 0.087

$$A = \frac{10^{-9}(m^2/s) \times 0.087(m^3/m^3)}{h(m)} = \left(\frac{8.7 \times 10^{-11}}{h}\right)(rm^3/s/m^2)$$

Forties area = 90 km^2 = 9×10^7 m^2. Hence bound on total diffusive flux is:

$$\left(\frac{8.7 \times 10^{-11} \times 9 \times 10^7}{h}\right)(rm^3/s) = \frac{7.83 \times 10^{-3}}{h}(rm^3/s) = \frac{2.5 \times 10^5}{h}(rm^3/year)$$

Take $h = 100$ m (for example), then the total vertical diffusive flux 100 m above the base of the cap rock is less than 2500 rm^3/year. Suppose that just 15% of Forties oil volume replaced with CO_2, then reservoir volume of CO_2 stored = 10^8 rm^3.

Lower bound on time to remove this amount of CO_2, by vertical diffusion through cap rock and overburden to a distance of at least 100 m above reservoir = 10^8 rm^3/(2500 rm^3/year) = 40,000 years. The proportion of stored CO_2 removed in 1000 years to at least 100 m above reservoir by vertical diffusion of dissolved CO_2 through the cap rock and overburden is at most 2.5%.

APPENDIX D: RISKS ASSOCIATED WITH WELL PATHWAYS

In this appendix we detail the key pathways associated with wells. A full assessment of risk for these pathways would require detailed information about the well design and abandonment strategy that has been adopted in the field to date, and may also require specification of well abandonment strategies for wells that are still operating. Some aspects of the risks have been quantified where appropriate information is available. Well pathways are identified as a key area requiring further study in order to gain a comprehensive and in-depth understanding of the risks.

Note that the parameter values listed in Appendix A have been used where appropriate in these calculations without further detailed justification; reference should be made as required to Appendix A for details of the parameter values that have been used.

Potential for Escape along Well Pathways
The cap rock has been penetrated many hundreds of times, in different locations, as a result of drilling activities associated with field development. A comprehensive risk assessment needs to consider the potential for CO_2 escape along each of the resulting well pathways at different stages in field life.

The long-term issues relate mainly to abandoned wells, since all wells will eventually be abandoned. In the short and medium terms, there are potential issues relating to operational wells (both producers and injectors), suspended wells and abandoned wells.

Once CO_2 has entered a well there are a range of pathways for transport to the surface. These include transport up the well bore followed by release into the formation at shallower depth, release at the seabed, or release at a platform. Note that release on the platform is only an issue for any particular well up to the time of final removal of casing strings down to the seabed. This is therefore also a well abandonment issue.

Number of Potential Well Pathways

The estimates of the number of potential well pathways shown in Table C1 are based on information provided by Apache North Sea (the operator of Forties) in December 2003.

Apache have estimated that they will drill a further 30 wells over the next 10 years (of which most are expected to be sidetracks, usually from dead or suspended wells) and that the program will result in a further 24 abandoned well bores over this period.

TABLE C1
ESTIMATES OF NUMBER OF POTENTIAL WELL PATHWAYS

Current well status	Well count
Producers	55
Water injectors	12
Dead (not used, but not formally suspended)	7
Suspended	27
Abandoned	89
Total	190

Well Abandonment Guidelines

The UK Offshore Operators Association has issued Guidelines for the Suspension and Abandonment of Wells. These guidelines require that two permanent barriers be set between the surface or seabed and any hydrocarbon-bearing permeable zone. A cement column of at least 100 ft measured depth of good cement is considered to constitute a permanent barrier; where possible 500 ft plugs are set. In addition, a single permanent barrier is required to isolate any water-bearing permeable zones from the seabed.

The guidelines recommend that the base of the first barrier be set across the top permeable zone of the reservoir or the top perforations, whichever is shallower, and should extend at least 100 ft above the highest point of potential inflow. The fracture pressure for the cement at the base of the first barrier should be in excess of the potential internal pressure (which is defined to be the maximum anticipated pressure that may develop below plugs in the well bore following abandonment).

On final abandonment it is good practice to retrieve all casing strings to a minimum of 10 ft below the seabed. In certain cases where large (e.g. concrete) structures remain permanently on the seabed, the requirement may be relaxed such that no casing strings may extend above the remaining structure.

Two categories of well pathways need to be considered in relation to abandoned wells.

(i) In the first group are the wells that have already been abandoned, where a detailed audit would be required to establish the criteria that have been used to design the abandonment strategy. It should be recognized that well abandonment design may have been carried out without anticipating the potential for leakage of CO_2 into the well bore, and that the abandoned well may not meet the same criteria that would apply to a future abandonment.

(ii) In the second group are all wells that have yet to be abandoned. The design of the abandonment strategy for these wells needs to take full account of the presence of CO_2 in the reservoir.

Effects of Increased Overpressure Due to CO_2 Filling Well Bore

The following sections consider some of the issues relating to the potential for leakage of carbon dioxide along well pathways and, in particular, the maximum level of overpressure that might arise below a permanent barrier in an abandoned well due to the accumulation of carbon dioxide below the plug and the possible consequences of that level of overpressure.

Calculation

The overpressure at base of concrete plug due to height h of CO_2 below it is given by

$$P_w = (\rho_b - \rho_c)gh$$

The maximum overpressure due to a height h of CO_2 is $5.8 \times 10^{-2}h$ bar (taking the minimum value for the density of CO_2 that has been estimated at 1000 m depth, corresponding to a pressure of 100 bar, and temperature 320 K). This is shown in Table D1 for various values of h.

We note the following.

(i) The Forties original oil–water contact is at 2217 m, corresponding to the spill point. Hence the maximum possible column of CO_2 below a plug set at 1000 m depth would be 1217 m, and the overpressure calculated for 1250 m represents an upper bound.

(ii) The effect of any overpressure would be to increase the average CO_2 density in the column and therefore to reduce the level of overpressure compared with the tabulated values.

TABLE D1
OVERPRESSURE AT BASE OF CONCRETE PLUG

h (m)	**Overpressure (bar)**	**Overpressure (psi)**
10	0.58	8.4
100	5.80	84.0
1000	58.00	840.0
1250	72.50	1050.0

Comparison with fracture pressure for rock

The fracture pressure of a rock can be estimated as a fraction (typically around 90%) of the overburden pressure; the overburden pressure gradient is approximately 1 psi/ft, so the fracture pressure can be related to the depth by

$$P_{frac} = 0.9d$$

where the fracture pressure is measured in psi and the depth is in ft. At a depth of 1000 m (= 3280 ft), the fracture pressure is around 2950 psi. This compares with the potential internal pressure below the cement plug of 2500 psi (equal to the sum of the hydrostatic pressure at 1000 m, which is 100 bar or 1450 psi, and the maximum overpressure of 1050 psi that results from a 1250 m column of CO_2 below the plug). This is still significantly lower than the fracture pressure.

Comparison with fracture pressure for cement plug
As discussed above, the UKOOA Guidelines for the Suspension and Abandonment of Wells require the strength of the cement to be such that the fracture pressure of the cement exceeds the potential internal pressure at the base of the plug. Two key questions are: what is the design strength of the concrete that has actually been used to date when abandoning wells, and how is this likely to degrade over long periods of time (e.g. over a 1000 year period) in the presence of CO_2?

Sensitivity to Depth of Placement for Cement Plug
The density of CO_2 changes very rapidly around 600–800 m depth (e.g. at 1000 m depth, 100 bar pressure the density is ~450 kg/m^3; at 600 m depth, 60 bar pressure, the density reduces to ~200 kg/m^3). If the plug is set shallower than 800 m, then we need to consider carefully the possible effects of fracturing of the cement plug. Below the plug, the pressure will depend on the thickness of the CO_2 column, and pressure will be higher than in surrounding formation and higher than that in the well bore above the plug. If the cement fractures, and CO_2 is able to escape, there will be a sudden drop in pressure and a corresponding (potentially large) rapid increase in gas volume. This might lead to an explosive blowout of the contents of the well bore.

If the cement plug is set closer to the surface, then the fracture pressure of the formation will also be correspondingly reduced. If the overpressure were sufficient to fracture the formation, then the same issues relating to sudden pressure release might apply, depending on the nature of the pathways that were formed. If the plug is set any higher than about 800 m depth, the risk is likely to be increased compared to a plug set at a deeper level. It will be necessary to consider the abandonment strategy and the depths at which the first and second permanent barriers are set and it would be prudent to ensure that second permanent barrier is set deeper than 800 m.

Leakage from Well into Formation Resulting from Overpressure
It has been shown above that fracturing of the formation is not likely to occur in the Forties scenario as a result of the level of overpressure that might arise from the accumulation of CO_2 in an abandoned well bore. Another issue that must be considered is the potential for leakage of CO_2 into permeable zones in the overburden. The UKOOA Guidelines for the Suspension and Abandonment of Wells require a permanent barrier to be set across any permeable water-bearing zones, the intention being to prevent leakage in either direction between the permeable layer and the well bore. A key question therefore is what permeability threshold has been applied in identifying a zone as permeable and what leakage might potentially occur into layers where the permeability was just below this threshold. The question of what leakage might potentially occur from the well bore into the formation, and what the fate of any such leakage would be (pathways to the seabed, timescales, etc.) is identified as an area for future investigation. This question can only be addressed through a detailed investigation of a range of specific examples of abandoned wells; this is beyond the scope of the present study.

Leakage from Well into the Sea
Risks associated with leakage from a well into the sea include the reduction in buoyancy due to gas bubbles; increased levels of dissolved CO_2 leading to an adverse impact on marine life; density-driven convection (reduced density resulting from dissolved CO_2) and a potential for instability with gas subsequently coming out of solution at reduced depths; and finally the effects of CO_2 release at sea surface.

Leakage Through Well with Release on Platform
Prior to the final removal of casing strings between platform and seabed there is a possibility of leakage back to the platform. Risks include CO_2 build-up in enclosed spaces on platform (e.g. sleeping areas); the build-up of a layer of dense CO_2 on platform (primarily of concern in calm conditions when it might lead to dangers of asphyxiation for workers.

Potential for Damage to Wells Due to Earthquake
Appendix A includes a discussion of frequency and magnitude of Earthquakes in the UK. Over a 22 year period the Forties area has had no seismic events above magnitude 4. There has been no reported damage to any North Sea wells due to earthquake over this period. Given the low frequency of earthquakes and their limited magnitude, this is considered to be a negligible risk.

NOMENCLATURE

BGS	British Geological Survey
EOR	enhanced oil recovery
EU	European Union
FEP	features, events and processes
GEUS	Geological Survey of Denmark and Greenland
GSGI	gravity stable gas injection
IFP	Institut Francais du Petrole
Ma	million years
mD	millidarcies
MMrb	million reservoir barrels
MMstb	million standard barrels
psia	pounds per square inch (absolute)
STOIIP	stock tank oil initially in place
WAG	water alternating with gas

REFERENCES

1. A.A. Espie, Options for establishing a North Sea geological storage hub, in: D.J. Williams, R.A. Durie, P. McMullan, C.A.J. Paulson, A.Y. Smith (Eds.), Greenhouse Gas Control Technologies: Proceedings of the Fifth International Conference on Greenhouse Gas Control Technologies, CSIRO, Collingwood, Australia, 2001, pp. 266–271.
2. M. Blunt, J.F. Fayers, F.M. Orr Jr., Carbon dioxide in enhanced oil recovery, *Energy Convers. Manage.* **34** (9–11) (1993) 1197–1204.
3. M. Simmonds, P. Hurst, M.B. Wilkinson, C. Watt, C.A. Roberts, A study of very large scale post combustion CO_2 capture at a refining and petrochemical complex, in: J. Gale, Y. Kaya (Eds.), Greenhouse Gas Control Technologies, vol. 1, Elsevier, Amsterdam, 2003, pp. 39–44.
4. M.A.E. Browne, S.L.B. Arkley, S. Holloway, The potential for carbon dioxide sequestration in the Midland Valley of Scotland and adjacent areas of the Forth Approaches, CO_2 Capture Project Technical Report, 2003, 74 pp.
5. R. Crawford, R.W. Littlefair, L.G. Affleck, The Arbroath and Montrose fields, blocks 22/17, 18, UK North Sea, in: I.L. Abbots (Ed.), United Kingdom Oil and Gas Fields, 25 Years Commemorative Volume, Memoir of the Geological Society of London, vol. 14, 1991, pp. 211–217.
6. B. Moss, D. Barson, K. Rakhit, H. Dennis, R. Swarbrick, Formation pore pressures and formation waters, in: D. Evans, C. Graham, A. Armour, P. Bathurst (Eds.), The Millenium Atlas: Petroleum Geology of the Central and Northern North Sea, The Geological Society of London, London, 2003, pp. 317–329.
7. F. Schneider, S. Wolf, L. Faille, D. Pot, A 3D basin model for hydrocarbon potential evaluation: application to Congo offshore, *Oil Gas Sci. Technol. Rev. IFP* **55** (2000) 3–13.
8. J.M. Wills, D.K. Peattie, The Forties Field and the evolution of a reservoir management strategy, in: A.T. Buller, E. Berg, O. Hjelmeland, J. Kleppe, O. Torsaeter, J.O. Aasen (Eds.), The North Sea Oil and Gas Reservoirs-II, The Norwegian Institute of Technology, Graham & Trotman, London, 1990, pp. 1–17.
9. M.R. Giles, Diagenesis: A Quantitative Perspective, Kluwer Academic Publishers, London, 1995, 526 pp.
10. D. Evans, C. Graham, A. Armour, P. Bathurst (Eds. and co-ordinators), The Millennium Atlas: Petroleum Geology of the Central and Northern North Sea, The Geological Society of London, London, 2003.
11. K. Pruess, A. Bielinski, J. Ennis-King, R. Fabriol, Y. Le Gallo, J. García, K. Jessen, T. Kovscek, D.H.-S. Law, Code Intercomparison Builds Confidence in Numerical Models for Geologic Disposal of CO_2, Sixth International Conference on Greenhouse Gas Control Technologies, IEA, Kyoto, Japan, 2002.

12. C. Oldenburg, D.H.-S. Law, Y. Le Gallo, S. White, Mixing of CO_2 and CH_4 in Gas Reservoirs: Code Comparison Studies, Sixth International Conference on Greenhouse Gas Control Technologies, IEA, Kyoto, Japan, 2002.
13. L.K. Thomas, D.L. Katz, et al., Threshold pressure phenomena in porous media, *SPE J.* **8** (2) (1968) 174–184.
14. M.Th. van Genuchten, A closed-form equation for predicting the hydraulic conductivity of unsaturated soils, *Soil Sci. Soc. Am. J.* **44** (1980) 892–898.
15. R.C. Reid, J.M. Prausnitz, B.E. Poling, The Properties of Gases and Liquids, McGraw-Hill, New York, 1987.
16. Z. Duan, N. Muller, J.H. Weare, An equation of state for the CH_4–CO_2–H_20 system: I. Pure systems from 50 to 1000 °C and from 0 to 8000 bar, *Geochim. Cosmochim. Acta* **56** (7) (1992) 2605–2617.

Carbon Dioxide Capture for Storage in Deep Geologic Formations, Volume 2
D.C. Thomas and S.M. Benson (Eds.)

Chapter 6

PREDICTING AND MONITORING GEOMECHANICAL EFFECTS OF CO_2 INJECTION

Jürgen E. Streit[1], Anthony F. Siggins[2] and Brian J. Evans[3]

[1]CRC for Greenhouse Gas Technologies, Australian School of Petroleum, The University of Adelaide, Adelaide, Australia
[2]CRC for Greenhouse Gas Technologies, CSIRO Petroleum, Perth, Australia
[3]CRC for Greenhouse Gas Technologies, Curtin University of Technology, Perth, Australia

ABSTRACT

Predicting and monitoring the geomechanical effects of underground CO_2 injection on stresses and seal integrity of the storage formation are crucial aspects of geological CO_2 storage. An increase in formation fluid pressure in a storage formation due to CO_2 injection decreases the effective stress in the rock. Low effective stresses can lead to fault reactivation or rock failure which could possibly be associated with seal breaching and unwanted CO_2 migration. To avoid seal breaching, the geomechanical stability of faults, reservoir rock, and top seal in potential CO_2 storage sites needs to be assessed. This requires the determination of in situ stresses, fault geometries, and frictional strengths of reservoir and seal rock. Fault stability and maximum sustainable pore fluid pressures can be estimated using methods such as failure plots, the FAST technique, or TrapTester (Badley Geoscience Ltd) software. In pressure-depleted reservoirs, in situ stresses and seal integrity need to be determined after depletion to estimate maximum sustainable pore fluid pressures. The detection of micro-seismic events arising from injection-induced shear failure of faults, fractures and intact rock is possible with geophone and accelerometer installations and can be used for real-time adjustment of injection pressures. In the event of injected CO_2 opening and infiltrating extensive fracture networks, this can possibly be detected using multi-component seismic methods and shear-wave splitting analysis.

INTRODUCTION

Underground storage of large quantities of anthropogenic CO_2 in geological formations is considered a viable option to significantly reduce greenhouse gas emissions [1,2]. One of the key objectives of geological CO_2 storage is the long-term underground containment of CO_2 in porous rock. To maximise storage quantities per unit volume of porous rock, CO_2 should be stored as a relatively dense phase in its supercritical state at depths below about 800 m [3]. Successful injection of CO_2 into a porous formation requires displacement or compression of the existing formation fluid and, thus, injection of CO_2 at pressures that exceed the formation pressure [4]. The excess pressure needs to be limited so it will not compromise the integrity of the reservoir seals.

It has been acknowledged by several authors that underground injection of CO_2 into porous rock at pressures higher than formation pressures can potentially induce fracturing and fault slip [5–7]. Indeed, it has been demonstrated that fluid injection into rocks can induce micro-seismic activity, as, e.g. in test sites such as the drill holes of the German continental deep drilling program (KTB) [8] or the Cold Lake oil field, Alberta [9]. Induced micro-seismicity is typically detected in the vicinity of the injector well within several hours to several days after fluid injection [8]. Deep well injection of waste fluids may even have induced earthquakes with moderate local magnitudes (M_L), as suggested for the 1967 Denver earthquakes ($M_L \leq 5.3$) [10] and

Abbreviations: FAST, fault analysis seal technology; MS, events, micro-seismic events; VSP, vertical seismic profiling.

the 1986/1987 Ohio earthquakes ($M_L \leq 4.9$) [11]. Seismicity that follows fluid injection is usually considered to result from increased pore fluid pressure in the hypocentral region of the seismic event [9,10].

Brittle failure of rocks and faults and associated micro-seismicity induced by overpressurisation would create or enhance fracture permeability while the formation of connected fracture networks and rough fault surfaces could provide pathways for unwanted CO_2 migration [12]. Thus, to avoid damage to top seal and fault seals due to injection-related pore pressure increase, maximum sustainable pore pressures need to be estimated for CO_2 injection. In addition, fluid pressures during injection as well as the CO_2 flow path should be monitored.

This chapter outlines key points of the geomechanical workflow that lead to assessments of fault stability and estimates of maximum sustainable fluid pressures in CO_2 storage sites. Monitoring of micro-seismic events, as well as seismic techniques suitable for the detection of unwanted CO_2 flow, are discussed as important methods for monitoring and controlling geomechanical effects of CO_2 injection.

STUDY METHODOLOGY

Some methods applicable to predict and monitor geomechanical effects of CO_2 injection were developed in the past to assist hydrocarbon exploration and production, while others are known from earthquake research and mining operations. Their application to CO_2 storage is new. The further development of such methods from different disciplines and their integration into a coherent and logical workflow is a key to facilitating sustainable storage of CO_2 in geological formations.

Predicting Geomechanical Effects of CO_2 Injection

The effect of increasing pore fluid pressure to decrease the strength of faults and rocks by decreasing effective stresses is extensively described in the geomechanical literature [13–15]. Geomechanical methods that are useful for estimating the stability of faults and maximum sustainable pore fluid pressures during CO_2 injection and storage are described below.

Effects of pore fluid pressure change on fault and rock stability

At depths greater than several hundred meters in the earth's crust, the maximum principal stresses usually are compressive. Where rocks contain pore fluid, the pressure of the pore fluid (P_f) acts in all directions and, thus, opposite to the compressive total stresses *(σ)* acting on the rock framework (Figure 1). Hence the pore fluid pressure combines with total stress to create a lower effective stress [13]:

$$\sigma' = \sigma - P_f \tag{1}$$

The effect of increasing pore fluid pressure on the state of stress on faults is schematically shown in Figure 2. When effective normal stresses ($\sigma_n - P_f$) are positive, they press opposing fault blocks together and resist sliding motion along the fault surface. Sliding can be induced by shear stresses (τ) acting parallel to the fault (Figure 2a). An increasing fluid pressure that decreases the effective normal stress therefore decreases the resistance to sliding. In a Mohr diagram, increasing pore fluid pressure shifts the Mohr circle towards the fault-failure envelope (Figure 2b). A relatively strong intact rock has a failure envelope further to the left from that of a relatively weak fault. Thus increasing fluid pressures often lead to fault failure before failure of relatively strong intact rock occurs. A failure envelope for a fault may be written in a general from as [14,18]

$$\tau_r = C + \mu(\sigma_n - P_f) \tag{2}$$

where τ_r is the shear stress that causes sliding and μ the coefficient of friction. *C* denotes an inherent shear strength of the fault which on cohesionless, gouge-lined fault surfaces is negligibly small [19]. On such faults, sliding occurs when the ratio of the shear stress to effective normal stress equals the coefficient of static friction of the fault:

$$\frac{\tau_r}{\sigma_n - P_f} = \mu \tag{3}$$

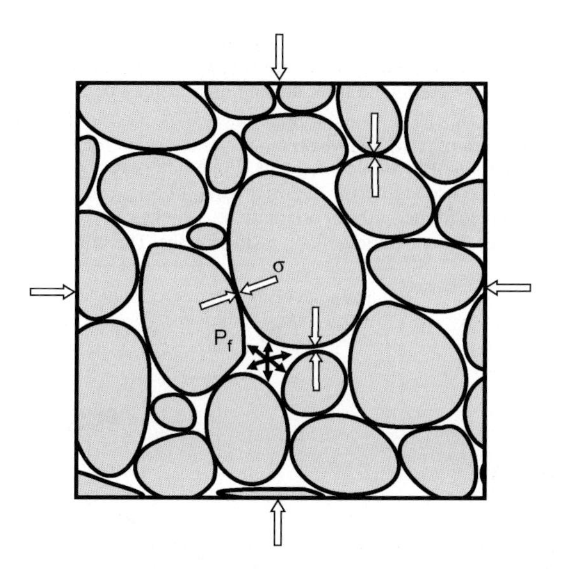

Figure 1: Sketch illustrating transmission of total stresses (σ) through grain boundaries and pore fluid pressure (P_f) acting in all directions, thus opposing total stresses. Diagram after Eisbacher [16].

The shear and effective normal stresses that act on a fault segment are a function of the fault geometry and are given in a two-dimensional form as

$$\tau = 0.5(\sigma_1 - \sigma_3)\sin 2\theta \text{ and } \sigma_n' = 0.5(\sigma_1' + \sigma_3') - 0.5(\sigma_1 - \sigma_3)\cos 2\theta \quad (4)$$

where σ_1 and σ_3 are the maximum and minimum principal stresses, respectively, and θ the angle between the fault and σ_1 (Figure 2a). Since the shear and normal stresses that act on a fault depend on the fault angle θ, some faults are more favourably oriented for slip than others within a homogeneous stress field. The analysis of fault stability thus requires knowledge of the in situ stress tensor and the geometry of pre-existing faults.

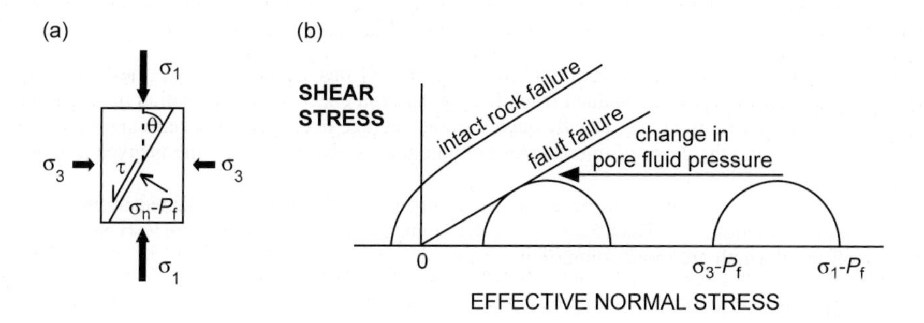

Figure 2: (a) Sketch showing the orientation of principal stresses, shear stress, and effective normal stress relative to a fault plane. (b) Mohr diagram showing shift of Mohr circle due to pore fluid pressure increase. Diagram from Streit and Hillis [17].

In situ stress determination
The orientation and magnitude of the vertical stress (S_v) and of the maximum (S_{Hmax}) and minimum horizontal stresses (S_{hmin}) can be determined from drilling data. It may be assumed that these stresses are principal stresses.

Stress orientation. The orientation of borehole breakouts (Figure 3a) which can be apparent on image logs and four-arm caliper logs can be used to derive the orientation of S_{Hmax} [21,22]. In cases where drilling-induced tensile fractures have formed, their orientation, which can be identified from image-log interpretation, directly indicates the orientation of S_{Hmax} (Figure 3b). The minimum horizontal stress and S_v are perpendicular to S_{Hmax}.

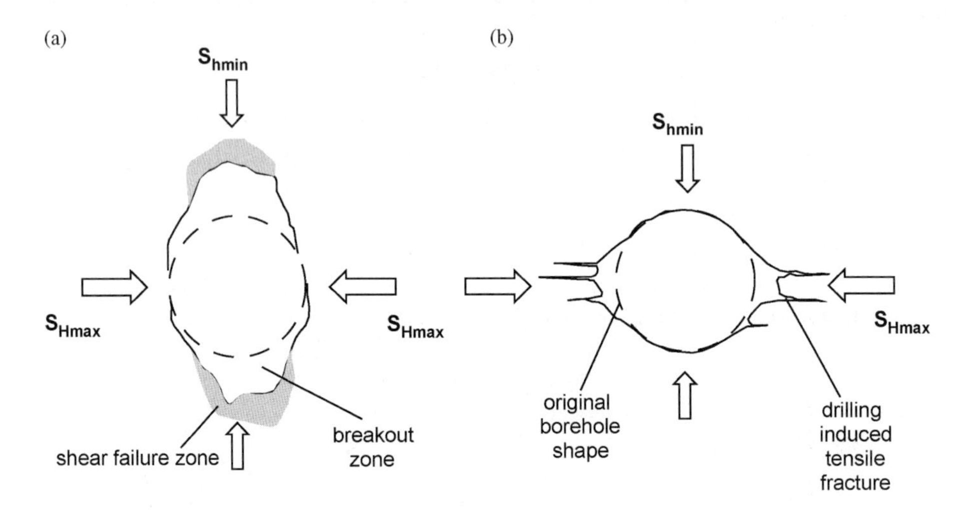

Figure 3: Schematic cross-section through borehole showing original circular borehole shape (broken line). (a) Borehole breakout due to spalling of borehole wall indicating the S_{hmin} direction. (b) Drilling-induced tensile fractures indicating the S_{Hmax} direction. Diagram modified from Dart and Zoback [20].

Stress magnitude. The overburden pressure at depth can be estimated by integrating the density of all overlying rocks and fluids over depth and calculating the resulting pressure [23]:

$$S_v = \int_z^0 \rho(z) g \, dz \tag{5}$$

S_v is the overburden pressure, g the gravitational acceleration (9.81 m/s^2), and z depth. ρ is the density of rocks and fluids. Rock densities can be obtained from density logs and check shot information on average sonic velocity. The overburden pressure is usually quoted as the average overburden (or vertical) stress gradient between the surface and the depth of interest. An example for a vertical stress profile is given in Figure 4.

The magnitude of S_{hmin} in wells can be estimated from fluid pressure levels attained during hydraulic fracturing of the formation [24]. Thus, S_{hmin} can be determined from hydraulic fracture tests and less ideally from leak-off tests, which are more commonly conducted.

The magnitude of S_{Hmax} can be constrained from the occurrence of borehole breakouts and drilling-induced tensile fractures, both of which can be interpreted on image logs [22,25]. Knowledge of the rock strength and the formation fluid pressure, as well as of the mud-weight during drilling and logging, is also required. However, estimates of S_{Hmax} are usually associated with relatively large uncertainty or cannot be obtained because the rock strength is not known.

In cases where S_{Hmax} cannot be determined, the stress regime may be constrained by applying a frictional limit calculation. This gives a crude upper bound on the magnitude of σ_1 based on the assumption that the strength of some optimally oriented faults within the area limits the magnitude of stresses that can accumulate. The limiting stress ratio for frictional sliding on optimally oriented faults can be written as [14]

$$\frac{\sigma_1 - P_f}{\sigma_3 - P_f} = ((\mu^2 + 1)^{1/2} + \mu)^2 \tag{6}$$

where P_f is the pore fluid pressure and μ the coefficient of static friction. An example for estimated frictional limits is given in Figure 4.

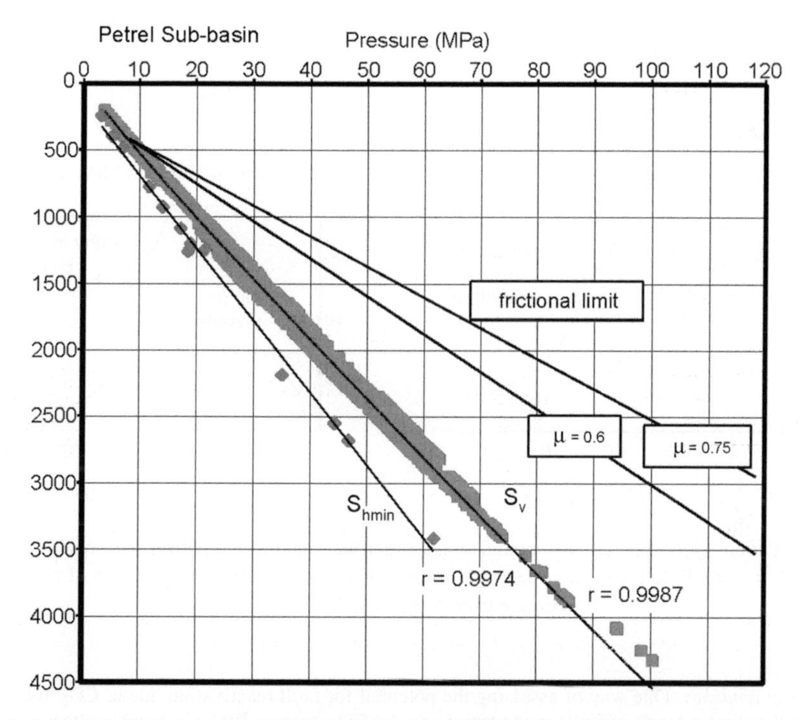

Figure 4: Stress profile for the Petrel Sub-basin based on drilling data. Estimates for S_{hmin} are based on pressures from leak-off tests; estimates for S_v were obtained by integrating density log data. Pearson correlation coefficients are indicated for curve fits. Frictional limits are indicated for different μ values. Diagram from Gibson-Poole et al. [6].

Fault stability and maximum sustainable pore fluid pressures in CO_2 storage sites

Assessments of fault stability require knowledge of fault geometries. Information on fault geometries in potential CO_2 storage sites is usually obtained from the structural interpretation of seismic data which need to be time-to-depth converted. Additional information on fracture geometry can be obtained from imaging fractures and faults that intersect wellbore walls. Such features may be imaged using, e.g. acoustic or ultrasonic scanners (borehole televiewers) or high-resolution resistivity imaging tools (FMS, FMI) in the wellbore.

Failure plots. Rock deformation experiments and field studies show that in some cases the formation of new fractures is more favourable than the reactivation of the pre-existing faults that have particular orientations

[26,27]. The orientation of relatively stable faults partly depends on the strength of their adjacent wall rock [14,28]. The orientation of relatively stable faults can be identified from failure plots [29,12], strictly for faults that contain the intermediate principal stress (σ_2) axis. It is further required that both fault and wall rock are subjected to the same stress and pore fluid pressure. This condition may hold for CO_2 storage scenarios.

The construction of failure plots for the identification of faults that are relatively stable is described in detail by Streit [29] for various rock types and fault strengths. Figure 5 shows failure plots for a hypothetical example in which faults are assumed to cut through Berea sandstone of the strength given by Handin et al. [13]. Figure 5 indicates that at a differential stress of 20 MPa, which is prevailing at approximately 2 km depth in some basins [30,31], faults with fault angles >65°–75° are relatively stable. The failure plot method has been applied in two study sites of the Australian GEODISC program for CO_2 storage [6,32]. However, the tendency for fault slip to occur should also be estimated using 3D methods, especially for faults that cannot be identified as relatively stable.

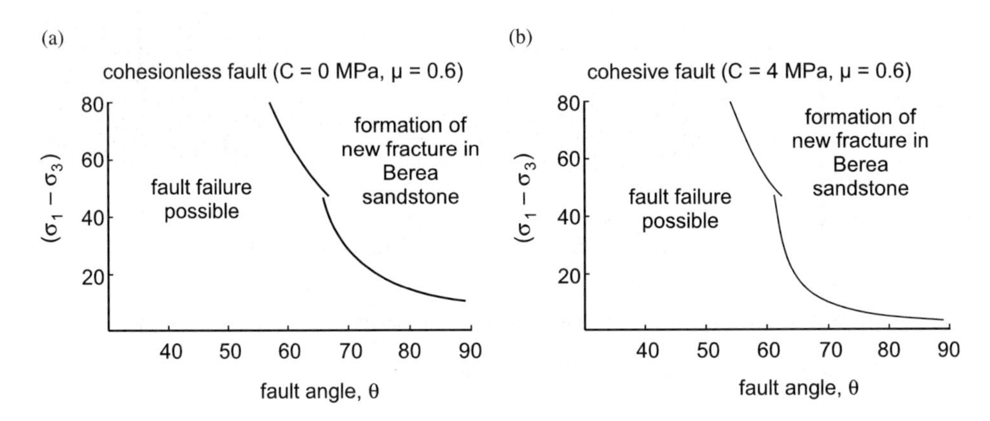

Figure 5: Failure plots showing differential stresses and fault angles that permit fault reactivation in the range $30 \leq \theta < 90$. Relatively stable faults fall in the fields that indicate the formation of new fractures. Figure from Streit and Hillis [12].

3D fault slip tendency. One way of assessing the potential for fault reactivation due to CO_2 injection is to determine the slip tendency of faults in the target area for CO_2 storage. By including the effect of pore fluid pressure, the slip tendency (T_s), which is defined as the ratio of resolved shear stress to normal stress acting on faults [33], may be expressed as:

$$T_s = \frac{\tau}{\sigma_n - P_f} \tag{7}$$

The slip tendency of a fault is evaluated by comparing the ambient stress ratio T_s to the stress ratio that would cause slip on a fault with no inherent shear strength ($C = 0$). Such a cohesionless fault is critically stressed when T_s equals the coefficient of static friction as shown in Eq. (3). Cohesionless faults are usually assumed to have Byerlee friction coefficients of $\mu = 0.6$–0.85 [22,34]. Where faults contain clay minerals, the friction coefficient can be less than $\mu = 0.6$ [18,35].

In cases where a 3D fault geometry can be constructed from the interpretation of depth-converted 3D seismic surveys or densely spaced 2D surveys, fault slip tendency can be calculated from Eq. (7) for each grid point on a fault. Figure 6 shows an example for the fault slip tendency computed from in situ stresses using commercially available software (TrapTester, Badley Geoscience Ltd, UK, http://www.badleys.co.uk).

Assuming that the two faults shown in Figure 6 have a coefficient of static friction of $\mu = 0.6$, their slip tendency is low to moderate. The maximum sustainable pore fluid pressure on these faults can be estimated by using progressively higher pore fluid pressures in Eq. (7) until the slip tendency becomes critically high.

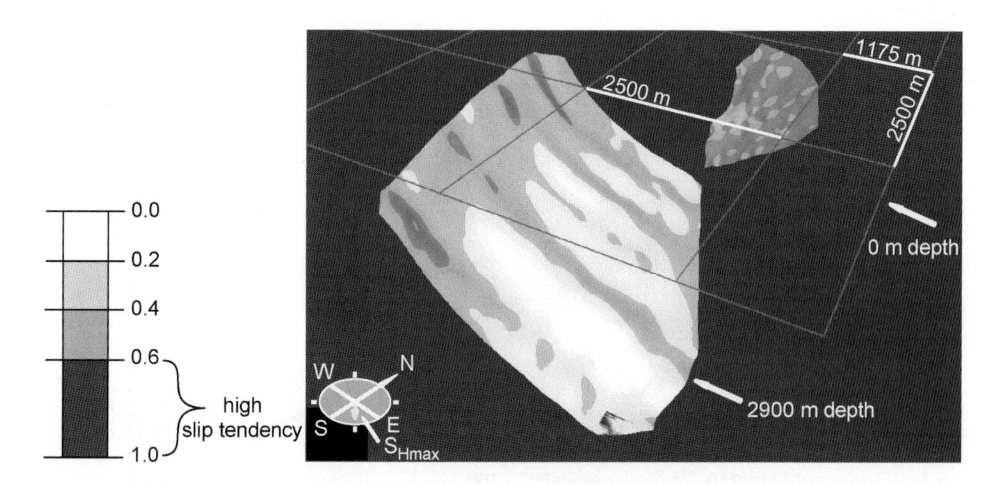

Figure 6: Slip tendency on two fault surfaces for $S_{\text{hmin}} = 15.2z + 0.5$, $S_{\text{v}} = S_{\text{Hmax}} = 22.5z + 0.5$, and $P_{\text{f}} = 9.7z + 0.5$ (z = depth in km); S_{Hmax} orientation is 116°N.

Critical pore fluid pressure increase. The likelihood for fault failure can be estimated by calculating the fluid pressure increase required to induce brittle failure (ΔP_{f}). This fluid pressure increase can be shown in a 3D Mohr diagram [14] for any fault angle (θ) as illustrated in Figure 7.

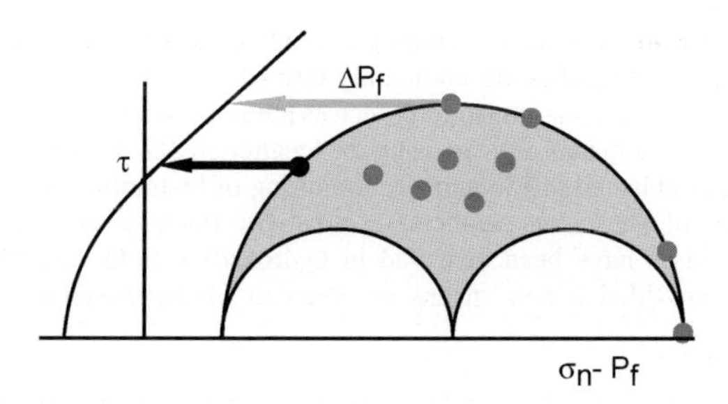

Figure 7: Illustration of state of stress on faults with different geometries in a 3D Mohr diagram and required pore pressure changes (ΔP_{f}) to reactive such faults. Diagram modified from Mildren et al. [31].

The FAST (Fault Analysis Seal Technology) technique of Mildren et al. [31] is used herein to illustrate the fluid pressure increase required for fault failure to occur (ΔP_{f}). This fluid pressure increase is calculated for a particular depth and shown in a southern hemisphere polar plot projection. The orientation of poles to fault planes in such a projection can then be attributed to the relevant ΔP_{f} value (Figure 8). Since the FAST technique can include an inherent shear strength of a fault or rock (C), it can also be used to estimate fluid pressures that induce failure in intact reservoir rock or seal [31]. The technique is thus suitable to calculate maximum sustainable pore fluid pressures on faults, in intact reservoir rock, and below top seals, given that the relevant rock or fault frictional strength is known. Application of the FAST technique on a study site for potential CO_2 storage is given by Gibson-Poole et al. [32].

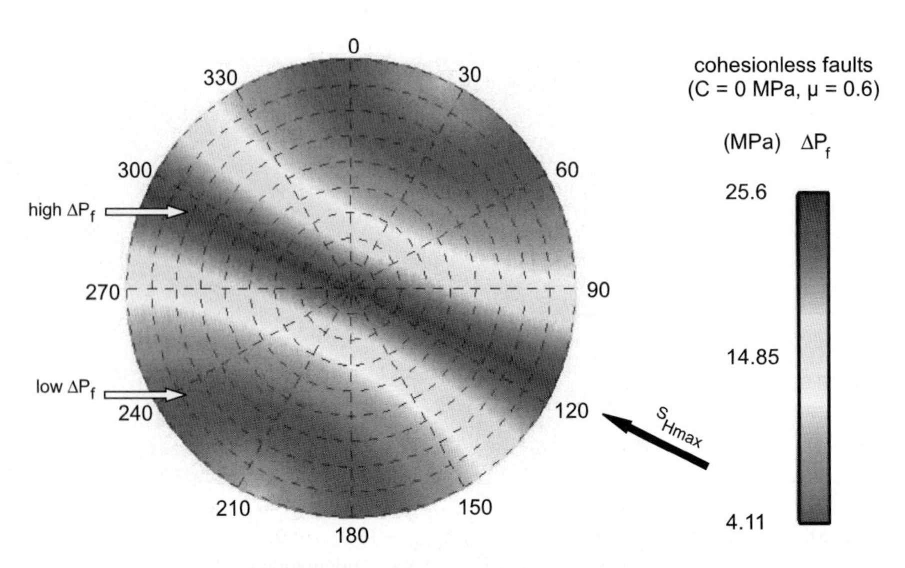

Figure 8: Polar plot projection showing pore fluid pressure increases required to cause failure for any fault orientation (poles to planes) at 2 km depth using the same stress tensor as in Figure 6. Faults are assumed to be cohesionless with $\mu = 0.6$.

Monitoring for Induced Shear Failure

Geophysics of micro-seismic events

In cases where fluid injection induces shear fracturing or fault slip at a seismic rate, micro-earthquakes can occur [36,37]. These micro-earthquakes are commonly termed "micro-seismic" (MS) events and can be readily monitored with geophysical instrumentation such as accelerometer, hydrophone or geophone arrays. Such instrumentation provides a means of visualising the location in 3D, the time, and the magnitude of the events. As has been well established in 15 years of monitoring of hydraulic fracturing experiments in Hot Dry Rock fields, the cloud of MS events can be associated with fractures associated with advancing fluid fronts [38]. Similar responses have been observed in hydrocarbon fields during production and water flooding. This then has provided a new means of reservoir characterisation termed seismicity-based reservoir characterisation.

Micro-seismic events arise when a sudden inelastic deformation occurs such as slip on a fracture or fault (Figure 2a). During slip the elastic strain energy stored in the rock is transformed into fracture surface energy, heat energy, and the radiation of seismic waves. The rate at which this transformation occurs determines the frequency of emission and efficiency of the radiating process. The micro-seismic event will be accompanied by stress release in the zone of the rupture [39].

Slip on a pre-existing fracture in a rock mass will generate a radiating seismic wavefield consisting of both compressional, P, and shear, S, wavelets. The bulk of the seismic energy generated will consist of S-waves (this will be manifested in the high S to P-wave amplitude ratio). Waveforms recorded will contain a superposition of primary P and S wavelets followed by secondary P and S components arising from reflections and refractions within the reservoir formations. Figure 9 illustrates the P-wave and S-wave radiation patterns from a double couple acting within an isotropic rock mass.

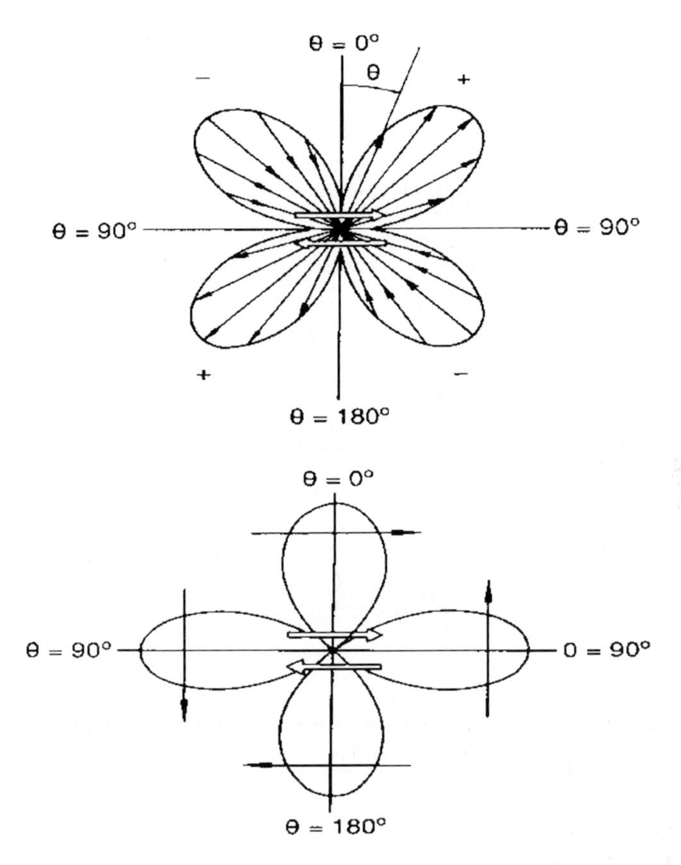

Figure 9: (a) Radiation pattern of the P-wave displacement at the source of a micro-seismic event. The P-wave lobes are shown in a plane of constant azimuth. The force couples are represented by central arrows. (b) Radiation pattern of the S-wave displacement at the source of a micro-seismic event shown in a plane of constant azimuth. The central arrows represent the force couple, giving rise to the event while the larger arrows represent the direction of particle displacement. After Aki and Richards [40].

Monitoring micro-seismic events

Most transducers used for monitoring micro-seismic activity are based on triaxial geophones, usually installed downhole. It is desirable to have at least six triaxial seismometers installed in a monitoring well in close proximity to the injection well. For long-term monitoring, some near-surface arrays are desirable although the geology of the field will influence the design of the arrays.

In recent times the "instrumented oil field" has moved from a concept to reality. It is rapidly gathering impetus with the development of a new generation of transducers. The aim of the instrumented field is to monitor changes in temperature, pressure, and seismic response over the life of a reservoir with arrays of permanently installed transducers. A recent trend has been to develop fibre-optics-based instrumentation that avoids the fragility of down-hole electrical connections over long periods. While temperature and pressure-measuring fibre-optics-based devices are relatively well established, a new type of seismometer that relies on optical diffraction has been developed and installed recently (Internet News Release—Weatherford.com). In the light of such technological advances, fibre-optics-based permanent installations for monitoring micro-seismic activity as well as the usual reservoir parameters, such as temperature and pressure, are strongly recommended for all subsurface CO_2 storage reservoirs.

Interpretation of transducer recordings
With sufficient number and distribution of receiver stations arranged in space around the source, it is theoretically possible to determine the radiation pattern, i.e. the P-wave and S-wave radiation pattern lobes, associated with a particular event and its orientation. This is usually represented graphically by equal area hemispherical projections. This allows the elements of the moment tensor to be calculated and consequently the fault plane orientation, as is illustrated in Figure 9. The procedure to determine the slip magnitude and orientation from an event is termed a "moment tensor inversion" and is described in depth in Ref. [41].

In the case of a small number of receivers, such as the arrays used in monitoring hydraulic fracturing, it is not possible to determine the full moment tensor. Instead, fault plane solutions derived from the first-motion polarities of P-waves and also S-waves can be determined by graphical means. This procedure consists of plotting the ray path vectors to each receiver station as points of compressional (P) or dilatational polarity (T) on to a hemispherical projection. It then becomes a matter of determining the best fit to the two orthogonal planes that separate the compressional and dilatational points. These two planes are termed

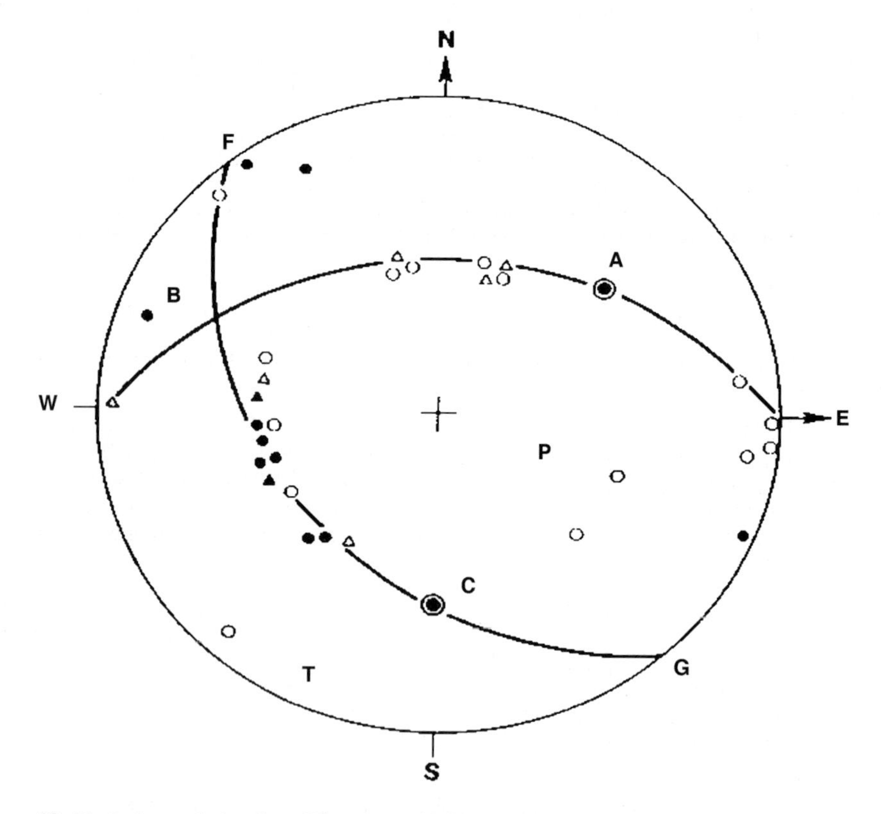

Figure 10: Fault plane solution from Gibowicz and Kijko [39] for a mining-related seismic event. A lower hemisphere equal area projection is used. Solid circles and triangles represent compressional arrivals while open circles and triangles represent dilatational arrivals. A and C are the poles of the two nodal planes. P and T are the axes of compression and tension, respectively.

the focal plane solution and comprise the fault plane and an equivalent orthogonal or auxiliary plane. Additional information such as structural geological information is needed to resolve the ambiguity. A typical fault-plane solution is presented in Figure 10.

Attempts to find fault-plane solutions are rarely reported in the literature that is related to micro-seismic monitoring during hydraulic fracturing. Instead the emphasis is placed on source location producing the conventional MS cloud. This is not surprising given the difficulty in achieving a spatial distribution of receivers that would give a fault-plane solution. However, if fault-plane solutions can be achieved, valuable information can be obtained concerning the orientation of the fault planes and the slip or shearing along those planes. A number of new analysis techniques have been developed which have allowed for very high spatial resolution of MS events. These techniques have greatly improved the visualisation of the intricate detail of subsurface fluid-flow associated with seismic slip [42–44]. Hence, the monitoring for MS events in CO_2 storage sites will be useful for detecting injection-induced slip on faults and fractures and for real-time adjustment of injection pressures.

Seismic Methods for Detecting CO_2 Migration into Fractures

Seismic methods

Surface seismic methods are the most dominant methods used to image underlying geological features, predict lithological variations, and detect the presence of hydrocarbons. Presently most of the surveys are 3D. Successful application of three-component (3C) surveys on land has led to similar, but far more complex, data acquisition procedures offshore. These are designed to record both pressure and particle velocities by utilising hydrophones and 3C geophones and are called 4C and are often 3D. Repeated surveys (usually 3D, 1C or 4C) are useful for monitoring hydrocarbon production and reservoir stimulation by detecting changes in fluid type, saturation, and pressure.

Seismic methods can also be applied in boreholes as, e.g. the vertical seismic profiling (VSP) method. The information obtained by this method is limited to a relatively small area around the borehole in comparison to an area typically covered by surface seismic methods. However, a VSP survey, unlike surface seismic surveys, is useful for recording transmitted waves, from which one can deduce rock properties.

Detection of fracture systems

In practice, a wide area around a borehole can be analysed through multi-azimuth, multi-offset 3C VSP measurements. However, over large areas, the detection of aligned fractures, their density and fill requires repeated 3D (time lapse 3D or "4D") surface seismic data, using single or multi-component recording and analysis methods.

One option for fracture detection and characterisation is the use of P-wave surveys. Many authors [45,46,47] have theoretically studied the behaviour of P-waves (amplitude, velocity, and frequency) propagating through fractured media, the results of which have been corroborated by researchers such as Nur and Simmons [48] and Sayers and Ebrom [49]. P-waves propagating parallel to fractures are subject to rock stiffness but across the fractures they encounter rock compliance (weakness). This results in azimuthally dependent P-wave velocities, amplitudes, and attenuation.

3D surface seismic data are suitable for the detection of P- and S-wave azimuthal anisotropies such as caused by fractures. A polar representation of 3D azimuthal normal move-out velocity will result in an ellipse, with semi-major axis being collinear with the fracture direction [50]. The elongation of the ellipse depends on fracture parameters such as fracture density, fracture aspect ratio, and fluid content. Figure 11 shows Thomsen's anisotropy parameters computed for dry and saturated fractures using different fracture densities. Thomsen's anisotropy parameters [46] are given by

$$\gamma \equiv \frac{c_{66} - c_{55}}{2c_{55}}; \ \varepsilon \equiv \frac{c_{11} - c_{33}}{2c_{33}}; \ \delta \equiv \frac{(c_{13} + c_{55})^2 - (c_{33} - c_{55})^2}{2c_{33}(c_{33} - c_{55})} \quad (8)$$

where c_{ij} are elastic stiffness values, ε and γ describe P- and S-wave anisotropies, respectively, through the differences between vertical and horizontal velocities. Parameter δ is considered to control the shape of the P-wave group-velocity surface away from normal incidence [47]. Simple models as presented in Figure 11 show that the difference in P-wave anisotropy (ε) between dry and fluid saturated fractures is considerable for high fracture densities. Azimuthal variations in P-wave amplitudes are also a tool for the detection and characterisation of fractures [51,52].

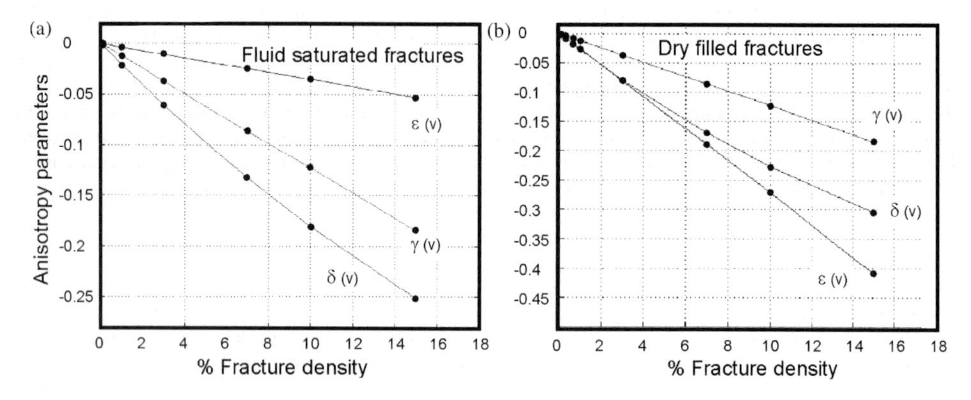

Figure 11: Variations of Thomsen anisotropy parameters $\varepsilon^{(V)}$, $\delta^{(V)}$ and $\gamma^{(V)}$ for horizontal transverse isotropic media with fracture density for a fixed aspect ratio: (a) fluid saturated fractures and (b) dry fractures. Note a high value of the fractional P-wave anisotropy $\varepsilon^{(V)}$ for a system of dry fractures of high density and $\varepsilon^{(V)} - \delta^{(V)} \cong 0$, which is the case of elliptical anisotropy. Also $\varepsilon^{(V)} - \gamma^{(V)}$ is positive for fluid saturated and negative for dry fractures.

Detection of CO_2-filled fractures and faults

P-waves can be utilised for the detection of fractures, their alignment, their properties, and their fluid contents through measurements of velocity, amplitude, attributes, and attenuation in different directions and incidence angles [50,53,54]. In practice, conventional 3D P-wave (particularly marine) surveys lack the range of azimuths and incidence angles required for comprehensive studies of fracture properties. In addition, a successful application of the P-wave methodology is strongly affected by seismic signal-to-noise ratio, and time-lapse P-wave imaging may not be effective at depths where CO_2 properties are similar to liquids. In such cases, the application of borehole time-lapse surveys using VSP and cross-well methods is useful. The frequency content of cross-well seismic and vertical sampling provides data having a vertical resolution comparable to that of wire-line logging. A disadvantage of the cross-well method is that only the 2D slice between the wells can be imaged and the areal extent of a VSP is also therefore limited.

Cross-well pre-stack depth migration using all body waves is useful for detecting small-scale faults and for detailed lithological interpretation. This method can be applied to detect CO_2 migration into vertical faults and fractures. Permanent seismic array deployment allows time-lapse cross-well seismic P- and S-wave recording methods to image-isolated fractures and faults during CO_2 flooding as shown by Wang et al. [55]. This technique works well in high-velocity, high-Q carbonates, but not in all formations.

A method that can be applied to detect aligned fractures is shear wave polarisation analysis (from 3C VSP surveys). This method provides clues on fracture alignment through variations in elastic properties, but this is non-unique due to the alignment of pore space by the in situ stress field. In aligned fracture systems, the shear wave splits into two modes—"fast S1" and "slow S2"—which are polarised along and perpendicular to fractures as suggested by Crampin [56]. The difference in travel time between these two modes is proportional to open fracture density.

Shear wave splitting away from symmetry directions that are parallel or perpendicular to the fracture planes is dependent on the nature of the fluid saturation [45,57]. Shear wave energy would be unaffected by the CO_2 state of phase, but shear wave polarisation, their velocities, and frequency content may change with saturation. Hence time-lapse multi-component VSP surveys using permanently installed 3C geophones may be most useful for the detection of CO_2 infiltration into fractures.

Field Examples

Multi-component seismic data recorded at the Vacuum Field, New Mexico, have shown changes in S-wave splitting within fractured carbonates in response to changes in pore pressure [58]. The interpretation of the results was that the opening of the fractures due to pore pressure increase at the injection well produces an increase in S-wave anisotropy. At the production well a decrease in pore pressure has the opposite effect on the split shear waves. Similar observations were reported during the Lost Hills CO_2 flood where CO_2 was observed to preferentially flow along fracture networks [59]. These examples show that multi-component seismic and in particular shear-wave polarisation analysis have a great potential for detecting migration of CO_2 into highly fractured zones. However, such analyses are unlikely to detect thin CO_2 accumulations in single fractures and faults with typical millimetre to centimetre-scale apertures.

DISCUSSION

This article focuses on the geomechanical effects of CO_2 injection that arise from the law of effective stress (see Eq. (1)). Other stress changes that can result from fluid pressure changes and which are not described by Eq. (1) are briefly discussed in this section. In addition, some key uncertainties that can affect the prediction of fault stability and maximum sustainable fluid pressures will be addressed.

Uncertainty in Rock Frictional Strength

The frictional strength of faults at depth in potential CO_2 storage sites, such as depleted gas reservoirs or saline formations, is difficult to determine. Core samples from faults are rarely available and unlikely to be representative for all faults in the vicinity of a CO_2 storage site. Thus geomechanical predictions, which are usually based on empirical frictional values, need to allow for variations in frictional properties of faults. The frictional strength of reservoir rock and top seal can be determined in laboratory measurements. This requires either rock samples from outcrops of the relevant lithological units of interest, or, even better, fresh core samples from wells in the relevant CO_2 storage site.

Aseismic Slip

While transducer recordings can be used to monitor for seismic slip on fractures and faults, other methods are required to detect aseismic fault slip or so-called fault creep. Appropriate methods include the installation of tiltmeters downhole and creepmeters across fault surface traces, as well as the repeated GPS surveying of reference stations in order to detect fault movement. However, to determine whether any detectable fault creep is related to fluid injection, pre-injection long-term monitoring would be required. Since fault creep may not significantly increase fault permeability and is not the primary study objective, it appears more practical to conduct seismic monitoring for CO_2 migration into faults and near-surface testing (e.g. soil gas testing) for excessive CO_2 accumulations.

Pore Pressure/Stress Coupling

Pressure depletion associated with production in hydrocarbon fields can be associated with a decrease in the total minimum horizontal stress [60,61]. While the vertical stress is usually assumed to remain essentially unaffected during pore pressure depletion, a change of only the horizontal total stresses can, in some tectonic settings, affect the shear stress acting on faults and rocks [61]. This is indicated in Figure 12 for normal fault stress regimes. Induced stress changes are thought to be the cause of faulting within and in the vicinity of reservoirs subjected to pore pressure depletion [60,62].

The effects of pore pressure/stress coupling are of relevance to geological CO_2 storage for a number of reasons. In cases where CO_2 storage is envisaged in pressure-depleted reservoirs, failure that was induced due to pore pressure/stress coupling during reservoir depletion can have compromised the integrity of seals and thus affect the suitability of the reservoir for CO_2 storage.

In addition, the compaction of reservoir rock that can occur due to severe pressure depletion can be partly elastic and also partly permanent [63]. In cases of permanent compaction (= pore collapse) the potential storage capacity for CO_2 would be diminished. Since pore pressure depletion can affect the in situ horizontal

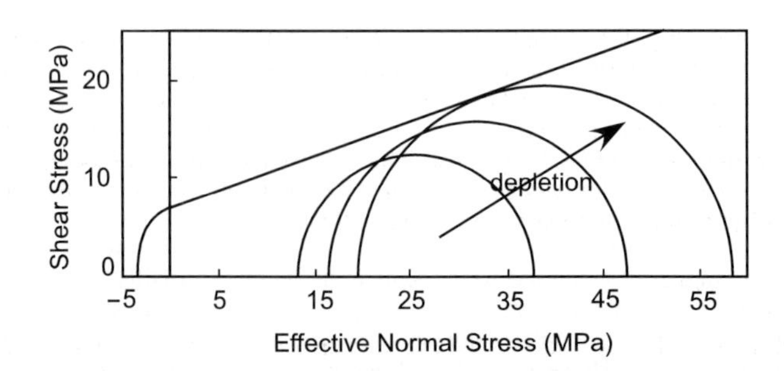

Figure 12: Mohr diagram showing the effect of pore pressure/stress coupling during pore pressure depletion in a normal fault stress regime. For the shown stress path, pressure depletion leads to failure due to decreases of S_{hmin}. Diagram from Hillis [61].

stresses, their orientation and magnitude need to be determined from post-production data in order to be useful for the evaluation of fault stability and maximum sustainable fluid pressures.

Pore pressure/stress coupling that is known to occur during pore pressure depletion may also occur during fluid injection. Since little is known about the poroelastic response of entire reservoirs to fluid pressure increase and the potential effects on total horizontal stresses this needs to be investigated during fluid injection in CO_2 storage sites [12,17].

CONCLUSIONS

This study has presented methods that can be used to predict and monitor geomechanical effects of underground CO_2 injection. The focus was the assessment of fault stability, the prediction and monitoring of maximum sustainable fluid pressures, and the application of seismic methods for the detection of CO_2 infiltration into fault-fracture networks.

The main conclusions of this study are:

- Low effective stresses can lead to fault reactivation at pore fluid pressures lower than those required to induce new fractures in intact rock, especially where faults are optimally oriented for reactivation.
- The effective stresses prevailing in potential CO_2 storage sites can be constrained from the interpretation of drilling data and the application of failure criteria for faults and wellbores.
- Utilising the information on the effective stresses in potential CO_2 storage sites and relevant rock strength data, the stability of faults and rocks and maximum sustainable fluid pressures can be estimated using techniques such as failure plots, FAST, and TrapTester software.
- For CO_2 storage in pressure-depleted reservoirs or fields, these need to be tested for depletion-related effects including damage to seals, permanent compaction of pore space, and stress changes.
- Reliable predictions of poroelastic responses of reservoir rocks to CO_2 injection-related pressure increases and any potentially related changes of total stresses need further studies.
- Seismometer monitoring of micro-seismic events in CO_2 storage sites is an ideal option for fast detection of induced faulting and fracturing related to CO_2 injection and associated effective stress changes.
- Active seismic monitoring methods (multi-component seismic methods and shear-wave splitting) are useful for detecting and monitoring CO_2 accumulations in porous reservoir rock and overburden, or in extensive fault-fracture networks, but may not be suitable for detecting the opening of isolated fractures and faults with millimetre to centimetre-scale widths.

RECOMMENDATIONS

Measures that should be taken for CO_2 storage include the assessment of fault and rock stabilities and the estimation of maximum sustainable fluid pressures in reservoir rock, on faults, and below top seals. Due to the usually sparse availability of testable rock samples, geomechanical and physical predictions need to allow for variations in rock and fault properties. Seismic methods should be applied to detect induced brittle failure that causes micro-seismic events and to identify significant CO_2 accumulations in extensive fracture networks. Thus, the permanent installation of acoustic transducers that record micro-seismicity is recommended for monitoring in CO_2 storage sites. Since the thickness of layered CO_2 accumulations that can be detected by active seismic monitoring methods is limited, monitoring for leak detection requires a combination of seismic and non-seismic monitoring methods.

ACKNOWLEDGEMENTS

This work is based on research undertaken for the CO_2 Capture Project (CCP) under subcontract T-1.1.4 and also based on research conducted by the Australian CRC for Greenhouse Gas Technologies (CO2CRC). The authors appreciate helpful discussions with R.F. Daniel, D.N Dewhurst, K. Dodds, R.R. Hillis, and M. Urosevic. Badley Geoscience Ltd. kindly provided *TrapTester* software. Comments on the manuscript by Curt Oldenburg (Associate Editor) and two anonymous reviewers are appreciated.

REFERENCES

1. S. Bachu, *Energy Convers. Manage.* **41** (2000) 953.
2. P.J. Cook, A. Rigg, J. Bradshaw, *APPEA J.* **40** (2000) 654.
3. S. Holloway, *Energy Convers. Manage.* **37** (1996) 1149.
4. S. Holloway, R. van der Straaten, *Energy Convers. Manage.* **36** (1995) 519.
5. P.D. Bergman, E. Winter, *Energy Convers. Manage.* **36** (1995) 523.
6. C.M. Gibson-Poole, S.C. Lang, J.E. Streit, G.M. Kraishan, R.R. Hillis, in: M. Keep, S.J. Moss (Eds.), The Sedimentary Basins of Western Australia 3: *Proceedings of the Petroleum Exploration Society of Australia Symposium*, Perth, WA, 2002, 2002, p. 439.
7. J. Sminchak, N. Gupta, C. Byrer, P. Bergman, *J. Energy Environ. Res.* **2** (2002) 32.
8. S.A. Shapiro, E. Huenges, G. Borm, *Geophys. J. Int.* **131** (1997) F15.
9. S. Talebi, T.J. Boone, J.E. Eastwood, *Pure Appl. Geophys.* **153** (1998) 95.
10. J.H. Healy, W.W. Ruby, D.T. Griggs, C.B. Raleigh, *Science* **161** (1968) 1301.
11. M.U. Ahmad, J.A. Smith, *Geology* **16** (1988) 739.
12. J.E. Streit, R.R. Hillis, in: J. Gale, Y. Kaya (Eds.), *Proceedings of the Sixth International Conference on Greenhouse Gas Control Technologies, 1–4 October 2002*, Kyoto, Japan, vol. I, Pergamon Press, Amsterdam, 2003, p. 495.
13. J. Handin, R.V. Hager Jr., M. Friedman, J.N. Feather, *Am. Assoc. Petrol. Geol. Bull.* **47** (1963) 718.
14. J. Jaeger, N.G.W. Cook, Fundamentals of Rock Mechanics, Chapman & Hall, London, 1979.
15. J.E. Streit, *J. Geophys. Res.* **102** (1997) 24619.
16. G.H. Eisbacher, Einführung in die Tektonik, Enke, Stuttgart, 1996.
17. J.E. Streit, R.R. Hillis, Energy, **29**(9–10) (2004), 1445–1456.
18. J. Byerlee, Friction of rocks, *Pure Appl. Geophys.* **116** (1978) 615.
19. T. Shimamoto, J.M. Logan, *J. Geophys. Res.* **86** (1981) 2902.
20. R.L. Dart, M.L. Zoback, *Log Analyst* **30** (1989) 12.
21. R.A. Plumb, S.H. Hickman, *J. Geophys. Res.* **90** (1985) 5513.
22. D. Moos, M.D. Zoback, *J. Geophys. Res.* **95** (1990) 9305.
23. A. McGarr, N.C. Gay, *Ann. Rev. Earth Planet. Sci.* **6** (1978) 405.
24. T. Engelder, Stress Regimes in the Lithosphere, Princeton University Press, Princeton, NJ, 1993.
25. P. Peska, M.D. Zoback, *J. Geophys. Res.* **100** (1995) 12791.
26. J. Handin, *J. Geophys. Res.* **74** (1969) 5343.
27. A. Nur, H. Ron, O. Scotti, *Geology* **14** (1986) 746.
28. R.H. Sibson, *J. Struct. Geol.* **7** (1985) 751.
29. J.E. Streit, *J. Geophys. Res.* **104** (1999) 17929.
30. A.D. Castillo, R.R. Hillis, K. Asquith, M. Fischer, in: P.G. Purcell, R.R. Purcell (Eds.), The Sedimentary Basins of Western Australia 2—Proceedings of Petroleum Exploration Society of Australia Symposium, Perth, WA, 1998, p. 325.
31. S.D. Mildren, R.R. Hillis, J. Kaldi, *APPEA J.* **42** (2002) 187.

32. C.M. Gibson-Poole, J.E. Streit, S.C. Lang, A.L. Hennig, C.J. Otto, *APPEA J.* **44** (2004) 653.
33. A. Morris, D.A. Ferrill, D.B. Henderson, *Geology* **24** (1996) 275.
34. R.H. Sibson, in: C.J. Marone, M.L. Blanpied (Eds.), Faulting, Friction, and Earthquake Mechanics, Part I Pageoph Topical Volumes, Birkhäuser Verlag, Basel, 1994, p. 645.
35. T. Shimamoto, J.M. Logan, *Tectonophysics* **75** (1981) 243.
36. S.A. Shapiro, *Geophys. J. Int.* **143** (2000) 931.
37. P. Talwani, S. Acree, *Pure Appl. Geophys.* **122** (1985) 947.
38. T. Wallroth, A. Jupe, R. Jones, *Marine Petrol. Geol.* **13** (4) (1996) 447.
39. S.J. Gibowicz, A. Kijko, An Introduction to Mining Seismology, Academic Press, New York, 1994.
40. K. Aki, P.G. Richards, Quantitative Seismology. Theory and Methods, Freeman, New York, 1980.
41. J.A. Snoke, in: W.H.K. Lee, H. Knanamori, P.C. Jennings, C. Kisslinger (Eds.), International Handbook of Earthquake Engineering Seismology, Academic Press, San Diego, 2003.
42. W.S. Phillips, L.S. House, M.C. Fehler, *J. Geophys. Res.* **102** (1997) 11745.
43. H. Moriya, H. Niitsuma, R. Baria, *Bull. Seism. Soc. Am.* **93** (1996) 1606.
44. R.A Jones, R.C. Stewart, *J. Geophys. Res.* **102** (1997) 8245.
45. J.A. Hudson, *Geophys. J. R. Astron. Soc.* **64** (1981) 133.
46. L. Thomsen, *Geophysics* **51** (1986) 1954.
47. L. Thomsen, *Geophysics* **53** (1988) 304.
48. A. Nur, G. Simmons, *J. Geophys. Res.* **74** (1969) 6667.
49. C.M. Sayers, D.A. Ebrom, *Geophysics* **62** (1997) 1570.
50. V. Grechka, I. Tsvankin, *Geophysics* **63** (1998) 1079.
51. W. Chen, AVO in azimuthally anisotropic media fracture detection using P-wave data and a seismic study of naturally fractured tight gas reservoirs, unpublished PhD thesis, Stanford University, 1995.
52. M. Luo, B.J. Evans, *64th Meeting: European Association of Geoscience Engineering*, 2002, p. C043.
53. C.M. Sayers, J.E. Rickett, *Geophys. Prosp. Eur. Assoc. Geosci. Eng.* **45** (1997) 165.
54. A. Bakulin, C. Slater, H. Bunain, V. Grechka, *70th Annual International Meeting, Society of Exploratory Geophysics*, 2000, p. 1405.
55. Z. Wang, M.E. Cates, R.T. Langan, *Geophysics* **63** (1998) 1604.
56. S. Crampin, *Wave Motion* **3** (1981) 343.
57. L. Thomsen, *Geophysics* **43** (1995) 805.
58. L. Duranti, T.L. Davis, R.D. Benson, *70th Annual International Meeting, Society of Exploratory Geophysics*, Expanded Abstracts, 2000, 1528.
59. P.R. Perri, DOE/BC/14938-12 (OSTI ID: 5127), *2001 Annual Report, National Petroleum Technology Office*, US Department of Energy, Tulsa, OK, 2002.
60. L.W. Teufel, D.W. Rhett, H.E. Farrell, in: J.-C. Roegiers (Ed.), Rock Mechanics as a Multidisciplinary Science: *Proceedings of the 32nd US Symposium*, Balkema, Rotterdam, 1991, p. 63.
61. R.R. Hillis, *Expl. Geophys.* **31** (2000) 448.
62. P. Segall, *Geology* **17** (1989) 942.
63. P. Johnson, D.W. Rhett, W.T. Siemers, *J. Petrol. Tech.* **July** (1989) 712.

Carbon Dioxide Capture for Storage in Deep Geologic Formations, Volume 2
D.C. Thomas and S.M. Benson (Eds.)

Chapter 7

GEOPHYSICAL AND GEOCHEMICAL EFFECTS OF SUPERCRITICAL CO_2 ON SANDSTONES

Hartmut Schütt, Marcus Wigand and Erik Spangenberg

Geo-Research-Center (GFZ) Potsdam, Potsdam, Germany

ABSTRACT

The overall objective of this laboratory study was to investigate the geophysical and geochemical effects of CO_2 storage in deep saline formations. We used a triaxial cell and autoclaves to reproduce reservoir pressure and temperature conditions that are representative of depths down to 2000 m. The CO_2 is in the supercritical state ($CO_{2,scr}$) at depths greater than approximately 800 m. We measured a number of geophysical parameters, such as seismic wave speeds and attenuation, and collected liquid samples that had been in contact with the rock. Geochemical reactions were studied in detail in autoclaves that are charged with either milled rock or mineral separates. We used three sandstone samples as reservoir rock, and 1 M NaCl solution in doubly deionized water as brine. The geophysical data showed that some effects were qualitatively predictable by standard models. The Gassmann model predicted the dependence of the saturating fluid on the bulk modulus, but underestimated the measured results by approximately 10%. This discrepancy may be due to the modulus dispersion between the low-frequency range of the Gassmann model and the ultrasonic laboratory frequency. The Voigt model reproduced the saturation dependence of v_p. Some experiments, however, indicated the existence of fluid front instabilities by reaching only 50% saturation. This corroborated the results of numerical modeling qualitatively. Unexpected was the increase of the compressional wave attenuation for $CO_{2,scr}$ saturation. Scattering can be excluded as a cause, and a local fluid flow model failed to predict the observed effect. Also unexpected and not predicted by the Gassmann equations was the dependence on the saturating fluid of the shear modulus, which is a few percent smaller for $CO_{2,scr}$ saturation than for brine saturation. This may be caused by fluid–mineral interactions. Mineralogical analysis of the rock before and after CO_2 flooding indicated that the concentration of major and trace elements decreased, whereas the Si content increased. The mobilization and removal of these elements was caused by the alteration of rock-forming minerals, e.g. biotite, plagioclase, alkali feldspar. Furthermore, we observed the mobilization of heavy metal cations. Precipitation of mineral phases (e.g. dawsonite) was not observed in the short-term experiments. We are still lacking a thorough understanding of the correlation between geophysical and geochemical data. Long-term experiments (duration of several weeks) and careful analysis on a smaller length-scale (individual grains and grain contacts) may help to address this issue.

INTRODUCTION

Geological storage of CO_2 has the potential to reduce the net CO_2 emissions of industrialized countries significantly, while still allowing for the use of fossil fuels. The storage of large amounts of CO_2 requires careful planning of the technical operations and prediction of the long-term behavior of the entire system (fluids, reservoir rock, cap rock, overburden) to minimize risks and to address public concerns about safety. Some key questions are given below.

- What is the long-term fate of the CO_2? How much will dissolve in brine, and how much will be trapped in newly precipitated minerals?
- Will the CO_2 dissolve and/or precipitate minerals in the reservoir, thereby changing the transport properties of the reservoir rock (sealing the reservoir or opening pathways)?
- Will the CO_2 affect the pressure in the reservoir and/or the stability of the reservoir (risk of subsidence)?

- Can geophysical methods (active and passive, surface and borehole) provide reliable data to assess the saturation and pressure state of the reservoir?
- Is there a correlation between geophysical data and geochemical processes in the reservoir?

Answering these questions requires—besides numerical modeling—careful experiments in the laboratory on reservoir and cap rock samples under realistic and controlled conditions. These measurements are needed to verify numerical results and calibrate geophysical and geochemical field measurements. The results may help to develop more realistic long-term storage and risk assessment scenarios.

CHEMICAL AND PHYSICAL PROPERTIES OF BRINE AND SUPERCRITICAL CO_2

The sodium chloride content of the brine was adjusted to saline formation compositions occurring in the North German basin [1,2] using sodium chloride (analytical grade) and deionized water. The composition of the initial 1 M brine (58.44 g/L NaCl) (L0), analyzed by inductively coupled plasma optical emission spectroscopy (ICP-OES) and ICP mass spectrometry (ICP-MS), is summarized in Table 6.

The critical temperature is the highest temperature at which a gas can be converted to a liquid by an increase in pressure; the critical pressure is the highest pressure at which a liquid can be converted to a gas by an increase in temperature [3], cf. Figure 1. A supercritical fluid possesses properties that are both gas and liquid-like. Figure 2a–c shows some isothermal ($T = 60$ °C) physical properties of CO_2 and brine that are relevant for seismic wave propagation in saturated porous rocks. The pressure and temperature conditions are representative of the experiments. The properties of the brine are virtually independent of pressure, while the properties of CO_2 change substantially. The density of CO_2 (300–600 kg/m^3) is of the order of the brine density (1000 kg/m^3), while the CO_2 viscosity and bulk modulus are 1–2.5 orders of magnitude smaller than the respective properties of brine (Figure 3). These pronounced viscosity and bulk modulus contrasts between brine and CO_2 can be expected to cause substantial changes in seismic attributes, such as seismic wave speeds and attenuation. This may allow for the detection of saturation as well as pore pressure changes and for monitoring the reservoir conditions during CO_2 storage using seismic surface and borehole methods.

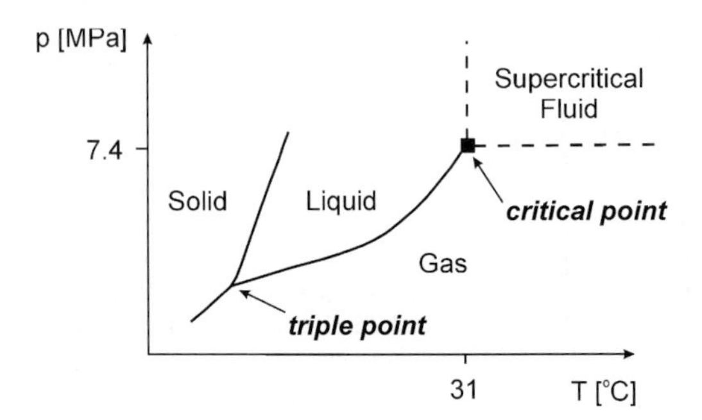

Figure 1: Phase diagram for CO_2 (modified from http://www.dekker.com/servlet/product/DOI/101081EECHR120005324).

SAMPLE DESCRIPTION

Three different types of sandstone were selected to represent reservoir rocks. They differ mainly by their porosity, permeability and grain size (Table 1). These sandstones are litharenites and sub-litharenites (Figure 4). They contain mainly quartz with variable amounts of feldspar (alkali feldspar, plagioclase) and clay minerals (mainly sericite, illite, kaolinite and glauconite). Chemical alteration of feldspars is common and typically involves replacement by clay minerals (e.g. sericite, kaolinite and illite) [4]. The common accessory minerals are zircon, apatite, tourmaline and rutile. Individual chlorite and muscovite crystals are also present in the sandstone samples BW3 and BW6. In contrast to sample H2, the samples BW3 and BW6

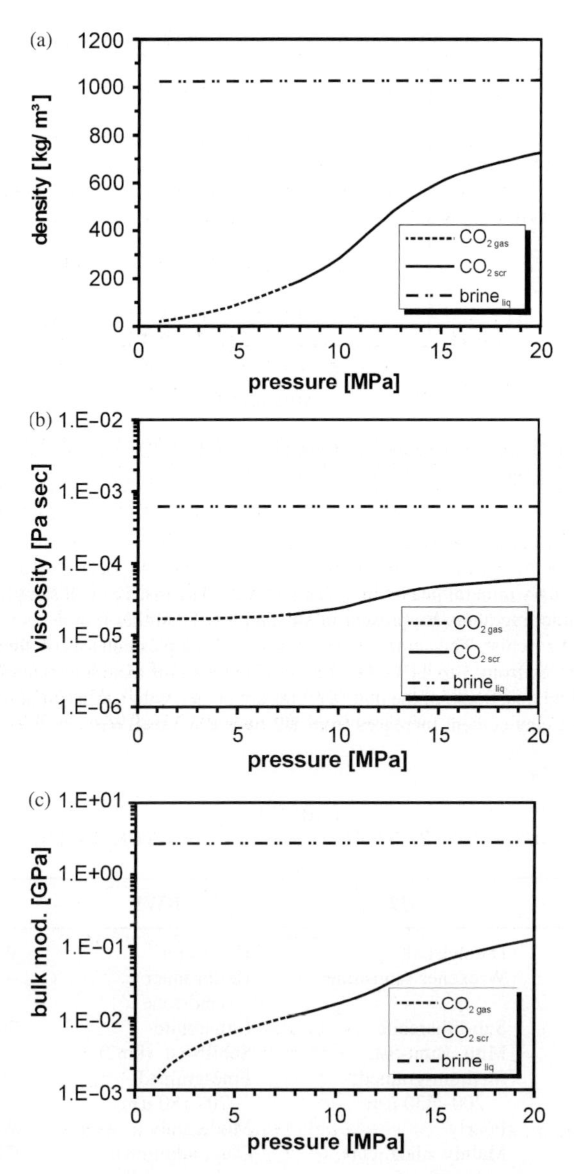

Figure 2: Physical properties of 1 M brine and CO_2 as function of pressure at 60 °C. The properties of brine are virtually constant over the entire pressure range, while the CO_2 properties change substantially. Data are from the NIST database, http://webbook.nist.gov/chemistry/.

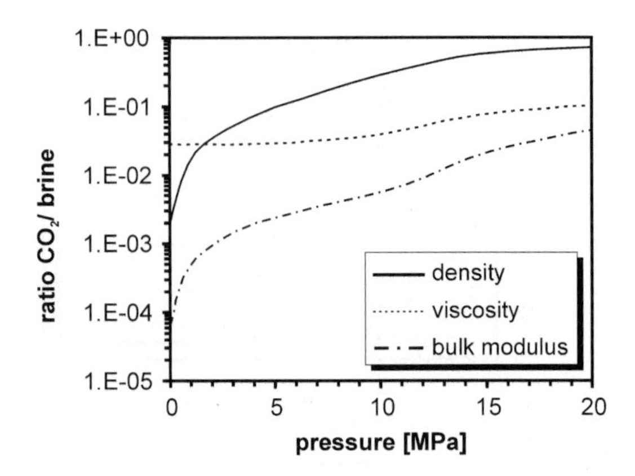

Figure 3: Ratio of the CO_2 and brine properties from Figure 2. The CO_2 density is similar to that of brine in this pressure range, while the viscosity and the bulk modulus are substantially smaller for CO_2 than for brine. Data are from the NIST database, http://webbook.nist.gov/chemistry/.

contain calcite as secondary mineral phases in minor amounts. The matrix of all samples contains abundant opaque iron oxide. Lithic fragments are present in all samples. The lithic fragments in BW3 and BW6 are commonly chert and clay schist. The cement of the sandstone samples consists of silica and clay, whereas the cement fraction varies from 3 to 11%. The quartz cement in all examined samples is in the form of overgrowths of detrital grains. Some of the quartz grains are coated with a thin rim of hematite and limonite cement. The amount of clay cement increases from H2 over BW3 to BW6.

TABLE 1
SAMPLE DESCRIPTION OF RESERVOIR ROCK SAMPLES

Sample	H2	BW3	BW6
Locality[a]	Diemelstadt	Heilbronn	Wertheim-Dietenhahn
Character[a]	Wrexener sandstone	Heilbronner sandstone	–
Classification	Sub-litharenite	Litharenite	Sub-litharenite
Formation[a]	Mittl. Bunt-Sst.	Schilf-Sst. (km2)	(sm1) sandstone
Grain size[a]	Medium-grained, 200–630 μm	Fine-grained, 70–180 μm	Fine-medium grained, 125–630 μm
Sorting[a]	Poorly	Moderately to well	Well
Cement[a]	Mainly silica, clay, growth of microgranular quartz and kaolinite, illite/sericite	Clay, additional silica and limonite	Clay, additional silica, hematite and limonite
Cement fraction (%)[a]	3	11	4
Porosity ϕ (%)[b]	20	22.24	16
Permeability K_{air} (mD)[b]	203.7–379.4	0.767	0.24–0.81

Sst.: sandstone.
[a] Ref. [5].
[b] U. Trautwein, personal communication, 2003. Classification after Ref. [6].

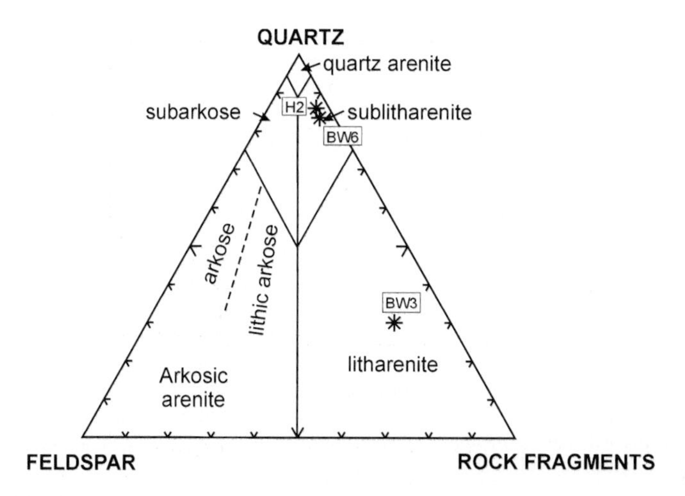

Figure 4: Classification of sandstone samples. The diagram is modified after Ref. [4]. The matrix fraction (i.e. grains $<30\ \mu m$) is $<15\%$ in all sandstone samples.

EXPERIMENTS IN THE TRIAXIAL CELL

A description of the triaxial apparatus (Figure 5) can be found in Ref. [7]. The triaxial cell can be heated externally. The fluids are pushed through the sample in the triaxial cell with two pairs of pumps. Each pair consists of an upstream and a downstream pump, which create a pressure difference between the opposite ends of the sample. While one upstream and one downstream pumps are connected to the sample, the other two pumps get fluid from a reservoir or dispose off fluid into a container. This ensures a continuous flow through the sample. The fluid reservoirs can be exchanged while the sample is in the triaxial cell. This allows for alternating brine and CO_2 floods (so-called water-alternating-gas (WAG) tests). During the experiment, liquid (brine) samples can be taken from the collecting reservoir. The liquid is analyzed for ions that were mobilized by chemical reactions between brine, CO_2 and the minerals.

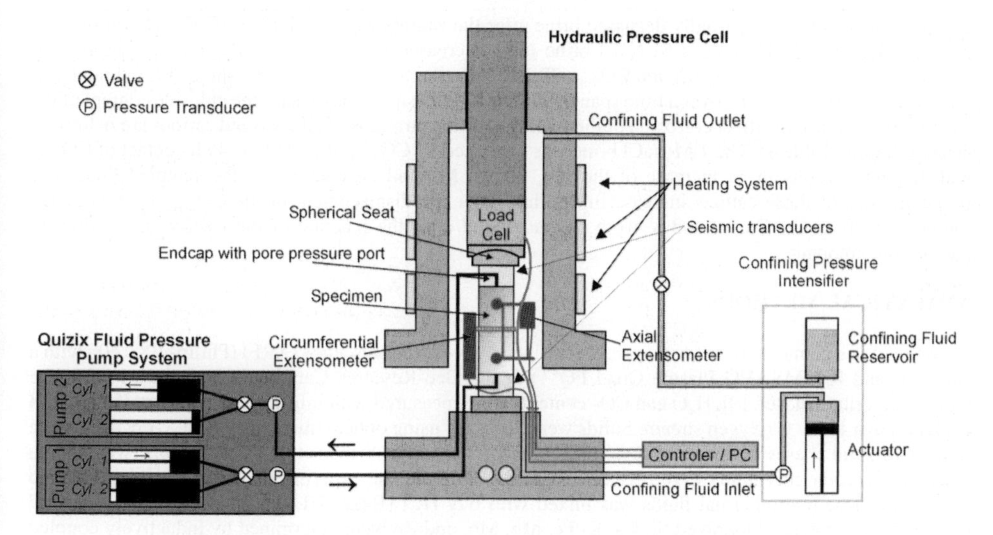

Figure 5: Sketch of the triaxial cell with the pump and heating systems.

The seismic compressional and shear wave speeds (v_p, v_s) are measured by passing an ultrasonic wave through the sample ($f \approx 0.5$ MHz). The waveforms are stored, and v_p and v_s are calculated from the travel times. The seismic wave attenuation is assessed using the spectral ratio method [8]. The waveforms of the dry samples are used as a low-attenuation reference. Lateral and axial strain data as well as temperature, stress and fluid pressure are measured at pre-selected time intervals.

The dry sandstone samples are of cylindrical shape, have a diameter of 50 mm and are 100 mm long. The shape of the samples is limited by the dimension and assembly of the triaxial cell. They are heated to 40 or 60 °C (Table 2) and the confining pressure is increased at a rate of 0.5 MPa/min to 30 MPa. The sample is allowed to settle at the new conditions for 1 day before v_p and v_s are measured. The sample is then saturated with brine and the pore pressure is raised in steps at a rate of 0.5 MPa/min. v_p and v_s are measured at several intermediate pore pressure levels after the sample has been allowed to settle for about 30 min. Brine is then passed through the sample at flow rates of 1 – 10 mL/min. The seismic wave speeds are measured at different pore pressure levels. In the next saturation step the brine is displaced by $CO_{2,scr}$. The measurements are equivalent to those conducted during brine saturation. After the last brine flood, the system is rinsed with deionized water to flush the salt out of the entire system. The duration of the saturation with each fluid corresponds to the porosity and permeability of each sample and is given in Table 2. In contrast to the other samples, BW3 was saturated with $CO_{2,scr}$ twice.

The in-vessel load cell of the MTS and QUIZIX pumps, which are in contact with the fluids during the experiments, are made of different types of stainless steel. The compositions of the steel types are given in Table 3.

The corrosion rate of stainless steel increases with increasing chloride ion content over the range from 10,000 to 100,000 ppm. The magnitude of this effect increases with increasing temperature over 60 °C. Corrosion intensity generally increases with CO_2 partial pressure. CO_2 is an acid gas, i.e. it has the ability to lower the pH when it is dissolved in an aqueous solution such as the 1 M NaCl brine. This increased aggressiveness results from the decrease of the pH of the aqueous phase as the partial pressure of CO_2 increases and leads to the mobilization of cations, which are components of stainless steel.

To avoid errors in the interpretation of the fluid data, a flow-through experiment under 60 °C and 15 MPa pore pressure was conducted on a chemically inert polytetrafluoroethylene (PTFE) cylinder with a centered drill-hole of diameter of 0.2 mm to evaluate the corrosion behavior of the steel. The PTFE cylinder has the same dimension and design as the sandstone samples. The results of this experiment are given in Table 4.

The pH was measured in partially degassed brine after the sampling of the fluid at 25 °C and atmospheric pressure. The initial pH of the 1 M NaCl brine (6.9) decreased to 5.9 and 5.5 after the saturation steps containing a mixture of 1 M NaCl and $CO_{2,scr}$ (Table 4). In the first saturation step, the PTFE cylinder was flushed with 1 M NaCl brine over a time span of 24 h. Under these circumstances, the chloride content of the aqueous phase does not affect corrosion of the steel. The concentrations of all analyzed cations are below the detection limit (Table 4). The 1 M NaCl brine was displaced by $CO_{2,scr}$ after 24 h. A 49 h contact of $CO_{2,scr}$ with the steel results in an increase of the Mn, Si, Ni, Mo and Fe contents of the sampled fluid. The concentrations of these cations increase in the fluid after the displacement of the $CO_{2,scr}$ by 1 M NaCl. Elements such as Cr, Co, W and V were not mobilized after the reaction of the steel with the fluid in measurable amounts.

ANALYTICAL METHODS

Major and trace element abundances were determined by X-ray fluorescence (XRF) (Philips PX 1400 with a Rh tube) and ICP-MS (VG Plasma Quad PQ^{2+}) at the Geo-Research-Center (GFZ) Potsdam using the method described in Ref. [9]. H_2O and CO_2 contents were measured with infrared spectrometry (LECO CH analyzer) in a 1000 °C oxygen stream. Solids were analyzed using optical microscopy and X-ray diffraction (XRD). Mineral phases were determined by XRD analysis using a Cu tube at 40 kV and 20 nA (wavelength Cu Kα). Hydrothermal fluids were periodically sampled during the experiments and analyzed. A defined volume of these hydrothermal fluids was mixed with 6 N HCl (Baker ULTREX) to dissolve oxide and hydroxide precipitates. Dissolved Si, Ca, K, Fe, Mg, Mn, and Zn were determined by inductively coupled plasma atomic emission spectrometry (Varian Liberty 200) at the Geo-Research-Center Potsdam.

TABLE 2
CONDITIONS OF THE FLOW-THROUGH EXPERIMENTS

Sample	T (°C)	Confining pressure (MPa)	Pore pressure (MPa)	Total duration of saturation of each sample (h)	Duration of saturation with 1 M NaCl (h)	Duration of saturation with $CO_{2,scr}$ (h)	Duration of saturation with 1 M NaCl (h)	Duration of saturation with $CO_{2,scr}$ (h)	Duration of saturation with deionized water (h)
T	60	30	15	149	24	49	72	–	4
H2	40	25	13	92	20	31	31	–	10
BW3	60	30	15	152	22	21	27	72	10
BW6	60	30	15	83	25	24	24	–	10

–, not performed.

TABLE 3
TYPE AND COMPOSITION OF THE STEEL TYPES, WHICH ARE IN CONTACT WITH THE FLUIDS DURING THE EXPERIMENT

(%)	X35CrMo17	X5CrNiMo1712.2	Hastelloy
C	0.38	0.08 max.	0.010
Mn	1.00 max	2.00 max.	1.00
P	–	0.045 max.	0.025
S	1.00 max	0.030 max.	0.010
Si	–	0.75 max.	0.08
Cr	16.00	16.00–18.00	15.50
Ni	0.80	10.00–14.00	57.00
Mo	1.00	2.00–3.00	16.00
Fe	Balance	Balance	5.50
Co	–	–	2.50
W	–	–	4.00
V	–	–	0.35

Balance; –, not given.

TABLE 4
CHEMICAL COMPOSITION OF FLUID SAMPLES AFTER FOUR STAGES OF SATURATION OF THE PTFE CYLINDER

Sample pH[b]	T (PTFE cylinder)		
	6.9 Saturation with 1 M NaCl	5.9 Saturation with supercritical CO_2	5.5 Saturation with 1 M NaCl
mg/l			
Mn	b.d.l.[a]	0.19	0.46
P	n.a.[c]	n.a.[c]	n.a.[c]
S	n.a.[c]	n.a.[c]	n.a.[c]
Si	b.d.l.[a]	0.2	2.1
Cr	b.d.l.[a]	b.d.l.[a]	b.d.l.[a]
Ni	b.d.l.[a]	2.1	8.2
Mo	b.d.l.[a]	0.1	0.5
Fe	b.d.l.[a]	1.0	45
Co	b.d.l.[a]	b.d.l.[a]	b.d.l.[a]
W	b.d.l.[a]	b.d.l.[a]	b.d.l.[a]
V	b.d.l.[a]	b.d.l.[a]	b.d.l.[a]

The cations were analyzed by ICP-MS and ICP-OES. The concentrations (mg/L) are normalized to the concentrations of initial 1 M NaCl brine.
[a] b.d.l., below detection limit.
[b] pH reported is stable value measured on sampled brine at standard temperature (25 °C) and at least partially degassed.
[c] n.a., not analyzed.

Concentrations of Al, Li, Co, Ni, Cu, Mo, Cd, Sn, Tl, Th, U and Pb were determined by ICP-MS (VG Plasma Quad PQ^{2+}) at the GFZ following the method described in Ref. [9].

RESULTS

Seismic Wave Speeds

The simultaneous use of compressional and shear waves allows for the assessment of both the bulk and shear moduli of the saturated rock [8]. We assume that full saturation is reached after several hours of pumping fluid through the sample, but we cannot evaluate intermediate saturation states directly. To estimate saturation effects, we calculate the Voigt and Reuss bounds of the compressional wave speed [8], for the system minerals/brine/$CO_{2,scr}$. The Voigt bound represents high-frequency laboratory data, while the Reuss bound represents low-frequency field data. We assume that the shear modulus is independent of the saturation state (Gassmann model, [8]). We can then predict the Voigt and Reuss bounds of the bulk modulus and the compressional wave speed for any saturation state. The result is shown in Figure 6 for sample BW3 at 60 °C and 15 MPa pore pressure.

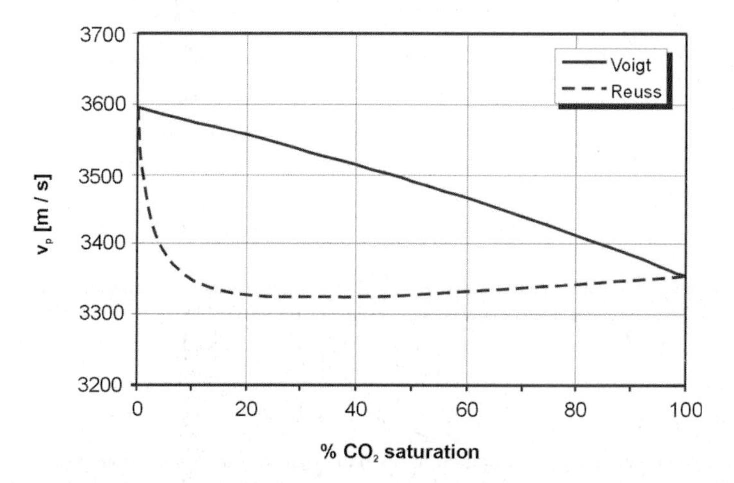

Figure 6: Predicted upper and lower bounds for the compressional wave speed as function of the CO_2 saturation for sample BW3. The Voigt bound represents the high-frequency laboratory data, the Reuss bound represents low-frequency field data. Note that both bounds are equal for full brine and full CO_2 saturation.

The wave speeds were measured at different pore pressures. They are shown in Figure 7 along with the Voigt/Reuss predictions for full brine and full $CO_{2,scr}$ saturation. At relatively low pressure, the velocity data fall along the upper bound in case of full brine saturation. For full CO_2 saturation, they fall along the lower bound. Superimposed is a pressure trend, indicated by the broken lines; the higher the pore pressure, the lower the v_p. This trend is expected, since increasing pore pressure opens small pores and cracks and reduces the moduli. The effect is more pronounced for CO_2 saturation than for brine saturation. The data at 15 MPa pore pressure in the middle between the data for full brine and full CO_2 saturation may reflect an intermediate saturation state, where both phases are present in the pores. We can now estimate the CO_2 saturation using the Voigt bound from Figure 6. It shows an almost linear decrease of v_p with increasing CO_2 saturation at a rate of 2.5 m/s per 1% CO_2. We assume that this rate is also applicable to seismic wave speeds that are measured at variable pore pressure. This approach suggests a partial CO_2 saturation of approximately 50% for the intermediate wave speeds at 15 MPa (Figure 7), since the corresponding data points fall in the middle between the upper and the lower Voigt bound under consideration of the pore pressure (broken lines in Figure 7). We can assess the approximate saturation resolution from the Voigt bound; if the wave speeds can be resolved in 1% steps, the corresponding CO_2 saturation can be resolved in 20% steps (cf. Figure 6). The variation with CO_2 content will be more pronounced in "softer" rock samples with lower moduli and wave speeds, i.e. seismic methods will be more sensitive to saturation changes in soft rocks and unconsolidated sands. Laboratory calibration of the saturation–velocity characteristics for each reservoir rock is a prerequisite for a successful application to seismic field data, and the velocity dispersion between laboratory and field seismics has to be accounted for.

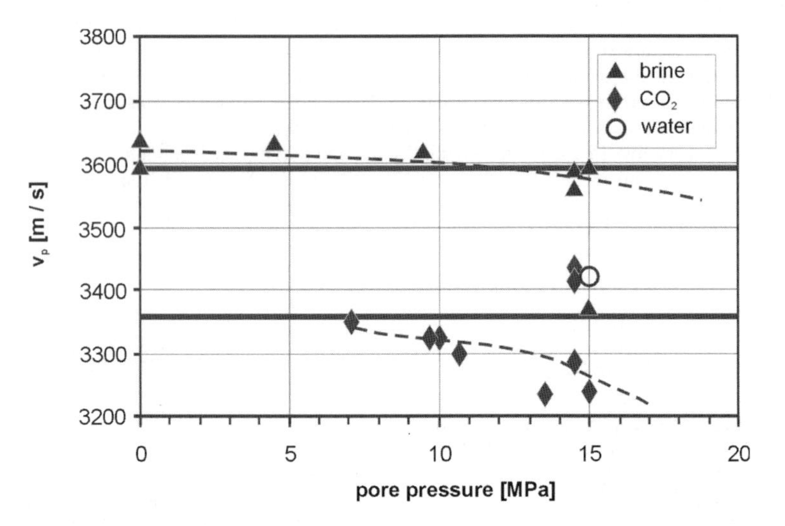

Figure 7: Compressional wave speed for sample BW3 at different saturation states and pore pressures. The heavy horizontal lines correspond to the end values (i.e. for 0 and 100% $CO_{2,scr}$ saturations, respectively) of the bounds in Figure 6. The broken lines are the trends of v_p with changing pore pressure: the higher the pore pressure, the lower the v_p.

CO_2 saturation

Ferer et al. [10] compute the CO_2 saturation at breakthrough in simple pore models for CO_2 displacing brine. They find breakthrough saturations of approximately 20% at a representative viscosity ratio M ($M = \eta_{CO_2}/\eta_{brine} = 0.05$) for a wide range of capillary numbers[1] Ca ($10^{-5} < Ca \leq 10^{-2}$). For the (unrealistic) viscosity ratio of 1.0, the CO_2 saturation at breakthrough increases from 20% at $Ca = 6 \times 10^{-6}$ to 50% at $Ca = 6 \times 10^{-3}$. These data suggest that fluid front instabilities may occur when low-viscosity and low-density supercritical CO_2 displaces brine. This may lead to both viscous destabilized and gravity destabilized flow [11], resulting in a reduced CO_2 sweep efficiency. We estimate the capillary number for our experiment with sample BW3 from the flow rate. The data fall into the transition zone between stable displacement, viscous fingering and capillary fingering when plotted in the M–Ca plane [12]. This suggests that some sort of fluid front instability will occur under the conditions of the simulations and experiments. This may lead to a final CO_2 saturation below 1. It is not clear, however, whether this will be a significant effect under field conditions, and why there is an inconsistency between the breakthrough saturation of Ferer et al. [10] (20–50%) and our data (50–100%).

Moduli

The seismic wave speeds depend on an appropriate modulus and on the density [8]. The equations suggest that the velocities decrease/increase with increasing/decreasing density. However, opposite velocity trends

[1] The capillary number, Ca, is equal to the ratio of the viscous force to the capillary force; $Ca = v\eta/\sigma$, where v is the average speed of the liquid, η the viscosity and σ the surface tension.

are frequently observed. This behavior is caused by the dependence of the moduli on the saturation state. Seismic wave speeds are usually dominated by modulus effects, while density effects are secondary.

The simultaneous use of compressional and shear waves allows for the decomposition of the measured seismic wave speeds into bulk and shear moduli [8]. The effects of saturation and pore pressure on the moduli can then be assessed. The standard model for the prediction and interpretation of fluid substitution effects are the Gassmann equations [8]

$$K_{sat} = K_{dry} + \frac{\left(1 - \dfrac{K_{dry}}{K_s}\right)^2}{\dfrac{\phi}{K_{fl}} + \dfrac{1-\phi}{K_s} - \dfrac{K_{dry}}{K_s^2}} = K_{dry} + \Delta K \tag{1}$$

$$\mu_{sat} = \mu_{dry} \tag{2}$$

where K_{sat} is the bulk modulus of the liquid-saturated rock, K_{dry} the bulk modulus of the dry[2] rock, K_s the bulk modulus of the solid matrix material, K_{fl} the bulk modulus of the fluid, ϕ the porosity, μ_{sat} and μ_{dry} the shear moduli of the fluid-saturated and the dry rocks, respectively.

The modulus increment, ΔK, is caused by the fluid, i.e. the fluid "stiffens" the dry rock and increases v_p. If the fluid is a gas with low bulk modulus, this effect is negligible. The Gassmann equations hold strictly only for static deformation. They are, however, routinely applied to data from field seismics or even to ultrasonic laboratory data. This may lead to erroneous results. According to the Gassmann equations, the shear modulus is independent of the saturating fluid. This may not be true if the liquid interacts physically or chemically with the minerals, e.g. through dehydration or mineral dissolution.

The bulk moduli of sample BW3 for different saturation conditions are shown in Figure 8 as a function of the differential pressure, $p_{diff} = p_{confining} - p_{pore}$ [13]. The confining pressure is constant (30 MPa), while the pore pressure is changed. The Gassmann prediction for brine saturation[3] is about 10% smaller than the measured value. This may be caused by the modulus dispersion, i.e. the increase of the modulus between the frequency range of the Gassmann equations (static) and the frequency range used in the laboratory (ultrasonic). The Gassmann prediction for CO_2 saturation is not shown here. It is practically identical to the dry modulus due to the high compressibility of the supercritical CO_2. It is not clear why some data at low differential pressure (i.e. high pore pressure) fall in the middle between CO_2 and brine saturation. It may be an indication of an intermediate saturation state with both brine and CO_2 present in the pores. The general picture is the same for samples H2 and BW6 in that the bulk modulus for $CO_{2,scr}$ saturation is identical to the dry bulk modulus and the moduli for brine saturation are larger than the predicted moduli by approximately 9 and 12%, respectively.

The Gassmann model predicts no effect of the saturating fluid on the shear modulus, i.e. $\mu_{sat} = \mu_{dry}$ for all fluids. Figure 9 shows the shear moduli of sample BW3 as a function of the differential pressure for different saturation conditions. The dry modulus is the highest. Once the sample is saturated with brine, the shear modulus drops by 3% at 0 pore pressure ($p_{diff} = 30$ MPa). The shear modulus depends almost linearly on the differential pressure and drops by 6% between 0 pore pressure and 15 MPa. The shear modulus drops by another 3% when brine is replaced by $CO_{2,scr}$. The shear modulus also depends linearly on the differential pressure for $CO_{2,scr}$ saturation. These trends are reproducible for alternating $CO_{2,scr}$ and brine saturation. The shear modulus decreases for sample H2 by 5% for the first brine saturation and 4% for the $CO_{2,scr}$ flood; the corresponding values for sample BW6 are 7 and 5%, respectively. The measurements show that, contrary to the Gassmann equations, the shear modulus does depend on the saturating fluid. It drops by a few percent when the dry sample is saturated with brine and by another few percent when the brine is displaced by supercritical CO_2. There seems to be a nearly linear dependence between the shear modulus and the differential pressure.

[2] Dry refers here to *drained* conditions, where the pore pressure is constant during the compression ("open" boundary condition).

[3] We assume a pure quartz matrix, $K_{quartz} = 37$ GPa [8].

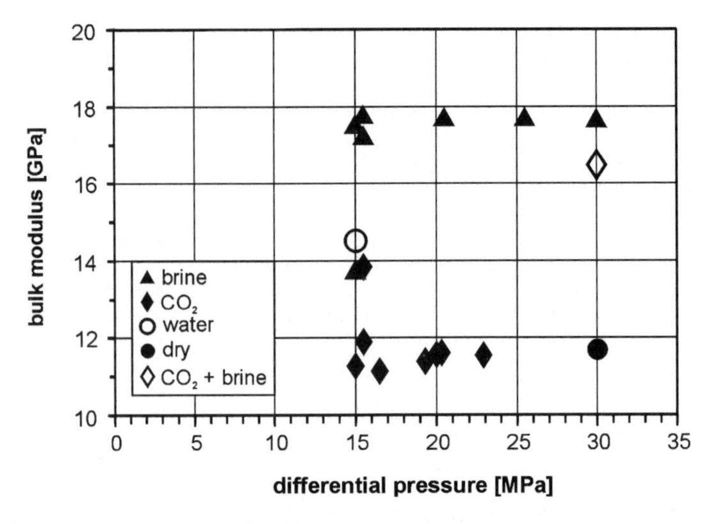

Figure 8: Measured bulk moduli for sample BW3 at different pressure and saturation conditions; the pressure is given as differential pressure $p_{\text{diff}} = p_{\text{conf}} - p_{\text{pore}}$, with $p_{\text{conf}} = 30$ MPa.

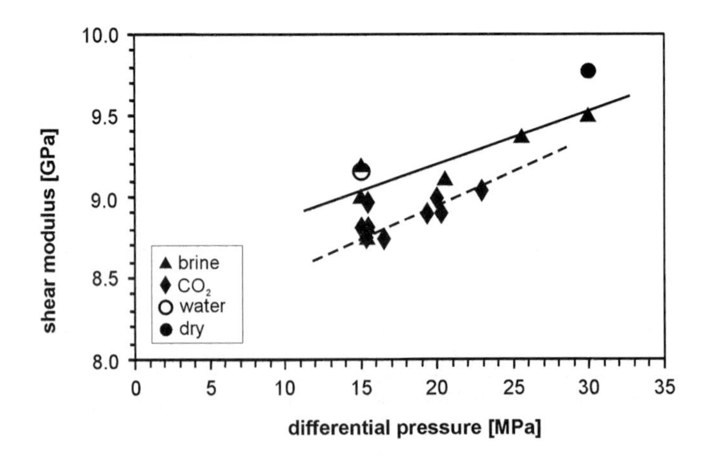

Figure 9: Measured shear moduli for sample BW3 as function of the differential pressure.

A comparison of Figures 8 and 9 shows that the bulk modulus depends mainly on the saturating fluid, while the pressure dependence is nearly negligible. On the other hand, the shear modulus depends on both the saturating fluid and the pressure. To quantify this dependence, we define the pressure sensitivity of the moduli with respect to the differential pressure, S_{pdiff}:

$$S_{\text{pdiff}} = \frac{1}{M} \frac{\text{d}M}{\text{d}p_{\text{diff}}} \tag{3}$$

where M is either the bulk modulus, K, or the shear modulus, μ; p_{diff} is the differential pressure, $p_{\text{diff}} = p_{\text{conf}} - p_{\text{pore}}$ [13].

The measured data suggest a linear dependence of the moduli on the differential pressure, i.e. dM/dp_{diff} is independent of p_{diff}. We can now easily evaluate the pressure sensitivities of both bulk and shear moduli for different saturation conditions using Eq. (3) and the data in Figures 8 and 9. We get approximate sensitivities of 4×10^{-3} and 3×10^{-3} MPa^{-1} for the shear modulus with brine and with $CO_{2,scr}$ saturation, respectively, while the corresponding values for the bulk modulus are 3×10^{-4} MPa^{-1} with brine and 1×10^{-3} MPa^{-1} with $CO_{2,scr}$. The shear modulus is more sensitive to the differential pressure than the bulk modulus. It is not clear why there is a difference between brine and CO_2 saturations. Neither is it clear why the bulk modulus is more sensitive for CO_2 saturation, while the shear modulus is more sensitive for brine saturation. The results demonstrate, however, that the different magnitudes of pressure and saturation sensitivities of the moduli may be used to discriminate between pressure and saturation effects using seismic v_p and v_s data. This corroborates the findings of Wang et al. [14]. However, the different frequency ranges of field and laboratory measurements have to be considered carefully before the laboratory results can be applied directly to the interpretation of field data.

Seismic Wave Attenuation

Seismic wave amplitudes decrease as the wave travels through a medium. This is due to the irreversible conversion of a fraction of the wave energy into heat.[4] Possible mechanisms of attenuation are scattering at small inhomogeneities, friction or fluid-related effects. Scattering can be excluded as a significant mechanism by applying the analysis of Aki and Richards [15]. Global fluid flow effects [16] cause significant attenuation in high-porosity and high-permeability media, such as unconsolidated sand [17], while they are often negligible in consolidated sedimentary rock [18]. Local fluid flow models [19] and squirt flow models [20] relate the attenuation to viscous dissipation in compliant pores, e.g. in penny-shaped cracks or grain contacts. This mechanism is found to cause substantial attenuation even at high hydrostatic pressure where many of the compliant pores can be expected to be closed [18].

The seismic wave attenuation can be expressed in terms of the quality factor Q, where lower Q implies larger attenuation. In terms of energies, the attenuation can be expressed as (cf. [8]):

$$\frac{1}{Q} = \frac{1}{2\pi} \frac{\Delta E}{E_{max}} \tag{4}$$

where ΔE is the energy dissipated per cycle and E_{max} the peak strain energy during the cycle.

A popular method to estimate $1/Q$ from both laboratory and field data is the spectral ratio method [8]. The logarithmic ratio of the spectral amplitudes of two signals, $\ln[A_2(f)/A_1(f)]$, is plotted as a function of the frequency, f. A_1 is a reference signal, measured in a medium with low attenuation. Aluminum serves often as the reference medium, while we use the dry sample as the reference. Dry rock usually exhibits an attenuation that is about one order of magnitude lower than saturated rock. The use of the dry sample as reference has the advantage that both signals are measured on the same piece of rock without changing the setup. This ensures identical source and receiver coupling conditions and thus very similar signal characteristics for both measurements.

Figure 10 shows the attenuation data for sample BW3. We can identify some characteristics despite the scatter in the data. The attenuation increases with increasing pore pressure (i.e. with decreasing differential pressure). This can easily be explained with opening micropores and microcracks that promote either friction along crack and grain surfaces or local fluid flow in compliant low-aspect ratio pores. Another feature is the dependence of the p-wave attenuation on the saturating fluid; it is significantly higher for $CO_{2,scr}$ saturation than for brine saturation. This dependence is not observed for the shear attenuation.

Figure 11 shows how the seismic wave attenuation may help to create a classification scheme for saturation assessment. Brine saturation is characterized by relatively high values of K/μ and Q_p/Q_s, while $CO_{2,scr}$ saturation is characterized by low values. This is true for all three sandstone samples. Note that the water-saturated sample BW3 (when deionized water is used to flush the salt out of the apparatus) lies in the domain

[4] Geometric spreading is negligible under laboratory conditions.

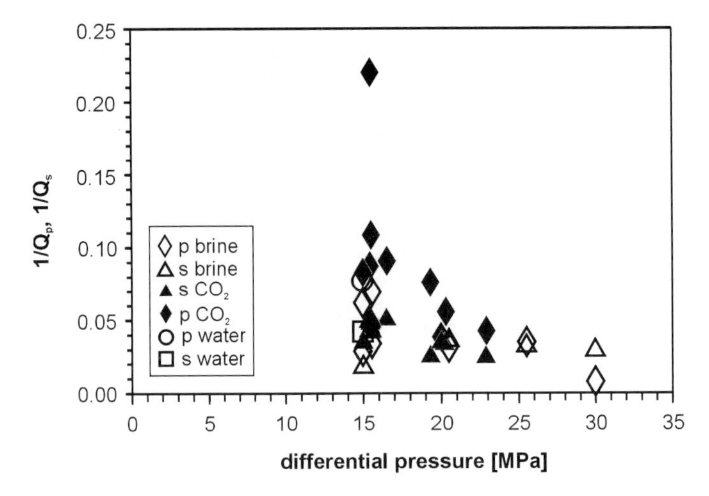

Figure 10: Seismic wave attenuation of sample BW3 as function of the differential pressure for various saturation conditions.

that is populated by $CO_{2,scr}$-saturated samples. This could be an indication of residual $CO_{2,scr}$ saturation or of alteration of the rock frame.

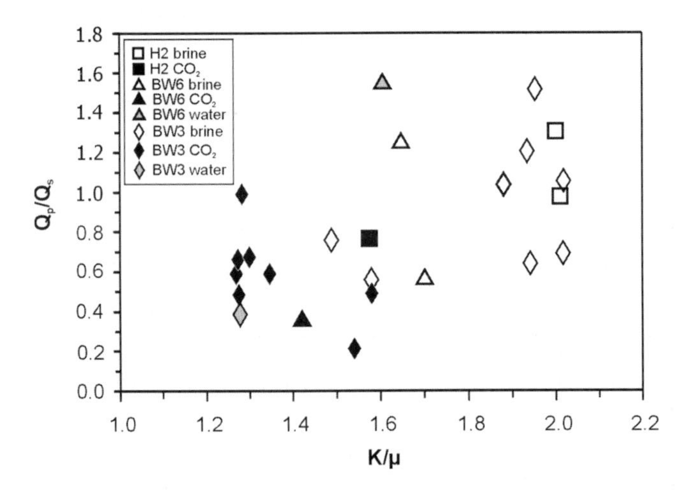

Figure 11: Classification of the saturation state using two seismic attributes. The upper right corner of the diagram is populated by data measured at brine saturation, while the lower left corner is mainly populated by data measured at $CO_{2,src}$ saturation.

Mineral Chemistry

During the experiment, fluid (brine) samples were taken from the collecting reservoir. Changes in the chemical composition of brine resulting from interaction with the sandstones H2, BW3 and BW6 are summarized in Figure 12A–C. The concentrations (mg/L) are normalized to the concentrations of initial 1 M NaCl brine. In contrast to the samples BW3 and BW6, aluminum and silicon were not analyzed in sample H2. The composition of the fluids before and after the reaction with sandstones is summarized in Tables 5–7. At similar duration, the contents of the analyzed cations in the fluids of the sandstone experiments are, with the exception of Fe and Ni, higher than the contents of the dummy experiment. This behavior suggests that particularly the alteration and dissolution of the rock-bearing mineral phases are responsible for increasing cation contents. Steel corrosion is in most cases irrelevant for geochemical data interpretation.

TABLE 5
CHEMICAL COMPOSITION OF FLUID SAMPLES AFTER FOUR STAGES OF SATURATION OF THE SANDSTONE H2

Sample	H2		
	Saturation with 1 M NaCl	Saturation with supercritical CO_2	Saturation with 1 M NaCl
Ca	31	134	38
K	10	15	16
Mg	4.9	54	8.8
Mn	0.13	103	12
Zn	0.2	40	19
Li	0.1	0.2	0.1
Co	b.d.l.[a]	0.3	0.1
Ni	0.01	14	4.1
Cu	0.2	36	11
Mo	b.d.l.[a]	0.2	0.1
Cd	b.d.l.[a]	0.16	0.03
Pb	b.d.l.[a]	0.5	0.3

The cations were analyzed by ICP-MS and ICP-OES. The concentrations (mg/L) are normalized to the concentrations of initial 1 M NaCl brine.
[a] b.d.l., below detection limit.

TABLE 6
CHEMICAL COMPOSITION OF FLUID SAMPLES AFTER FOUR STAGES OF SATURATION OF THE SANDSTONE BW3

Sample	L0 Initial brine	BW3			
		Saturation with 1 M NaCl	Saturation with supercritical CO_2	Saturation with 1 M NaCl	Saturation with supercritical CO_2
Na	22,989[c]	n.a.[b]	n.a.[b]	n.a.[b]	n.a.[b]
Cl	35,453[c]	n.a.[b]	n.a.[b]	n.a.[b]	n.a.[b]
Ca	b.d.l.[a]	64	93	305	43
Fe	b.d.l.[a]	4.0	6.8	21	11
K	b.d.l.[a]	6.1	7.4	12	10
Mg	b.d.l.[a]	7.7	21	42	11
Mn	b.d.l.[a]	3.9	4.7	1.4	25
Si	b.d.l.[a]	6.1	10	24	14
Al	0.05	0.06	0.2	0.2	2.4
Zn	0.1	1.9	0.43	0.91	2.5
Co	b.d.l.[a]	b.d.l.[a]	0.41	0.38	2.7
Ni	b.d.l.[a]	0.8	5.0	2.3	14.6
Cu	b.d.l.[a]	0.1	3.0	2.5	4.4
Rb	<0.1	<0.1	0.1	<0.1	0.1
Sr	<0.1	1.2	1.4	0.9	0.7

(continued)

TABLE 6
CONTINUED

Sample	L0 Initial brine	BW3			
		Saturation with 1 M NaCl	Saturation with supercritical CO_2	Saturation with 1 M NaCl	Saturation with supercritical CO_2
Mo	b.d.l.[a]	0.1	0.2	<0.1	0.3
Cd	<0.01	0.02	0.10	0.02	0.04
Pb	b.d.l.[a]	b.d.l.[a]	0.41	0.50	1.75
Sn	b.d.l.[a]	b.d.l.[a]	b.d.l.[a]	b.d.l.[a]	b.d.l.[a]
Cs	b.d.l.[a]	b.d.l.[a]	b.d.l.[a]	b.d.l.[a]	b.d.l.[a]
Tl	b.d.l.[a]	b.d.l.[a]	b.d.l.[a]	b.d.l.[a]	b.d.l.[a]
Th	b.d.l.[a]	b.d.l.[a]	b.d.l.[a]	b.d.l.[a]	b.d.l.[a]
U	b.d.l.[a]	b.d.l.[a]	b.d.l.[a]	b.d.l.[a]	b.d.l.[a]

The cations were analyzed by ICP-MS and ICP-OES. The concentrations (mg/L) are normalized to the concentrations of initial 1 M NaCl brine.
[a] b.d.l., below detection limit.
[b] n.a., not analyzed.
[c] Na and Cl contents of the initial brine are calculated.

TABLE 7
CHEMICAL COMPOSITION OF FLUID SAMPLES AFTER FOUR STAGES OF SATURATION OF THE SANDSTONE BW6

Sample	BW6		
	Saturation with 1 M NaCl	Saturation with supercritical CO_2	Saturation with 1 M NaCl
Ca	65	77	123
Fe	45	b.d.l.[a]	2.1
K	7.6	6.1	8.9
Mg	8.7	7.7	9.3
Mn	4.6	0.45	0.19
Si	21	4.0	8.4
Al	0.3	0.1	0.1
Zn	3.9	1.8	0.4
Co	0.65	0.16	b.d.l.[a]
Ni	51	11.4	0.3
Cu	5.8	3.8	0.7
Rb	0.2	0.2	0.2
Sr	0.5	0.6	0.4
Mo	1.5	0.2	<0.1
Cd	0.02	0.03	0.01
Pb	0.12	0.02	0.01
Sn	b.d.l.[a]	b.d.l.[a]	b.d.l.[a]
Cs	b.d.l.[a]	b.d.l.[a]	b.d.l.[a]
Tl	b.d.l.[a]	b.d.l.[a]	b.d.l.[a]
Th	b.d.l.[a]	b.d.l.[a]	b.d.l.[a]
U	b.d.l.[a]	0.02	b.d.l.[a]

The cations were analyzed by ICP-MS and ICP-OES. The concentrations (mg/L) are normalized to the concentrations of initial 1 M NaCl brine.
[a] b.d.l., below detection limit.

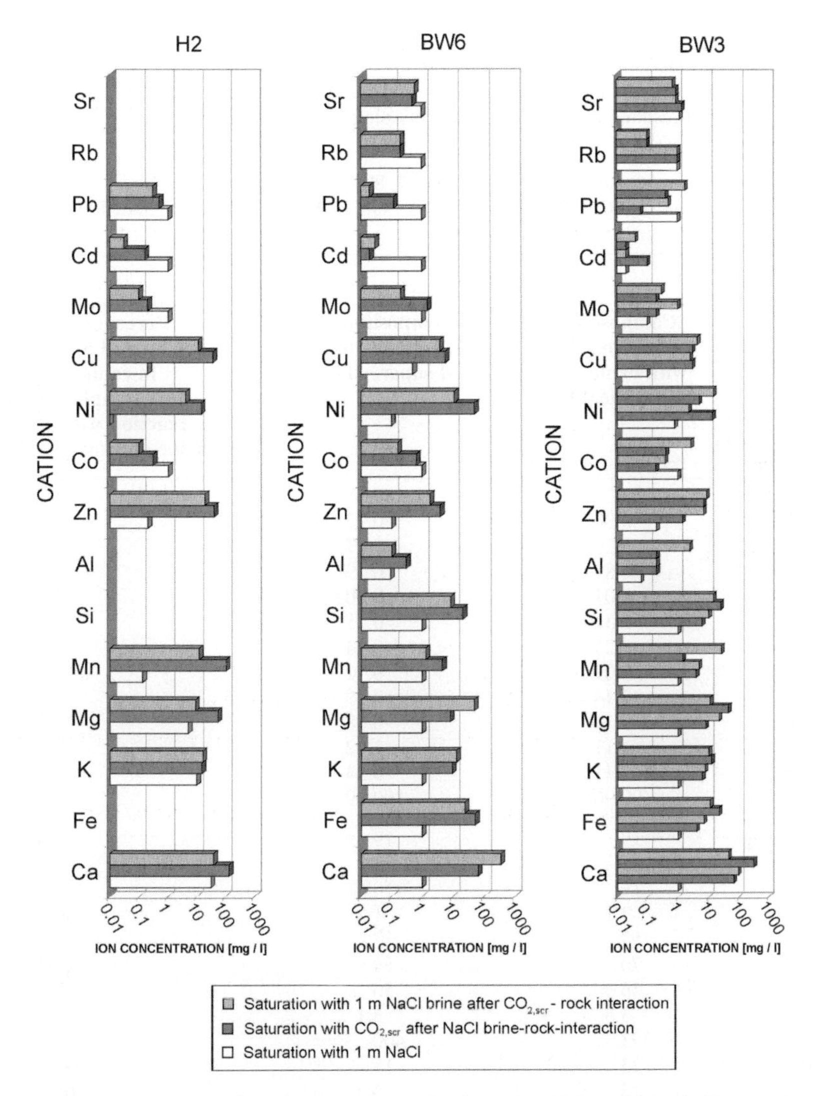

Figure 12: Bar graphs of the samples H2, BW6 and BW3, which indicate changes in the element composition of brine resulting from interaction with the sandstones. Fe, Si, Al, Rb and Sr contents were not analyzed for sample H2.

Figure 12A–C display the stages (steps) of saturation. As a result of the reaction with the brine, Ca, Fe, K, Mg, Mn, Si, Al, Zn, Co, Ni, Cu, Rb, Sr, Mo, Cd and Pb were leached in measurable amounts (see Tables 5–7 and Figure 12A–C). While all sandstone samples also contain Sn, Cs, Tl, Th and U (Table 8 and authors' unpublished data), there is no significant indication for mobilization of these cations during the reaction with 1 M NaCl brine and/or $CO_{2,scr}$. The leaching behavior of the cations is uneven and depends mainly on the following attributes:

- mineral composition of the rock,
- duration of fluid contact,
- pH,
- concentration and composition of the brine.

The highest leaching rate was observed during the rock–$CO_{2,scr}$ interaction in nearly all cases (Figure 12A–C and Tables 5–7). As observed during the dummy experiment and described in Ref. [21], there was a noticeable drop in pH after the injection of $CO_{2,scr}$ as a result of an increase in total HCO_3^- concentration. Such a decrease in pH resulted in primary mineral dissolution, e.g. carbonates.

TABLE 8
MAJOR AND TRACE ELEMENT DATA FOR RESERVOIR ROCK SAMPLE H2

Sample	**H2**		
	Composition before reaction with brine and supercritical CO_2	**Composition after reaction with brine and supercritical CO_2**	**Percentage increase (+) and decrease (−) in composition of sample H2 after reaction with 1 M NaCl brine and supercritical CO_2**
SiO_2 (wt%)	88.2	89.8	+1.8
TiO_2 (wt%)	0.88	0.22	−75.3
Al_2O_3 (wt%)	5.26	5.07	−3.6
Fe_2O_{3tot}(wt%)	0.87	0.59	−31.7
MnO (wt%)	0.013	0.002	−82.1
MgO (wt%)	0.31	0.25	−18.8
CaO (wt%)	0.25	0.14	−45.0
Na_2O (wt%)	0.09	0.11	+23.2
K_2O (wt%)	2.44	2.49	+2.2
P_2O_5 (wt%)	0.14	0.08	−45.2
H_2O^+ (wt%)	1.11	1.02	−8.6
CO_2 (wt%)	0.41	0.34	−16.7
Total	99.96	100.09	–
Cr (ppm)	19	19	+1.3
Ni (ppm)	11	21	+99.4
V (ppm)	14	10	−27.8
Co (ppm)	2.1	1.3	−38.1
Cu (ppm)	57	37	−35.1
Zn (ppm)	70	54	−23.4
Sr (ppm)	56	50	−10.7
Cs (ppm)	3.3	3.1	−6.1
Ba (ppm)	393	400	+1.8
Pb (ppm)	23	15	−34.8
Th (ppm)	13	4.2	−67.7
U (ppm)	4.4	1.3	−70.5
Li (ppm)	18	16	−11.1
Sn (ppm)	3.0	2.0	−33.3
Mo (ppm)	1.0	0.3	−70.0
Cd (ppm)	0.07	b.d.l.[a]	–
Sb (ppm)	0.8	0.5	−37.5
Tl (ppm)	0.6	0.6	+/−0.0
Bi (ppm)	0.03	0.01	−66.7

The major elements were analyzed by X-ray fluorescence (XRF) and all trace elements were analyzed by inductively coupled plasma mass spectrometry (ICP-MS). H_2O and CO_2 contents were measured with infrared spectrometry.

[a] b.d.l., below detection limit.

XRD analyses indicate that the duration of the experiments was too short to form measurable amounts of secondary mineral phases. For this reason the whole rock composition of sample H2 was analyzed by XRF and ICP-MS before and after the experiment to get an indirect indication on the changes in mineral composition. The whole rock data of sample H2 together with the relative changes in composition are summarized in Table 8.

In comparison with the initial composition of sample H2, most of the major elements and trace elements have lower concentrations (Table 8). However, SiO_2, Na_2O and K_2O content increases during the reaction with brine and $CO_{2,scr}$. Furthermore, the carbonate contents of reservoir rock H2 decreased after the reaction with brine and supercritical CO_2. As a result, dissolution prevails over precipitation of rock-bearing carbonates in this initial pre-steady-state experimental investigation. Decreasing H_2O^+ (structural water) contents suggest reaction of OH-bearing minerals (e.g. sheet silicates) with $CO_{2,scr}$.

CONCLUSIONS

The geochemical data show that the exposure of sandstone to brine and $CO_{2,scr}$ under temperature and pressure conditions that are representative for depths below 1000 m leads to the mobilization of a number of different ions from the rock's mineral framework. This may be seen as an indication for mineral dissolution. Mineral precipitation has not been found for the duration of our experiments (up to 10 days). These findings suggest an increase of the porosity and consequently of the hydraulic permeability. These effects are probably too small to be detected using seismic methods, at least during short-term experiments.

The seismic data show a dependence of both the bulk and the shear moduli on the saturation state as well as on the (differential) pressure. The Gassmann model underestimates the fluid substitution effect by an amount that is representative for the modulus dispersion between the static regime (Gassmann model) and the ultrasonic frequency range (laboratory data). The dependence of the shear modulus on the saturating fluid is not predicted by the Gassmann model. The shear modulus shows a higher pressure sensitivity than the bulk modulus. This may lead to the discrimination of pressure and saturation effects through the simultaneous use of compressional and shear waves. The seismic wave attenuation shows the potential to contribute useful information about the saturation state. The experiments corroborate numerical studies that predict fluid front instabilities—and thus breakthrough saturation below 1—when $CO_{2,scr}$ displaces brine. The application of geophysical methods for in situ reservoir monitoring requires the calibration of all effects in laboratory experiments for each reservoir rock, taking into account the different length scales and frequency ranges. The assessment of the reservoir state will certainly benefit from the incorporation of further geophysical methods, such as electrical resistivity and gravity methods.

RECOMMENDATIONS

We recommend effort and funding be directed toward the following laboratory research:

- long-term experiments (several months),
- investigation of fluid-phase behavior (e.g. solubility of $CO_{2,scr}$ in brine),
- pH and electrical resistivity measurements,
- porosity and permeability evolution in triaxial cell, mercury porosimetry,
- assessment of fluid displacement process (fluid front instabilities, fingering),
- cap rock experiments.

ACKNOWLEDGEMENTS

The contributions of the GFZ Potsdam to the Carbondioxide Capture Project (CCP) were initiated and directed by Prof. Dr G. Borm, Prof. Dr J. Erzinger and Dr E. Huenges.

The authors wish to thank Rudolf Naumann for performing the XRF analyses. We are grateful to Knut Hahne, Heike Rothe (ICP-MS) and Sabine Tonn (ICP-OES) for their help with analyses at the GeoForschungsZentrum Potsdam. The sandstone samples along with poro-perm data were provided by Ute Trautwein. Liane Liebeskind and Olaf Ryll helped with the experiments in the triaxial cell. Many thanks to

Siegfried Raab for his patience in endless discussions concerning the improvement of the experimental setup.

We are very grateful for the comments and suggestions by Curt Oldenburg and two anonymous reviewers.

REFERENCES

1. A. Seibt, T. Kellner, P. Hoth, in: P. Hoth, A. Seibt, T. Kellner, E. Huenges (Eds.), *Geothermie Report 97-1: Geowissenschaftliche Bewertungsgrundlagen zur Nutzung hydrogeothermaler Ressourcen in Norddeutschland*, Scientific Technical Report STR97/15, 1997, pp. 134–146.
2. D. Naumann, Salinare tiefenwässer in Norddeutschand, Scientific Technical Report STR00/21, 2000, p. 116.
3. T. Clifford, Fundamentals of Supercritical Fluids, Oxford University Press, New York, 1998.
4. R.L. Folk, *Am. J. Sci.* **245** (1947) 388–394.
5. W.D. Grimm, Bildatlas Wichtiger Denkmalgesteine der Bundesrepublik Deutschland, Bayerisches Landesamt für Denkmalpflege, Arbeitsheft 50, 1990.
6. F.J. Pettijohn, P.E. Potter, R. Siever, Sand and Sandstone, Springer, New York, 1987.
7. M. Alber, J. Heiland, *Rock Mech. Rock Eng.* **34** (2001) 167–186.
8. G. Mavko, T. Mukerji, J. Dvorkin, The Rock Physics Handbook, Cambridge University Press, New York, 1998.
9. H.G. Plessen, H. Rothe, M. Zimmer, J. Erzinger, in: K. Govindaraju, P.J. Potts, P.C. Webb, J.S. Watson (Eds.), *Geostandard Newsletters*, 18, 1994, pp. 211–300.
10. M. Ferer, G.S. Bromhal, D.H. Smith, *J. Energy Environ. Res.* **2** (2001) 120–132.
11. R.J. Glass, M.J. Nicholl, *Geoderma* **70** (1996) 133–163.
12. R. Lenormand, E. Touboul, C. Zarcone, *J. Fluid Mech.* **180** (1988) 165–187.
13. Y. Guéguen, V. Palciauskas, Introduction to the Physics of Rocks, Princeton University Press, Princeton, NJ, 1994.
14. Z. Wang, M.E. Cates, R.T. Lagan, *Geophysics* **63** (1998) 1604–1617.
15. K. Aki, P.G. Richards, Quantitative Seismology: Theory and Methods, W.H. Freeman and Co., New York, 1980.
16. M.A. Biot, *J. Acoust. Soc. Am.* **28** (1956) 168–191.
17. R.D. Stoll, Sediment Acoustics, Springer, Berlin, 1989.
18. A.-M. Wulff, H. Burkhardt, *J. Geophys. Res.* **102** (1997) 3043–3050.
19. W.F. Murphy, K.W. Winkler, R.L. Kleinberg, *Geophysics* **51** (1986) 757–766.
20. J. Dvorkin, R. Nolen-Hoeksema, A. Nur, *Geophysics* **59** (1994) 428–438.
21. J.P. Kaszuba, D.R. Janecky, M.G. Snow, *Appl. Geochem.* **18** (2003) 1065–1080.

Carbon Dioxide Capture for Storage in Deep Geologic Formations, Volume 2
D.C. Thomas and S.M. Benson (Eds.)

Chapter 8

REACTIVE TRANSPORT MODELING OF CAP-ROCK INTEGRITY DURING NATURAL AND ENGINEERED CO_2 STORAGE

James W. Johnson, John J. Nitao and Joseph P. Morris

Environmental Sciences Division, Lawrence Livermore National Laboratory, Livermore, CA, USA

ABSTRACT

Long-term cap rock integrity represents the single most important constraint on the long-term isolation performance of natural and engineered CO_2 storage sites. CO_2 influx that forms natural accumulations and CO_2 injection for EOR/storage or saline-aquifer disposal both lead to geochemical alteration and geomechanical deformation of the cap rock, enhancing or degrading its se
al integrity depending on the relative effectiveness of these interdependent processes. Using our reactive transport simulator (NUFT), supporting geochemical databases and software (GEMBOCHS, SUPCRT92), and distinct-element geomechanical model (LDEC), we have shown that influx-triggered mineral dissolution/precipitation reactions within typical shale cap rocks continuously reduce microfracture apertures, while pressure and effective-stress evolution first rapidly increase then slowly constrict them. For a given shale composition, the extent of geochemical integrity enhancement in the cap rock is nearly independent of key reservoir properties (permeability and lateral continuity) that distinguish EOR/sequestration and saline formation settings and of CO_2 influx parameters (rate, focality, and duration) that distinguish engineered disposal sites and natural accumulations, because these characteristics and parameter have negligible (indirect) impact on mineral dissolution/precipitation rates. In contrast, the extent of geomechanical integrity degradation is highly dependent on these reservoir properties and influx parameters because they effectively dictate magnitude of the pressure perturbation. Specifically, initial geomechanical degradation has been shown inversely proportional to reservoir permeability and lateral continuity and proportional to influx rate. Hence, while the extent of geochemical alteration is nearly independent of filling mode, that of geomechanical deformation is significantly more pronounced during engineered storage. This suggests that the currently secure cap rock of a given natural CO_2 accumulation may be incapable of providing an effective seal in the context of an engineered injection, a potential discrepancy that limits the extent to which natural CO_2 reservoirs and engineered storage sites can be considered analogous. In addition, the pressure increase associated with CO_2 accumulation in any compartmentalized system invariably results in net geomechanical aperture widening of cap-rock microfractures. This suggests that ultimate restoration of pre-influx hydrodynamic seal integrity—in both EOR/storage and natural accumulation settings—hinges on ultimate geochemical counterbalancing of this geomechanical effect. To explore this hypothesis, we have introduced a new conceptual framework that depicts such counterbalancing as a function of effective diffusion distance and reaction progress. This framework reveals that ultimate counterbalancing of geochemical and geomechanical effects is feasible, which suggests that shale cap rocks may in fact *evolve* into effective seals in both natural and engineered storage sites.

Abbreviations: CCP, CO_2 Capture Project; GEMBOCHS, Geologic and Engineering Materials: Bibliography of Chemical Species (Thermodynamic/kinetic database and software library [24,25]); LDEC, Livermore Distinct Element Code (geomechanical modeling software [32,33]); NUFT, Non-isothermal Unsaturated Flow and Transport (reactive transport software [19,20]); SUPCRT92, SUPerCRiTical (geochemical modeling software and database: [26]).

INTRODUCTION

Successful engineered CO_2 storage in geologic formations hinges on our ability to identify optimal sites and forecast their long-term security. This ability, in turn, relies upon predictive models for assessing the relative effectiveness of CO_2 migration and storage processes (isolation performance) as a function of key target-formation and cap-rock properties (screening criteria). It also relies on detailed knowledge of naturally occurring CO_2 reservoirs and clear understanding of the extent to which they represent natural analogs to engineered storage sites. In the opinion of the author, among key screening criteria, long-term cap rock integrity is the most important constraint on the long-term isolation performance of both natural and engineered CO_2 storage sites. And among predictive methodologies, the reactive transport modeling approach is uniquely well suited to quantify this fundamental constraint.

In this study, we have extended and applied our computational toolbox to address this central issue of long-term hydrodynamic seal capacity. In the development phase, we first interfaced our existing reactive transport and geomechanical modeling capabilities to facilitate assessment of stress–strain evolution along and above the reservoir/cap-rock contact during and after CO_2 influx. We then constructed a new conceptual framework for evaluating the net impact on long-term cap rock integrity of influx-triggered geochemical alteration and geomechanical deformation processes.

In the application phase, we have used our modeling capabilities to address two fundamental questions. First, what is the evolution of cap-rock integrity during engineered CO_2 storage—and does this evolution vary significantly between EOR/storage and saline aquifer settings? This work builds directly upon our earlier modeling studies, which demonstrated enhanced hydrodynamic seal capacity of shale cap rocks as a function of injection-triggered geochemical processes during saline aquifer disposal [1–4]. Here, these earlier analyses have been extended to include explicit account of the concomitant geomechanical processes, and to assess dependence of this coupled geochemical–geomechanical evolution on key reservoir properties (permeability and lateral continuity) that distinguish typical oil reservoirs and saline aquifers [5,6].

We then address a closely related key issue: is the predicted evolution of cap-rock integrity for engineered CO_2 disposal sites similar to or appreciably different from that of natural CO_2 accumulations, i.e. what is the dependence of this evolution on the rate, duration, and focality of CO_2 influx? The widely espoused natural analog concept implicitly assumes a lack of such dependence; however, this assumption—upon which strict validity of the concept hinges—may be invalid in some cases. For example, a given reservoir/cap rock system that now holds a natural CO_2 accumulation may be incapable of doing so in the context of an engineered injection owing to significant differences in the magnitude and style of CO_2 influx. Further, the currently secure cap rock of a given natural accumulation may have *evolved* into an effective hydrodynamic seal following geochemical alteration that attended some degree of CO_2 migration through it. To address these issues, we have conducted and compared reactive transport simulations of a representative generic natural CO_2 reservoir for natural and engineered "filling" modes [7,8].

Because cap-rock integrity represents the ultimate constraint on the long-term isolation performance of geologic CO_2 storage sites, our reactive transport modeling analysis is linked to a number of additional CCP-funded studies presented in this volume [9–14]. There are potential direct links to three studies: the SAMCARDS analysis of Wildenborg [9], into which our simulation results could be directly incorporated, and the natural analog and experimental studies of Stevens [10] and Borm et al. [11], respectively, with which future coordinated efforts might provide field- and laboratory-scale "proof of concept" for our modeling capabilities. In addition, the reactive transport modeling approach used here could be employed to simulate the advective and diffusive migration of imposed anomalies in noble gas isotope ratios, as measured in the field by Nimz and Hudson [12]; to generate the fluid-phase pressures, saturations, densities, and viscosities required to predict dependent geophysical properties, as discussed by Hoversten and Gasperikova [13]; and to predict the migration paths of CO_2-charged fluids within magma-hydrothermal systems, as inferred from field measurements by Evans et al. [14].

METHODOLOGY

Reactive transport modeling is a computational method for quantitatively predicting the long-term consequences of natural or engineered perturbations to the subsurface environment [15,16]. Because these predictions typically involve space, time, and system complexity scales that preclude development of direct analytical or experimental analogs, they often provide a unique forecasting tool. The necessary point of departure for predictive investigations of this kind is established by successful application of the method to simulate well-constrained laboratory experiments [17,18].

The method is based on mathematical models of the integrated thermal, hydrological, geochemical, and geomechanical processes that redistribute mass and energy in response to the disequilibrium state imposed

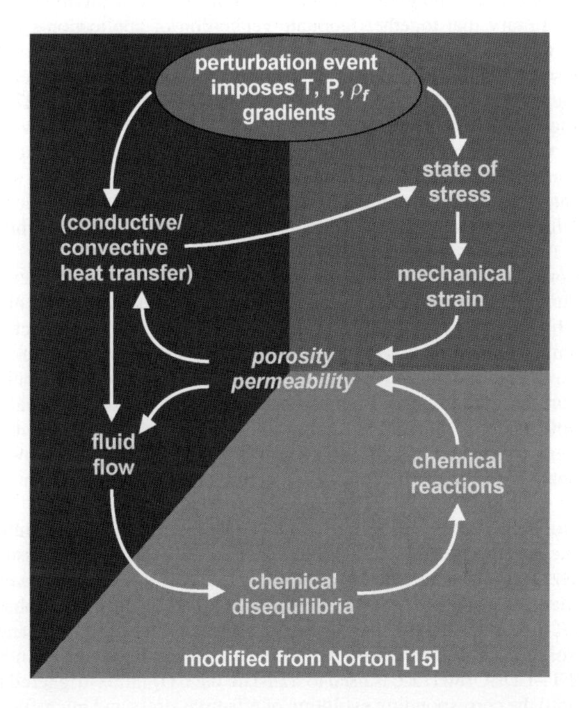

Figure 1: Schematic depiction of coupled subsurface processes that redistribute mass and energy in response to natural or engineered perturbation events. Porosity and permeability are the key variables that link hydrological, geochemical, and geomechanical sectors of the diagram.

by perturbation events such as magmatic intrusion or CO_2 influx (Figure 1). Traditionally, such models have been developed as separate entities and applied as such to address specific issues relevant to their individual scope. The fundamental advance embodied in reactive transport modeling is its explicit integration of these conceptually distinct process models. In practice, however, present-day simulators address and couple various subsets of these models, while the ultimate simulation tool—one that implements and explicitly couples all of the relevant processes—remains on the horizon.

We have developed a computational capability that integrates a state-of-the-art reactive transport simulator (NUFT), comprehensive supporting geochemical software and databases (SUPCRT92, GEMBOCHS), and a versatile distinct-element geomechanical model (LDEC). NUFT [19,20] is a software package that simulates non-isothermal multiphase/multicomponent flow and reactive transport within a wide range of subsurface environments characterized by multiscale physical and compositional heterogeneity. NUFT uses an integrated finite-difference spatial discretization to solve the flow and reactive-transport equations, with the Newton–Raphson method to solve the resulting non-linear systems at each time step. Explicit account is taken of multiphase advection, diffusion, and dispersion; of relative permeability and capillary pressure using an extended Van Genuchten formulation [21]; and of kinetically controlled fluid–mineral reactions, using rate laws from transition state theory [22]. Moreover, explicit account is also taken of *coupling between* these transport and geochemical processes through the dependence of permeability on porosity changes due to mineral precipitation/dissolution, using a normalized Kozeny equation [23], and through the dependence of fluid-phase volumetric saturations on gas (e.g. CO_2(g)) generated or consumed by fluid–mineral reactions.

The GEMBOCHS system [24,25] integrates a comprehensive relational thermodynamic/kinetic database and dedicated software library that together facilitate generation of application-specific thermodynamic/kinetic datafiles for use with a variety of geochemical modeling codes and reactive transport simulators. The thermodynamic database covers about 3200 distinct chemical species, spanning 86 elements of the periodic table; its core component is the current version of the SUPCRT92 database [26,27], which covers about 1550 species, spanning 82 elements. Custom datafiles are generated using Jewel [24], a GUI-driven software package that extrapolates reference-state properties to elevated $P-T$ conditions using a number of standard algorithms, the core set of which are those encoded with the SUPCRT92 software package [26]. These include global- and critical-region equations of state and a dielectric formulation for H_2O [28] that are explicitly integrated with equations of state for both aqueous solutes [29,30] and minerals/gases [31].

LDEC [32,33] is a geomechanical model that implements the distinct element method, which facilitates representation of fractured rock mass using arbitrary polyhedra, detection of new contacts between blocks resulting from relative block motion using the "Common-Plane" approach [34], exact conservation of linear and angular momentum, and simplified tracking of material properties as blocks move. Use of an explicit integration scheme allows extreme flexibility with respect to joint constitutive models, which here include effects such as cohesion, joint dilation, and friction angle. Both rigid and deformable approximations to block response are implemented. The rigid block approximation assumes that the compliance of fractured rock mass is closely approximated by lumping all compliance at the joints alone; however, this formulation also includes an optional second joint stiffness term that approximates deformation of the rock matrix.

The current method for one-way coupling between NUFT and LDEC is our integrated model's key approximation. Specifically, the NUFT–LDEC interface facilitates mapping pressure evolution into the corresponding effective stress, fracture aperture, and permeability history; however, at present, this geomechanical-dependent evolution (LDEC) is not back-coupled into the multiphase flow and reactive transport model (NUFT). As a result, the dependence of permeability, fluid flow, and pressure (including capillary pressure) evolution on concomitant geomechanical aperture history is not represented. In the present study, the NUFT–LDEC interface is used to translate the CO_2 influx-triggered pressure perturbation within basal cap rock into the corresponding evolution of effective stress and microfracture apertures, which permits first-order assessment of influx-induced geomechanical deformation. It is important to recognize that in the context of a bi-directionally coupled multiphase flow and geomechanical model, the magnitude of this pressure perturbation would likely be diminished—through concomitant evolution (initial widening) of cap-rock microfracture apertures—relative to that predicted here. Owing to their functional dependence on such magnitude, both the extent of CO_2 migration into undeformed cap rock and that of aperture widening predicted by the present one-way coupled model should be viewed as upper-limit values; on the other hand, likely enhanced advective CO_2 migration through initially widened microfractures is not accounted for here.

In order to evaluate the net impact on long-term cap rock integrity of concomitant geochemical and geomechanical processes, we introduce a new conceptual model that depicts geochemical counterbalancing of geomechanical aperture evolution as a function of effective diffusion distance and reaction progress. This model provides a theoretical framework for assessing the extent to which cap-rock integrity will ultimately be enhanced or degraded in specific reservoir/cap-rock systems in the context of specific CO_2 influx scenarios.

RESULTS AND DISCUSSION

Predicting long-term permeability evolution within the cap-rock environment of CO_2 storage sites requires first identifying, then quantifying its functional dependence on key system parameters and dynamic processes. The most important factors influencing this evolution are conveniently subdivided into three groups: intrinsic cap-rock properties, chemical conditions at the reservoir/cap-rock interface, and the CO_2 influx-triggered pressure perturbation.

Relevant cap-rock properties include geomechanical parameters, such as fracture normal stiffness, and geochemical characteristics, such as bulk concentrations of carbonate-forming cations—principally Fe, Mg, Ca, Na, and Al. These cation concentrations represent the primary control on geochemical alteration processes, while chemical conditions at the reservoir/cap-rock interface, which are determined by reservoir compositions and CO_2 waste-stream impurities (e.g. CH_4, H_2S, SO_x, NO_x concentrations), exert a secondary control. Magnitude, duration, and focality of the injection-induced pressure perturbation—which depend on these same characteristics of CO_2 influx as well as on reservoir permeability, lateral continuity, compartment height (for laterally confined settings), depth, and thickness—represent the fundamental controls on geomechanical deformation processes.

In the context of these dependencies, long-term enhancement or degradation of cap-rock integrity hinges on the relative contributions of geochemical alteration, which tends to reduce microfracture apertures in typical shale, and geomechanical deformation, which widens them (Figure 2). As a result, long-term performance forecasting of potential CO_2 storage sites requires a predictive capability that quantifies this pivotal interplay of geochemical and geomechanical processes. Previously, we have modeled the geochemical contribution within a full system analysis of coupled hydrological and geochemical processes [1–4]. Here, we first assess the geomechanical contribution—through analysis of its dependence on hydrological processes, key reservoir properties, and CO_2 influx parameters—then evaluate the ultimate net effect of opposing geochemical and geomechanical contributions to cap-rock integrity for both natural and engineered storage scenarios.

In describing this work, we begin with a review of subsurface CO_2 migration and storage processes, which provides not only the geochemical contribution to long-term cap rock integrity, but also full-system context for the subsequent analysis, which focuses on the cap-rock environment.

Subsurface CO_2 Migration and Storage Processes

Our previous modeling studies [1–4] have been largely based on simulating CO_2 injection at Statoil's North-Sea Sleipner facility—the world's first commercial saline-aquifer storage site. Here, CO_2-rich natural gas is produced from 3500 m below the seabed. Excess CO_2 is removed by amine absorption on the platform, then stripped from the amine, and finally injected—at the rate of one million tons per year since 1996—into the Utsira formation 2500 m above the hydrocarbon reservoir [35]. The 200-m-thick Utsira is a highly permeable fluid-saturated sandstone capped by the Nordland Shale. Hydrologic and compositional properties of the Utsira are relatively well characterized [1,4,36], while those of the Nordland Shale are virtually unknown, and must be estimated [1,4].

All of our Sleipner simulations have been carried out within a common $600 \times 250\ m^2$ spatial domain, which represents the near-field disposal environment, and over a single 20-year time frame, which encompasses equal-duration prograde (active-injection) and retrograde (post-injection) phases. The domain includes a 200-m-thick saline aquifer (35% porosity, 3-darcy permeability), 25-m-thick shale cap rock (5% porosity, 3-microdarcy permeability), and an overlying 25-m-thick saline aquifer. Its lateral boundaries are open to multiphase flow and mass transfer, while its top and bottom boundaries are not. During the prograde phase, pure CO_2 is injected at a rate of 10,000 ton/yr into the basal center of this domain (37 °C, 111 bar), which therefore corresponds to a 1-m-thick cross-section through the actual 100 m screen length at Sleipner.

Within the common domain, we have evaluated three distinct injection scenarios—models XSH, CSH, and DSH [1,4]. Model XSH examines CO_2 injection into a shale-capped homogeneous sandstone aquifer. Models CSH and DSH impose into XSH four thin (3-m thick) intra-aquifer shales, which are separated from the cap rock and each other by 25 m. Model CSH examines the effect of imposing laterally continuous

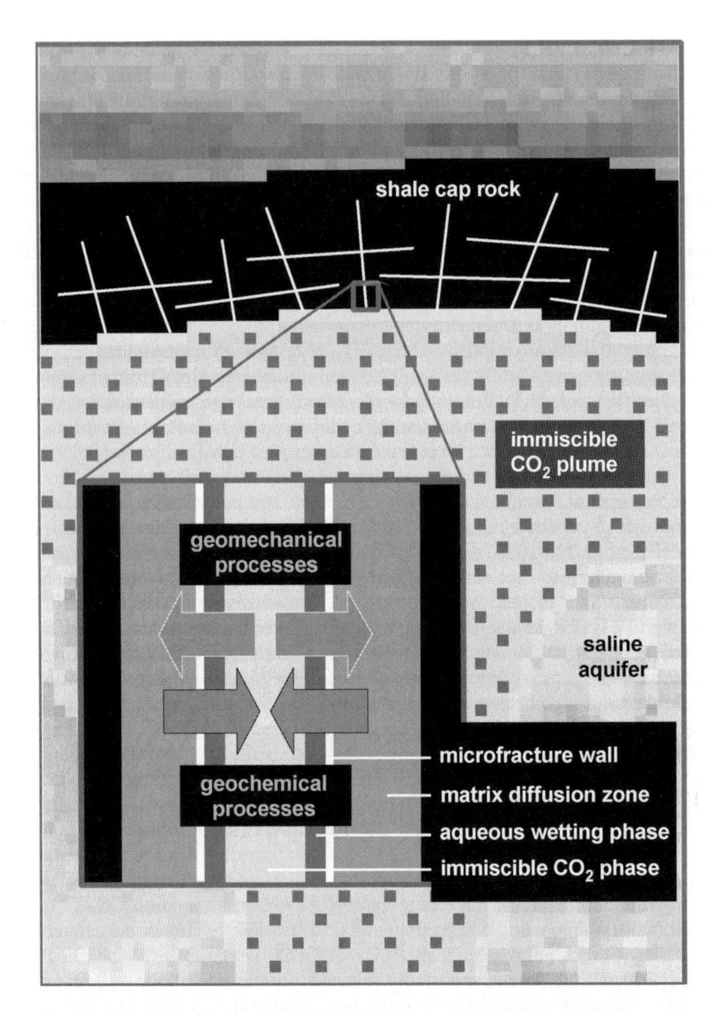

Figure 2: Schematic depiction of concomitant CO_2 influx-triggered geochemical and geomechanical processes within shale cap rock microfractures. Mineral dissolution/precipitation reactions tend to continuously reduce microfracture apertures for typical shale compositions, while pressure evolution initially widens then reduces them (net widening).

microfractured shales having assigned permeability (3 md) that equates to a continuum representation of 100 μm fractures spaced roughly 30 m apart. Model DSH examines the effect of imposing laterally discontinuous shales, which are bridged by lateral facies change to sandstone. Assigned permeability of these shales (3 μd; same as the cap rock) reflects typical shale integrity.

Compositionally, the well-characterized saline aquifers are represented as impure quartz sand: 80% quartz, 10% K-feldspar, 5% plag-ab_{80}, 3% muscovite, and 2% phlogopite [1,4]. The virtually uncharacterized shale cap rock is estimated to contain 60% clay minerals (50% muscovite, 10% Mg-chlorite), 35% quartz, and 5% K-feldspar; this mineralogy and bulk $K_2O/(FeO + MgO)$ ratio closely approximate those of typical (non-carbonaceous) shales, while permitting avoidance of more realistic illite, smectite, and montmorillonite solid solutions, for which thermodynamic and kinetic data are currently lacking [1,4]. Mg end-member components are used to represent Fe/Mg solid solutions because *in situ* oxidation states are unknown. The saline aquifers and shale are all saturated with an aqueous phase of near-seawater composition [1,4,36].

Our Sleipner simulations suggest that the ultimate fate of CO_2 injected into saline aquifers is governed by three interdependent yet conceptually distinct processes: CO_2 migration as a buoyant immiscible fluid phase, direct chemical interaction of this rising plume with ambient saline waters, and its indirect chemical interaction with aquifer and cap-rock minerals through the aqueous wetting phase. Each process is directly linked to a corresponding trapping mechanism: immiscible plume migration to hydrodynamic trapping, plume–water interaction to solubility trapping, and plume–mineral interaction to mineral trapping.

Immiscible plume migration and hydrodynamic trapping

Intra-aquifer permeability structure controls the path of prograde immiscible CO_2 migration, thereby establishing the spatial framework of plume–aquifer interaction and the potential effectiveness of solubility

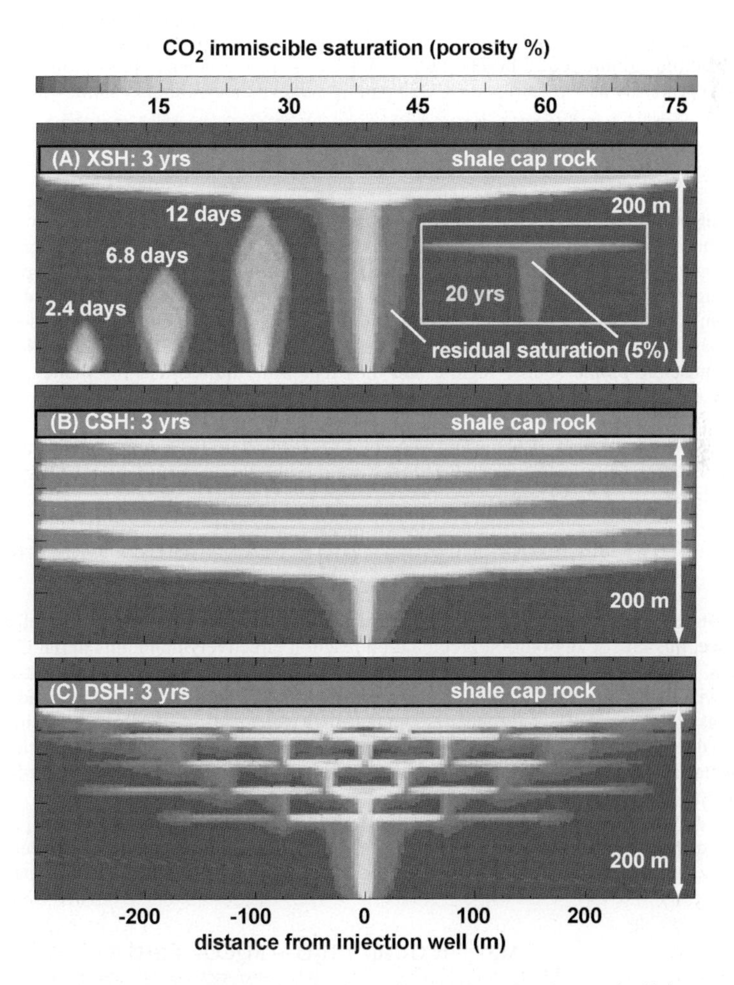

Figure 3: Immiscible plume migration and hydrodynamic trapping after 3 years in Sleipner models XSH, CSH, and DSH; interbedded thin shales not shown for CSH and DSH.

and mineral trapping. Actual efficacy of these trapping mechanisms is determined by compositional characteristics of the aquifer and cap rock. By retarding vertical and promoting lateral plume mobility, interbedded thin shales significantly expand this framework (i.e. CO_2 storage capacity), enhance this potential, and delay outward migration of the plume from the near-field environment (Figure 3). Seismic data strongly suggest that the Utsira formation combines elements of models CSH and DSH [1,3,4].

In all three models, steady-state configuration of the immiscible CO_2 plume is realized within 1 year. During the prograde phase, a residual saturation zone marks the wake of initial plume ascent to the cap rock or deepest interbedded shale (e.g. Figure 3A, left insets). During the retrograde phase, this zone encompasses virtually the entire prograde steady-state plume (e.g. Figure 3A, right inset)—effectively maintaining the prograde extent of solubility trapping and continually enhancing that of mineral trapping, as described below for model DSH. In the near-field environment of Sleipner-like settings, 80–85% by mass of injected CO_2 remains and migrates as an immiscible fluid phase ultimately subject to hydrodynamic trapping beneath the cap rock, which represents an effective seal in these models [1–4], where geomechanical processes are not accounted for.

Geochemical trapping mechanisms

As the immiscible plume equilibrates with saline formation waters, intra-plume aqueous CO_2 concentrations (primarily as $CO_2(aq)$ and HCO_3^-) rapidly achieve their solubility limit, while pH decreases [1–4]:

$$CO_2(g) + H_2O = CO_2(aq) + H_2O = HCO_3^- + H^+ \tag{1}$$

For the chemical system and $P-T$ conditions that characterize the Utsira formation at Sleipner, equilibrium aqueous CO_2 solubility is 1.1–1.2 m, accounting for 15–20% by mass of injected CO_2 (Figure 4A). Owing to residual saturation of immiscible CO_2, this degree of solubility trapping is virtually constant throughout the prograde and retrograde phases. The initial pH drop caused by solubility trapping—from 7.1 to 3.4—catalyzes silicate dissolution, which after 20 years has increased pH from 3.4 to 5.3. This dissolution hydrolyzes potential carbonate-forming cations (here, primarily Na, Al, and Mg) within the immiscible-plume source region, and thus represents the critical forerunner of all mineral-trapping mechanisms.

We have identified four distinct mechanisms whereby CO_2 precipitates as carbonate minerals. Intra-plume dawsonite cementation (Figure 4B) is catalyzed by high ambient Na^+ concentration, CO_2 influx, and acid-induced K-feldspar dissolution [1–4].

$$\underset{\text{K-feldspar}}{KAlSi_3O_8} + Na^+ + CO_2(aq) + H_2O \leftrightarrow \underset{\textit{dawsonite}}{NaAlCO_3(OH)_2} + \underset{\textit{silica}}{3SiO_2} + K^+ \tag{2}$$

The volume of co-precipitating dawsonite and silica polymorphs slightly exceeds that of dissolving K-feldspar. Hence, this kinetic dissolution/precipitation reaction effectively maintains initial CO_2 injectivity; after 20 years, porosity has decreased by a factor of less than 0.1% (Figure 5A). Pervasive dawsonite cementation will likely be characteristic of saline aquifer storage in any feldspathic sandstone. In fact, natural analogs for this process have been documented: widespread dawsonite cement in the Bowen–Gunnedah–Sydney Basin, Eastern Australia, which has been interpreted to reflect magmatic CO_2 seepage on a continental scale [37], and sporadic dawsonite cement in the clastic Springerville-St. Johns CO_2 reservoir [38].

Calcite-group carbonate rind (here, magnesite) forms along—and therefore effectively delineates—both lateral and upper plume boundaries (Figure 4C). Genetically distinct, these two processes can be described by [1–4]:

$$Mg^{+2} + CO_2(aq) + H_2O \leftrightarrow \underset{\textit{magnesite}}{MgCO_3} + 2H^+ \tag{3}$$

As intra-plume formation waters, progressively enriched in Mg^{+2} from phlogopite dissolution, migrate

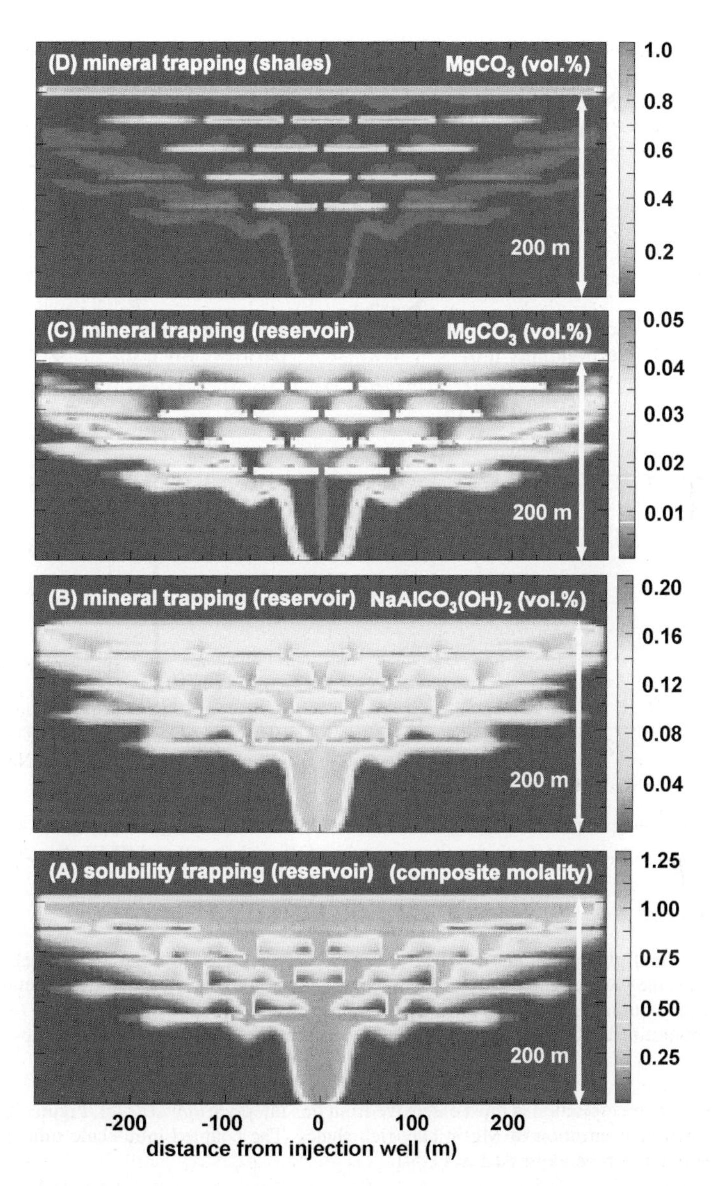

Figure 4: Geochemical trapping mechanisms after 20 years in model DSH: (A) solubility trapping (composite molality of all carbon-bearing aqueous species), (B) intra-plume dawsonite cementation, (C) plume-bounding magnesite precipitation (shales shown in white [off-scale high]), and (D) intra-shale magnesite precipitation.

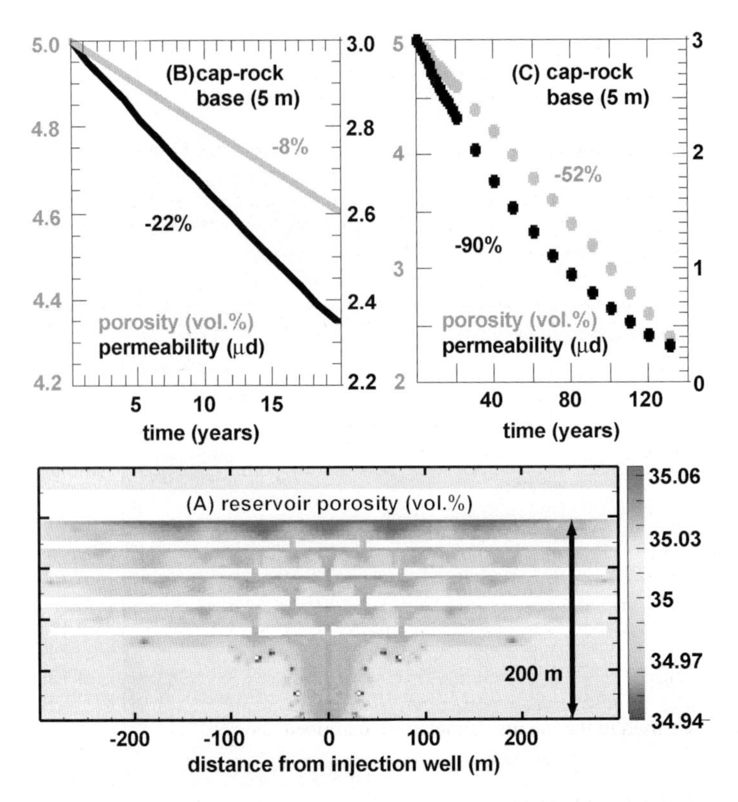

Figure 5: Porosity and permeability reduction in model DSH due to mineral trapping (A) after 20 years in the reservoir (initial porosity: 35%; shales shown in white [off-scale low]) and (B,C) after 20 and 130 years in the cap rock.

outward across lateral plume boundaries, they traverse steep gradients in CO_2(aq) and pH; the net effect strongly promotes magnesite precipitation. Along upper plume boundaries, CO_2(aq) concentration and pH are nearly constant, but aqueous Mg^{+2} concentration increases most rapidly here because formation-water saturation is minimized; this leads to magnesite cementation from the reservoir/cap-rock interface downward.

However, magnesite precipitation is most extensive from this interface *upwards* (cf. Figure 4C,D), owing to the relatively high concentration of Mg in clay-rich shales. The coupled intra-shale mineral dissolution/ precipitation reaction can be expressed as [1–4]:

$$\underset{\textit{K-feldspar}}{KAlSi_3O_8} + 2.5\underset{\text{Mg-chlorite}}{Mg_5Al_2Si_3O_{10}(OH)_8} + 12.5CO_2(aq) \leftrightarrow \underset{\text{muscovite}}{KAl_3Si_3O_{10}(OH)_2}$$

$$+ 1.5\underset{\text{kaolinite}}{Al_2Si_2O_5(OH)_4} + 12.5\underset{\text{magnesite}}{MgCO_3} + 4.5\underset{\text{silica}}{SiO_2} + 6H_2O \tag{4}$$

This kinetic reaction proceeds to the right with an increase in solid-phase volume of 18.5% (magnesite accounting for 47 vol% of the product assemblage). After 20 years, porosity and permeability of the 5-m-thick cap-rock base have been reduced by 8 and 22%, respectively, by this process (Figure 5B), which upon hypothetical completion at 130 years would reduce initial porosity by half and initial permeability by an order of magnitude (Figure 5C), thereby significantly improving cap-rock integrity. A natural analog to reaction (4) has recently been documented in the Ladbroke Grove natural gas field, where post-accumulation CO_2 influx has converted Fe-rich chlorite to Fe-rich dolomite (ankerite), kaolinite, and silica [39].

Although composite mineral trapping accounts for less than 1% by mass of injected CO_2 in our models of the near-field disposal environment at Sleipner, it has enormous strategic significance: it maintains initial CO_2 injectivity (reaction (2)), delineates and may partially self-seal plume boundaries (reaction (3)), and—most importantly—reduces cap-rock permeability (reaction (4)), thereby enhancing hydrodynamic containment of immiscible and solubility-trapped CO_2 [1–4].

The CO_2 migration and storage processes reviewed above in the context of engineered saline-aquifer storage are equally applicable to CO_2-flood EOR operations in shale-capped water-wet oil reservoirs, which are primarily distinguished by the presence of a hydrocarbon phase and lateral confinement, and the formation of natural CO_2 reservoirs, which are fundamentally distinguished by the rate, focality, and duration of CO_2 influx. However, in all of these settings the effect of geochemical alteration to improve the seal integrity of typical (non-carbonaceous) shale cap rocks may be counterbalanced or even overwhelmed by concomitant geomechanical deformation, which initially acts in opposition. Hence, in evaluating long-term hydrodynamic sealing capacity, explicit account must be taken of both processes.

Pressure Evolution and Geomechanical Deformation

A first-order assessment of cap-rock geomechanical deformation can be obtained from evaluating the dependence of microfracture aperture evolution on the influx-triggered pressure perturbation. In a new series of NUFT/LDEC simulations, we have assessed this dependence, first as a function of reservoir permeability and lateral continuity—two key parameters that typically distinguish saline-aquifer disposal sites and oil reservoirs, and second, as a function of CO_2 influx rate—the fundamental parameter that distinguishes engineered and natural storage scenarios. Within these new models, the values adopted for other important parameters that influence geomechanical response to CO_2 injection (e.g. reservoir depth and thickness) are those used in the Sleipner simulations described above.

In the Sleipner models, we addressed coupled hydrological and geochemical processes. In the following simulations, we explicitly address only the effect of hydrological (multiphase flow) processes. However, this approximation has negligible impact for impure sandstone reservoirs (such as the Utsira formation), where reservoir porosity and permeability—and thus the injection-induced pressure perturbation—are not modified appreciably by geochemical alteration, as demonstrated above (Figure 5A).

Dependence on reservoir properties: saline aquifer versus EOR settings

In this analysis, four distinct simulations have been carried out within two spatial domains (Figure 6). Reservoir permeability and lateral continuity are varied from 3000 md and infinite in model UHP (laterally-unconfined, high permeability), which represents desirable saline-aquifer storage sites, to 300 md and 2000 m in model CLP (laterally-confined, low permeability), which represents a typical compartmentalized EOR setting. Models ULP and CHP represent cross-combinations of these values, which facilitate evaluation of specific dependence on reservoir permeability and lateral confinement. In both laterally confined models, compartment height—itself a parameter that exerts second-order influence on the injection-induced pressure perturbation—is 150 m. In all four models, supercritical CO_2 is injected at the rate of 10,000 ton/yr during the prograde event.

Magnitude of the influx-triggered pressure perturbation within basal cap rock varies significantly with (and inversely proportional to) reservoir permeability and lateral continuity (Figures 7–10), although the general style of its evolution during prograde and retrograde phases of the influx event does not (Figure 11). For highly permeable, laterally extensive reservoirs (model UHP), this perturbation follows a characteristic three-stage evolution: (1) rapid increase to maximum pressure as the aqueous phase is displaced upwards during initial ascent of the immiscible CO_2 plume to the cap rock, (2) rapid asymptotic decrease to a near steady-state value intermediate to ambient and maximum pressures that is maintained thereafter during the prograde regime, and (3) a second rapid asymptotic decrease towards the ambient value, which is triggered by onset of the retrograde regime (Figure 7). This pressure evolution suggests that the potential for dependent geomechanical deformation events is maximized during three very brief, distinct episodes that

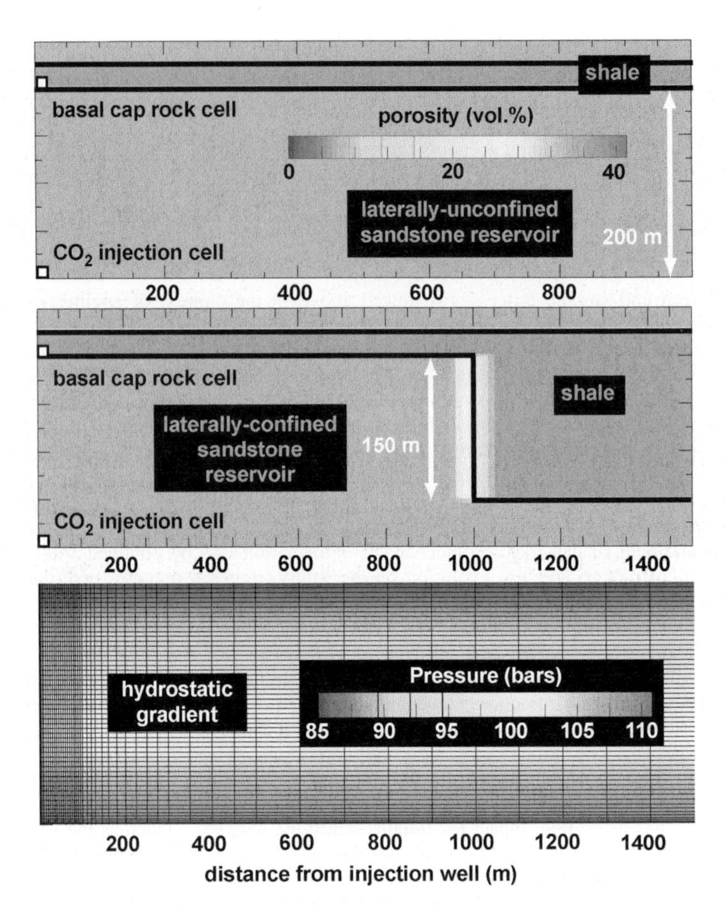

Figure 6: Schematic depiction of the laterally unconfined and laterally confined simulation domains used for models UHP/ULP and CHP/CLP, respectively. Illustrated basal cap rock and CO_2 injection cells are not drawn to scale. Actual cell granularity is overlain upon the hydrostatic gradient plot, which is identical in the two systems (shown for models CHP/CLP).

occur during the earliest stages of prograde and retrograde storage. Note that for this Sleipner-like setting, the range of injection-induced pressure variation is small—on the order of 3 bar.

Decreasing reservoir permeability from 3000 to 300 md without imposing lateral confinement (i.e. model ULP) significantly increases magnitude of the pressure perturbation—from roughly 3 to nearly 22 bar—without altering the three-stage evolution described above (cf. Figures 7 and 8). Also noteworthy from this comparison is the inverse dependence of CO_2 storage capacity on reservoir permeability, which suggests that for pure-storage scenarios the additional energy cost of exploiting less permeable reservoirs—which require higher injection pressures—may be partially offset by the benefit of increased storage and delayed migration into the far-field environment, providing cap-rock performance is not significantly compromised.

The influence of reservoir compartmentalization on the influx-triggered pressure perturbation within basal cap rock is examined in models CHP and CLP (Figures 9 and 10). Although the functional form of pressure evolution in these models is analogous to that described above for laterally unconfined reservoirs, three

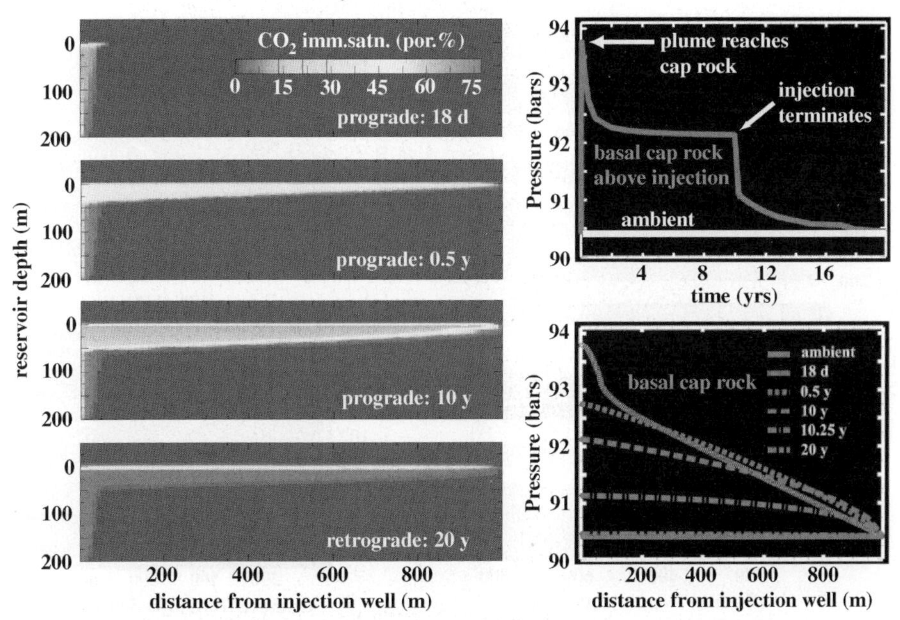

Figure 7: NUFT simulation of immiscible CO_2 migration together with associated pressure evolution within the basal cap rock directly above and as a function of distance from CO_2 injection for the laterally unconfined 3000 md reservoir.

significant variations are introduced by compartmentalization. First, the magnitude of initial pressure increase during plume ascent to the cap rock is significantly enhanced—reaching 60 bar in model CLP—owing to the restricted lateral flow (increased flow resistance) of displaced formation water. Second, a permeability-dependent fourth stage of pressure evolution—one that bridges cap-rock and spillpoint plume arrival times—is introduced that either causes a secondary pressure increase (CHP) or slows prograde decrease (CLP) of the initial pressure anomaly. Third, owing to presence of the accumulated CO_2 column, during the retrograde phase pressure decays asymptotically toward a steady-state value that exceeds hydrostatic and whose magnitude is proportional to column height. This final variation is extremely significant because it imposes a long-term pressure increase at and above the cap-rock interface, which does not occur in unconfined reservoirs.

Propagation of the injection-triggered pressure perturbation from the well to and above this interface effects CO_2 migration into undeformed cap rock in cases where its magnitude—more specifically, that of the difference between increased gas and liquid pressures—is sufficient to overcome capillary forces, i.e. capillary entry pressure is exceeded, which permits increased CO_2 saturation within the cap rock as a function of further increased capillary pressure. CO_2 migration into the 25-m-thick 3-μd cap rock through this process is minimized in model UHP, where after 20 years CO_2 saturations of roughly 1% are obtained for a penetration distance of only 5 m, and maximized in model CLP, where CO_2 saturations of about 10% are achieved within this basal 5 m, and penetration distance actually breeches the overlying reservoir, although here CO_2 saturations of $<1\%$ are realized (Figure 12).

The injection-triggered pressure perturbation also leads to geomechanical deformation of the cap rock, through dependent changes in effective stress and microfracture apertures. Here, we adopt a simplified form

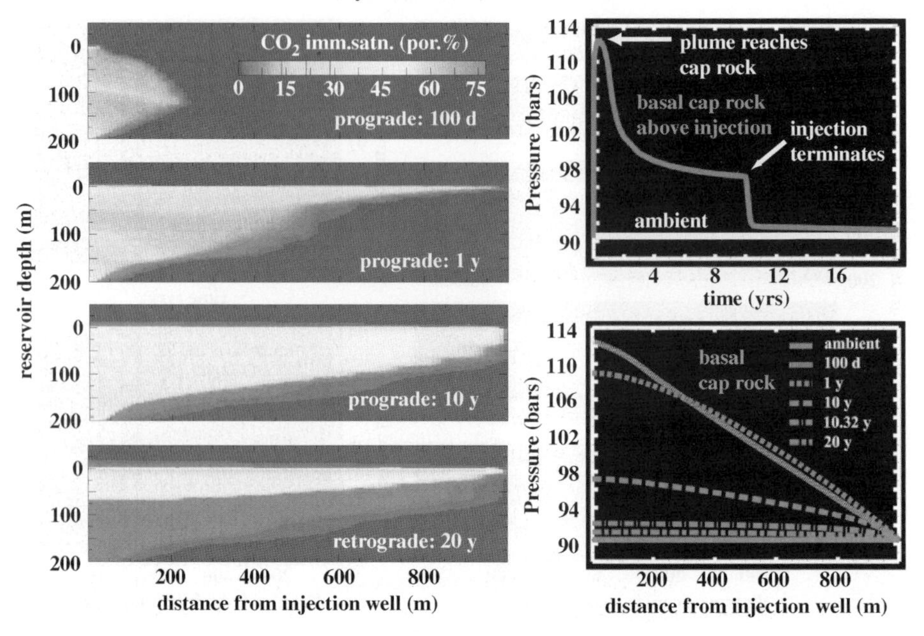

Figure 8: NUFT simulation of immiscible CO_2 migration together with associated pressure evolution within the basal cap rock directly above and as a function of distance from CO_2 injection for the laterally unconfined 300 md reservoir.

of the constitutive relationship between effective stress (σ_E), total stress (σ_T), and pressure (P_f):

$$\sigma_E = \sigma_T - P_f, \tag{5}$$

where σ_T is assumed to be constant ($\Delta\sigma_E = -\Delta P_f$). By further neglecting the non-linear aperture dependence of fracture normal stiffness (K_N), normal aperture displacement due to reduced effective normal stress (Δa_N) can be expressed as

$$\Delta a_N = (\Delta P_f / K_N). \tag{6}$$

Using Eqs. (5) and (6) together with an estimated normal stiffness for shale fractures at depth [40], we first translate the maximum injection-induced pressure perturbation within basal cap rock for each of the four models (Figure 11) into the corresponding maximum aperture normal displacement in order to gauge relative scale (Figure 13). As can be seen, the potential maximum aperture increase due to reduced effective normal stress is on the order of 100–1000 μm. Because attainment of this pressure maximum coincides with arrival of the CO_2 plume at the cap rock—after only 15–100 days in all four models—the potential for geomechanical deformation is maximized very early during the prograde phase.

Simulating long-term aperture evolution requires use of the NUFT–LDEC interface, which facilitates translation of pressure evolution within a given reservoir cap-rock system into the dependent evolution of effective stress and microfracture apertures—here cast within the simplifying context of Eqs. (5) and (6). In this application, the interface is applied to a representative sub-grid from our NUFT domains: a 60 m-by-50 m half-space that encompasses the uppermost 10 m of the lower reservoir (two NUFT grid cells), the 25 m-thick shale cap rock (five cells), and the 25-m-thick upper reservoir (five cells).

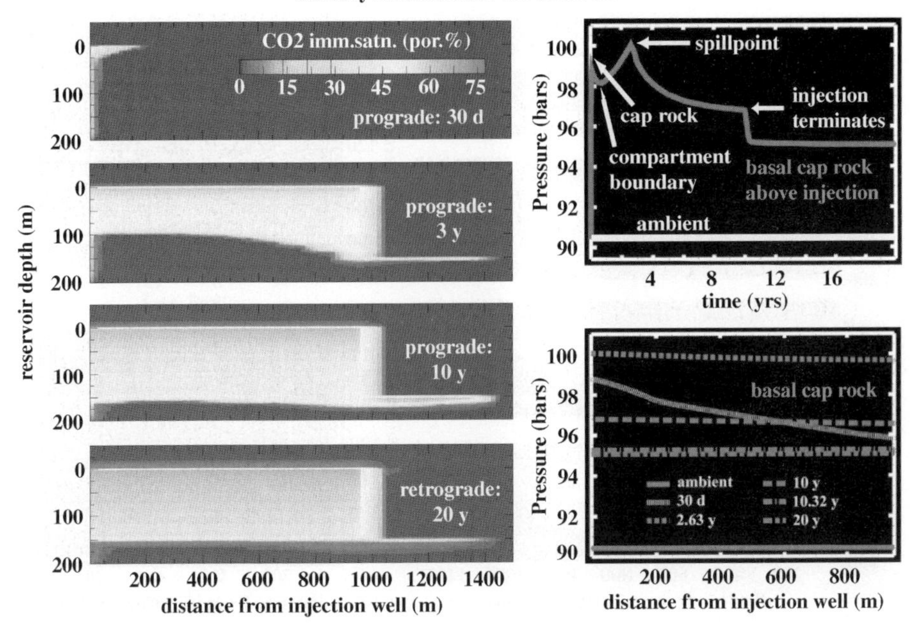

Figure 9: NUFT simulation of immiscible CO_2 migration together with associated pressure evolution within the basal cap rock directly above and as a function of distance from CO_2 injection for the laterally confined 3000 md reservoir.

The functional form of aperture evolution within basal cap rock is directly analogous to that described above for pressure, as exemplified by LDEC simulation of such evolution for model CLP (Figure 14). Here, during the prograde phase apertures rapidly increase by roughly 1000 μm during initial plume ascent, then asymptotically decrease to a steady-state value that reflects net widening of about 400 μm. During the retrograde phase, they first rapidly decrease from this prograde steady state, then continue to decrease asymptotically towards a final steady-state value that reflects ultimate net widening of roughly 100 μm per the approximate 5-bar net pressure increase associated with CO_2 accumulation. Hence, geomechanical deformation degrades cap-rock integrity only during the earliest stages of the prograde phase, after which it continuously self-mitigates this initial degradation event.

Unless counterbalanced by geochemical effects, ultimate net aperture widening through geomechanical deformation could facilitate long-term CO_2 migration into the cap rock. Moreover, although maximum prograde and ultimate net aperture increases of 1000 and 100 μm, respectively, occur just above the reservoir interface, concomitant increases of 200–900 and a few 10 s of μm, respectively, are realized *throughout* the lowest 20 m of the 25-m-thick shale cap rock (Figure 15). Such pervasiveness suggests the potential development of microfracture continuity sufficient to permit CO_2 migration into and perhaps completely through relatively thin shale cap rocks in certain influx settings.

Dependence on influx parameters: engineered versus natural storage

In this analysis, three distinct simulations have been carried out within a single spatial domain (Figure 16) that represents a confined sandstone reservoir whose compartment width (10 km), height (100 m), and width:height aspect ratio (100:1) typify those of natural CO_2 reservoirs [41]. In all three models, reservoir and shale cap rock permeability are 300 md and 3 μd, respectively. The models are distinguished primarily

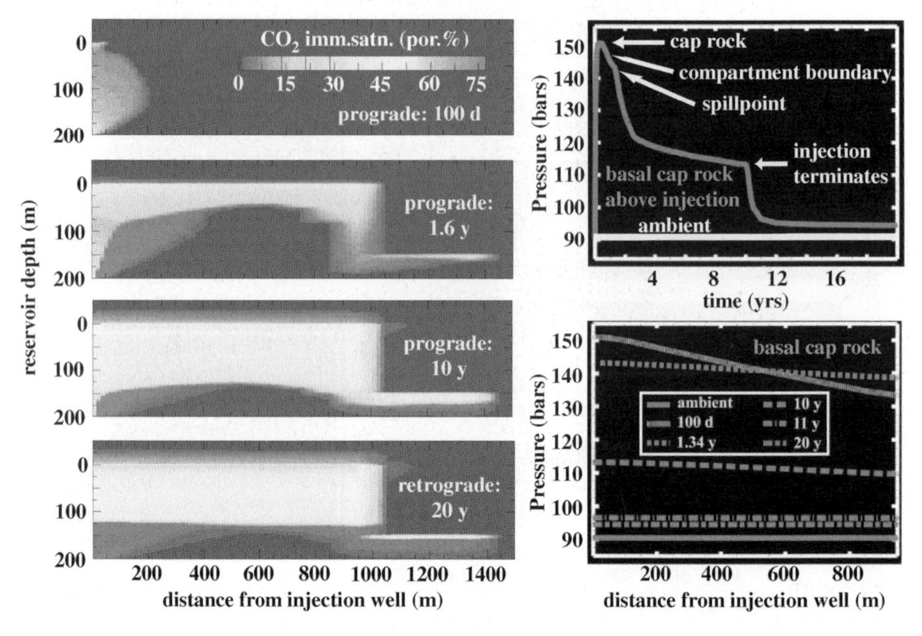

Figure 10: NUFT simulation of immiscible CO_2 migration together with associated pressure evolution within the basal cap rock directly above and as a function of distance from CO_2 injection for the laterally confined 300 md reservoir.

by prograde CO_2 influx rate, which is varied from 10^4 to 10^3 to 10^2 ton/yr, representing engineered injection, "fast" natural accumulation, and "slow" natural accumulation, respectively. The engineered injection rate is that used in all of the preceding simulations, while the two values adopted for natural accumulation rates—which are presently unknown [41]—are rough estimates. A secondary difference is duration of the prograde and retrograde events, both of which span 10 years for the engineered injection, but are extended to 40 and 20 years in both natural accumulation models.

Because the engineered-injection model adopts the same injection rate used in the preceding set of simulations, it illustrates dependence of the pressure perturbation on compartment width and aspect ratio, while providing a baseline for evaluating its dependence on influx rate per comparison with the two natural accumulation models (Figure 17). Increasing compartment width from 2 to 10 km causes pressure to increase even after the plume has reached the cap rock, owing to the increased volume of formation water that must be displaced. Hence, while pressure increases from 90 to 150 bar during initial plume ascent in both models CLP and here (cf. Figures 10 and 17), in this case pressure ultimately reaches 250 bar before declining after the plume reaches the lateral compartment boundary. Subsequent asymptotic pressure decline during the post-spillpoint prograde and retrograde phases is dampened by increased compartment width.

When influx rate is reduced by one and two orders of magnitude, migration of the plume is retarded and the pressure perturbation is reduced proportionately, while its functional form remains unchanged (Figures 18 and 19). In the "fast" natural accumulation model, the immiscible plume does not reach the lateral compartment boundary until just before termination of the 40-year prograde event, while the maximum pressure perturbation (about 22 bar) is a factor of 7–8 less than that for the engineered injection model. In the "slow" natural accumulation model, the plume has not quite advanced halfway to the compartment

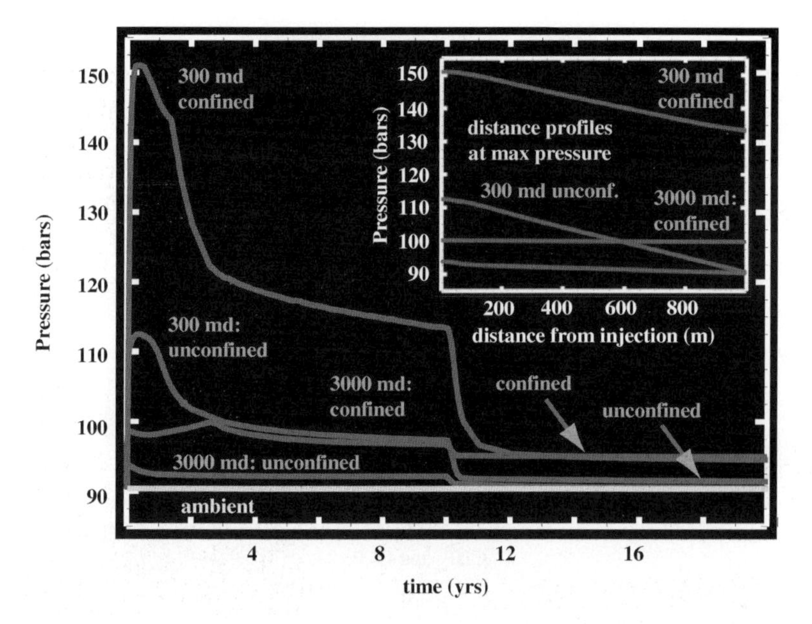

Figure 11: NUFT simulation of pressure evolution within the basal cap rock directly above CO_2 injection for models UHP, ULP, CHP, and CLP together with the distance profiles associated with attainment of pressure maxima (inset).

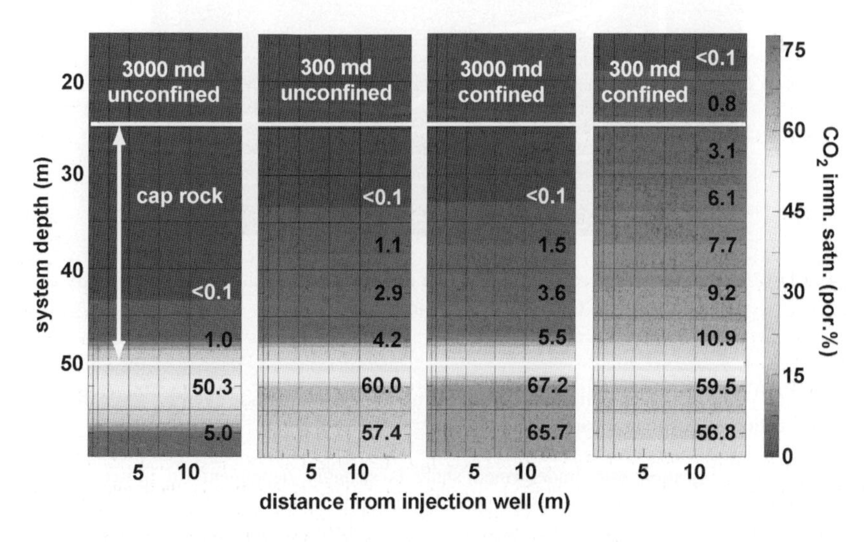

Figure 12: NUFT simulation of CO_2 migration into geomechanically undeformed cap rock as a function of the CO_2 influx-induced pressure perturbation, which in models UHP, ULP, CHP, and CLP (shown from left to right) is sufficient to overcome resistive capillary forces.

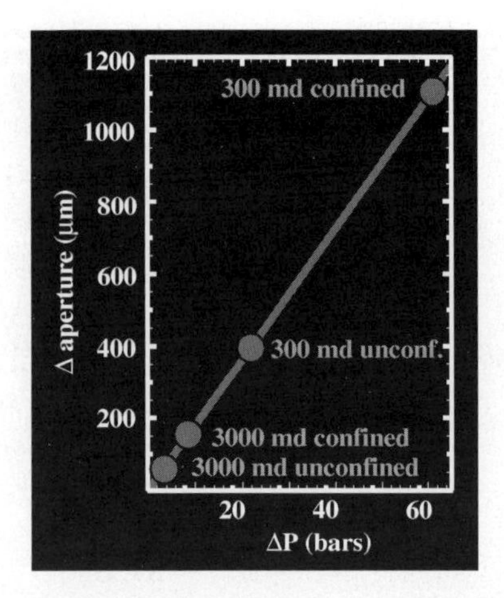

Figure 13: Maximum potential aperture increase of basal cap rock microfractures as a function of the CO_2 influx-triggered initial pressure increase and reduced effective normal stress in models UHP, ULP, CHP, and CLP.

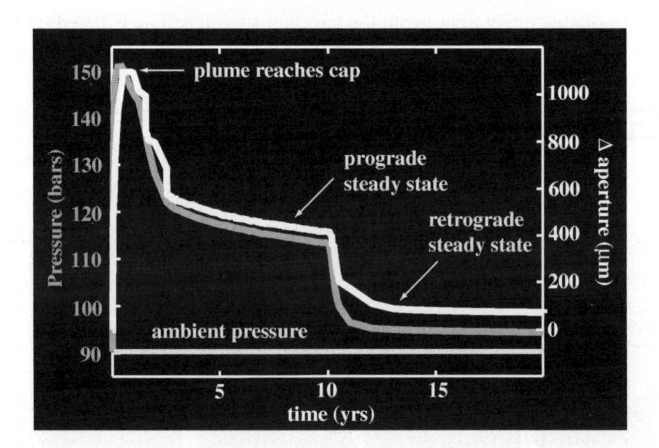

Figure 14: LDEC simulation of microfracture aperture evolution in the basal cap rock directly above the CO_2 injection well in model CLP.

boundary after 60 years (which encompasses both the prograde and retrograde events), while the maximum pressure perturbation is less than 3 bar.

The extent of CO_2 migration into undeformed shale is strongly dependent on influx rate, through dependence of the injection-triggered pressure perturbation on this rate. Such migration extends halfway through the 25-m-thick shale in the "slow" accumulation model (intra-shale saturations approaching 8%), completely through this shale and halfway through the overlying 25-m-thick reservoir in the "fast" accumulation model (upper reservoir saturations approaching 12%), and completely through this upper

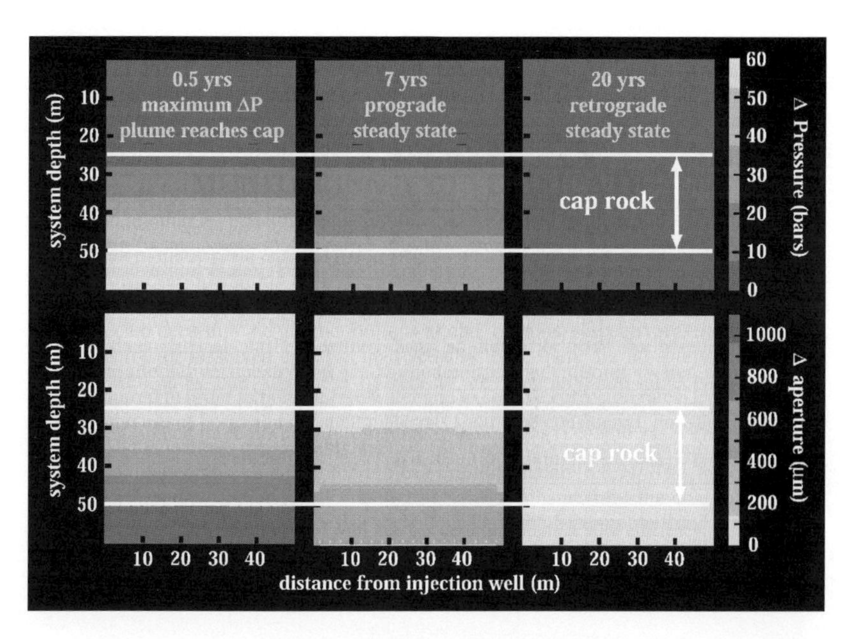

Figure 15: LDEC simulation of aperture evolution within and immediately surrounding the cap rock in model CLP.

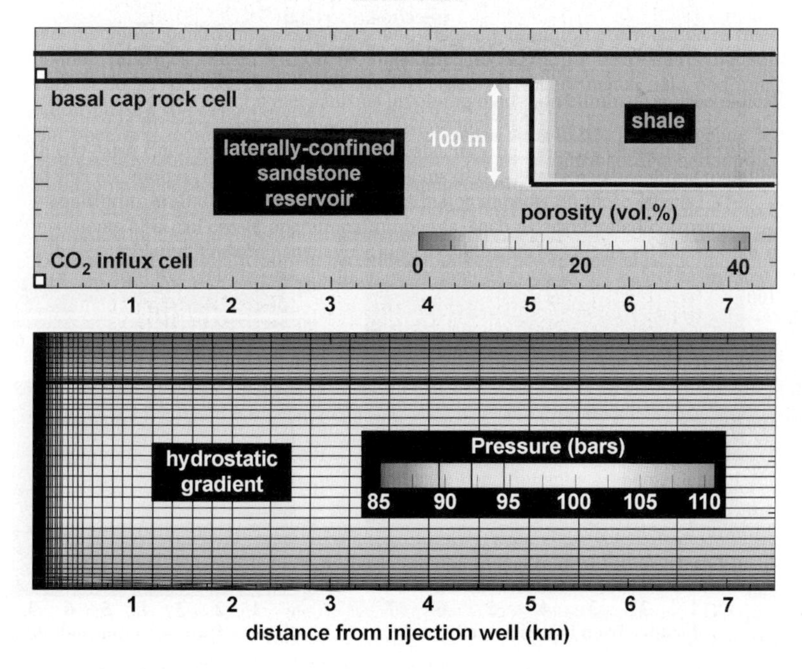

Figure 16: Schematic depiction of the laterally confined simulation domain used to represent natural CO_2 reservoirs. Illustrated basal cap rock and CO_2 influx cells are not drawn to scale. Actual cell granularity is overlain upon the hydrostatic gradient plot.

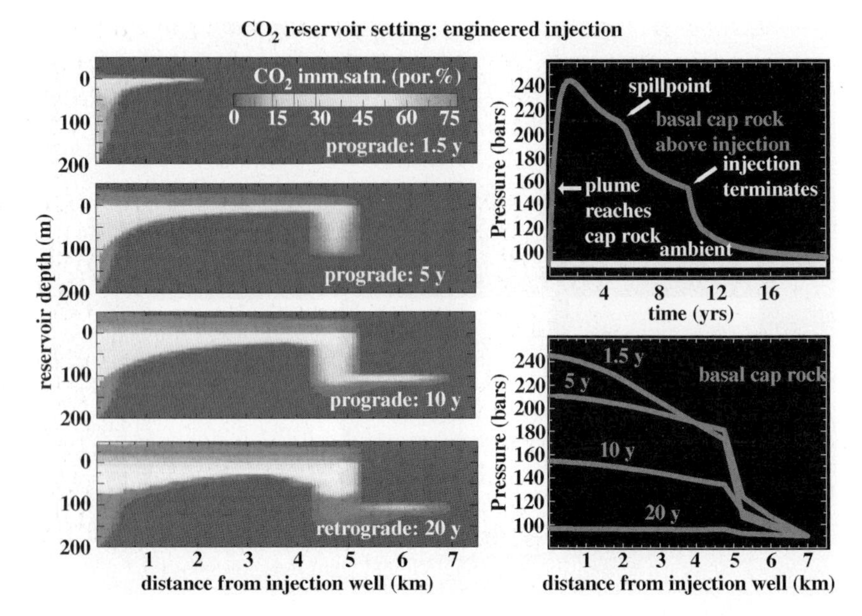

Figure 17: NUFT simulation of immiscible CO_2 migration together with associated pressure evolution within the basal cap rock directly above and as a function of distance from CO_2 injection for engineered injection.

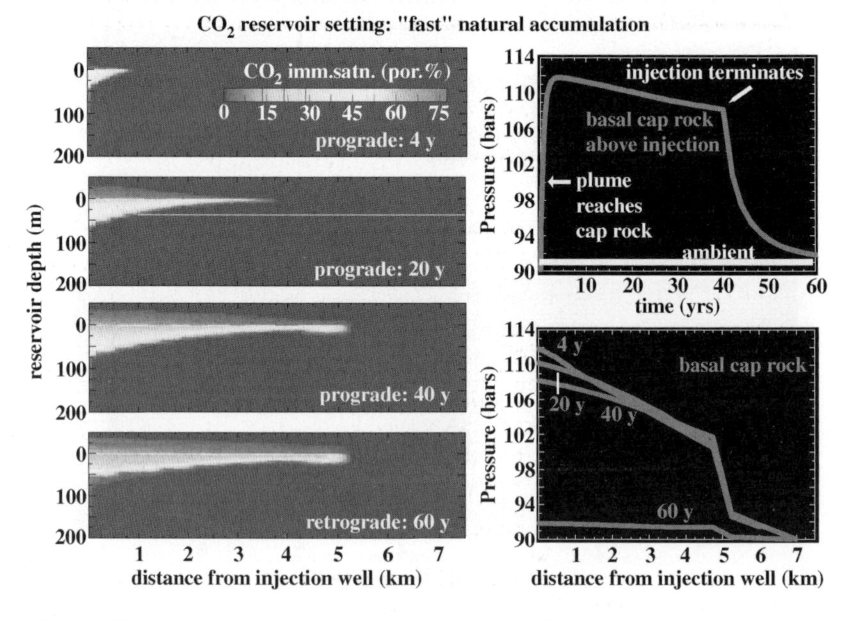

Figure 18: NUFT simulation of immiscible CO_2 migration together with associated pressure evolution within the basal cap rock directly above and as a function of distance from CO_2 influx for "fast" natural accumulation.

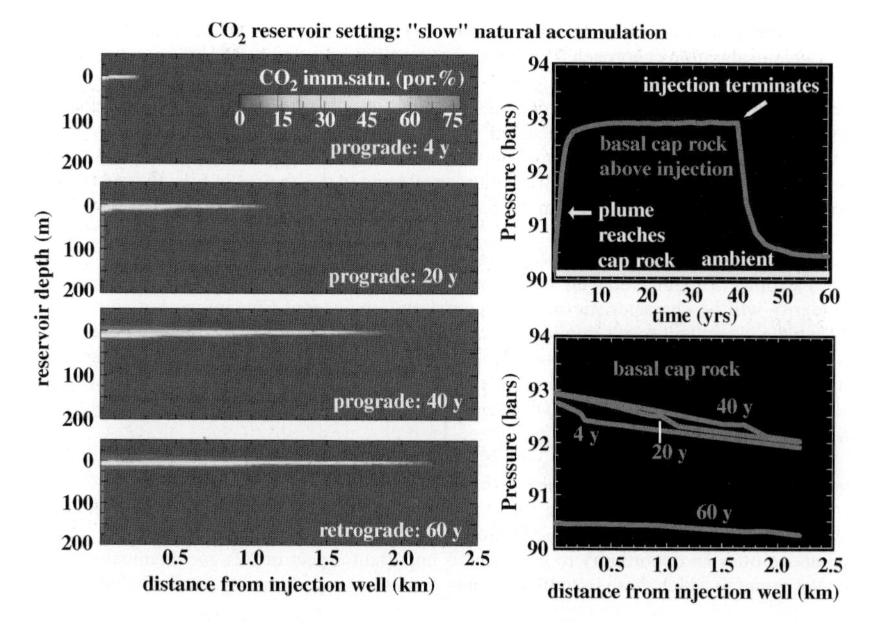

Figure 19: NUFT simulation of immiscible CO_2 migration together with associated pressure evolution within the basal cap rock directly above and as a function of distance from CO_2 influx for "slow" natural accumulation.

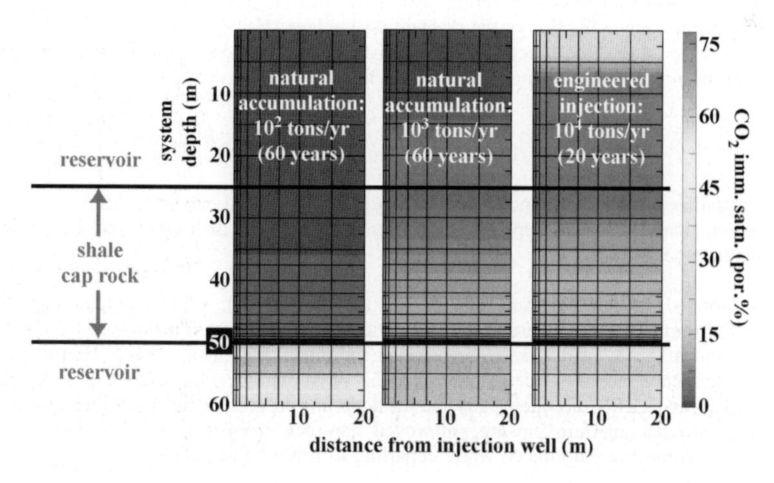

Figure 20: NUFT simulation of CO_2 migration into geomechanically undeformed cap rock as a function of the influx-triggered pressure perturbation, which in the "slow" and "fast" natural accumulation and engineered injection models (shown from left to right) is sufficient to overcome resistive capillary forces.

reservoir to form a laterally-restricted (see Figure 17) accumulation zone beneath the upper domain boundary (where saturations approach 25%) in the engineered injection model (Figure 20).

The extent of geomechanical cap-rock deformation through changes in effective stress and dependent aperture evolution is also strongly dependent on influx rate. As the maximum pressure perturbation realized within basal cap rock increases from 3 to 22 to 160 bar with a 10- to 100-fold increase in influx rate (Figures 17–19), the dependent aperture widening—evaluated in the context of Eqs. (5) and (6)—increases from approximately 50 to 350 to 2900 μm.

The three simulations described above address a fundamental question regarding natural CO_2 reservoirs: are they natural analogs to engineered CO_2 storage sites? The models suggest that geomechanical degradation of seal integrity will be characteristic of both natural and engineered CO_2 influx, but significantly more severe during the latter. This result implies that cap-rock isolation performance may vary considerably as a function of filling mode, which further suggests that the currently secure cap rock of a given natural CO_2 accumulation may be incapable of providing an effective seal in the context of an engineered injection. This potential discrepancy limits the extent to which natural CO_2 reservoirs can be considered directly analogous to engineered CO_2 storage sites.

Geochemical Counterbalancing of Geomechanical Effects

Long-term enhancement or degradation of shale cap-rock integrity ultimately hinges on the relative effectiveness of concomitant geochemical alteration and geomechanical deformation. The analyses presented above offer an opportunity to evaluate an important aspect of this geochemical/geomechanical interplay: the extent to which these initially opposing processes may ultimately counterbalance one another.

This cross-comparison requires a common reference frame, the choices for which are changes in porosity or fracture aperture, which have been used above to represent the respective contributions of geochemical and geomechanical effects. Converting aperture change into the corresponding porosity change requires an initial aperture or fracture density (neither of which are known here), while the aperture change associated with matrix expansion due to a specific mineral dissolution/precipitation reaction can be represented as a function of the dependent variables. Hence, we adopt the latter approach and translate the geochemical contribution into the aperture-change reference frame.

For a given dissolution/precipitation reaction within the matrix, the associated aperture change (Δa) depends on the initial volume fraction of the reactant assemblage (V_R/V_T), standard molal volume change of the reaction ($\Delta V_r^0 = V_P^0 - V_R^0$), effective diffusion distance (L_D, how deep into matrix blocks the reaction occurs), and reaction progress (C, the extent to which the reaction proceeds towards completion) [5]:

$$\Delta a = -2[(V_R/V_T)(\Delta V_r^0/V_R^0)L_D C] \tag{7}$$

All of these variables are typically known or can be closely estimated except for diffusion distance and reaction progress. Hence, it is both appropriate and convenient to plot Δa isopleths as a function of these latter two parameters.

We have constructed such a diagram for reaction (4) (Figure 21), where the Δa-isopleths plotted correspond to the range of geomechanical aperture widening—from initial maximum to final net values (roughly 1000 and 100 μm, respectively)—predicted for model CLP (Figure 14). Hence, they can be viewed as geochemical counterbalance isopleths for this most extreme case of the four systems modeled (CLP, CHP, ULP, and UHP), i.e. along any curve, departing to greater diffusion distances or reaction progress equates to net aperture closure (improved cap-rock integrity) as a function of combined geochemical and geomechanical effects, while departing to lesser values equates to net aperture opening (degraded integrity).

This diagram reveals that ultimate geochemical counterbalancing of initial maximum aperture widening (1000 μm) requires diffusion distances of 3–6.5 cm for reaction progress of 30–60%. Moreover, such counterbalancing of the final net widening (100 μm) requires <0.5 cm diffusion distance for the same range

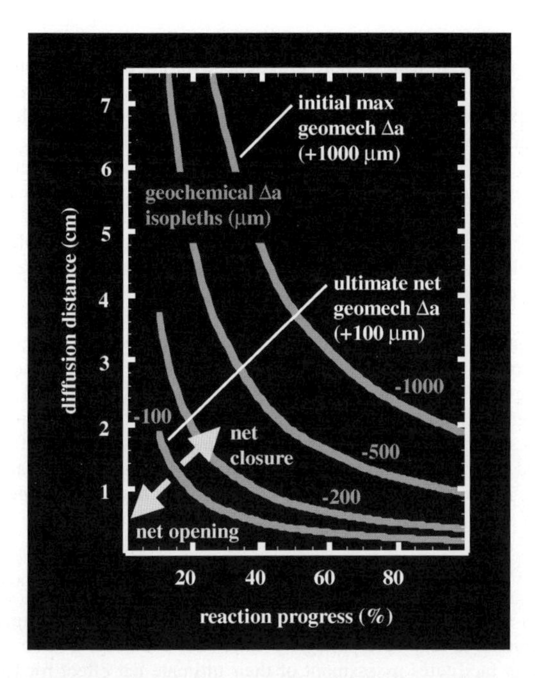

Figure 21: Conceptual framework for assessing potential long-term geochemical counterbalancing of geomechanical effects: geochemical Δaperture isopleths plotted as a function of diffusion distance and reaction progress for mineral dissolution/precipitation reaction (4), where $\Delta V_r^0/V_R^0$ is 0.185 and V_R/V_T is 0.1425, constructed using reaction (7) over the range of geomechanical Δaperture defined by initial maximum and ultimate net widening for model CLP (Figure 14).

of reaction progress; this diffusion length scale and extent of reaction progress—both of which are commonly observed in natural systems—strongly suggest that CO_2 influx-triggered geomechanical deformation may be ultimately counterbalanced by long-term geochemical alteration. This raises the distinct possibility that currently secure shale cap rocks in natural CO_2 reservoirs may have *evolved* into effective seals following some degree of CO_2 migration through them. Careful mineralogical and petrographic analyses of these shale cap rocks may shed light on this important concept.

CONCLUSIONS

Reactive transport and geomechanical models have been interfaced and a new conceptual framework developed to evaluate long-term cap rock integrity in natural and engineered CO_2 storage sites. For typical (non-carbonaceous) shale compositions, influx-triggered geochemical alteration and geomechanical deformation act in opposition to enhance and degrade hydrodynamic seal capacity through aperture narrowing and widening of cap-rock microfractures; hence, net impact of these concomitant processes hinges on their relative effectiveness. The extent of geochemical enhancement is largely independent of reservoir characteristics that distinguish saline-aquifer from EOR/storage settings and influx parameters that distinguish engineered disposal sites from natural accumulations, because such characteristics and parameters have negligible (indirect) effect on mineral dissolution/precipitation rates. In contrast, the extent of geomechanical degradation is highly dependent on these reservoir characteristics and influx parameters, because they effectively dictate magnitude of the pressure perturbation. Specifically, it is has been shown inversely proportional to reservoir permeability and lateral continuity and proportional to influx rate.

As a result, while the extent of geochemical alteration is nearly independent of filling mode, that of geomechanical deformation is significantly more pronounced during engineered storage. This suggests that the currently secure cap rock of a given natural CO_2 accumulation may be incapable of providing an effective seal in the context of engineered injection, a potential discrepancy that limits the extent to which natural CO_2 reservoirs and engineered storage sites can be considered analogous. In addition, the pressure increase associated with CO_2 accumulation in any compartmentalized system invariably results in net geomechanical aperture widening of cap-rock microfractures. This suggests that ultimate restoration of pre-influx hydrodynamic sealing capacity—in both EOR/storage and natural accumulation settings—hinges on ultimate geochemical counterbalancing of this geomechanical effect, which further suggests that the well documented leaky-to-secure character of fossil CO_2 reservoirs may reflect the incomplete-to-complete nature of such restoration.

To explore these hypotheses, a new conceptual framework has been introduced that depicts ultimate geochemical counterbalancing of geomechanical aperture evolution as a function of effective diffusion distance and reaction progress. This framework reveals diffusion length scales and reaction progress extents consistent with those observed in nature, which suggests that ultimate counterbalancing of geochemical and geomechanical effects is feasible, and, therefore, that shale cap rocks may in fact *evolve* into effective seals—in both natural and engineered storage sites. Further, it provides a theoretical model for assessing the extent to which cap-rock integrity will ultimately be enhanced or degraded in specific reservoir/cap-rock systems in the context of specific engineered injection scenarios.

RECOMMENDATIONS

The present contribution can be viewed as a scoping study in which influx-triggered geochemical and geomechanical contributions to cap-rock integrity have been modeled, then merged within a new conceptual framework that facilitates assessment of their ultimate net effect for CO_2 storage sites whose compositional and influx parameters can be well characterized. As such, it provides a unique computational methodology for addressing two central issues for geologic storage—long-term prediction of isolation performance and the extent to which natural and engineered sites are analogous. A number of model development and application activities are immediately posed by this inaugural work.

In terms of important technological advances, there is a pressing need to develop a simulation capability that fully integrates reactive transport and geomechanical processes, which we have merely interfaced here. There are many ways to accomplish this, ranging from, ideally, a global-implicit approach to, perhaps more realistically in the short-term, bi-directional coupling of distinct models. Equally pressing is the need for improved kinetic descriptions of mineral dissolution and (especially) precipitation processes as well as more accurate and comprehensive databases of the associated species-specific parameters; these developments will lead to improved predictive capabilities. Also very important is the need to develop methodology for assessing the specific rates and time frames of geochemical counterbalancing that involves multiple dissolution/precipitation reactions; here, we have addressed this concept only in a time-integrated sense and for a single representative reaction.

In parallel with such development activities, several key applications could provide critical benchmarking, validation, and refinement for both the simulation capabilities and new hypotheses described above. For example, detailed reactive transport modeling of well-characterized fossil or active CO_2 reservoirs—ideally, a suite of leaky-to-secure systems for which cap-rock core is available—would provide a crucial field-scale test bed for the incomplete-to-complete geochemical counterbalancing concept. Similarly detailed modeling of carefully designed and precisely characterized batch and plug-flow reactor experiments would provide an analogous laboratory-scale test bed for this theory—as well as the ideal means of benchmarking simulation capabilities for all mineral trapping mechanisms.

Closely integrated modeling/experimental studies such as these—on both the field and laboratory scale—also provide an effective methodology for evaluating key compositional dependencies of long-term cap rock (and reservoir) integrity. Such dependencies include those associated with formation waters (e.g. salinity, specific cation/anion concentrations) and waste-stream impurities (e.g. CH_4, H_2S, SO_x, NO_x concentrations) as well as the effect of lithologic diversity, ranging from the influence of carbonate cements on the shale-capped sandstone systems addressed here to a dramatic shift from such environments into, for example, anhydrite-capped carbonate reservoirs.

Finally, for a suite of well-characterized potential CO_2 disposal sites, reactive transport and geomechanical modeling could be used to identify and evaluate the volume change associated with key injection-triggered mineral dissolution/precipitation reactions, to assess concomitant pressure-dependent geomechanical deformation, and to determine net impact of these interdependent processes on long-term cap rock integrity (e.g. Figure 21). It would be particularly instructive and useful to carry out this modeling study for a suite of prospective sites that spans the broad range of potential reservoir/cap-rock lithologies—well beyond the single sandstone/shale combination examined here. Such an analysis would provide a unique means of quantitatively ranking long-term isolation performance as a function of important lithologic and other dependent variations.

ACKNOWLEDGEMENTS

This work was performed under the auspices of the U.S. Department of Energy by University of California, Lawrence Livermore National Laboratory under Contract W-7405-Eng-48. Funding support was provided by the JIP CO_2 Capture Project and the U.S. Department of Energy (Office of Fossil Energy) through a CRADA (TC-02038). It is a pleasure to thank Associate Editor Curt Oldenburg and two anonymous reviewers for their critiques of the original manuscript; each contributed a number of comments that were invaluable in preparing the revised contribution.

REFERENCES

1. J.W. Johnson, J.J. Nitao, C.I. Steefel, K.G. Knauss, Reactive transport modeling of geologic CO_2 sequestration in saline aquifers: the influence of intra-aquifer shales and the relative effectiveness of structural, solubility, and mineral trapping during prograde and retrograde sequestration, *Proc. First Natl Conf. Carbon Sequestration*, Washington, DC, May 14–17, 2001, p. 60.
2. J.W. Johnson, J.J. Nitao, C.I. Steefel, Fundamental elements of geologic CO_2 sequestration in saline aquifers, *Am. Chem. Soc., Fuel Chem. Div. Prepr.* **47** (1) (2002) 41–42.
3. J.W. Johnson, J.J. Nitao, Reactive transport modeling of geologic CO_2 sequestration at Sleipner, Proc. *Sixth Int. Conf. Greenhouse Gas Control Technologies (GHGT-6)*, Kyoto, Japan, October 1–4, 2002, vol. 1, pp. 327–332.
4. J.W. Johnson, J.J. Nitao, K.G. Knauss, Reactive transport modeling of CO_2 storage in saline aquifers to elucidate fundamental processes, trapping mechanisms, and sequestration partitioning, in: S.J. Baines, R.H. Worden (Eds.), *Geologic Storage of Carbon Dioxide*, Geological Society, London, Special Publication, **233** (2004) 107–128.
5. J.W. Johnson, J.J. Nitao, J.P. Morris, S.C. Blair, Reactive transport modeling of geohazards associated with offshore CO_2 injection for EOR and geologic sequestration, *Proc. Offshore Technology Conf.*, Houston, TX, May 5–8, 2003a, p. 9.
6. J.W. Johnson, J.J. Nitao, J.P. Morris, Reactive transport modeling of long-term cap rock integrity during CO_2 injection for EOR or saline-aquifer storage, *Second Natl Conf. Carbon Sequestration*, Alexandria, VA, May 5–8, 2003b.
7. J.W. Johnson, J.J. Nitao, J.P. Morris, S.C. Blair, CO_2 reservoirs: are they natural analogs to engineered geologic storage sites? *AAPG Annual Meeting*, Salt Lake City, UT, May 11–14, 2003c.
8. J.W. Johnson, J.J. Nitao, J.P. Morris, Reactive transport modeling of cap rock integrity during natural and engineered CO_2 sequestration, *Am. Chem. Soc. Natl Meeting*, New York, NY, September 7–11, 2003d.
9. A.F.B. Wildeborg, A.L. Leijnse, E. Kreft, M.N. Nepveu, A.N.M. Obdam, B. Orlic, E.L. Wipfler, B. van der Grift, W. van Kesteren, I. Gaus, I. Czernichowski–Lauriol, P. Torfs, R. Wójcik, Risk Assessment Methodology for CO_2 Storage—the Scenario Approach, CCP Summary, vol. 2, 2004, 1293–1316.
10. S.H. Stevens, Natural CO_2 Fields as Analogs for Geologic CO_2 Storage, CCP Summary, vol. 2, 2004, 687–697.

11. H. Schütt, M. Wigand, E. Spangenberg, Geophysical and Geochemical Effects of Supercritical CO_2 on Sandstones, CCP Summary, vol. 2, 2004, 767–786.
12. G.J. Nimz, G.B. Hudson, The Use of Noble Gas Isotopes for Monitoring Leakage of Geologically Stored CO_2, CCP Summary, vol. 2, 2004, 1113–1128.
13. G.M. Hoversten, E. Gasperikova, Non-seismic Geophysical Approaches to Monitoring, CCP Summary, vol. 2, 2004, 1071–1112.
14. Z.K. Shipton, J.P. Evans, B. Dockrill, J. Heath, A. Williams, D. Kirchner, P.T. Kolesar, Natural Leaking CO_2-Charged Systems as Analogs for Failed Geologic Sequestration Reservoirs, CCP Summary, vol. 2, 2004, 699–712.
15. D. Norton, Theory of hydrothermal systems, *Annu. Rev. Earth Planet. Sci.* **12** (1984) 155–177.
16. J.W. Johnson, J.J. Nitao, A.F.B. Tompson, C.I. Steefel, 21st-century tools for modeling reactive transport in dynamic geologic systems of economic and environmental significance, *Earth and Environmental Sciences 1998 Annual Report*, 1999, pp. 7–11, LLNL, UCRL-LR-126434-98.
17. C. Bertrand, B. Fritz, J.F. Sureau, Hydrothermal experiments and thermo-kinetic modelling of water–sandstone interactions, *Chem. Geol.* **116** (1994) 189–192.
18. J.W. Johnson, K.G. Knauss, W.E. Glassley, L.D. DeLoach, A.F.B. Tompson, Reactive transport modeling of plug-flow reactor experiments: quartz and tuff dissolution at 240°C, *J. Hydrol.* **209** (1998) 81–111.
19. J.J. Nitao, Reference Manual for the NUFT Flow and Transport Code, Version 2.0: LLNL, UCRL-MA-130651, 1998a, p. 55.
20. J.J. Nitao, User's Manual for the USNT Module of the NUFT Code, Version 2.0 (NP-Phase, NC-component, Thermal): LLNL, UCRL-MA-130653, 1998b, p. 76.
21. J.C. Parker, R.J. Lenhard, T. Kuppusamy, A parametric model for constitutive properties governing multiphase flow in porous media, *Water Resour. Res.* **23** (4) (1987) 618–624.
22. A.C. Lasaga, Kinetic Theory in the Earth Sciences, Princeton University Press, Princeton, NJ, 1998, p. 811.
23. A.E. Scheidegger, The Physics of Flow Through Porous Media, third ed., University of Toronto Press, Toronto, 1974, p. 353.
24. J.W. Johnson, S.R. Lundeen, Jewel: A Graphical–User Interface for Generating Custom GEMBOCHS Thermodynamic Datafiles for Use with Geochemical Modeling Software: LLNL-YMP Milestone Report MOL63, 1994a, p. 23.
25. J.W. Johnson, S.R. Lundeen, GEMBOCHS Thermodynamic Datafiles for Use with the EQ3/6 Software Package: LLNL-YMP Milestone Report MOL72, 1994b, p. 99.
26. J.W. Johnson, E.H. Oelkers, H.C. Helgeson, SUPCRT92: a software package for calculating the standard molal thermodynamic properties of minerals, gases, aqueous species, and reactions from 1 to 5000 bars and 0 to 1000°C, *Comp. Geosci.* **18** (7) (1992) 899–947.
27. E.L. Shock, An updated and augmented version (slop98.dat) of the original SUPCRT92 database (sprons92.dat) is available on Dr. Shock's website: http://geopig.asu.edu, 1998.
28. J.W. Johnson, D. Norton, Critical phenomena in hydrothermal systems: state, thermodynamic, electrostatic, and transport properties of H_2O in the critical region, *Am. J. Sci.* **291** (1991) 541–648.
29. J.C. Tanger IV, H.C. Helgeson, Calculation of the thermodynamic and transport properties of aqueous species at high pressures and temperatures: revised equations of state for the standard partial molal properties of ions and electrolytes, *Am. J. Sci.* **288** (1) (1988) 19–98.
30. E.L. Shock, E.H. Oelkers, J.W. Johnson, D.A. Sverjensky, H.C. Helgeson, Calculation of the thermodynamic properties of aqueous species at high pressures and temperatures: effective electrostatic radii, dissociation constants, and standard partial molal properties to 1000°C and 5 kb, *J. Chem. Soc. (London) Faraday Trans.* **88** (6) (1992) 803–826.
31. H.C. Helgeson, J.M. Delany, H.W. Nesbitt, D.K. Bird, Summary and critique of the thermodynamic properties of rock-forming minerals, *Am. J. Sci.* **278-A** (1978) 229.
32. J.P. Morris, L.A. Glenn, S.C. Blair, The distinct element method—application to structures in jointed rock, *Lect. Notes Comput. Sci. Eng.* **26** (2002) 291–306.
33. J.P. Morris, M.B. Rubin, S.C. Blair, L.A. Glenn, F.E. Heuze, Simulations of underground structures subjected to dynamic loading using the distinct element method, *Eng. Comput.* **21** (2–4) (2004) 384–408.

Gas storage technologies can therefore make a significant contribution to the technology needs of the geologic CO_2 storage industry. In particular, the significant technology development that occurred during the early stages of development of natural gas aquifer storage projects should be relevant to geologic CO_2 storage. These technologies are unique to the natural gas storage industry and are not generally practiced by the oil and gas exploration and production industry.

RESULTS AND DISCUSSION

The review of gas storage technologies throughout this report focuses on three major areas:

- gas storage field integrity determination,
- gas storage field monitoring and leak detection techniques, and
- gas storage field operator response to leaks and gas leak mitigation.

Discussion of relevance to geologic CO_2 storage has been integrated throughout the report. Particular emphasis has been placed on the technologies developed by the aquifer gas storage industry. Although the hydrocarbons trapped in depleted fields by definition demonstrate the natural storage integrity of these fields, the operators of these projects have developed and utilized monitoring technologies as well that are included in our analysis.

Injection into depleted oil or gas reservoirs is the most widely utilized method of storing natural gas in geologic formations. This is due to the fact that these reservoirs have effective seals that have prevented the escape of hydrocarbons for thousands of years so that the risk of leakage is minimal. However, there are not enough depleted hydrocarbon fields in areas where natural gas storage fields are needed. The same is also true for geologic CO_2 storage for which the sites are needed in the industrial and highly populated areas where depleted oil and gas fields are rare if present at all.

The gas storage industry has overcome this obstacle in part by creating storage fields in aquifers. The same process is an obvious choice for storage of CO_2 in many of the industrial and high-population regions of the United States and around the world. Storage of natural gas in aquifers is the process of injection of gas into an aquifer under structural conditions that mimic natural oil and gas reservoirs, e.g. anticlinal high or up-dip pinch-outs. In addition, the target aquifer must be free of transmissive faults so that stored gas will not leak through faults. Many fault systems are comprised of sealing faults that provide effective containment of fluids as evidenced by the accumulation of oil and gas within these systems. The challenge for aquifer projects is to prove that a fault system has sealing faults.

The keys to the success of storing natural gas and/or CO_2 in geologic formations are proper site selection and accurate delineation of the host formations to ensure that they are continuous and extend over a wide area without encountering faults or other features that could allow escape of the injected gas.

The storage zone must be contained below impermeable beds, preferably structurally undisturbed, and laterally continuous to allow storage of a large quantity of gas injected continuously over months or years. In addition, for any method of gas storage or geologic CO_2 storage to have value, a reliable monitoring procedure must be available to ensure that the process is following the projected path and to implement early remedial action when required.

A number of technologies developed by the gas storage industry in the United States and Europe have been identified as having potential application to geologic CO_2 storage. Table 1 identifies these technologies.

Migration and Leakage of Injected Gas in Underground Natural Gas Projects

An important finding of this study is that only 10 of the approximately 600 storage reservoirs operated in the United States, Canada, and Europe have been identified to have experienced leakage; four due to cap rock issues, five due to wellbore integrity, and one due to reservoir selection (too shallow). All observed leaks through cap rocks have occurred in aquifer storage fields. Table 2 lists the reported incidents of leaks in gas storage fields, the type of leak, and the mechanism or procedure implemented for control of the leak.

TABLE 1
GAS STORAGE TECHNOLOGIES WITH POTENTIAL APPLICATION TO CO_2 STORAGE

Inventory verification
Pressure–volume techniques
Reservoir simulation
Volumetric gas in place calculations
"Watching the barn doors"
Gas storage monitoring techniques
Vegetation monitoring and surface observations
Shallow water wells
Gas storage observation wells
Well logging
Seismic monitoring
Gas metering
Gas sampling and analysis
Tracer surveys
Production testing
Remote sensing
Leak mitigation techniques
Shallow gas recycle
Aquifer pressure control
Caprock sealing (not proven technique)
Caprock integrity techniques
Geologic assessment
Threshold pressure
Production/injection tests (pump test)
Flow/shut-in pressure tests
Air/CO_2 injection

It should be noted that this list might not include all leaks that have occurred, but is as complete as a literature search and interviews of storage operators (as conducted through this study) could provide.

Many of these gas migration incidents have been discovered by state-of-the-art monitoring technologies utilized by the gas storage industry, and in most cases the gas migration has been successfully controlled. Given the number of gas storage reservoirs in the world, the gas storage industry has an excellent record for the safe and effective storage of natural gas.

The gas storage industry has developed a series of actions to be taken when a leak in a storage field occurs. Emphasis is placed on mitigation techniques for cap rock leaks in particular, as the oil and gas industry has a great deal of experience and capability for addressing well workovers and handling of wellbore leaks.

Mitigation of gas leakage from underground natural gas storage projects
In the case of a leak in an aquifer gas storage field, the following mitigation steps are taken. Many of these steps will apply to leaks from any type of storage field.

1. When the gas leakage is first observed and reported, the geographic area of the leak is surveyed for homes, farms, businesses or other entities that may be endangered by the leak. Local and state officials are notified as necessary to protect the public.

TABLE 2
REPORTED INCIDENTS OF LEAKS, TYPE OF LEAK AND REMEDIATION EFFORTS TAKEN IN GAS STORAGE FIELDS

Field type and location	Type of leak	Remedial action taken
Aquifer Storage Field, Galesville Formation, Midwestern US	Caprock leak	Gas recycle from shallow zones followed by water removal from storage zone for pressure control
Aquifer Storage Field, Mt. Simon Formation, Midwestern US	Caprock leak	Gas recycle from shallow zones above aquifer
Aquifer Storage Field, Mt. Simon Formation, Midwestern US	Caprock leak	Field abandoned after small volume of gas stored
Aquifer Storage Field (Leroy), Thaynes Formation, Uinta County, Wyoming, US	Wellbore leak	Wellbore remediation
Salt Cavern Field (Yaggy), Shallow Salt Zone, Kansas, US	Wellbore leak	Wellbore remediation/abandonment
Aquifer Storage Field, St. Peter Sandstone, Midwestern US	Caprock leak	Zone abandoned, deeper formation developed
Depleted Oil and Gas Reservoir Ontario, Canada	Wellbore leak	Wellbore remediation
Depleted Gas Reservoir, Multiple Formations, West Virginia, US	Casing leaks	Rework/recompletion of wells. Casing defect repair
Depleted Oil and Gas Reservoir, West Montebello, California, US	Improperly plugged old well	Proper plugging of old well
Aquifer Storage Field, Shallow Sand, Northern Indiana, US	Reservoir selected too shallow	Abandon field
Russian Fields		No data available

2. If gas is being injected into the gas reservoir, injection may be temporarily halted or injection into wells in the suspected vicinity of the leak discontinued.
3. If gas leakage is observed during the gas withdrawal season, scheduling of gas withdrawal from the storage field may be accelerated. This can be done in the vicinity of the leak and/or can include the entire gas storage field.
4. An investigation into the source of the leak begins immediately. Wellbores in the suspected area are checked for anomalous pressures. Well logs such as temperature and neutron logs may be run in suspect wells. The neutron logs in particular are useful for determining the presence of shallow natural gas accumulations albeit only in the wellbore vicinity. (Note: Neutron logs detect hydrogen densities in the nearby formation, and thus will not be useful for the direct detection of CO_2. Neutron logs may be useful for detecting CO_2 gas through the displacement of water in some cases.)
5. In the case of a cap rock leak, the local geology is reviewed for the most likely area of gas accumulation above the storage zone. Ideally this geologic assessment has been done previously and is available. The shallow zones to be investigated for accumulations of leaking gas are those that are porous and permeable with some type of cap rock just above to slow down or trap a significant accumulation of

gas. Dr Donald Katz in his gas storage research coined the phrase "the cats (Katz) and the doors" to illustrate the most likely location for accumulation of migrating gas [3]. Figure 2 illustrates the concept. In the analogy, the gas storage reservoir is a large room full of cats (the gas) trying to escape. The door leads to a series of rooms connected by doors of various sizes. If the doors leading from rooms 2, 3, 4, are larger than the door from the main storage room, no cats will accumulate in the intermediate rooms since all the cats passing the first door can pass through the larger ones. The cats will accumulate only in room 5, which has a door smaller than the door of the first room. Similarly, the accumulation of gas above a leaky cap rock will occur only when a cap rock is reached which does not leak, or if it leaks more slowly than the primary cap rock. It is believed that gas migrating through the first, second, and third observation zones might well give a significant pressure perturbation at the start of leakage even without significant gas accumulation. This may not always be the case, however, as leakage has occurred to the surface in some storage fields without any observable pressure change in shallow observation zones [3]. In some cases it is possible for gas to leak from the storage reservoir and accumulate in shallow zones without any gas escaping to the atmosphere. (Note: All fields with leakage reviewed within this report experienced gas escaping to the atmosphere.)

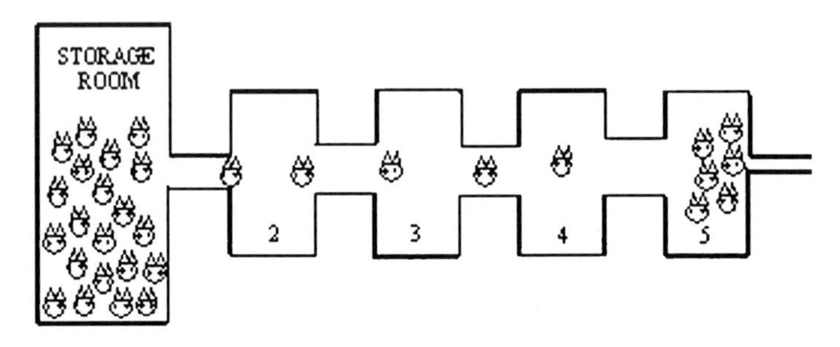

Figure 2: Analogy of "Cats and Doors" to migration and accumulation of gas above a storage reservoir [3].

6. Once the shallow geology is reviewed, a study is conducted integrating all the information on hand. This includes the surface location of the leak in comparison to structural high points in shallow zones and the relative existence and location of permeable zones and cap rocks. Seismic data may be reviewed or new data obtained. From this information, shallow wells are drilled to attempt to locate and produce the gas as it accumulates on its migration path to the surface.
7. Shallow wells that encounter shows of gas after drilling are completed in the gas-bearing zones and production of the shallow gas begins. The process of production lowers the pressure in the zone and helps mitigate further gas movement to the surface. Control of migrating gas has been accomplished at two aquifer fields in the midwestern United States. Locating a shallow zone that is well connected to a significant volume of the leaking gas is an important accomplishment in controlling leakage. This may require the drilling of several wells. Advanced seismic techniques available today may assist with locating shallow gas accumulations.
8. Once shallow wells are drilled and completed, an ongoing gas recycle program is initiated and performed for the remaining life of the storage field or until the leak is located and stopped. In the case of a leak in the cap rock, the recycle goes on for the life of the storage field. Figure 3 illustrates the possible pathway for gas leakage, its accumulation in a shallow zone, and the completion of a shallow well to recover and recycle the leaking gas.
9. Another technique used for control of leaking gas is the continuous withdrawal of water below the gas storage bubble. The removal of a sufficient volume of water lowers the pressure in the gas storage zone to near or below original aquifer pressure. This in turn reduces the volume of gas that leaks through the cap rock, thus controlling the leak. This practice has been put in place at one midwestern gas storage field and continues to be utilized (Midwestern US gas storage operator, personal communication). In this case, the water withdrawn from the zone below the gas bubble in the storage reservoir is injected into shallow zones above the gas storage field.

10. After implementation of a gas recycle program or pressure control procedure via water withdrawal, the injection-withdrawal schedule for the entire storage field may be modified. In particular, the injection season may be delayed as late into the year as possible and withdrawal commenced as early as possible. This has the overall effect of minimizing the time the cap rock experiences high pressure. Another step implemented is to withdraw enough gas every year such that the reservoir pressure is taken below its original pressure each year. This is essential in the case of an aquifer storage project. Yet another step taken to assess the leakage problem is a field-wide shut-in of wells with pressures monitored on each well. The objective is to observe anomalous pressures that may indicate the area of gas leakage.
11. The last mitigation step for leaking storage fields is to identify the location and source of the leak and plug the leak. When the leak occurs due to a mechanical problem in a storage well, repair is accomplished through well workover procedures such as casing patches, squeeze cementing, installation of liners or other accepted practices available from the oil and gas industry. If the leak is through a flaw in the cap rock the problem is much more difficult (see below).

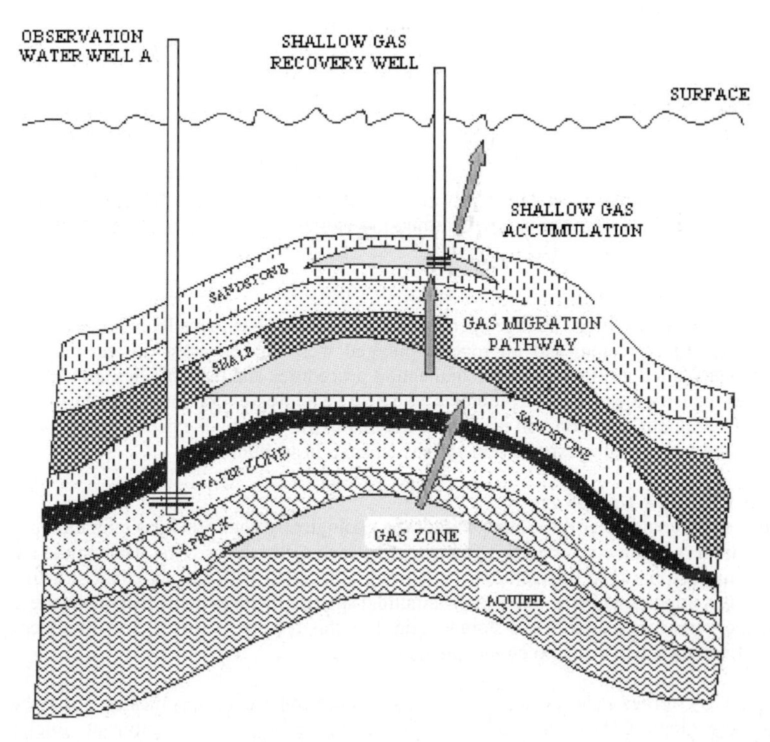

Figure 3: Pathway of migrating gas from storage reservoir, accumulation in shallow zone and recovery and recycle by shallow gas well.

Geologic CO_2 storage needs. Throughout the study, more than 40 participants (which were predominantly gas storage field operators) were asked where they felt the greatest technology needs reside with respect to geologic CO_2 storage. The top 10 needs are listed in Table 3 along with the percentage of respondees selecting each need. The majority of the suggested technological advances involve injection well cementation, completion, inventory verification, and risk analysis operations. Major research efforts are also needed in the development of hardware and software for testing, monitoring, and modeling/simulation.

Important findings for geologic CO_2 storage from gas storage operations. Gas storage operators have accumulated a significant knowledge base for the safe and effective underground storage of natural gas.

TABLE 3
TECHNOLOGY AREAS IN NEED OF IMPROVEMENT FOR CO_2 SEQUESTRATION AS DETERMINED BY SURVEY RESPONSES

Technology	% of Responses
Injection well completion	54
Inventory verification	53
Injection well cementation	39
Risk analysis	39
Storage performance	31
Monitoring cap rock leaks	31
Monitoring gas location	31
Simulation	31
Leak response	23
Leak mitigation	23

While unwanted gas migration has occurred both due to mechanical problems with wells and geologic factors, overall gas storage has been effectively and efficiently performed. The following topics are felt to be the most relevant findings from the study regarding gas storage technology application to CO_2 geological storage.

Wellbore gas leakage
Most gas leakage incidents in gas storage operations have occurred due to wellbore integrity problems. Poor cement jobs, casing corrosion, and improperly plugged wells in converted oil and gas fields have all contributed to unwanted gas leakage. Remedial action procedures and technologies are well established in the oil and gas industry to solve these problems, and new technologies continue to be developed to address these issues. Continuous attention will need to be applied to this area by the geologic CO_2 storage industry but practices and technologies exist to remedy gas leakage in wellbores.

Geologically controlled gas leakage
As far as this study could determine, almost all of the geologically controlled gas migration problems have occurred in aquifers being converted to gas storage. In each of these cases the flaws in the cap rock were most likely due to some type of fracturing or faulting associated with the anticlinal structure of the gas storage field. It is important to note that a large anticlinal structure with as many feet of closure as possible is an important criteria for an aquifer gas storage field. It is this feature, however, that introduces the greater possibility of cap rock flaws and potential leakage.

The geologic CO_2 storage industry may find this experience important. Specifically, it may be in the best interest of the geologic CO_2 storage industry to avoid aquifer areas with significant structural features. Gently sloping structure and cap rock formations may be preferable for long-term CO_2 storage.

"Significant structural features" are those with significant structural relief which increases the possibility of faulting or fracturing that may lead to leakage situations through the cap rock.

It is of special note that leakage of gas has occurred (unobserved until significant gas release to the atmosphere was observed) in at least one field despite application of practically all available technology and integrity determination techniques. Accordingly, the caution directed at the gas storage industry by Dr Donald Katz, a pioneer in natural gas engineering and gas storage, in the 1960s is likely to be appropriate for the newly developing CO_2 storage industry today. The caution is quoted below:

> Caution must be exercised in claiming that no gas will ever be found outside the intended well—gathering line—reservoir system. If any gas is found outside the intended system, it is possible that it can

be handled so as to cause little or no harm, and should be no cause for calling a halt to the operations. Therefore, it is necessary in any full description of a fifty-year life for a storage operation to admit that, on occasions, some gas will enter the waters and even the soil, but that mechanical repairs are available so that the leak can be halted [3]

Testing the integrity of cap rocks above storage zones

The gas storage industry has successfully used several cap rock integrity testing techniques, which are included in Table 1. Each of these techniques can be used individually or combined with other techniques to assure safe storage conditions exist.

The issue of cap rock integrity is where the "rubber meets the road" with regard to storage of gases both for natural gas to be utilized for deliverability needs and for long-term geologic CO_2 storage. The necessity of cap rocks for trapping hydrocarbons is well understood within the oil and gas industry. The gas storage industry has performed research and studied the issue of cap rocks in particular in the area of aquifer gas storage. The interest and need are greatest for aquifer gas storage as there is no natural occurrence of oil or gas to test the integrity or sealing capability of the cap rock.

Potential for assessing field integrity with pilot storage of CO_2 or air

While the natural gas storage industry is required to perform expensive tests to assess field integrity, the geologic CO_2 storage industry is dealing with a noncombustible gas and may not have the need to withdraw gas from storage. This presents the opportunity to test a potential storage site by simply injecting CO_2 and monitoring for pressure disturbances above the zone of interest. If CO_2 is not available at a given site, consideration can be given to injecting air. Air injection is not feasible at a potential natural gas site as the subsequent storage of gas in the presence of air creates the obvious problem of potential unwanted combustion.

A possible procedure is to deploy a portable compressor and one or two wells, one injection well and one observation well, above the storage zone. CO_2 or air could be injected into the potential storage zone creating an over-pressure situation against the potential cap rock. Careful measurements in the observation well above the cap rock can assist with cap rock integrity determination. This type of test could provide significant insight into the integrity and quality of a potential storage site. If air is utilized, it should be kept in mind that CO_2 and air have quite different physical and chemical properties. Air may be very useful for assessing cap rock integrity and basic reservoir properties but would act quite differently than CO_2 in the reservoir, especially deep formations.

Leak mitigation possibilities

The gas storage and oil and gas industries have been successful in repairing wellbore leaks but there is no known case where a geologic leak through a confining layer or cap rock has been sealed. In the case of the oil and gas industry, the need is usually not present, as any cap rock flaw would have precluded the trapping of commercial volumes of hydrocarbons. Without the commercial potential the oil and gas industry has neither interest in these features nor any incentive to investigate cap rock seals. The gas storage industry does have interest in cap rock seals, especially in the aquifer storage area, and has performed limited research.

In the case of aquifers with gas leakage, there have been attempts to determine the location and type of the leak. Tracer surveys, seismic and well tests have been used in this regard. Most of these efforts were undertaken in the 1970s shortly after the development of many of the aquifer storage projects. Little has been done since then due to a lack of new storage development and the application of leak mitigation techniques, primarily gas recapture in shallow horizons or pressure control techniques.

There have been significant advances in recent years in many areas that may allow for the successful sealing of a cap rock leak in the future. Seismic technology has advanced significantly to include 3D and 4D seismic, high-resolution crosswell and vertical seismic profiling (VSP). The technology to carefully drill and steer a wellbore to a given location is available today with a precision unprecedented relative to 1970s technology. Research has been performed on using foams and other materials to control the flow of fluids within reservoirs and wellbores that may eventually lend themselves to the sealing of a geologic fault or fracture. Again, while there is no known successful or attempted geologic fault/cap rock flaw-sealing project, new technologies may open this door in the future where and if it is required. This is an area where the CO_2 storage industry may wish to perform additional research.

Matrix of gas storage technology with applications to geologic CO_2 storage
Table 4 lists the 24 gas storage technologies discussed in the report and notes the application of these technologies for geologic CO_2 storage.

TABLE 4
GAS STORAGE TECHNOLOGIES AND APPLICATION TO GEOLOGIC CO_2 SEQUESTRATION

Gas storage technology area	Gas in place determination	Leak detection	Leak control	Gas movement monitoring	Caprock integrity determination	Reservoir suitability for storage
Pressure–volume techniques	X	X				
Reservoir simulation	X			X		X
Volumetric techniques	X	X				X
"Watching the Barn Doors"	X	X		X	X	
Surface observations		X	X	X		
Change in vegetation		X	X	X		
Shallow water wells		X	X	X		
Gas storage observation wells	X	X	X	X	X	X
Well logging	X	X	X	X		X
Seismic monitoring		X		X		X
Gas metering	X					
Gas sampling and analysis		X		X		
Tracer surveys		X	X	X		
Production testing	X	X			X	
Remote sensing		X	X	X		
Shallow gas recycle			X			
Aquifer pressure control			X			
Caprock sealing techniques		X	X			
Geologic assessment	X	X	X	X	X	X
Threshold pressure					X	X
Pump tests		X			X	X
Flow/shut-in pressure tests		X			X	X
Air/CO_2 injection		X			X	X
Over pressuring					X	X

CONCLUSIONS

Our study resulted in the following conclusions and recommendations:

- The best "early warning signals" for leak detection are observation wells and surface monitoring techniques.
- Control technology for leaking gases from storage operations exists (shallow gas recycle and pressure control). These techniques require continuous, expensive operations and may not be feasible for long-term CO_2 storage.
- All "geologic" cap rock leaks are related to the gas storage need for "steep" structural closure. The geologic CO_2 storage industry (particularly in aquifers) can learn from this experience and significantly mitigate risk.
- Cap rock leak "sealing", while not successful to date, has significant potential through application of newer seismic and well steering for locating and accessing the leak zone. New fluids such as foams and other materials to control fluid flow in the storage zone and the overlying cap rock could then be applied.
- Field-integrity testing should include all available techniques. The design of a pilot test for storage field integrity testing, utilizing the principles of the gas storage industry "pump tests" has potential. Utilization of CO_2 and/or air could provide significant savings.

- Successful monitoring of geologic CO_2 storage projects, as with gas storage, requires a combination of techniques (observation wells, pressure–volume studies, remote sensing). These technologies are available.
- The fact that only 10 gas migration incidents have been reported from operation of approximately 600 storage fields over 90 years of history suggests that natural gas can be safely stored.
- Issues that operators face for geologic CO_2 storage facilities are similar to what natural gas underground storage project operators experience. Both are concerned about:
  - the migration of injected gas over time,
  - technologies for monitoring the location of the injected gas,
  - integrity of cap rocks, and
  - monitoring of zones above cap rocks for leakage.
- Significant technology development has occurred within the natural gas storage industry, especially for aquifer storage, which will have direct applicability to CO_2 storage. The five most relevant technologies/techniques are
  - "Watching the Barn Doors" (application of all available techniques),
  - gas storage observation wells,
  - pump testing techniques,
  - cap rock sealing, and
  - surface monitoring
- Small volumetric release rates can manifest themselves at the surface (crop damage, visible bubbling in streams, water wells, etc.) giving the perception of a very significant leak.
- Pressure–volume, reservoir-simulation, and volumetric inventory verification techniques are not always precise enough to identify vertical gas migration during early stages (possibly years) of gas storage.

RECOMMENDATIONS

The geologic CO_2 storage industry should:

1. Further the "science of observation wells" through additional research.
2. Investigate the integration of new seismic, well steering, and fluid control technologies to pinpoint, locate, and seal a geologic leak.
3. Investigate the design of a custom test for field integrity based on gas storage industry pump testing (high rates of fluid withdrawal while monitoring pressure) techniques.
4. The CO_2 storage industry should heed the caution directed at the gas storage industry during its infancy: "Caution must be exercised in claiming that gas will never be found outside the intended area" [3].

ACKNOWLEDGEMENTS

The author and Gas Technology Institute would like to thank the Research Council of Norway/Klimatek, Statoil, and the Carbon Capture Project (CCP) for funding support of this project. Many gas storage companies in the United States, Europe, and Canada participated in the study by responding to surveys, providing data and offering advice and suggestions through personal communications. Their contributions were invaluable and much appreciated. I would like to thank the reviewers of the document for their recommendations and valuable edits. Finally, the editors of the publication deserve special recognition for their foresight and efforts in compiling information on a timely and important topic.

REFERENCES

1. K.F. Perry, *Gas Storage Technology Applicability to CO_2 Sequestration*, Gas Technology Institute, 2003, Des Plaines, Illinois.
2. Energy Information Administration, EIAGIS—NG Geographic Information System, Underground Storage Database, as of June 2002.
3. D.L. Katz, K.H. Coats, Underground Storage of Fluids, Ulrich's Books, Inc., 1968, Ann Arbor, Michigan.

Carbon Dioxide Capture for Storage in Deep Geologic Formations, Volume 2
D.C. Thomas and S.M. Benson (Eds.)

Chapter 10

LEAKAGE OF CO_2 THROUGH ABANDONED WELLS: ROLE OF CORROSION OF CEMENT

George W. Scherer[1], Michael A. Celia[1], Jean-Hervé Prévost[1], Stefan Bachu[2], Robert Bruant[1], Andrew Duguid[1], Richard Fuller[1], Sarah E. Gasda[1], Mileva Radonjic[1] and Wilasa Vichit-Vadakan[3]

[1]Department of Civil & Environmental Engineering, Eng. Quad. E-319, Princeton University, Princeton, NJ 08544, USA
[2]Alberta Geological Survey, Alberta Energy and Utilities Board, Edmonton, AB, T6B 2X3, Canada
[3]Department Civil Engineering and Geological Science, University of Notre Dame, 160 Fitzpatrick Hall, Notre Dame, IN 46556, USA

ABSTRACT

The potential leakage of CO_2 from a geological storage site through existing wells represents a major concern. An analysis of well distribution in the Viking Formation in the Alberta basin, a mature sedimentary basin representative of North American basins, shows that a CO_2 plume and/or acidified brine may encounter up to several hundred wells. A review of the literature indicates that cement is not resistant to attack by acid, but little work has been reported for temperatures and pressures comparable to storage conditions. Therefore, an experimental program has been undertaken to determine the rate of corrosion and the changes in properties of oil well cements exposed to carbonated brine. Preliminary results indicate a very high rate of attack, so it is essential to have accurate models of the composition and pH of the brine, and the time that it will remain in contact with cement in abandoned wells. A model has been developed that incorporates a flash calculation of the phase distribution, along with analysis of the fluxes and pressures of the liquid, solid and vapor phases. A sample calculation indicates that wells surrounding the injection site may be in contact with the acidified brine for years.

INTRODUCTION

Possible leakage of injected CO_2, from the formation into which it is injected to other subsurface formations or to the atmosphere, constitutes a major concern associated with geological storage of captured CO_2 because it may contaminate existing energy, mineral and/or groundwater resources, pose a local hazard at the ground surface, and contribute to increased concentrations of CO_2 in the atmosphere. Possible pathways for leakage include diffuse leakage across cap rock formations, concentrated leakage through natural features such as faults and fractures, and leakage through human-made features such as wells. In areas where little exploration for, or production of, hydrocarbons has occurred, there are few existing wells, and potential leakage through them is not a major concern. (Although old exploration wells might leak, it would not be prohibitively expensive to repair a small number of wells.) However, in mature sedimentary basins, such as those found in North America, more than a century of exploration and production has resulted in a very large number of wells. For example, in the state of Texas in the United States, more than 1 million wells have been drilled [1], while in the Province of Alberta in Canada, more than 350,000 wells have been drilled, with approximately 15,000 new wells currently being drilled annually (www.eub.gov.ab.ca). A significant fraction of these wells are abandoned [2], and information about abandonment practices and general record keeping are of variable quality, especially for older wells. Because of the large number of wells in locations such as these, the potential for leakage through existing wells is an important concern that requires quantitative investigation. For context, a schematic of a possible well-leakage scenario is shown in Figure 1, where an injected CO_2 plume moves under the influence of both pressure drive and buoyancy, and then some of the injected CO_2 moves vertically upward upon encountering a preferential flow path that corresponds to an abandoned well. It is this kind of scenario that requires quantitative analysis.

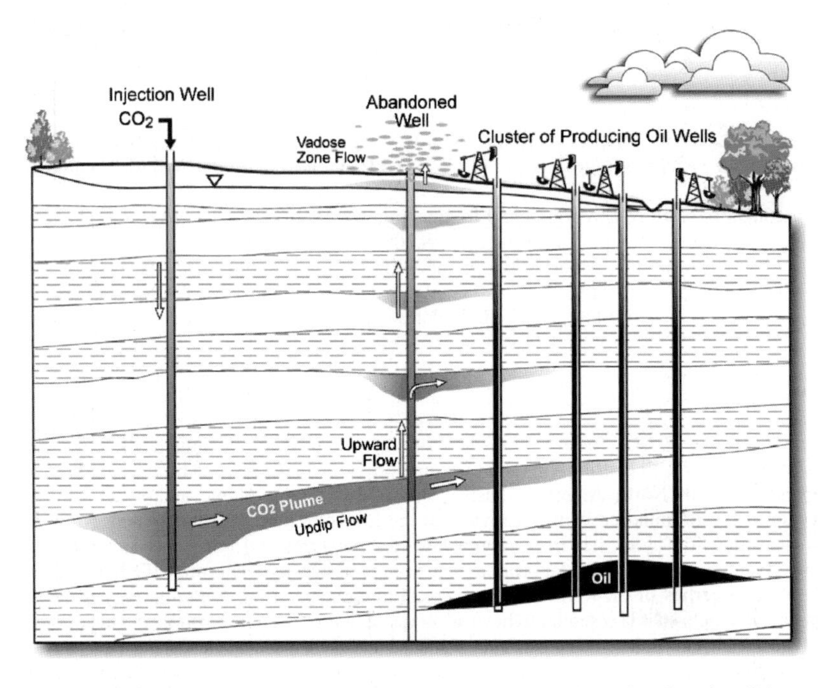

Figure 1: Schematic of injection site and leakage of CO_2 plume through an abandoned well that penetrates a formation in the sedimentary succession. From Ref. [2]. Reproduced with permission from Environmental Geology; copyright Springer, Berlin.

If an exploration well is drilled and the operator decides to abandon the well without further development, the open hole would typically be filled with a series of cement plugs. If the well is developed for production, then a casing would be inserted into the hole, and cement would be emplaced along a portion of the annular space between the casing and the rock. Possible leakage pathways along an existing well are shown schematically in Figure 2, and include preferential flow pathways along the rock–cement interface, the casing–cement interface, and through degraded materials. Because well-formed cement has very low permeability, of the order of 10^{-20} m^2 [3], no significant flow of CO_2 can occur unless there are preferential flow paths, or the material has degraded, or the material was not emplaced properly. If such preferential flows occur, then the overall well materials need to be assigned a quantitative measure of flow potential, which we might take to be the effective permeability of the composite materials associated with the well. Assignment of this composite measure requires estimation of cement degradation under in situ conditions, including possible contact with CO_2-rich fluids, some assessment of the initial emplacement of the cement, including its location along the well and the quality of the emplacement procedure, and knowledge of the location of wells in the vicinity of the injection operation. The first of these requires careful laboratory and modeling studies, which are the primary subject of this chapter, while the second and third require examination of historical records and detailed analysis of existing wells.

To place the problem in some context, consider a simple simulation in which one injection well and one possible leaky well are modeled, with injection and formation parameters as shown in Figure 3a. If we define the leakage fraction to be the rate of leakage along the leaky well (mass of CO_2 per time) divided by the CO_2 injection rate (mass of CO_2 per time), then the leakage fraction is a function of both the distance

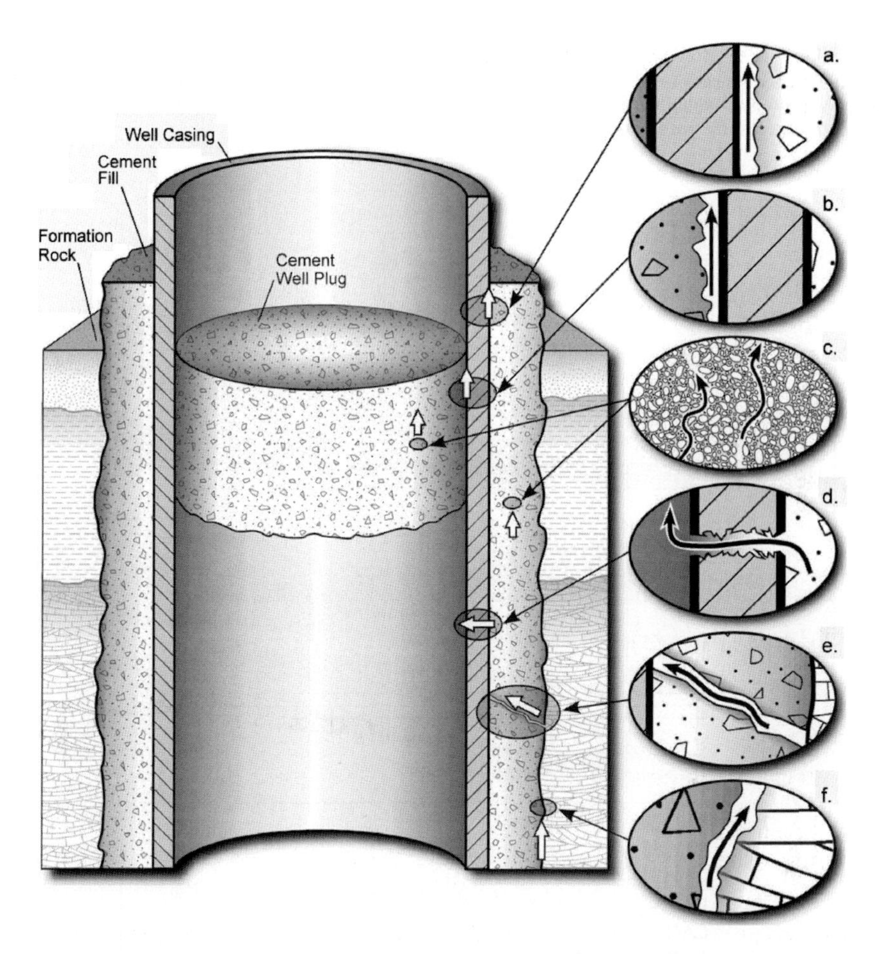

Figure 2: Schematic representation of possible leakage pathways through an abandoned well: (a) between casing and cement; (b) between cement plug and casing; (c) through the cement pore space as a result of cement degradation; (d) through casing as a result of corrosion; (e) through fractures in cement; and (f) between cement and rock. From Ref. [2]. Reproduced with permission from Environmental Geology; copyright Springer, Berlin.

between the two wells and the effective permeability of the leaky well. This relationship is captured in Figure 3b, which shows leakage fraction as a function of distance between the wells and abandoned-well permeability. This figure, taken from Ref. [4], shows that a very large increase in effective permeability is required to produce significant leakage: for a leakage fraction $> 1\%$ at a well spacing of 500 m, the effective permeability associated with the abandoned well must increase to about 10^{-10} m^2. This is many orders of magnitude larger than the permeability of intact cement, showing clearly that well-formed cement will not leak any CO_2. However, this value of effective permeability also corresponds to the effective permeability of an annular opening between the rock and cement that is 1 mm thick. So a thin (1 mm) degraded zone of cement, with very large permeability in the degraded zone, can lead to large effective permeabilities if the annular opening is continuous along the well. Therefore, while the material (cement) emplaced along the well has properties that can suppress all leakage, the system is also extremely sensitive to small irregularities in the system structure. Clearly this problem requires detailed studies of cements, on very small length scales, to properly capture possible small-scale system irregularities that can lead to significant leakage rates. In addition, these detailed small-scale studies must ultimately couple to analyses of injection and leakage at the field scale. This very large range of length scales over which the leakage analysis must be performed, from millimeters to kilometers, is one of the features that makes this a challenging and scientifically interesting problem.

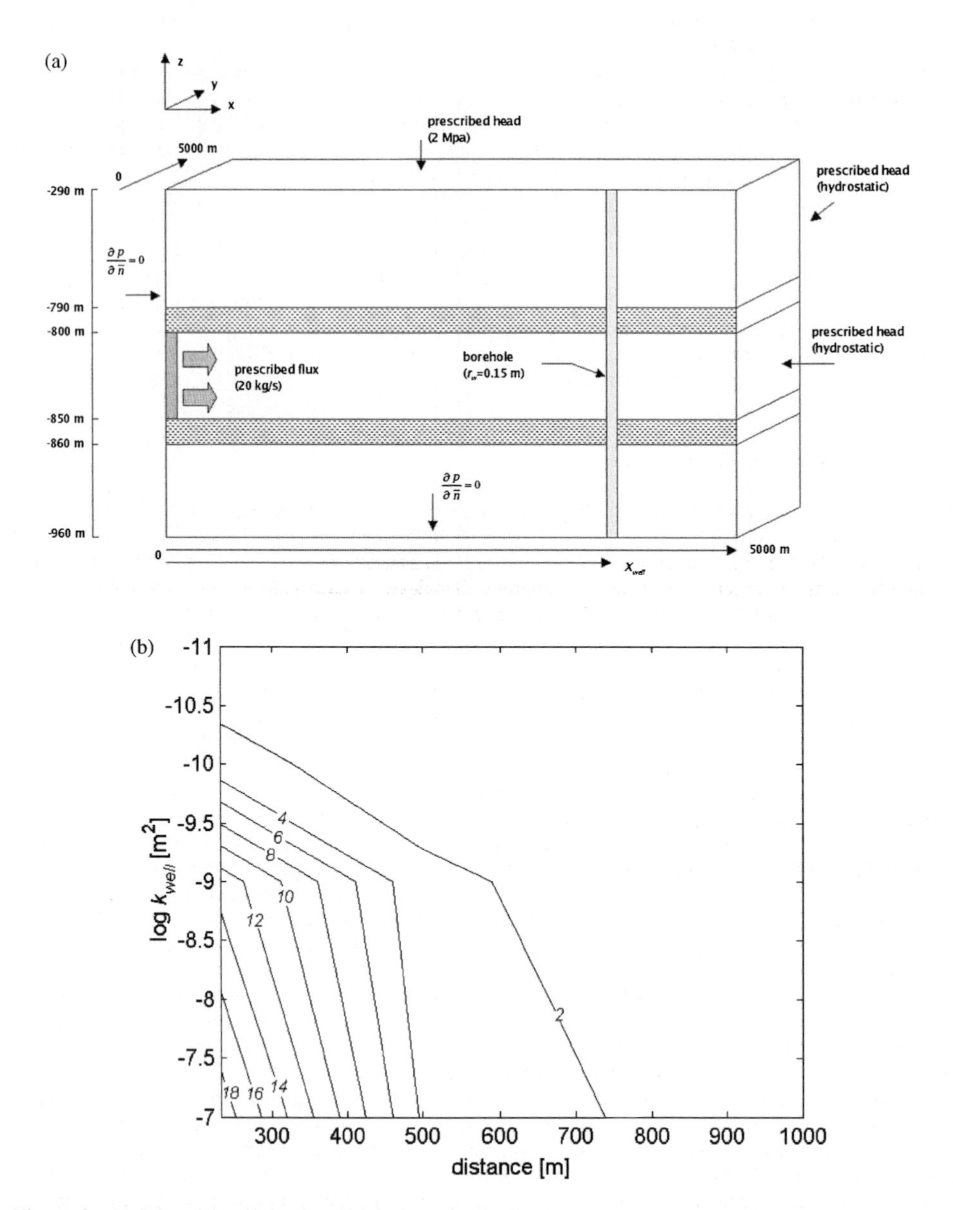

Figure 3: (a) Schematic of injection with leakage from a single passive well; (b) total leakage as a percentage of total CO_2 injected after 4 years at 20 kg/s in 2D parameter space. Leakage is a function of borehole location from injector well (x-axis) and the borehole effective permeability, k_{well} (y-axis). From Ref. [2].

Figure 3b provides the context for studying well leakage. If we can identify the values to be used on the two axes of the plot, we can estimate leakage along a well. Of course, the difficulty is in identification of these values along the axes. Consider first the distance. For some locations, all well locations are known and records are available; for others, this is not the case. In addition, we know that more than one well can be impacted by an injection operation. Therefore, in order to characterize the "distance" axis, we actually need to know many distances, associated with fields of existing wells. On the "permeability" axis, we need to have information about the well cements used, and we need to estimate degradation rates and the properties of degradation products. We then need a set of modeling tools that can integrate these data to produce meaningful estimates of leakage fractions associated with an individual injection operation or with a series of operations. Our research has focused on each of these three aspects of the well-leakage problem: spatial statistics of well locations in mature basins, cement degradation dynamics and small-scale geochemical modeling and large-scale modeling including many wells and uncertainties in their properties.

In this chapter, we review some of our work on spatial statistics of wells, and then present both experimental and modeling work related specifically to cement degradation. This work complements other ongoing work within our extended group that includes larger scale modeling of plume evolution and leakage [5–7], CO_2 transport through shallow unsaturated soils [8,9], geochemical responses and possible water quality changes in shallow aquifers due to introduction of leaked CO_2 [10], and analysis of plume evolution and extent in ongoing acid–gas injection operations [11].

SPATIAL ANALYSIS OF WELLS

Because the Alberta Basin has an outstanding database with a wide range of information on oil and gas wells, we conducted a study of well locations in a formation in Alberta in order to determine spatial characteristics of oil and gas well patterns in a mature basin. We analyzed all wells that penetrate the Viking Formation, which is an areally extensive formation that contains numerous oil and gas pools. Both a cross-section of the basin and the spatial location of all wells that penetrate the Viking Formation are shown in Figure 4. The well locations show clear clustering, which is expected given the nature of oil and gas pools in the formation and in the overall basin. In order to characterize the number of wells that one would expect to be impacted by an injection operation in the Viking Formation, we performed a cluster analysis and separated spatial regions into "high-density", "medium-density", and "low-density" areas. High-density areas, typically associated with oil production, constitute about 3% of the area while accounting for about 30% of the wells, and have mean well density of close to 4 wells per square kilometer. Medium-density wells account for another 30% of the wells, and have a density of about 1 well per square kilometer, and correspond roughly to gas-producing clusters. The low-density background wells cover close to 90% of the area, correspond to a bit more than one-third of the wells, and have a density of 0.15 wells per square kilometer. The low-density background regions also have the highest fraction of abandoned wells. Summary statistics are shown in Table 1.

These numbers can be translated into number of wells that would be impacted by a typical injection scenario. If we estimate a typical CO_2 plume to evolve radially on the order of 5 km, based on solutions in Lindeberg [12], Xu et al. [13] and others or Nordbotten et al. [5,6,14], then we can analyze the spatial data to determine the number of wells impacted by an injection. Results of such an analysis, taken from Gasda et al. [2], are shown in Figure 5, where for each of the three density classes we present a histogram based on a discretization of the data, showing how the number of wells varies for different points within clusters. We present these histograms for both the total number of wells (left column of three figures) and for only the abandoned wells within the specific density class (right column of figures). We see that in high-density areas, the number of wells impacted by a modest plume size of 5 km is several hundred; the mean is 240, and the largest value is greater than 700. For injections into the low-density background regions, the numbers are much more modest, with a mean of about 18 and a maximum number of 130; about 35% of the bins give an adjacent well count of zero. These numbers indicate that in the Viking Formation, injection operations should be expected to contact a significant number of existing wells, up to many hundreds per injection operation. Because the Viking Formation is characteristic of North America's onshore sedimentary basins, we expect these statistics to apply to other mature sedimentary basins.

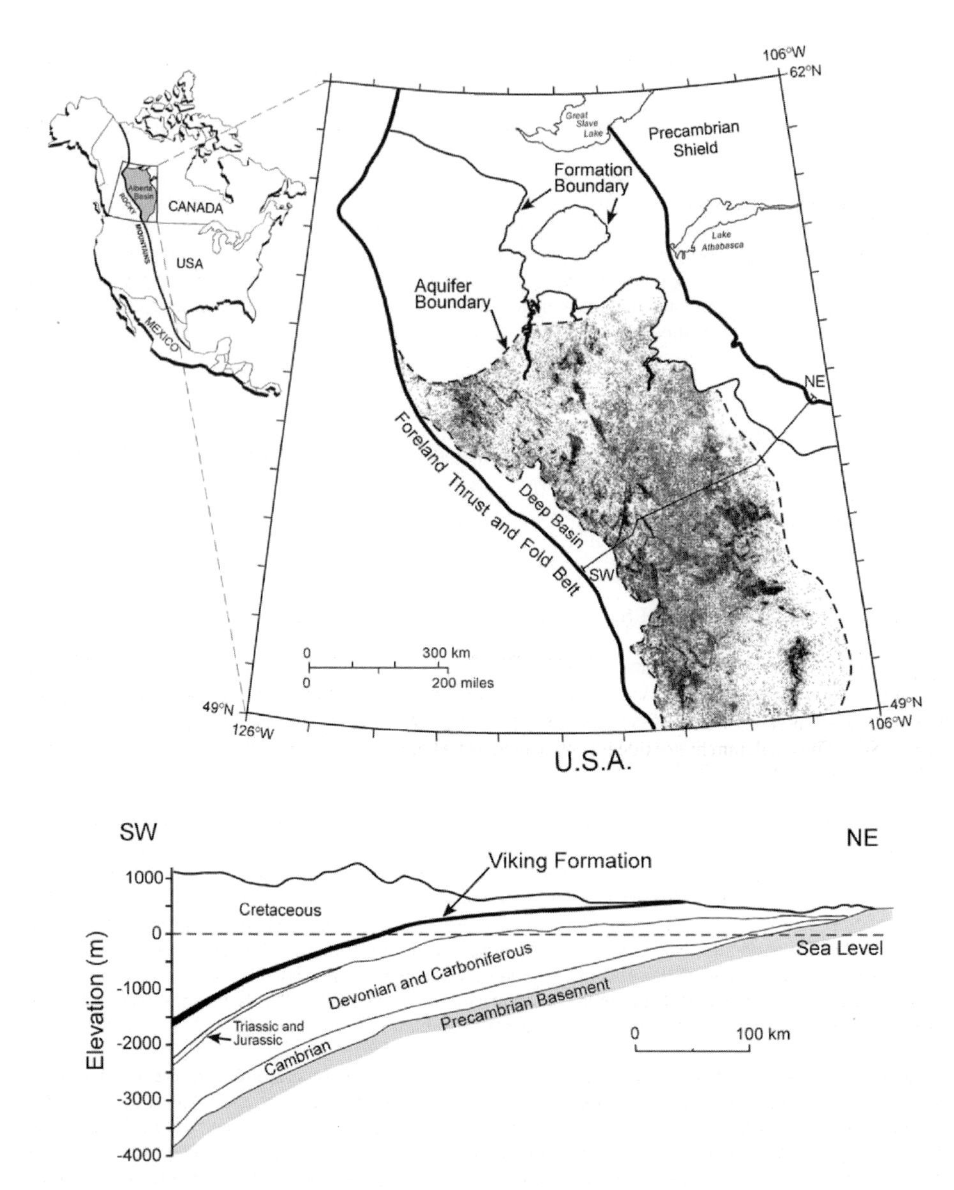

Figure 4: Location of the Viking aquifer in the Alberta basin, Canada: (a) plan view, showing also the location of all wells that penetrate the aquifer; (b) cross-section of the Alberta basin showing the location of the Viking Formation. From Ref. [2]. Reproduced with permission from Environmental Geology; copyright Springer, Berlin.

CEMENT DURABILITY

Cement is used to seal the annulus between the casing and the formation, as shown in Figure 2. Cement powder is mixed with water and various additives to control the density and rheology of the slurry,

TABLE 1
STATISTICS OF DISTRIBUTION OF WELLS IN THE VIKING AQUIFER OF THE ALBERTA BASIN

	High density	**Medium density**	**Low density (background)**
Number of clusters	268	963	–
Number of wells (% total)	28.0	28.6	38.2
Area (% total)	2.7	10	87.2
Mean intrinsic density (wells/km^2)	3.75	1.13	0.15
Fraction of wells abandoned (%)	28.9	45.0	50.0

pumped down through the casing and up the annulus. The rate of the hydration reaction must be carefully controlled, so that the cement paste does not harden prematurely. Once in place, the hardened cement paste must have strength, comparable to that of the surrounding formations, and low enough permeability to provide zonal isolation and to protect the steel casing [3]. Unfortunately, cement is not resistant to acids, so it will be attacked by the carbonated brines produced by storage of CO_2 in saline aquifers. Studies in the literature indicate that the rate of deterioration of the cement may be problematic, but there are no data specifically applicable to the range of temperature and pressure relevant for storage. In the following, we will briefly review the chemistry of cement and then summarize what is known about reaction of cement with carbon dioxide. Finally, we will describe a research program in our lab that is designed to provide quantitative information about the rate of reaction of cement under storage conditions, and the effect of the reaction on the relevant physical properties of the cement paste.

Cement Chemistry

In the following discussion we will use ordinary chemical notation, set in italics, and the shorthand notation commonly used in cement chemistry: $C = CaO$, $S = SiO_2$, $A = Al_2O_3$, $F = Fe_2O_3$, $H = H_2O$, $\bar{C} = CO_3$, and $\bar{S} = SO_4$. Thus, calcium hydroxide is represented as CH or $Ca(OH)_2$ and tricalcium aluminate as C_3A or $3CaO{\cdot}Al_2O_3$ or $Ca_3Al_2O_6$.

Ordinary Portland cement (OPC) is made by grinding calcium carbonate from a natural deposit of limestone, mixing it proportionally with clay, and firing the mixture in a rotary kiln at 1450 °C to form clinker [15]. Once clinker cools, it is ground to a mean particle size of approximately 30 μm. OPC is composed primarily of four compounds: C_3S, C_2S, C_3A, and C_4AF, which are defined in Table 2. The various types of cements shown in Table 3 differ in the ratios of the four compounds and in the fineness of the grind. Oil wells are generally made with Class G or H, which are very similar chemically to ASTM Type I (OPC), which is the most widely used cement in ordinary construction. Class H differs from Type I in that H has a larger particle size and its aluminate is primarily in the form of C_4AF, with little C_3A; both these factors increase the setting time of Class H relative to Type I.

The hydration of OPC at atmospheric temperature and pressure yields several products, but the one that is responsible for the strength of the hardened paste is colloidal calcium silicate hydrate (C-S-H), often called the "gel phase". C-S-H has no fixed composition (which is why the dashes are included in the abbreviation), but the average ratio of calcium to silicon is approximately 1.7 [16]. The building blocks are believed to be particles approximately 2.2 nm in diameter [17] having a semi-crystalline layered structure, with siloxane chains attached to sheets of Ca–O; variations in chain length and substitutions in the Ca–O sheets result in a range of stoichiometry and lattice spacing in the crystallites. Both C_2S and C_3S hydrate into C-S-H, but C_2S reacts much more slowly than C_3S, and C_3S produces three times as much CH as C_2S:

$$2C_3S + 6H \rightarrow C_3S_2H_3 + 3CH \tag{1}$$

$$2C_2S + 4H \rightarrow C_3S_2H_3 + CH \tag{2}$$

Figure 6 is an SEM picture of C-S-H with CH embedded. It is the CH that will play a significant role in the wet and dry carbonation processes. Although the amount of C_3A is small, it reacts violently with water and

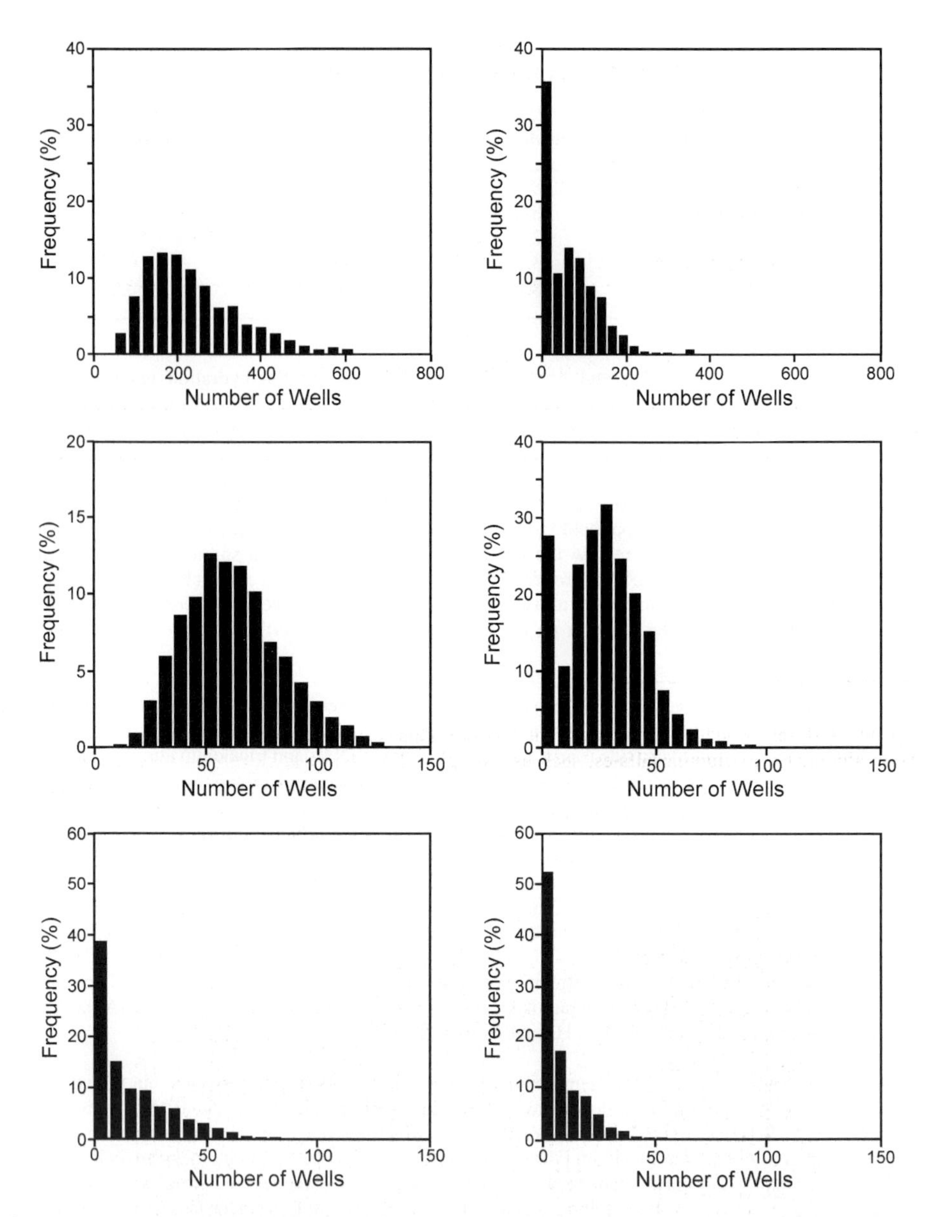

Figure 5: Frequency distributions for the number of wells within an approximate 5 km radius of a potential injection well—all wells in (a) high-density clusters, (b) medium-density clusters, (c) background;—and only abandoned wells in (d) high-density clusters, (e) medium-density clusters, (f) background. From Ref. [2]. Reproduced with permission from Environmental Geology; copyright Springer, Berlin.

TABLE 2
MAIN COMPOUNDS OF PORTLAND CEMENT [15]

Name of compound	Oxide composition	Abbreviation
Tricalcium silicate	$3CaO{\cdot}SiO_2$	C_3S
Dicalcium silicate	$2CaO{\cdot}SiO_2$	C_2S
Tricalcium aluminate	$3CaO{\cdot}Al_2O_3$	C_3A
Tetracalcium aluminoferrite	$4CaO{\cdot}Al_2O_3{\cdot}Fe_2O_3$	C_4AF

TABLE 3
TYPES AND CLASSES OF CEMENT

API class	ASTM type	C_3S (%)	C_2S (%)	C_3A (%)	C_4AF (%)
A	I	53	24	8 +	8
B	II	47	32	5 −	12
C	III	58	16	8	8
D		26	54	2	12
E		26	54	2	12
F		–	–	–	–
G		50	30	5	12
H		50	30	5	12

can create "flash set", a sudden hardening of the mixture. This is avoided by the addition of gypsum, which converts aluminates to aluminosulfates, such as ettringite. The combination of the calcium silicate and aluminum hydrates and CH form the gel phase of hardened cement paste. The formation of a percolated gel phase is called "setting", and it marks the transition from a fluid slurry to an elastic solid. The setting process is governed by C_3S and C_3A because these two have high rates of reaction. In contrast, C_2S governs the subsequent hardening process, which occurs over a period of weeks.

Figure 6: SEM photo showing large polyhedral crystals of CH embedded in C-S-H matrix.

At the setting point, the cement grains are lightly bonded by C-S-H, and the micron-scale interstices (called capillary pores) between the grains are filled with water. As the reaction proceeds, the water is consumed and the capillary pores are gradually filled with the gel phase, which has approximately 28% porosity and pores ranging from approximately 2 to 100 nm in diameter. The network of capillary pores initially controls the permeability of the paste, but the permeability drops drastically if the quantity of hydration products is sufficient to interrupt the capillary pores and force flow of pore liquid to pass through the gel. The pore structure is strongly affected by the particle size of the cement: the finer the grind, the smaller the interstices, and the easier they are to fill with hydration products. However, the most important factor is the water/cement ratio, w/c, used in hydration. If $w/c > 0.4$, then the volume of hydration products is not sufficient to fill the capillary pores; however, the capillary pore network can be blocked (depercolated) by gel at higher w/c. When the permeability is high, chemical attack on the hydrated paste can be very fast, because its surface area is quite high (approximately 50–300 m^2/g).

The exposure of hydrated cement to high temperature and pressure results in the transformation of the amorphous C-S-H gel into other crystalline forms, as shown in Figure 7. The transformation is gradual and depends on the temperature and pressure of the system. Xonotlite (C_6S_6H), a substantially weaker and more porous material than C-S-H, is commonly found in geothermal wells [16]. Silica-rich materials, called pozzolanic admixtures, prevent or delay the strength retrogression by reacting with CH to form more C-S-H. The delay in strength retrogression is partly due to an increased quantity of C-S-H for conversion. In addition, the presence of additional silica shifts the ratio of CaO/SiO_2 down, which can cause other crystal structures to form, as shown in Figure 7.

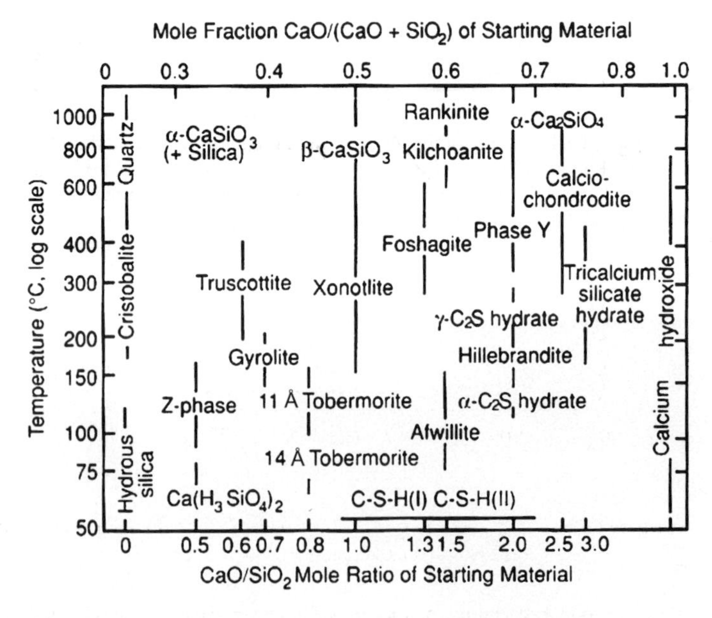

Figure 7: High-temperature phases of calcium silicate [3].

Reaction with CO_2

When moist OPC is exposed directly to dry CO_2, the calcium hydroxide is carbonated to form calcium carbonate ($CaCO_3$ or $C\bar{C}$):

$$Ca(OH)_2 + CO_2 \rightarrow CaCO_3 + H_2O \tag{3}$$

The increase in molar volume from 26.2 cm^3/mole for CH to 33.1 cm^3/mole for $C\bar{C}$ makes the cement stronger and less permeable [15]. The effect of carbonation on C-S-H is more complicated, as explained below.

Exposure to supercritical CO_2 has been shown to increase the strength and reduce the permeability of concrete [18], so exposure to dry CO_2 during injection is not expected to harm cement, However, when CO_2 is introduced into an aquifer, the chemistry changes drastically, because dissolution of CO_2 in water creates carbonic acid (H_2CO_3). Hydrated cement is a highly alkaline material that is chemically stable only when $pH > 10$ [16]. Therefore, the introduction of large quantities of CO_2 into an aquifer will make the downhole conditions extremely aggressive against the existing cement.

The capacity for water to dissolve CO_2 increases with pressure and decreases with rising temperature. Therefore, the depth of storage will play a significant role in the capacity of the aquifer to hold CO_2, as indicated in Figure 8, which shows how the pH of the aquifer varies according to the depth of injection. In a quartzitic sandstone formation, the pH is about 3 over the range of depth of interest; even in equilibrium with limestone, the pH is about 5. Solutions with such low pH can attack cement rapidly.

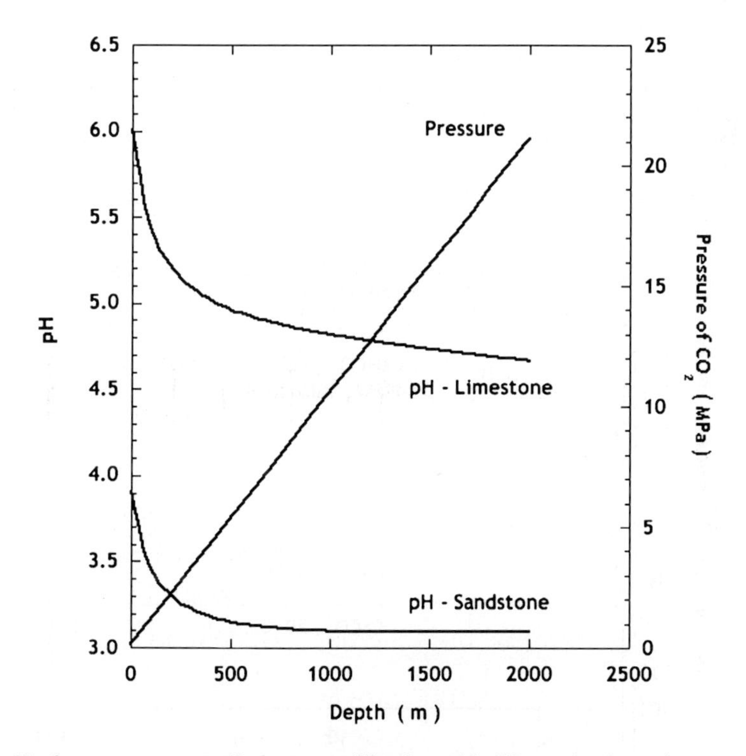

Figure 8: pH of a pure water aquifer upon equilibration with CO_2 at the formation temperature and pressure, calculated using EQ3/6 v. 7.2.

Calcium hydroxide is a product of cement hydration, as shown in Eqs. (1) and (2), and constitutes the alkaline reserve to provide acidic resistance [19]. Although the solubility of CH is quite low (approximately 1–2 g/kg of water), leaching of CH from cement by water has been well documented in the literature [20–26]. The consequences of the removal of CH include lower pH, higher porosity, higher permeability, and lower strength. A lower pH allows steel to corrode and oxidize, and the stability of the C-S-H gel is compromised when the pH drops below 10. Increased porosity and permeability allow greater influx of contaminants and aggressive agents, so the corrosion process accelerates.

Powers et al. [20] found that leaching of CH occurred during permeability measurement. Continuing hydration of the cement offsets the leaching effect during the first 600 days, but once the cement was fully hydrated, the effect of leaching as indicated by the increasing permeability became obvious. More recently, Carde et al. [23] performed a series of experiments on pure OPC paste and paste with silica. They concluded that the macroporosity created through the leaching of CH leads to decreasing strength; since macroporosity controls transport properties, the permeability would also increase, but that property was not measured. Progressive leaching by flowing water, which increases permeability, leads to self-accelerating attack; this mode of deterioration probably constitutes the greatest threat to cement in abandoned wells.

Far more serious than leaching by water is the reaction of CH with carbonic acid to create calcium bicarbonate, $Ca(HCO_3)_2$:

$$Ca(OH)_2 + 2H_2CO_3 \rightarrow Ca(HCO_3)_2 + 2H_2O \quad (4)$$

Calcium bicarbonate is two orders of magnitude more soluble than CH, as shown in Figure 9, and the solubility increases as the pressure of CO_2 rises, and the pH drops.

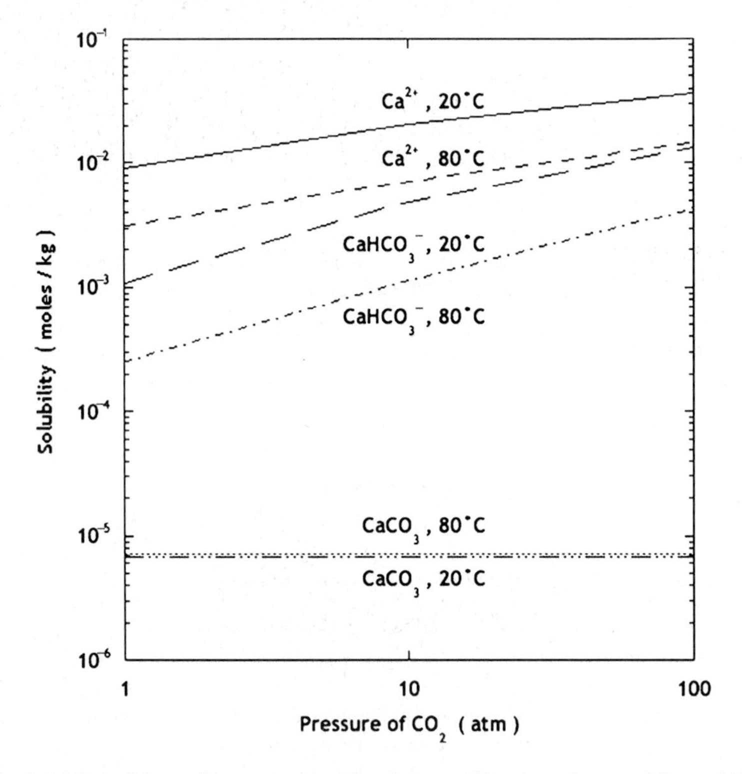

Figure 9: Solubility of the major species resulting from equilibration of water, CO_2, and limestone.

Although the carbonation of CH is favored, C-S-H can also be decomposed by carbonic acid. This reaction can be significantly accelerated by the increased porosity and permeability of the cement due to the absence of CH. The decomposition of C-S-H by H_2CO_3 can be approximated as [27]:

$$CaO{\cdot}SiO_2{\cdot}H_2O + H_2CO_3 \rightarrow Ca^{2+} + CO_3^{2-} + SiO_2{\cdot}H_2O \quad (5)$$

This produces soluble species and silica gel, which is highly porous and has very poor mechanical properties.

There are numerous studies in the literature on the degradation of cement in acidic water. Some have been on the degradation at elevated temperature and pressure, which is typically found in geothermal wells [28–34]. Others have been tested at atmospheric pressure and temperature but with varying pH [19,35–38]. Milestone et al. [31,32] investigated the failure of geothermal wells in the Broadlands, at temperatures of 150 and 250 °C. The primary focus was on the added silica, which, under ambient temperature and pressure, tends to drive down the permeability by reacting with the CH and creating more C-S-H. The hypothesis was that, since CH was depleted by the pozzolanic reaction of silica and CH, aggressive CO_2 immediately attacked the C-S-H, which accelerated the degradation process. Strength retrogression was observed, and the depth of carbonation increased with silica content to a maximum depth of 3 mm. The permeability decreased after exposure to carbonic acid, when the amount of added silica was less than 20%; however, as the amount of added silica exceeded 20%, the permeability increased by 50–3000% after only 2 weeks exposure to CO_2.

Bruckdorfer [29] varied the size (and thus the surface to volume ratio) of his samples and tested them after 3 and 6 weeks after exposure at 79.4 °C and 20.68 MPa. After 6 weeks of exposure, both Classes C and H cement exhibited 80% strength loss; by decreasing the w/c from 0.53 to 0.42, the strength loss was reduced by 25%.

Apparently the only study performed on cement recovered from a well was done by Shen and Pye [33], who obtained samples from the intermediate casing annulus, the production casing annulus, and the re-drilling casing annulus. Most samples had permeability below the API recommended value of 0.2 mD. Attempts to correlate permeability and strength with $C\bar{C}$ content and in-service time were unsuccessful. However, the cement started to lose strength and gain permeability after nine production shutdowns, an effect attributed to thermal stresses during the shutdowns. Fissures of 0.1–5 mm were observed extensively, and some of the finer fissures were filled with $C\bar{C}$, but it could not be determined whether these fissures were due to the coring process or the chemical attack.

Revertegat et al. [37] published a paper on the effect of pH on the durability of paste immersed in a bath of water. At a pH of 4.6, obtained by controlling the partial pressure of CO_2 over the bath, 70–75% decalcification of samples occurred after 3 years of exposure, leaving a weak and highly permeable gel of silica and alumina. Electron microprobe analysis revealed a sharp concentration gradient, indicating a deterioration front between the corroded and uncorroded regions. As the deterioration progressed, cracks large enough to be visible under optical microscopes were observed within the corroded region. The results from X-ray fluorescence showed that the decrease in CaO over 3 years obeyed Fick's Law, indicating diffusion-controlled deterioration.

These studies have yielded useful indications of the severity of attack to be expected under conditions of storage, but none have comprehensively looked at performance parameters, such as strength and permeability, along with microstructure and chemistry, such as the degree of hydration and crystal structure. There are no data available for transport rates in cement subjected to attack by carbonated brine, and essentially nothing is available within the temperature and pressure range of interest for CO_2 storage. Therefore, we are undertaking a comprehensive study of the mechanical and transport properties of cement subjected to the range of pH and temperature expected to exist in carbonated brine.

EXPERIMENTAL PROGRAM

As indicated in Figure 8, if injection is done at a depth of 1–2 km, the pH of the brine will fall in the range of 3 in purely quartzitic sandstone formations and 5 in limestone formations; assuming a temperature gradient of 30 °C/km, the temperature will be about 50–85 °C at that depth. Therefore, we are studying the durability of cement exposed to brine (3 wt% NaCl) in that range of pH and temperature. The samples are prepared from Class H cement (Lafarge; 61% C_3S, 16.3% C_2S, 16.6% C_4AF, 0 C_3A) with additions of 0, 6, or 12 wt% bentonite. The cement pastes are mixed at room temperature and cured in brine at 20, 50, or 85 °C for 28 days, prior to exposure to acidic brine. One set of samples consists of a cylinder of stone (5.5 cm diameter by 10 cm high) with a 2.5 cm hole drilled parallel to the axis, but off center, as shown in Figure 10. Cylinders have been prepared using Salem limestone (13.2% porosity, density 2.33 g/cm^3) and Berea sandstone (19.1% porosity, density 2.14 g/cm^3). The varying thickness of the stone results in a difference in the time to transport acid to the cement, so that the depth of attack varies around the perimeter. The diffusion coefficient of water in the pores of the sandstone, measured using ^{1}H NMR, is $D = 6 \times 10^{-10}$ m^2/s (which is approximately four times slower than the self-diffusion coefficient in bulk water, owing to the tortuosity of the pore network). The time to diffuse a distance x is approximately $t = x^2/D$, and the thickness of the stone around the cement core is between 5 and 25 mm, so the diffusion time ranges from approximately 12 to 290 h. To investigate the effect of exposure to the acified brine, the cylinder is cut into slices approximately 1 cm thick and the faces are sealed between sheets of Teflon and stainless steel, so that the brine can only enter radially through the stone.

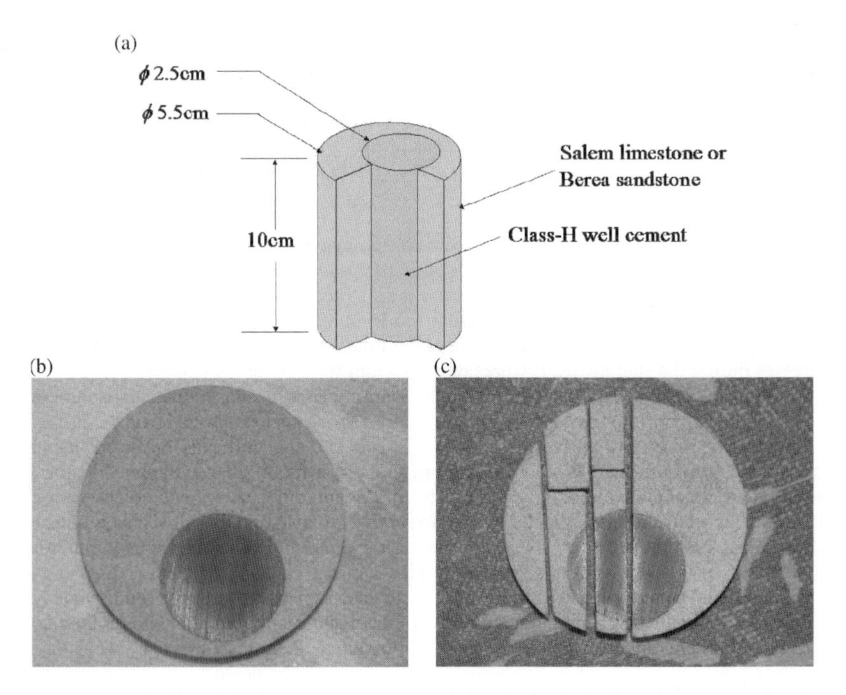

Figure 10: (a) Design of sample consisting of cylinder of stone with off-center hole parallel to the axis, filled with cement paste; (b) slice from cylinder is exposed to acidic brine, then cut into sections for analysis of structure, composition, and properties. From Ref. [45].

The slices are immersed in a static bath with a volume of brine about 35 times greater than that of the samples, at a temperature of 20, 50, or 85 °C. The pH is controlled by saturating the brine with carbon dioxide at ambient pressure, which produces pH 4, then adjusting with NaOH or HCl. If the formation used for storage is primarily quartzitic rock, there will be negligible dissolution of the stone in the brine. To simulate that situation, the samples made with the Berea sandstone are exposed to a brine containing 3% NaCl that is saturated with silica and CO_2 at atmospheric pressure; the pH is adjusted to 3, 4, or 5. On the other hand, if the formation is limestone, there will be substantial dissolution near the point of injection, so we expect that the brine will be saturated with the components of calcium carbonate by the time it reaches any cement-filled wells. Therefore, the samples made with Salem limestone are exposed to a brine saturated with calcium carbonate, with a pH of 4, 5, or 6.

At appropriate intervals of time (depending on temperature and pH), samples are removed for analysis. The slice is cut into several pieces, as shown in Figure 10b, and subjected to the following examinations: (a) the composition is profiled along the radius of the cement core using an electron microprobe; (b) the structure (porosity, mineral distribution, cracking) is examined using an environmental scanning electron microscope (ESEM), which permits imaging without drying of the sample; (c) the hardness is profiled along the radius of the cement core using Vickers indentation, to reveal changes in mechanical integrity of the cement.

Another set of cement samples is cast in the form of cylinders with diameter of 8 or 12 mm and length of about 250 mm. This form is chosen to permit measurement of the permeability by the beam-bending method recently developed in this lab [39]. The saturated cylinder of cement paste is subjected to a sudden deflection in three-point bending, which causes compression of the sample above the midplane and tension below. The liquid in the pores in the upper half of the sample is compressed, while tension is created in the pores in the lower half, resulting in flow of the liquid within the cylinder to eliminate the gradient in pore pressure. The force required to sustain a constant deflection of the cylinder changes as the pore pressure equilibrates, and the rate of equilibration depends on the permeability, so the permeability of the sample can be found by analyzing the force exerted by the cylinder against the pushrod. This method has been applied to cement paste to determine the permeability of mature paste [40,41] and the evolution of permeability and viscoelastic properties in young paste [42]). A typical result is shown in Figure 11, for a sample of Class H cement prepared with a water/cement ratio of 0.38, then aged in 3% brine at neutral pH for 28 days at 50 °C. The permeability is found to be 7.4×10^{-20} m^2 (equivalent to 7.4×10^{-13} m/s); although the permeability is quite low, the measurement was completed in about 1 h.

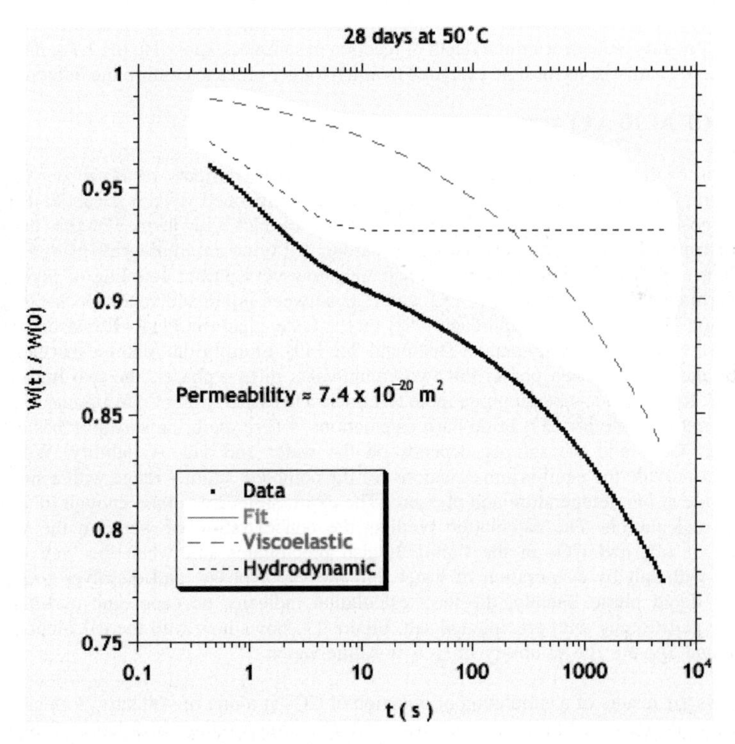

Figure 11: Normalized force, W, on saturated cylinder of Type H cement paste (water/cement ratio = 0.38, aged 28 days at 50 °C in 3% NaCl solution) versus time under load in three-point bending. The data are the symbols and the fit to the theoretical curve is invisible under the data points. The relaxation of the pore pressure is described by the Hydrodynamic relaxation curve (short dashes) and the viscoelastic stress relaxation is described by the Viscoelastic curve (long dashes); the total relaxation is the product of the two. The plateau in the Hydrodynamic curve occurs when the pore pressure reaches atmospheric pressure. Analysis of the kinetics of relaxation using the theory in Ref. [39] yields a permeability of 7.4×10^{-20} m^2.

Cylindrical samples of cement paste containing 0, 6, and 12% bentonite, with dimensions suitable for beam-bending, are exposed to flowing brine with pH values of 3 or 4 at temperatures of 20, 50, and 85 °C. The composition of the effluent is monitored using ICP analysis to determine the rate of attack on the cement. Samples are periodically removed from the bath and subjected to three-point bending to determine changes in permeability and viscoelastic properties. Small sections are then cut from the end of the rod for analysis of compositional and structural changes, using ESEM and microprobe.

These experiments on stone and cement are performed at ambient pressure, because we expect the chemical reactions to be strongly affected by temperature and pH, but weakly dependent on pressure [43,44]. To test this assumption, some experiments are performed at elevated pressure using apparatus developed by Bruant et al. [43]. Figure 12 shows an example of two samples of Class H cement paste that were exposed to brine under 10 MPa CO_2 at 50 °C for 9 days; one sample had been cured at 21 °C and the other at 50 °C for 3 months prior to exposure [45]). There is a reaction rim about 1 mm deep that has turned from gray to red, apparently owing to the change in oxidation state of the iron; within the reaction zone there are several rings with slightly different colors. The ESEM reveals that the structure of the outer layer is more porous, and chemical analysis by energy dispersive X-ray spectroscopy (EDX) reveals extensive removal of calcium. It is expected (e.g. Ref. [46]) that the acid will dissolve the CH in the paste, and then leach calcium from the C-S-H, and that process is evident in this sample. In the sample that had been cured at 50 °C, the Ca/Si ratio in the reacted zone is 1.25, whereas the ratio in the unreacted zone is about 3. The ratio would be about 1.7, if only C-S-H were present, so the leaching has decalcified the C-S-H, as well as removing the CH. Preliminary results indicate a similar depth of reaction in samples exposed to pH 2.7 at 53 °C at ambient pressure, but these comparisons must be extended to much longer times to confirm the influence of pressure.

MODELING OF ACID ATTACK

The rapid attack of the sample in Figure 12 is consistent with preliminary results of our experiments at ambient pressure. Clearly the potential risk to the cement in abandoned wells is great, so it is essential to know the composition of the solution that will come into contact with them. For this reason, detailed simulations of the composition and flow rate of the carbonated brine are in progress using Dynaflow [46], which is a finite element program capable of analyzing flow with proper coupling of pore pressure and strain in the formation. The partitioning of CO_2 and H_2O between liquid and vapor species is achieved in a "flash calculation" based on a new approach [47]) that is faster than the Peng–Robinson [48] approach, and is consistent with the solubility data of Duan and Sun [49]. Equilibrium requires that the fugacity of a component be the same in each phase. For two components in two phases, the two fugacity equations alone determine the two independent phase mole fractions. This separation of partitioning and equilibrium calculations is not possible for CO_2 brine flash calculations, where the brine's salinity has a strong effect on solubility of CO_2, and the salinity depends on the water and salt availability. We use fugacity expressions that provide the equilibrium equations for the complete salinity range with a single fit to CO_2 solubility in brine at one temperature and pressure. The expressions are simple enough to allow a new fit for each flash calculation. The calculation predicts the concentration of water in the vapor and the concentrations of salt and CO_2 in the liquid. It also precipitates salt when the water-rich phase is supersaturated with salt by evaporation of water into the vapor phase, and dissolves solid salt into an undersaturated liquid phase. Finally, the flash calculation indicates no-vapor and no-liquid conditions along with any partitioning with precipitated salt. Figure 13 shows how well the calculation matches the data of Duan and Sun for the solubility of CO_2 in saline water.

Figure 14 shows the results of a simulation of injection of CO_2 at a rate of 100 kg/s (3 megatons/yr) into a formation at a pressure of 10 MPa at temperature 60 °C, corresponding to a depth of about 1 km; the formation is assumed to have a permeability of 10^{-13} m^2 (100 mD) and porosity of 10%. The total dissolved solids in the brine is assumed to be 5 wt%. The saturation of supercritical fluid is near unity at the injection site, but drops rapidly with distance; the advancing front has a vapor saturation that ranges from about 25 to 0% over a distance of about 75 m. The liquid saturation rapidly drops near the site of injection, reaching zero in about 2.5 years, at which point the leading edge of the plume has advanced approximately 1200 m; thereafter, the profile becomes self-similar, so that it is a function only of the Boltzmann variable, $R/\sqrt{t}$, where R is the distance from the injection site and t the time. This profile, which is shown in Figure 15, can be used to determine the size of the plume at any subsequent time or position surrounding the injection site.

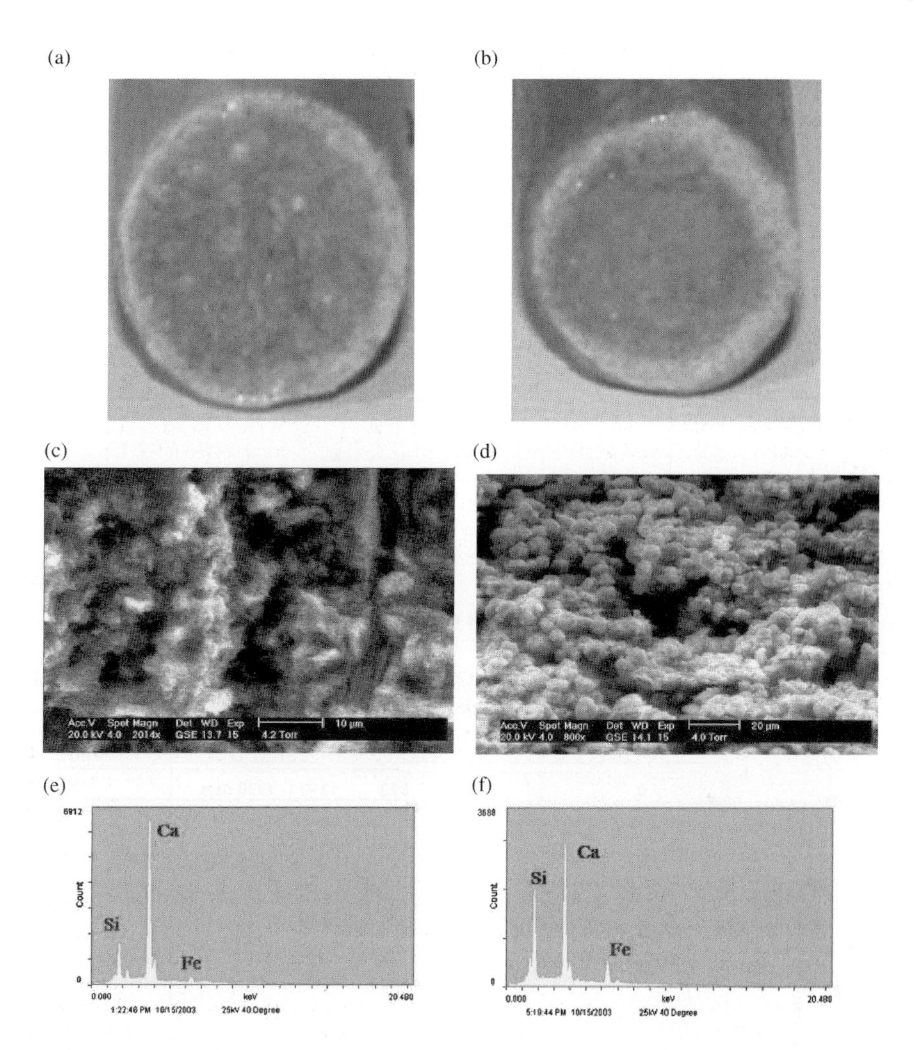

Figure 12: The effect of CO_2-rich water at 10 MPa and 50 °C for 9 days on cement paste samples. (a) Reaction rims on a cement sample cured at 23 °C prior to CO_2 exposure. (b) Reaction rims/color change on a cement sample cured at 50 °C prior to CO_2 exposure. (c) ESEM micrograph of the middle part of cement sample, with a typical open texture CSH due to curing at 50 °C. (d) ESEM micrograph of cement paste subjected to high pressure carbonation, (e) and (f) show EDX composition of cement paste prior and after CO_2 exposure showing a Ca depletion due to the carbonic acid attack, observed as different Si/Ca ratios. From Ref. [45].

To get an idea of the duration of exposure of a well to acid, we multiply the flux of liquid by the mole fraction of dissolved CO_2 in the brine, with the result shown in Figure 16. This plot indicates that a well located approximately 800 m from the injection site would be exposed to the acidified brine after about a year. The peak of the acid flux corresponds to the edge of the plume, where the liquid saturation is high ($>75\%$) and the liquid is saturated with CO_2. At the peak, the flux of aqueous carbonate species is roughly 100 kg/m^2 yr for a period of about a month, then drops to approximately 10 kg/m^2 yr for approximately 10 years. Therefore, as the carbonated brine passes by an abandoned well, the cement will suffer the most

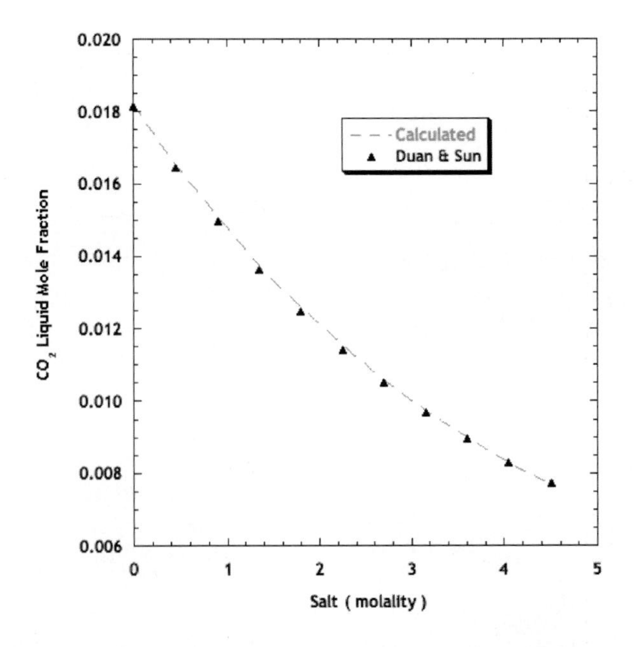

Figure 13: Flash calculation of mass fraction of (a) CO_2 dissolved in liquid phase, along with data of Duan and Sun [49] and (b) water in vapor phase, as functions of salt content.

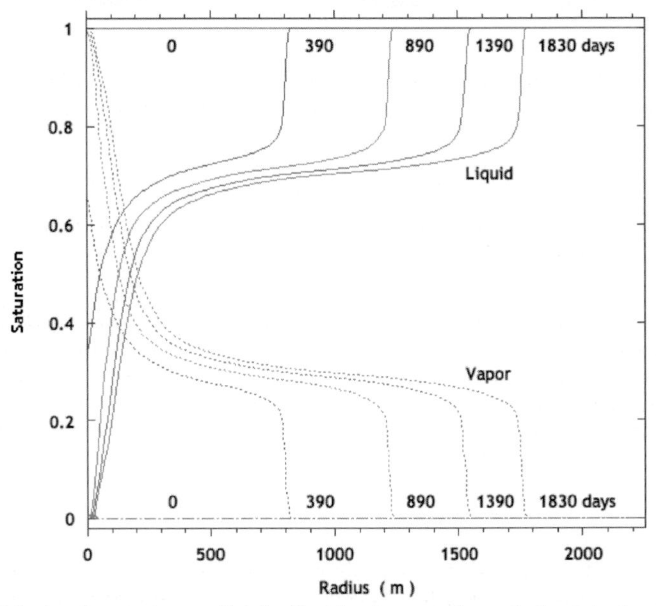

Figure 14: Calculated saturation profiles for liquid and vapor phases during injection of CO_2 at a rate of 100 kg/s (3 megatons/yr) into a formation at a pressure of 10 MPa at temperature 60 °C; the formation is assumed to have a permeability of 10^{-13} m^2 (100 mD) and porosity of 10%. The total dissolved solids in the brine is assumed to be 5 wt%.

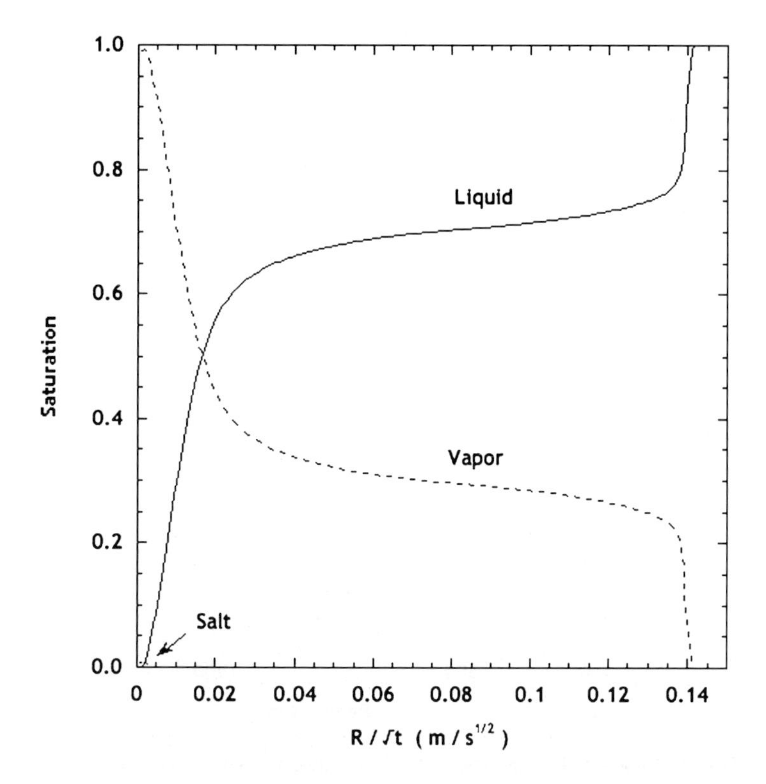

Figure 15: Saturation of pore space with liquid and vapor phases as a function of radial distance from the injection site (*R*) divided by the square root of time; these curves apply from the time when the liquid saturation near the injection site drops to zero (viz., ~900 days, under the conditions of this simulation, which are the same as in Figure 14).

aggressive attack over a period of a few months, but will continue to be exposed to a flow of acidic brine for a decade (under the conditions of this simulation).

The corrosion experiments will provide quantitative information about the depth of attack that could occur during that period. The greatest risk of leakage would occur if there were an annular gap between the cement and the cap rock, or a region of permeable cracks from drilling damage near the well, as shown in Figure 2. In that case, the acidified brine would flow through the annulus for a period of months or years, and could turn a small leak into a large one by dissolving the cement. The kinetics of this process will be explored by simulating flow through an annulus using Dynaflow, together with experimental data on the corrosion rate of cement.

CONCLUSIONS

The potential leakage of CO_2 from a geological storage site through existing wells represents a major concern. An analysis of well distribution in the Viking Formation in the Alberta basin, a mature sedimentary basin representative for North American basins, shows that a CO_2 plume and/or acidified brine may encounter up to several hundred wells. If carbon dioxide is geologically stored in regions, such as this, that have experienced intensive exploration for petroleum products, the acidified brine will come into contact with numerous abandoned wells. Corrosion of the cement that seals the well could lead to rapid leakage, so

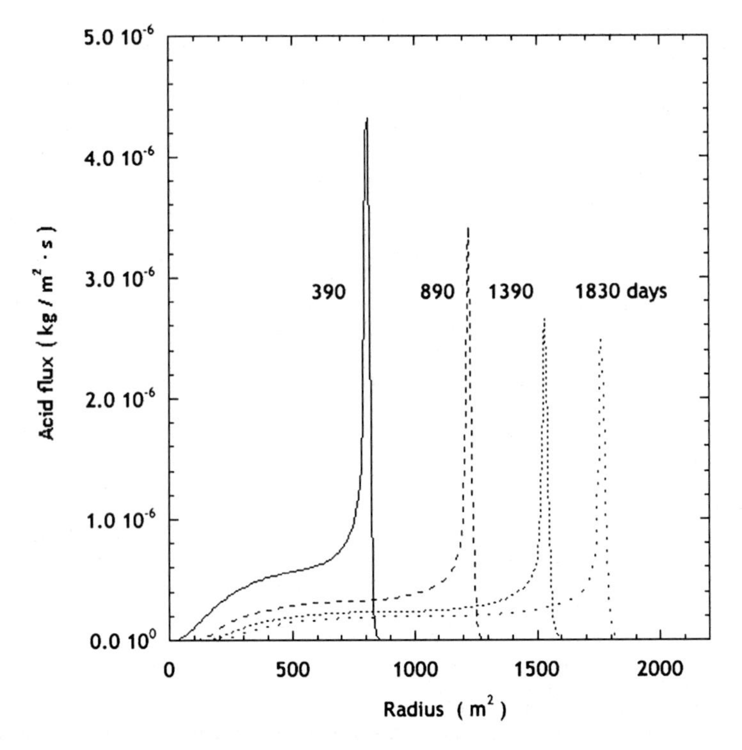

Figure 16: Flux of aqueous carbon species, under the conditions specified in Figure 14. The acidic brine would reach a well 800 m from the injection site in about 1 year; the edge of the plume has a high saturation and acid concentration, but it passes quickly. The acid flux continues for years at a lower intensity (owing to reduced saturation of the pore space with liquid) as the plume expands.

it is essential to determine the duration and intensity of exposure to the acid. Detailed numerical simulations with Dynaflow, incorporating a flash calculation to find the phase distribution and speciation in the brine, indicate that the carbonated brine may spend years in contact with the cement in abandoned wells. Preliminary results from an ongoing experimental study of cement corrosion indicate that the rate of attack is rapid, when the pH of the solution is low, so the risk of leakage will be high if the acidic brine can flow through an annulus and bring fresh acid into contact with the cement.

ACKNOWLEDGEMENTS

This work was supported by BP and Ford Motor Company through the Carbon Mitigation Initiative at Princeton University.

REFERENCES

1. Railroad Commission of Texas, Current and historical oil and gas wells and other existing wells, 1999, http://www.rrc.state.tx.us/index.html.
2. S.E. Gasda, S. Bachu, M.A. Celia, Spatial characterization of the location of potentially leaky wells penetrating a geological formation in a mature sedimentary basin, *Environ. Geol.* (2004) in press.
3. E.B. Nelson, *Well Cementing*, Schlumberger Educational Services, Sugar Land, TX, 1990.
4. S.E. Gasda, CO_2 sequestration into a mature sedimentary basin: determining the capacity and leakage potential of a saline aquifer formation, Master's Thesis, Princeton University, 2004.

5. J.M. Nordbotten, M.A. Celia, S. Bachu, Analytical solutions for leakage rates through abandoned wells, *Water Resour. Res.* **40** (2004) W04204, doi 10.1029/2003WR002997.
6. J. Nordbotten, M.A. Celia, S. Bachu, H.K. Dahle, Analytical solution for CO_2 leakage between two aquifers through an abandoned well, *Environ. Sci. Technol.* (2004) in press.
7. S.E. Gasda, M.A. Celia, Upscaling relative permeabilities in a structured porous medium, in: C.T. Miller, M.W. Farthing, W.G. Gray, G.F. Pinder (Eds.), *Proc. XVth CMWR Conf.*, 13–17 June 2004, Chapel Hill, NC, New York: Elsevier (2004) pp. 793–804.
8. A.S. Altevogt, M.A. Celia, Numerical modeling of carbon dioxide in unsaturated soils due to deep subsurface leakage, *Water Resour. Res.* **40** (2004) W03509: doi: 10.1029/2003WR00284.
9. A.S. Altevogt, P.R. Jaffe, Modeling the effects of gas-phase CO_2 intrusion on the biogeochemistry of variably saturated soils, in: C.T. Miller, M.W. Farthing, W.G. Gray, G.F. Pinder (Eds.), *Proc. XVth CMWR Conf.*, 13–17 June 2004, Chapel Hill, NC, New York: Elsevier (2004) 817–826.
10. S. Wang, P.R. Jaffe, Dissolution of trace metals in potable aquifers due to CO_2 release from deep formations, *Energy Convers. Manage.* **45** (18–19) (2004), 2833–2848.
11. S. Bachu, J.M. Nordbotten, M.A. Celia, Evaluation of the spread of acid gas plumes injected in deep saline aquifers in western Canada as an analogue for CO2 injection in continental sedimentary basins, *Proceedings, Seventh International Greenhouse Gas Technologies Conference,* Vancouver, BC, September 5–9, 2004.
12. E. Lindeberg, Escape of CO_2 from aquifers, *Energy Convers. Manage.* **38S** (1997) S235–S240.
13. T.F. Xu, J.A. Apps, K. Pruess, Reactive geochemical transport simulation to study mineral trapping for CO_2 disposal in arenaceous formations, *J. Geophys. Res. Solid Earth* **10** (B2) (2003) 2071.
14. J.M. Nordbotten, M.A. Celia, S. Bachu, Injection and storage of CO_2 in deep saline aquifers: analytical solution for CO_2 plume evolution during injection, *Transport Porous Media* (2004) in Press.
15. A.M. Neville, *Properties of Concrete*, fourth ed., Wiley, New York, NY, 1997.
16. H.F.W. Taylor, *Cement Chemistry*, second ed., Thomas Telford, London, 1997.
17. H. Jennings, Colloid model of C-S-H and implications to the problem of creep and shrinkage, *Mater. Struct. Concr. Sci. Eng.* **37** (2004) 59–70.
18. T. Hartmann, P. Paviet-Hartmann, J.B. Rubin, M.R. Fitzsimmons, K.E. Sickafus, The effect of supercritical carbon dioxide treatment on the leachability and structure of cemented radioactive waste-forms, *Waste Manage.* **19** (1999) 355–361.
19. A. Hidalgo, C. Andrade, C. Alonso, Role of alkaline reserve in the acidic resistance of cement pastes, in: J. Skalny, J. Gebauer, I. Odler (Eds.), *Material Science of Concrete: Calcium Hydroxide in Concrete*, The American Ceramic Society, Westerville, OH, 2001, pp. 93–111.
20. T.C. Powers, L.E. Copeland, H.M. Mann, Capillary continuity or discontinuity in cement pastes, *J. Portland Cem. Assoc. Res. Dev. Lab.* (1959) 38–48. May.
21. F. Adenot, M. Buil, Modelling of the corrosion of the cement Paste by deionized water, *Cem. Concr. Res.* **22** (2/3) (1992) 489–496.
22. D.P. Bentz, E.J. Garboczi, Modeling the leaching of calcium hydroxide from cement paste: effects on pore space percolation and diffusivity, *Mater. Struct.* **25** (1992) 523–533.
23. C. Carde, R. Crancois, J.-P. Ollivier, Microstructural changes and mechanical effects due to the leaching of calcium hydroxide from cement paste, in: K.L. Scrivener, J.F. Young (Eds.), *Mechanisms of Chemical Degradation of Cement-Based Systems*, E&FN Spon, London, England, 1997, pp. 30–37.
24. B. Delagrave, J. Gerard, Marchand, Modelling the calcium leaching mechanisms in hydrated cement pastes, in: K.L. Scrivener, J.F. Young (Eds.), *Mechanisms of Chemical Degradation of Cement-Based Systems*, E&FN Spon, London, England, 1997, pp. 38–49.
25. B. Kienzler, P. Vejmelka, H.-J. Herbert, H. Meyer, C. Altenhein-Haese, Long-term leaching experiments of full-scale cemented waste forms: experiments and modeling, *Nucl. Technol.* **129** (1) (2000) 101–118.
26. J. Marchand, D. Bentz, E. Samson, Y. Maltais, Influence of calcium hydroxide dissolution on the transport properties of hydrated cement systems, in: J. Skalny, J. Gebauer, I. Odler (Eds.), *Material Science of Concrete: Calcium Hydroxide in Concrete*, American Ceramic Society, Westerville, OH, 2001, pp. 113–129.
27. J. Cowie, F.P. Glasser, The reaction between cement and natural waters containing dissolved carbon dioxide, *Adv. Cem. Res.* **4** (15) (1991) 119–134.
28. D.D. Onan, Effects of supercritical carbon dioxide on well cements, *SPE Technical Paper No. 12593*, Society of petroleum Engineers, Richardson, TX, 1984, pp. 161–172.

29. R.A. Bruckdorfer, Carbon dioxide corrosion in oilwell cements, *SPE Technical Paper No. 15176*, Society of Petroleum Engineers, Richardson, TX, 1986, pp. 531–539.
30. N.B. Milestone, D.A. St John, J.H. Abbott, L.P. Aldridge, CO_2 corrosion of geothermal cement grouts, *Proceedings from the Eighth International Congress on the Chemistry of Cement in Rio de Janeiro,* Brazil, September, 1986, pp. 141–144.
31. N.B. Milestone, T. Sugama, L.E. Kukacka, N. Carciello, Carbonation of geothermal grouts—part 1: CO_2 attack at 150 °C, *Cem. Concr. Res.* **16** (6) (1986) 941–950.
32. N.B. Milestone, T. Sugama, L.E. Kukacka, N. Carciello, Carbonation of geothermal grouts—part 2: CO_2 attack at 250 °C, *Cem. Concr. Res.* **17** (1) (1987) 37–46.
33. J.C. Shen, D.S. Pye, Effects of CO_2 attack on cement in high-temperature applications, *SPE Technical Paper No. 18618,* Society of Petroleum Engineers, Richardson, TX, 1989, pp. 19–28.
34. J.W. Hedenquist, M.K. Stewart, Natural CO_2-rich steam-heated waters in the Broadlands–Ohaaki Geothermal System, New Zealand: their chemistry, distribution and corrosive nature, Transactions from the geothermal resources council, vol. 9, 1985, Part II, August, pp. 245–250.
35. J. Jambor, V. Zivica, Porosity of mortar and its influence on resistance against corrosion caused by aggressive carbon dioxide, *Proceedings from RILEM/IUPAC International Symposium on Pore Structure and Properties of Materials,* Prague, September 18–21, 1973, pp. F83–F93..
36. Y. Ballim, M.G. Alexander, Carbonic acid water attach of Portland cement based matrices, in: R.K. Dhir, J.W. Green (Eds.), *Proceedings from the International Conference on the Protection of Concrete*, September 11–12, 1990, pp. 93–104.
37. E. Revertegat, C. Richet, P. Gegout, Effect of pH on the durability of cement pastes, *Cem. Concr. Res.* **22** (1992) 259–272.
38. V. Zivica, A. Bajza, Acidic attack of cement-based materials—a review: part 2. Factors of rate of acidic attack and protective measures, *Construction Building Mater.* **16** (2002) 215–222.
39. G.W. Scherer, Measuring permeability of rigid materials by a beam-bending method: I. theory, *J. Am. Ceram. Soc.* **83** (9) (2000) 2231–2239.
40. W. Vichit-Vadakan, G.W. Scherer, Measuring permeability of rigid materials by a beam-bending method: III. Cement paste, *J. Am. Ceram. Soc.* **85** (6) (2002) 1537–1544.
41. J.J. Valenza II, G.W. Scherer, Measuring permeability of rigid materials by a beam-bending method: V. Cement paste plates, *J. Am. Ceram. Soc.* (2004) in press.
42. W. Vichit-Vadakan, G.W. Scherer, Measuring permeability and stress relaxation of young cement paste by beam-bending, *Cem. Concr. Res.* **33** (2003) (2003) 1925–1932.
43. D.E. Giammar, R.G. Bruant Jr., C.A. Peters, Forsterite dissolution and magnesite precipitation at conditions relevant for deep saline aquifer storage and sequestration of carbon dioxide, *Chemical Geology*, in press.
44. Y. Soong, A.L. Goodman, J.R. McCarthy-Jones, J.P. Baltrus, Experimental and simulation studies on mineral trapping of CO2 with brine, *Energy Convers. Manage.* **45** (2004) 1845–1859.
45. A. Duguid, R. Bruant, M. Radonjic, G.W. Scherer, M.A. Celia, C. Christopher, The effect of sequestered CO_2 on well cement, presented at Water–Rock Interactions Induced by Reservoir Exploration, CO_2 Sequestration, and Other Geological Storage, Paris, France, November 18–20, 2003.
46. J.-H. Prévost, DYNAFLOW: A Nonlinear Transient Finite Element Analysis Program, Princeton University, Princeton, New Jersey, NJ, 2003, last revision (2004).
47. R. Fuller, J. Prévost, A new approach to flash calculation and its implementation for three phase carbon dioxide, water, and salt mixtures, to be submitted.
48. D.-Y. Peng, D.B. Robinson, A new two constant equation of state, *Ind. Eng. Chem. Fundam.* **15** (1) (1976) 59–64.
49. Z. Duan, R. Sun, An improved model calculating CO_2 solubility in pure water and aqueous NaCl solutions from 273 to 533 K and from 0 to 2000 bar, *Chem. Geol.* **193** (2003) 257–271.

SECTION 3:
STORAGE OPTIMIZATION

STORAGE OPTIMIZATION PREFACE

Jos Maas,

Shell International Exploration and Production
B.V. Kessler Park 1, 2288 GS Rijswijk, The Netherlands

SMV "Optimization" studies have been grouped into three sections. The first two, "Industry Experience with CO_2 Injection" and "Simulation of CO_2 Injection", focus on subsurface aspects of CO_2 storage, monitoring and verification. The tens of years of industry experience with CO_2 injection have led to significant advances in predictive modelling capabilities.

The CO_2 collected is rarely a pure component stream. This has important implications for capture, compression, transport and geological sequestration itself. The section on "Cost Reduction" covers these issues both for the subsurface and surface equipment.

Industry Experience with CO_2 Injection
Geological sequestration of CO_2 has basically four options: injection into

- oil reservoirs,
- gas reservoirs,
- aquifers,
- coal beds.

Of these, the first two can be considered relatively mature with a history of 30 years CO_2 EOR and 15 years of acid gas disposal. CO_2 injection for EOR roughly is applied in two forms: miscible and immiscible projects. In miscible projects, the injected CO_2 becomes miscible with the oil in the reservoir and displaces the oil very efficiently. Residual oil saturations may be very low to close to zero. In immiscible projects, the injected CO_2 dissolves into the oil and oil production is enhanced due to reduced viscosity and oil swelling.

Grigg (Chapter 11) presents a survey of Permian Basin reservoirs used, or considered, for CO_2 injection. Both miscible and immiscible projects are considered. It is of interest to note that over 90% of all crude produced in CO_2 EOR projects globally, is produced in West Texas. Consequently, Grigg's survey can be deemed representative. The summary conclusion is that over the 30 years of CO_2 EOR, the reservoir seals are maintaining their integrity and perform as expected in retaining CO_2. Monitoring and verification of CO_2 flow in geological formations is critical to verification of sequestration, but technical development is in its infancy.

Bachu and Haug (Chapter 12) discuss in situ characteristics of acid gas injection operations in the Alberta Basin. This technology development was driven by the need to remove hydrogen sulphide from gas produced from deep sour gas reservoirs in the Alberta Basin in West Canada. The acid gas that was generated in the process contains H_2S and CO_2. Canadian acid gas is injected at 27 sites into deep saline aquifers and at 21 sites into depleted oil and gas fields. The acid gas injection operations may serve as an analogue for future large-scale CO_2 geological sequestration operations. No safety or leakage incidents have been reported in 15 years since the first acid gas injection in the world started in Alberta.

Simulation of CO_2 Injection
Simulation technology for oil and gas field engineering literally exists for as long as computers have been used in the industry. Development is through an iterative process: building on data from process studies in the laboratory and from field operations, new numerical models are constructed and used to predict future

field behaviour over tens of years. With fields indeed producing tens of years, the models are further improved. This provides a good starting point to study new applications like injection of CO_2 in saline aquifers and coal beds.

Kumar et al. (Chapter 13) present a study of a prototypical CO_2 sequestration project in a deep saline aquifer. They show that the storage capacity of CO_2 in residual gas may well surpass storage as brine-dissolved CO_2 or CO_2 locked into minerals. Moreover, they argue that CO_2 storage in residual gas may mitigate CO_2 escape pathways, because residual gas is not mobile.

Laboratory experiments have shown that coal preferentially adsorbs CO_2 and some other gases over CH_4. Therefore, coal beds that have a natural content of CH_4 may see an enhanced production of CH_4 when other gases are injected. While coal bed methane (CBM) operations are mature technology with today's natural gas in the USA consisting for 7% of CH_4 from coal beds, enhanced CBM (ECBM) operations are still experimental. Presently, only four or five pilots are active in the world with two in the US and one in Canada.

Wo and Liang (Chapter 14) present a simulation study on CO_2 storage in coal beds. Apart from pure CO_2, also a mixture of CO_2 and N_2 as flue gas is considered. Focus was on modelling an actual pilot and interpreting field behaviour. The pilot showed early N_2 breakthrough. Model results suggested that CO_2 seepage could occur if the injection wells were closer than 2 miles from the outcrop.

Cost Reduction

Optimized design for compression and the pipeline system requires accurate predictions of fluid properties, in particular density and water solubility. Thermodynamic models and tools for calculating properties for CO_2 and CO_2-rich mixtures including CH_4, N_2, H_2S and amines, have been verified by Heggum et al. (Chapter 15) as discussed in their contribution on CO_2 conditioning and transportation. The study also addresses gas condensation and prediction of hydrate formation.

Dry and pure CO_2 is essentially non-corrosive. However, CO_2 is rarely produced as a pure component stream. A first concern is water content and associated corrosion, because corrosion issues impact costs. Materials selection for capture, compression, transport and injection of CO_2 is addressed in the contribution by Seiersten and Kongshaug (Chapter 16). Main focus of the work was to determine the corrosion rate of carbon, 13% chromium and duplex steels in liquid/supercritical CO_2 as function of water content at high pressure. It is shown that 20 ppm CO_2 corrosion inhibitor is sufficient to lower the corrosion rate below 0.1 mm/yr at temperatures up to 30 °C and CO_2 pressures up to 72 bar.

The two last contributions in this section address the impact of impurities of SO_x and NO_x in flue gas. The papers are complementary: one looks at surface facility engineering and the other, the subsurface.

The three main components of the study by Sass et al. (Chapter 17) are impact of these impurities on the performance of amine CO_2 separation systems; the evaluation of the phase behaviour of these mixtures on multi-stage compressors; and a literature review of compressed gases to determine corrosivity of these mixtures in the presence of moisture. It is also noted that metals such as arsenic and mercury are derived from coal and oil and may hinder downstream processes.

Bryant and Lake (Chapter 18) discuss the effect of SO_x and NO_x impurities on subsurface CO_2 processes, notably on well injectivity and on the efficiency of CO_2 EOR. They conclude that well injectivity is probably insensitive to possible geochemical alterations prompted by SO_x or NO_x. Moreover, based on correlations developed over several decades of field experience, the impact of these impurities at the levels typical of flue gases are unlikely to affect oil recovery of miscible and immiscible CO_2 EOR floods.

Carbon Dioxide Capture for Storage in Deep Geologic Formations, Volume 2
D.C. Thomas and S.M. Benson (Eds.)

Chapter 11

LONG-TERM CO_2 STORAGE: USING PETROLEUM INDUSTRY EXPERIENCE

Reid B. Grigg

New Mexico Petroleum Recovery Research Center, New Mexico Institute of Mining and Technology, Socorro, NM, USA

ABSTRACT

This study comprised a survey of Permian Basin reservoirs where CO_2 is being injected for enhanced oil recovery, or where CO_2 injection was seriously considered. The focus was the assessment of successes and problems in these projects.

There is significant experience and knowledge in the oil and gas industry to separate, compress, transport, inject, and process the quantities of CO_2 that are envisioned for CO_2 storage. Improvements will occur as incentives, time and fluid volumes increase.

In some cases, certain phenomena that had been noted during waterflood were not included in simulating CO_2 processes—an omission that can prove, and has proven in some cases to be detrimental to the success of the project. When the reservoir is well understood, CO_2 has performed as expected. Also, the thermodynamic phase behavior of CO_2 must be honored in predictive models. High-pressure CO_2 performs as expected: it mobilizes oil, dissolves into brine, and promotes dissolution of carbonates. Brine can become supersaturated with dissolved solids; when pressure drops as it advances through the reservoir, precipitants can form. However, the kinetics of dissolution and precipitation under many reservoir conditions requires further study.

In the time frame wherein CO_2 has been actively injected into geological formations, seals appear to have maintained their integrity and retained CO_2. Monitoring and verification of CO_2 flow in geological formations is critical to verification of storage, but additional research and monitoring demonstration are needed.

INTRODUCTION

The petroleum industry has been injecting carbon dioxide (CO_2) into geological formations for about 50 years. The bulk of this injection, taking place over the last two decades, has not been for storage, but to displace/dissolve oil for increased oil production. Currently, about 39 Mt of CO_2 is being injected into geological formations for the purpose of improving oil recovery (IOR). Though most of the injected CO_2 remains in oil reservoirs, the majority of the floods cannot be considered storage projects because the CO_2 source is from naturally occurring CO_2 reservoirs. Geological formations presently producing high-purity CO_2 for IOR are located in southwest Colorado (McElmo Dome), southeast Colorado (Sheep Mountain), northeast New Mexico (Bravo Dome), and Mississippi (Jackson Dome). Combined, these produce about 29 Mt of CO_2 annually. There are a number of notable exceptions in which the CO_2 source is an industrial by-product. Industrial projects such as the coal gasification plant in North Dakota, fertilizer plants in Oklahoma and Michigan, and hydrocarbon gas purification plants in Texas (Val Verde gas plants) and Wyoming (La Barge gas plant) supply CO_2 to a number of IOR field projects. These can also be considered CO_2 storage projects. These projects are supplying about 10 Mt of CO_2 annually. The experience that operators have obtained from injecting CO_2 in diverse oil-bearing reservoirs and the potential storage capacity of oil reservoirs are resources that ought to be tapped for CO_2 storage knowledge and future storage potential.

During this study, we identified over 135 reservoirs in the United States (USA) into which CO_2 is being injected or has been injected, or the operating company has indicated that there would be a future CO_2 miscible flood. These include:

- 70 field projects that are currently operating.
- 47 terminated projects, of which at least 20 were field demonstration pilots. Most of the others are field projects that have been completed or abandoned.
- 18 projects that have not been started. Of these, about 10 are still listed as future projects and the remainder were announced in the past as future projects but for one reason or another (mergers, changes in company philosophy, downturn in oil prices) were not.

These projects are distributed throughout the continental USA. Table 1 summarizes the number of total and active projects by region and state. In addition, about 25 immiscible CO_2 projects have been initiated in the USA; most began and terminated in the 1980s. Only a few projects persisted into the 1990s. Thus, there are around 160 projects on record that have been studied as prospects for CO_2 injection with about 140 having actually had CO_2 injected into a geological formation. Figure 1 shows the approximate density and location of these projects on a USA map. The injection time varied from a few months for some pilots to about 30 years for some field projects. These numbers do not include fields considered for CO_2 injection but never announced outside the company as an imminent project.

TABLE 1
CO_2 MISCIBLE PROJECT LOCATIONS IN THE UNITED STATES WITH THE NUMBER OF TOTAL AND ACTIVE PROJECTS LISTED BY STATE

Region	State	Total projects	Active projects
East	Pennsylvania	2	0
	West Virginia	2	0
Midwest	Kansas	1	1
	Michigan	2	2
	North Dakota	1	0
South	Alabama	1	0
	Louisiana	10	0
	Mississippi	4	3
Southwest	New Mexico	8	3
	Oklahoma	6	5
	Texas	80	47
West	California	2	0
	Colorado	2	1
	Montana	1	0
	Utah	3	3
	Wyoming	11	6

Of the miscible tests, about 65% of the total projects and 70% of the current operating projects are located in the Permian Basin. At least 30 different organizations have operated CO_2 projects in the Permian Basin. Projects have been performed in sandstone, limestone, and dolomite reservoirs, with more than half being located in San Andres formation. The other projects are found in more than a dozen different formations. Because of the concentration of CO_2 projects in the Permian Basin, this region was the focus of this.

This type of study becomes more difficult to conduct as time progresses, because of mergers, property sales, and personnel changes that will result in lost or limited access to valuable information. Several fields have

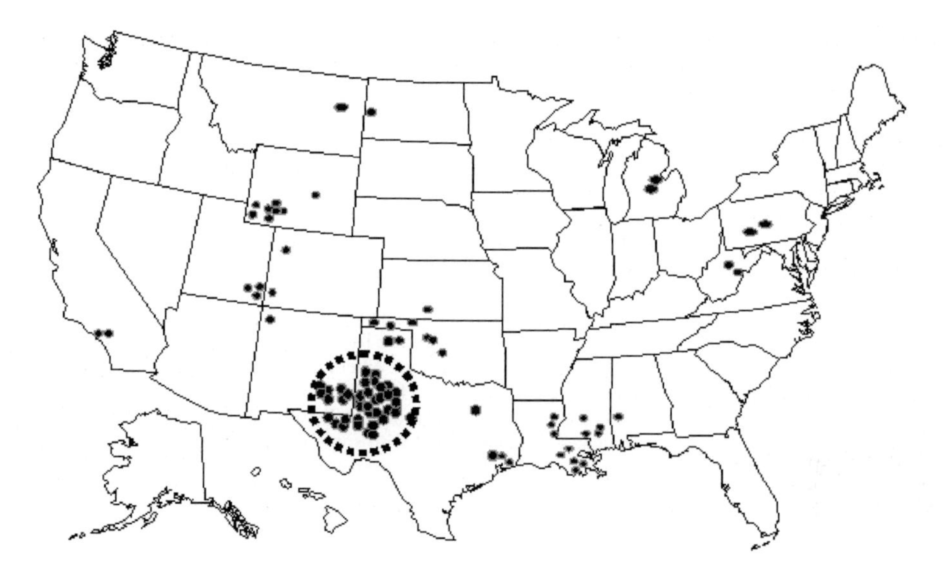

Figure 1: Map of the USA with black dots indicating location and approximate density of CO_2 injection projects for IOR in the USA. The study area, Permian Basin, is indicated by the circle.

changed operators since termination and often the new operators have little incentive to relay information on previous operations. In some cases information was obtained from earlier publications and interaction with engineers from before the operators were changed.

This study was not carried out as a simple survey, but included visits to the engineering center sites and archives of the appropriate operating companies to gather information and obtain clarifications. The goal was 100% coverage, with a minimum goal of 75% since it was not assured that all operators would participate. This survey had 80% participation from the operators that cover about 60% of the fields. Two operators that did not participate have considerable holdings.

STUDY METHODOLOGY

Steps that were taken to identify and analyze CO_2 injection project in the Permian Basin included:

1. Identification of CO_2 field projects from the biannual EOR Survey published in the *Oil & Gas Journal* in each even year since 1978 [1–13]. These surveys always list present projects, including pilot and full-scale projects and often mention announced future projects and projects terminated since the last publication.
2. Identification of those projects in the lists mentioned above which are within the Permian Basin (the defined study area).
3. A literature search on the projects identified above, most of which was available from the Society of Petroleum Engineer conferences and publications.
4. Selection of a number of parameters, items, and questions to answer for each project.
5. Gathering information from the literature of the items listed in "4" and entering them into spreadsheets. Each spreadsheet was then sent to a representative of the operating company, usually the field or project engineer, for review and additions.
6. A facility visit with each project engineer that could accommodate the survey team.
7. Analysis of information in hand in order to aid those considering CO_2 injection into a geological formation.

8. Finally, interpretation of the information obtained from each engineer and literature source. Note that this information was based on data gathered from reliable sources; it cannot be construed as an official stance or opinion of the production company.

RESULTS AND DISCUSSION

Below is a summary of the data we have in hand.

1. Over 160 CO_2 projects were initially identified in the United States in 16 states.
2. Over 100 projects were identified in Texas and New Mexico. Among these we found some that had not been CO_2 flooded, nor did the operator ever intend it to be a CO_2 project, as in a number of early projects outside the Permian Basin. We also combined some pilot project with a later field projects or several pilot projects in the same field into a single one. Table 2 contains a list of projects that were considered in this study. Among these, some had little available information. Where present project operators declined to participate, results from earlier work were considered [14]. Also listed in Table 2 are the state and operating status of projects. Found in an earlier publication is a list of Society of Petroleum Engineer published papers related to the indicated reservoir, most with some mention of CO_2 injection [15].
3. A spreadsheet of two to four pages for each reservoir was prepared, though not included in this paper.

Listed below are some general observations from this study. Some of these probably seem intuitive. More details are provided in the following sections.

1. Many of the problems that have been encountered could have been avoided or at least anticipated and minimized with better reservoir characterization. Such problems could become more severe when CO_2 is injected into a geological formation that had not been flooded and/or studied extensively previously. Generally, produced petroleum reservoirs are extensively studied formations with a fair amount of detail developed from their production history. These reservoirs still present challenges when starting injection of a fluid such as CO_2.
2. The flow paths of the CO_2 are not always well understood.
3. Retention of CO_2 is significant in most reservoirs.
4. CO_2 injectivity is often lower than expected and in many cases is a critical parameter when considering economics.
5. In one reservoir that has been CO_2 flooded and is about to be plugged and abandoned, the produced CO_2 is being injected into a brine aquifer.
6. In many cases, CO_2-saturated water seems to be reacting with formation rock and might be at least part of the cause of significant formation injectivity changes.
7. Reservoir engineers working on these projects believe that there is still much to learn with regard to the long range implications of CO_2 injection and storage in geological formations.

The following subsections summarize responses to questions on parameters that were included in the survey sent to engineers for each CO_2 injection project and from subsequent discussions. Very few respondents answered all questions.

TYPES OF RESERVOIR ROCK

Table 3 lists the rock types with the number of reservoirs reporting the indicated rock type(s). For example, out of 81 reservoirs reporting rock types, 43 reported dolomite only as a rock type and 17 others had a mixture of dolomite and one of the other rock types. Thus, dolomite is the principal reservoir type being flooded in the Permian Basin CO_2 floods. Limestone and sandstone are about equal. Of the 81, 72 are all or partly carbonate (dolomite, limestone, tripolite). Thus, the general statements in this report are for carbonate reservoirs.

Types of Seals

The number of responses to this inquiry was relatively low. Of the 12 responses to the question on type of seals, four indicated that the seal was structural, two seals were salt barriers, and six seals were evaporites or anhydrites. The integrity of the seal is vital for long-term storage. Reservoir engineers were the principal

TABLE 2

CO_2 FIELD PROJECTS IN THE PERMIAN BASIN, USA LISTED BY UNIT NAME, STATE, CURRENT OPERATING STATUS, AND RESERVOIR FLUID VOLUME

Unit name	State	Current status	Total reservoir fluid (vol. $\times 10^6$ m^3)
Adair San Andres	Texas	Operating	46.4
Anton Irish	Texas	Operating	134.8
Bennett Ranch	Texas	Operating	143.1
Brahaney	Texas	Future	4.0
Brahaney Plains	Texas	Future	4.0
Cedar Lake	Texas	Operating	49.1
Central Vacuum	New Mexico	Operating	13.0
Cogdell	Texas	Operating	16.1
Cordona Lake	Texas	Operating	38.2
Dollarhide (Clearfork "AB")	Texas	Future	39.4
Dollarhide (Devonian)	Texas	Operating	49.4
East Ford	Texas	Operating	11.1
East Huntley	Texas	Terminated	6.7
East Penwell (SA)	Texas	Operating	3.3
East Vacuum	New Mexico	Operating	72.0
El Mar	Texas	Operating	80.6
Ford Geraldine	Texas	Terminated	26.6
Garza	Texas	Terminated	20.0
GMK South	Texas	Operating	7.0
Goldsmith	Texas	Field demonstration	4.8
Hanford	Texas	Operating	7.6
Hanford East	Texas	Operating	2.2
Hansford Marmaton	Texas	Terminated	8.9
Jess Burnes	Texas	Never started	1.3
Kingdom Abo	Texas	Terminated	19.7
Leamex	New Mexico	Pilot terminated	2.4
Levelland	Texas	Pilots terminated	205.6
Levelland	Texas	Never started	26.4
Loco Hills	New Mexico	Pilot terminated	14.5
Mabee	Texas	Operating	92.5
Maljamar Pilot & Field	New Mexico	Terminated	44.5
McElroy	Texas	Terminated	22.9
McElroy	Texas	Field demonstration	1073.3
Means (San Andres)	Texas	Operating	89.7
Mid Cross-Devonian	Texas	Operating	14.8
North Cowden	Texas	Pilots terminated	1.7
North Cross (Crossett)	Texas	Operating	27.0
North Dollarhide	Texas	Operating	17.3
North El Mar	New Mexico	Never started	24.6
North Farnsworth	Texas	Terminated	3.5
North Hansford Cherokee	Texas	P&A	13.5
North Hobbs	New Mexico	Future	61.7
North Van Rueder	Texas	Never started	7.9
North Ward Estes	Texas	Terminated	596.8
Philmex	New Mexico	Pilot terminated	3.2
Ranger Lake	New Mexico	Never started	4.0

(*continued*)

TABLE 2
CONTINUED

Unit name	State	Current status	Total reservoir fluid (vol. × 10^6 m^3)
Rankin	Texas	Pilot Terminated	1.0
Reeves	Texas	Never started	63.0
Reinecke	Texas	Operating	4.6
Robertson (Central and N.)	Texas	Future	21.5
Russell	Texas	Never started	59.5
Sable	Texas	Terminated	4.3
SACROC	Texas	Operating	795.2
Salt Creek	Texas	Operating	177.6
Seminole-Main Pay	Texas	Operating	274.2
Seminole-ROZ Phase 1	Texas	Operating	14.6
Sharon Ridge	Texas	Operating	136.2
Slaughter (started June 1989)	Texas	Operating	6.2
Slaughter (started May 1985)	Texas	Operating	63.3
Slaughter Alex Estate	Texas	Operating	9.9
Slaughter Central Mallet	Texas	Operating	18.6
Slaughter Estate & Pilot	Texas	Operating	54.4
Slaughter Frazier	Texas	Operating	3.5
Slaughter HT Boyd Lease	Texas	Operating	96.5
Slaughter Sundown	Texas	Operating	86.2
South Cowden	Texas	Operating	3.5
South Cowden (Emmons)	Texas	Future	15.6
South Cross (Crossett)	Texas	Operating	15.6
South Huntley	Texas	Terminated	11.8
South Welch & Pilots	Texas	Operating	58.3
Spraberry Trend	Texas	Pilot	2654.5
State 35 Unit (Hale Mable)	New Mexico	Operating	5.1
T-Star	Texas	Operating	4.3
Twofreds-East & West	Texas	Operating	21.3
University Waddell	Texas	Terminated	6.8
VGSAU	New Mexico	Future	13.0
Wasson	Texas	Operating	8.7
Wasson Cornell	Texas	Operating	40.9
Wasson Denver	Texas	Operating	564.7
Wasson ODC & Pilot	Texas	Operating	173.1
Wasson South	Texas	Operating	70.4
Wasson Willard & Pilot	Texas	Operating	166.3
Wellman	Texas	Terminated	33.1
West Brahaney	Texas	Terminated	2.5
West Welch	Texas	Operating	4.8

respondents to this study. Many do not worry about the reservoir seal as long as it is sufficient to trap crude oil. It seems to be assumed it will trap CO_2. If the oil contains significant amounts of methane and the lighter hydrocarbons it is expected to trap the CO_2, which is similar in molecular size. It is concluded that for at least the foreseeable future, or life of the CO_2 project, that the seal will be maintained. CO_2 IOR projects consider decades of containment compared to a minimum of hundreds or preferably thousands of years when considering long-term storage.

TABLE 3
INVENTORY OF ROCK TYPES IN CO_2 IOR OPERATIONS IN THE USA PERMIAN BASIN

Rock type	Dolomite	Sandstone	Limestone	Tripolite
Dolomite	43			
Sandstone	6	9		
Limestone	10	1	7	
Tripolite	1	1	0	3

Injectivity
In many IOR injection projects, injectivity is a key parameter dictating the success or failure of the process. In many reservoirs, injectivity has been lower than expected. When injecting water alternating with gas (WAG), brine and/or CO_2 injectivities are often lower than the waterflood injectivity. This decrease in injectivity is more dramatic and persistent as predicted when considering relative permeability effects of multiphase flow. As shown in Table 4, the majority of operators indicated changes in injectivity after CO_2 injection. For those that changed, most of them decreased. There were no reports of water injectivity increasing once CO_2 injection occurred. The decreases ranged from 10 to 100% decrease. In one case after CO_2 injection, no brine could be injected during the water half-cycle. The problems seemed to be greater in the carbonates, especially dolomite. The average decrease was in the 40–50% range.

TABLE 4
INJECTIVITY CHANGES AFTER START OF WAG, COMPARED TO WATERFLOOD INJECTIVITIES IN IOR CO_2 MISCIBLE FLOOD OPERATIONS IN THE PERMIAN BASIN, USA

Injectivity changes	Brine	CO_2
None noted	4	5
No comment	7	9
Changed (decreased for all brine and about half the CO_2)	16	13

During the CO_2 half-cycle the change from waterflood injectivity was not as severe as during brine half-cycles. Because of the lower viscosity of CO_2 (at reservoir conditions at least 90% less than the brine) one might expect the injectivity during the CO_2 half-cycle to be much higher than the waterflood injectivity. In most cases brine saturation remains sufficient to reduce the relative permeability close to that of waterflood injectivity, but even with this, CO_2 injectivity is expected to be higher than brine. For the projects reporting CO_2 injectivity changes, the changes ranged from a decrease of 40% to an increase of 30% with an average near-zero change from waterflood injectivity. This result is disappointing when an increase was generally expected. Seven projects reported a decrease without indicating the magnitude.

One might ask what it means when a respondent indicates no injectivity change was noted or had no comment. In discussions with engineers, this generally meant that the desired injection rates were maintained, whether or not injectivity changed. Thus, there could be a significant decrease in injectivity that was not noted because injectivity was still sufficient to achieve desired injection rate.

In one reservoir there were no injection problems in one area of the field, but in another area the brine injectivity decreased, and in the third area both CO_2 and brine injectivity decreased. The difference among the three areas of the reservoir was that they had relatively high, medium, and low permeability, respectively. This is an indication that if a reservoir is operating a waterflood near the injection limit and it is converted to a CO_2 flood, there is a high probability that the project will be injection limited.

CO_2 Reservoir Retention

Reservoir CO_2 retention is a key storage parameter. In an IOR project, CO_2 retention is the quantity of the purchased CO_2 that remains in the reservoir at the present time and ultimately remaining in the reservoir at the time the reservoir is plugged and abandoned. One has to be careful not to include recycled gas when determining the retention quantities. The objective of IOR is not to maximize reservoir CO_2 retention rates, but to maximize profit. The maximum retention might correspond to the maximum sweep efficiency and thus maximum oil production, but often this is not the optimum economical scenario. In several reservoirs that were relatively homogeneous, the sweep was too efficient and the production rate was too slow and/or the timing of significant oil production increases took too long to obtain the desired rate of return on the capital investment. It appears that sufficient heterogeneity in the reservoirs is necessary for some relatively early oil recovery to recoup investment. Then, after breakthrough, action can be taken to mitigate the early breakthrough caused by heterogeneity and continue oil recovery while minimizing CO_2 production.

As we look at CO_2 storage in depleted petroleum reservoirs, heterogeneity in both producing petroleum reservoirs and aquifers will have a similar effect. A need for the economy of high injectivity over maximum storage efficiency of the reservoir may be an important trade-off.

Many of the floods in the Permian Basin are not mature enough to predict final retention. Retention was reported for eight reservoirs and ranged from 38 to 100% with an average of 71%. The reservoir that had 100% retention was a pilot. Respondents speculate that insufficient CO_2 was injected and insufficient time was allowed to detect CO_2 breakthrough. After 10 years they have not seen CO_2 in the produced gas above background concentrations. In mature reservoirs retention was listed as low as 38% of the total CO_2 injected, including recycled volumes. This is the estimated total amount of CO_2 that does not return to the surface once injected, thus not recycled. Essentially 100% of the purchased CO_2 is still in the system. Practically, 100% of the fluid will be stored in the reservoir unless a reservoir blowdown is instigated. To date, six other projects reported retentions in the range of 60–90% of the CO_2 remaining in the reservoir, with an average of 71% retention. These estimates were from reservoirs that had been undergoing CO_2 injection from 5 to 30 years. Most of the projects are early in their lifecycles and thus not reporting ultimate retention.

CO_2 Distribution

In some cases CO_2 is not going where it had been expected to go and engineers made statements such as

1. CO_2 left the intended target area.
2. CO_2 went into upper and lower zones with much of the reservoir in between untouched. Sweep efficiency was less than what had been expected.
3. CO_2 was not detected at a producer after 2 years of injection. It is believed CO_2 had greater sweep—both vertical and horizontal—than expected; thus not enough time and insufficient injection occurred for a successful project.

Each of the three comments above demonstrates that a better understanding of the reservoir would improve predictions, and the project's technical and economic success.

Monitoring/Detection Methods

The most common method used to determine CO_2 movement in IOR projects is tracking produced gas composition. Logging of pilot project observation wells has also been one of the more successful methods used to detect CO_2 movement and saturation changes. Monitoring tools being considered for widespread monitoring are seismic methods that include crosswell tomography, 3D and 4D seismic, and microseismic. Each method has been used with varying levels of success. Cases of the successful use of seismic tools were cited, but respondents were not sure if the signal changes were activated by fluid saturation changes or formation deformation. If the formation deformation tracks fluid movement, it will not be of consequence, but if the deformations do not track fluid movement, it will be difficult to interpret. In one test, seismic changes were noted in a formation several hundred meters above the injection zone. It was feared that CO_2 was flowing into a higher zone that could potentially cause problems. Perforations into the zone in question found no CO_2, no compositional changes in reservoir fluids, or any pressure changes. The reason for the anomaly is unknown. Thus, more work is required in the area of seismic monitoring.

Losses Out of Zone

It is desirable to know how successfully CO_2 is delivered to the intended zone. Generally, CO_2 was retained in the formation intended and could be accounted for within engineering accuracy. Many respondents noted that CO_2 was going into zones that were in communication with the injection zone. Generally, CO_2 is less dense than liquids in the reservoir and might be expected to migrate upwards in the formation, but CO_2 has been found migrating into water or residual oil zones below the zone of interest. This is probably caused by several phenomena, e.g. diffusion and brine density caused by dissolved CO_2. Diffusion is thought to be slow compared to injection fluid flow rates, but especially for long-term storage, diffusion may be important.

Unexpected fractures, thief zones, and loss out of the flanks of the structure have been suspected as culprits of CO_2 loss. However, often the ratio of injection to production fluid has not been tracked as closely as it could be and water production is not tracked as closely as oil or gas, resulting in mass balance uncertainty.

What has Gone Well?

To provide some idea of what petroleum producers look for when considering success of a project, engineers were asked, "What had gone well in the project?" The foremost concern was the timing of the oil response (see Table 5). Most modeling and engineering studies center on optimizing and predicting oil response. Respondents mentioned injectivity in a couple of cases, confirming this as a concern in many projects.

TABLE 5
WHAT HAS GONE WELL IN IOR CO_2 MISCIBLE FLOOD OPERATIONS IN THE PERMIAN BASIN, USA

Response	Number
Oil response at or above that predicted	20
Project performed well (usually oil response was at or above expectations)	5
Injectivity is sufficient	2
Gas production within designed limits	4
Other: minimum asphaltene deposit, cost in line with predictions, lower corrosion than expected, acceptable well failure rate	4

What has not Gone Well?

The question of what has not gone as well as expected in the project was also asked. The answers again provide some idea of parameters to consider when designing a project. Oil response time and magnitude were premier among concerns and were disappointing in a number of CO_2 miscible floods. The second most undesirable situation (Table 6) often occurs with low oil response, i.e. early CO_2 breakthrough and high gas production. At essentially the same level of negative response was low injectivity that also resulted in a low or late oil response. Scaling/deposition was identified in a number of responses. Deposition in the reservoir can result in increased CO_2 retention as well as modified injectivity.

TABLE 6
WHAT HAS NOT GONE WELL IN IOR CO_2 MISCIBLE FLOOD OPERATIONS IN THE PERMIAN BASIN, USA

Response	Number
Low and/or late oil response	19
CO_2 early breakthrough or high cycling, high GOR, conformance	13
Low injectivity	12
Scaling	7
Other: corrosion, cost too high, completion problems, old wellbores	6

Attempted Remediation and Success Rates

Methods employed to remedy problems mentioned in the previous sections are listed in Table 7. Most of the remediation methods were used to reduce CO_2 production, to improve on CO_2 reservoir sweep efficiency, or to increase injectivity. WAG management to control/improve (decrease) CO_2 production while maintaining or increasing oil production has generally made improvements. Control of conformance with gels, foams, or squeeze jobs has had fair technical success, but with a concern for expense. Attempts to improve injectivity have met with temporary or no success.

TABLE 7
REMEDIATION ACTIONS IN IOR CO_2 MISCIBLE FLOOD OPERATIONS IN THE PERMIAN BASIN, USA

Responses[a]	Number
WAG management	8
Conformance control (foam, gel, etc.)	7
Cement squeeze	4
Acid stimulation	5
Scale inhibitor	2
Other: horizontal well, infill drilling, increase reservoir pressure, increase production	5

[a] Remediation actions listed above may not increase storage, but some would be effective in increasing injectivity and thus might improve on the economics of CO_2 storage.

What would You do Differently if Starting Over, or for Another Flood?

Hindsight does not profit a company unless it is used to improve subsequent projects. Today, almost without exception, new CO_2 floods start with a large CO_2 slug (Table 8) and do not switch to WAG before CO_2 breakthrough or a targeted amount such as a 20% pore volume slug size has been injected. The large CO_2 slug has the advantage of minimizing the time of the first occurrence of a significant oil response, as well as reducing the impact of reduced injectivity in the brine half-cycle by delaying brine injection as long as practical. Additionally, possible reduction in CO_2 injectivity is delayed in subsequent CO_2 half-cycles. Again, reservoir characterization was near the top of concerns by a number of individuals.

Mysteries of the System

Project engineers were asked if they had any unresolved problems (Table 9). These are items that, if better understood, would improve the project. This could mean improved profits and in some cases a modification

TABLE 8
FUTURE IMPROVEMENTS SUGGESTED BY OPERATORS OF IOR CO_2 MISCIBLE FLOOD OPERATIONS IN THE PERMIAN BASIN, USA

Responses	Number
Lower CO_2 and lower surface facilities cost, and effective government incentives	12
Start with a larger CO_2 slug, more aggressive with CO_2	10
Better reservoir characterization or honor waterflood characterization	9
Start CO_2 earlier in waterflood	2
Conformance control	2
Horizontal and infill wells, patience, lower reservoir pressure, stimulate early	5

of the project area. Thirty-one of the responses (~90% of the total) indicated a desire to better understand fluid flow patterns in the reservoir, reservoir characterization, and injectivity, which all concern the interconnection of reservoir petrophysics, fluid flow, and fluid-reservoir rock interactions.

TABLE 9
UNSOLVED ISSUES IN IOR CO_2 MISCIBLE FLOOD OPERATIONS IN THE PERMIAN BASIN, USA

Responses	Number
Fluid flow patterns in the reservoir	12
Reservoir characterization	12
Injectivity	7
Scaling, asphaltenes, conformance, equipment	4

Research Focus

Petroleum producers want improved sweep and productivity/injectivity to increase reservoir efficiency. The first three items in Table 10 are relevant to long-term storage of CO_2. First, an understanding of the fluid flow patterns in the reservoir is critical. This is connected to the second response of injection and production rates. The third response shows that, even though it is important for IOR to monitor CO_2, the ability to monitor the CO_2 plume is essential to understanding and predicting long-term CO_2 storage.

TABLE 10
RESEARCH FOCUS SUGGESTED BY OPERATORS OF IOR CO_2 MISCIBLE FLOOD OPERATIONS IN THE PERMIAN BASIN, USA

Responses	Number
Sweep/profile/conformance	10
Productivity/injectivity	8
Monitoring	3
Predictions, mechanism, improve economics of known technology	8

Safety

Safety is an item that was not mentioned in the discussions. In the author's experience, more than 20 years in the area of CO_2 production, transportation, and injection into geological formations of significant quantities of CO_2 have passed without a fatality. Since CO_2 is not flammable and is much less toxic than many other fluids that are transported in great quantities and at high pressure, it is well within the capability of the industry to separate, compress, transport, inject, and process enormous quantities of CO_2 at acceptable safety levels for the public.

CONCLUSIONS/RECOMMENDATIONS

Listed below are major lessons from CO_2 injection into geological formation for IOR that are most applicable to CO_2 storage.

1. Significant experience and knowledge in the industry exists to separate, compress, transport, inject, and process the quantities of CO_2 that are envisioned for CO_2 storage. As the volume of injected CO_2 increases, significant technological improvements are expected.
2. Monitoring and verification of CO_2 flow in geological formations is in the infancy of its development.
3. Experience has shown that CO_2 goes where expected. The challenge is developing detailed reservoir characterizations and honoring them. In some cases, phenomena have been noted during waterflood, but not included when simulating the CO_2 oil recovery process, resulting in surprises during the project that could have been avoided. The phase behavior of CO_2 must be honored also.
4. CO_2 does what is expected: mobilizes oil, dissolves in brine, and promotes dissolution of carbonates. Saturated brine will become supersaturated as it flows away from the injector, dropping the pressure and resulting in precipitation. The kinetics of these processes under a wide range of reservoir conditions requires further studies.
5. In the short geological timeframe that CO_2 has been actively injected into geological formations for IOR, seals generally are retaining the CO_2 subsurface. Oil reservoir seals, to date are generally performing as expected, but it must be remembered that a maximum of several decades is short compared to the longer time periods required for effective CO_2 storage.

ACKNOWLEDGEMENTS

We would like to express our gratitude for the support of the members of the CO_2 Capture Project, for their financial support and the opportunity to work with them. We express our thanks also for the support of the State of New Mexico for its financial support through the New Mexico Petroleum Recovery Research Center, a research division of New Mexico Institute of Mining and Technology. Finally we appreciate the effort of the many project engineers from the various CO_2 injection projects that took their time to review their projects, answer many questions, and offer their support to this effort.

REFERENCES

1. G. Moritis, Worldwide EOR Survey, *Oil Gas J.* **100** (15) (2002) 45.
2. G. Moritis, Worldwide EOR Survey, *Oil Gas J.* **98** (12) (2000) 39–61.
3. G. Moritis, Worldwide EOR Survey, *Oil Gas J.* **96** (16) (1998) 49–77.
4. G. Moritis, Worldwide EOR Survey, *Oil Gas J.* **94** (16) (1996) 39–61.
5. G. Moritis, Production report, *Oil Gas J.* **92** (39) (1994) 51–69.
6. G. Moritis, Annual production report, *Oil Gas J.* **90** (16) (1992) 51–79.
7. G. Moritis, Annual production report, *Oil Gas J.* **88** (17) (1990) 49–82.
8. L.R. Aalund, Annual production report, *Oil Gas J.* **86** (16) (1988) 33–73.
9. J. Leonard, Production/enhanced recovery report, *Oil Gas J.* **84** (15) (1986) 71–101.
10. J. Leonard, Annual production report: enhanced oil recovery, *Oil Gas J.* **82** (14) (1984) 83–105.
11. J. Leonard, Annual production survey, *Oil Gas J.* **80** (14) (1982) 139–159.

12. S.L. Matheny Jr., Production report, *Oil Gas J.* **78** (13) (1980) 79–124.
13. D. Noran, Annual production issue, *Oil Gas J.* **76** (13) (1978) 113–140.
14. R.B. Grigg, D.S. Schechter, State of the industry in CO_2 floods, *Paper SPE 38849 Presented at the 1997 SPE Annual Technical Conference and Exhibition* held in San Antonio, October 5–8.
15. R.B. Grigg, Improving CO_2 efficiency for recovering oil in heterogeneous reservoirs, *First Annual Technical Report*, DOE Contract No. DE-FG26-01BC15364, January 2003.

Carbon Dioxide Capture for Storage in Deep Geologic Formations, Volume 2
D.C. Thomas and S.M. Benson (Eds.)

Chapter 12

IN SITU CHARACTERISTICS OF ACID-GAS INJECTION OPERATIONS IN THE ALBERTA BASIN, WESTERN CANADA: DEMONSTRATION OF CO_2 GEOLOGICAL STORAGE

Stefan Bachu and Kristine Haug

Alberta Energy and Utilities Board, Edmonton, AB, Canada

ABSTRACT

Acid-gas injection in the Alberta basin in western Canada occurs over a wide range of subsurface characteristics, acid gas compositions, and operating conditions. The subsurface characteristics of the injection sites are representative for compacted continental sedimentary basins, like those in the North American mid-continent. No safety or leakage incidents have been reported in the 15 years since the first acid-gas injection operation in the world started in Alberta, and this record indicates that acid-gas injection is a mature technology that can be applied elsewhere in the world. Furthermore, these acid-gas injection operations constitute a commercial-scale analogue for future large-scale CO_2 geological storage operations to reduce CO_2 emissions into the atmosphere from large CO_2 point sources. This review of the subsurface characteristics of the acid-gas injection operations in western Canada provides data and information that can be used in future studies for site selection.

INTRODUCTION

Deep sour-gas reservoirs in the Alberta basin in western Canada contain hydrogen sulphide (H_2S) that has to be removed from the produced gas to meet pipeline and market specifications, generating in the process acid gas (a mixture of H_2S and CO_2). Since 1989, regulatory agencies in western Canada require that gas plants with a sulphur throughput of more than 1 t/d recover the sulphur from the acid-gas stream rather than burn it in flare stacks or incinerators, as previously done if sulphur-recovery technology could not economically remove the sulphur. Because desulphurization is uneconomic in a weak market dominated by recovered sulphur, and the surface storage of the produced sulphur constitutes a liability, increasingly more operators in the Alberta basin are turning to acid-gas disposal through injection into deep geological formations. Compared to other options, acid-gas injection has less environmental consequences than sulphur recovery (where leaching of sulphur piles can lead to groundwater contamination) or flaring (which essentially substitutes SO_2 for H_2S in the atmosphere, as well as releasing CO_2). In addition, although the purpose of acid-gas injection is to dispose of H_2S, significant quantities of CO_2 are being injected at the same time because of the cost involved in separating the two gases, thus reducing the release of CO_2 into the atmosphere.

Forty-eight injection sites have been approved since the start of the first acid-gas injection operation in 1989, of which 41 are currently active. One operation was not implemented, three were rescinded after a period of operation, either because injection volumes reached the approved limit, or because the gas plant producing the acid gas was decommissioned, and three sites have been suspended by the regulatory agency because of reservoir overpressuring. The annual injection rate in 2003 varied between 0.5 and 280 kt/yr, with an average of 25 kt/yr. By the end of 2003, approximately 2.5 Mt CO_2 and 2 Mt H_2S have been injected into deep saline formations and depleted hydrocarbon reservoirs in western Canada.

These acid-gas injection operations constitute a commercial-scale analogue to geological storage of CO_2, with a 15-year track record of industrial implementation and regulatory stewardship. Because acid-gas injection occurs over a wide range of characteristics in the subsurface environment, acid gas compositions and operating conditions, these operations are truly representative of the geological media that most likely will be the target for large-scale CO_2 geological storage, particularly for continental sedimentary basins like the ones in North America situated between the Rocky and Appalachian mountains.

REGULATORY REQUIREMENTS

In Alberta, the Oil and Gas Conservation Act requires that operators apply for and obtain approval to dispose of acid gas from the Alberta Energy and Utilities Board (EUB), a provincial regulatory agency. Similarly, in British Columbia (BC) operators have to apply to the BC Oil and Gas Commission for approval. The regulatory agencies review applications to maximize conservation of hydrocarbon resources, minimize environmental impact and ensure public safety. To adequately address these matters, the regulators require that the applicants submit information regarding surface facilities, injection well configurations, characteristics of the injection reservoir or saline formation, and operations. Approvals set limits for the maximum H_2S mole fraction in the injected acid-gas stream, maximum wellhead injection pressure and rate, maximum volume, and the size of the Emergency Protection Zone (EPZ) in the case of an atmospheric release of H_2S. No application has been rejected to date; however, in some cases the operator had to provide additional information and/or had to make changes to satisfy requirements and requests from the regulatory agencies. After approval for acid-gas injection is granted, the operators have to submit to the regulatory agencies (bi)-annual progress reports. These progress reports usually contain information about the actual composition of the injected acid gas, and wellhead injection pressure, temperature, volume and rate.

The selection of an acid-gas injection site needs to address various considerations that relate to (1) proximity of the injection site to the sour oil and gas facility that is the source of acid gas; (2) confinement of the injected gas; (3) effect of acid gas on the rock matrix; (4) protection of energy, mineral and groundwater resources; (5) equity interests; and (6) wellbore integrity and public safety [1–3]. The specific location is based on a general assessment of the regional geology and hydrogeology, which is designed to evaluate the potential for leakage [3]. The injection wells are considered as Class III disposal wells, unless the acid gas is dissolved in produced water prior to injection, in which case the well is designated as either Class Ib or Class II, depending on the produced-water designation.[1] Completion and logging requirements are similar for Class II and III wells, and include: (1) identification of all geological zones using logs and/or cores; (2) isolation by cement of all potential hydrocarbon-bearing zones and shallow potable groundwater aquifers; (3) confirmation of hydraulic isolation and cement integrity by a full-length cement bond log; (4) injection through tubing, and filling of the annulus with a corrosion-inhibiting fluid; and (5) installation of safety devices both above the ground and in the well bore to ensure that failure of any component in the system does not result in environmental damage and risk to life.

The integrity of the acid-gas disposal zone is critical. To optimize disposal and minimize risk, advantage is taken of the properties of the acid gas [5,6], which is injected: (1) in a dense-fluid phase (liquid or supercritical), to increase storage capacity and decrease buoyancy; (2) at bottom-hole pressures greater than the formation or reservoir pressure; (3) at temperatures in the system generally greater than 35 °C to avoid hydrate forming, which could plug the pipeline and disposal well; and (4) with water content below the saturation limit, to avoid corrosion. Because of a water-solubility minimum in the 3–5 MPa pressure range that depends on the acid gas composition, dehydration is naturally supplied in most cases by the compression cycle [7,8]. Only in a few cases triethylene glycol, refrigeration, or a desiccant is used.

To avoid gas migration through the cap rock, the difference between the pressure at the top of the disposal formation and the pressure in the confining layer must be less than the cap rock threshold displacement pressure (the pressure needed for the acid gas to overcome the capillarity barrier and displace the water that saturates the cap rock pore space). The injection zone must be free of natural fractures, and the bottom hole injection pressure (BHIP), although higher than the formation or reservoir pressure, must be below 90% of the rock-fracturing threshold, to avoid inducing fractures. Lately, if acid gas is injected into a depleted oil

[1] Well classification in western Canada is different from EPA classification in the United States [4].

or gas reservoir, regulatory agencies set the maximum BHIP at the initial reservoir pressure or at 10% lower to ensure reservoir integrity.

A historical review of acid-gas injection operations, based on published literature to date, is provided in Ref. [9].

OPERATING CHARACTERISTICS

Acid gas is injected in the Alberta basin in free phase (*dry acid gas*), dissolved in, or mixed with water. Two originally water disposal operations have been subsequently approved to co-inject acid gas dissolved at surface at very low concentrations in very large volumes of water, resulting in a weak acidic solution (*sour water*) that is injected deep into the ground through 17 and 49 wells, respectively. Because the water has a much larger capacity for dissolved acid gas than actually used, there are no safety issues relating to the possibility of a well blow out, and these operations are generally not subject to the same level of requirements as the other operations. A third sour water injection operation has also been implemented. At seven other sites, of which three have been rescinded, wet acid gas (i.e. acid gas with free water present) is injected. The free water is present in these cases as a result of mixing at surface. Dry acid gas is injected at all other sites (i.e. no free water is present). Figure 1a shows the location of the various acid gas sites in the Alberta basin and the type of injection.

The approved H_2S mole fraction of the injected acid gas varies between 5 and 97%. The rest comprises mostly CO_2, with a few percent C^+ gases for the balance, except for the cases where the acid gas is dissolved in or contains free water. Table 1 shows the operating range of the licensed acid-gas operations and of the actually injected acid gases. Averages of the actual operating characteristics were calculated on the basis of the progress reports submitted by the operators to the regulatory agencies (Table 1 and Figure 2a). Based on the estimated total injection volume and capacity of the injection reservoir or saline formation, the acid-gas injection sites are planned to operate for periods of 10–25 years.

Usually four-stage electric or diesel compressors are used to bring the acid gas to the wellhead pressure needed for injection. Compressor power varies between 50 and 900 hp (horse power). Dehydration of the acid gas is achieved in most cases naturally through the compressing cycle. In a few cases refrigeration or dry desiccation is used. Pipelines from the gas plant to the injection well are on the order of several hundred meters, with the shortest at 130 m and the longest at 20 km. Pipeline diameter ranges from 48.3 to 168.3 mm, and pipeline wall thickness varies between 3.2 and 11.1 mm.

Injection takes place usually through a single well, although in several cases more than one well is used. The well consists of a central steel tubing string with an outer annulus bounded by a steel casing that is cemented to the subsurface formations. Well casing varies in diameter from 114 to 244 mm, and the diameter of the well tubing ranges from 60.3 to 178 mm. The wells are protected against corrosion with inhibited crude oil, inhibited fresh or produced water, or diesel in the annulus. The casing is isolated by installing a packer, which is pressure tested for integrity once a year, in the annulus between the casing and the tubing string just above the disposal formation. A down-hole safety valve or a check valve is incorporated in the tubing string so that, if equipment fails at surface, the well is automatically shut-in to prevent acid gas backflow. The wellhead of the injection well is similarly protected with valves. The surface facilities and injection well are monitored for leaks, but no in situ monitoring is performed.

IN SITU CHARACTERISTICS

In their pure state, CO_2 and H_2S have similar phase equilibria, with CO_2 condensing at lower temperatures than H_2S [10]. The critical points are $T = 31.1\,°C$ and $P = 7380\,kPa$ for CO_2 and $T = 100.2\,°C$ and $P = 8963\,kPa$ for H_2S (Figure 2b). The phase behavior of the acid-gas system is represented by a continuous series of two-phase envelopes separating the liquid and gas phases, located between the CO_2 and H_2S bounding systems in the pressure–temperature space. The in situ temperature and pressure position of the injected acid gas is located in the P–T space mostly between the supercritical points for CO_2 and H_2S (Figure 2b). Phase calculations [11] indicate that the acid gas will be mostly in liquid phase as a result of gas

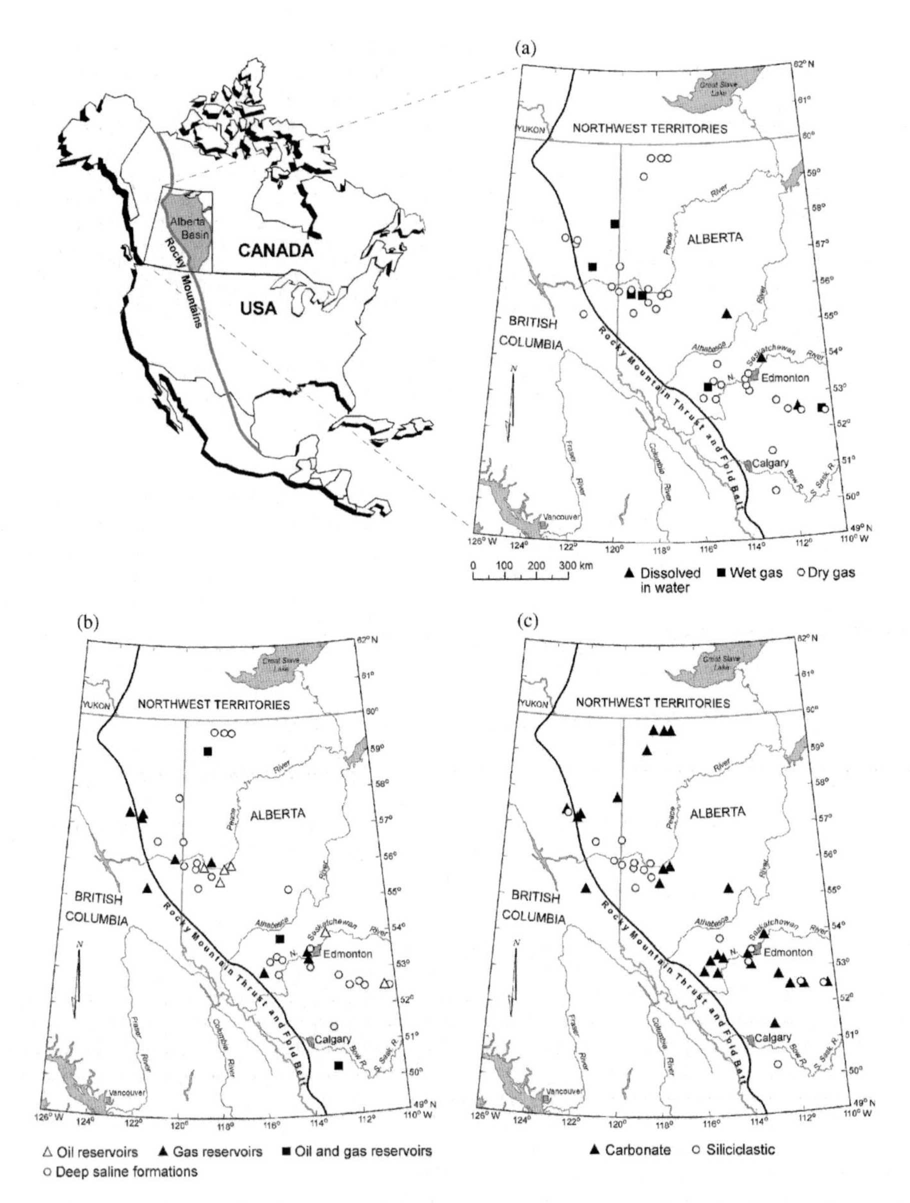

Figure 1: Location of acid-gas injection operations in western Canada and relevant characteristics: (a) injected acid-gas stream; (b) host unit; and (c) rock type.

composition and in situ conditions, and only in six cases it will be either in gaseous or supercritical phase. The density of the acid gas at in situ conditions varies between 205 and 728 kg/m^3 [11].

At 26 sites the acid gas is injected into a deep saline formation, at 18 sites it is injected into a depleted oil or gas reservoir, and at four sites it is injected in the water leg underlying a reservoir (Figure 1b). The top of the most shallow injection zone is at 705 m depth, and the bottom of the deepest one reaches 3478 m, but

TABLE 1
OPERATING RANGE OF ACID-GAS INJECTION SCHEMES IN WESTERN CANADA

Characteristic	**Minimum**	**Maximum**
Licensed H_2S (mol fraction)	0.05	0.97
Actual injected H_2S (mol fraction)	0.02	0.83
Actual injected CO_2 (mol fraction)	0.14	0.95
Maximum well head pressure (kPa)	3750	19,000
Maximum injection rate (10^3 m^3/day)	2	900
Actual average injection rate (10^3 m^3/day)	0.84	500.7
Maximum injection volume (10^6 m^3)	6	1876

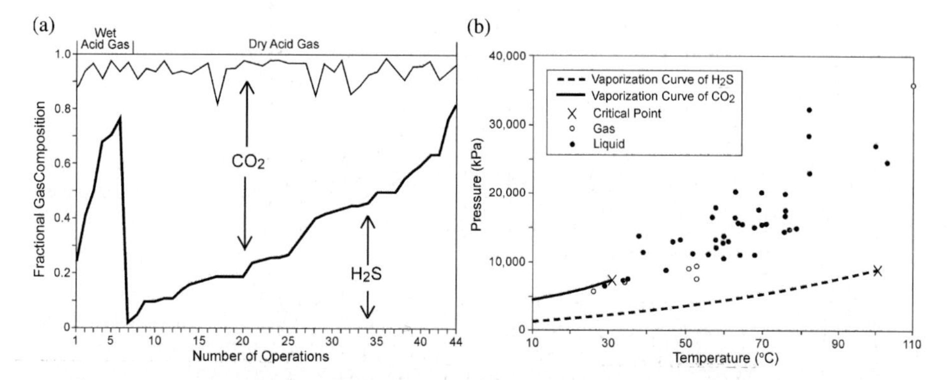

Figure 2: Characteristics of acid-gas injection operations in western Canada: (a) average acid-gas composition and (b) position of the $P-T$ space at in situ conditions.

most injection zones vary in depth between 1100 and 2300 m. The average depth of the actual injection interval varies between 824 and 3432 m.

The thickness of the injection formation, as defined geologically, varies between 4 and 276 m; however, the actual net pay, defined by layers with porosity and permeability adequate for injection, reaches only a maximum of 100 m. Of these, 29 are in carbonate rocks (limestone and dolostone) and 19 are in siliciclastics (sandstone and quartz arenites) (Figure 1c). In most cases shales and shaly siliciclastics constitute the overlying confining unit, the remainder of the injection zones is confined by tight limestones, evaporites and anhydrites. The cap rock thickness varies between 2 and 270 m.

The porosity of the injection zone varies between 4 and 30%, but in most cases is less than 12% (Figure 3a). The carbonate rocks generally have low porosity (less than 10%), except for carbonate reefs where porosity is as high as 22%. There is no trend in porosity for carbonate rocks. Only the porosity in siliciclastics displays a general trend of decreasing porosity with increasing depth (Figure 3a). Rock permeability varies from as low as 1 mD to as high as 4250 mD, although most values are of the order of 10^1-10^2 mD (Figure 3b). As expected, there is no trend in permeability for carbonate rocks, but the siliciclastic rocks exhibit a trend of decreasing permeability with decreasing porosity.

The original formation pressure in the disposal zones is generally slightly subhydrostatic, which is characteristic of the Alberta basin. Two cases of above-hydrostatic pressures correspond to isolated reefal gas reservoirs. The only overpressured case corresponds to injection into a deep structural trap in the thrust

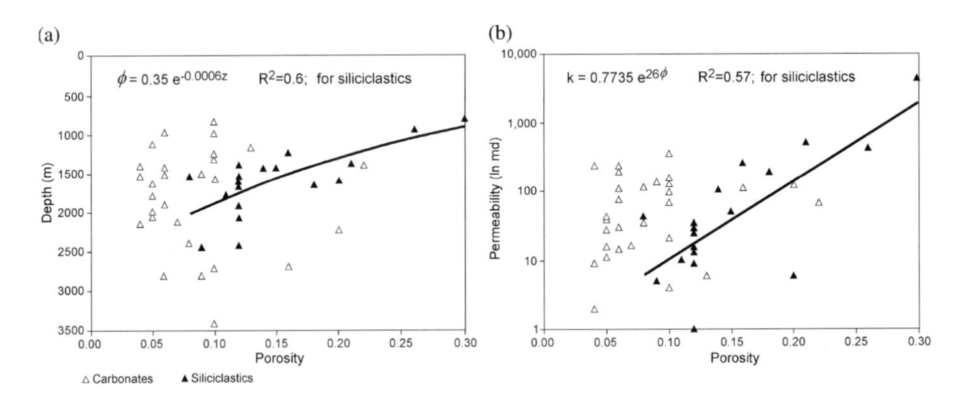

Figure 3: Characteristics of rock matrix at acid-gas injection sites in Western Canada: (a) porosity and (b) permeability.

and fold belt of the Rocky Mountains. In the case of acid-gas injection into depleted oil or gas reservoirs, the original reservoir pressure has been drawn down as a result of production, such that formation pressure at the start of acid-gas injection was less than the original formation pressure, sometimes significantly. From this point of view, injection into a depleted oil or gas reservoir has the advantages of injection pressures being low and of wells and pipelines being already in place [2].

Acid-gas injection could increase oil or gas recovery if it is injected to maintain reservoir pressure [12,13]. Generally, CO_2 can be used for enhanced oil recovery of light and medium oils (gravity greater than 27 °API, [14]), and the presence of H_2S has the effect of decreasing the minimum miscibility pressure [15]. The oil gravity in the oil reservoirs used for acid-gas injection in western Canada varies between 16 and 68 °API, and the specific gravity of the native gas in gas reservoirs used for injection varies between 0.573 and 1.121. The density of the native oil or gas in these reservoirs at initial reservoir conditions (prior to production) was calculated on the basis of shrinkage factor and gas–oil ratio (GOR) for oils, and compressibility (*Z* factor) for gases [16]. The injected acid gas is lighter than the original oil (Figure 4a), but heavier than the original gas (Figure 4b). Since these reservoirs have been produced, the drop in pressure results in the remaining oil losing some of the gas in solution, therefore becoming heavier (denser), while the remaining gas became lighter than when the pools were discovered. Given the properties of the acid gas with respect to the native oil or gas, acid-gas injection can be used for enhanced oil or gas recovery, including heavier oils than in the case of pure CO_2 floods, and this was the case for a few years for one of the acid-gas injection operation in northern Alberta [17]. If acid-gas injection is applied in conjunction with enhanced oil or gas recovery, the produced oil and/or gas has to be desulphurized. Thus, acid-gas injection for enhanced recovery is more suitable for sour oil and gas pools that already contain H_2S and have the desulphurization infrastructure already in place, but the economics still needs to be established on a case-by-case basis.

Formation temperature varies between 26 °C at 843 m depth and 110 °C at 3432 m depth. Figure 5a shows the in situ temperature for the cases of acid-gas injection into deep saline formations. The spread in the variation of temperature with depth is due to the variability in geothermal gradients across the Alberta basin, which exhibits a trend of increasing gradients from the south, where they are as low as 20 °C/km, to the north, where they reach more than 50 °C/km. Formation waters are generally very saline, with salinity varying in a very wide range, from ~20,000 to ~341,000 mg/L (Figure 5b). Cases of relatively low-salinity water encountered at great depths are due to the influx of fresher meteoric water in recharge areas. The cases of very high salinity encountered at relatively shallow depths correspond to injection into saline formations in the vicinity of salt beds. The density of formation water, calculated on the basis of in situ pressure, temperature, and salinity [18], varies between 1007 and 1273 kg/m^3 (Figure 5c). The strong resemblance between Figure 5b and c illustrates the strong dependency of water density on salinity [18]. In most cases of acid-gas injection into deep saline formations, the density of the acid gas at in situ

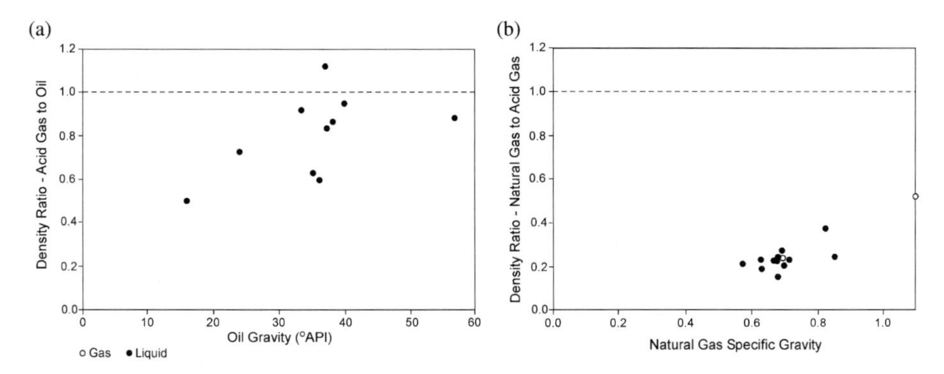

Figure 4: Comparison at in situ conditions between the density of acid-gas injected in oil and gas reservoirs in western Canada, and the native reservoir fluid: (a) oil and (b) gas.

conditions is approximately 50–60% of the density of the resident formation water (Figure 5d). However, in a few cases, particularly for the acid gas in gaseous phase, the density of the acid gas is as low as 10–20% of the water density. This indicates that the buoyancy force acting on the injected acid gas is quite strong, and that acid-gas migration will be mostly updip, regardless of the direction of the natural flow of formation water.

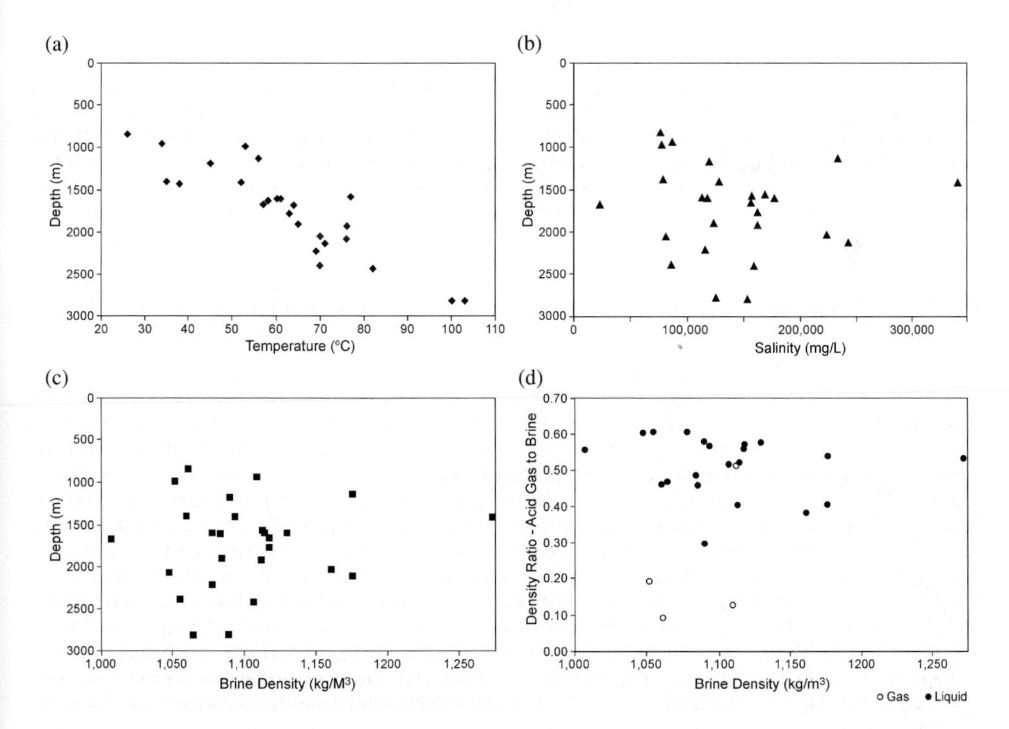

Figure 5: Characteristics of host aquifers for acid-gas injection operations in western Canada: (a) temperature; (b) salinity; (c) formation water density; and (d) in situ acid gas density in relation to formation water density.

If the acid gas is injected into depleted oil or gas reservoirs, the maximum volume allowed for injection is established on the basis of the respective reservoir volume, such that the acid gas will be contained within the reservoir and will not spill out. Furthermore, the pressure in the reservoir is not allowed to increase past the initial pressure, and in three cases where the operator has overpressured the reservoir, the regulatory agency suspended the operations. Thus, in the case of injection into a reservoir, the concern about potential leakage is limited only to other wells that penetrate that reservoir.

In the case of injection into deep saline formations, the pressure around the injection well is higher than the initial formation pressure. Furthermore, the plume of the injected acid gas is no more contained physically, as in the case of reservoirs. The issue of plume spread and migration becomes very important in the context of determining what existing wells may be encountered by the acid gas. Given the geological and operational complexity of these acid-gas injection operations, only detailed numerical modeling of multi-phase, multi-component flow in heterogeneous porous media can provide a prediction of the fate of the injected acid gas, particularly after cessation of injection. However, simple analytical solutions can provide an estimate of the plume spread during injection [19].

The flow of a plume of a fluid injected into a horizontal formation, immiscible with and lighter than the formation water, is driven by the bottom-hole pressure differential and buoyancy, but the plume spread and evolution are controlled mainly by the viscosity difference between the two fluids [19]. Assuming a constant injection rate Q, and a sharp front between the injected acid gas and the formation water, the maximum radial extent, $r_{\max}$, of the plume during injection is given by [19]

$$r_{\max}(t) = \sqrt{\frac{\lambda_c Q t}{\lambda_w \varphi \pi B}}$$

In the above relationship φ and B are, respectively, formation porosity and thickness, and λ is the individual phase mobility, defined as the ratio $\lambda_\alpha = k_{r\alpha}/\mu_\alpha$ of relative permeability k_r to fluid viscosity μ, where α identifies each fluid ($\alpha = w$ for formation water and $\alpha = c$ for acid gas). For the cases of acid-gas injection into saline formations in western Canada, the ratio of brine to acid gas viscosities, calculated at in situ conditions [11,18], varies between 5.97 and 30.38, and the estimated radii of plume spread to date around the injection well vary between 147 and 2070 m [19]. In reality, other processes operate simultaneously, such as dissolution of the acid gas into formation water at the contact between the two, residual acid gas and water saturation, capillarity, and buoyancy in dipping strata, and some of these processes have the effect of slowing the plume spread, while other have the effect of speeding it. While better estimates for plume spread could be obtained using complex numerical simulations, this analytical solution provides a first-order estimate for identifying wells that have been or will be reached by injected acid gas during injection.

CONCLUSIONS

Acid-gas injection in the Alberta basin in western Canada occurs at 41 sites over a wide range of formation and reservoir characteristics, acid gas compositions, and operating conditions. Injection rates in 2003 varied between 0.5 and 280 kt/yr, with an average of 25 kt/yr and a cumulative total of 1 Mt. To the end of 2003, approximately 2.5 Mt CO_2 and 2 Mt H_2S have been successfully injected into deep hydrocarbon reservoirs and saline formations. The size of these operations is smaller by one to two orders of magnitude than of Sleipner West in the North Sea, where CO_2 is injected into the Utsira Formation deep under the sea bottom and which is currently the only greenhouse gas storage operation in the world, and of planned future operations for CO_2 storage in geological media. However, the number of the acid-gas injection operations, cumulative injection rate, diversity in injection conditions and length of operations provide valuable information that may serve as a guide for site selection and implementation of large-scale geological storage of greenhouse gases.

The subsurface characteristics of the injection sites are representative for low-porosity and low-permeability strata found usually in compacted continental sedimentary basins that have been subjected to burial and uplift, such that those in the North American mid-continent. This is in contrast to weakly compacted offshore sedimentary basins that are currently undergoing compaction and that are characterized generally by higher porosity and permeability. In the 15 years since the first acid-gas injection operation in the world started on the outskirts of the city of Edmonton, Alberta, no safety or leakage incidents have been reported. Together with the approximately 16 acid-gas injection operations in the United States, these acid-gas injection operations indicate that acid-gas injection is a mature technology that can be applied elsewhere in the world as increasingly more sour gas is produced from deep gas reservoirs.

These acid-gas injection operations constitute a commercial-scale analogue for future large-scale CO_2 geological storage operations to reduce CO_2 emissions into the atmosphere from large CO_2 point sources. Given that H_2S is more toxic and corrosive than CO_2, the success of these acid-gas injection operations indicates that the technology and engineering experience developed at these operations (i.e. design, materials, leakage prevention and safety) can be easily adopted for large-scale operations for CO_2 geological storage. The major issues that need addressing in the near future are the long-term containment of the injected gases in the subsurface, and the safety of large-scale operations. This review of acid-gas injection operations in western Canada may help in addressing these issues.

ACKNOWLEDGEMENTS

The work reported in this paper was supported by a consortium of Canadian federal and provincial government departments and agencies, whose contribution is gratefully acknowledged. The authors wish to express their gratitude for the opportunity to publish this work in this volume.

REFERENCES

1. Alberta Energy and Utilities Board, Guide 65: Resources Applications for Conventional Oil and Gas Reservoirs, Alberta Energy and Utilities Board, Calgary, AB, 2000, pp. 113–136, DOI, URL eub.gov.ab.ca/bbs/products/guides/g65.pdf.
2. H. Keushing, Hydrogen sulphide—if you don't like it, put it back, *J. Can. Petrol. Technol.* **34** (6) (1995) 18–20.
3. H.L. Longworth, G.C. Dunn, M. Semchuk, Underground disposal of acid gas in Alberta, Canada: regulatory concerns and case histories, SPE Paper 35584, *Proc. Gas Technology Symposium,* Calgary, AB, Canada, 28 April–1 May 1996, SPE, pp. 181–192.
4. Alberta Energy and Utilities Board, Guide 51: Injection and Disposal Wells, Alberta Energy and Utilities Board, Calgary, AB, 1994, DOI, URL eub.gov.ab.ca/bbs/products/guides/g51-1994.pdf.
5. J.J. Carroll, D.W. Lui, Density, phase behavior keys to acid gas injection, *Oil Gas J.* **95** (25) (1997) 63–72.
6. H.-J. Ng, J.J. Carroll, J.R. Maddocks, Impact of thermophysical properties research on acid-gas injection process design, *Proc. 78th Gas Processors Association Annual Convention,* Nashville, TN, March 1–3, 1999, Tulsa, OK, pp. 114–120.
7. M.A. Clark, W.Y. Syrek, W.D. Monnery, A.K.M. Jamaluddin, D.B. Bennion, F.B. Thomas, E. Wichert, A.E. Reed, D.J. Johnson, Designing an optimized injection strategy for acid gas disposal without dehydration, *Proc. 77th Gas Processors Association Annual Convention*, Dallas, TX, March 16–18, 1998, Tulsa, OK, pp. 49–56.
8. E. Wichert, T. Royan, Acid gas injection eliminates sulfur recovery expense, *Oil Gas J.* **95** (17) (1997) 67–72.
9. S. Bachu, W.D. Gunter, Acid gas injection in the Alberta basin, Canada: a CO_2 storage experience, in: S.J. Baines, R.H. Worden (Eds.), *Geological Storage of Carbon Dioxide for Emissions Reduction: Technology*, Special Publication of the Geological Society, Bath, UK, 2004, in press.
10. J.J. Carroll, Phase diagrams reveal acid-gas injection subtleties, *Oil Gas J.* **96** (9) (1998) 92–96.
11. S. Bachu, J.J. Carroll, In-situ phase and thermodynamic properties of resident brine and acid gases (CO_2 and H_2S) injected in geological formations in western Canada, in: E.S. Rubin, D.W. Keith, and C.F. Gilboy (Eds.), *Proc. 7th International Greenhouse Gas Technologies Conference (GHGT-7),* Volume 1: Peer-reviewed papers and plenary sessions September 5–9, 2004, Vancouver, Canada, in press.

12. H.L. Longworth, G.C. Dunn, M. Semchuk, Underground disposal of acid gas in Alberta, Canada: regulatory concerns and case histories, SPE Paper 35584, *Proc. of the Gas Technology Symposium*, Calgary, AB, Canada, 28 April–1 May 1996, SPE, pp. 181–192.
13. L. Connock, Acid gas injection reduces sulphur burden, *Sulphur* **272** (2001) 35–41.
14. J.C. Shaw, S. Bachu, Screening, evaluation, and ranking of oil reservoirs suitable for CO_2-flood EOR and carbon dioxide sequestration, *J. Can. Petrol. Technol.* **41** (9) (2002) 51–61.
15. H.M. Sebastian, R.S. Wenger, T.A. Renner, Correlation of minimum miscibility pressure for impure CO_2 streams, *J. Petrol. Technol.* **37** (12) (1985) 2076–2082.
16. M. Vasquez, H.D. Beggs, Correlations for fluid physical property prediction, *J. Petrol. Technol.* **32** (6) (1980) 968–970.
17. R.J. Davison, A. Mayder, D.W. Hladiuk, J. Jarrell, Zama acid gas disposal/miscible flood implementation and results, *J. Can. Petrol. Technol.* **38** (2) (1999) 45–54.
18. S. Bachu, J.J. Adams, Equations of state for basin geofluids: algorithm review and intercomparison for brines, *Geofluids* **2** (2002) 257–271.
19. S. Bachu, J.M. Nordbotten, M.A. Celia, Evaluation of the spread of acid gas plumes injected in deep saline aquifers in western Canada as an analogue for CO_2 injection in continental sedimentary basins, in: E.S. Rubin, D.W. Keith, and C.F. Gilboy (Eds.), *Proc. 7th International Greenhouse Gas Technologies Conference (GHGT-7),* Volume 1: Peer–reviewed papers and plenary sessions September 5–9, Vancouver, Canada, 20004, in press.

Carbon Dioxide Capture for Storage in Deep Geologic Formations, Volume 2
D.C. Thomas and S.M. Benson (Eds.)

Chapter 13

SIMULATING CO_2 STORAGE IN DEEP SALINE AQUIFERS

Ajitabh Kumar, Myeong H. Noh, Gary A. Pope, Kamy Sepehrnoori, Steven L. Bryant and Larry W. Lake

University of Texas, Austin, TX, USA

ABSTRACT

We present the results of compositional reservoir simulation of a prototypical CO_2 storage project in a deep saline aquifer. The objective was to better understand and quantify estimates of the most important CO_2 storage mechanisms under realistic physical conditions. Simulations of a few decades of CO_2 injection followed by 10^3-10^5 years of natural gradient flow were done. The impact of several parameters was studied, including average permeability, the ratio of vertical to horizontal permeability, residual gas saturation, salinity, temperature, aquifer dip angle, permeability heterogeneity and mineralization. The storage of CO_2 in residual gas emerges as a potentially very significant issue meriting further study. Under some circumstances this form of immobile storage can be larger than storage in brine and minerals.

INTRODUCTION

Geological Storage

Geological storage of CO_2 is one of the few ways to remove combustion emissions in sufficient volumes [1] to mitigate the greenhouse effect. Several groups have reported aquifer-scale simulations of the storage process, usually in order to estimate the volume that can be stored [1–14]. Most schemes that have been put forward depend on storing CO_2 in the supercritical state. In these schemes, buoyancy forces will drive the injected CO_2 upward in the aquifer until a geological seal is reached. The permanence of this type of storage depends entirely on the integrity of the seal over very long periods of time. Assuring such integrity in advance is very difficult.

Our study focuses on three modes of CO_2 storage that avoid this concern: (1) pore-level trapping of the CO_2-rich gas phase within the geologic formation; (2) dissolution into brine in the aquifer; and (3) precipitation of dissolved CO_2 as a mineral, e.g. calcite. All three modes are familiar, though to date not much attention has been paid to the first in the context of CO_2 trapping mechanisms. Each of these modes is permanent for the time frame of interest in CO_2 storage. The key issues then become (1) how to maximize these three highly desirable forms of storage so that very large volumes of CO_2 can be permanently stored in aquifers, without the need for ensuring long-term seal integrity and (2) how long it takes for the injected CO_2 to migrate into these modes of storage.

The principal petrophysical parameters influencing storage as an immobile gas phase (in this chapter, we use the term "gas" as shorthand for "supercritical fluid") are relative permeability, including hysteresis, and the residual saturation of a nonwetting phase. Both depend on the rock making up the aquifer and thus can vary with location. The phase behavior of the CO_2/brine mixture controls storage in solution, and this depends upon brine salinity, temperature, and pressure. The principal geochemical driver accompanying storage is the acidification of the brine resulting from dissociation of dissolved CO_2. Low pH brine 10 in turn induces several reactions with minerals in the formation. An obvious example is the dissolution of carbonate cements. Other reactions are analogous to weathering, in which the acid extracts cations from aluminosilicates (feldspars, clays, etc.). The released cations may form relatively insoluble carbonate precipitates such as siderite. The competition between these reactions will determine the potential for additional storage by mineralization.

The time scales for these processes vary widely. Once CO_2 injection ends, the fluid displacement leading to residual saturations depends on absolute and relative permeabilities, hysteresis, buoyancy forces, the potential gradient caused by dip of the formation, and the magnitude of the residual saturation. Dissolution of CO_2 into brine is rapid, but the overall rate of mass transfer depends on contact between the phases. This is a complicated function of time, especially after injection stops, controlled by the same parameters as the post-injection fluid displacement. Geochemical reactions (mineral dissolution and precipitation) are typically slow [1,10] though under some conditions the rate may be comparable to other mass transport processes [4,14].

EXPERIMENTAL/STUDY METHODOLOGY

To study these processes, their dependence on aquifer parameters, and their characteristic time scales, we conducted a large set of two- and three-dimensional simulations with fully coupled reactive flow and transport. The Computer Modeling Group's GEM simulator was used in this study [8]. Base case simulations were conducted for aquifer storage times of 1000 years. Some simulations were continued for up to 100,000 years.

Because this is a generic study of CO_2 storage in deep, saline aquifers rather than the study of a specific aquifer, the goal was to select representative characteristics for the aquifer as a base case for a systematic parameter study. This provides insight into the potential for CO_2 storage in forms that have minimal tendency to escape from the aquifer.

The input parameters for the base case simulation are summarized in Table 1. The simulated aquifer is 53,000 ft (16,154.4 m) long, 53,000 ft (16,154.4 m) wide and 1000 ft (304.8 m) thick. Constant pressure wells are used along all boundaries to model an open aquifer, while the injector is in the center of the aquifer. These wells are all at the same pressure, so only gravity-driven flow occurs after CO_2 injection. The relative permeability curves are shown in Figure 1.

Pure supercritical CO_2 is injected into the aquifer for 10 years. The injector is then shut in, and the simulation continues with only density differences driving the flow. Having established the base case, we conducted several simulations to study the effect of the parameters influencing the distribution of CO_2 in the aquifer. These parameters include permeability, the ratio of vertical to horizontal permeability, residual gas saturation, salinity, temperature, and dip. Table 2 summarizes the different runs made. These runs did not include geochemical reactions.

This study assumed no conductive faults and no leaky wellbores in the aquifer. Such features would provide a potential escape route for mobile CO_2-rich gas, but not for CO_2 trapped as a residual phase, dissolved in brine, or precipitated as minerals. They would introduce a critical length scale—distance from injector to the potential leak—that would influence the design of strategies to permanently store CO_2. If the injected CO_2 is transformed into trapped forms before it reaches conductive vertical pathways, then risk of escape is small.

RESULTS AND DISCUSSION

Phase Behavior

The calibration of the fluid property models with experimental data is a very important first step in establishing the input to the simulator for this problem. CO_2 solubility is of obvious importance in evaluating storage in brine. Critical to evaluating the permanence of this mode of storage is the brine density: it increases with CO_2 content, hence brine will sink relative to other fluid phases in the aquifer. Thus, CO_2 solubility, brine density and brine viscosity models were calibrated against experimental data as a function of salinity, temperature, and pressure. The brine density and viscosity also depend on the CO_2 concentration.

We made an extensive literature search to find the best sources of experimental data. Table 3 lists the different sources of solubility data for CO_2 in brine [15–19]. These sources give similar trends over a wide range of temperature and salinity.

TABLE 1
SIMULATION INPUT FOR BASE CASE SIMULATION

Aquifer properties		
Length (m)	16,154.4	
Width (m)	16,154.4	
Thickness (m)	304.8	
Depth at top of formation at injection well (m)	1615.44	
Temperature (°C)	60	
Initial pressure (MPa)	15.6	
Dip (degree)	1	
Salinity (ppm)	100,000	
Dykstra–Parsons coefficient	0.7	
Vertical to horizontal permeability ratio	0.001	
Mean permeability (md)	100	
Horizontal permeabilities of each layer,[a] (md)		
Layers 1–4	89	
Layers 5–8	65	
Layers 9–12	46	
Layers 13–16	30	
Layers 17–20	15	
Layers 21–24	120	
Layers 25–28	165	
Layers 29–32	235	
Layers 33–36	840	
Layers 37–40	370	
Porosity	0.25	
Residual water saturation	0.25	
Residual gas saturation	0.25	
Gas end point relative permeability	1.0	
Water end point relative permeability	0.334	
Grid	40 × 40 × 40	
Maximum injection pressure (MPa)	22.75	
Maximum injection rate (MMSCM/D)	1.416	
Description of components		
Component	CO_2	H_2O
Critical pressure (MPa)	7.38	22.06
Critical temperature (°C)	30.98	373.94
Critical volume (l/gmole)	0.094	0.056
Molecular weight (g/gmole)	44.01	18.015
Acentric factor (dimensionless)	0.22394	0.344
Parachor (dimensionless)	78	52

[a] Layer 1 is the top layer.

We tuned the Peng–Robinson equation-of-state [20,21] to fit available experimental data on the solubility of CO_2 in brine and the density of brine [22–28] as a function of CO_2 concentration in the brine, brine salinity, temperature, and pressure. Flash calculations are done in the compositional simulator each time step to calculate the phase behavior of the CO_2 and H_2O mixtures in each grid block as well as the density of both the gas and aqueous phases. The binary interaction parameter between the CO_2 and H_2O was adjusted to fit the CO_2 solubility data and the volume shift parameter for H_2O was adjusted to fit the aqueous phase density. The computed curves for CO_2 solubility as a function

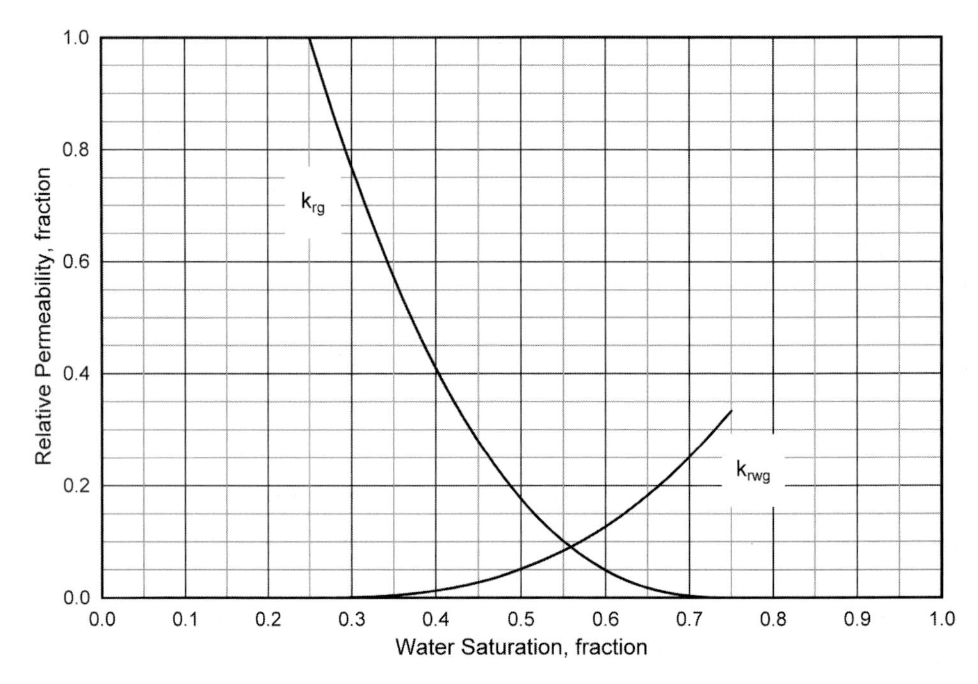

Figure 1: Water–gas relative permeability curves.

TABLE 2
SUMMARY OF SIMULATIONS MADE FOR SENSITIVITY ANALYSIS

Parameter varied	Results/comments
Layered permeability—injection for 10 years	
Temperature = 43.33 °C	Increase in temperature leads to increased dissolution of gas into brine
Temperature = 60 °C[a]	
Temperature = 76.66 °C	
Temperature = 93.33 °C	
Temperature = 110 °C	
Mean permeability = 10 md	Increase in mean permeability leads to greater injectivity as well as greater migration of CO_2
Mean permeability = 100 md[a]	
Mean permeability = 1000 md	
Salinity = 0 ppm	Increase in salinity leads to decreased dissolution of gas into brine
Salinity = 50,000 ppm	
Salinity = 100,000 ppm[a]	
Salinity = 200,000 ppm	
Salinity = 300,000 ppm	

(*continued*)

TABLE 2
CONTINUED

Parameter varied	Results/comments
$k_v/k_h = 0$ $k_v/k_h = 0.001$[a] $k_v/k_h = 0.01$ $k_v/k_h = 0.1$ $k_v/k_h = 1$	Increase in k_v/k_h value leads to upward migration of gas and finally its migration along seal
$S_{gr} = 0.05$ $S_{gr} = 0.15$ $S_{gr} = 0.25$[a] $S_{gr} = 0.35$ $S_{gr} = 0.5$	Low value for S_{gr} leads to increased gas migration and dissolution in brine, while high value leads to increased trapping as residual gas
Dip = 0° Dip = 1°[a] Dip = 2.5° Dip = 5°	Increase in dip leads to increased gas migration and dissolution into brine
Stochastic permeability—injection for 50 years (correlation lengths: $\Delta x = \Delta y = 127, \mathrm{m}$, $\Delta z = 1.27, \mathrm{m}$)	
Mean permeability = 10 md (other properties correlated) Mean permeability = 1000 md (other properties correlated)	Increase in mean permeability leads to increased injectivity and dissolution into brine

[a] Base case.

of salinity and pressure are shown in Figure 2 along with selected experimental data points. Similar agreement occurred at temperatures ranging from 68 to 212 °F (20–100 °C). Using the available solubility data, the binary interaction coefficient was correlated linearly with temperature and salinity for a temperature range of 68–212 °F and salinity range of 0–350,000 ppm of NaCl. We tuned the Pedersen correlation for brine viscosity [27].

Density data for pure water was taken from Ref. [26]. This source was preferred because it is based on the IAPWS-95 formulation adopted by International Association for the Properties of Water and

TABLE 3
EXPERIMENTAL DATA FOR CO_2-SOLUBILITY IN BRINE

Source	Temperature range (°C)	Pressure range (MPa)	Salinity range (ppm total dissolved solids)
15	40–160	0.69–9.65	230,000–350,000
16	48.9–150	10–40	0
17	20–100	0.1–60	0
18–19	4.85–19.85	930–4280	0–31,000

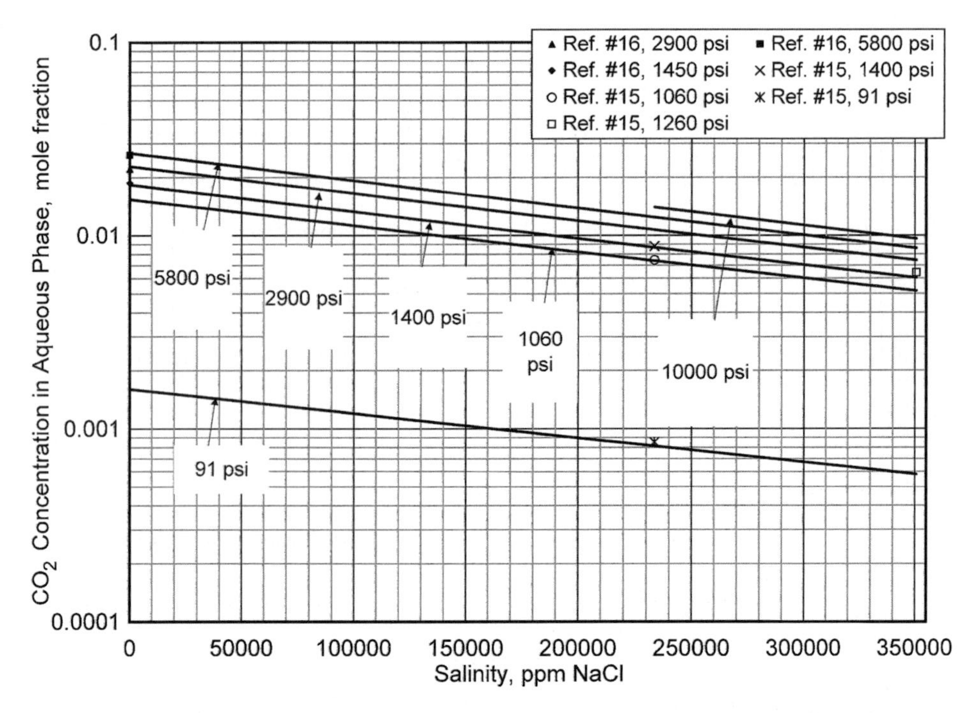

Figure 2: Effect of brine salinity on CO_2 solubility in the aqueous phase at 140 °F (60 °C).

Steam (IAPWS). Density data for pure brine have been taken from Ref. [25] for a wide range of temperature (77–477 °F), pressure (1030–5830 psi), and salinity (30,000–300,000 ppm of NaCl). Unfortunately, there are very few experimental data in the temperature and pressure range of interest for the density of brine saturated with CO_2. Parkinson and Nevers [28] give density values for CO_2–H_2O mixtures for pressures less than 500 psia and temperatures less than 105 °F (40.5 °C). Teng et al. [18,19] give density values of CO_2–brine mixtures for temperatures less than 68 °F (20 °C). Data from Ref. [24] were used to verify density trends. Those few density data that could be found were used to develop a correlation for the volume shift parameter of H_2O used in the Peng–Robinson EOS over the same range of temperature and salinity. Figure 3 shows an example of the predicted density of both brine and brine saturated with CO_2 as a function of salinity at 140 °F (60 °C) and 5830 psia (40.2 MPa). The density of brine saturated with CO_2 is slightly greater than that of brine without CO_2. However, the differences decrease as salinity increases.

Effect of Aquifer Properties

Table 2 summarizes the results when a wide range of aquifer properties were varied individually. Less CO_2 is stored in the 10-year injection period when the formation permeability is small. This is because the simulation includes a maximum bottom hole pressure for the injector, which limits its injection rate.

The effects of temperature and salinity reported in Table 2 reflect the changes in CO_2 solubility and in density of CO_2-saturated brine. The solubility of CO_2 in brine and the viscosity of brine both decrease with an increase in temperature. The former tends to lessen dissolution of CO_2 in brine, while latter increases the same due to increased contact of injected CO_2 with brine. The second phenomenon is more prominent hence at higher temperatures a greater percentage of injected CO_2 goes into aqueous phase. Similarly smaller salinity corresponds to more dissolution because of increased solubility. Larger values for dip lead to greater

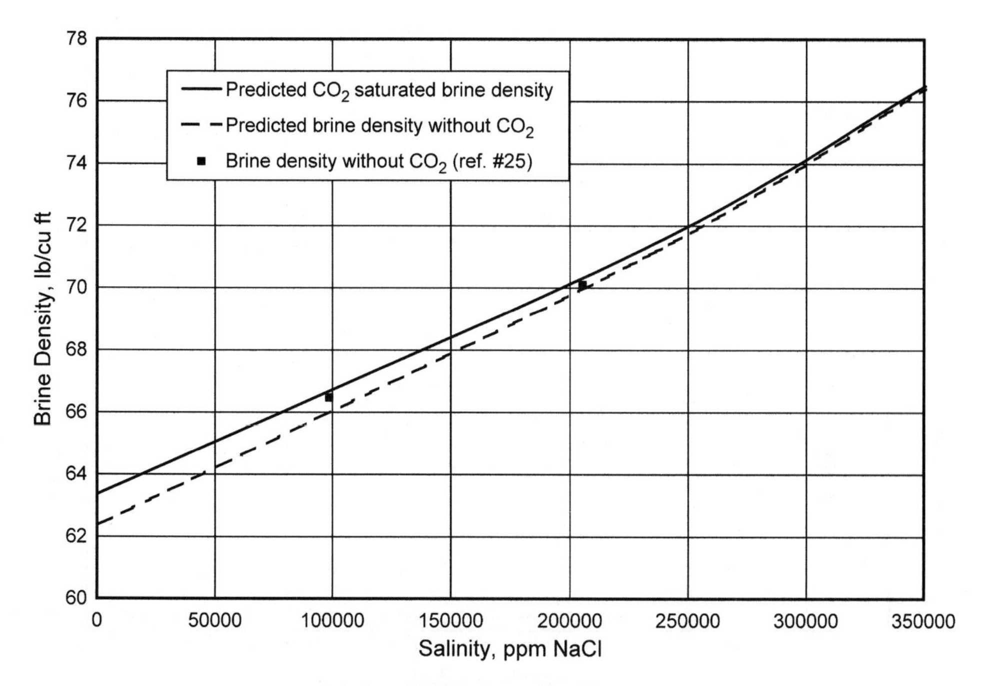

Figure 3: Effect of CO_2 on brine density at 122 °F and 5830 psi.

lateral movement of CO_2, which in turn leads to increased dissolution. Although the vertical to horizontal permeability ratio (k_v/k_h) does not affect the distribution of CO_2 among phases significantly, it does affect spatial distribution. At small values of k_v/k_h, there is more horizontal movement of the CO_2 in the layers into which injection occurred. At larger values, there is more vertical migration followed by movement along the top seal.

Residual gas saturation has the greatest effect on the distribution of CO_2 among the three modes of storage (Figure 4). For small values of residual gas saturation, nearly 20% of the CO_2 is still mobile after 1000 years. Thus, there is greater movement of the CO_2-rich gas phase in the post-injection period. This increases the extent of contact between CO_2 and brine, which in turn leads to increased dissolution of CO_2 in brine. On the other hand, this also permits migration of CO_2 to the top seal of the aquifer. As illustrated later, it also leads to considerable migration along the top of the aquifer in the up-dip direction. In contrast, at larger values of residual gas saturation, most of the CO_2 is trapped as residual gas. There is correspondingly less CO_2 dissolved in brine. Most importantly, the amount of CO_2 that is still mobile after 1000 years is very small.

The strong influence of residual gas saturation on CO_2 storage in aquifers is one of the most important findings of this study. The simulations discussed above assume a single value of residual gas saturation for the entire aquifer. In general, this parameter will vary with rock type [29]. For example, data suggest a correlation between residual gas saturation and porosity [30]. To examine the implications of this variability, we conducted a second set of simulations with stochastic porosity/permeability realizations (Table 2). The porosity values for each block were then calculated using the following correlation [30].

$$\phi = \left(\frac{k}{7 \times 10^7}\right)^{1/9.606}$$

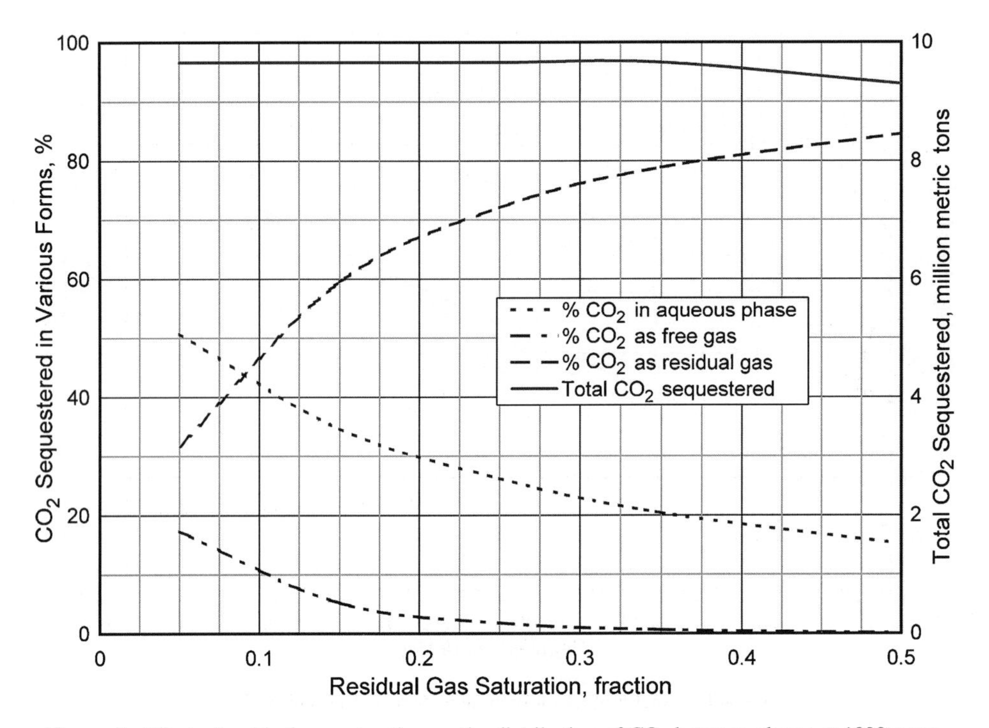

Figure 4: Effect of residual gas saturation on the distribution of CO_2 between phases at 1000 years.

Based on the values of porosity for each grid, maximum residual gas saturation and residual water saturation values were found using following correlations [30].

$$S_{gr}^{max} = 0.5473 - 0.9696\phi; \qquad S_{wirr} = 5.6709 \times (\text{Log}(k)/\phi)^{-1.6349}$$

We also accounted for the fact that the relative permeability of the gas phase depends on whether it is displacing or being displaced by water. GEM models hysteresis with the following equations:

$$k_{rg}(S_g) = k_{rg}(\text{Drainage};\ S_g)\text{during drainage}; \qquad k_{rg}(S_g) = k_{rg}(\text{Drainage};\ S_g(\text{shifted}))\text{during imbibition};$$

where

$$S_g(\text{shifted}) = \frac{(S_g - S_{grh})(S_{gh})}{(S_{gh} - S_{grh})} \qquad \text{and} \qquad \frac{1}{S_{gr}^{max}} - 1 = \frac{1}{S_{grh}} - \frac{1}{S_{gh}}$$

S_{gh} is the value of S_g when the shift to imbibition occurs, S_{grh} is the value of S_{gr} corresponding to S_{gh} via Land's equation, and S_{gr}^{max} has the value of the user-entered parameter S_{gr}^{max}.

In these simulations, a set of 10–15 intervals of porosity values was defined. Each interval was assumed to represent a single rock type and hence was assigned a different relative permeability curve and a different value of S_{gr}^{max} and S_{wirr}. The latter were calculated using the average porosity value for the interval. Figure 5 shows the correlation between different aquifer properties plotted with actual values used in simulations. An example relative permeability curve is shown in Figure 6.

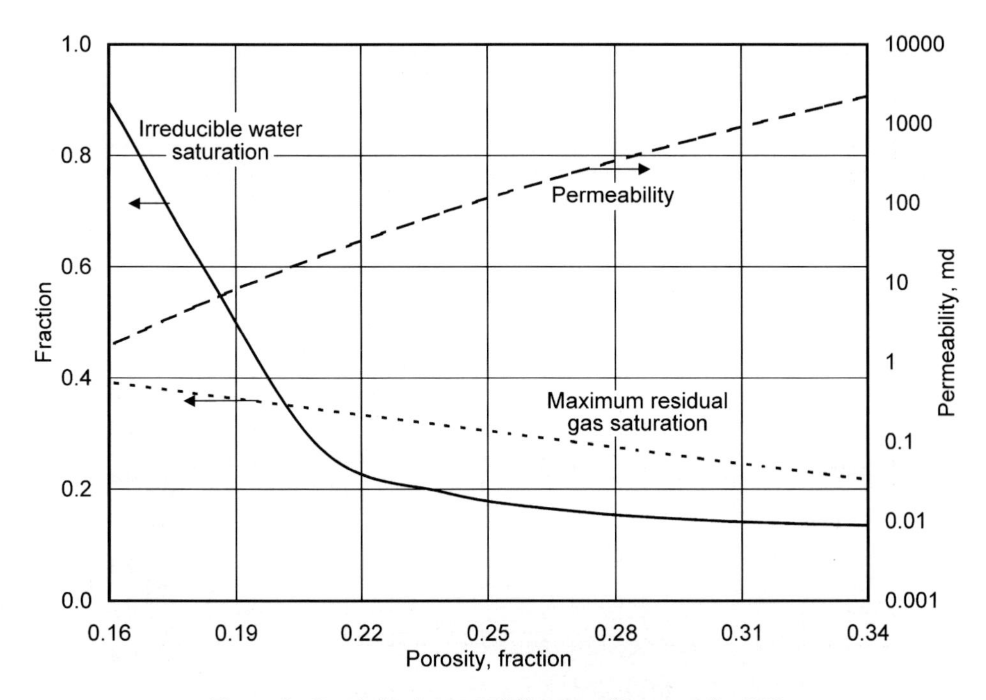

Figure 5: Correlation between different aquifer properties [30].

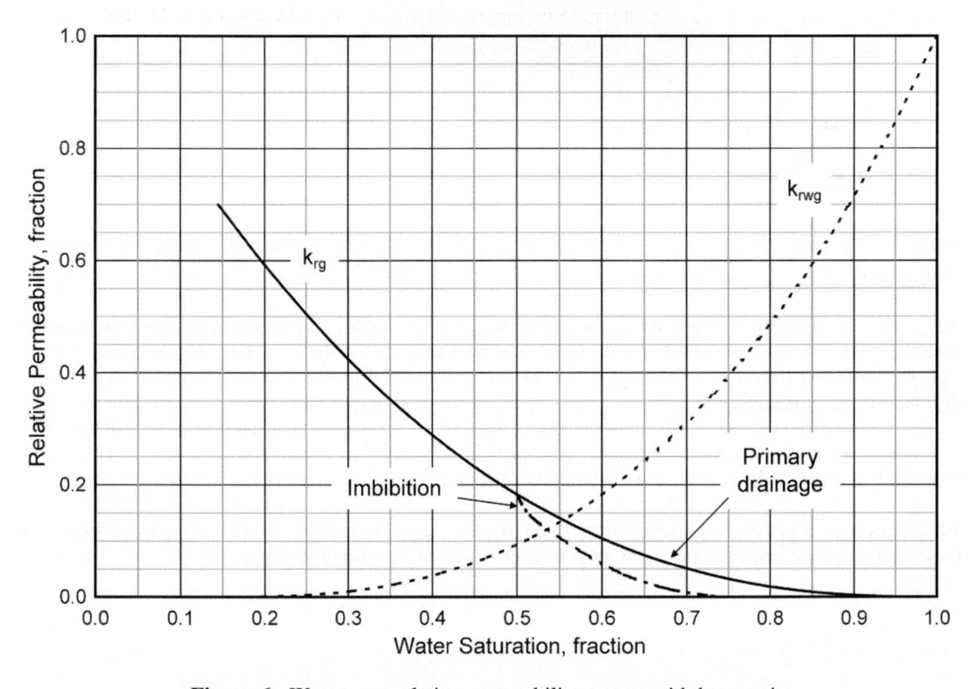

Figure 6: Water–gas relative permeability curves with hysteresis.

To study the effect of the injector completion, CO_2 was injected only in the bottom half of the aquifer. Also, CO_2 was injected for 50 years, rather than 10 years, to investigate how the much larger volume of CO_2 would affect storage.

The simulations conducted with partial well completion in stochastic porosity/permeability realizations with hysteretic relative permeability and rock-type-dependent residual gas saturation indicate that with time all the gas will be trapped in various forms and will never reach the top seal of the aquifer. Figure 7 shows the gas injection profile at 50 years for a vertical $x-z$ cross-section through the injector. Figure 8 shows the same profile after 1000 years. Figure 9 shows the CO_2 mole fraction in the aqueous phase for the same cross-section after 1000 years.

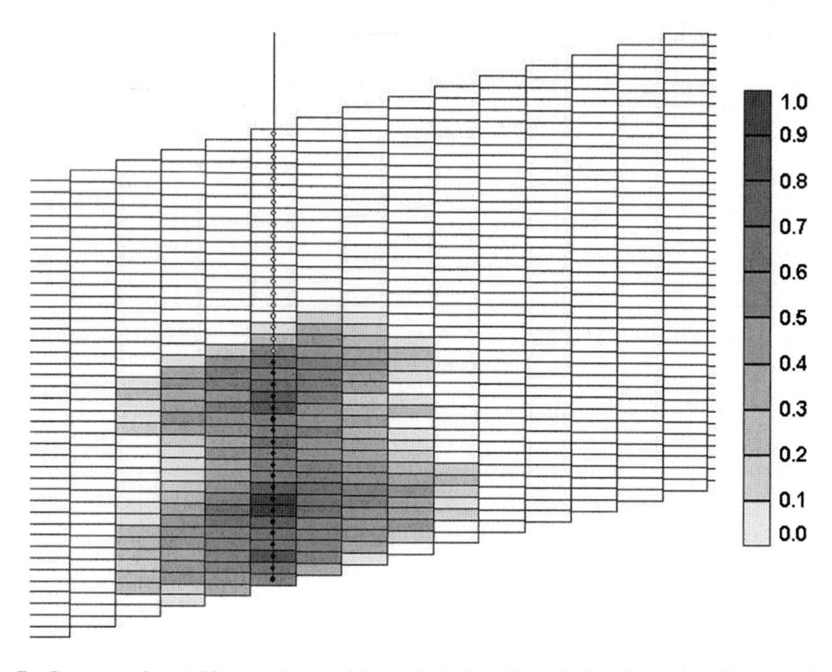

Figure 7: Gas saturation at 50 years (zoomed-in vertical slice through the injection well in $x-z$ direction).

Some 25% of the injected CO_2 exists as a mobile CO_2-rich gas phase at the end of the 50-year injection period. Figure 10 shows an important consequence of buoyancy-driven fluid movement after injection ends: CO_2 is transferred from the mobile phase into permanently stored forms. The time scale for this transfer depends strongly on aquifer properties, including dip; for this example the transfer is essentially complete within 1000 years. This simulation shows the benefit of CO_2 movement after injection ends, but this movement also presents a potential disadvantage. Figure 11 shows the gas saturation profile at 1000 years when CO_2 is injected through the entire interval of the well, rather than the bottom half. Migration of CO_2 up dip along the top seal is evident. This result emphasizes the importance of engineering design in an aquifer storage scheme. A good understanding of the target formation, of the key physicochemical phenomena, and of classical reservoir engineering concepts will be prerequisite for ensuring long-term storage.

Influence of Mineralization

To study the possible contribution of mineralization to CO_2 storage, we performed a third set of simulations in a one-dimensional tilted aquifer (1° dip) derived from the base case described above. This is shown

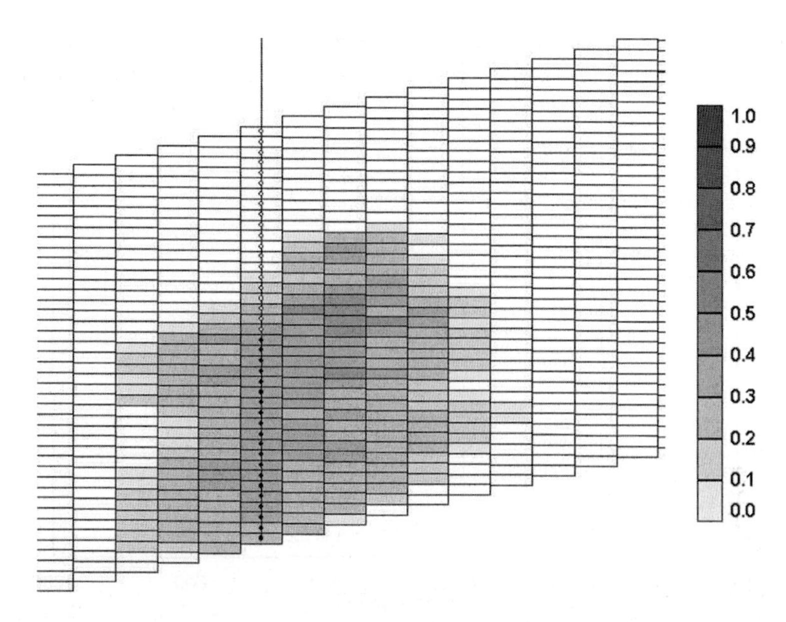

Figure 8: Gas saturation at 1000 years (zoomed-in vertical slice through the injection well in x–z direction).

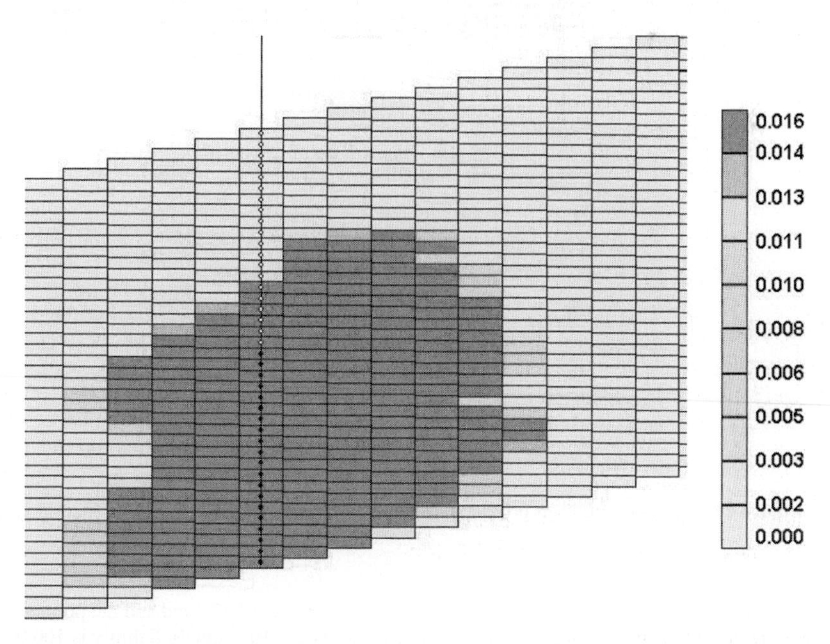

Figure 9: CO_2 mole fraction in aqueous phase at 1000 years (zoomed-in vertical slice through the injection well in x–z direction).

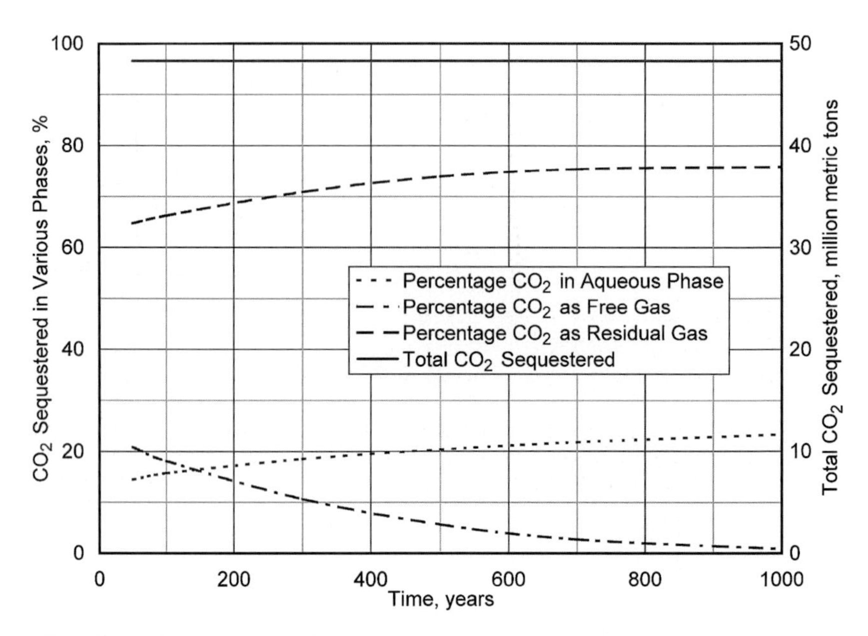

Figure 10: Effect of gravity-driven fluid migration on the distribution of CO_2 between phases after injection for 50 years (at 1000 years).

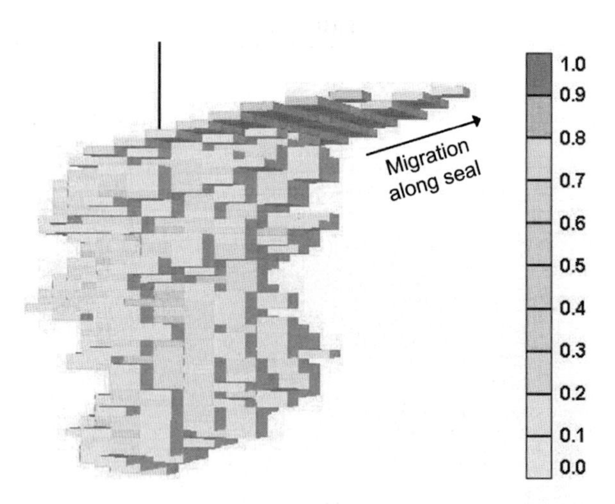

Figure 11: 3D gas saturation profile at 1000 years for injection along whole interval.

schematically in Figure 12. The homogeneous horizontal permeability is 197.5 md and the porosity is 0.25. The reservoir temperature is 60 °C and the diffusion coefficient is 2×10^{-5} cm^2/s. Salinity is 100,000 ppm. For simplicity, CO_2 solubility was modeled with Henry's law [8,20], using a constant of 3.85×10^5 kPa. Relative permeability curves are shown in Figure 1 and capillary pressure is ignored.

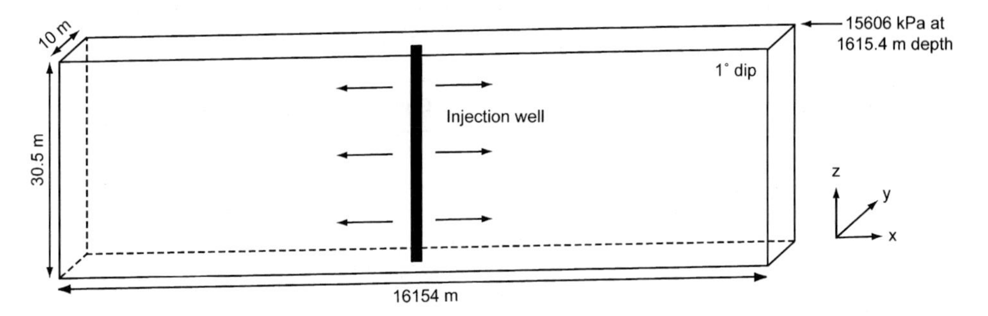

Figure 12: Schematic of 1D flow field used for simulations that account for mineralization.

The three aqueous reactions and five mineral reactions, described in Tables 4 and 5, were used in all simulations. The mineral properties and compositions are based on a glauconitic sandstone aquifer in the Alberta Sedimentary Basin, Canada [4–8]. The reaction equations for the five minerals are as follows:

$$\text{Calcite} + H^+ \leftrightarrow Ca^{2+} + HCO_3^-$$

$$\text{Anorthite} + 8H^+ \leftrightarrow 4H_2O + Ca^{2+} + 2Al^{3+} + 2SiO_2(aq)$$

$$\text{Kaolinite} + 6H^+ \leftrightarrow 5H_2O + 2SiO_2(aq) + 2Al^{3+}$$

$$\text{Siderite} \leftrightarrow Fe^{2+} + CO_3^{2-}$$

$$\text{Glauconite} + 14H^+ \leftrightarrow 1.5K^+ + 2.5Fe^{3+} + 0.5Fe^{2+} + Mg^{2+} + Al^{3+} + 7.5SiO_2(aq) + 9H_2O$$

TABLE 4
AQUEOUS REACTIONS

Reaction	Equilibrium constant, $\log_{10} K$
$H_2O \leftrightarrow H^+ + OH^-$	−13.2631
$CO_2(aq) + H_2O \leftrightarrow H^+ + HCO_3^-$	−6.3221
$CO_2(aq) + H_2O \leftrightarrow 2H^+ + CO_3^{2-}$	−16.5563

TABLE 5
MINERAL REACTIONS

Mineral	$\text{Log}_{10} K_{sp}$	$\text{Log}_{10} k_\beta$ (mole/m² s)	$\hat{A}_\beta$ (m²/m³)	Ea_β (J/mole)
Calcite	1.36	−8.8	88	41,870
Anorthite	−8	−12	88	67,830
Kaolinite	5.47	−13	17,600	62,760
Siderite	10.7	−9.35	88	41,870
Glauconite	−8.6	−14	4400	58,620

Table 6 shows the initial concentrations for aqueous components and the mineral properties and initial volume fractions are shown in Table 7. In this example, we set the residual gas saturation to 0.25 and the initial gas saturation to be zero. Supercritical CO_2 is injected for 10 years with the rate of 100 m³/day. A production

TABLE 6
INITIAL CONCENTRATIONS FOR AQUEOUS COMPONENTS

Aqueous species	Concentration, mole/g H_2O
$H+$	1.0×10^{-10}
Ca^{2+}	9.12×10^{-8}
$SiO_2(aq)$	2.35×10^{-11}
Al^{3+}	2.32×10^{-14}
Fe^{2+}	3.22×10^{-9}
Fe^{3+}	4.99×10^{-8}
Mg^{2+}	5×10^{-10}
K^+	5×10^{-10}
OH^-	5.46×10^{-10}
CO_3^{2-}	2.49×10^{-5}
HCO_3^-	1.17×10^{-8}

TABLE 7
MINERAL PROPERTIES

Mineral	Molecular weight	Density (g/m^3)	Initial volume fraction
Calcite	100.1	2.71	0.0088
Anorthite	278.2	2.74	0.0088
Kaolinite	258.16	2.41	0.0176
Siderite	115.86	3.96	0.0088
Glauconite	426.93	2.67	0.044

well is placed at each boundary to maintain constant far-field pressure. The total amount of CO_2 injection is 9.2×10^9 gmole. Then we stop the CO_2 injection and continue the simulation for 10,000 years.

The average abundances of calcite and siderite for case 1 are shown in Figure 13. During the CO_2 injection period, the calcite initially present in the aquifer starts to dissolve because the dissolved CO_2 perturbs the initial aqueous phase composition so that it becomes undersaturated with respect to calcite. Since the average water saturation decreases during the first 10 years, mineral abundances increase even though mineral dissolution occurs. Figure 13 shows that the mineralization (precipitation of calcite) starts after the injection stops. The siderite curve does not show significant responses after 10 years.

Anorthite and calcite average abundances are presented on a linear time axis in Figure 14. The calcite abundance increases nonlinearly and stabilizes at 1.62×10^3 gmole/kg water. Calcite precipitation requires a source of calcium cations, which provided in this example by the dissolution of anorthite. Thus, the calcite precipitation is symmetric with the anorthite dissolution. Because very little fluid migration occurs after injection ends, the perturbation of the aqueous phase composition is limited to the region contacted by CO_2 during injection. This defines the mineralization region. The anorthite abundance in Figure 14 becomes constant when most of the anorthite in the mineralized region has dissolved, after 10,000 years. In this example, 90.8% of injected CO_2 remains as a gas phase and 6.4% dissolves into water. About 2.7% of the CO_2 is mineralized into calcite. A relatively small amount of CO_2 stays as the bicarbonate ion (HCO_3^-) and the amounts of the siderite precipitation and the carbonate ion are negligible. Even though the residual gas saturation is a modest 0.25, the residual saturation trapping is 46.8 and 44% of total CO_2 is still mobile.

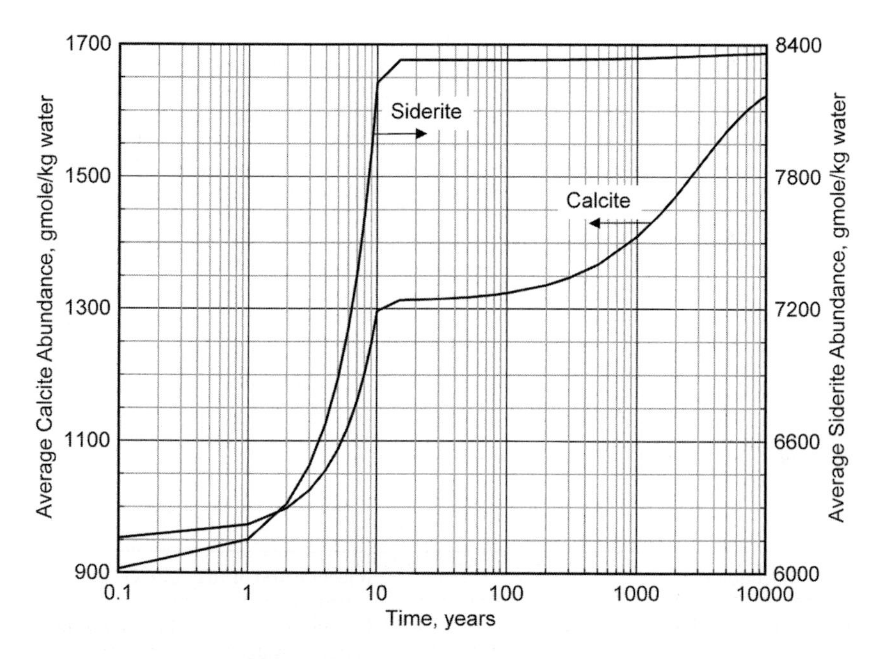

Figure 13: Mineral abundances for case 1.

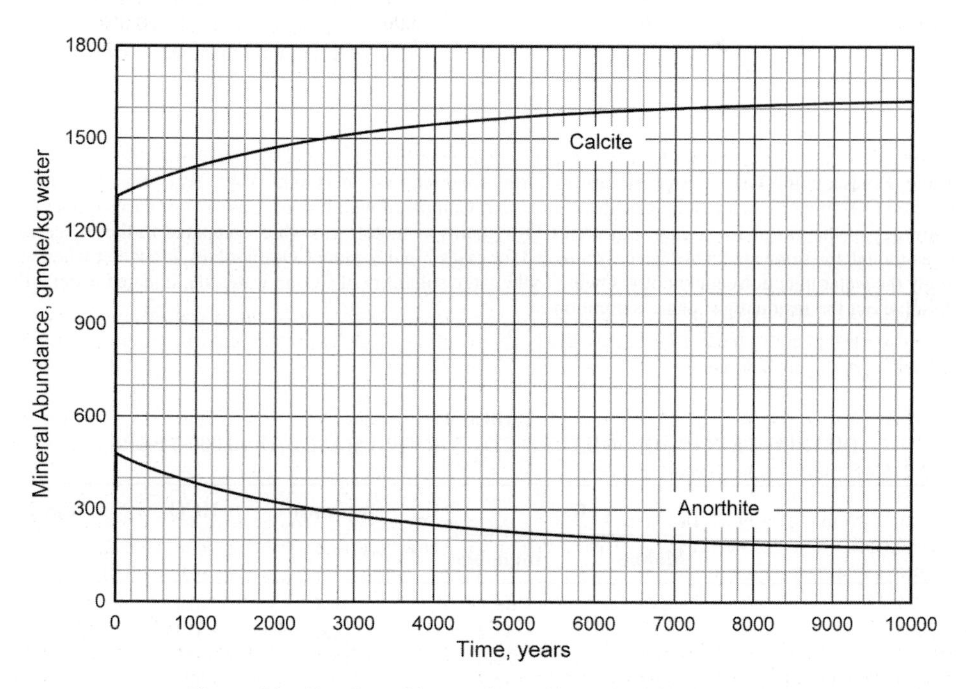

Figure 14: Abundance history of anorthite and calcite for case 1.

In the case described above (case 1), we consider only CO_2 injection, and the majority of CO_2 remains in the gas phase. To evaluate the potential for reducing the amount of mobile gas in the aquifer, we simulated the injection of water simultaneously with the CO_2 injection (case 2). We also simulated the injection of the same amount of the water as in case 2, but immediately after the CO_2 injection (case 3). As was mentioned previously, mineral precipitation depends highly on the amount and type of the source minerals, e.g. the anorthite dissolution as a precursor for calcite precipitation. If we inject CO_2 in an anorthite-rich aquifer (case 4), more calcite precipitation will occur. In case 4, we increase the initial volume fraction of anorthite to 0.088, which is 10 times larger than case 3, and the sequential water injection is also applied. Table 8 summarizes the formulation of simulation runs. The injection of water causes the gas saturation to decrease in the region around the injector because the CO_2 is displaced, and because the CO_2 remaining dissolves into water. Saturation fronts for cases 1 and 2 are the same because the same amount of CO_2 is injected for 10 years for both cases. When CO_2 and water are injected sequentially, water pushes the gas saturation front and there is less mobile gas than the simultaneous injection case because CO_2 has more contact with the formation water. Only 10% of injected CO_2 remains mobile after 10,000 years in case 3.

TABLE 8
SUMMARY OF SIMULATIONS WITH MINERAL REACTIONS

Case 1: CO_2 injection only	Injection 100 m^3/day of CO_2 for 10 years and shut-in
Case 2: simultaneous water injection	Co-injection 100 m^3/day of CO_2 and 100 m^3/day of water for 10 years and shut-in
Case 3: sequential water injection	Sequential injection 100 m^3/day of CO_2 for 10 years, then 100 m^3/day of water for another 10 years and shut-in
Case 4	Increase initial anorthite abundance to 10 times more than that of case 3

Table 9 presents the CO_2 storage in various forms for each case at 10,000 years. Forty-four percent of injected CO_2 remains as a mobile gas phase in case 1. Compared with Figure 4, as 1D test cases ignore the buoyancy of the gas phase so more injected CO_2 remains as mobile gas when compared to the 3D cases. Even though the same amount of water is injected for cases 2 and 3, more CO_2 dissolves into water when we apply the water injection sequentially. Owing to the large solubility of CO_2 in water, the injected water will dissolve out the residual gas phase saturation.

TABLE 9
DISTRIBUTION [%] OF INJECTED CO_2 FOR TEST CASES AT 10,000 YEARS

	Gas		**Aqueous**	**HCO_3^-**	**Calcite**
	Mobile	**Immobile**			
Case 1	44.0	46.8	6.4	0.1	2.7
Case 2	31.9	55.2	9.4	0.1	3.4
Case 3	10.0	70.6	14.7	0.2	4.5
Case 4 (70,000 years)	2.7	43.3	10.3	0.1	43.6

Figure 15 compares the calcite precipitation between case 3 and 4. In case 4, the calcite precipitation occupies 43.6% of CO_2 for 70,000 years and keeps increasing thereafter. Compared with case 3, about 22% of CO_2 in gas phase is precipitated as calcite and the CO_2 dissolution in the aqueous phase is slightly decreased. If all the anorthite in the aquifer were converted to calcite, the theoretical potential of mineral trapping would be 46.2% of the injected CO_2.

Figure 15 shows that mineralization is negligible over the time scales considered in Figures 7–9, i.e. over the span of 1000 years. The fraction of injected CO_2 stored as calcite begins to increase after a few thousand years. The transfer of CO_2 from the gas phase to the mineral phase (mediated by the aqueous phase dissolution of anorthite) is limited by the rate of anorthite dissolution. Given enough time and a sufficient supply of calcium ion, however, this mechanism substantially decreases the amount of CO_2 stored as a mobile gas phase.

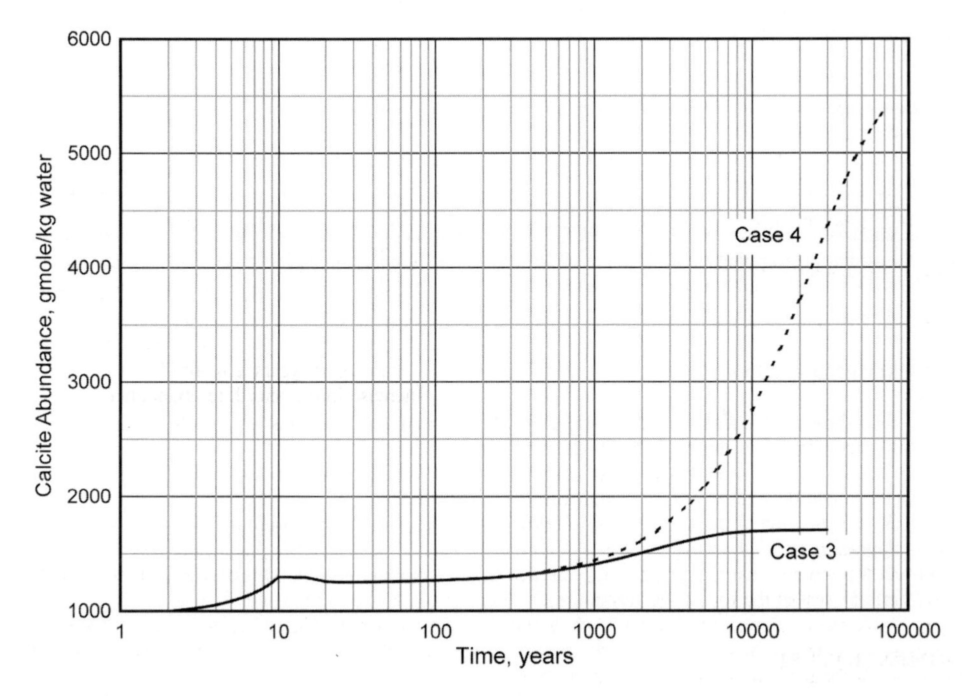

Figure 15: Comparison of calcite precipitation histories for cases 3 and 4. Case 4 has 10 times more initial anorthite than case 3.

CONCLUSIONS

The concerns about CO_2 escape pathways from aquifers used for storage can be considerably mitigated if all or almost all of the CO_2 were stored in the immobile forms of residual gas, dense brine, and minerals. We simulated CO_2 injection in deep, saline aquifers with emphasis on those mechanisms that would immobilize (store) the CO_2. The most significant conclusion from this scoping study is that the effect of residual gas on CO_2 storage can be very large, even more significant than storage in brine or minerals. Potentially all of the CO_2 can be stored in an immobile form when advantage is taken of this well-known phenomenon of capillary trapping. Therefore, the magnitude and variation of residual gas saturation as a petrophysical property merit further study. Both aquifer dip and vertical to horizontal permeability ratio have a significant effect on gas migration, which in turn affects CO_2 dissolution in brine and mineralization.

Well completions play an important role in deciding the fate of CO_2 after injection. When the supercritical CO_2 enters the aquifer near the top seal, it is likely to continue to migrate up dip for long distances and thus may eventually find an escape path. In contrast, when the CO_2 is injected in the bottom half of the aquifer, gravity-driven flow steadily reduces the amount of mobile gas before it can migrate to the top of the aquifer. The time scale for reduction of mobile gas to insignificant values strongly depends on the petrophysical parameters of the aquifer. Over the range of parameters investigated in this scoping study, very little mobile gas remained in the aquifer after a few hundred years.

For the cases studied, mineralization (conversion of dissolved CO_2 into carbonate minerals) occurs over a much longer time scale, on the order of 10^4 years, primarily because of the slow reaction rates of the chemical reactions. However, if the rate of gravity-driven gas movement is sufficiently small, mineralization could play a significant role in immobilizing injected CO_2.

Injecting water after the CO_2 injection period increases the storage capacities of solubility and mineral trapping. The amount of the mobile gas phase drops significantly because the gas phase is displaced by the injected water and spreads out. This effect would be attenuated if the injected water were saturated with CO_2.

For the cases studied here, the capacity of CO_2 storage by mineral trapping is relatively small compared to residual saturation trapping or mobile gas. The amount of minerals containing divalent cations initially present in the aquifer, and the rate at which they dissolve, control the relative amounts of carbonate minerals precipitated.

RECOMMENDATIONS

1. As study shows, significant amount of injected CO_2 (around 75%) remains as trapped gas at the end of 1000 years, hence it is important to model residual gas saturation correctly.
2. The possibility that mobile CO_2-rich gas could reach conductive fractures/faults before becoming trapped should be studied in more detail for any particular formation.
3. Accurate estimation of dip and vertical to horizontal permeability ratio would help predict the extent of gas migration.
4. Proper well completion may significantly reduce chances of CO_2 leakage. This may obviate the need for a "perfect" seal at the top of the formation.

NOMENCLATURE

k permeability, md
k_{rg} gas relative permeability
k_{rw} water relative permeability
S_g gas saturation, fraction
S_{gr}^{max} maximum residual gas saturation, fraction
S_{wirr} irreducible water saturation, fraction

Greek Symbols
ϕ porosity

ACKNOWLEDGEMENTS

This research was supported by fellowships and grants from the National Energy Technology Laboratory of the U.S. Department of Energy and the CO_2 Capture Program. We especially wish to thank Duane Smith at

NETL for his support of the initial CO_2 research by Vikas and A. Kumar at UT as part of their MS theses. We also want to thank CMG Ltd. for making available the GEM simulator used in this research. Larry W. Lake holds the W. A. (Monty) Moncrief Centennial Chair, Kamy Sepehrnoori holds the Bank of America Centennial Professorship, and Gary A. Pope holds the Texaco Centennial Chair at The University of Texas at Austin.

REFERENCES

1. K. Pruess, T. Xu, J. Apps, J. Garcia, Numerical modeling of aquifer disposal of CO_2, *SPE J., SPE 83695* (2003) 49–60. March.
2. S. Bachu, W.D. Gunter, E.H. Perkins, Aquifer disposal of CO_2: hydrodynamic and mineral trapping, *Energy Convers. Manage.* **35** (1994) 269–279.
3. C. Doughty, K. Pruess, Modeling supercritical CO_2 injection in heterogeneous porous media, *Paper Presented at TOUGH Symposium*, California, May 2003.
4. W.D. Gunter, B. Wiwchar, E.H. Perkins, Aquifer disposal of CO_2-rich greenhouse gases: extension of the time scale of experiment for CO_2-storing reactions by geochemical modeling, *Mineral. Petrol.* **59** (1997) 121–140.
5. B. Hichon, W.D. Gunter, T. Gentzis, The serendipitous association of sedimentary basins and greenhouse gases, *Proceedings, American Chemical Society Symposium on CO_2 Capture*, Utilization and Disposal Orlando, Florida, 1996, pp. 25–29.
6. N.J. House, D.D. Faulder, G.L. Olson, J.R. Fanchi, Simulation study of CO_2 storage in a North Sea Formation, *Paper SPE 81202 Presented at the SPE/EPA/DOE Exploration and Production Environmental Conference*, San Antonio, March 2003.
7. J. Ennis-King, Role of convective mixing in the long-term storage of carbon dioxide in Deep Saline Formations, *Paper SPE 84344 Presented at SPE Annual Technical Conference and Exhibition*, October 2003.
8. L. Nghiem, Compositional Simulator for Carbon Dioxide Storage, Computer Modeling Group Ltd., 2002.
9. S.M. Pasala, C.B. Forster, S.J. Lim, M.D. Deo, Simulating the impact of faults on CO_2 storage and enhanced oil recovery in Sandstone Aquifers, *Paper SPE 84186 Presented at the SPE Annual Technical Conference and Exhibition*, Denver, October 2003.
10. K. Pruess, A. Bielinski, J. Ennis-King, J. Fabriol, Y.L. Gallo, J. García, K. Jessen, T. Kovscek, D.H.S. Law, P. Lichtner, C. Oldenburg, R. Pawar, J. Rutqvist, C. Steefel, B. Travis, C.F. Tsang, S. White, T. Xu, Code Intercomparison Builds Confidence in Numerical Models for Geologic Disposal of CO_2, http://www-esd.lbl.gov/GEOSEQ/index.html.
11. J.G. Seo, D.D. Mamora, Experimental and simulation studies of storage of supercritical carbon dioxide in depleted gas reservoirs, *Paper SPE 81200 Presented at the SPE/EPA/DOE Exploration and Production Environmental Conference*, San Antonio, March 2003.
12. T. Xu, J.A. Apps, K. Pruess, Analysis of Mineral Trapping for CO_2 Disposal in Deep Aquifers, *Report LBNL-46992*, Lawrence Berkeley National Laboratory, Berkeley, CA, 2001.
13. Vikas, Simulation of CO_2 storage, MS thesis, University of Texas at Austin, 2002.
14. T.P. Wellman, R.B. Grigg, B.J. McPherson, R.K. Svec, P.C. Lichtner, Evaluation of CO_2–brine–reservoir rock interaction with laboratory flow tests and reactive transport modeling, Paper SPE 80228 *Presented at the SPE International Symposium on Oilfield Chemistry*, Houston, February 2003.
15. B. Rumpf, H. Nicolaisen, C. Ocal, G. Maurer, Solubility of carbon dioxide in aqueous solutions of sodium chloride: experimental results and correlation, *J. Solution Chem.* **23** (3) (1994) 431–438.
16. P. Scharlin, Carbon Dioxide in Water and Aqueous Electrolyte Solutions, Solubility Data Series, vol. 62, Oxford University Press, *International Union of Pure and Applied Chemistry*, Oxford, 1996.
17. N. Spycher, K. Pruess, J. Ennis-King, CO_2–H_2O Mixtures in the Geological Storage of CO_2. I. Assessment and Calculation of Mutual Solubilities from 12 to 100 °C and up to 600 bar, *Report LBNL-50991*, Lawrence Berkeley National Laboratory, 2002.
18. H. Teng, A. Yamasaki, Solubility of liquid CO_2 in synthetic sea water at temperatures from 278 K to 293 K and pressures from 6.44 MPa to 29.49 MPa, and densities of the corresponding aqueous solutions, *J. Chem. Eng. Data* **43** (1998) 2–5.

19. H. Teng, A. Yamasaki, M.K. Chun, H. Lee, Solubility of liquid CO_2 in water at temperatures from 278 K to 293 K and pressures from 6.44 MPa to 29.49 MPa and densities of the corresponding aqueous solutions, *J. Chem. Eng. Data* **29** (1997) 1301–1310.
20. Y.K. Li, L.X. Nghiem, Phase equilibria of oil, gas and water/brine mixtures from a cubic equation of state and Henry's law, *Can. J. Chem. Eng.* **64** (1986) 486–496.
21. A. Firoozabadi, R. Nutakki, T.W. Wong, K. Aziz, Predictions of compressibility and phase behavior in systems containing water, hydrocarbons, and CO_2, *SPE Reservoir Eng., SPE 15674* **3** (2) (1988).
22. J.E. Garcia, Density of Aqueous Solutions of CO_2, *Report LBNL-49023*, Lawrence Berkeley National Laboratory, 2001.
23. U. Grigull, J. Straub, P. Schiebener, Steam *Tables in SI-Units*, third ed., Springer, Berlin, 1990.
24. L. Hnědkovský, R.H. Wood, V. Majer, Volumes of aqueous solutions of CH_4, CO_2, H_2S and NH_3 at temperatures from 298.15 K to 705 K and pressures to 35 MPa, *J. Chem. Thermodyn.* **28** (1996) 125–142.
25. J.M. Simonson, C.S. Oakes, R.J. Bodnar, Densities of NaCl(aq) to the temperature 523 K at pressures to 40 MPa measured with a new vibrating-tube densitometer, *J. Chem. Thermodyn.* **26** (1994) 345–359.
26. W. Wagner, A. Pruß, The IAPWS formulation 1995 for the thermodynamic properties of ordinary water substance for general and scientific use, *J. Phys. Chem. Ref. Data* **31** (2) (2002) 387–535.
27. I.D. Zaytsev, G.G. Aseyev, Properties of Aqueous Solutions of Electrolytes, CRC Press, Boca Raton, FL, 1992.
28. W. Parkinson, N.D. Nevers, Partial molal volume of carbon dioxide in water solutions, *Ind. Eng. Chem. Fundam.* **8** (4) (1969) 709–713.
29. L.W. Lake, Enhanced Oil Recovery, Prentice-Hall, New Jersey, 1989.
30. H.M. Holtz, Residual gas saturation to aquifer influx: a calculation method for 3-D computer reservoir model construction, *Paper SPE 75502 Presented at SPE Gas Technology Symposium*, April–May 2002.

Carbon Dioxide Capture for Storage in Deep Geologic Formations, Volume 2
D.C. Thomas and S.M. Benson (Eds.)

Chapter 14

CO_2 STORAGE IN COALBEDS: CO_2/N_2 INJECTION AND OUTCROP SEEPAGE MODELING

Shaochang Wo[1] and Jenn-Tai Liang[2]

[1]Institute for Enhanced Oil Recovery and Energy Research University of Wyoming, 1000 E. University Ave., Dept 4068, Laramie, Wyoming 82071
[2]University of Kansas, Lawrence, KS 66045, USA

ABSTRACT

Methane (CH_4) production from coalbeds can be enhanced by injection of carbon dioxide (CO_2), nitrogen (N_2), or a mixture of both (flue gas) to accelerate methane production at sustained or increased pressures. Coal has the capacity to adsorb considerably more CO_2 than either methane or nitrogen. However, the actual field performance of enhanced methane recovery processes, wherein CO_2 is concurrently stored, is largely dictated by how effectively injected gases contact and interact with coalbeds over the active project lifetime. By history matching the early nitrogen breakthrough time and nitrogen cuts in BP's Tiffany Unit, simulation indicated that the injected N_2 may only contact a small portion of the total available pay, which was evidenced by the spinner surveys conducted in some of the N_2 injectors. As a possible explanation, the elevated pressure affected by N_2 injection may expand the coal fractures on the preferential permeability trends in the Tiffany Unit. Simulation prediction of CO_2–N_2 mixed gas injections was performed following the history matching in the pilot area. Methane seepage has already been observed from many locations along the north and west Fruitland outcrops in the San Juan Basin. The concern is that injected CO_2 could likely follow the methane seepage paths and leak from the outcrops. Based on the geological setting of the Fruitland coal outcrop, a representative seepage model was used to simulate the effects of CO_2 contact volume (net pay interval) in coal and the injection distance from the outcrop on methane and CO_2 seepage. Under certain conditions, simulation predicted that a large volume of methane and CO_2 breakthrough could occur if the CO_2 injection wells are placed too close to the outcrop.

INTRODUCTION

There is a growing consensus in the international community that CO_2 emission from burning fossil fuels plays an important role in global climate change. Of the storage options currently under consideration, geologic storage of CO_2 in coal formations is considered to be one of the methods with significant short-term potential. A recent report by Reeves [1] estimates that the total storage potential in unmineable coalbeds in the US alone is about 90 gigatonnes for CO_2 storage, with an additional benefit of 152 trillion cubic feet of methane recovery.

Quantitative modeling is necessary to estimate storage capacity, in situ concentration, transport velocity, CO_2 sweeping volume, and the timeframe for filling, monitoring, and storage. The actual CO_2 storage capacity of coal is largely determined by how effectively injected gases contact and interact with the reservoir over the active project lifetime. The economic limit for methane recovery and CO_2 storage is usually dictated by CO_2 breakthrough, poor injectivity or a variety of other factors that make further operation economically prohibitive. Obvious factors, which may control contact and interaction, include gas adsorption isotherms, reservoir heterogeneity, respective roles of convective and diffusive transports in a fractured medium, CO_2 dissolution in water, and the effect of CO_2 adsorption on coal permeability. In this study, the focus was placed on an actual field case (Tiffany Unit), the sensitivity study of critical coal reservoir properties, and CO_2 seepage from outcrops. This approach establishes a link between the first-hand knowledge from an actual field performance and a more realistic CO_2 seepage forecast. A compositional model, BP-Amoco's GCOMP [2], was used in the simulation of the history match and CO_2-N_2 mixed gas injections in the pilot area. The sensitivity study and outcrop seepage modeling were performed on the COMET2 [3,4] CBM simulator developed by the Advanced Resources International. COMET2 can only model single gas or binary gas mixtures (CH_4-N_2 or CH_4-CO_2) but provides more coalbed-specified features, such as coal matrix shrinkage/swelling, which GCOMP does not provide.

Nitrogen Injection in the Tiffany Unit

In the San Juan Basin, two commercial demonstration projects of enhanced coalbed methane recovery (ECBM) by gas injection have been implemented at the Allison and Tiffany Units [5,6] (Figure 1). Carbon dioxide is being injected into the Fruitland coal in the Allison Unit, operated by Burlington Resources, while nitrogen injection into the same coal formation is being tested at the Tiffany Unit, operated by BP America Inc. The field performance of N_2-ECBM not only provides valuable knowledge of how the coal formation interacts with injected N_2 while the coal swelling due to CO_2 injection is absent, but also has important implications for CO_2 storage via flue-gas injection.

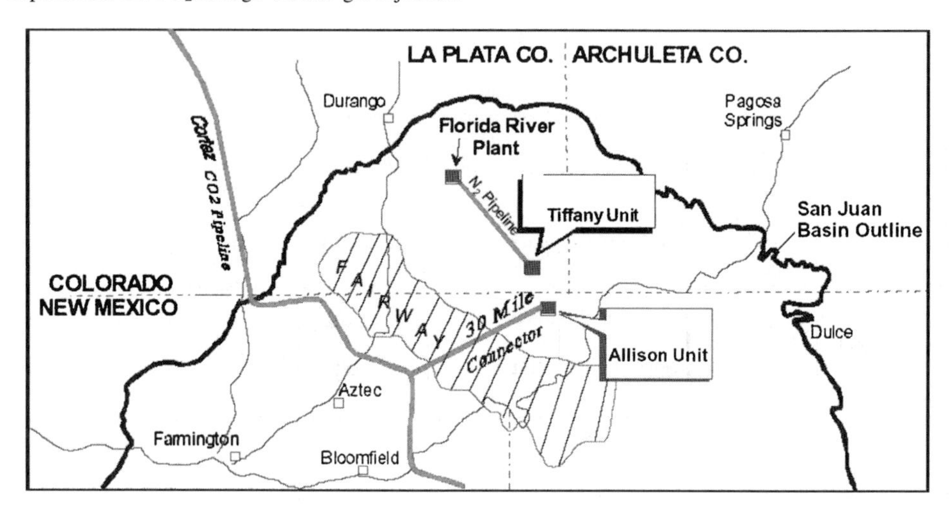

Figure 1: Locations of Tiffany and Allison Units, San Juan Basin [5].

The Tiffany Unit is located in the southern Colorado portion of the San Juan Basin (Figure 1). The pilot area for nitrogen injection is about 10,000 acre and consists of 36 production wells and 12 nitrogen injection wells with a mix of 320 and 160-acre well spacing (Figure 2). Methane is being produced from five Upper Cretaceous Fruitland Formation coal seams, named A, B, C, D, and E (from shallowest to deepest) [6]. A summary of basic coal reservoir properties is provided in Table 1. Note that the reported coal permeability of 1–3 md [6] appears much lower than the permeability of 3–8 md obtained from the history match of primary production in the Tiffany Unit.

Of the 12 N_2 injection wells, 10 were drilled directionally from existing production well pads. The remaining two injection wells were converted production wells. The directional wells were realigned vertically before penetrating the coal horizons. All injection wells were cased, perforated in the coal seams, and hydraulically fractured. To avoid the potential connection with N_2 injection into non-coal strata the wells were not intentionally hydraulically fractured. The production wells were completed with casing and then perforated and simulated by hydraulic fracturing. After the water production declined to a low rate, the wells were configured with a tubing/packer arrangement and produced on natural flow [5].

The source of the injected nitrogen is a cryogenic air separation plant located at BP's Florida River gas processing facility (Figure 1). Injection operations at the field began in February 1998 and continued intermittently until January 2002. Because generation costs become prohibitively high when the ambient

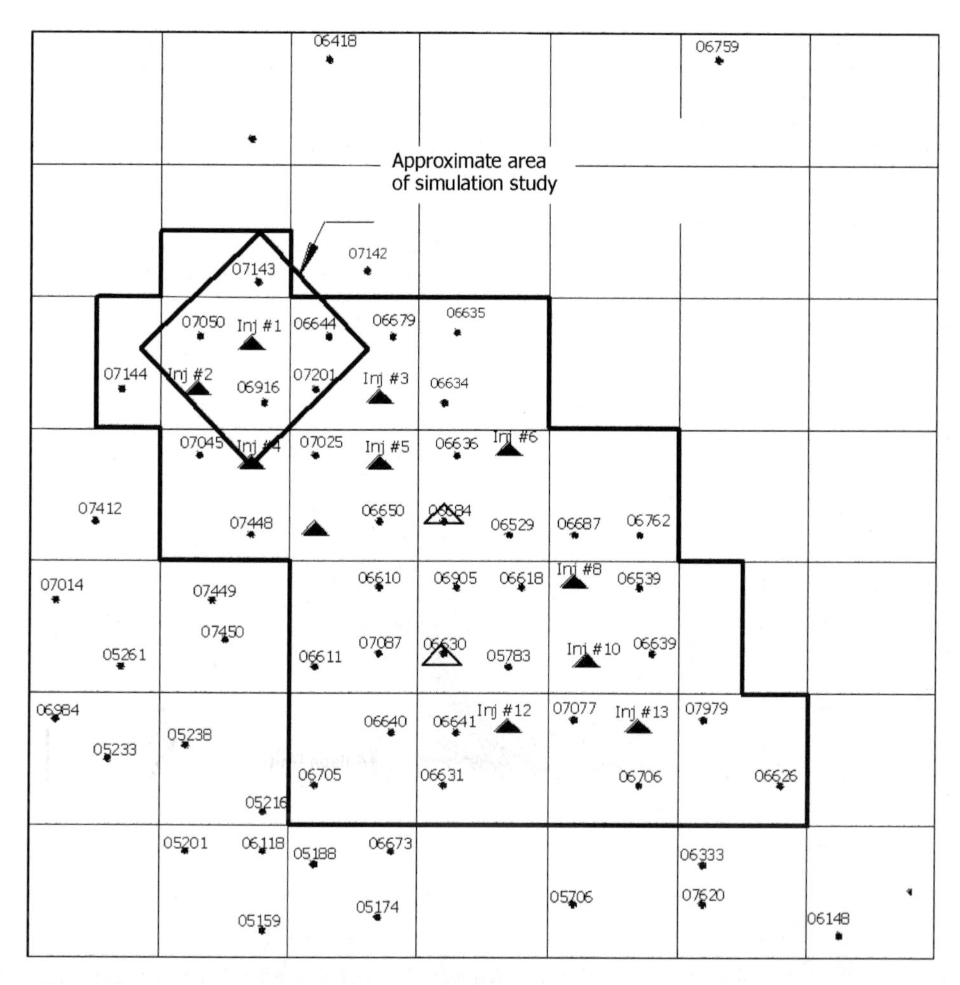

Figure 2: Injection/production well configurations and the area of simulation study, Tiffany Unit.

TABLE 1
TIFFANY UNIT BASIC COAL RESERVOIR PROPERTIES

Property	Value
Number of coal seams	5 (A, B, C, D, and E)
Total coal thickness	40–60 ft
Approximate depth to coal	3200 ft
Original reservoir pressure	1620 psi
Original reservoir temperature	120 °F
Coal seam porosity	0.01–0.02
Coal seam permeability	1–3 md

temperature was greater than 65 °F, BP adopted the strategy of injecting primarily during the cooler (winter) months. Nitrogen injection was suspended after January 2002. The injection of N_2 resulted in a 5-fold increase in methane production [6].

Early N_2 breakthrough was observed from many producing wells. Figure 3 shows the injection history of four injection wells in comparison to the N_2 breakthrough time and N_2 cut responses from the five production wells in the simulation study area. N_2 cuts from all wells except Well 6644 reached 20% in about 1 year after the beginning of N_2 injection. Simulation has shown that Well 6644 is not aligned to any injector on the preferential permeability trends. In an internal report by Raterman [7], two distinct kinds of breakthrough were identified. The first type is characterized by a strong methane response. This behavior is consistent with a homogeneously fractured coal description wherein volumetric sweep of the target coals are largely unaffected. The second type of breakthrough is not associated with coal but rather a distinct thief zone or fracture network.

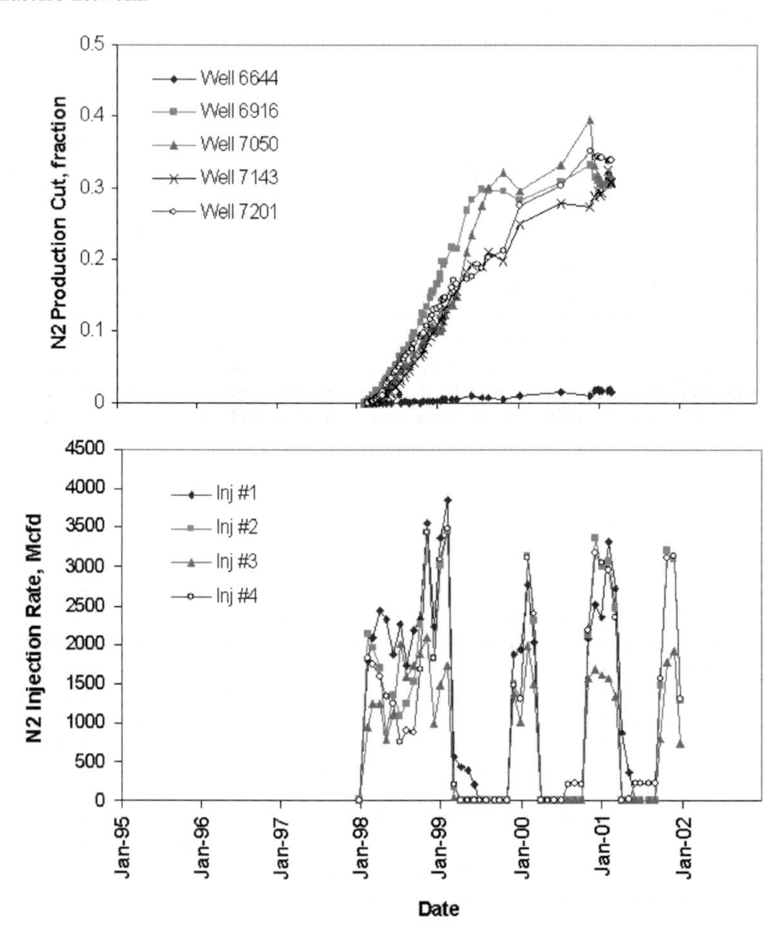

Figure 3: Nitrogen breakthrough time and nitrogen cut responses to nitrogen injection, Tiffany Unit.

In October 1996, a single well injectivity test was conducted in the Southern Ute Gas Unit "U" #1 producer [7]. The test was designed to specifically assess the potential for poor N_2 sweep at Tiffany field. Initially, perforation and fracture integrity were evaluated by breakdown test that consisted of isolating 3 ft sections of the perforated interval, injecting a small water volume, and recording the threshold pressure at which flow was initiated. The testing data indicated that over 95% of the 54 ft interval, including all the five coal seams, in the well was open. Within the open interval, fluid entry pressures appeared relatively uniform. Following the breakdown test the well was placed on production to remove the injected water. The well was then reconfigured for N_2 injection. An analysis of the spinner survey, conducted in the well during N_2 injection, revealed that about 75% gas flow entered approximately 25% of the perforated interval. The highly conductive zone is mostly associated with coal seam B. Similar results were later observed from spinner surveys conducted in other N_2 injectors including Injector #1 and #4 in the simulation study area (Figure 2).

With BP's proposal to supplement the nitrogen injection with the CO_2 captured from its gas processing plant, the effectiveness of combined CO_2 storage and ECBM recovery was assessed including a full-field simulation modeling. The model provided a good history match of the primary production but was unable to predict N_2 breakthrough time and N_2 cut responses at the majority of the responding producers. The proposed injection of CO_2 was postponed due to economic considerations.

ECBM Modeling

Coal has the capacity to hold considerably more CO_2 than either methane or nitrogen in the adsorbed state, in an approximate ratio of 4:2:1 for typical Fruitland coal [6,8,9]. The injected CO_2 becomes preferentially adsorbed onto the coal and thereby displaces methane from the coal matrix. On the other hand, the injection of N_2 will decrease the partial pressure of gaseous methane in the cleat system. As a result, methane desorbs and is pulled into the gaseous phase to achieve partial pressure equilibrium. The N_2-ECBM process is generally referred to as methane stripping. However, the actual field performance of enhanced methane recovery processes is largely dictated by how effectively injected gases contact and interact with the coalbed over the active project lifetime. As observed from spinner surveys, it is likely that a highly conductive coal zone may exist within the Tiffany field. The elevated pressure by N_2 injection could expand the coal fractures on the preferential permeability trends and result in poor N_2 sweep. Early N_2 breakthrough and high N_2 cuts suggest that the permeability on the preferential trends appears much higher than initially assessed even in the low-pressure regions near the producers. Consequently, simulation models that can provide good historical matches of primary productions are often proven inadequate in many aspects to accurately match field performances during the gas injection phase [10–12].

The porous structure of coal is normally described using the Warren and Root [13] concept, wherein the coal matrix blocks are considered to be rectangular parallelepipeds or cubes, and the fractures are considered to be parallel cleats between the matrix blocks. The two orthogonal cleat sets, perpendicular to bedding, are commonly referred to as face (dominant) and butt (subordinate) cleats. Permeability is essentially negligible in the matrix of coal. The aspect ratio of face cleat permeability to butt cleat permeability and cleat orientations largely dictate the preferential permeability trends of coal. The factors that control the permeability of cleats are frequency, connectivity, and aperture width. Gas movement in coal is controlled by diffusion in the coal matrix and the water–gas transport through the cleat system is described by Darcy's law for two-phase flow. Conventional compositional reservoir models, such as GCOMP, have successfully been used to model the primary methane production [10,11,14] and have been attempted to simulate the ECBM process. In this approach, coal is treated as immobile oil and instantaneous gas diffusion is assumed in the coal matrix. The sorption of gas mixtures is described by equilibrium K-values. GCOMP also provides a coal degasification option, in which the multi-component gas sorption is modeled by the extended Langmuir model. The extended Langmuir model is used by most CBM simulators, such as the COMET2/3. In addition, CBM simulators provide more coalbed-specified features that are lacking in conventional models, such as dual porosity/dual permeability, Fick's law for gas diffusion in coal matrix, and coal shrinkage (swelling) due to gas desorption (adsorption).

Methane production rates are commonly used as the well constraint in the history match of the primary production recovery process, while reservoir and well parameters are tuned to achieve a match on water production rates and bottomhole pressures. During the ECBM phase, CO_2 or N_2 is injected by either gas rate or pressure control. However, this simulation approach may encounter difficulty in matching the bottomhole producing pressures for both phases. As observed in Tiffany Unit, initial methane producing rates are usually low even though under low bottomhole producing pressures. The slow release of methane is due to the slow drawdown of coal potentiometric surface. The drawdown could take several months until a sizeable quantity of CBM water has been produced. In order to match both the initial low gas rates and the low bottomhole flowing pressures, a lower permeability often has to be set near the producers. In contrast, during the gas injection phase, the early N_2 breakthrough time and high N_2 cuts indicate the existence of high-permeability trends linking injectors to producers. In other words, a reservoir model resulting from the history match of primary production may not be adequate in simulating the gas injection phase if coal reacts differently to the pressure increase by gas injection.

TIFFANY UNIT SIMULATION STUDY

Previously, a full-field simulation model was developed by BP-Amoco's engineers, which incorporates the full geologic description. The description consists of the five coal seams, some of which do not extend throughout the unit. Coal continuity and thickness are greatest in the northern portion of the field. The model provided good historical matches of the field performance during the primary production period. During the subsequent enhanced recovery phase, N_2 was injected into the field to accelerate methane recovery. However, the field model was unable to predict nitrogen breakthrough time and nitrogen cut responses at the majority of the responding producers. The actual N_2 breakthrough time was much earlier than that predicted by the field model. As evidenced by spinner surveys, the nitrogen injection would have to be restricted into one geological layer, i.e. coal seam B, which accounts for only 25% of the total pay but extends throughout the unit. However, the injectivity tests, such as conducted in the Southern Ute Gas Unit "U" #1 [7], showed nearly uniform fluid entry pressures at most perforated intervals. For a more meaningful history match of the gas injection phase, instead, we developed a 3-layer mechanistic model specific to CO_2 storage in the Fruitland coal of the Tiffany Unit. The simulation area is a five-spot pattern in the northern part of the field where BP planned to conduct a micro-pilot test of CO_2 injection. Figure 2 shows that the pattern consists of one in-pattern and three off-pattern injectors as well as four in-pattern and one off-pattern producers.

Model Description

To match the field performance during the enhanced recovery phase, we assumed that the high-permeability streaks or conduits such as fractured and well-cleated coal within each geologic layer contributed to the early nitrogen breakthrough. Although the high-permeability pay dominates early production response, the long-term response is mostly dictated by the amount of gas exchanged between high and low-permeability packages. Instead of dividing each geologic layer into a fast and a slow component, we modified the model to include a high-permeability fast layer sandwiched between two low-permeability slow layers. In this mechanistic model, the fast layer represents well-cleated and fractured coal from all geological layers while the slow layers represent coal with little or no fracture development from the same geological layers. Initially, a northwest–southeast permeability trend was assumed and the simulation grid blocks were rotated 45° counter-clockwise to match the field permeability trend. However, later from history matching of N_2 injection, it was found that the preferential permeability trend orients roughly along the north–south direction in the simulation area.

History Matching

During history matching, layer thickness, permeability, and vertical transmissibility between layers were adjusted to control N_2 breakthrough time and N_2 cut response. Figure 4 shows that the mechanistic model matched the nitrogen breakthrough time and nitrogen cut reasonably well for all in-pattern producers. The total gas production rate was used as the producing control for all in-pattern producers. As shown in Figure 5, the model resulted as a good match for all producers. However, in order to match nitrogen breakthrough time and nitrogen cut, the vertical transmissibility had to be set to zero. This means that there was no communication between the fast and the slow layers. In this model, nitrogen was allowed to enter all three layers, not just the high-permeability fast layer. However, because the permeabilities of layers 1 and 3 were low and there is no communication between the fast and the slow layers, most of the injected nitrogen entered the high-permeability fast layer. Figures 6–8 show the nitrogen saturations at the end of the nitrogen injection for the high-permeability fast layer (Layer 2) and the two low-permeability slow layers (Layers 1 and 3), respectively. From Figure 6, we can clearly see the preferential permeability trends between the injectors and the producers. A comparison between Figure 6 and Figures 7 and 8 shows that at the end of the nitrogen injection, the nitrogen saturations were very high in the fast layer (Layer 2) and very low in the slow layers (Layers 1 and 3). This is consistent with the observation from spinner surveys and implies that the nitrogen injection and enhanced methane recovery were mostly restricted to only about one-third of the available pay.

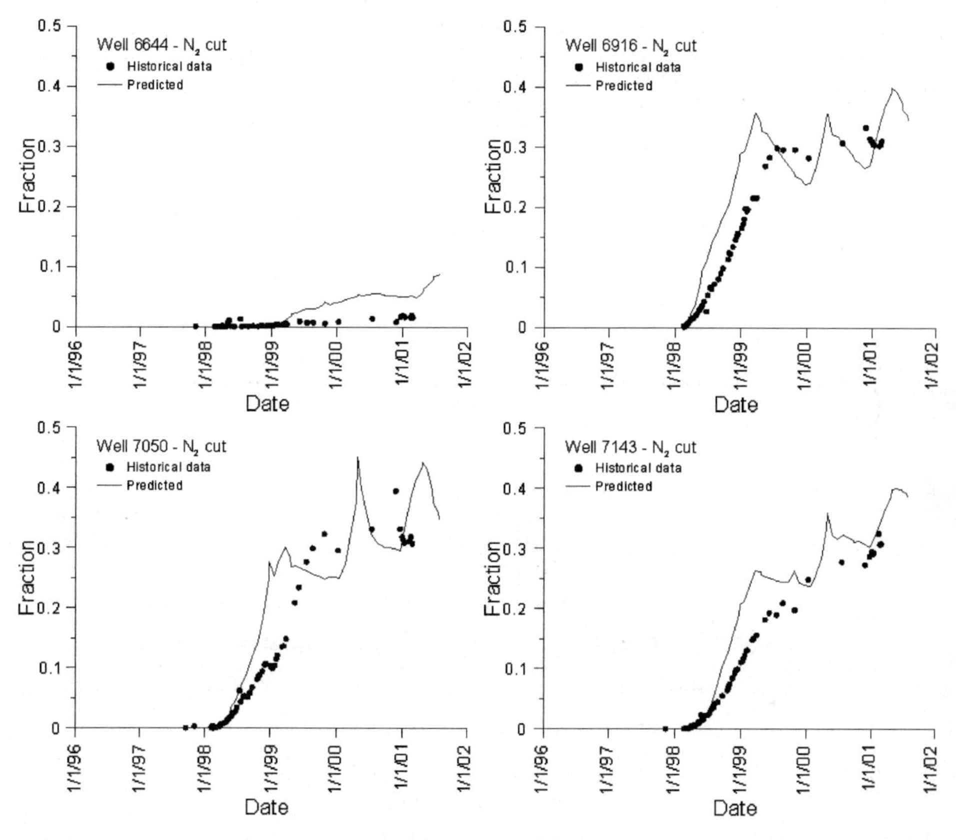

Figure 4: Nitrogen production cut.

Figure 9 shows that the mechanistic model did a reasonable job, matching the bottomhole flowing pressures of all in-pattern producers during the enhanced recovery phase. However, it overestimated the bottomhole flowing pressures during the primary production period for all but one producer. As shown in Figure 9, the mechanistic model matched the pressure responses of Well 6644 reasonably well during both the primary, except in the initial producing period, and the enhanced recovery phases. As discussed before, the difficulty in matching the early bottomhole flowing pressures is because a large pressure drawdown due to a low bottomhole pressure will instantaneously desorb a large volume of methane from coal matrix in the grid block where a producer is placed. The instantaneous gas release does not represent the actual behavior of typical CBM wells during the initial producing period.

Figure 6 shows that unlike other producers, Well 6644 is not linked to any injector on the preferential permeability trends in the simulation area. In other words, the well is least affected by the pressure increase during the gas injection. These findings suggest that the coal formation along the preferential permeability trends in the simulation area reacted differently to pressure depletion during the primary production period and gas injection during the enhanced recovery phase. During nitrogen injection, the elevated pressure may cause coal fractures along a highly conductive zone not only to expand but also to extend from injectors to producers, which was indicated from spinner surveys conducted in some of the N_2 injectors. This permeability enhancement may be additionally supported by matrix shrinkage caused by a lower

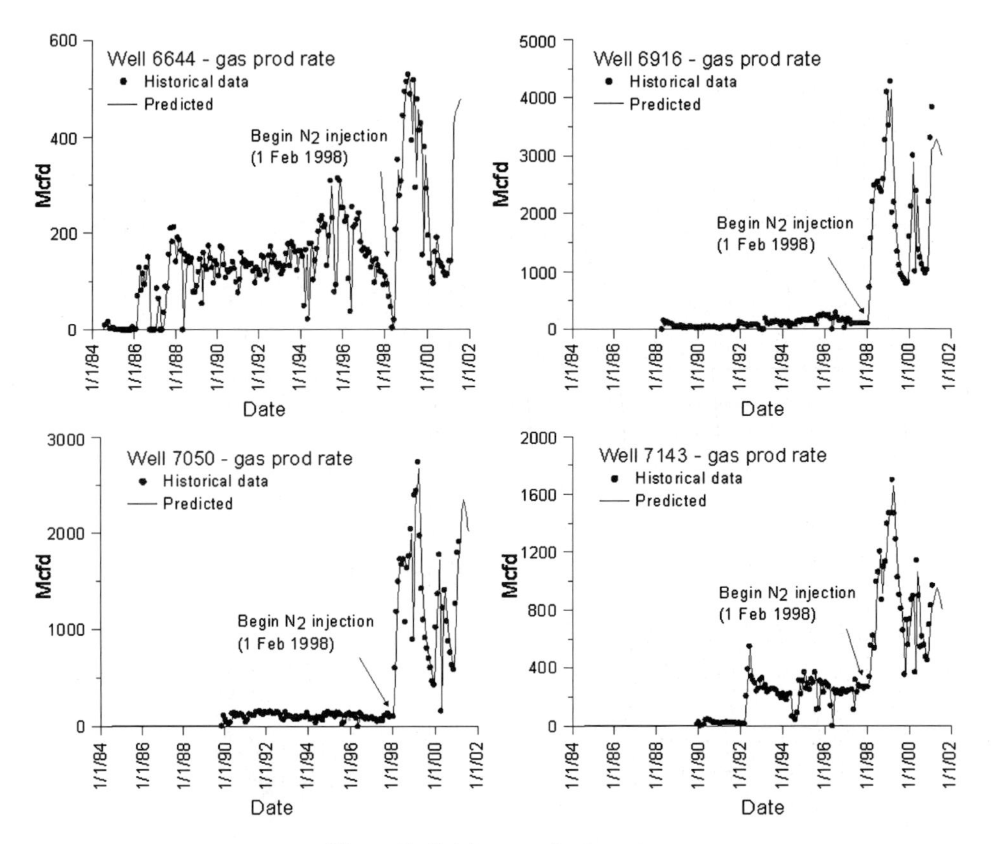

Figure 5: Total gas production rate.

equilibrium adsorbed nitrogen concentration (phase volume) vs. methane. One possible way to satisfactorily simulate both the primary and enhanced recovery phases is to apply negative skin factors to wells on the preferential permeability trends during nitrogen injection but not during the primary production period. Another way is to use one stress–permeability relationship during primary production and a different one during enhanced recovery with gas injection. Also, different stress–permeability relationships might be required for different injector/producer pairs with different degrees of connectivity. Unfortunately, no such specific experimental data are available. Since the mechanistic model is based on field performance during the enhanced recovery phase with N_2 injection, it should be adequate in predicting the field performance during the subsequent CO_2 and N_2 injections.

Model Predictions

The important factors that control the lifetime of an ECBM project are the inert gas (CO_2 and N_2) production and the inert gas cut with time. While methane production represents the income potential, it is the amount of inert gas reprocessed that actually determines the economic limit for an ECBM project. The injection of different mixtures of CO_2 and N_2 was simulated to evaluate their effects on inert gas production and retained CO_2 in coal. The same model settings from the history matching were used except the well controls in the injectors and producers during the injection period from 2/26/1998 to 1/1/2010. In all cases, a continuous injection was assumed with a constant total injection rate of CO_2 and N_2 mixtures. Figure 10 shows the effect of CO_2 content on the cumulative methane, CO_2, N_2, and total gas productions. With an increase in CO_2 percentage in the injected mixture, the cumulative methane production shows an increasing trend while the total cumulative gas production

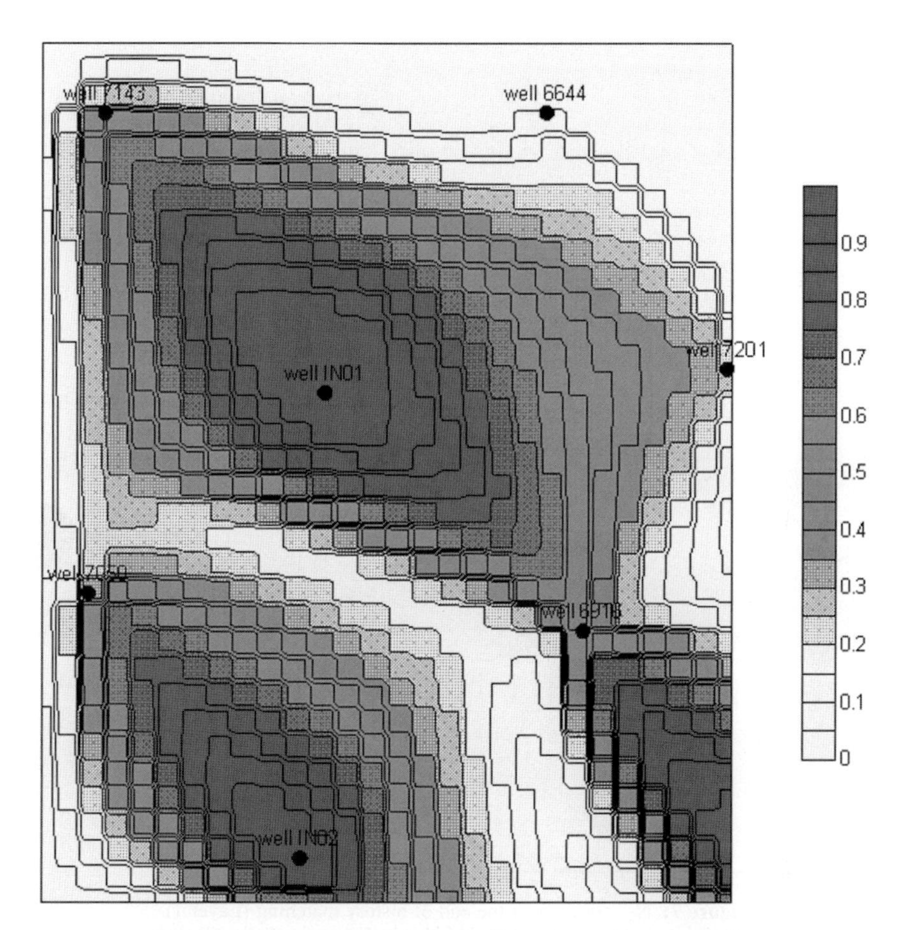

Figure 6: N_2 saturation at the end of history matching (Layer 2).

decreases. Because the coalbed gas in the Tiffany Unit contains about 2–6% CO_2, a certain amount (680 mmscf) of CO_2 was produced when only N_2 was injected as shown in Figure 10. Figure 11 shows the estimated retaining percentage of injected CO_2 in coal. The CO_2 retaining percentage increases as the CO_2 content in the injected gas mixture increases, and reaches to about 44% under 100% CO_2 injection. The estimation was made by subtracting the produced CO_2 and the amount of CO_2 produced under 100% N_2 injection from the total injected CO_2. Coal swelling and permeability reduction due to CO_2 adsorption, which was not considered in this modeling, could significantly increase the CO_2 sweeping volume. Therefore, CO_2 retaining percentage in coal could be much higher for the actual field performance of CO_2-ECBM processes.

EFFECTS OF COALBED PROPERTIES

By virtually reducing the coal thickness, the mechanistic model achieved the history matching of the actual N_2 breakthrough time and production cut. The question is what are the effects of other coalbed properties. To identify dominant reservoir factors, a sensitivity study was performed. Here the COMET2 CBM simulator was used to provide a comparison with GCOMP. Based on the Tiffany field data, a single-well model was used for matching the primary production and a dual-well model was used for matching

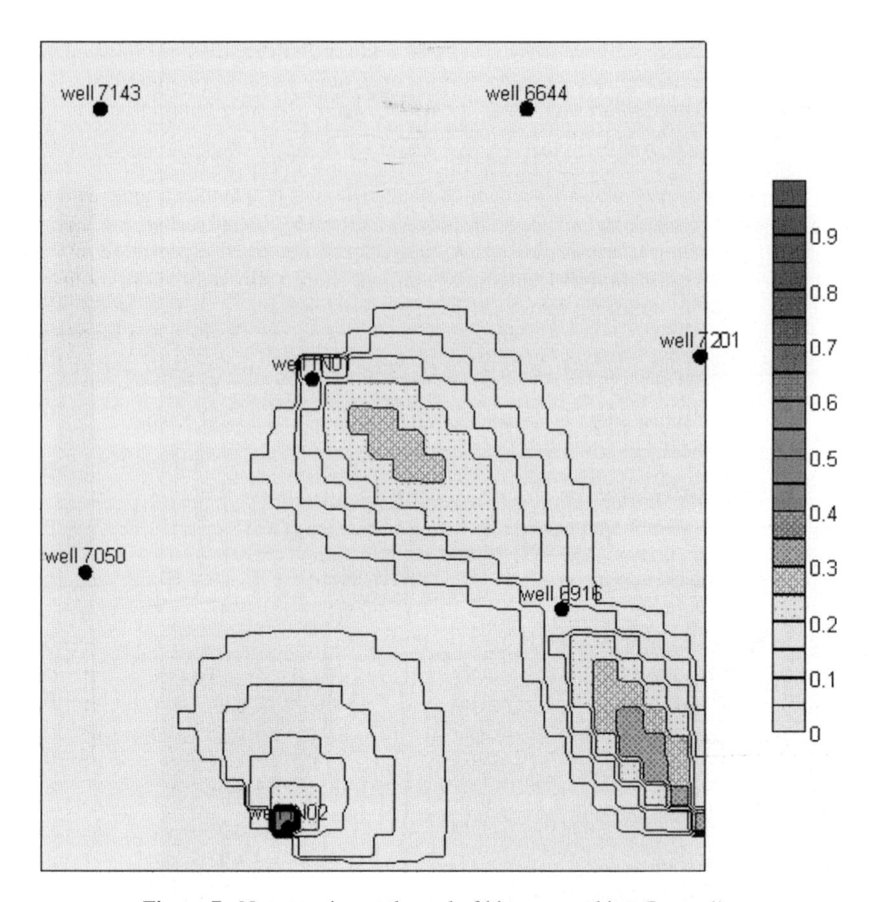

Figure 7: N_2 saturation at the end of history matching (Layer 1).

the performance of N_2 injection. In all cases, no CO_2 was initially assumed in coalbed gas. For comparison, CO_2 injections were also simulated under same model settings and assumptions. Since wells produced on natural flow, fixed bottomhole pressures were used as the producing control. The findings from this sensitivity study are summarized below.

Isotherms

For a pure gas (CH_4, CO_2, or N_2), laboratory-measured isotherm data of Fruitland coal can usually be described by the Langmuir adsorption isotherm model, given by Eq. (1)

$$C = \frac{V_L P}{P_L + P} \tag{1}$$

where C is the adsorbed gas content, P the coal formation pressure, and V_L and P_L the two Langmuir constants. Simulations show that isotherms are the most dominant factor affecting gas production. Laboratory-measured isotherms (CH_4, CO_2, and N_2) on dry coal are available from the Tiffany field [6]. However, the gas content in dry coal (at any given pressure) is significantly higher than that in wet coal as in the reservoir condition. The simulated methane production rates appeared much higher than the actual rates when the methane isotherm on dry coal was used. Instead, the methane isotherm used in simulation was

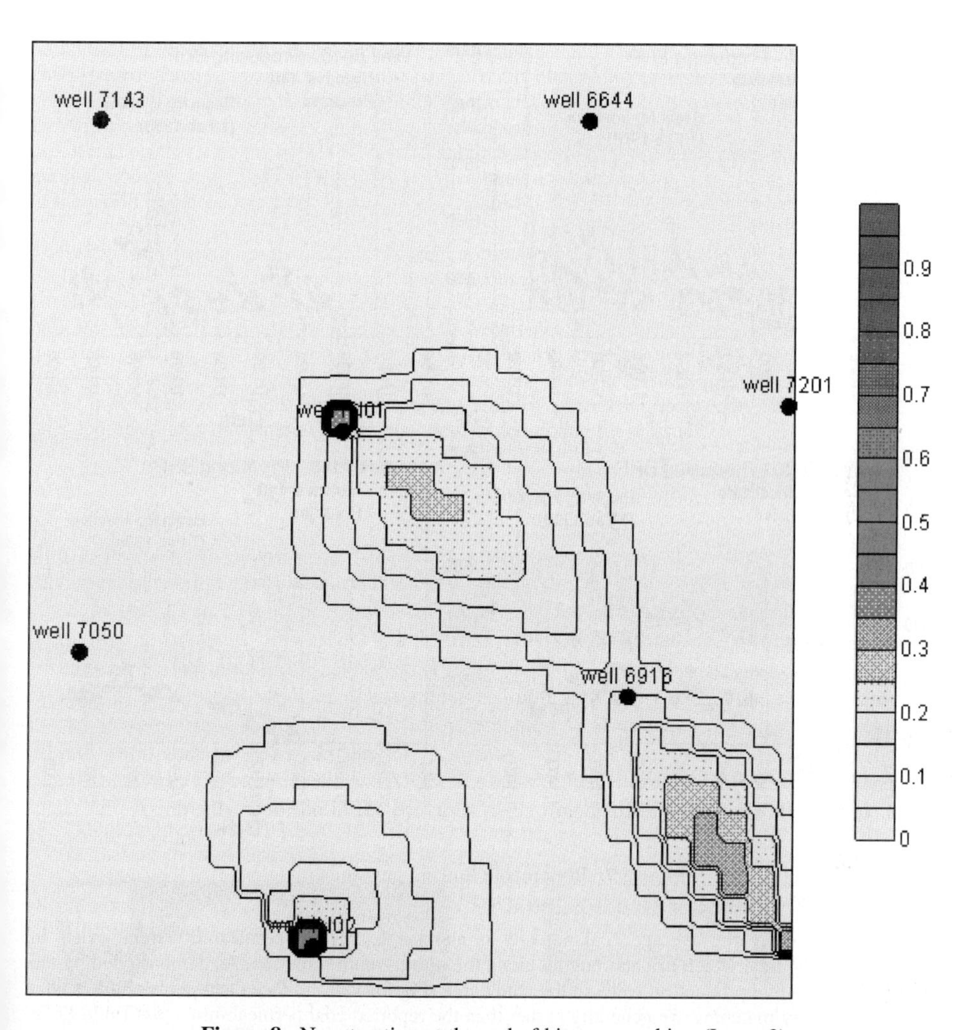

Figure 8: N_2 saturation at the end of history matching (Layer 3).

obtained from matching the primary production. CO_2 and N_2 isotherms were accordingly rescaled using the ratio between the field and laboratory methane isotherms.

Initial Methane in Place

The initial methane in place consists of free gas in the cleat system and the adsorbed gas on the coal matrix. The adsorbed gas (initial gas content) can be estimated from the net pay coal volume and the initial reservoir pressure via Eq. (1) if the methane isotherm is available, either from laboratory or from history matching. Measured initial reservoir pressures are usually available and regarded as reliable data. When the initial pressure is high enough, e.g. greater than 1200 psi in the Tiffany Unit, coal becomes nearly fully saturated with methane. In that case, the initial gas content is usually not very sensitive to the initial pressure.

Porosity and Permeability

In matching the primary production, the gas to water production ratio was found to be very sensitive to cleat porosity. The coal porosity (mainly cleat porosity) is usually very small and initially filled with water, such

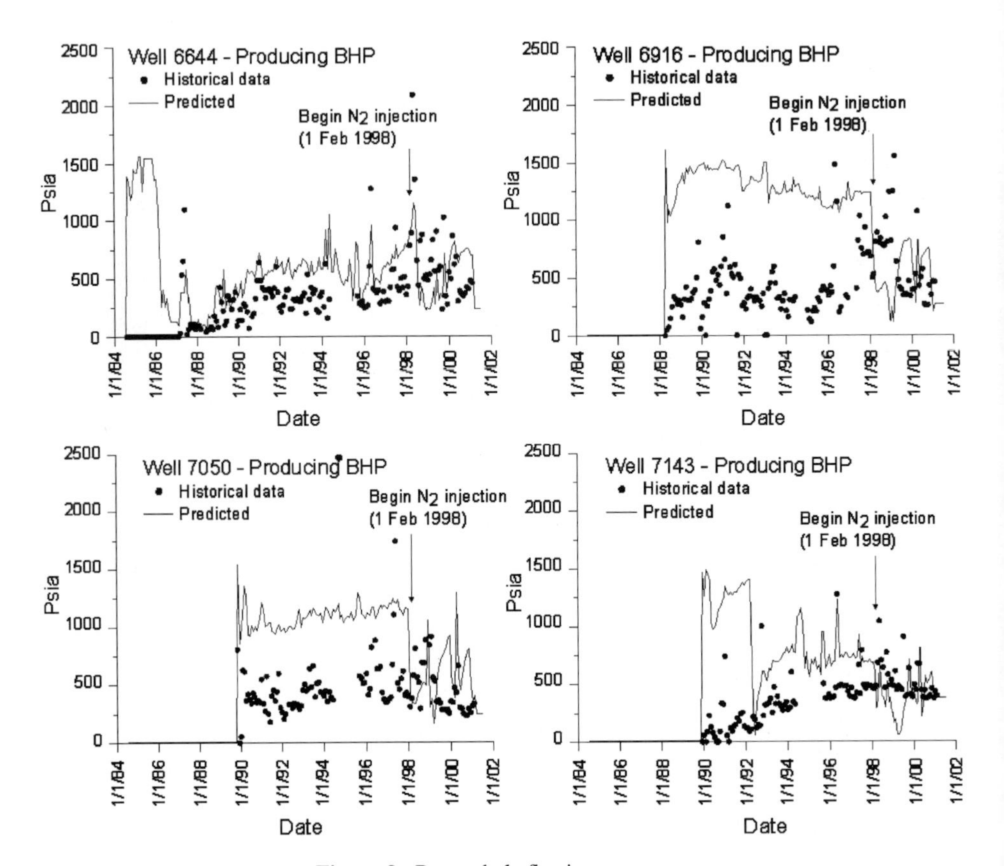

Figure 9: Bottomhole flowing pressures.

as in the Tiffany where the average coal porosity is about 1%. A field permeability trend exists in the simulation study area, which orients roughly along the north–south direction. As demonstrated by the history match of the five production wells in the study area (Figure 12), the face cleat permeabilities (K_x) obtained from history matching are generally higher than the reported coal permeabilities [6] (Table 1). In addition, Figure 12 also shows that the permeability aspect ratio of face cleat permeability to butt cleat permeability (K_y) could have significant effect on gas and water production rates, and an acceptable historical match can be achieved by adjusting the butt cleat permeability (and therefore the permeability aspect ratio). As shown in Figure 12, the actual methane and water production trends generally fall between the curves simulated with the permeability aspect ratio of 2:1 and 3:1.

Relative permeabilities
As shown in Figure 12, simulations predicted much higher initial gas rates than the actual gas rates. This is due to the low bottomhole pressure control, close to 1 atm, set in the production wells. When a simulation begins a large pressure drawdown instantaneously occurs in the grid block in which the well is placed and causes a large volume of methane to desorb from the coal matrix. This behavior does not represent the actual field case. The actual pressure (potentiometric surface) drawdown in coalbeds is usually much slower than that simulated and so is the methane release. This again explains the difficulty in matching the bottomhole pressure when the methane production rate is used as the well control (Figure 9). Tuning relative permeabilities was proven insignificant when a large pressure drawdown becomes the dominant factor of methane release.

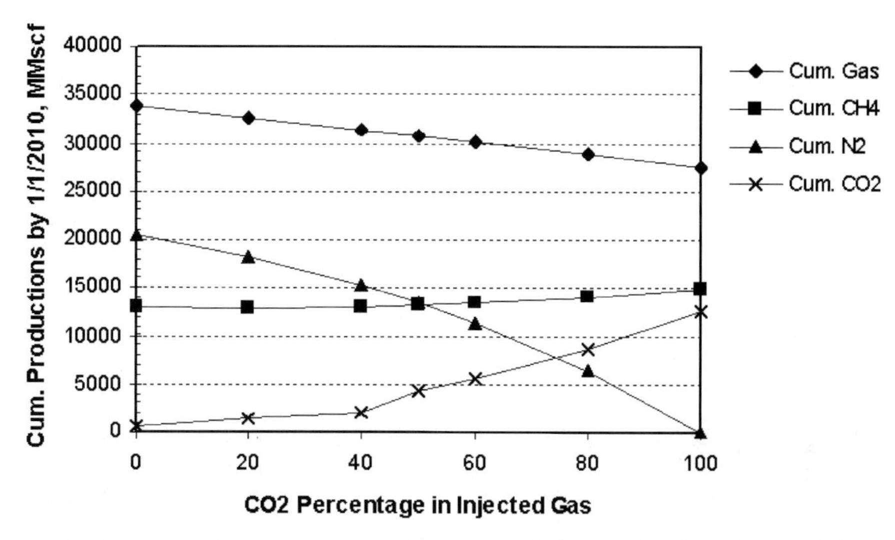

Figure 10: Predicted performance of N_2–CO_2 mixed injections (Well 7201 is excluded).

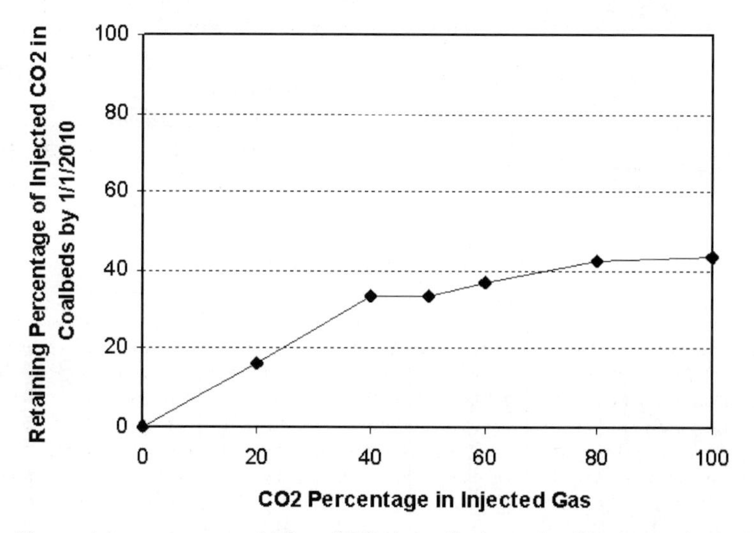

Figure 11: The retaining percentage of injected CO_2 in coalbeds vs. the CO_2 content in the injected gas mixture (Well 7201 is excluded).

For water or gas flooding in conventional oil and gas reservoirs, relative permeabilities are among the most important reservoir properties. A change in relative permeabilities could significantly affect the simulation prediction of water or gas producing rates. However, in coalbeds, injected CO_2 or N_2 could be entirely adsorbed by the coal before reaching a production well if a large CO_2/N_2–coal contact volume (or a large coal thickness) is assumed. To verify the assumption, a dual model consisting of a pair of injection–production wells on a 160-acre well spacing was used to simulate nitrogen injections. If a net pay thickness of 50 ft (the average coal thickness in the Tiffany Unit) is used, Figure 13 shows that little difference resulted even with a large variety of gas relative permeability sets, where the same water relative

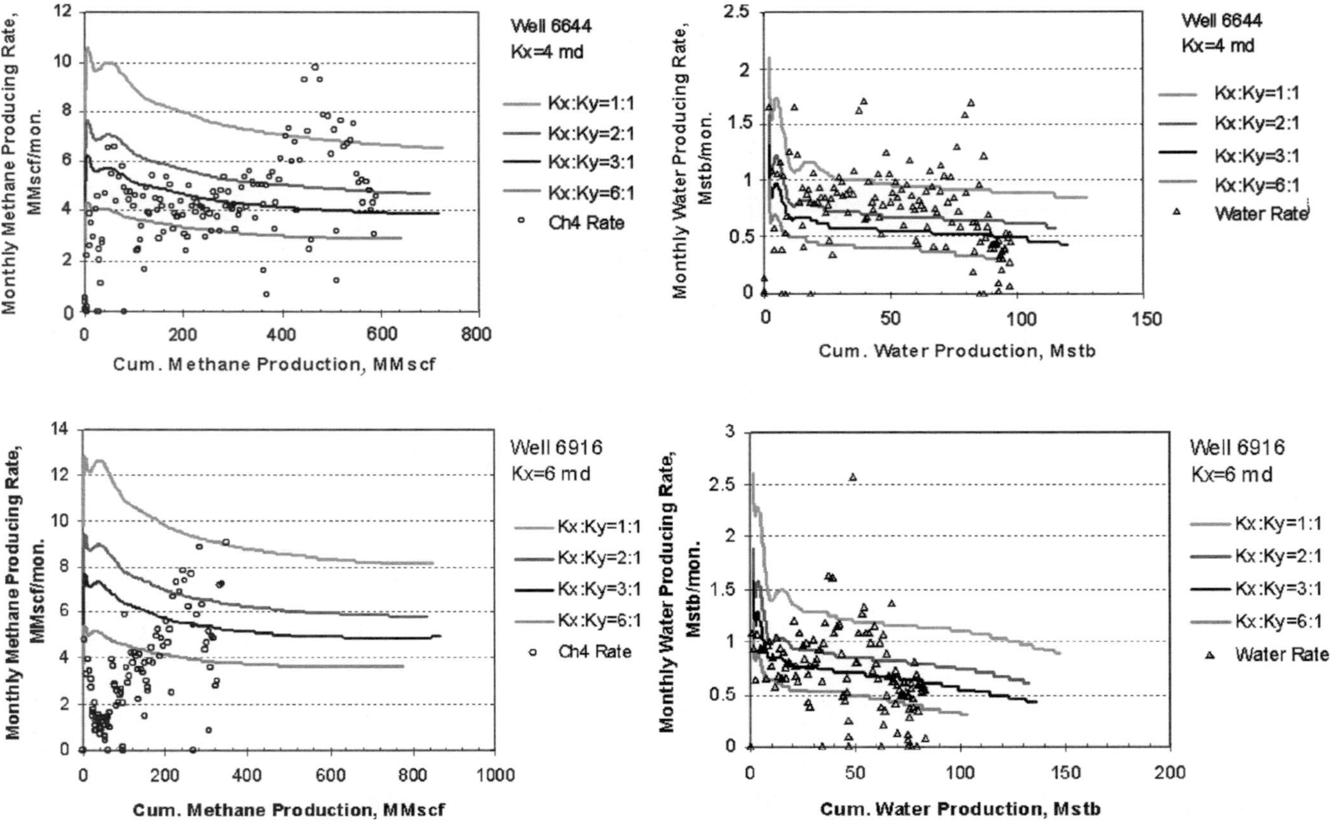

Figure 12: Effect of the permeability aspect ratio on methane (left) and water (right) production rates. In all figures, simulated curves from high to low appear in ascending order of the permeability aspect ratios.

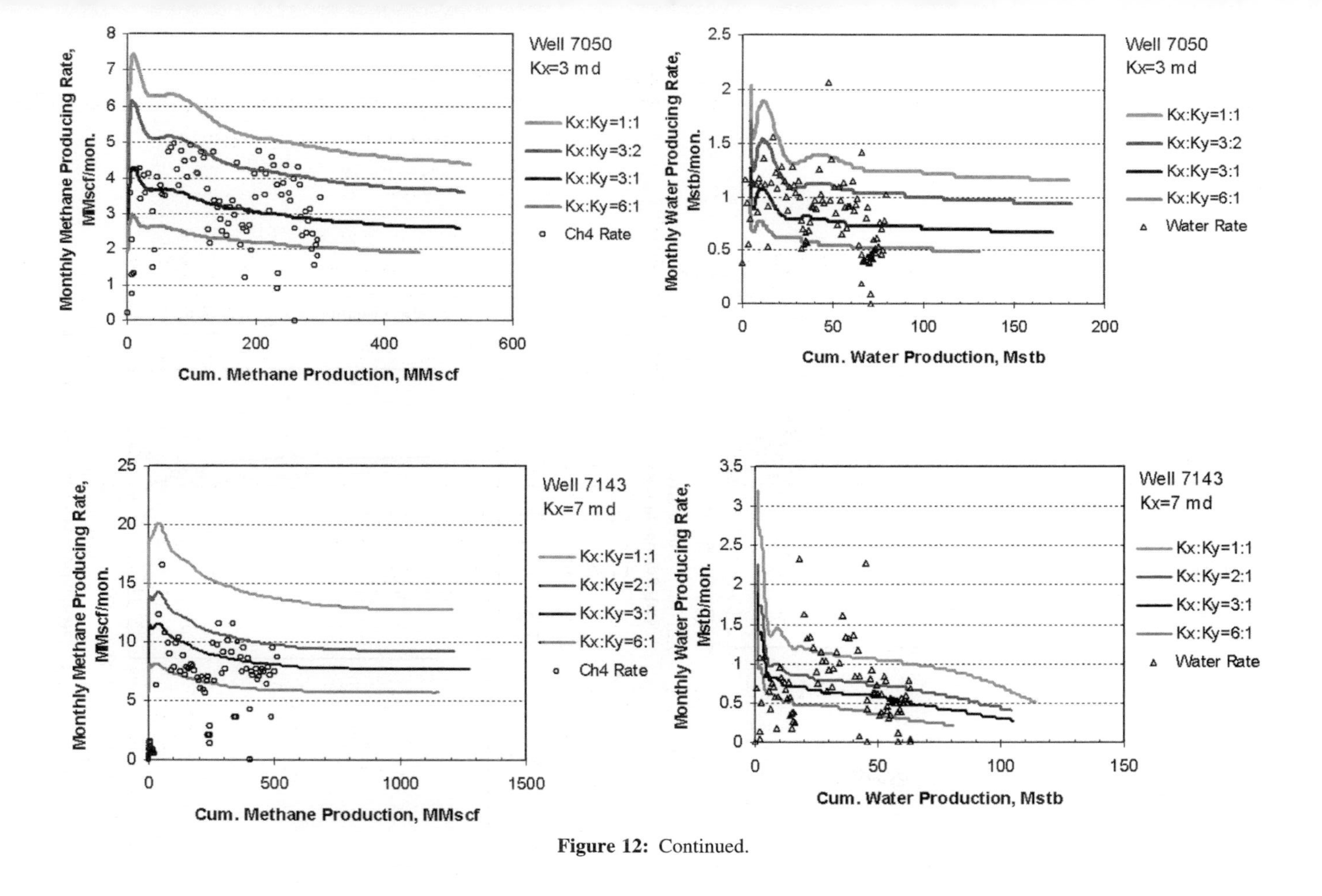

Figure 12: Continued.

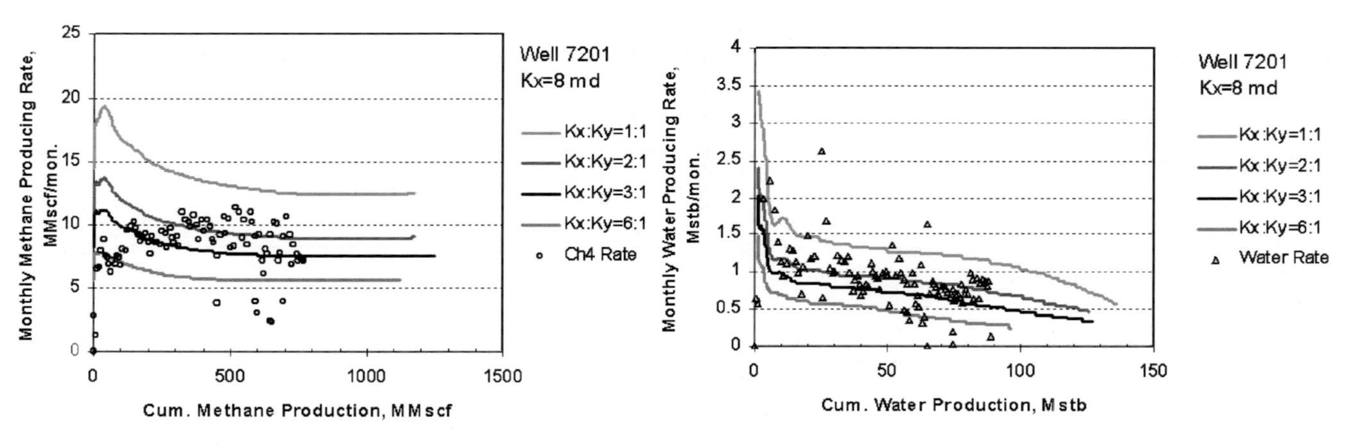

Figure 12: Continued.

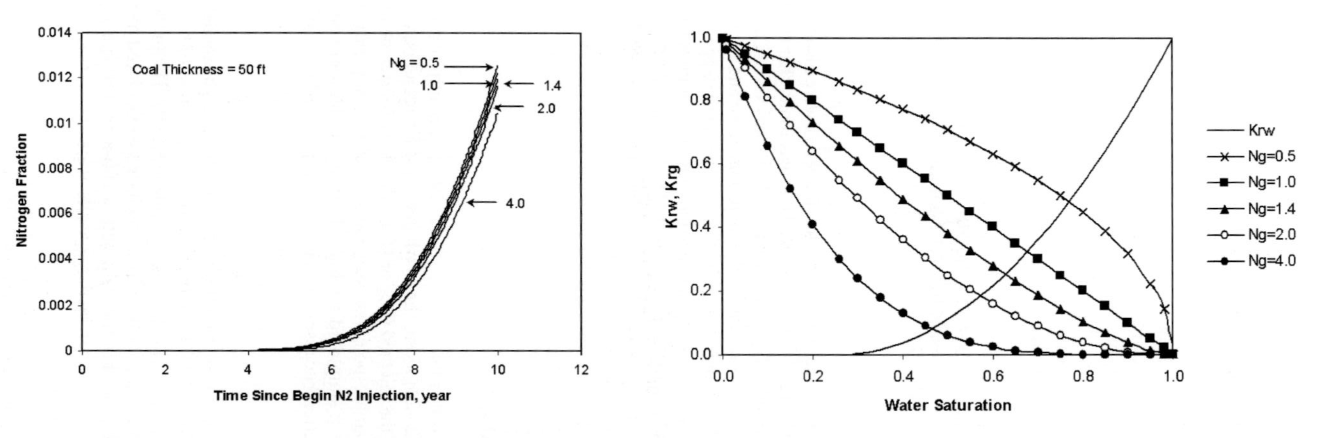

Figure 13: Nitrogen production cuts (left) simulated with different relative permeability curves (right). In the left figure, simulated curves are in ascending order of N_g from left to right.

permeability used for primary production was assumed. In Figure 13, N_g is the parameter used to define a gas relative permeability curve by Eq. (2):

$$K_{rg} = (1 - S_w)^{N_g} \tag{2}$$

Coal Matrix Shrinkage and Swelling

Cleat permeability is directly dependent on the width of the cleats and the cleat frequency. Cleat frequency is generally assumed to be constant, but cleat width is dependent on the in situ stress, the coal properties, and the gas content of the coal. Coal shrinks on desorption of gas and expands again upon readsorption, which changes the cleat width as well as permeability [3,15–18]. The matrix shrinkage (volumetric strain) due to the release of the adsorbed gas can be modeled with a Langmuir curve analogous to the adsorbed gas isotherm [18]. The coal shrinkage and permeability model developed by Sawyer et al. [3] was applied in the COMET2 simulator. No injectivity loss due to N_2 injection was observed in the Tiffany Unit. Because no laboratory data of coal shrinkage/swelling were available from the Tiffany field, different parameter settings were tested for CO_2 injection. In some cases, the increase of bottomhole pressure caused by permeability reduction became too high to sustain the injection rate because of the restricted injection pressure. More importantly, when a total net pay of 50 ft was used simulations failed to predict any CO_2 breakthrough even for a simulated time of more than 100 years. This is not consistent with what observed from the Allison Unit [6,9].

CO_2/N_2 Contacted Volume in Coal

Besides the gas relative permeability, other key reservoir parameters were also tuned in an attempt to match the early N_2 breakthrough time and high N_2 cuts (Figure 3). It was found that an acceptable match could be achieved only if a significant reduction in N_2–coal contact volume was assumed. The left figure of Figure 14 shows the effect of the net pay thickness on the N_2 breakthrough time and N_2 cut. In comparison with the actual field performance (Figure 3), it suggests that only about one-tenth to one-fifth of the total pay interval may be contacted by the injected N_2. The result is consistent with the findings from the mechanistic model. Under the same model settings and assumptions, the effect of CO_2–coal contact volume was also examined. A much-delayed CO_2 breakthrough was predicted as illustrated in the right figure of Figure 14. In comparison to an N_2 breakthrough time of about 2 years, the predicted CO_2 breakthrough time may occur about 20 years after the CO_2 injection on a 160-acre well spacing in the Tiffany Unit. The CO_2 breakthrough time is also much later than that predicted by the mechanistic model (Figure 10).

OUTCROP SEEPAGE MODELING

Methane seepage has been observed from the Pine River [19–21], South Texas Creek, Valencia Canyon, Soda Springs, and other areas [22–24] along the north and west Fruitland outcrops. If injection wells are placed too close to seepage sites, the injected CO_2 or N_2 could likely follow the methane seepage paths and seeps from the outcrops. To examine potential seepage scenarios, a representative seepage model was developed. The model represents a simplified geological setting of the north and west Fruitland outcrops [24]. The focus was on CO_2 injection because potential CO_2 seepage paths must be assessed for any large-scale CO_2 storage in the basin.

Model Configuration

Figure 15 shows the configuration of the representative seepage model. The model is a 2-layer, 1.25-mile by 12-mile strip with a downward dip of 2.92° from the outcrop to the bottom of the basin. There are a total of 240 (5 by 48) grids in each layer with a grid size of 0.25 mile (1320 ft). The model consists of two seepage wells to represent the 1.25-mile outcrop and three water recharge wells placed just below the water table. A total of 28 production wells were placed in the strip with a 160-acre well spacing. Production wells were perforated only in the top layer, and water recharge wells were opened only to the bottom layer. The thickness ratio between top and bottom layers was set to 10:1.

Groundwater Recharge

The annual precipitation in the Colorado portion of San Juan basin varies from 10 to 30 in. per year [25,26]. Along the Fruitland outcrop, an average precipitation of 22 in. per year was used in this study. The recharge

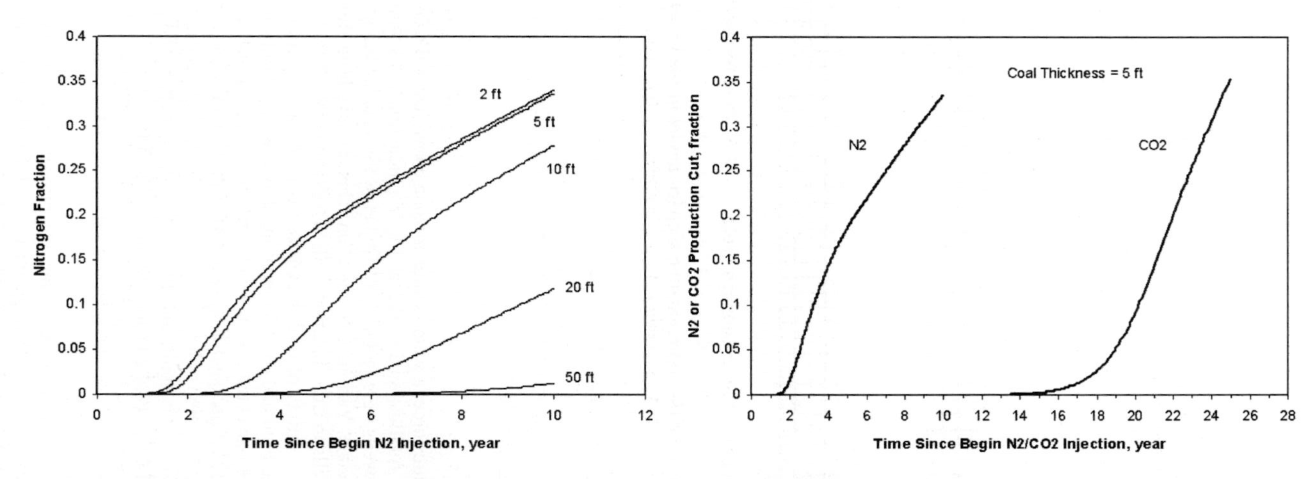

Figure 14: Nitrogen production cuts simulated with different coal net pay thicknesses (left). The comparison between nitrogen and CO_2 breakthrough time and cuts with a net pay interval of 5 ft (right).

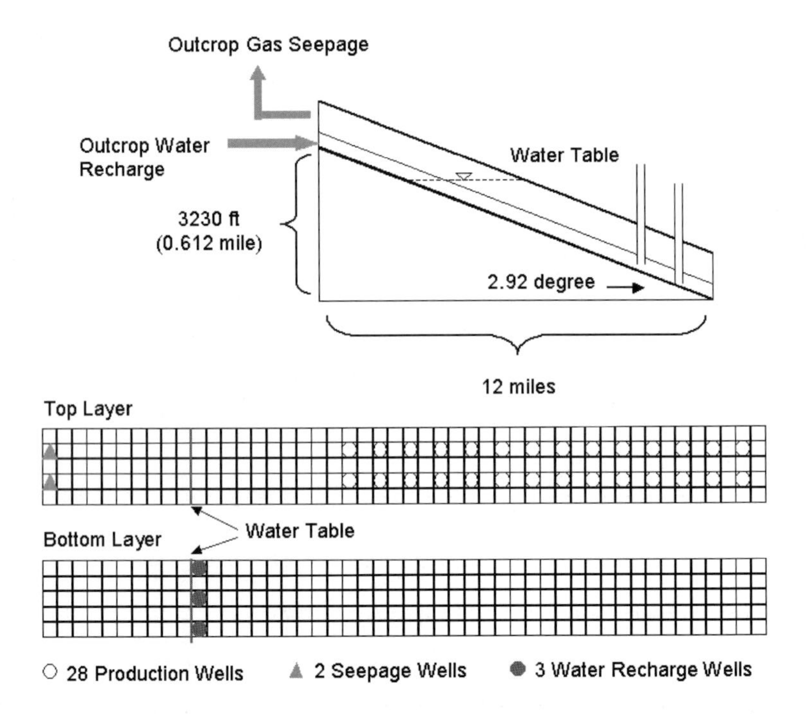

Figure 15: The configuration of the representative seepage model of the Fruitland coal outcrops.

rate is only about 1% of the precipitation [25,26]. Most recharge water migrates to adjacent rivers and creeks. An estimated 15% of the recharge water actually enters the basin. Based on above statistics, an estimated outcrop recharge rate of about 100 barrels per mile per day was calculated and used in the representative seepage model.

Preferable Scenarios

Because of the capillary pressure force, a water-saturated zone above injected CO_2 could help to prevent CO_2 migrating up to outcrops. As illustrated in Figure 15, 28 production wells were placed on a 160-acre well spacing where the top two wells were vertically 673 ft below the water table and horizontally more than 2.5 mile away from the water table. Various CO_2 injection schemes have been simulated, which includes converting 2–14 production wells to CO_2 injection wells. To examine the effect of CO_2 sweeping volume on methane and CO_2 seepage, a variety of coal thickness was used, ranging from 2 ft to an approximate average thickness of 52 ft of the Fruitland coal. Simulations started with a stabilization period of 100 years to stabilize the methane seepage rate at the current level. Carbon dioxide was then injected in the converted injection wells at a rate of 3200 Mcf per day for 30 years. After the CO_2 injection simulations continued for another 200 years without any production or injection. For all cases, no CO_2 seepage was predicted from the outcrop. Also as shown in Figure 16, no significant change in methane seepage was predicted even for cases with small pay intervals (small CO_2–coal contact volumes) of 2–5 ft.

Extreme Scenarios

When methane recovery reaches an economic limit, the priority objective will change to effectively store the injected CO_2. The extraction of a large quantity of CBM water that is required to release methane from coal surface usually causes a large drawdown of the potentiometric surface of depleted coalbeds. Consequently, it may result in a drawdown of the water table in coal seams and increase

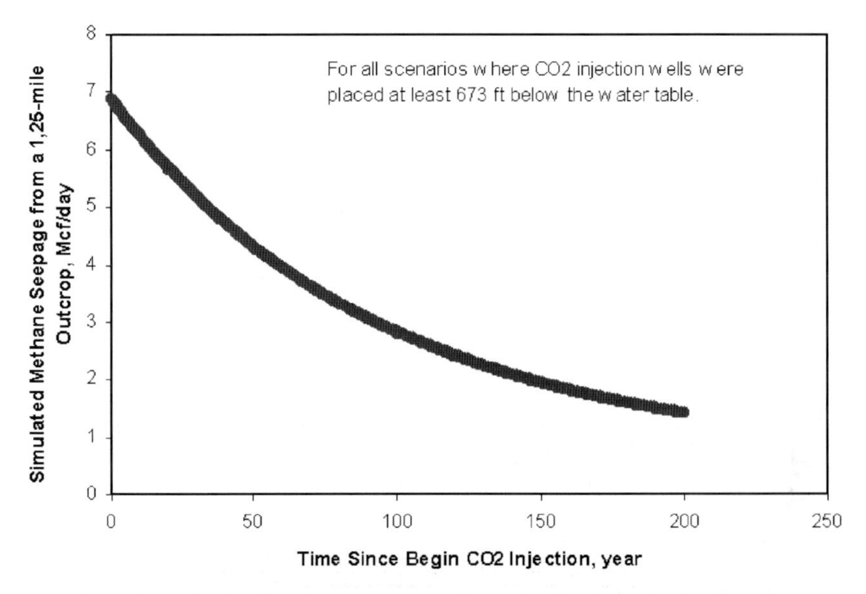

Figure 16: Simulated methane seepage under preferable CO_2 injection scenarios.

the risk of CO_2 migrating above the water table. To simulate the worst-case scenarios, two CO_2 injection wells were placed above the water table. The same injection rate and simulation scheme used for the preferable scenarios were used in simulating the extreme scenarios. Cases with various combinations of coal thickness, between 2 and 50 ft, and the distance of injection wells to the outcrop, from 1 to 5 mile, have been simulated. Figures 17 and 18 show that a large CO_2 and methane breakthrough may occur if the CO_2 injection wells are placed too close to the outcrop (within 2 mile). Figures 19 and 20 show that CO_2 and methane seepage rates reduced significantly when the injection wells were located more than 2 mile away from the outcrop.

DISCUSSIONS

The heterogeneity of the Fruitland coal in both its distribution and composition [27–29] strongly affects the effectiveness of the gas injection for ECBM and CO_2 storage. Methane recovery efficiency is on a well-by-well basis as observed in the Tiffany and Allison Units. Critical factors include cleat permeability, coal seam continuity, CO_2/N_2 sweeping volume, coal shrinkage/swelling, and seal integrity to prevent leakage of injected gas.

A good understanding of the sorption of CH_4, CO_2, N_2, and water mixtures on coal is essential for a credible modeling of the gas injection processes. Given the complexity of mobile gas mixtures of unequally sized molecules with different interactions adsorbed on the heterogeneous surface of coal matrices, the prediction of multi-component adsorption equilibriums on wet coal from single-component data is one of the most challenging problems in ECBM simulation. Simple analytical models, such as Langmuir, Gibbs, or potential theory based approaches [30,31] often show difficulty in accurately predicting the sorption behaviors of a mixture that contains three or more components [8,32].

Accordingly, further improvement to CBM simulation model is needed, especially in modeling coal structure reactions to gas injection and the multiple component adsorption/desorption processes.

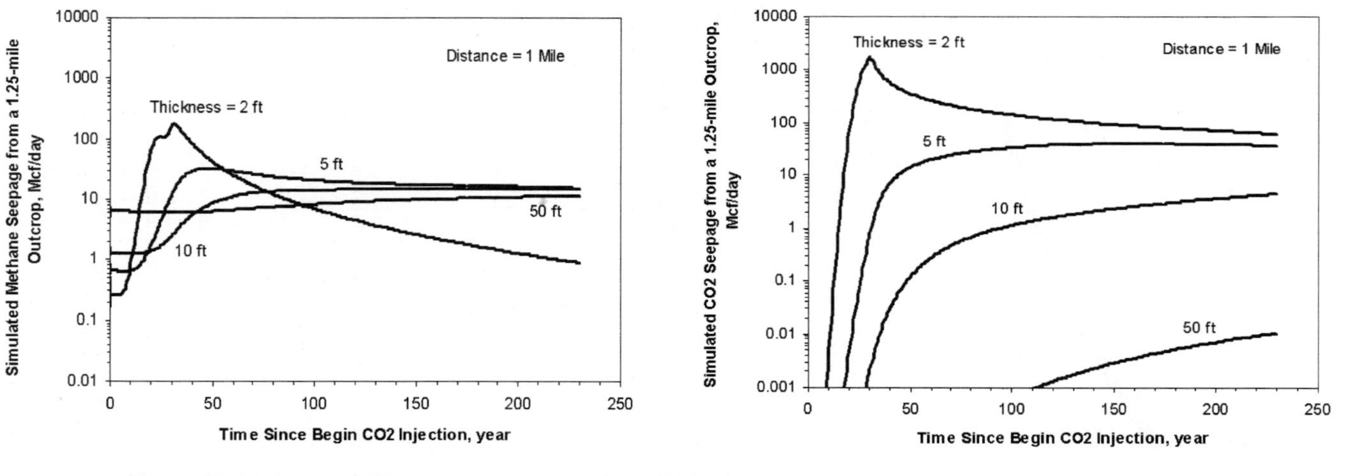

Figure 17: Methane and CO_2 seepage rates vs. coal net pay thickness where the injection is 1 mile from the outcrop.

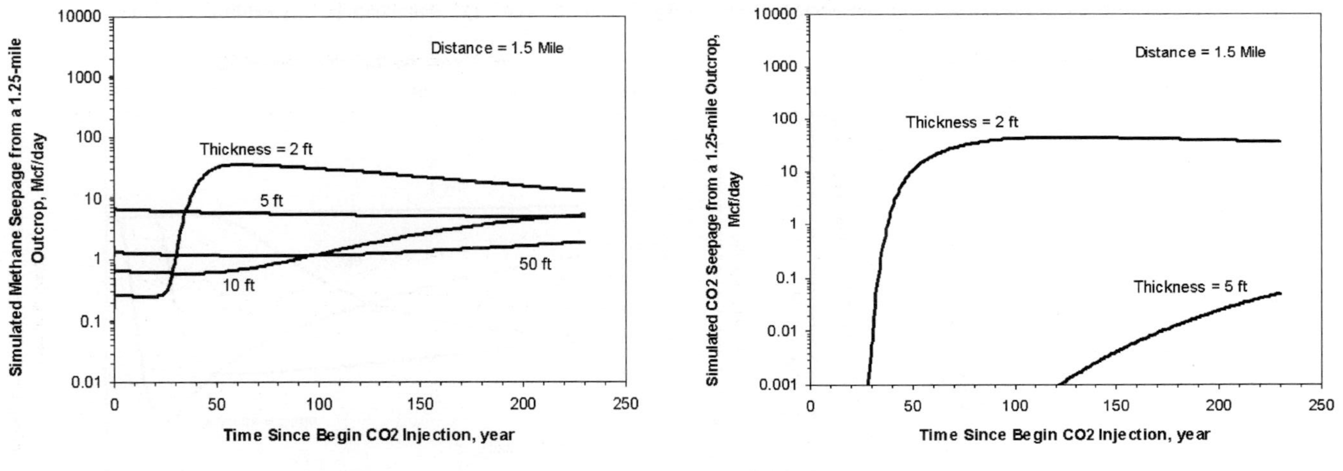

Figure 18: Methane and CO_2 seepage rates vs. coal net pay thickness where the injection is 1.5 mile from the outcrop.

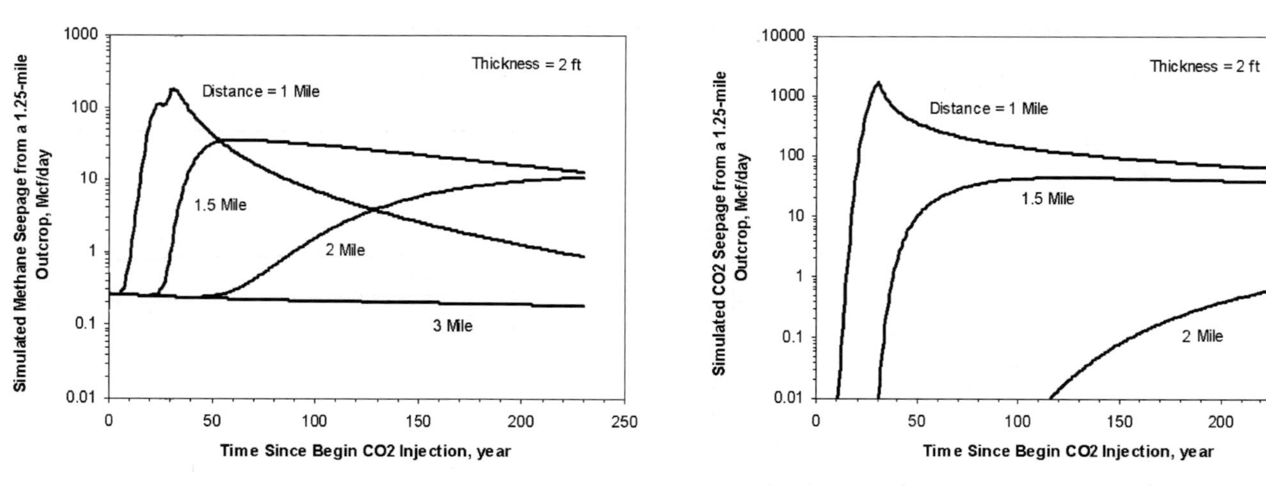

Figure 19: Methane and CO_2 seepage rates vs. injection distances using a net pay thickness of 2 ft.

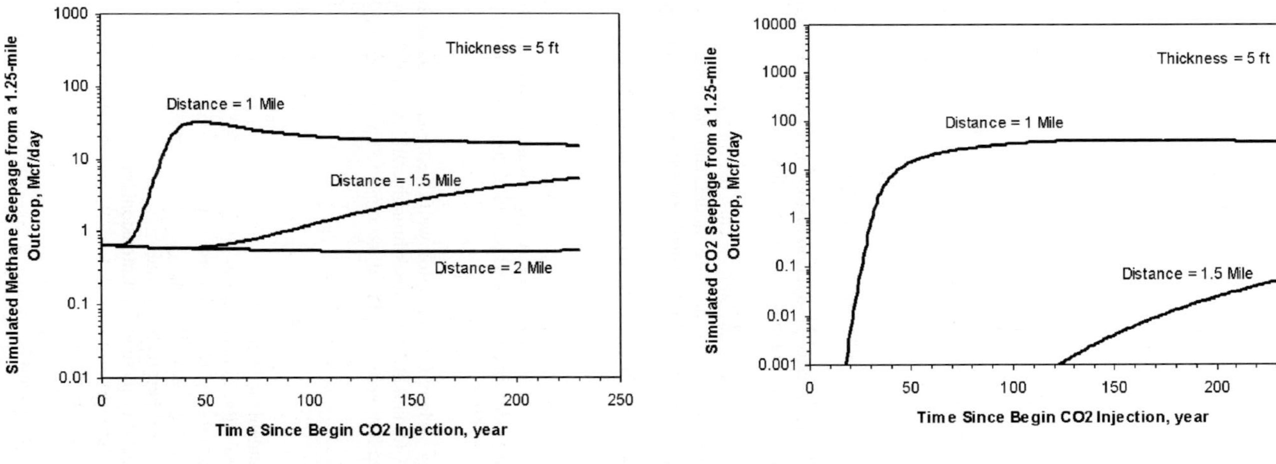

Figure 20: Methane and CO_2 seepage rates vs. injection distances using a net pay thickness of 5 ft.

CONCLUSIONS

Conventional compositional reservoir simulators, such as GCOMP, and currently available CBM simulators, such as COMET2/3, are generally capable of modeling the primary methane production in coalbeds but may encounter more difficulties in the history match and prediction of gas (CO_2, N_2, or CO_2–N_2 mixture) injection processes. With the limitations discussed in the chapter and the specific data set from the Tiffany Unit, the following conclusions have been drawn from this simulation study.

Simulations should use sorption isotherms measured under the actual reservoir conditions. Laboratory-measured isotherms on dry coals need to be rescaled by matching field history performance. Without rescaling, the simulation forecast of CO_2 or N_2 injection may not be accurate.

During the primary production, the gas to water production ratio is very sensitive to cleat porosity in low porosity coalbeds, such as in the Tiffany Unit.

Based on the history match, simulation verifies that the field permeability aspect ratio in Tiffany Unit is approximately in the range of 2:1 to 3:1.

During nitrogen injection, the elevated pressure caused the coal fractures on the preferential permeability trends not only to expand but also to extend from injectors to producers. Even in the low-pressure regions near the producers, the permeabilities were higher than expected.

Simulation models that match the primary production history may not be accurate in forecasting CO_2 or N_2 injection due to the heterogeneity of coalbeds and the reaction of coal structure to gas injection.

To match the early N_2 breakthrough time and high N_2 cuts, the coal thickness had to be reduced to one-third of the average total pay (50 ft) for the mechanistic model, and one-tenth (5 ft) for the dual model. This suggests that the injected N_2 may only contact a small portion of the available coal volume.

In matching the gas (CH_4, CO_2, or N_2) production cut, it may not be effective to tune the gas relative permeability while gas–coal contact volume and gas adsorption/desorption are the more dominant factors.

Under preferable scenarios, if CO_2 injection wells are placed below the water table, vertically more than 673 ft below the water table in the simulated cases, no significant change in methane seepage from outcrop was predicted by the seepage model. In a simulated period of 200 years, no CO_2 seepage from outcrop was predicted after 30-year CO_2 injection.

Under the worst case scenario, where CO_2 injection wells were placed above the water table, the seepage model predicted that a large CO_2 and methane breakthrough could occur if the sweeping volume of injected CO_2 is limited and CO_2 injection wells are placed too close to an outcrop, e.g. within 2 mile.

NOMENCLATURE

C	coal matrix gas content, scf/ton coal
K_{rg}	gas relative permeability, dimensionless
K_{rw}	water relative permeability, dimensionless
K_x	face (dominant) cleat permeability, md
K_y	butt (subordinate) cleat permeability, md
N_g	gas relative permeability parameter, dimensionless
P	coal reservoir pressure, psi
P_L	Langmuir pressure constant, psi
S_g	gas saturation, dimensionless
S_w	water saturation, dimensionless
V_L	Langmuir volume constant, scf/ton coal
BP	British Petroleum

BLM	Bureau of Land Management
CCP	CO_2 Capture Project
CBM	coalbed methane
CRADA	Cooperative Research and Development Agreement
DOE	Department of Energy
ECBM	enhanced coalbed methane recovery
INEEL	Idaho National Engineering and Environmental Laboratory
JIP	Joint Industry Program
NETL	National Energy Technologies Laboratory

ACKNOWLEDGEMENTS

We are grateful to Daryl Erickson of BP America Inc. for providing helpful insights and the Tiffany field data. This work was supported in part by a Cooperative Research and Development Agreement (CRADA) between BP America Inc., as part of the CO_2 Capture Project (CCP) of the Joint Industry Program (JIP), and the US Department of Energy (DOE) through the National Energy Technologies Laboratory (NETL) under contract DE-AC07-99ID13727.

REFERENCES

1. S.R. Reeves, Enhanced CBM recovery, coalbed CO_2 sequestration assessed, *Oil Gas J.* (2003) July 14.
2. Amoco Production Company, Computer User's Instructions for GCOMP, May 16, 1998.
3. W.K. Sawyer, G.W. Paul, R.A. Schraufnagle, Development and application of a 3D coalbed simulator, Paper CIM/SPE 90-119, *Proceedings of the Petroleum Society CIM*, Calgary, June 1990.
4. Advanced Resources International, Inc., COMET2 Version 2.11 Users Guide, December 11, 2001.
5. S. Reeves, C. Clarkson, D. Erickson, Selected field practices for ECBM recovery and CO_2 sequestration in coals based on experience gained at the Allison and Tiffany Units, San Juan Basin, *DOE Topical Report*, December 2002.
6. D. Erickson, Overview of ECBM commercial demonstration pilot at Tiffany Unit, *Internal Report*, BP America Inc., October 2002.
7. K.T. Raterman, Assessing reservoir heterogeneity from a single well injection test, *Internal Report*, Amoco Production Company, November 1996.
8. K.A.M. Gasem, R.L. Robinson Jr., S.R. Reeves, Adsorption of pure methane, nitrogen, and carbon dioxide and their mixtures on San Juan Basin coal, *DOE Topical Report*, May 2002.
9. S.R. Reeves, Geologic sequestration of CO_2 in deep, unmineable coalbeds: an integrated research and commercial-scale field demonstration project, SPE 71749, *Proceedings of the SPE Annual Technical Conference and Exhibition*, New Orleans, September 30–October 3, 2001.
10. J. Liang, K.T. Raterman, E.P. Robertson, A mechanistic model for CO_2 sequestration in Tiffany coal bed methane field, *2003 International Coalbed Methane Symposium*, Tuscaloosa, AL, May 7–8, 2003.
11. D.H.S. Law, L.G.H. van der Meer, M.J. Mavor, W.D. Gunter, Modeling of Carbon Dioxide Sequestration in Coalbeds: A Numerical Challenge, Alberta Research Council, 2002.
12. S. Reeves, A. Taillefert, L. Pekot, C. Clarkson, The Allison Unit CO_2-ECBM pilot: a reservoir modeling study, *DOE Topical Report*, February 2003.
13. J.E. Warren, P.J. Root, The behavior of naturally fractured reservoirs, *Trans. AIME* **228** (1963) 245–255.
14. J.P. Seidle, L.E. Arri, Use of conventional reservoir models for coaled methane simulation, CIM/SPE Paper 90-118, presented at the *International Technical Meeting*, Calgary, Alberta, Canada, June 10–13, 1990.
15. I. Palmer, J. Mansoori, How permeability depends on stress and pore pressure in coalbeds: a new model, *Paper SPE 52607, SPEREE*, December 1998, pp. 539–544.
16. L.J. Pekot, S.R. Reeves, Modeling the effects of matrix shrinkage and differential swelling on coalbed methane recovery and carbon sequestration, Paper 0328, *Proceedings of the 2003 Coalbed Methane Symposium*, Tuscaloosa, AL, May 5–7, 2003.
17. J.Q. Shi, S. Durucan, Changes in permeability of coalbeds during primary recovery—part 1: model formulation and analysis, *Proceeding CBM Symposium*, Tuscaloosa, AL, May 5–9 (2003).

18. S. Harpalani, G. Chen, Estimation of changes in fracture porosity of coal with gas emission, *Fuel* **74** (1995) 1491.
19. Advanced Resources International, Inc., Gas Seepage in the Pine River Area, Colorado, prepared for the Geological and Reservoir Engineering Subcommittee of the Pine River Fruitland Coal Investigative Team, November 1994.
20. Amoco, Pine River Fruitland Coal Outcrop Investigation: Southern Rockies Business Unit, Amoco Production Company, Denver, CO, 1994.
21. P. Oldaker, Monitoring Data Review, Pine River Ranches, prepared for Colorado Oil & Gas Conservation Commission and Amoco Production Company, USA, 1999.
22. Bureau of Land Management (BLM), San Juan Field Office, Coalbed Methane Development in The Northern San Juan Basin of Colorado, December 1999.
23. Questa Engineering Corporation, The 3M Coalbed Methane Reservoir Model, Prepared for the Southern Ute Indian Tribe, Ignacio, CO, the Colorado Oil and Gas Conservation Commission, Denver, CO, and the Bureau of Land Management, Durango, CO, May 26, 2000.
24. L.L. Wray, Late Cretaceous Fruitland Formation geologic mapping, outcrop measured sections, and subsurface stratigraphic cross sections, Northern La Plata County, CO, *Colorado Geological Survey Open File Report 00-18*, Denver, CO, 2000.
25. Applied Hydrology Associates, Inc., 3M Project, San Juan Basin, Colorado and New Mexico, Hydrologic Modeling Report, prepared for the Southern Ute Indian Tribe, the Colorado Oil and Gas Conservation Commission, and the Bureau of Land Management, 2000.
26. J.A.M. Thomson, Modeling groundwater–surface water interaction, presented at the *Conference of CBM Water Management Strategies*, Durango, CO, February 10–11, 2003.
27. D. Bland, Coalbed methane from the Fruitland Formation, San Juan Basin, New Mexico, *North Mexico Geological Society 43rd Annual Conference*, 1992.
28. C.M. Tremain, S.E. Laubach, N.H. Whitehead III, Fracture (cleat) patterns in Upper Cretaceous Fruitland Formation coal seams, San Juan basin, in: W.B. Ayers Jr., W.R. Kaiser (Eds.), *Coalbed Methane in the Upper Cretaceous Fruitland Formation*, San Juan Basin, New Mexico and Colorado: New Mexico Bureau of Mines and Mineral Resources, Bulletin 146, 1994, pp. 87–102.
29. C.R. Nelson, Geologic controls on effective cleat porosity variation in San Juan basin Fruitland Formation coalbed reservoirs, *Proceedings, International Coalbed Methane Symposium*, Paper 108, Tuscaloosa, AL, 2001, pp. 11–19.
30. D.M. Ruthven, *Principle of Adsorption and Adsorption Processes*, Wiley, New York, 1984.
31. R.T. Yang, Gas Separation by Adsorption Processes, Butterworths Publishers, London, 1987.
32. F.R. Siperstein, A.L. Myers, *Mixed-Gas Adsorption*, *AIChE J.* **47** (5) (2001).

Carbon Dioxide Capture for Storage in Deep Geologic Formations, Volume 2
D.C. Thomas and S.M. Benson (Eds.)

Chapter 15

CO_2 CONDITIONING AND TRANSPORTATION

Geir Heggum[1], Torleif Weydahl[2], Roald Mo[1], Mona Mølnvik[2] and Anders Austegaard[2]

[1]Reinertsen Engineering, Trondheim, Norway
[2]SINTEF Energy Research, Trondheim, Norway

ABSTRACT

The aim of the CO_2 Conditioning and Pipeline Transportation project is to advance the development of cost effective and safe methods for CO_2 compression and pipeline transportation. Optimized design for the compression process and pipeline system requires accurate and reliable predictions of fluid properties, particularly density and water solubility.

Existing CO_2 pipeline transportation systems (onshore USA and Canada; offshore Norway) are reviewed in terms of operational parameters, particularly drying specifications. Based on calculations of water solubility for a selected case, it is found that the most stringent drying requirements (e.g. 50 ppm proposed for Hammerfest LNG) may be relaxed to ~600 ppm (present USA Kinder Morgan specification). Today there is little experience with subsea pipelines for CO_2 transportation, particularly in deep waters and over long distances. The intension of this study is to build up confidence in the technology and save costs for future projects.

Thermodynamic models and tools for calculating properties for CO_2 and CO_2-rich mixtures have been verified against experimental data. For CO_2 density the Lee–Kesler model is in satisfactory agreement with National Institute of Standards and Technology (NIST) data both in gas and liquid phase. For solubility of water in pure CO_2, the Soave–Redlich–Kwong equation of state with adjusted binary coefficient to 0.193 in van der Waals mixing rule can be applied, and gives a good approximation to the data collected from literature. Adding impurities as CH_4, N_2, H_2S and amines to the CO_2 mixture will affect the solubility of water, e.g. adding 5% methane lowers the water solubility in the liquid phase considerably. However, very little experimental data on water solubility in these mixtures is available in the literature.

In order to inhibit hydrate formation and prevent excessive corrosion rates for carbon steel, no free water should be allowed in the pipeline. Thus, water removal is usually required upstream of the pipeline inlet. For a typical case, theoretical calculations show that the limit for free water precipitation at supercritical conditions in the pipeline averages ~1300 ppm. This suggests that water content requirements might be relaxed and opportunities for alternative, more cost-effective water removal solutions are provided.

INTRODUCTION

Project Background

The overall goal of this study is to support the development of cost effective and safe methods for CO_2 compression and transportation. The combined effort by SINTEF, Reinertsen and IFE is aimed at qualifying state-of-the-art process engineering, material selection and pipeline engineering methods to ensure optimal design of CO_2 compression and pipeline systems.

Unlike natural gas, CO_2 will be in the liquid or supercritical state, quite close to the critical point in the pipeline. In this state, the CO_2 density may be three times higher than the corresponding density of natural gas (molar densities) at same temperature and pressure. The density is an important property with respect to the prediction of the static pressure gain in the pipeline flow.

The water solubility in CO_2 is dependent on the temperature and pressure, and these characteristics can be utilized in the drying process. Free water in the pipeline will cause hydrate formation and corrosion problems under certain conditions, and must hence be avoided. Accurate estimations of the water solubility are consequently an important issue, which is addressed in the present work. The density and the solubility of water in CO_2 will also vary as a function of other gases in the CO_2 such as hydrocarbons, nitrogen and other impurities.

Experience with CO_2 Transportation

Onshore transportation and injection of CO_2 for enhanced oil recovery (EOR) has been going on for three decades in the United States, providing much experience on issues regarding on-shore facilities. Conversely, offshore experience is very limited.

Onshore

Kinder Morgan (KM) is a US based company with more than 30 years of experience in CO_2 transportation and injection for EOR. In 2002, 17 Mt CO_2 was transported in their pipeline system [1].

The Cortez pipeline is the largest example, with a diameter of 30 in. and a capacity of 12.2 Mt CO_2/yr. The CO_2 for this pipeline is produced from the McElmo field, which contains 97 mole% of pure CO_2. Before the CO_2 is pumped down the pipeline, it is cleaned, dehydrated and compressed to supercritical pressure (145 bar). To achieve the required pressure for CO_2 transportation, KM makes use of pumps rather than compressors, which reduces operational costs. No chemicals are necessary in the transportation systems. Dehydration is necessary to obtain the requirements for using carbon steel. According to KM, the requirement for CO_2 pipelines, used for EOR in the US, is maximum 600 ppm water [2]. Due to the rather relaxed water requirement, diethylene glycol (DEG) can be used for dehydration, instead of TEG (triethylene glycol, since the boiling point is lower, DEG is less effective than TEG). The transportation of CO_2 is in the liquid phase, below the supercritical point for CO_2 (below critical temperature, but above critical pressure).

In Western Canada, acid gas (a mixture of H_2S and CO_2) is removed from the natural gas. Typically, a four-stage compression process from 0.80 to 60 barg is used with cooling below 20 °C. The acid gas is then in liquid phase, provided the methane content is no greater than 1–2%. Selection of material for the acid gas injection line between the plant and the injection well is generally related to whether or not the acid gas has been dehydrated. For dehydrated gas, sour service carbon steel materials could be used, such as CSA-Z662 Grade 359 Category II [3]. If dehydration by glycol is used to ensure that no free water drops out, it should be installed after the second or third compression stage, since the solubility of glycol is lowest in the pressure range of 40–55 bar [4].

Offshore

At Sleipner Vest, operated by Statoil in the North Sea, CO_2 removal and injection is done offshore. Amine separation is used to remove CO_2 from the gas stream. The CO_2 is injected into a saline aquifer, called the Utsira formation, 800 m below seabed. The purpose of the CO_2 removal system on Sleipner Vest is to reduce the CO_2 content in the export gas from maximum 8.95 to 2.5 vol% to meet sales gas specifications and to avoid taxed emissions to the atmosphere [5]. The gas injection system is designed for a gas rate of 1.7 MSm^3/d. CO_2 is separated by an amine absorber. The separated CO_2 is saturated with water at operating conditions (0.1 barg at 70 °C). Water is condensed in a CO_2 water wash column operating directly counter-current with cooled re-circulated water. No further drying is implemented before the CO_2 compressor train.

The CO_2 gas is compressed to 80.0 barg in four stages. An aftercooler is installed downstream the compressor on stage 1–3, which cools the gas to 30 °C. The condensed liquid (mainly water) is removed in a suction scrubber on each stage. The water content is 3.9 mole% for the 1st stage compressor and 0.3 mole% for the 3rd stage. The solubility of water in CO_2 is lower at the third stage pressure of 32 barg than at the wellhead pressure of 80 barg. This ensures robustness with respect to hydrate formation.

EXPERIMENTAL/STUDY METHODOLOGY

In the present work both commercial and in-house calculation tools have been used to calculate thermodynamic properties of CO_2 and CO_2 mixtures. The commercial tools are the flowsheet programs

PRO II 6.0 [6] and HYSYS 3.0.1 [7]. In addition some in-house tools to calculate solubility in binary and ternary mixtures have been developed. These tools solve the general equations for predicting solubility and density of CO_2 mixtures. The reader is referred to Refs. [9,10] for a comprehensive description of the equations of state (SRK, LK and BWRS) and mixing rules presented in this chapter. This study is limited to the simple van der Waals mixing rule.

The CO_2 compression and drying process is calculated in HYSYS. For CO_2 pipeline transport, a program that calculates the wellhead pressure for a mixture consisting mainly of CO_2 has been developed. The program solves the distributed mass, momentum and heat balance equations for one or two-phase stationary flow in a one-dimensional pipeline. A more detailed description of the program is found in Ref. [8]. The output from the calculations are the stationary temperature, pressure, density and void fraction profile in the pipeline in addition to the temperature distribution in the pipeline material.

RESULTS AND DISCUSSION

Properties for CO_2 Pipeline Transportation

In this section, the thermodynamic models and tools are discussed through comparison with available experimental data.

Properties for pure CO_2

Regarding thermodynamic properties of pure CO_2, much research has been carried out and accurate measurements are available. The National Institute of Standards and Technology (NIST) [11] provides properties such as densities, enthalpy and viscosity for CO_2, from 216 to 1100 K, and for pressures up to 8000 bar. These data are used as reference densities when comparing computations with experiments.

Figure 1 shows the comparison between NIST data and various models. The sudden increases in density with increasing pressure are due to the phase transition from gas to liquid. As seen in the figure, all models compare well with the NIST data in the gas phase, but not as well as in the liquid phase, especially at temperatures close to the critical point ($T_c = 31\,°C$, $P_c = 73.8$ bar). The Benedict–Webb–Rubin–Starling

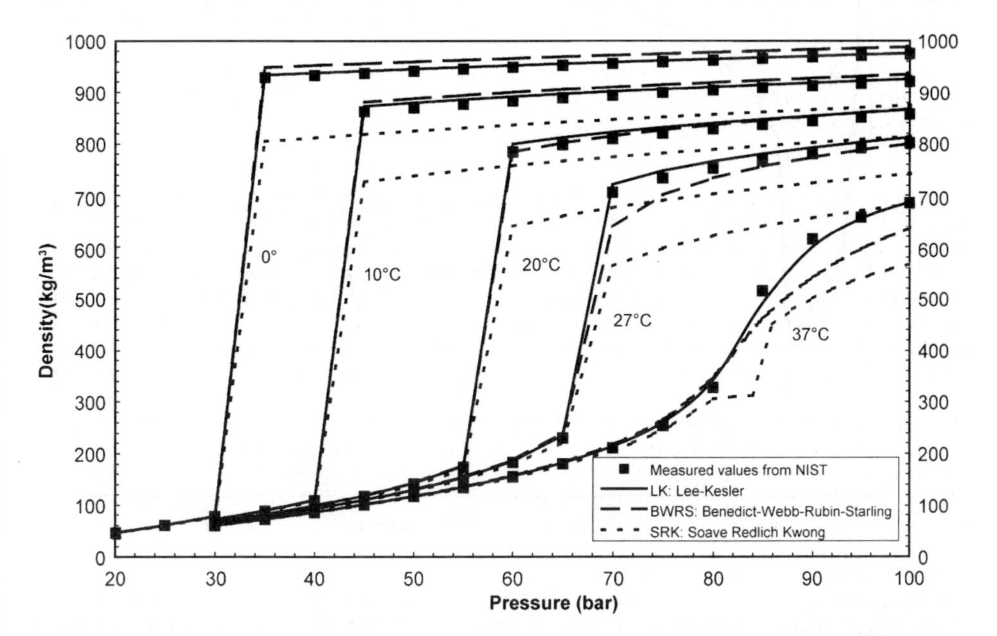

Figure 1: Calculations and experiments for CO_2 density as a function of pressure at selected temperatures.

(BWRS) model is calculated with PRO II, the Soave–Redlich–Kwong (SRK) and Lee–Kesler (LK) models are calculated with in-house codes. The LK model gives the most accurate result with a maximum error of 4.7% for the results in figure, even when CO_2 is close to the critical point. For these results the SRK and the BWRS model have a maximum error of 26 and 12%, respectively.

Water solubility in pure CO_2

The solubility of water in CO_2 vapor at a given temperature decreases to a minimum as pressure is increased (see Figure 2). When the pressure is further increased, a phase transition to liquid occurs and the solubility increases again. These physical properties of CO_2 are important to consider when dehydrating the CO_2, as the minima in water solubility is the best operating point when "knocking out" water.

The binary coefficient between water and CO_2 is adapted to 0.193 for the standard van der Waals mixing rule with SRK equation of state. With this adaptation of the model the mean absolute error between calculations and experiments is 6.3%. The results in the liquid phase are much more sensitive to the binary interaction parameter than the results in the gas phase as shown in Figure 2. Three different binary coefficients are plotted in the figure, where 0.0392 and 0.23 are the default values for HYSYS and PRO II, respectively. It is important to note that the optimum binary coefficient for solubility of water in CO_2 is different from the coefficient for solubility of CO_2 in water. Because the commercial programs only operate with a single default binary coefficient for the CO_2–H_2O system, the calculations will not fit the results properly, unless the parameter is adjusted (Figure 2).

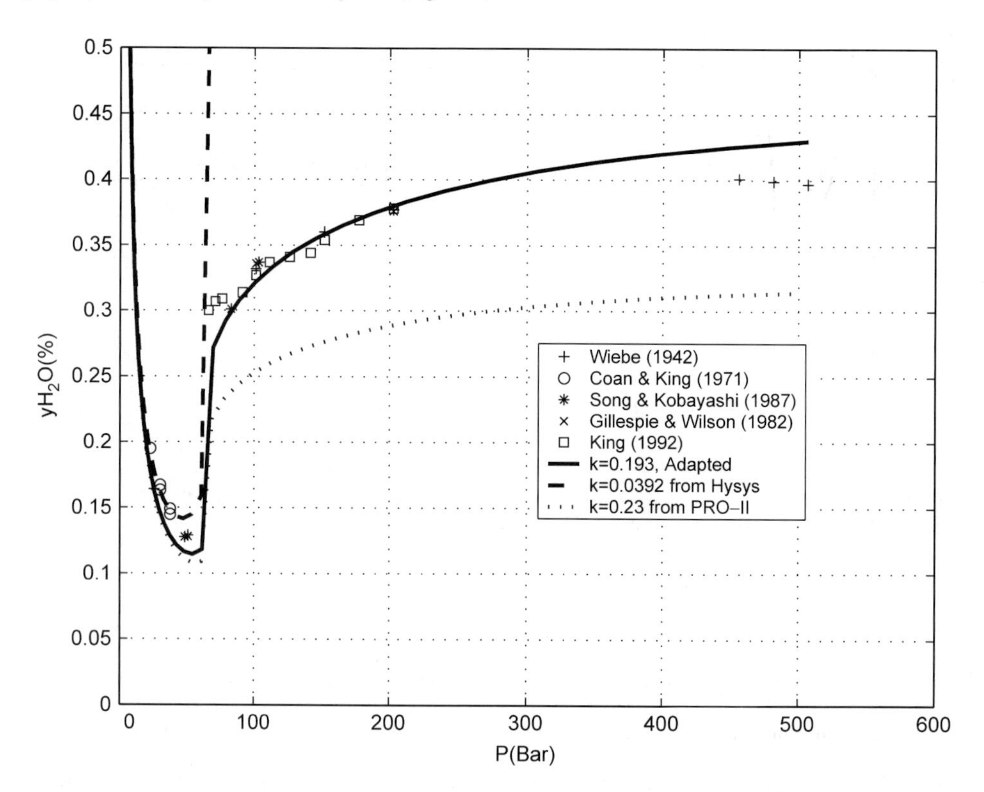

Figure 2: Comparison of model calculations using SRK with van der Waals mixing rule and various binary coefficients (k) at 26 °C with measurement data (from 24 to 28 °C) for solubility of water in CO_2. Given in molar percent (1% = 10,000 ppm). The experimental data are found in Refs. [12–16].

Solubility of water in CO_2 mixtures
Figure 3 shows experiments and calculations for solubility of water in pure CO_2 (solid line) and in CO_2 with 5.3% intermixture of methane (dotted line and the experiments of Song and Kobayashi [17]). The solubility of water in CO_2 and CO_2-CH_4 mixtures is about the same in the gas phase. The difference is however much larger in the liquid phase, where the solubility of water in a mixture of CO_2 and CH_4 is much lower than in pure CO_2. A practical consequence is that, to avoid free water precipitation, the gas must be dried to a lower water level, which again increases the cost. As can be seen from Figure 3, the pressure where liquid appears is also higher in a mixture of CO_2-CH_4 (71 bar) than in pure CO_2 (65 bar).

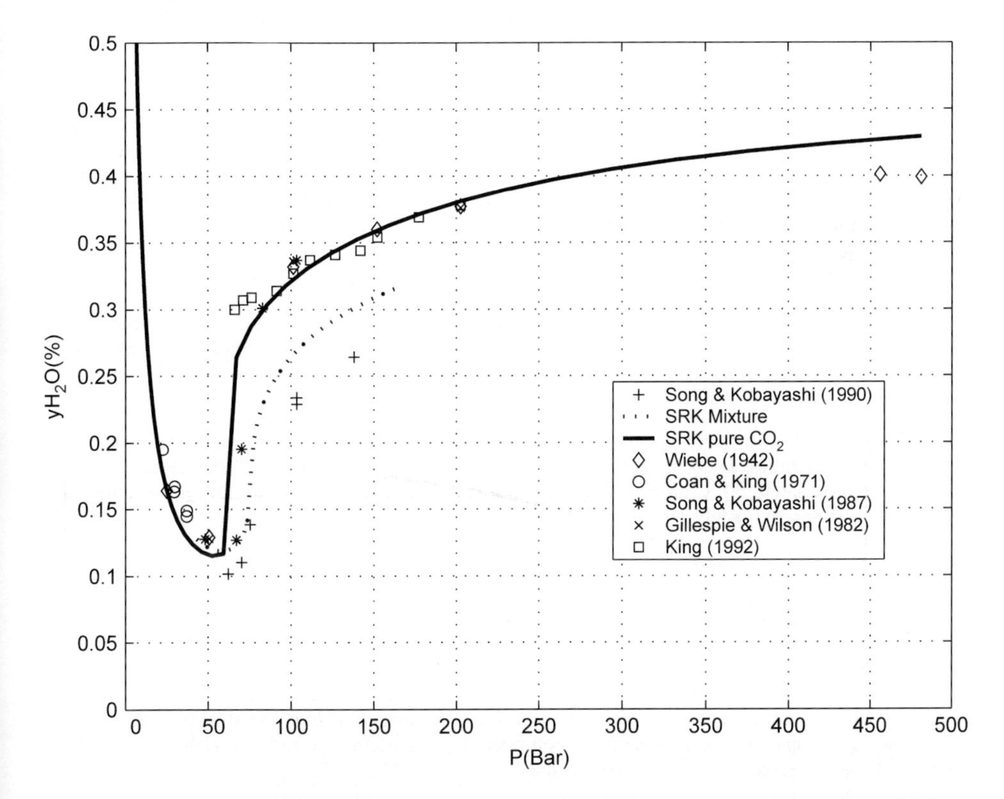

Figure 3: Comparison of model calculations at 26 °C with measurement data (from 24 to 28 °C) for solubility of water in CO_2 (solid line and Refs. [12–16]) and in a mixture of CO_2 and 5.3% CH_4 (dotted line and Ref. [17]).

A binary coefficient of 0.52 is used for the methane–water system. With this binary coefficient, the water solubility in the ternary mixture is overestimated in the liquid phase according to the experimental data. The coefficient can be further adjusted to fit the data better, but generally the amount of data is too sparse to draw any conclusions regarding model performance. The SRK with Huron Vidal mixing rule (Huron et al. [18]) is a better model for this mixture. Work in progress is to apply more advanced models to this mixture.

The effect of water solubility with intermixture of N_2 in CO_2 has not been properly verified due to lack of experimental data. A paper by Seo et al. [19] describes the three phase equilibrium conditions (aqueous liquid-hydrate-vapor) for binary mixtures of CO_2 and N_2, but does not measure the water solubility in this mixture. According to Carroll [20] no data exist for water solubility in a mixture of CO_2 and H_2S. However, we can assume that very small amounts of H_2S will not have a major effect on the water solubility in a CO_2 mixture. The phase behavior of H_2S and CO_2 is qualitatively similar [21]. The solubility in water is slightly larger for H_2S than for CO_2. Amines, from an amine separation process, will mainly occur in a liquid or aqueous phase (high boiling point), and will only affect the liquid/water phase in the mixture and not the gas phase.

System Design

The transportation system proposed includes a process facility for conditioning the CO_2 gas mixture, primarily compression, and pipeline(s) for transportation to the injection/deposition point. Typically, CO_2 is separated from hydrocarbon gas by amine absorption and may contain up to 5% CH_4, 5% N_2, 0.5% water, 100 ppm H_2S and a small amount of amines as it is fed to the compression process.

Compression process

In order to inhibit hydrate formation and prevent excessive corrosion rates for carbon steel, no free water should be allowed in the pipeline. Thus, water removal usually is required upstream the pipeline inlet.

The base case for the compression process includes compressors with coolers and scrubbers between each compression stage in order to reduce the gas temperature and knock out free water. In order to increase the pressure of CO_2 from 1 to 150 bar (pipeline requirement), four compression stages are considered. The "pure compression process" is shown schematically in Figure 4.

Water removal. By using coolers with seawater at 9 °C as a cooling medium, the gas temperature can be reduced to approximately 15 °C between each compression stage. For this case, theoretical calculations show that the water content in the CO_2 mixture can be reduced to approximately 600 ppm (mole), only by compression, intermediate stage cooling and scrubbing for dehydration.

Additional drying. The water content in the gas depends on the available cooling media temperature. In areas with air temperature at 20 °C as the only available cooling media, the CO_2 temperature could be reduced to approximately 30 °C after the coolers. In this case, the lowest theoretically achievable water content in CO_2 is approximately 1600 ppm [2].

If precipitation of free water in the pipeline is possible and/or likely, additional drying may be required. In general, the following type of dryers can be utilized for water removal in CO_2:

- Adsorption units using, e.g. molecular sieves (MSA).
- Absorption with TEG or DEG as an absorber medium.

Molecular sieve adsorption is the recommended drying method due to low investment costs, compact design, low maintenance rate and generally, good operating experience. In order to reduce size, a MSA downstream the 2nd stage scrubber is recommended.

Gas condensation. Typically, compression to a point above the bubble point pressure is required, prior to transportation. Hence, in the compression process, the gas is condensed (to liquid phase). For pure CO_2, the bubble and dew point curves coincide and the pressure during condensation is constant (at constant temperature). Intermixture of CH_4 in CO_2 leads to a higher bubble point pressure, compared to pure CO_2, and the bubble point pressure is above the dew point pressure. Between these two pressures, is the two-phase region, with gas and liquid in equilibrium. As can be seen from Figure 5 (95% CO_2 and 5% CH_4), if the temperature is kept constant at 15 °C and the gas is pressurized, droplets will start to condense out at 50 bar and complete condensation is reached at 62 bar.

Because N_2 is more volatile than CH_4, introduction of N_2 leads to an even higher bubble point pressure. For a mixture of 95% CO_2, 2% CH_4 and 3% N_2, complete condensation at 15 °C requires a pressure of approximately 70 bar [2].

If the content of volatile components (N_2 and CH_4) in the gas is low, typically less than 5%, the pressure fluctuation during condensation is moderate and condensation of the CO_2 mixture is feasible. In this case,

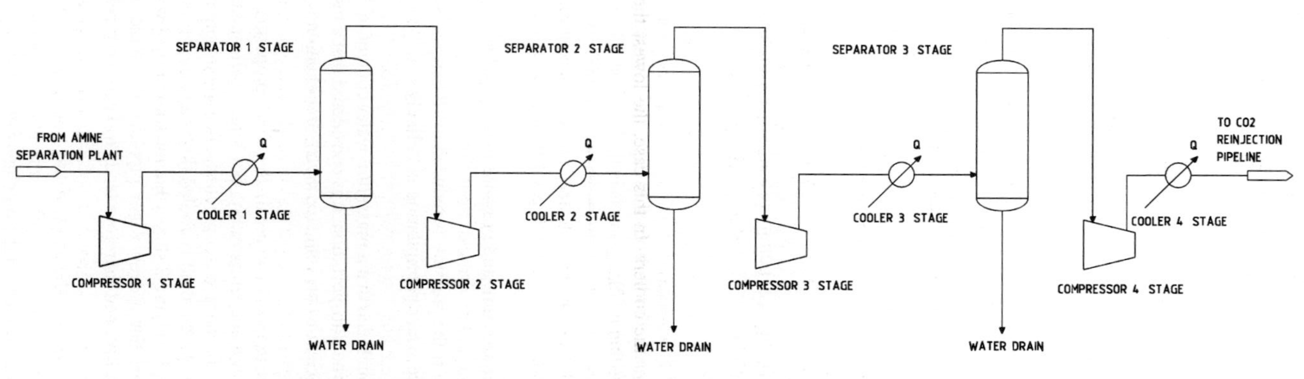

Figure 4: Four stage compression process with interstage cooling and scrubbing.

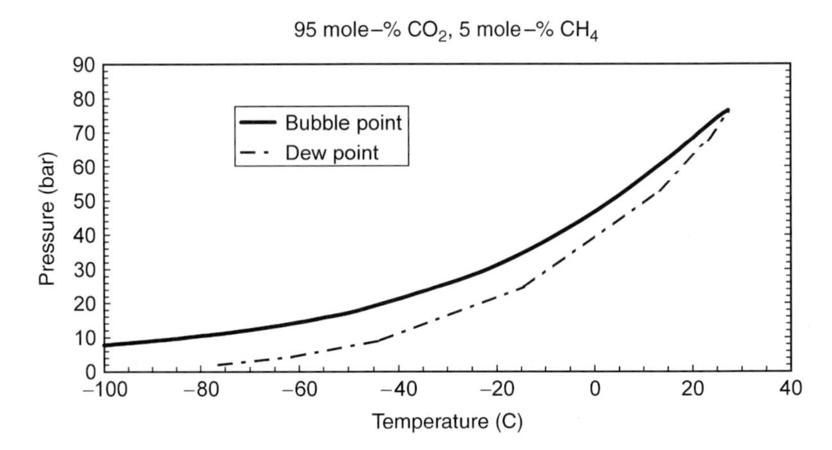

Figure 5: Dew point and bubble point curve for CO_2 mixed with CH_4, calculations in HYSYS with SRK equation of state.

the last compression stage may include a pump instead of a compressor. By pumping, the power consumption and hence, the operational costs will be reduced in the order of 10% compared to a pure compression process.

Pipeline transportation

Water content and drying requirements. For hydrocarbon pipelines, a requirement to dry the gas to 50 ppm water is often used in order to ensure that no free water is in the pipeline. The requirement for CO_2 pipelines, used for EOR in the United States (New Mexico), is maximum 600 ppm water (Kinder Morgan) [2]. Experiments performed at IFE, with CO_2 and carbon steel, show low/insignificant corrosion rates for water content below 600–700 ppm (mole), over a wide range of temperatures [22].

For a typical case, the minimum temperature and pressure in the pipeline is 5 °C and 85 bar, respectively. Theoretical calculations indicate a lower critical limit for free water precipitation of approximately 1300 ppm (mole) in this case. Even if the pressure is reduced to 70 bar, the water solubility is considerably higher than 600 ppm [2]. Thus, free water precipitation is unlikely.

Based on this consideration, it is concluded that a maximum water content of 50 ppm may be too stringent a requirement. For typical CO_2 pipeline transportation, it is considered that a maximum water requirement of 600 ppm may be suitable and sufficient to prevent free water precipitation (see Seiersten, this volume).

Onshore/offshore pipeline transportation. In general, the fluid properties, such as density and water solubility, depend on pipeline pressure and temperature. For long pipelines the fluid will be cooled down to ambient temperature, typically 5 °C for deep water pipelines. For an onshore pipeline, the fluid temperature will be close to the air temperature. As the fluid is cooled down and the pressure is reduced, water tends to precipitate out of the gas. For example, Figure 2 shows that the solubility of water in CO_2 is at a minimum at a pressure of approximately 50 bar. For a deep water pipeline, the static head contributes to increased pressure and increased water solubility, and therefore tends to keep the pipeline dry (no free water).

The compression/pumping requirement at the inlet of the transportation pipeline depends not only on the reservoir pressure, but also on the reservoir depth at the injection point. Deep water combined with a deep reservoir results in a large static pressure in the pipeline/well, which again contributes to obtaining the required injection pressure. Thus, if the water and reservoir depths are large, the boosting requirement at the pipeline inlet is reduced, resulting in reduced capital and operational costs for compression/pumping.

Distance. Knowledge about the distance from the capture plant to the injection point is needed in order to establish the pressure and boosting requirements at the pipeline inlet. Increased pipeline length results in higher frictional pressure loss and increased costs for compression/pumping. Alternatively, the pipe diameter can be increased in order to reduce the flow velocity and thus, reduce the pressure drop. In this case, the capital costs for the pipeline will increase.

In principle, high water content in the CO_2 mixture can be handled either by

- sufficient drying (water removal) downstream in the capture process, or
- using corrosion resistant materials for pipeline and process equipment.

The distance from capture to injection may influence the strategy for handling high water content, i.e. drying vs. corrosion resistant materials. If the pipeline is long, an ordinary carbon steel pipeline is considered to be the only realistic alternative due to cost. In this case, sufficient drying (water removal) upstream of the pipeline is required in order to prevent free water and excessive corrosion rates. On the other hand, if the pipeline is short, the use of corrosion resistant materials should be considered. For short pipelines, the costs for corrosion resistant materials may be less than the costs for installing and operating a separate drying unit.

Materials evaluation and corrosion protection

The corrosion rates in carbon steel pipelines strongly depend on the water content and the water solubility in the CO_2 mixture (see Seiersten [22] and this volume). If free water exists in the pipeline, it will be saturated with CO_2 and the corrosion rate will be significant for carbon steel. For a free water phase without inhibitor the corrosion rate due to CO_2 may be several mm/yr. If free water is expected to occur frequently or normally, MEG or a commercial corrosion inhibitor like, e.g. "Dynea KI-350", may be used to obtain acceptably low corrosion rates (less than 0.1 mm/yr). If MEG or corrosion inhibitor is not used, and free water is present, corrosion resistant alloys will be required.

In general, it is recommended to dry the gas sufficiently in order to inhibit precipitation of free water in the pipeline. For long pipelines, water removal is considered to be the most cost-effective solution, since ordinary CMn steel (e.g. API X65) can be used in this case.

The minimum operating temperature for ordinary CMn steel is -46 °C. If the pipeline leaks and the ambient pressure is low/atmospheric, the liquid CO_2 may be transformed to dry ice (solid CO_2), with a temperature of approximately -79 °C and low temperature steel may be required to avoid further failure.

In general, equipment in the compression train can be made from carbon steel. However, for the following components, corrosion resistant alloys may be required:

- piping at the inlet of the compressors as well as critical components in the compressors,
- coolers and piping just downstream of the coolers, and upstream of the scrubbers, and
- scrubbers.

Alternatively, if acceptable from a mechanical point of view, such components can be made from carbon steel with a certain internal corrosion allowance added to the required wall thickness.

CONCLUSIONS

Documentation about existing CO_2 pipeline transportation systems is provided. Today's onshore systems are primarily CO_2 pipelines used for EOR in the US and acid gas removal in Western Canada. The only existing offshore injection is at Sleipner Vest in the North Sea, operated by Statoil. Drying requirements for the CO_2 pipelines for EOR, operated by Kinder Morgan, is 600 ppm. At the LNG plant at Hammerfest in Norway, CO_2 will be removed from the natural gas. The drying requirement for this plant is 50 ppm water, which is the engineering practice for transportation of natural gas. Based on theoretical calculations as well as experimental data from IFE, it is concluded that a maximum water content of 50 ppm may be a too stringent requirement. A maximum of 600 ppm water may be a suitable and sufficient requirement for CO_2 pipelines.

Typically, the gas mixture out of the capture plant may contain up to 5% CH_4, 5% N_2, 0.5% water, 100 ppm H_2S and an unknown amount of amines. The pure compression process, as opposed to less expensive pumping, is considered to be robust with respect to handling likely composition ranges. High water content in the gas is handled by the scrubbers. Because there is no phase transition, pressure fluctuation due to condensation is eliminated. The water content in the CO_2 mixture can be reduced to approximately 600 ppm (mole), by compression, intermediate stage cooling and scrubbers for dehydration. In this case, it is assumed that seawater at approximately 10 °C is available as a cooling medium. Theoretical calculations indicate a lower/critical limit for free water precipitation of approximately 1300 ppm (mole) in the pipeline. Thus, precipitation of free water is not likely in this case.

In order to inhibit hydrate formation and prevent excessive corrosion rates for carbon steel, sufficient water removal is required upstream the pipeline inlet. The CO_2 pipeline can be made from CMn steel (e.g. API X65), provided that no free water is present and that the minimum operation temperature will not drop below −46 °C. If free water precipitation in the pipeline is possible/likely, additional drying may be required, preferably by molecular sieve adsorption.

Accurate and reliable predictions of fluid properties, particularly density and water solubility, are critical with respect to process and pipeline design. Thermodynamic models and tools for calculating properties for CO_2 and CO_2-rich mixtures have been verified against experimental data. For CO_2 density the Lee–Kesler model is in satisfactory agreement with NIST data both in gas and liquid phase.

RECOMMENDATIONS

For solubility of water in pure CO_2, the SRK model with adjusted binary coefficient to 0.193 in van der Waals mixing rule gives the best approximation to the data collected from literature. Adding impurities as CH_4, N_2, H_2S and amines to the CO_2 mixture will affect the solubility of water, i.e. adding 5% methane lowers the water solubility in the liquid phase considerably. However, very little experimental data is available in the literature for these mixtures. More experimental data is needed in order to verify the model performance.

In the case of a sudden pressure drop to atmospheric pressure (worst case), liquid CO_2 may transform to dry ice (solid CO_2), with a temperature of approximately −79 °C. Thus, low temperature steel materials should be considered for onshore and shallow water pipelines. In future work, criteria for minimum design temperature and material selection for CO_2 pipelines should be established.

Little data exists on the behavior of wet CO_2 gas in scrubbers. In the present work, the process considerations for CO_2 dehydration are based on theoretical calculations. The results of the calculations have not been verified by operational or experimental data. In order to clarify this uncertainty, it is recommended to collect data from existing CO_2 facilities and compare the data against model simulations.

NOMENCLATURE

T_c	Critical temperature
P_c	Critical pressure
K	Kelvin
Pa	Pascal (Unit for pressure)
M	Mega (one million)
SRK	Soave–Redlich–Kwong (equation of state)
LK	Lee–Kesler (equation of state)
BWRS	Benedict–Webb–Rubin–Starling
NIST	National Institute of Standards and Technology
EOR	Enhanced oil recovery
TEG	Triethylene glycol
MSA	Molecular sieve absorption

DEG Diethylene glycol
CMn Carbon manganese
API American Petroleum Institute
IFE Institute for Energy Technology
NorCap The Norwegian part of CCP (CO_2 Capture Project)

ACKNOWLEDGEMENTS

The authors are grateful to CCP and the Norwegian Research Council for financial support, as well as Institute for Energy Technology (IFE) for providing experimental data and expertise. The authors also would like to thank Bjørn Berger, Geir Owren, Trond Soligard, Lars Volden and Rune Meidal for important contributions to the project.

REFERENCES

1. Kinder Morgan homepage: http://www.kindermorgan.com/.
2. T. Weydahl, A. Austegaard, M.J. Mølnvik, R. Mo, G. Heggum, CO_2 Capture Project—An Integrated, Collaborative Technology Development Project for Next Generation CO_2 Separation, Capture and Geologic Sequestration—CO_2 Conditioning and Pipeline Transportation, CCP Report 50058, 2004.
3. G. Moritis, EOR dips in U.S. but remains a significant factor, *Oil Gas Journal*, September 26, 2004.
4. O. Bolland, S. Sæter, M. Hyllseth, O. Lunde, Gas Fired Power Plant with Reduced Emissions of Carbon dioxide, *SINTEF Report*, December 1991.
5. A. Baklid, R. Korbøl, G.A. Owren, Sleipner Vest CO_2 injection in shallow underground aquifer. Society of Petroleum Engineers, *71st Annual Technical Conference and Exhibition*, October 6–9, 1996, Denver, Colorado, USA.
6. PRO II, Version 6.0, http://www.simsci.com/.
7. HYSYS, Version 3.0.1, http://www.hyprotech.com/.
8. A. Austegaard, User Manual—CO_2 Wellhead Pressure Calculator, Version 2.1. *SINTEF-Report TR F5698*, Trondheim, Norway, 2002.
9. R.C. Reid, J.M. Prausnitz, B.E. Poling, The Properties of Gases and Liquids, fourth ed., New York: McGraw-Hill, 1987 (ISBN: 0-07-051799-1).
10. J.M. Prausnitz, R.N. Lichtenthaler, E.G. de Azevedo, Molecular Thermodynamics of Fluid-Phase Equilibria, third ed., Prentice Hall International Series in the Physical and Chemical Engineering Sciences, 1999.
11. National Institute of Standards and Technology (NIST), http://webbook.nist.gov/chemistry/.
12. R. Wiebe, The Binary System Carbon Dioxide—Water Under Pressure, *Chem. Rev.* **29** (1942) 475–481.
13. C.R. Coan, A.D. King Jr., Solubility of Water in Compressed Carbon Dioxide, Nitrous Oxide and Ethane. Evidence of Hydration of Carbon Dioxide and Nitrous Oxide in the Gas Phase, *J. Am. Chem. Soc.* **93** (8) (1971).
14. K.Y. Song, R. Kobayashi, Water Content of CO_2 in Equilibrium with Liquid Water and/or Hydrates, SPE *Formation Evaluation*, December 1987.
15. P.C. Gillespie, G.M. Wilson, Vapor–Liquid and Liquid–Liquid Equilibria: Water–Methane, Water–Carbon Dioxide, Water–Hydrogen Sulfide, Water–*n*Pentane and Water–Methane–*n*Pentane, *Research Report RR-48*, Wiltec Research Co., Inc., Provo, UT, 1982.
16. M.B. King, A. Mubarak, J.D. Kim, T.R. Bott, The mutual solubilities of water with supercritical and liquid carbon-dioxide, *J. Supercrit. Fluids* **5** (4) (1992) 296–302.
17. K.Y. Song, R. Kobayashi, The water content of a CO_2-rich gas mixture containing 5.31 mol% methane along three-phase and supercritical conditions, *J. Chem. Eng. Data* **35** (3) (1990).
18. M.-J. Huron, J. Vidal, New mixing rules in simple equations of state for representing vapour–liquid equilibria of strongly non-ideal mixtures, *Fluid Phase Equilib.* **3** (1979) 255–271.
19. Y.T. Seo, S.P. Kang, H. Lee, C.S. Lee, W.M. Sung, Hydrate phase equilibria for gas mixtures containing carbon dioxide: a proof-of-concept to carbon dioxide recovery from multicomponent gas stream, *Korean J. Chem. Eng.* **17** (2000) 659–667.

20. J.J. Carroll, Phase equilibria relevant to acid gas injection: part 1—non-aqueous phase behaviour, *J. Can. Petrol. Technol.* **41** (2002) 25–31.
21. J.J. Carroll, Phase equilibria relevant to acid gas injection: part 2—aqueous phase behaviour, *J. Can. Petrol. Technol.* **41** (2002) 39–43.
22. M. Seiersten, K.O. Kongshaug, Material selection for capture, compression, transport and injection of CO_2, Technical Report, Institute for Energy Technology, 2003.

Carbon Dioxide Capture for Storage in Deep Geologic Formations, Volume 2
D.C. Thomas and S.M. Benson (Eds.)

Chapter 16

MATERIALS SELECTION FOR CAPTURE, COMPRESSION, TRANSPORT AND INJECTION OF CO_2

Marion Seiersten[1] and Kjell Ove Kongshaug[2]

[1]Institute for Energy Technology, Kjeller, Norway
[2]University of Oslo, Oslo, Norway

ABSTRACT

The principal alternative for long-distance transportation of CO_2 from source to storage site is in pipelines. To a large extent pipelines can be made in carbon steel as pure, dry CO_2 is essentially non-corrosive. More corrosion-resistant materials or corrosion inhibition must be considered when the CO_2 contains water that condenses out during transportation. This will occur where it is impossible to dry CO_2 to a dew point well below the ambient temperature. Water-saturated CO_2 is corrosive when water precipitates, but experiments show that corrosion rates at high CO_2 pressures in systems containing only water or water/MEG (monoethylene glycol) mixtures are considerably lower than predicted by corrosion models. This applies particularly at low temperatures that are typical for sub-sea pipelines in northern waters. In our previous study, it has been demonstrated that 20 ppm CO_2 corrosion inhibitor is sufficient to lower the corrosion rate below 0.1 mm/y at temperatures up to 30 °C and CO_2 pressures up to 72 bar.

The present study focuses on determining the corrosion rate as function of CO_2 pressure up to 80 bar. The results are compared to existing corrosion models that have been developed to cover a pressure range relevant for oil and gas transportation, i.e. pressures up to 20 bar. The objective of the present study was to verify or extend the use of corrosion models at CO_2 pressure above 20 bar. The experiments show that the models overestimate the corrosion rate when they are used above their CO_2 partial pressure input limit. At low temperature the models predict more than 10 times the measured corrosion rate. Furthermore, the results indicate that the corrosion rate has a maximum as function of CO_2 pressure at 40 and 50 °C. The maximum is at 30–50 bar depending on temperature.

Part of the present study was devoted to determine the solubility of water in CO_2 containing up to 5% CH_4 at high pressure. The results show that CH_4 lowers the water solubility and hence increases the risk of free water in liquid or supercritical CO_2.

INTRODUCTION

Choice of materials for transportation and storage of CO_2 is a critical issue although the oil industry has re-injected CO_2 for the purpose of enhanced oil recovery (EOR) for decades with little or no problems related to corrosion. Low alloy carbon steel pipelines have been used for transportation of liquid CO_2 at high pressure, but in all these cases, drying the CO_2 to less than 100 ppm water and thus removing free water in the pipeline has eliminated the corrosion risk. Drying the CO_2 increases the handling costs especially at offshore installations.

There are several alternatives for CO_2 transportation. Pipelines are the most realistic alternative to bring CO_2 from the source to the storage site. If the transportation distance is more than a few km, carbon steel will be the most cost-effective alternative. In some cases, it may also be practical to reuse old pipelines or co-feed the CO_2 in existing multiphase pipelines. The latter has already been considered in the North Sea [1].

Abbreviations: CR, corrosion rate; MEG, monoethylene glycol; LPR, linear polarization resistance.

More such cases will probably arise when the use of depleted oil fields or aquifers for final storage of CO_2 increases. When CO_2 is transported in existing pipelines from old platforms, it will be impossible or very costly to dry the CO_2 and avoid all free water in the pipeline.

The use of carbon steel either requires that the CO_2 is dried to eliminate free water in the pipeline or that corrosion caused by free water is inhibited. Reliable corrosion data and prediction models are needed in order to evaluate the inhibition and estimate corrosion allowance.

Carbon dioxide has been utilized for EOR for 30 years and there are more than 100 installations worldwide. Most of these use carbon steel pipelines for CO_2 transportation. Despite this, there are few thorough investigations on the corrosion of steels and other materials in CO_2 at pressures above 50 bar [2]. The reason is that there have been few problems with the recovery and transportation systems. The CO_2, in this case, is pure and is dried to a dew point well below the ambient temperature before transportation. At the Sleipner Field, wet CO_2 is injected into the Utsira aquifer. The transportation distance is short and the use of corrosion-resistant duplex steel is therefore cost effective.

Table 1 lists candidate steels for CO_2 processing and transport. It is evident from the table that little is known about the performance of steels in these environments. The transportation costs can be considerable when the CO_2 storage sites are located at some distance from the source. For a CO_2 storage scenario with a 200 km transport line, the transportation costs have been estimated to 20–40% of the total costs [3]. It is not known what the materials costs amount to, but the cost figures in Table 1 clearly show that carbon steel is the most attractive alternative for long pipelines and that 13% Cr steels can be considered for shorter lines. Earlier studies have indicated that the corrosion rate of pipeline steel in wet CO_2 is less than anticipated, and that some water wetting of the pipeline may be allowed for a limited period of time [4]. More data for the corrosion of pipeline steels will be needed to be able to specify CO_2 quality and set limits for trace chemicals and free water. It should also determine the possible extent to which carbon steels can be used with corrosion and hydrate inhibitors. Furthermore, little is known on the corrosion of 13% Cr steel in liquid or supercritical CO_2 with free water. It is a candidate material if water wetting is anticipated, especially for shorter pipelines.

TABLE 1

CORROSION RATES AND EXPERIENCES REPORTED IN THE LITERATURE FOR CANDIDATE STEELS FOR CO_2 PIPELINES AND PROCESS EQUIPMENT

Environment	**Quantitative measurements or reported experience**		
	Carbon steel	**13% Cr steel**	**Duplex and other high-alloy steels**
Dry pure CO_2	Good	–	Good
Wet pure CO_2	Some investigations indicate corrosion rate > 10 mm/y	Not investigated	Most are resistant; corrosion rate ~ 1 μm/y
Dry CO_2 with traces of chemicals from the separation process and hydrocarbons	Few investigations, probable limits for trace elements	Not investigated	Depending on the trace elements (stress corrosion cracking, SCC, must be considered)
Wet CO_2 with traces of chemicals from the separation process and hydrocarbons	Not investigated, corrosion rate probably high	Not investigated	Depending on the trace chemicals (SCC must be considered)
Cost factor for piping material	1	2	≥4

The table summarizes investigations at CO_2 pressure above 70 bar. A cost factor for piping materials (with carbon steel as 1) is indicated in the bottom row.

The possibility of free water will determine the materials selection and thus costs. Depending on its origin, CO_2 for injection will contain other substances that may reduce the water solubility in the fluid. CO_2 separated from natural gas may for instance contain up to 5% CH_4. While the solubility of water in pure CO_2 (liquid or supercritical) is well known as function of pressure and temperature, few data are available for the effect of trace chemicals on solubility. It is known, e.g. that CH_4 lowers the solubility of water substantially, but the solubility as function of composition has only been measured for a few compositions and the applicable pressure and temperature range is not extensive [5]. The data available are therefore inadequate for use as design parameter for CO_2 injection pipelines and there is a need of accurate solubility limits in actual mixtures.

Project Objectives

The objective of the study was to establish a basis for materials selection for the processes of CO_2 capture, compression, transportation, and injection. The project was coordinated with the CCP Transportation project run by Reinertsen Engineering and SINTEF. The sub-goals were:

1. To quantify the amount of water that can be dissolved in CO_2-NGL (max 5%) mixtures at 50–500 bara and temperatures up to 30 °C. The effect of trace components from the separation process will also be addressed.
2. To provide the data needed by the Reinertsen/SINTEF project in the development of the guidelines to be used for cost-effective development of CO_2 transportation systems.
3. To determine the corrosion rate of carbon, temperature and pressure and to clarify if it is possible to extend the use of carbon steels with corrosion inhibitors.

EXPERIMENTAL

Water Solubility in CO_2 and in Mixtures of CO_2 and CH_4

The experiments were based on the use of tritium-labeled water. The water phase was allowed to equilibrate with a gaseous or liquid CO_2 phase at a given temperature. Samples of the gas phase were washed out with water to pick up the tritium-labeled water in the gas phase and the resulting water was analyzed for tritium.

The experimental apparatus shown in Figure 1 is a modification of the one described by Song and Kobayashi [5]. It consists of a titanium grade 2 autoclave with gas inlet and outlet and a sampling cylinder in stainless steel AISI 316L. The volume of the autoclave is 1000 mL. The autoclave is completely submerged in a thermostatic bath. The water in the bath circulates continuously and is controlled to ±1 °C. The autoclave and the sampling system can be evacuated to 0.1 bar.

Tritium-labeled water was obtained from the nuclear reactor at IFE, Kjeller. It was thinned 10 times to obtain an activity of 6 Mbq (Mega Becquerel) in the test solution.

Experiments were started by filling the autoclave with 100 mL tritium-labeled water. The autoclave was evacuated and the gas or fluid phase was added through the bottom inlet and bubbled through the water phase. Measurements on CH_4 and CO_2 mixtures were carried out by adding CH_4 and let it equilibrate with the water phase at the correct partial pressure before CO_2 was added and the pressure was stabilized at the correct total pressure. In this way, it was easy to obtain correct gas mixtures and avoid uncertainty due to the high solubility of CO_2 in the water phase. Liquid CO_2 was be pumped in for high-pressure studies.

The autoclave was decoupled after filling and shaken in the thermostatic bath before it was coupled to the sampling system. It was equilibrated for 4–24 h before sampling started. The sampling system was thoroughly dried and evacuated before sampling began.

The volume of the sampling cylinder was 150 mL and before sampling, it was filled with ca. 100 mL distilled water. The water content was accurately determined by weighing. The cylinder was evacuated before sampling. The pressure change in the autoclave during the operation was less than 1 bar.

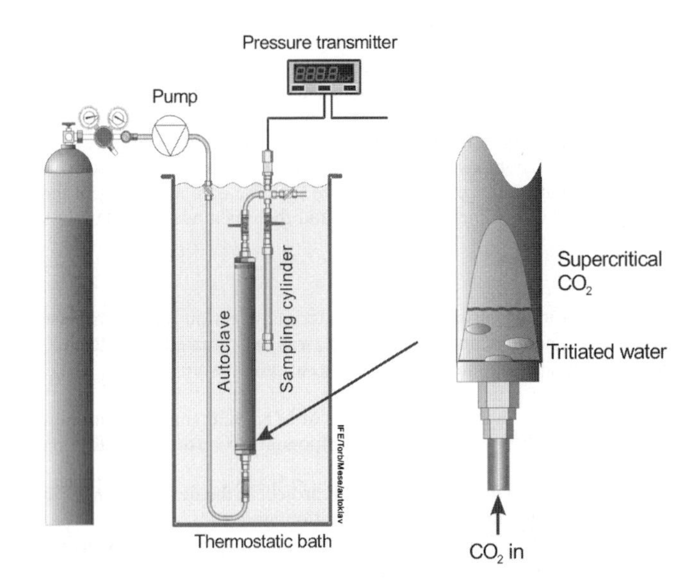

Figure 1: Apparatus for measurements of water solubility in liquid/gaseous CO_2.

The sample amount was determined by weighing and the sampling cylinder was thoroughly shaken before the tritium content of the water was measured. The analysis was carried out on a Quantulus low background level liquid scintillation counter. The analysis samples were diluted 1 to 10 by a low-level tritium Ultra Gold scintillation liquid.

Set-Up for Studies of Corrosion of Carbon Steel in CO_2 and CO_2-Saturated Water

The experimental equipment is shown in Figure 2. It consists of an autoclave with a circulation pump. All exposed materials are made of Hastelloy C. The autoclave has both a cooling and heating system. It can be operated from 0 to 100 °C and is certified for a pressure of 80 bar. The total volume of the autoclave is 8.9 L. The autoclave can be evacuated with a pressure-driven vacuum pump. When both water and liquid CO_2 are added, there is a phase separation with the water phase at the bottom. Corrosion coupons can be inserted both in the CO_2 phase at the top of the autoclave and in the CO_2-saturated water phase at the bottom. In most of the present experiments, there were corrosion coupons in the water phase only.

The test specimens were machined from X65 low-carbon steel. The composition of this steel is given in Table 2. The specimens were ground with 1000 mesh SiC paper wetted with isopropanol, cleaned with technical acetone in ultrasonic bath and flushed with ethanol. The specimens were blow dried before they were mounted on the specimen holder.

The test solutions were prepared from technical or analytical grade chemicals and distilled water. Some of the experiments were performed with MEG as hydrate-preventing agent. When MEG was applied, the concentration was always 50% by weight in the aqueous phase. The solutions were deaerated by CO_2 bubbling for at least 4 h. Oxygen was removed from the autoclave by repeated evacuation and CO_2 flushing. The test solution was transferred to the autoclave by vacuum suction. The autoclave was not filled completely; a 0.3 L gas cap was left to ensure CO_2 gas/liquid equilibrium. For experiments below/above room temperature the test solution was cooled/heated to the experimental temperature before the CO_2 was let in. To obtain pressures above the saturation pressure of CO_2 at room temperature (~58 bar), the test solution was saturated with liquid CO_2 at 15 °C. The autoclave was then heated to the experimental temperature, and the experimental pressure was obtained by venting off CO_2 during heating. The experiments were typically run for 5–7 days.

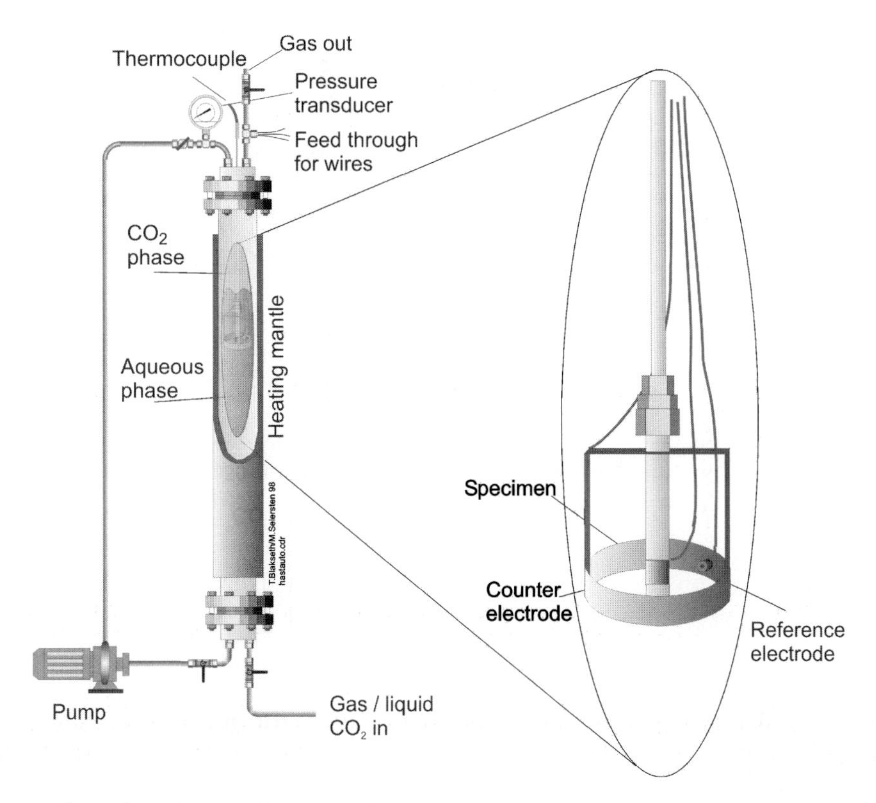

Figure 2: Schematic illustration of the test autoclave used for corrosion experiments.

TABLE 2
ELEMENT ANALYSIS (WT%) FOR THE CARBON STEEL USED IN THE TESTS

Steel	C	Si	Mn	S	P	Cr	Ni	V	Mo	Cu	Al	Sn	Nb
#57	0.08	0.25	1.54	0.001	0.019	0.04	0.05	0.095	0.01	0.02	0.038	0.001	0.043

The measurements are at 100 bara.

The corrosion rate was measured by the linear polarization resistance (LPR) technique in the three-electrode configuration. The specimens were small cylinders with surface area of 3.14 cm^2. The counter electrode was a Pt-foil mounted around the specimen. The reference electrode was a 1 mm Ag rod mounted in a PTFE bar in the counter electrode. Even though the electrode was anodized in 0.1 M HCl before each experiment, it did not remain stable during the experiments. A new agar-based Ag/AgCl reference electrode was therefore constructed (Figure 3) and used in the late experiments (CCP_K11–K20). The electrode was made with standard Swagelock fittings, and was composed of an Ag rod covered with AgCl, 3 M KCl in agar and a porous ceramic plug. The electrode was placed in the bypass line. Even this electrode did not provide sufficient stability and it had to be regenerated before each experiment following a procedure described in Ref. [6].

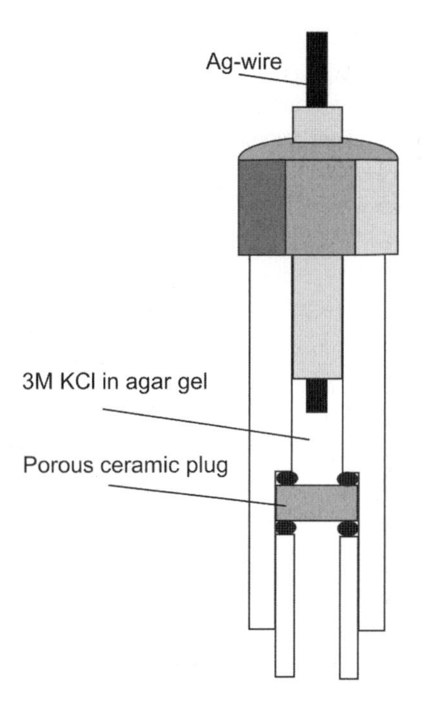

Figure 3: Schematic representation of the Ag/AgCl reference electrode.

RESULTS AND DISCUSSION

Water Solubility in CO_2 and in Mixtures of CO_2 and CH_4

Verification test

A series of experiments was carried out to verify the experimental set-up. The test condition of 25 °C and 100 bar was chosen, as this was the parameter set where most data could be found in the literature. An additional series was carried at 36.5 °C and 100 bar to study the effect of temperature on the time needed to reach equilibrium. The results are given in Table 3. The results show that 8 h are needed to reach equilibrium at 25 °C, while equilibrium is obtained after 2 h at 36 °C.

Measurements on mixtures of CO_2 and CH_4

A series of experiments was conducted to measure the solubility of water in pure CO_2 and in mixtures of CO_2 and CH_4. The CH_4 concentration of the mixtures was 5%. Figure 4 gives the results at 25 °C. The spread in the measured values is considerable, but that is also the case for literature data. Measurements at 14 °C were not successful, as equilibrium could not be established in a reasonable time. The conclusion from these experiments is that the method is best suited for high-density CO_2 fluids and that it is difficult to obtain reliable results for gaseous CO_2. The reason is that it is more difficult to sample the gaseous phase as only small temperature gradient leads to water condensation in the tubing and the valves.

Corrosion of Carbon Steel in CO_2 and CO_2-Saturated Water

Previous work

Pure, dry CO_2 is essentially non-corrosive. Experimental studies indicate this [10–12] and, in addition, field experience shows few problems with transportation of high-pressure dry CO_2 in carbon steel pipelines [13–15]. More than 3000 km of pipelines carrying CO_2 are in operation worldwide, most of these in the USA.

TABLE 3
WATER SOLUBILITY IN PURE CO2 AT 100–103 BAR

Temperature (°C)	Mole fraction water in CO_2/ppm ($X \times 10^6$)	Comments
25	2743	Equilibrium time 2 h
25	2909	Equilibrium time 4 h
25	3381	Equilibrium time 8 h
36.5	4145	Equilibrium time 2 h
36.5	4312	Equilibrium time 4 h
36.5	4199	Equilibrium time 8 h
25	3539	101 bar, Data from Wiebe and Gaddy [7]
25	3374	103 bar, Data from Song and Kobayashi [5]
25	3270	101 bar, Data from King et al. [8]
25	3360	101 bar, Data from Dewan [9]
35	4070	101 bar, Data from King et al. [8]

Designation: API 5L X65, microstructure: ferrite–perlite.

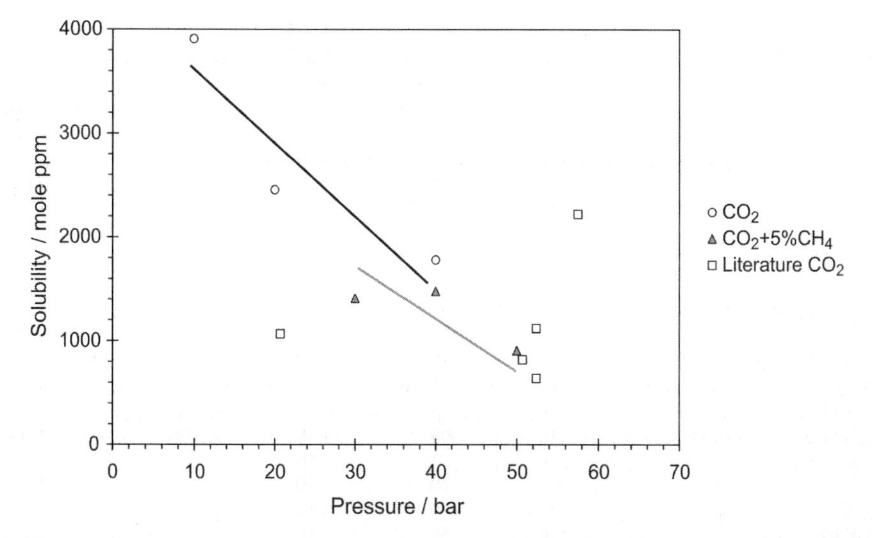

Figure 4: Water solubility in pure CO_2 and in a 5% CH_4 in CO_2 mixture at 25 °C as function of pressure.

CO_2 in the presence of water, however, will form carbonic acid which is corrosive. The impact of CO_2 corrosion on carbon steel has been studied extensively at pressures relevant for oil and gas transport (up to 20 bar). At higher pressures experimental data are sparse. Studies of CO_2 corrosion of carbon steel conducted at 170 °C and 100 bar CO_2 for 200 days [10] and at 50 °C and 240 bar CO_2 for 24 h [16] provide qualitative evidence for corrosive attacks on carbon steel. In both these cases CO_2 contains water just above the solubility limit. A study conducted in a 0.8 L autoclave filled with 1.0 M NaCl solution at 80 °C and CO_2 pressures up to 50 bar [16] showed that the pH change of the test solution, during the experiments introduced by the corrosion process, affects the corrosion rates at high CO_2 pressures. Experiments performed under "floating pH" conditions showed small differences in corrosion rates at 5 and 50 bar CO_2, whereas experiments performed at constant pH showed 1.5–3 times higher corrosion rate at 50 bar than at 5 bar. For instance, the corrosion rate at pH 3.5 was about 10 mm/y at 5 bar and 15 mm/y at 50 bar.

Generally, the corrosion rates decreased with increasing pH, and the experiments performed under "floating pH" conditions gave the lowest corrosion rates (~2.5 mm/y). Formation of a protective carbonate scale under "floating pH" conditions explains this observation. Within the project "Large Scale CO_2 Transportation and Deep Ocean Storage" sponsored by US Department of Energy (DOE) and performed by McDermott International, corrosion experiments were conducted in a high-pressure autoclave at conditions simulating deep ocean environment (i.e. 300 bar and 2 °C) [18]. Corrosion rates on carbon steel were measured by LPR for 0–30% CO_2 in sea water. The relevant DOE report has not been made public.

Reported field experiences with wet CO_2 at high pressures are also sparse. Accumulation of corrosion products due to insufficient drying and a leak at a low point due to water build up are reported from the start up of one CO_2 pipeline [19]. Failures of carbon steel pipe fittings with high-pressure CO_2 line tapping were reported in a urea plant. The CO_2 pressure in the pipeline was 156 bar and the temperature was 104 °C [20].

Experimental results
Table 4 summarizes the corrosion rate for the experiments CCP_K1–K21. The reported rates are the average rates during the last 24 h of the experiments. In systems containing 50 wt% MEG, experiments were run at three different temperatures 5, 25 and 50 °C, respectively.

TABLE 4
CORROSION RATES FOR THE EXPERIMENTS CCP_K1–K21

Eksp. No.	Temperature (°C)	MEG concentration (wt%)	CO_2 pressure (bar)	NaCl (g/kg)	Precorrosion	Average corrosion rate (mm/y)
CCP_K1	5	50	44	10		0.1
CCP_K2	5	50	5	10		0.04
CCP_K3	25[a]	50	1–58	10		2.3
CCP_K4	25[a]	0	58	10		2.4
CCP_K5	25[a]	50	58	10		2.3
CCP_K6	25[a]	90	58	10		0.2
CCP_K7	25[a]	50	30	10		1.3
CCP_K8	5	50	10	10		0.06
CCP_K9	5	50	20	10		0.075
CCP_K10	5	50	1–44	10		0.045
CCP_K11	25	50	64	10	yes	0.6
CCP_K12	25	0	64	10	yes	3.6
CCP_K13	50	50	1	10		1.5
CCP_K14	50	50	10	10		2.3
CCP_K15	50	50	30	10		2.7
CCP_K16	50	50	55	10		2.5
CCP_K17	50	50	80–64	10		1.7
CCP_K18	50	0	80–63	10		4.6
CCP_K19	50	0	40	10		6.9
CCP_K20	50	0	20	10		4.3
CCP_K21	50	0	5–60	10		2.3

The reported corrosion rates are the average of the last 24 h.
[a] Temperature control failed, and the actual temperature was slightly higher than 25 °C.

Figures 5 and 6 illustrate the pressure dependence of the corrosion rates at the different temperatures. At 5 °C the corrosion rates increase with increasing pressure, and the maximum rate is 0.1 mm/y at 44 bar. Increasing the temperature to 25 °C results in a large increase in the corrosion rate (0.1 mm/y at 5 °C to 2.3 mm/y at 25 °C). Also at this temperature there seems to be an increase in corrosion rates with increasing pressure. However, just two experiments are performed at this temperature. Increasing the temperature further to 50 °C has only a limited effect on the corrosion rates (2.3 mm/y at 25 °C to 2.7 mm/y at 50 °C). In contrast to the behavior at the 5 and 25 °C, the corrosion rates do not increase with increasing pressure. Instead the corrosion rate reaches a maximum of 2.7 mm/y at 30 bar (Figure 6). A similar behavior with respect to the influence of pressure increase on the corrosion rates is observed in systems containing only water at 50 °C (Figure 7). Here the corrosion rate has a maximum of 6.9 mm/y at 40 bar. Similar behavior of decreasing corrosion rates with increasing pressure has also been reported previously at 40 °C [21].

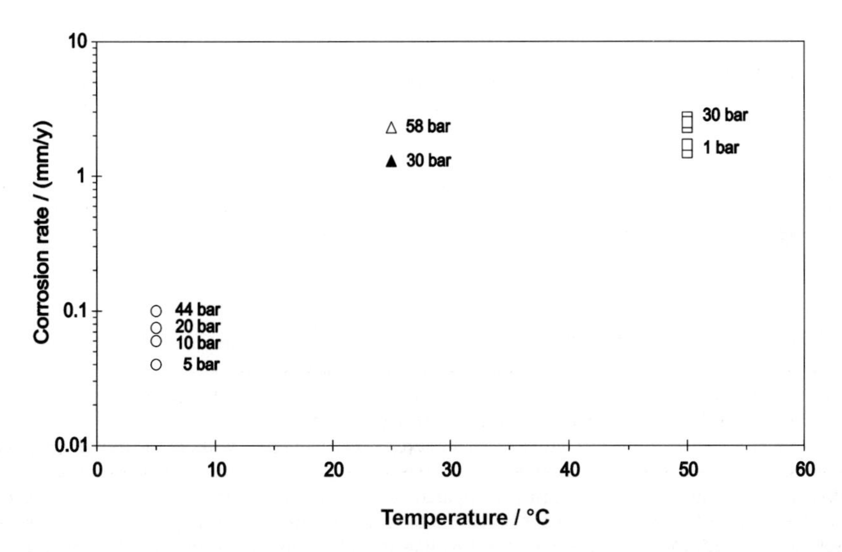

Figure 5: Pressure dependence of the final corrosion rates for experiments performed at 5, 25 and 50 °C, 50 wt% MEG, 10 g/L NaCl, semi-stagnant conditions, floating pH.

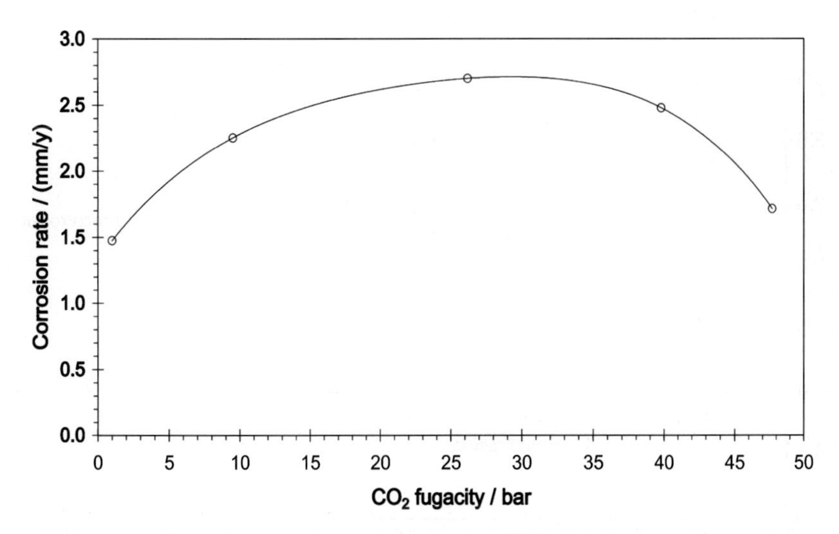

Figure 6: Final corrosion rates in 50 wt% MEG solution (10 g/L NaCl) at 50 °C as a function of CO_2 fugacity at semi-stagnant conditions with floating pH.

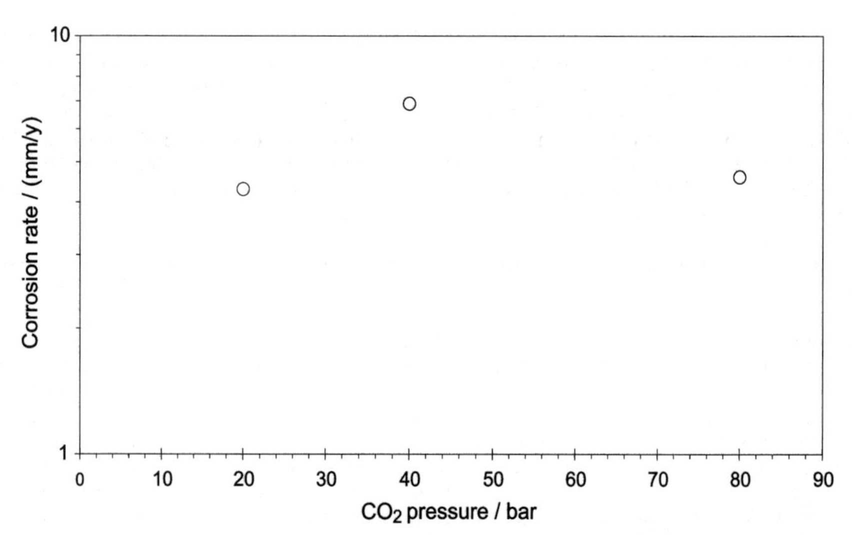

Figure 7: Final corrosion rates in water (10 g/L NaCl) at 50 °C as a function of CO_2 pressure fugacity at semi-stagnant conditions with floating pH.

Two experiments were run for corrosion at low pressure (1 bar) for a period of 24 h (CCP_K11 and K12). These experiments can be compared with almost identical experiment runs without the low-pressure period, see Figure 8. In the system containing 50 wt% MEG the corrosion rate is lower than in a similar experiment without a low-pressure period (0.6 mm/y compared to 2.3 mm/y). On the other hand, in the system containing water only, the corrosion rate is highest in the experiments with a low-pressure period (2.4 mm/y compared to 3.6 mm/y). From these data it seems that in water only precorrosion at low pressure might "activate" the carbon steel and give a higher corrosion rate at high pressure, but the data are too sparse to draw any firm conclusion.

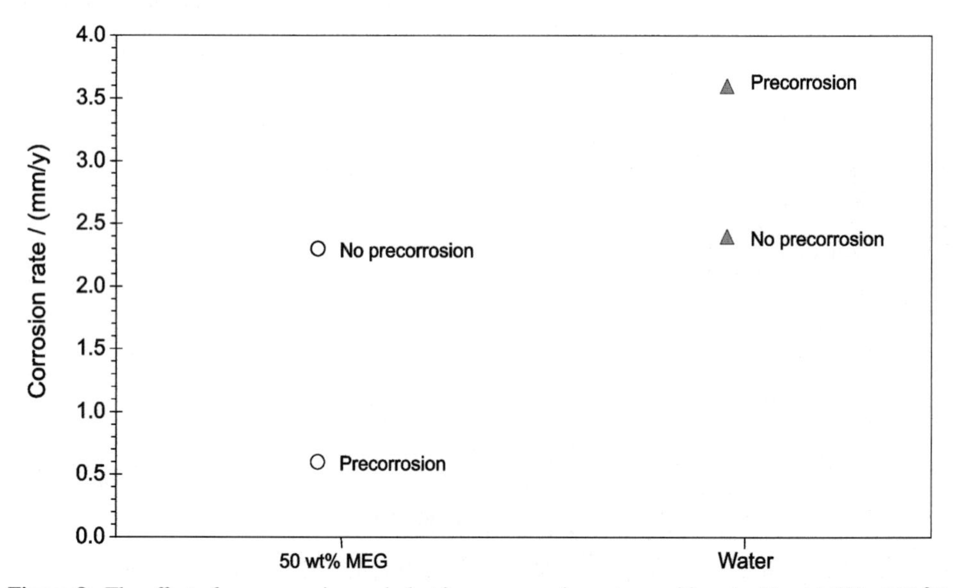

Figure 8: The effect of a precorrosion period at low pressure in systems with and without MEG at 25 °C.

In an actual high-pressure pipeline, MEG or another hydrate inhibitor will be injected to prevent hydrate formation if a temperature of less than 15–20 °C is expected. From studies conducted at lower CO_2 pressures, it is well known that MEG, in addition to preventing hydrates, reduces CO_2 corrosion [22]. In the 50 wt% MEG solution in water applied in this work, the corrosion rate is reported to reduce by a factor of 0.33 [23]. In two experiments conducted at 25 °C (Table 4, CCP_K4 and K5) under the same experimental conditions, but with and without MEG, respectively, the corrosion rates are almost the same. But at 50 °C (Table 4, CCP_K17 and K18) the correction factor of 0.33 fits the data very well. It is difficult to draw clear conclusions from these data on the inhibiting effect of MEG at high CO_2 pressures.

Corrosion coupons mounted in the CO_2 phase at the top of the autoclave did not show any signs of corrosion attacks.

Measured corrosion rates compared to model predictions

There exist a number of corrosion prediction models that can be used to assess CO_2 corrosion rates (Table 5) [24]. However, they have been developed to cover a pressure range relevant for oil and gas transportation, i.e. pressures up to 20 bar. When CO_2 pressure above 10 bar is used as input in these models, they tend to predict corrosion rates that are higher than the experimental rates that are reported in this study (Figures 9–12). Some of the models do not have an MEG correction factor incorporated, and the corrosion rates predicted in 50 wt% MEG solutions are obtained by multiplying corrosion rates predicted in pure water systems with a factor of 0.33 [23]. It should also be noted that not all models are applicable at temperatures below 20 °C.

TABLE 5
SURVEY OF PREDICTION MODELS FOR CO_2 CORROSION WITH APPLICATION LIMITS

Model	**Developed by**	***T* (°C)**		***P* (bar)**	**pCO_2 (bar)**		**pH**	
		Min	**Max**	**Max**	**Min**	**Max**	**Min**	**Max**
de Waard[a]	de Waard and coworkers (Shell, IFE), published	0	140			10		
HYDROCOR	Shell	0	150	200		20		
Cassandra 98[b]	BP		140	200		10		
NORSOK[c]	Hydro, Saga, Statoil (IFE data)	20	150	1000		10	3.5	6.5
CORMED[d]	Elf		120					
LIPUCOR	Total	20	150	250		50		
KSC model[e]	IFE (JIP)	5	150	200	0.1	20	3.5	7
Tulsa model[f]	University of Tulsa	38	116			17		
PREDICT[g]	InterCorr International	20	200			100	2.5	7
Ohio model[h]	Corrosion in Multiphase Systems Center at Ohio University	10	110	20				
SweetCor[i]	Shell	5	121		0.2	170		

[a] None of the de Waard papers give application limits. Maximum values in nomogram shown.
[b] Accepts input outside these values but displays a warning.
[c] Wall shear stress between 1 and 150 Pa. Will be extended down to 5 °C.
[d] CORMED accepts higher temperatures and ionic strengths but displays a warning, as the pH calculation becomes uncertain. The corrosion risk prediction is still valid.
[e] Flow velocity between 0.2 and 30 m/s.
[f] Recommends these limits, but accepts input outside these values.
[g] PREDICT does not give any limits, either in the software or in the manual.
[h] Minimum 10% water cut. Can be used at higher pressures with fugacity coefficient input.
[i] Analysis of CO_2 corrosion by managing a large database of corrosion data from laboratory experiments and field data.

A partial explanation for the low experimental rates compared to model predictions is that the models are developed for flowing conditions. Although water solution in the autoclave is continuously pumped from the bottom to the top, the flow rate at the specimen site is low. Based on flow loop experiments at lower CO_2 pressures, it is expected that the corrosion rate may increase by a factor 2–3, at normal pipeline flow rates. A discrepancy in the order of 10 is difficult to explain as a flow factor as long as the solution is undersaturated with iron.

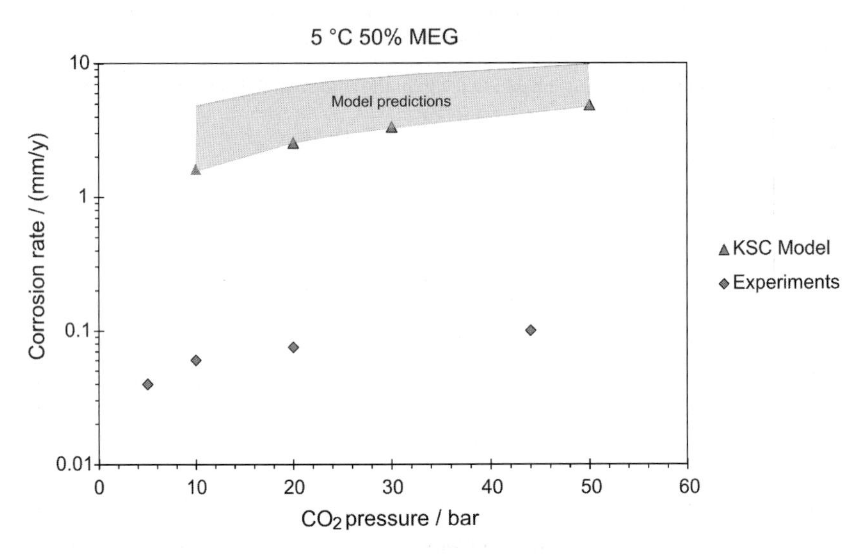

Figure 9: Experimental corrosion rates compared to model predictions in 50 wt% MEG solutions at 5 °C. The model predictions fall in the shaded area and only the values calculated by the KSC model are shown.

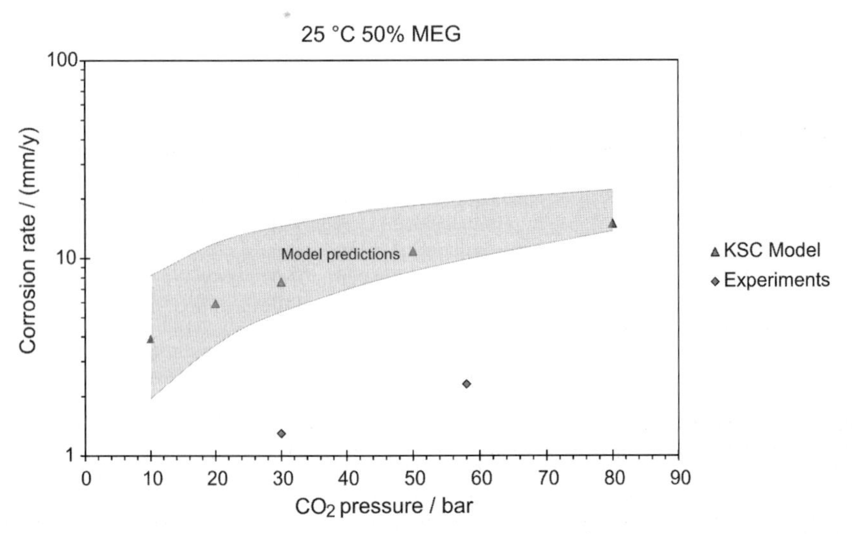

Figure 10: Experimental corrosion rates compared to model predictions in 50 wt% MEG solutions at 25 °C. The model predictions fall in the shaded area and only the values calculated by the KSC model are shown.

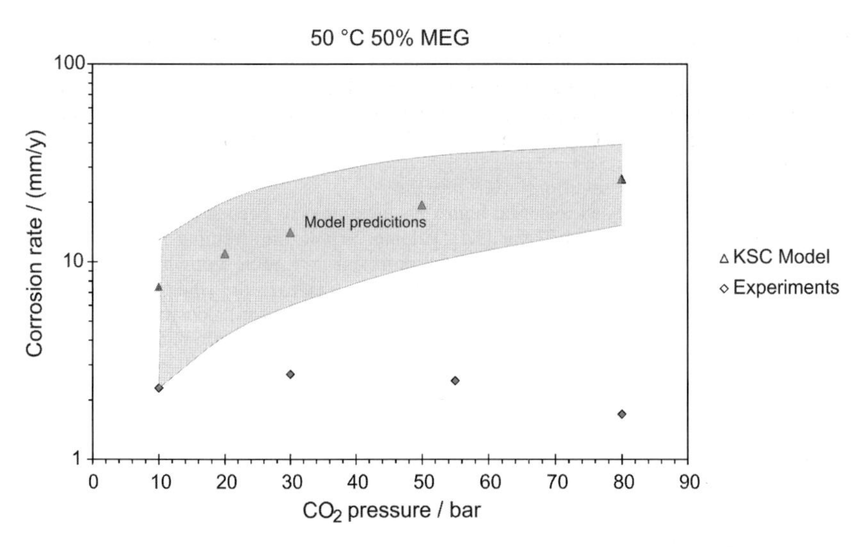

Figure 11: Experimental corrosion rates compared to model predictions in 50 wt% MEG solutions at 50 °C. The model predictions fall in the shaded area and only the values calculated by the KSC model are shown.

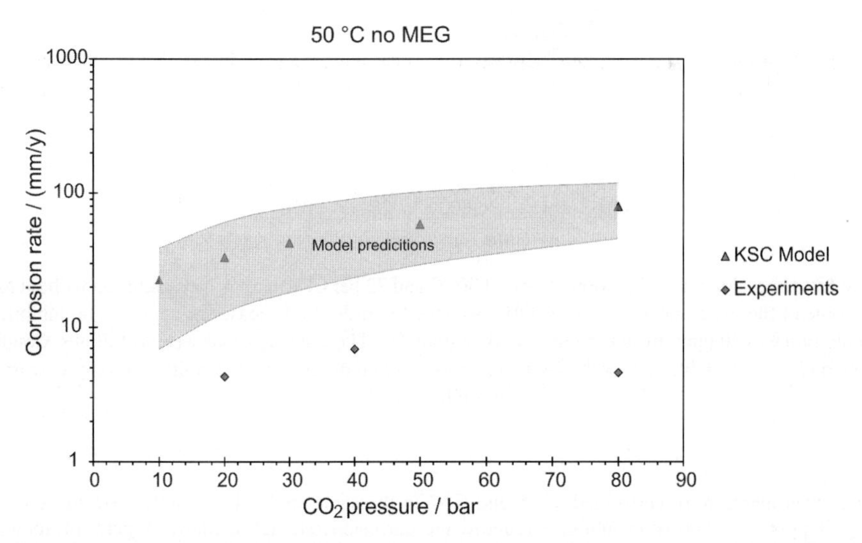

Figure 12: Experimental corrosion rates compared to model predictions in water at 50 °C. The model predictions fall in the shaded area and only the values calculated by the KSC model are shown.

A possible explanation for the large inconsistency between experimental and model values could be the formation of a protective $FeCO_3$ film at higher pressures. However, the pH in these experiments will be in the area 3–3.5, and thus the iron solubility limit will be above 200 ppm. Such iron concentrations were not reached in these experiments. The iron concentration at the end of the experiments was in the range

10–100 ppm. Examination of the specimens after the tests showed only patches covered by corrosion products, too thin to be analyzed. It is thus unlikely that $FeCO_3$ films cause the low corrosion rates. Another possible explanation for the low corrosion rates is that there is a change in the CO_2 corrosion mechanism at high pressures. This will be explored further in forthcoming experiments with a more reliable reference electrode.

Inhibiting corrosion of carbon steel at high CO_2 pressures

The possibility of inhibiting CO_2 corrosion at high CO_2 pressures has been studied earlier [1]. Figure 13 summarizes the findings at 30 °C and 72 bar CO_2. Adding 20 ppm film-forming inhibitor in addition to MEG decreases the corrosion rate markedly and the corrosion rate soon drops below the target value: 0.1 mm/y. In these experiments, 14 g/L NaOH was added to decrease the acidity. This does not have a beneficial effect on the inhibited corrosion rate as it decreases slower and does not reach the same low value as without NaOH.

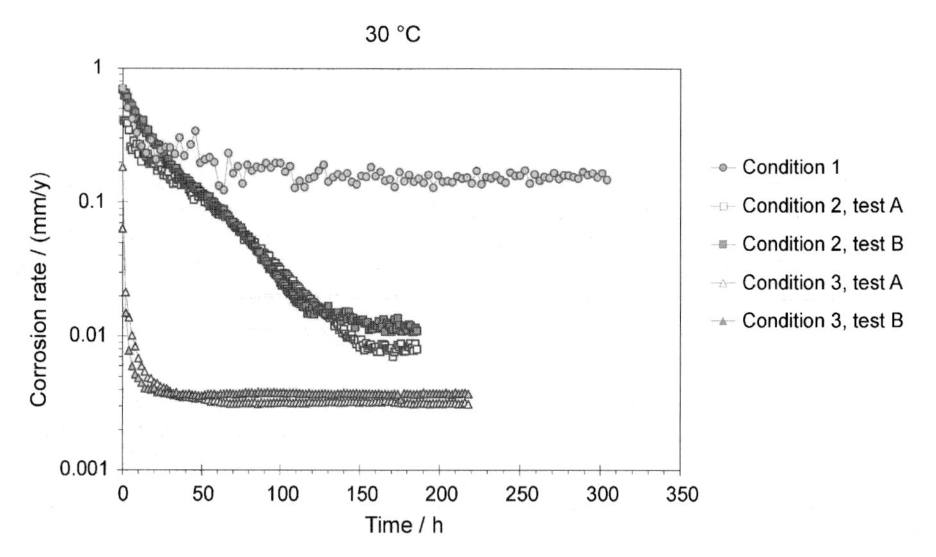

Figure 13: Corrosion rate as function of time at 30 °C and 72 bar CO_2 in a 1% NaCl solution without MEG. In some of the experiments 14 g/L NaOH was added in order to decrease the acidity. The inhibitor concentration was 20 ppm. Results from an earlier study [1]. The test conditions were as follows: Condition 1: no additions; Condition 2: both NaOH and corrosion inhibitor added; Condition 3: only corrosion inhibitor added.

Similar experiments were performed at 15 and 5 °C in the same study. The results were the same, i.e. adding 20 ppm CO_2 corrosion inhibitor reduced the corrosion rate substantially. Figure 14 shows the results at 5 °C and 44 bar CO_2. The corrosion rate without inhibitor is 0.2–0.3 mm/y independent of NaOH addition. Adding 20 ppm inhibitor to the solution lowers, also in this case, the corrosion rate well below 0.01 mm/y.

CONCLUSIONS

The literature survey and experiments performed in this study show that dry CO_2 and CO_2 that is not saturated with water is non-corrosive to carbon steel at transportation pipeline operation conditions.

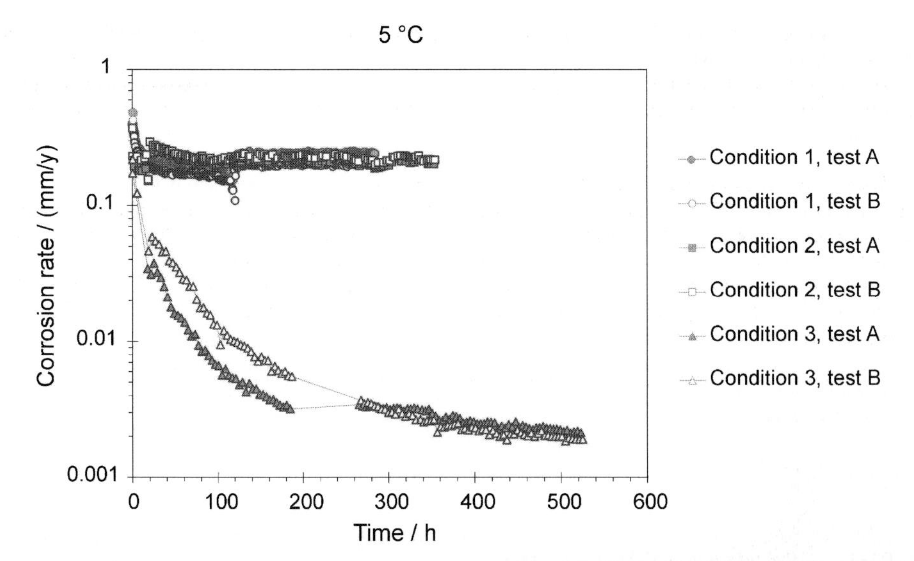

Figure 14: Corrosion rate as function of time at 5 °C and 44 bar CO_2 in a 1% NaCl solution without MEG. In some of the experiments 14 g/L NaOH was added in order to decrease the acidity. The inhibitor concentration was 20 ppm. Results from an earlier study [1]. The test conditions were as follows: Condition 1: no additions, CO_2 pressure 35 bar; Condition 2: only NaOH added, CO_2 pressure 44 bar; Condition 3: NaOH and corrosion inhibitor added, CO_2 pressure 44 bar.

CO_2 with water content above saturation is corrosive when water precipitates, but the experiments show that the corrosion rates at high CO_2 pressures in systems containing only water and that containing water/MEG mixtures are considerably lower than predicted by existing corrosion models. This applies especially at low temperatures typical for sub-sea pipelines in northern waters. In a previous study it was demonstrated that 20 ppm CO_2 corrosion inhibitor is sufficient to lower the corrosion rate below 0.1 mm/y at temperatures up to 30 °C and CO_2 pressures up to 72 bar [1].

More work is needed in order to understand the apparent change in corrosion mechanism at high CO_2 partial pressure. It should also be emphasized that the present investigations have been performed at non-flowing conditions in autoclaves with gentle agitation. Before the results can be utilized for pipeline designs, the effect of flow should be investigated in flow loop experiments.

RECOMMENDATIONS

For the transportation of CO_2 in carbon steel pipelines it is recommended that more work be conducted in the following areas.

1. Verification of water solubility as function of temperature and pressure for actual fluid compositions. For corrosion evaluations this information will be needed to be able to predict the amount of water precipitation, if any. This information is also necessary in order to optimize compression cycles and water removal during compression.
2. Determination of corrosion risks at realistic flowing rates with actual fluid composition should be made. The effect of flowing conditions should be investigated before the corrosion results are utilized for design. This should also include inhibitor evaluation. The consequences of other fluid components than

CO_2 should also be explored, especially the effect of H_2S in reducing conditions or sulfur or nitrogen oxides at oxidizing conditions.
3. More work on the mechanisms of CO_2 corrosion at high CO_2 partial pressures will be needed in order to extend present corrosion models.

When stainless steels are used it will be necessary to evaluate the corrosion risks if the fluids are more aggressive than pure CO_2. In addition, the integrity of steels with sealants (such as might be used in abandoned wells) requires special experiments.

ACKNOWLEDGEMENTS

The funding of this work by the CO_2 Capture Project (CCP) and the Research Council of Norway/Klimatek under Award No. 144737/228 is greatly acknowledged. Any opinions, findings, conclusions or recommendations expressed herein are those of the authors and do not necessarily reflect the views of the CCP or Klimatek.

REFERENCES

1. S.M. Hesjevik, S. Olsen, M. Seiersten, Corrosion at high CO_2 pressure, CORROSION/2003, Paper No. 03345, *NACE International*, Houston, 2003.
2. M. Seiersten, Corrosion of pipeline steel in supercritical CO_2/water mixtures, *Proceedings of EuroCorr2000*, London, September 11–14, 2000.
3. T. Wildenborg, Costs of CO_2 sequestration by underground storage, *Greenhouse Issues* (2000) 47.
4. M. Seiersten, Vurdering av rørledning for CO_2 injeksjon, Report to Statoil, IFE/KR/F-97/122.
5. K.Y. Song, R. Kobayashi, Water content of CO2 in equilibrium with liquid water and/or hydrates, SPE Formation Evaluation, 1987, pp. 500–508.
6. A.W. Hassel, K. Fushimi, M. Seo, An agar-based silver/silver chloride reference electrode for use in micro-electrochemistry, *Electrochem. Commun.* **1** (1999) 180.
7. R. Wiebe, V.L. Gaddy, *J. Am. Chem. Soc.* **63** (1941) 475–477.
8. M.B. King, A. Mubarak, J.D. Kim, T.R. Bott, *J. Supercrit. Fluids* **5** (1992) 296–302.
9. A.K.R. Dewan, Water saturation prediction for CO_2 rich mixtures containing traces of CH_4 and N_2, Paper 27B, *AIChE National Meeting*, Houston, March 24–28, 1985.
10. W.A. Propp, T.E. Carleson, C.M. Wai, P.R. Taylor, K.W. Daehling, S. Huang, M. Abdel-Latif, Corrosion in supercritical fluids, *US Department of Energy Report* DE96014006, Washington, DC, 1996.
11. F.W. Schremp, G.R. Roberson, Effect of supercritical carbon dioxide on construction materials, SPE Paper No. 4667, 1973.
12. F.W. Schremp, G.R. Roberson, Effect of supercritical carbon dioxide on construction materials, *Soc. Petrol. Eng. J.* **15** (1975) 227.
13. J.M. West, Design and operation of a supercritical CO_2 pipeline-compression system Sacroc Unit, Scurry County, Texas, SPE Paper No. 4804, 1974.
14. L.E. Newton, CORROSION/84, Paper No. 67, *NACE International*, Houston, 1984.
15. T.E. Gill, Canyon Reef Carriers, Inc. CO_2 pipeline: description and 12 years of operation, *ASME Energy-Source Technology Conference*, Pipeline Engineering Symposium, 1985, p. 59.
16. E.M. Russick, G.A. Poulter, C.L.J. Adkins, N.R. Sorensen, Corrosive effects of supercritical carbon dioxide and cosolvents on metals, *J. Supercrit. Fluids* **9** (1996) 43.
17. G. Schmitt, M. Krieck-Defrain, pH-static experiments in high pressure CO_2 corrosion of steel, Effect on scale morphology and chemical composition, corrosion rate, pitting susceptibility and inhibitor performance, *11th International Corrosion Congress*, Paper No. 480, 1990, p. 529.
18. H. Sarv, Personal communication.
19. L.E. Newton, R.A. Mcclay, Corrosion and operational problems, CO_2 Project, Sacroc Unit, *SPE Paper* No. 6391, 1977.
20. R. Kaul, N.G. Muralidharan, K.V. Kasiviswanathan, Failure analysis of carbon dioxide branch line in a urea plant, *Pract. Metallogr.* **32** (1995) 633.

21. M. Seiersten, CORROSION/2001, Paper No. 01042, *NACE International*, Houston, 2001.
22. E. Gulbrandsen, J.H. Morard, Why does glycol inhibit CO_2 corrosion, CORROSION/98, Paper No. 221, *NACE International*, Houston, 1998.
23. C. De Waard, U. Lotz, Prediction of CO_2 corrosion of carbon steel, CORROSION/93, Paper No. 69, *NACE International*, Houston, 1993.
24. R. Nyborg, Overview of CO_2 corrosion models for wells and pipelines, CORROSION/2002, Paper No. 02233, *NACE International*, Houston, 2002.

Carbon Dioxide Capture for Storage in Deep Geologic Formations, Volume 2
D.C. Thomas and S.M. Benson (Eds.)

Chapter 17

IMPACT OF SO_x AND NO_x IN FLUE GAS ON CO_2 SEPARATION, COMPRESSION, AND PIPELINE TRANSMISSION

Bruce Sass, Bruce Monzyk, Stephen Ricci, Abhishek Gupta, Barry Hindin and Neeraj Gupta

Battelle, 505 King Avenue, Columbus, OH 43201, USA

ABSTRACT

This study is an assessment of the effects of impurities in CO_2 streams on aboveground processing equipment. It is primarily a literature review that focuses on SO_x and NO_x impurities in flue gas. The three main components of the data analysis include:

1. Impact of impurities on the performance of amine separation systems.
2. Evaluation of the phase behavior of multi-component gas mixtures on multi-stage compressors.
3. Literature review of compressed gases to determine the corrosivity of pipeline materials in contact with CO_2, SO_x, and NO_x species with moisture present.

Flue gas impurities, such as SO_x, NO_x, other trace gases, and volatile metals have the potential of interacting unfavorably with capture, compression, and pipeline transmission of CO_2. Absorption and regeneration characteristics of amines and other solvents used to separate CO_2 are affected adversely by acid gas impurities, as their amine salts form essentially irreversibly. Compression of gas mixtures is subject to condensation of the higher boiling constituents, which may limit the ability to achieve adequate interstage cooling and may damage the compressor and other related processing equipment. Materials used in separation, compression, and transmission are subject to corrosion by acids formed from hydrolysis of SO_x and NO_x species in the presence of water. Finally, metals such as arsenic and mercury are accumulated from the coal and oil, and may hinder downstream processes.

INTRODUCTION

Flue gas produced by combustion of carbon-rich fuels consists mainly of carbon dioxide (CO_2), inert nitrogen (N_2), and excess oxygen (O_2) from the combustion air. Nitrogen oxides NO, NO_2, and NO_3 (collectively NO_x) form due to reactions between available nitrogen and oxygen. Compounds such as sulfur oxides SO_2 and SO_3 (collectively SO_x), hydrogen sulfide (H_2S), arsenic, and mercury occur from the coal combustion process, as well as, to a smaller extent, from oil and natural gas combustion processes. Ash and other particulate also are present in combustion gases.

These various acid gas impurities have the potential to interact unfavorably with capture, compression, and pipeline transmission of CO_2. The impurities have an adverse impact on absorption and regeneration characteristics of amines and other solvents used to separate CO_2 from flue gas. Compression of gas mixtures is subject to condensation of the higher boiling constituents, which may limit the ability to achieve adequate interstage cooling and may damage the compressor and other related processing equipment. Finally, materials used in separation, compression, and transmission are subject to corrosion by acids formed from hydrolysis of SO_x and NO_x species in the presence of water.

This article reviews the current status of existing technologies used for CO_2 gas separations. It addresses major issues affecting the industry and suggests potential research areas where further advancement is required. Most of the discussion in this chapter is limited to information obtained from published literature.

However, some thought has been given to devising an alternative scrubbing system, which to the authors' knowledge, has not been discussed elsewhere.

Chemical absorption by aqueous solutions of amine solvents or alkaline salt solutions is regarded as the most efficient CO_2 scrubbing process for low pressure, dilute gas streams. At higher pressure, or when the feed is more highly concentrated, physical absorption or a mixture of chemical and physical solvents (hybrid systems) may be cost effective. Other methods such as pressure swing adsorption, membrane separation, and cryogenic separation also are possible options for capturing CO_2 from flue gas but at higher cost [1]. The characteristics of CO_2 capture methods are summarized in Table 1. Prior cleanup to remove SO_2, NO_x, or H_2S also may be needed for many of these methods.

CO_2 can be produced as a by-product of many natural and chemical processes, which makes it somewhat unique in relation to the industrial gas market. However, CO_2 derived from combustion gases may contain sulfur oxides, nitrogen oxides, several different low molecular weight hydrocarbons, carbon monoxide, and mercury. The concentrations of these impurities may vary greatly in individual processes; also, the variety of possible CO_2 sources is responsible for a large number of potential impurities in the produced CO_2. Examples of possible impurities in several typical sources of CO_2 are listed in Table 2.

Successful implementation of a CO_2 storage system requires capture, compression, pipeline transmission, and injection steps that provide reliable and cost-effective operation. Impurities in the flue gas are potentially able to move through each step of the process and may interact adversely with equipment used in unit operations, as well as with the geologic media. A preliminary survey of some potential effects is shown in Table 3.

Chemistry of Amine Absorption Systems

Amine absorption is a chemical absorption process that involves capturing CO_2 using a reversible reaction between CO_2 and an aqueous solution of an amine giving either the carbonate ion salt or the carbonate as illustrated by the following reactions [3]:

$$2HOCH_2CH_2NH_2 + H_2O + CO_2 \leftrightharpoons 2HOCH_2CH_2NH_3^+ + CO_3^{2-} \qquad (1)$$

$$2HOCH_2CH_2NH_2 + CO_2 \leftrightharpoons HOCH_2CH_2NHCOO^- + HOCH_2CH_2NH_3^+ \qquad (2)$$

$$HOCH_2CH_2NH_2 + H_2O + CO_2 \leftrightharpoons HOCH_2CH_2NH_3^+ + HCO_3^- \qquad (3)$$

The reaction proceeds to the right at low temperature [25–65 °C (77–149 °F)] allowing absorption from the gas stream. The solvent can be regenerated, and the CO_2 recovered as a concentrated stream, by heating the solvent solution into the temperature range of 100–150 °C (212–302 °F) to reverse the absorption chemistry. Examples of solvents used for CO_2 recovery are summarized in Table 4.

CO_2 capture with solvent absorption is conducted using a pair of contacting columns typically referred to as the absorber and the reactivator (see Figure 1). Flue gas containing CO_2 enters the bottom of the absorber where it contacts amine solution flowing down through the column. High temperature in the absorber increases the rate of reaction between the CO_2 and the solvent, but also decreases the affinity of the solvent for CO_2. The competing effects of reaction rate and absorption affinity limit the optimum operating temperature in the absorber to about 50–60 °C (122–140 °F) [5]. The absorber typically operates at a pressure in the range from 30 to 45 kPa (207–310 psi). The amine solution, laden with CO_2, needs to retain fluidity so it can exit the bottom of the absorber, where it passes through a heat exchanger to recover some of the heat from, and cool, the reactivated amine, and then enters the top of the reactivator. The reactivator is equipped with a reboiler circulating and heating solution at the bottom of the tower such that steam rises up through the column, stripping CO_2 out of the amine solution thereby shifting Reactions (1)–(3) to the left. The reactivator typically operates at about 100 °C (212 °F) and 150–175 kPa (22–25 psi). The stripped CO_2 and steam exit the top of the reactivator and pass through a condenser to remove water vapor and produce a concentrated CO_2 stream containing about 6–8 v/o % water and traces of N_2, O_2, and NO as impurities. The reactivated amine solution exits the bottom of the reactivator and passes through the recovery heat exchanger, where it gives up some of its heat to the CO_2-loaded solution, thereby recovering its CO_2 sorptive property, and then it passes on to the top of the absorber.

TABLE 1
SUMMARY OF CO_2 CAPTURE METHODS

Capture process type	Description	Example separation materials	Comments
Amine solution absorption	Process involves capturing CO_2 using a reversible reaction between CO_2 and an aqueous solution of an amine. The amine is regenerated (by pressure reduction and heating) and recirculated	Monoethanolamine (MEA), diethanolamine (DEA), diglycolamine (DGA), methyldiethanolamine (MDEA), sterically hindered amines	Used at the commercial scale to remove low concentrations of acid gases (e.g. CO_2) from natural gas or breathing air Solution tends to saturate with high CO_2 loading, so the process is more efficient for lower CO_2 concentrations
Alkaline salt solution absorption	CO_2 captured using a reversible reaction between CO_2 and an aqueous solution of an alkaline salt. The salt solution is regenerated (by pressure reduction and heating) and recirculated	Potassium carbonate with additives such as boric acid or glycine to increase the solution capacity for CO_2	Used at the commercial scale to remove low concentrations of acid gases (e.g. CO_2) from natural gas Solution tends to saturate with high CO_2 loading so the process is more efficient for lower CO_2 concentrations CO_2 capacity of salt solution (even with additives) is lower than that of amine solutions
Physical absorption	CO_2 captured using physical dissolution in an absorption fluid. The fluid is regenerated (by pressure reduction and moderate heating) and recirculated	Propylene carbonate, *N*-methyl-2-pyrrolidone, methanol, dimethyl ether of polyethylene glycol, methyl isopropyl ether of polyethylene glycol	Used at the commercial scale to remove high concentrations of acid gases (e.g. CO_2) from natural gas More efficient for high CO_2 partial pressure (i.e. concentration and/or pressure) Does not typically remove acid gases as completely as chemical or hybrid absorption

(*continued*)

TABLE 1
CONTINUED

Capture process type	Description	Example separation materials	Comments
Hybrid absorption	CO_2 captured using a combination of chemical absorption and physical dissolution. The fluid is regenerated (by pressure reduction and moderate heating) and recirculated	Sulfolane (tetrahydrothiopene 1,1-dioxide) (physical solvent) and diisopropanolamine (DIPA) or MDEA (chemical solvent), sterically hindered amines, MDEA plus proprietary solvents	Used at the commercial scale to remove intermediate concentrations of acid gases (e.g. CO_2) from natural gas
Pressure swing adsorption (PSA)	Process involves using the intermolecular forces between gases and the surfaces of solid sorbent materials to capture CO_2. The sorbent is loaded at high pressure and regenerated by pressure reduction and, in some cases, heating	Molecular sieves, activated alumina, zeolites, activated carbon	Used at the commercial scale to remove CO_2 and other impurities from H_2. Some hydrogen gas cleanup processes also produce high purity CO_2
Gas separation membrane	Process involves pressurizing the flue gas and separating CO_2 from other gases by preferential permeation through a membrane. CO_2 is collected near atmospheric pressure as a permeate	Semipermeable membranes made of polyphenylene oxide, cellulose acetate, polysulfone, or polyamide	Used at the commercial scale to recover CO_2 used for enhanced oil recovery (EOR) (i.e. high CO_2 concentration) Requires two or more separation stages to reach a CO_2 removal of 90% and purity of 99%. Each stage requires compression, which increases cost, so the process typically is used for gas with high CO_2 content (e.g. pulverized coal/O_2 plants) Membranes are very sensitive to particulate fouling

Gas absorption membrane	The process involves using a semipermeable membrane as a barrier between the flue gas and an absorption fluid. Preferential removal of CO_2 from the gas stream occurs because the fluid (e.g. MEA) selectively absorbs CO_2	Microporous membranes made of Teflon®, polyphenyleneoxide, or polydimethylsiloxane	Innovative process The membrane allows a high surface area for transfer between the gas and liquid phases without requiring the two streams to mix. As a result the gas separation unit is more compact than the tall towers needed for chemical or physical absorption Membranes are very sensitive to particulate fouling
Cryogenic separation	Flue gas is cooled and compressed to condense CO_2 which can then be captured and purified by distillation	Not applicable	Used at the commercial scale to recover CO_2 used for EOR (i.e. high CO_2 concentration) Gas fed to the cryogenic separation unit must be dehydrated to prevent formation of solids (e.g. ice and CO_2 clathrates) Due to energy needed to reach cryogenic conditions, cryogenic separation typically is used for gas with high CO_2 content (e.g. pulverized coal/O_2 plants)

TABLE 2
POSSIBLE TRACE LEVEL IMPURITIES BY SOURCE TYPE (EXCLUDING AIR GASES AND WATER)

Component	Combustion	Wells/ Geothermal	Fermentation	Hydrogen or ammonia	Phosphate rock	Coal gasification	Ethylene oxide	Acid neutralization
Aldehydes	X	X	X	X		X	X	
Amines	X			X				
Benzene	X	X	X	X		X	X	X
Carbon monoxide	X	X	X	X	X	X	X	X
Carbonyl sulfide		X	X	X	X	X		X
Cycloaliphatic hydrocarbons	X	X		X		X	X	
Dimethyl sulfide		X	X		X	X		X
Ethanol	X	X	X	X		X	X	
Ether		X	X	X		X	X	
Ethyl acetate		X	X			X	X	
Ethyl benzene		X		X		X	X	
Ethylene oxide						X	X	
Halocarbons	X					X	X	
Hydrogen cyanide	X					X		
Hydrogen sulfide	X	X	X	X	X	X	X	X
Ketones	X	X	X	X		X	X	
Mercaptans	X	X	X	X	X	X	X	
Mercury	X					X		
Nitrogen oxide	X		X	X		X	X	X
Phosphine					X			
Radon		X			X			X
Sulfur dioxide	X	X	X	X	X	X		X
Toluene		X	X	X		X	X	
Vinyl chloride	X					X	X	
Volatile hydrocarbons	X	X	X	X		X	X	
Xylene		X	X	X		X	X	

Source: Ref. [2].
Note: The source types are generic sources, and there are variations in individual processes.

TABLE 3
SUMMARY OF IMPURITIES AND POTENTIAL EFFECTS ON SEQUESTRATION OPERATIONS

Impurity	Potential effects on capture by amines	Potential effects on compression	Potential effects on pipeline transmission	Potential effects on injection
N_2	None identified	Will increase compression energy consumption	Will increase transmission energy consumption	Will occupy space in the reservoir
O_2	None identified	Will increase compression energy consumption	Will increase transmission energy consumption	Will occupy space in the reservoir
Ar	None identified	Will increase compression energy consumption	Will increase transmission energy consumption	Will occupy space in the reservoir
H_2O	None identified	Condenses creating corrosive environment in presence of acid gases (e.g. CO_2 and SO_2)	Condenses creating corrosive environment in presence of acid gases (e.g. CO_2 and SO_2)	None
SO_2	Reacts irreversibly with some amine absorbents	Dissolves in water to form corrosive acid	Dissolves in water to form corrosive acid	May help form stable compounds to provide a long-term benefit to sequestration
SO_3	Reacts irreversibly with some amine absorbents	Dissolves in water to form corrosive acid	Dissolves in water to form corrosive acid	None identified
N_2O	Not strongly absorbed and tends to remain in flue gas	Not applicable	Not applicable	Not applicable
NO	Not strongly absorbed and tends to remain in flue gas. Readily converts to NO_2 with O_2	Not applicable	Not applicable	Not applicable
NO_2	Reacts irreversibly with some amine absorbents	Dissolves in water to form corrosive acid	Dissolves in water to form corrosive acid	None identified
Hydrocarbons	Increased toxicity (e.g. dioxins and furans)	Some hydrocarbons may increase compression energy consumption Increased toxicity (e.g. dioxins and furans)	Some hydrocarbons may increase compression energy consumption Increased toxicity (e.g. dioxins and furans)	Potential for unfavorable interaction with aquifer rock or fluid Increased toxicity (e.g. dioxins and furans)
Metals	Increased toxicity	Increased toxicity	Increased toxicity	Increased toxicity

TABLE 4
MAJOR CHEMICAL ABSORPTION SOLVENTS [4]

Chemical absorbent solvent	Solution strength (% amine)	Acid gas loading (mole gas/mole amine)	Amine heat of reaction with CO_2 (kJ/kg)
Monoethanolamine (MEA)	15–20	0.3–0.35[a]	1917
Diethanolamine (DEA)	25–35	0.3–0.35[a]	1517
Diglycolamine (DGA)	50–70	0.3–0.35[a]	1975
Methyldiethanolamine (MDEA)	20–50	>0.3	1394

[a] Limited by corrosion of carbon steel unless a corrosion inhibitor is used.

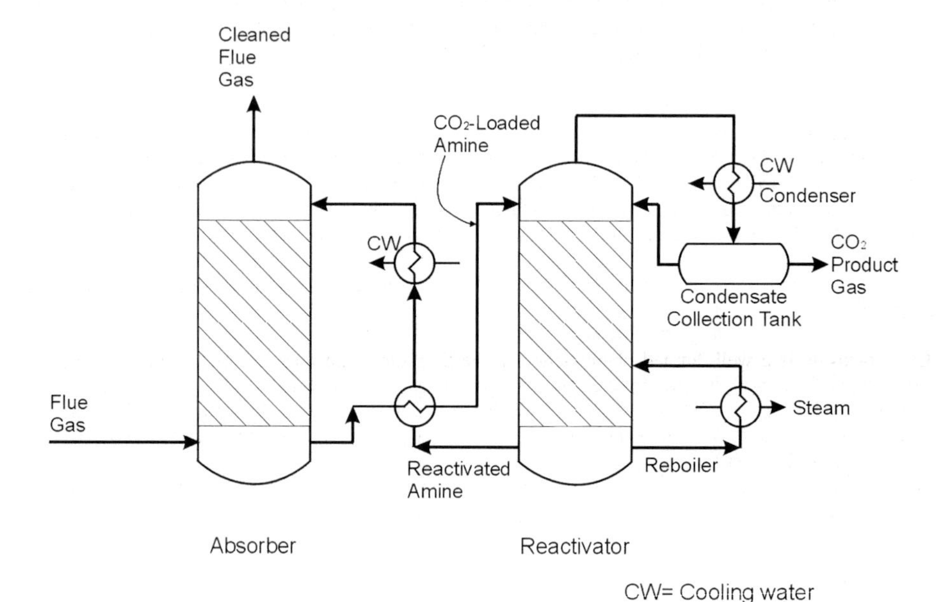

Figure 1: Recovery of CO_2 using solvent absorption.

As suggested by Reactions (1)–(3), the solubility of CO_2 in ethanolamines is influenced by temperature, amine solution strength, and pressure. Increasing the pressure of the flue-gas stream increases the potential concentration of CO_2 in the amine solution, which reduces the volume of solution circulating through the absorber and reactivator. The reduced volume decreases the required size and thus the capital cost and energy consumption of the process. However, a higher CO_2 concentration also increases the corrosiveness of the CO_2-rich amine solution. The CO_2 concentration typically is limited to 20% unless a solvent with corrosion inhibitors is used [3].

MEA used to separate CO_2 from flue gas also removes nearly all of the SO_2 and some of the NO_2 but very little NO and N_2O [6]. The SO_2 and NO_2 react with the amine to form stable salts that cannot be regenerated by heating in the reactivator and so represents a loss of solvent from the system, while increasing viscosity undesirably. The limits on SO_2 and NO_2 concentration in the flue gas being treated for CO_2 removal by MEA absorption are recommended to be in the range from 10 [6] to 50 ppmv [7].

Compression of CO_2 with Impurities Present

The importance of fluid properties, notably the phase behavior, and the associated impact on pipelines and compressors is well established. Applicable equations of state are known and numerical methods for predicting fluid properties, phase behavior, and sizing of equipment are available. However, very little specific information concerning the properties and phase behavior of $CO_2/SO_x/NO_x$ mixtures was found in published literature. The bulk of the information available discusses either pure CO_2 or mixtures of hydrocarbons and CO_2. This section discusses what is known about the impact of impurities on the compression of CO_2 and recommends a path forward for research and development.

Injection of CO_2 for enhanced oil recovery (EOR) or disposal requires processing, compression, and transmission of very large fluid volumes. CO_2 can be transmitted as a gas, liquid, or supercritical dense phase depending on the characteristics of the source, ambient temperature along the route, and other process and economic considerations. Transmission in the gas phase is not cost effective because of the volume requirement. Liquid transmission requires high purity and refrigeration, and may require heating for vaporization upon use. It is now accepted that high-pressure supercritical pipelines are the most cost-effective mode of CO_2 transmission. In this mode, the gas must be compressed to pressures well above the critical pressure, usually in the range of 8.3–15.9 MPa (1200–2300 psia).

Corrosive Effect of Impurities in CO_2 Stream

A literature search was carried out to identify available information related to the effect of impurities on corrosion properties of CO_2 pipeline. The impurities of interest to corrosion are SO_x and NO_x. Various resources covered for the search included technical publications, journals, conference proceedings, Internet searches, and industry reports. The scope of the search was limited to information related to the pipeline industry and to a lesser extent atmospheric pollution control of SO_x and NO_x.

Based on the analysis of the information from various resources, the following observations can be highlighted:

- CO_2 corrosion is a well-defined phenomenon in the literature. The mechanism, factors affecting, and prevention of corrosion related to CO_2 pipelines are covered extensively.
- The presence of H_2S in CO_2 and its effect on CO_2 corrosion is also well documented in the literature.
- Very little information is available on the effect of SO_x and NO_x on the corrosion properties of CO_2-carrying pipelines.
- No guidelines are available on maximum allowed concentrations of SO_x and NO_x to protect pipeline steel against corrosion.

RESULTS AND DISCUSSION

Chemistry of Amine Unit Operations

A substantial amount of worldwide activities, both recently completed and still under way, have targeted the development of CO_2 recovery technologies from power plant flue gas. In support of these efforts to achieve operability and cost-effectiveness, information is needed that will lead to an understanding of the complex physicochemical mechanisms within the proposed CO_2 capture concepts. To date, all such capture concepts appear very costly from capital and operating cost perspectives.

Nearly all proposed CO_2 sorption process schemes indicate a need to remove certain contaminants (notably dust, NO_x, and SO_x) prior to CO_2 scrubbing. The following unit operations are involved:

- A caustic wash of the flue is recommended to remove particulate and water by cooling/condensation because they can cause excessive foaming when alkyl amine solutions are used for capturing CO_2.
- The gas is dried may be sent through a catalytic converter to remove remaining O_2 because O_2 can oxidize the amine, most likely to N-oxides, olefinic compounds, aldehydes, carboxylic acids, etc.; the latter form stable emulsifying salts with the amine or metal ions (from corrosion), i.e. $RCOO^-M^+$ or $(RCOO^-)_2M^{+2}$ (where M represents a monovalent or divalent metal ion).

- The stream is sent through a desulfurization plant for sulfur removal to <100 ppm, and preferably <20 ppm(v/v), to minimize ammonium-SO_x^{2-} salt formation in the CO_2 scrubber, which does not thermally regenerate easily and in high yield.
- The gas is compressed prior to CO_2 capture by MEA or other amine technology to reduce needed equipment size and to increase sorption yield.

These unit operations each handle the full flue-gas stream and, therefore, are very capital intensive. These operations are required for performing CO_2 capture with minimal degradation of the amine. With respect to the specific CO_2 sorption unit operation, vendors have and are optimizing amine structures and formulations to minimize this degradation, to maximize CO_2 loading capacity, and to reduce the energy requirements for amine regeneration. What is lacking is a sufficient understanding of the mechanisms for amine degradation, and loss of CO_2 throughput capacity, in the face of natural contaminants contained in the flue gas, at flue gas scrubber conditions, and the conditions needed to minimize them. This section offers a preliminary examination of the impact of impurities on CO_2 capture by amines to illustrate where useful areas for mechanistic research exist.

Flue gas compositions for CO_2 scrubber feeds

Flue gas compositions are determined by several factors, including

- Fossil fuel composition (reservoir-specific).
- Fuel beneficiation process.
- Combustion process conditions, including O_2/fuel ratio, operating temperatures, and other process parameters.
- Pretreatment of the gas prior to CO_2 scrubbing (for example, manner and operational details of dust removal, N_2 conversion to NO_x, SO_x scrubbing and other components of the feed).

Specific compositions of flue gas are provided in the references cited throughout this chapter, with theoretical values for coal, oil, and natural gas provided in Table 5.

TABLE 5
MAJOR COMPONENTS OF FLUE GASES (MOLE %) FROM THREE HYDROCARBON FUEL TYPES BURNED WITH 110% THEORETICAL AIR [8]

Fuel	CO_2	H_2O	O_2	N_2
Coal	15.4	6.2	1.8	76.6
Oil	12.9	10.3	1.8	74.9
Natural gas	8.7	17.4	1.7	72.1

The level of NO_x in flue gas is determined by the fuel-bound nitrogen content, where increasing nitrogen concentration (normally 0.1–2% for certain coals) results in lower conversion efficiencies to NO_x [1]. Fuel nitrogen oxidation is controlled kinetically by the physical make-up of the feed and the chemical form of the nitrogen. In contrast to nitrogen, where only about half the fuel-bound N is converted to NO_x, essentially all the sulfur present in the fuel is converted to SO_x. Because CO_2 is a product of air/O_2 combustion of fossil fuels, and because carbon represents substantial portions of fossil fuels, especially coals, enormous amounts of CO_2 are present in the flue gas. Much more CO_2 is present in the flue gas than NO_x and SO_x. For example, flue gases from pulverized coal combustion are listed in Table 6, and compared to the postcapture product.

Pertinent amine chemistry relative to CO_2 scrubbing

Amines have a demonstrated ability to capture CO_2 from many types of gas streams, ranging from such extremes in application as removal from breathing air and flue gas from large power plants [9].

TABLE 6
TYPICAL COMPOSITION OF FLUE GAS AND CAPTURED CO_2 FROM A PULVERIZED COAL COMBUSTION POWER PLANT

Component	Flue gas composition (vol%)	Captured gas composition[a] (vol%)
CO_2	13.5	93.2
N_2	74.7	0.17
O_2	4.0	0.01
Ar	Trace	Trace
H_2O	7.7	6.5
SO_2	0.016[b]	Trace
SO_3	Trace	Trace
NO_x	0.06	Trace
Hydrocarbons	Trace	Trace
Metals	Trace	Trace[c]

[a] Assuming amine absorption.
[b] Assumes 2.2 wt% sulfur coal with flue gas desulfurization at 90% efficiency.
[c] Depends on fuel source and number of process factors; there is potential for volatile metals to be present in trace quantities.

The following list summarizes the key information regarding flue gas composition and the effectiveness of primary, secondary, and tertiary amines to scrub CO_2 from this environment [9–11].

- O_2, SO_x, and NO_x in flue gases can reduce the operational capacity of amines to scrub (absorb) CO_2 emissions by several chemical mechanisms.
- Amines can be lost by degradation mechanisms involving oxidation elimination, free radical C–N bond cleavage, catalyzed by trace emission species.
- Too much CO_2 also can diminish the operational capacity of amines by changing the chemical form of the sorbed species; more is not always better. Problems of this nature can occur due to buffer effects and solubility equilibria. Amine chemical structure has major impacts on these effects.
- Trace constituents also can create operational problems, such as accelerated corrosion.
- To avoid these problems, contaminant levels must be reduced by means of appropriate pretreatment methods.
- Amine tolerance levels are reported to be 90 ppm O_2, 10 ppm SO_2, and 20 ppm NO_x. These parameter values do not prevent the above degradation effects, but simply reduces them to economically and operationally acceptable rates.

Not much is known or understood about these amine decomposition reaction mechanisms, let alone the conditions that enhance or discourage them. Therefore, controlled and systematic laboratory and on-site testing are needed that uses actual flue gas slipstreams and commercial-scale supplied chemicals to elucidate these chemical reactions and the impact of trace contaminants on them. General amine reactive mechanisms have been extensively studied in other areas of technology which might offer some fundamental science from which candidate mechanisms and amine stabilizer chemistries could be drawn.

Flue-gas cleanup technologies
Future CO_2 removal technology research is being performed in several areas, including

- Improving amine scrubbing technology, where absorption is performed using "formulated amines", which include corrosion inhibitors, amine stabilizers, and other additives.
- Dry scrubbing technologies (adsorption media, membranes).
- Electrochemical capture of CO_2.

The impact of contaminants on these new and novel technologies remains to be determined and presents an important area of needed research. Potential problems may exist for membranes, for which adsorption sites are notoriously sensitive to fouling by low-level contaminants that bind surface reaction sites and small pores, causing accumulation, scale buildup, or direct chemical reactions in unfavorable directions. For example, poisoning of hydrogenation catalysts by low levels of sulfur is well known to occur. Reduction of cathodic over-potential in electrochemistry is another example.

More examples of flue-gas cleanup technologies, both commercially available and undergoing laboratory development, have been reported [12–63]. One such development is Ammogen™, a product that generates ammonia from hydrolyzing aqueous urea under pressure, but also produces CO_2. Ammogen™ is used to reduce flue gas levels of NO_x, SO_x, and other impurities.

Formulated amines (FA) are expected to offer substantial advantages over MEA alone. For example, improvements can include oxidation resistance, reduction in regeneration energy requirements, faster kinetics, higher selectivity, and less degradation. However, it is important to realize that impurities/contaminants may not react uniformly with each ingredient in the formulation. Therefore, proper amine sorbent make-up procedures are essential for these formulated products, as well as an understanding of how the various components of the FA react with the expected contaminants, concentration levels, and physical conditions of the CO_2 scrubber. Formulations also allow continuous improvements to be made to incrementally improve operability and reduce cost of overtime. Making such improvements is a typical challenge faced by formulations chemists.

Flue-gas cleanup chemistries and the impact of impurities on CO_2 removal

This section describes CO_2 scrubber chemistry in more detail. The following section describes the impact of other flue gas components, especially the NO_x and SO_x impurities, on the CO_2 removal process in terms of physical behavior, process operability, and economics.

Two major chemical products of CO_2 are formed during sorption: carbamate and bicarbonate ion. Although carbamate formation has advantages over bicarbonate in terms of selectivity for CO_2, the reverse reaction (regeneration) involves breaking covalent bonds that require large amounts of energy and can be slow. In contrast, bicarbonate ions form more slowly than carbamate since the carbamate is formed as an intermediate first, but requires significantly lower energy for regeneration. (Reaction kinetics are attributed to lower free OH^- ion activity than the free amine activity requires the carbamate to form first rather than forming HCO_3^- directly; therefore, the regeneration of CO_2 from the ionic species (e.g. $MDEA^+$, HCO_3^-) is a slower reaction.) Interaction of contaminants can take place at many points along the chemical pathway, including impacting the solubilization rate of CO_2 in the aqueous phase, the formation of OH^- from amine/H_2O reaction, and the breakup of bicarbonate ions into water, CO_2, and amine.

The degree to which carbamate forms depends primarily on the amine used in the sorber. Carbamate formation is decreasingly favored in the series primary > secondary ≫ tertiary (with tertiary amines, carbamate is not observed as the intermediate because it is unstable with respect to bicarbonate salt formation). Bicarbonate salts represent "ionic" capture of CO_2 as HCO_3^-, which is efficient but slow for all aliphatic amines. In scrubbing operations, carbamate formation is preferred due to its fast CO_2 sorption chemistry:

$$2RR'NH_{(aq)} + CO_{2(aq/g)} \leftrightarrows RR'N{-}COO^- + RR'NH_2^+$$

For carbonate formation, no water is required for CO_2 sorption, so the reaction also can occur in the gas phase. However, specific amine groups (identified by R and R′) generally are chosen to be very water soluble and having low volatility so that the amine is not lost to the large volume of flue gas passing through the contactor or vaporized with the CO_2 during regeneration. When impurities are present, their impact on water-soluble amines needs to be considered, for both cost and performance reasons.

Because a number of proprietary amine formulations exist, it is important for fundamental chemical mechanism studies to focus on understanding the interactions between other components of the flue gas, particularly SO_x and NO_x, dust, trace metal catalysts, and mercury. (Note that thiols and amines both tend to be strong sorbers for Hg.)

Typically it is necessary to accommodate the effects of flue gas contamination, resulting from upstream processing (dust, SO_x, and NO_x). For example, SO_x scrubbers using limestone or dolomite slurries will contribute additional CO_2, where water is both consumed and released in the reaction:

$$CaCO_3 + SO_3 + H_2O \rightarrow CaSO_4 + H_2O + CO_2$$

$$H_2O + CaCO_3 \leftrightarrows Ca^{2+} + HCO_3^- + OH^-$$

$$SO_3 + H_2O \leftrightarrows HSO_4^- + H^+$$

$$H^+ + OH^- \leftrightarrows H_2O$$

$$Ca^{2+} + HSO_4^- + HCO_3^- \leftrightarrows H_2O + CO_2 + CaSO_4$$

Whether sulfate (as shown) or sulfite is the main product depends upon scrubber pH, amount of excess O_2 present, combustion process temperatures, and presence of catalysts.

MEA is both low cost per pound, and nonvolatile, however it is expected to be particularly sensitive to contaminants because it readily degrades, especially by catalytic oxidation to compounds such as $^-OOCCH_2NH_3^+$. Inhibitors are sacrificial in nature and require replacement, and free radical inhibitors are also particularly sensitive to catalyzed oxidation of the inhibitors (as part of their beneficial action in protecting the amines from degradation). Hence consumption rates of the amine and stabilizers depend strongly upon contaminant effects, as does process complexity needed to deal with maintaining these reagents at operable levels.

Contaminant effects on CO_2 sorption chemistry

A number of process configurations exist for purification of flue gases. Detailed impacts of each possible process scenario are possible but outside of the scope of this chapter. An overview of the contaminant impact on CO_2 recovery processing is appropriate and illustrates the challenges, difficulties, and opportunities. Given the large size of the flue gas flow stream and, therefore, the high capital cost for handling such a large stream inherent in any such process, there will always be a strong incentive to minimize the number of gas treatment unit operations. Since gas decontamination is the objective, it is not necessary that the captured contaminants also be separated from each other as such operations would be more cost effective on small condensed product streams. However, separation of contaminants may be required at the large scale in order to accommodate different contaminant chemistries. For example, amines will effectively capture acid gases in one unit operation at the appropriate scrubber conditions (NO_x, SO_x, CO_2, and any HCl) but not low reactivity, volatile gases and vapors, e.g. NO or N_2O. Storage of toxic metals (Hg, etc.), if suitable sorbent chemistry and temperature are applied and NO oxidation to NO_x might be accomplished during dust and ash recovery in an initial treatment step, leaving N_2O in N_2, with some O_2 (added for NO oxidation) to be dealt with at the end of the process. Some of the water also would be captured, but could be minimized by operating the $NO_x/SO_x/CO_2$ scrubber at elevated temperature.

A key contaminant effect control consideration during CO_2 absorption is the prevention of premature hydrolysis of the carbamate driven by the strong acid (H^+) effect of the HNO_x (nitric and nitrous) and H_2SO_x (sulfuric and sulfurous) acids, i.e.

$$R_2N{-}COO^- + H^+ \leftrightarrows R_2NH + CO_2$$

This reaction, facilitated by moisture, would lower CO_2 absorption capacity. To prevent this sorption reversal reaction requires maintaining a high pH of the scrubber fluids ($pH > 7$). Although pH control would normally be accomplished with slaked or unslaked lime, the presence of SO_x would give rise to solids formation ($CaSO_x \cdot y\,H_2O$ and $CaCO_3$), if the pH is sufficiently high (>7), thereby requiring the recovery of alkyl amine from a solid/aqueous slurry. This mixture normally leads to unacceptable losses of amine through adsorption onto solid particles and requiring make-up.

Instead, water-soluble sulfates and carbonates could be produced using low-cost, water-soluble alkaline raw materials, such as NaOH and KOH, which can be used to prevent the solids from forming. These water-soluble alkalis are commodity priced, but still more costly than lime—and especially limestone. Hence, to improve the economics, NaOH would have to be regenerated from lime on site.

Using a water insoluble amine concept absorption occurs by the following set of reactions:

$$
\begin{gathered}
2R_2NH + CO_2 \leftrightarrows R_2NCOOH^- + R_2NH_2^+ \\
R_2NH + CO_2 + H_2O \leftrightarrows R_2NH^+ + HCO_3^- \\
R_2NH + NO_x + H_2O \leftrightarrows R_2NH_2^+ + NO_x^- \\
2R_2NH + SO_x + H_2O \leftrightarrows 2R_2NH_2^+ + SO_x^{2-} \\
R_2NCOO^- + H^+ \leftrightarrows R_2NCOOH(pH < 7)
\end{gathered}
$$

If caustic soda is provided to maintain the pH at approximately 6–8:

$$R_2NCOOH + NaOH \rightarrow R_2NCOO^- + Na^+ + H_2O$$

The use of water in insoluble amine leads to advantages in lower CO_2 regeneration temperatures, better water balance control, higher solubility of CO_2 in the organic phase, and ease of separations. The involvement and role of the major contaminants in flue gases (dust, NO_x, and SO_x) on the CO_2 recovery process using amines can be appreciated from the hypothetical process illustrated in Figure 2.

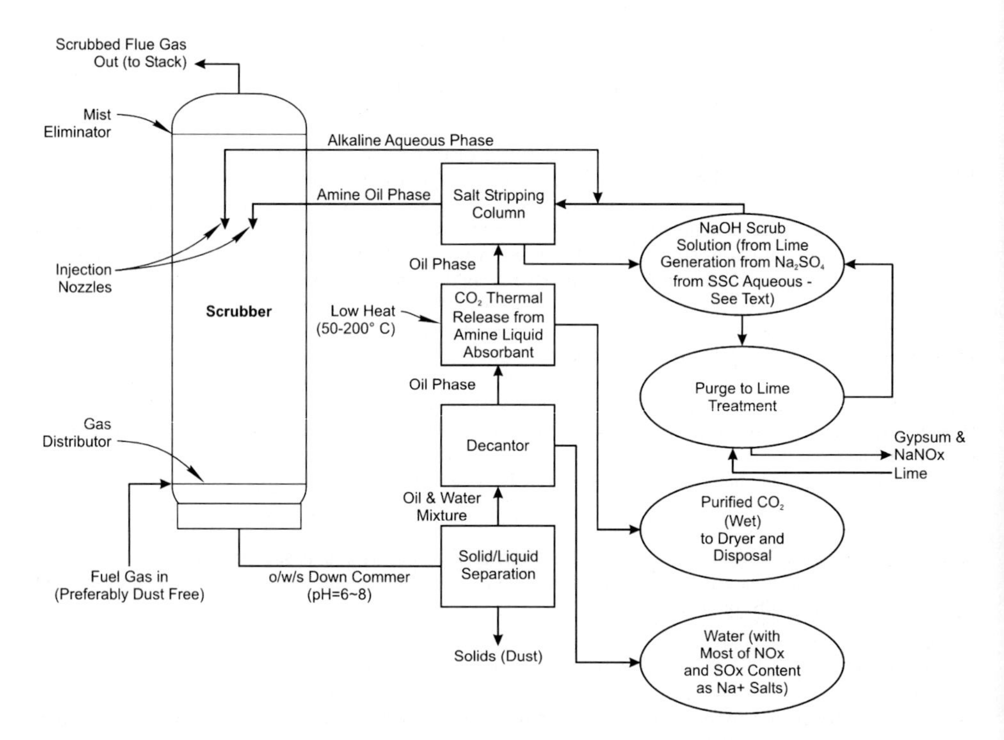

Figure 2: One-step flue gas cleanup concept using a water-insoluble amine.

Some scenarios can be envisaged where the organic amine losses are minimized by avoiding excessive heating, using mild pH, decreasing the maximum temperature for CO_2 release, and recovering amine from

NO_x and SO_x amine salts. Costs are further minimized by reducing the equipment size and complexity by a small inventory requirement for the flue gas by designing fast mixing conditions and short reaction times. Process reliability is enhanced by maximizing the use of proven, large-scale unit operations. All of these opportunities represent needed areas of research, development, and/or demonstration.

Another significant impact of these contaminants on the CO_2 capture process is that the high heat of neutralization of the NO_x and SO_x acid compounds by the amine causes high local heating at the absorption sites in the scrubber. This heat reverses the CO_2 sorption reactions, which occur in the 100–200 °C region, depending on the amine used. Due to the enormous amount of CO_2 present, such losses in capacity tend to be very adverse to process economics.

Chemistry of amine sorbent degradation chemistry related to contaminants
Because amines are reactive organic compounds, they can degrade by changing into other compounds as process feed compositions and conditions change. Since such reactions represent losses of throughput capacity and increase raw material costs, such degradation reactions and conditions are minimized. Therefore, although MEA is readily available and of low cost relative to current CO_2 separation processes serving the specialty chemical market, substantial improvements in amine chemical structures with respect to stability and regenerability are needed to cost-effectively concentrate CO_2 waste from power plant flue gas. Some new structures and supporting formulations development are in progress. Amine decomposition reactions to be avoided are given in the three reaction sequences:

Decomposition reactions of primary amines

$$\mathrm{RNH_2 + NaOH \rightarrow RNC\ (isocyanides)}$$
$$\mathrm{RNH_2 + oxidant \rightarrow RNO_2}$$
$$\mathrm{RNH_2 + aldehydes\ and\ formic\ acid \rightarrow secondary\ and\ tertiary\ amines}$$
$$\mathrm{R_2NH + aldehydes \rightarrow imines}$$

Decomposition reactions of secondary amines

$$\mathrm{R_2NH + esters \rightarrow amides}$$
$$\mathrm{R_2NH + H_2\ on\ metal\ surface \rightarrow tertiary\ amine\ (R_2NCH_2R')}$$
$$\mathrm{R_2NH + aldehydes/ketones \rightarrow keto\text{-}amines\ (branched)}$$
$$\mathrm{R_2NH + ketone + H^+ \rightarrow enamines}$$

Decomposition reactions of tertiary amines

Far fewer degradation reactions exist for tertiary amines than for secondary and primary amines. One such process involves alpha elimination to form an enamine, followed by oxidation and bond cleavage of the olefinic bond:

$$\mathrm{R_3N \xrightarrow{OH^-} R_2N{-}CH{=}CHR' \xrightarrow{oxidation} aldehyde + degraded\ amine}$$

Oxidation to the amine N-oxide ($R_3N + O_2 + catalyst \rightarrow R_3N \rightarrow O$).

Many of these reactions involve other components, which can be contaminants formed in the process or which enter with the feed. Nitrous acid (HNO_2) readily reacts with amines to form compounds with N–N bonds, which often readily decompose by hydrolysis to N_2 and N_2O, organic acids, ketones, aldehydes, and alcohols. Nonvolatile, water-insoluble amines may allow better control and purging of such degradation products. Therefore, amine structure is key to its performance and manufacturing cost, so further research is needed to optimize the amine structure best suited for CO_2 storage from power plant flue gas.

CO_2 product exit stream purity projections
As previously noted, CO_2 is released from the amine thermally during the regeneration step. Because the CO_2-loaded amine is in liquid form, it can be sharply separated from the inlet flue gas phase. Hence, the feed

gas "as is" can be cleanly prevented from carrying over into the CO_2 product gas. On the other hand, the sorbent amine solution is a concentrated aqueous solution of amine organic, so this stream is capable of scrubbing many other contaminants from the feed gas. Due to the alkalinity of the amine solution, SO_x (as $H_2SO_{3/4}$ and NO_x (as $HNO_{2/3}$) will be absorbed quantitatively (i.e. theoretically to $\ll 1$ ppm) provided adequate spray tower contact mixing in the scrubber and mist elimination is provided, and the amine is maintained in excess. In fact, the levels of passage of HNO_x or especially H_2SO_x is a direct measure of contactor stage efficiency. Therefore, unless these gases are removed in earlier separate steps, they will be completely absorbed into the amine solution during CO_2 sorption. However, because the CO_2 is desorbed from the amine at such mild conditions, and the nitrate and sulfate salts are not, these acids are not to be expected in the CO_2 product stream provided mist elimination is provided for in the CO_2 product gas as it exits the amine regeneration tower. However, for H_2SO_3 and HNO_2, the indications are much different. These acids break down at low temperatures back into SO_2 and NO_x, respectively. Hence, if sulfur and nitrogen oxidation was not complete, then some sulfur (S) and nitrogen (N) are expected to report to the CO_2 stream. These S and N products are expected to be quite corrosive, especially because moisture also is present (see below).

Water vapor is expected to be both absorbed by the amine scrubbing solution, as it occurs at < 100 °C, and evaporated with the CO_2 during amine sorbent regeneration as this occurs at > 100 °C. Hence, the CO_2 will be saturated with a high level of water vapor. Since water vapor will condense if compressed, leading to corrosion conditions and other problems, it is advisable to chill out most of the water vapor prior to CO_2 compression and/or use corrosive resistant compressor materials of construction.

With respect to toxic metal ions, mercury, lead, cadmium, and many other volatile metals (e.g. Zn, Ag, etc.), these ions form strong water-soluble complexes with amines

$$Hg^{2+} + 2RNH_2 \leftrightarrows Hg(RNH_2)_2^{2+}; \ \log K_f^{25\,°C} = 17.32$$

and so would absorb along with the CO_2 if not already removed at an earlier stage (e.g. as dust or with gypsum during SO_2 scrubbing). It is not expected that these nonvolatile metal ion complexes will follow the CO_2 concentrate stream during regeneration, other than by physical entrainment as aerosols, which can be eliminated by an in-line mist eliminator stage.

Because the CO_2 will be released from the amine under low to moderate pressure conditions, care will be necessary to avoid even small air leaks that will allow noncondensable gases from the air to enter the CO_2 product stream (N_2, O_2, or Ar). Noncondensable gases will not liquefy or dissolve in downhole storage of CO_2 and therefore will result in undesirable back pressure and downhole gas pockets.

As the CO_2 sorbent is a reactive organic phase, which degrades at a very slow rate if selected and used properly, some low levels of organics, mostly derivatives and fragments of the amine used to sorb the CO_2, will be sufficiently volatile to follow the CO_2 stream. Some of these organics will report to the dry condenser condensate, and the balance would proceed to accompany the CO_2 through to deep well disposal. Very low levels of these organics are expected so it should not create a disposal problem. Higher levels of organics are to be avoided for economic reasons, and also to avoid supplying nutrients for microbial growth once the CO_2 has been disposed of downhole.

Effect of Impurities on Compression

A literature search to determine thermodynamic properties of gas mixtures resulted in a compilation of sources, which are listed in the references section [64–76].

Impact of impurities on compression of CO_2

Pressurizing the CO_2 to the dense phase can be accomplished using a number of thermodynamic paths. A low-compressibility path involves both pumps and compressors because the fluid is taken thermodynamically through a two-phase region during the cooling cycle. In this path, the fluid is compressed, cooled to a liquid, and then pumped to its final pressure. The advantages of such a system are that refrigeration is not required and pressurization of a liquid requires less work. A high-compressibility thermodynamic path involves compression only and requires that the pressure and temperature of the fluid be maintained above the critical point to prevent two-phase conditions. Both centrifugal and reciprocating compressors are used for pressurization. Centrifugal machines are best suited for the initial stages, where large volumes are compressed from low pressure to a certain intermediate pressure. Multi-stage reciprocating compressors then are used to bring the fluid to its final pressure [69].

In the supercritical state, CO_2 behaves partly as a liquid and partly as a gas. In the normal pressure and temperature ranges of an operating pipeline, the presence of even small amounts of impurities can have a significant and nonlinear effect on density and compressibility and, therefore, flow behavior [64]. Table 7 shows the effect on flow capacity from the addition of small amounts of methane and nitrogen. Loss of flow capacity and increased flowing pressure drop increase the power requirements of the compression system. Figure 3 shows a pressure–enthalpy phase diagram for pure CO_2 and Figure 4 shows a similar plot for CO_2

TABLE 7
EFFECT OF IMPURITIES ON CO_2 PIPELINE CAPACITY [64]

Composition of flowing fluid	Flow velocity at design pressure drop (m^3/s)	Flow relative to pure CO_2
Carbon dioxide	98.3	1.00
Methane	90.8	0.92
Nitrogen	63.9	0.65
Carbon dioxide plus 5% methane	89.5	0.91
Carbon dioxide plus 10% methane	82.3	0.84
Carbon dioxide plus 5% nitrogen	85.9	0.87
Carbon dioxide plus 10% nitrogen	77.0	0.78

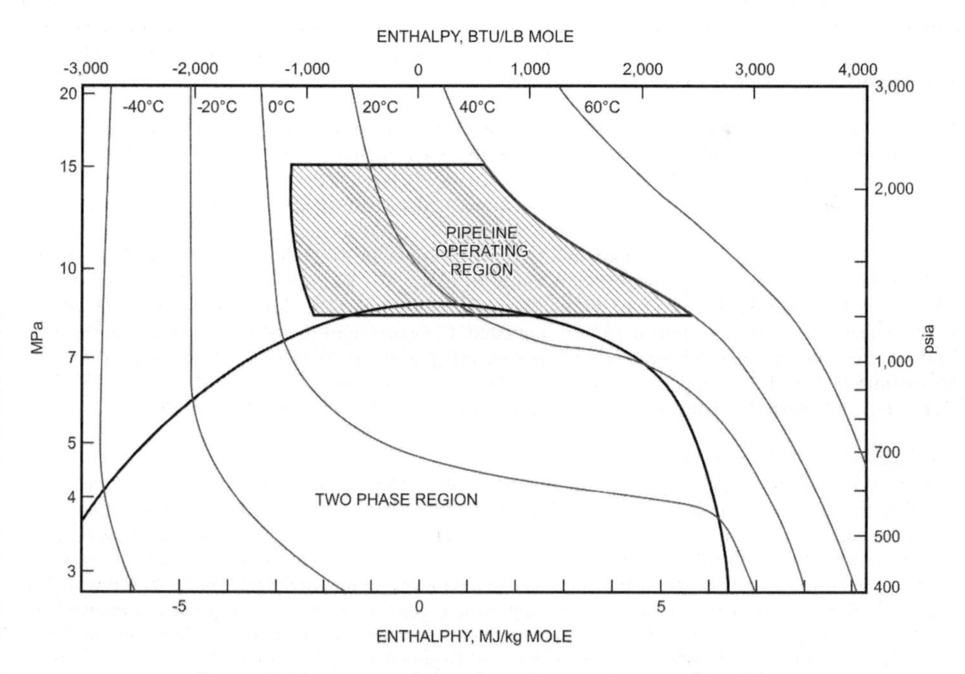

Figure 3: Pressure–enthalpy phase diagram for pure CO_2 [64].

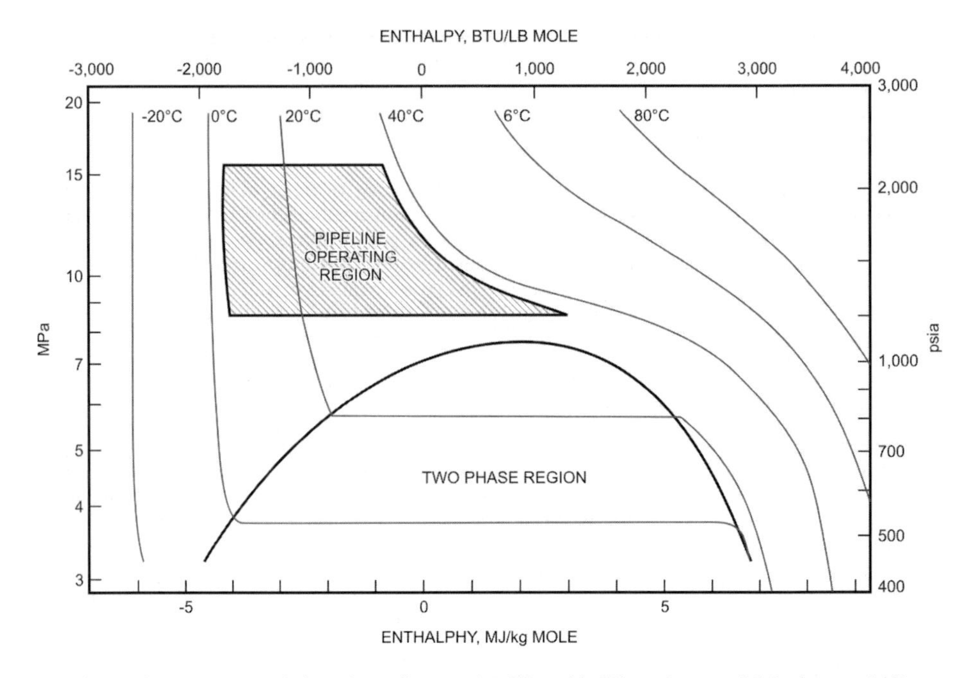

Figure 4: Pressure–enthalpy phase diagram for CO_2 with 5% methane and 5% nitrogen [64].

with 5% methane and 5% nitrogen. The operating range of a pipeline is shown on the plots. Comparison of the two diagrams shows that the operating range of a pipeline extends into the two-phase region when methane and nitrogen are present.

The effect of small quantities of SO_2 and NO_2 on the properties and behavior of CO_2 in a pipeline and compression system is uncertain. Although CO_2 and SO_2 have similar critical pressures (7.3 MPa for CO_2 and 7.9 MPa for SO_2), their compressibilities at 150 °C (300 °F) and 6 MPa (880 psia) are 0.91 [71] and 0.60 [74], respectively. Compressibility is the ratio of actual gas volume to ideal gas volume. The lower the compressibility, the easier the gas is to compress. Furthermore, SO_2 condenses at a higher temperature than CO_2. Based on the comparison of compressibility, it appears possible to compress a mixture of CO_2 and SO_2 without substantially increasing pressure or compressor energy per unit of fluid volume. Figure 5 shows a pressure–enthalpy phase diagram for CO_2 containing 2.5% SO_2 by weight [65]. The plot was computed from simulation, but the equations used were not cited. Comparison of the bubble pressures on the 20 °C isotherm from Figures 3 and 5 reveals some important differences. At 20 °C, CO_2 vaporizes at a pressure of approximately 5.5 MPa (800 psi), whereas the CO_2–SO_2 is shown to vaporize at 5.3 MPa (770 psi). Apparently, a distinct difference in the phase behavior exists, even with less than 5% SO_2.

Cost factors

Compression usually is the highest cost component in a CO_2 transmission system. In some cases, compressor equipment costs are as high as 2/3 the capital costs for the transmission system [68]. Equipment selection, energy requirements, number of stages, and interstage process components all must be given careful consideration in the design. All these design decisions depend on the composition and properties of the fluid being processed. It is essential to either maintain operating conditions above the critical point to prevent phase separation or tailor process steps where generation of multiple phases is unavoidable. If phase separation is unavoidable, it may be more economical to choose a low-compressibility thermodynamic path for compression (where the fluid is pumped to its final pressure as a liquid). Phase behavior of mixtures can be estimated using a number of well-known correlations and equations of state

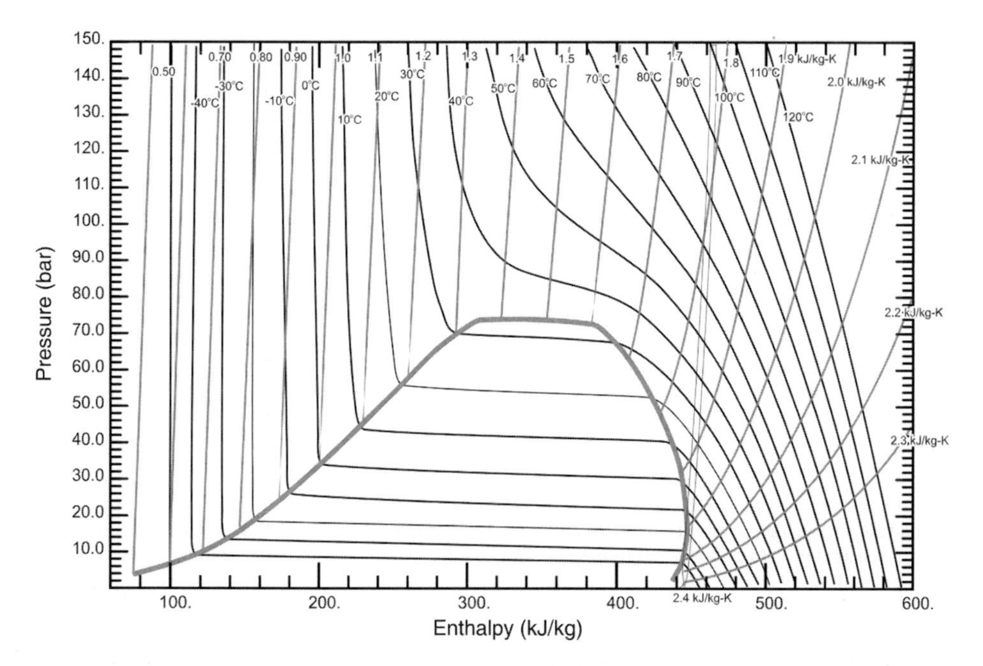

Figure 5: Pressure–enthalpy phase diagram for 97.5% CO_2 and 2.5% SO_2 [65].

such as the Benedict–Webb–Rubin–Starling and Wilson–Redlich–Kwong equations [64,75]. Numerical techniques and tools for computing bubble curves and other thermodynamic data for mixtures are also available [67,72].

Characterization of CO_2 streams

The composition of the flue-gas process stream depends on a number of factors, which include the source fuel for power generation, power plant operating conditions, and the nature and efficiency of flue-gas processing steps prior to compression. Gas composition from applications of interest needs to be identified as the first step of the development process.

Once characterized, various available techniques for property and thermodynamic state estimation should be applied to generate phase diagrams, thermodynamic properties, fluid and flow properties, and heat transfer characteristics for the mixtures of interest. It is expected that equations available in the literature for high-ratio CO_2 mixtures can be suitably modified for this development effort [75].

Compression equipment selection and design

The property and phase data will enable precise estimation of limits and specification of requirements for the compression system. The conditions required to avoid phase separation will be known. The final phase of development would involve a review of the state of the art in compression technology, along with the associated applications and costs. A design optimization effort can then be performed, taking into account the most appropriate thermodynamic paths, compression equipment and size, and interstage processing requirements.

Effect of Impurities on Corrosion

A significant amount of effort has been expended to control external corrosion of buried pipeline and, in fact, an entire industry has developed around its corrosion prevention and control. The most common practice is the use of coatings in conjunction with cathodic protection. Cathodic protection involves the use of either sacrificial anodes or impressed current, with the latter technique being the most common. Unfortunately, the corrosion control techniques used to protect the outside of pipelines are ineffective for protecting the pipeline from internal corrosion. This situation is often exacerbated by the fact that detection of internal corrosion is sufficiently difficult that problems may not be identified until an actual failure occurs. Perhaps the most common detection method for internal corrosion is the use of "pigs", which are mechanical devices used to clean and inspect the internal surfaces of pipelines.

Failure to adequately monitor and prevent internal corrosion can be disastrous. An example was the rupture of a natural gas transmission pipeline in Carlsbad, NM, in August 2000. The pipeline failure killed 12 people and property loss approached $1 million. The National Transportation Safety Board (NTSB) concluded that the pipeline failure was due to "significant reduction in pipe wall thickness due to severe internal corrosion" [77]. This report also suggests that internal corrosion was likely caused by a combination of microbes and such contaminants as moisture, chlorides, oxygen, carbon dioxide, and hydrogen sulfide. Cleaning pigs were not used in the section of pipe that failed because that section could not accommodate pigs. Among the recommendations of the NTSB was the request by the National Association of Corrosion Engineers (NACE) to update or replace guidelines for internal corrosion inspection of pipelines [78].

Role of contaminants and pH in corrosion of pipeline steel

SO_2 and NO_2 have high solubility in water/moisture. Similar to CO_2, SO_2 and NO_2 will not cause corrosion in the absence of moisture. It was mentioned in the literature that impurities in CO_2 tend to increase the rate of corrosion, but no specific details were cited [79]. The mechanism of CO_2 corrosion, in the presence of impurities, is also not understood properly. Van der Meer [80] has mentioned the increase of CO_2 corrosion due to the presence of SO_2, but no data or explanation was provided.

In the case of CO_2 transmission, the most important parameter determining possible corrosion damage will be the amount of water and dissolved O_2 present in the gas. The role of water in corrosion is its ability to act as an electrolyte and solvent for dissolved gases such as O_2 and other contaminants. O_2 is important in corrosion mechanisms because it provides several of the cathodic reaction paths for corrosion to proceed. Removing O_2 or de-aerating the solution will significantly reduce corrosion rates. According to Farthing [81] the internal corrosion rate of the pipe containing CO_2 is directly related to its concentration (when H_2O is present) and the system's temperature and pressure.

The maximum allowed moisture content in the pipeline should be evaluated and the tradeoffs for the dehydration systems examined. Experience with existing CO_2 compression and pipeline systems has established the requirement to reduce the moisture sufficiently to prevent condensation of liquid water. Dry CO_2 is generally considered noncorrosive. However, when combined with H_2O, carbonic acid is produced, which will corrode carbon steel. As the partial pressure of the CO_2 increases, the corrosivity of carbonic acid increases. At a CO_2 partial pressure greater than 207 kPa (30 psi), the carbonic acid formed is sufficiently strong to rapidly corrode carbon steel [82]. Therefore, CO_2 entering the pipeline must be dehydrated to minimize the potential for the formation of liquid water. Water concentrations between 288 and 480 $kg/10^6\ m^3$ (18–30 $lb/10^6$ scf) typically are adequate for CO_2 transmission in carbon steel pipe [83].

A preliminary analysis indicates that the compression, pipeline, and injection systems designed for CO_2 also should accommodate SO_2. Prior to the introduction of chlorofluorocarbons, SO_2 was the working fluid of choice for large commercial refrigeration systems. This application provides background data about compressing and handling SO_2. Performance in these systems indicates that corrosion is acceptable as long as the moisture content is $<0.005\%$ [70]. This dehydration requirement is similar to that for CO_2.

When H_2O is present, the main factor affecting the corrosion rate can be related to pH. In the absence of CO_2, the corrosion rate of iron and steel is fairly independent of pH between 4 and 10 (see Figure 6), but increases significantly below 3.5. The presence of CO_2 can complicate the dependency of corrosion rate on pH. De Waard et al. [89] indicated that the pH will largely be controlled by the concentration of dissolved iron carbonate ($FeCO_3$). Figure 7 is a graph showing the calculated corrosion rate, based on De Waard's

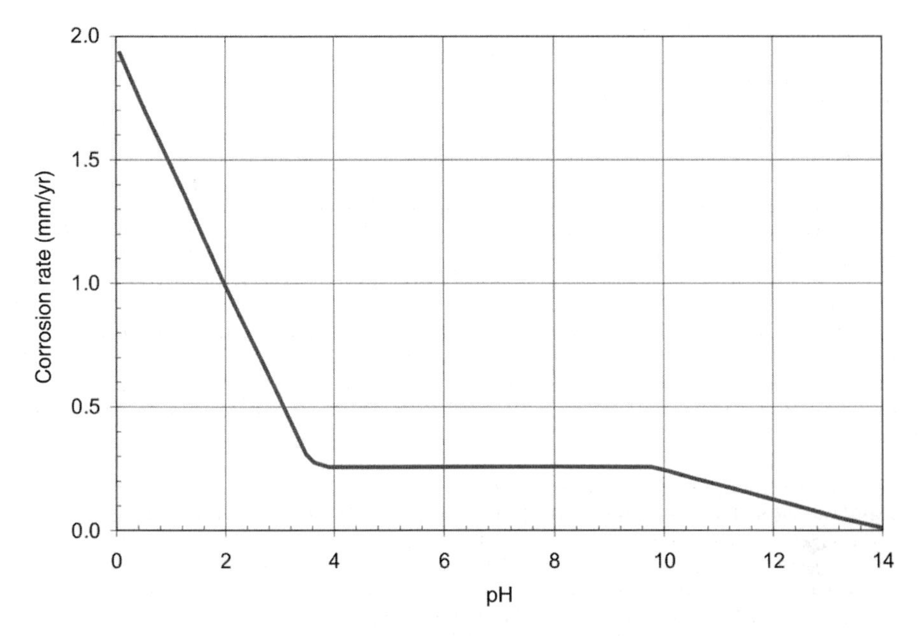

Figure 6: Corrosion of steel as a function of water pH [85].

semi-empirical model, as a function of both pH and liquid flow rate at a temperature of 40 °C and partial pressure of CO_2 of 1 bar (14.5 psi).

Studies on flue-gas condensate corrosion for residential high-efficiency furnaces [84] have shown that low concentrations of dissolved chlorides, nitrates, and sulfates can lead to dilute mixed acid solutions

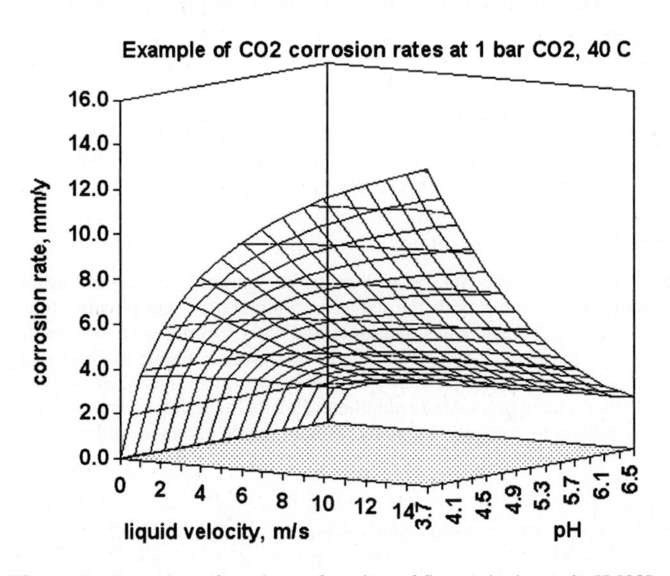

Figure 7: Corrosion of steel as a function of flow velocity and pH [89].

having a pH less than 2.5. It is not entirely clear to what degree NO_x and SO_x will lower the pH of any aqueous phase in pipeline condensate, but this is an area that should be studied to further understand its effect on internal corrosion rates.

Water is typically removed from a system using various devices such as separation equipment, dryers, and dehydrators. Water present in the system might be removed using drip pots, pig launders, valves, and separators. Addition or injection of compatible corrosion inhibitors to the pipeline stream should be considered when H_2O cannot be reduced below an acceptable concentration. Amines are often used as corrosion inhibitors, but other commercially available compounds exist that may improve on their effectiveness.

Pipeline alloy selection

Pipeline steels are typically made from high-strength American Petroleum Institute (API) grade material. Common grades are X52, X60, and X80 where the number following the "X" refers to the minimum yield strength in ksi or thousands of pounds per square inch. These alloys do not have an inherently good corrosion resistance, particularly to aqueous solutions with low pH. It is imperative, therefore, that an internal corrosion control program be implemented for pipelines carrying CO_2. Alloys that have greater corrosion resistances are not likely to be a cost-effective alternative. Chemical process industries, which often use stainless steel (300 series) alloys to minimize corrosion, would be at least three times more expensive than the carbon steels used in gas transmission pipelines.

Corrosion by acid gases

CO_2 corrosion or "sweet corrosion" is a common phenomenon for carbon and low-alloy steels. Pure, dry CO_2 is essentially noncorrosive except at temperatures exceeding 400 °C [85]. However, CO_2 in the presence of H_2O will form carbonic acid, which will corrode the pipeline steel. The corrosiveness of the carbonic acid increases with an increase in CO_2 partial pressure. CO_2 corrosion occurs primarily in the form of general corrosion but also as localized corrosion (e.g. pitting, mesa attack). CO_2 corrosion is affected by a number of parameters [86], including environmental, physical and metallurgical variables. All parameters are independent and can influence the CO_2 corrosion in many different ways.

Options for corrosion control

In the absence of data related to the effect of SO_x and NO_x on CO_2 corrosion, the effect of the presence of SO_x and NO_x on variables affecting CO_2 corrosion (as mentioned in "Introduction" section) should be considered. One of the factors affected by the presence of SO_x and NO_x will be the pH of the corroding fluids. Based on the concentration of SO_x, NO_x, and moisture in CO_2, the pH of the corroding fluid will change. The effect of pH on CO_2 corrosion is defined in the literature and, therefore, can be used to define the change in corrosion properties. However, a number of mechanisms were cited in the literature [87] to correlate the effect of pH on CO_2 corrosion depending on variables such as type of steel, composition of gasses, temperature, and pressure of gases. Sridhar [88] mentions that the increase in pH in $CO_2/H_2S/O_2$ systems results in the decrease of the corrosion rate.

CONCLUSION

Flue-gas impurities such as SO_x and NO_x, as well as arsenic and mercury present in solid fuels, have the potential of interacting unfavorably with the capture, compression, and pipeline transmission of CO_2. Absorption and regeneration characteristics of amines and other solvents used to separate CO_2 are affected adversely by acid gas impurities. Compression of gas mixtures is subject to condensation of the higher boiling constituents, which may limit the ability to achieve adequate interstage cooling and may damage the compressor and other related processing equipment. Finally, materials used in separation, compression, and transmission are subject to corrosion by acids, which include carbonic, sulfuric, sulfurous, nitric, and nitrous acids.

CO_2 Capture Chemistry

With respect to the specific CO_2 sorption unit operations, amines and amine formulations have been developed commercially to minimize degradation, maximize CO_2 capacity, and reduce the energy requirements for amine regeneration. What is lacking is a sufficient understanding of the mechanisms for amine degradation and energy balance control factors in the face of natural contaminants contained in the flue gas, at flue gas scrubber conditions, and the conditions needed to minimize them. The important research and development needs in the capture area are as follows:

- Identification of degradation-resistant amines and associated stabilizers with low volatilities. Generally, these amines will have branched structures of deactivated "R" groups and be essentially nonvolatile.
- Identification of synergistic formulations for CO_2 absorbents that will enable low-energy regeneration and release and that are stabilized against degradation by O_2 oxidation. Note that, as the regeneration temperature is lowered, substantial reduction in degradation is to be expected.
- Identification, and validation at continuous, flow, and sorbent recycle conditions, of a "one stage of contact" technology for scrubbing all of the needed contaminants (especially CO_2, NO_x, SO_x, dust, and toxic metals) from flue gas, with the objective of achieving substantial decreases in the capital cost of flue-gas cleanup.
- Characterization of the mechanisms for the reactions between NO_x and amine, with the objective of limiting interactions that lead to loss of the amine.
- Evaluation of the oil/water concept for "one step" flue-gas cleanup (including CO_2) and removal proposed in the "Results and Discussion" section of this chapter (Figure 2).

Compression

The effect of small quantities of SO_2 and NO_2 on the behavior of CO_2 in a pipeline and compression system has not been well established due to lack of data about specific systems of interest. It is known that CO_2 and SO_2 have similar critical pressures (7.3 and 7.9 MPa, respectively), but their compressibilities at 150 °C and 6 MPa are substantially different (0.91 and 0.60, respectively).

There is a clear need for more detailed information regarding CO_2 streams and the influence of NO_x and SO_x impurities on fluid properties and compressor performance. It is not possible to conclude from this study whether multi-phase conditions should be expected during compression stages or whether the impurities would cause substantial increases in energy consumption, other operating costs, or capital costs. It is recommended that research and development be undertaken in this area to clarify these issues. An experimental program may be needed to measure all important properties as a function of temperature and pressure, including

- thermodynamic properties, including enthalpy, compressibility, and boiling points (bubble curve);
- fluid flow properties, especially viscosity and density data; and
- heat transfer properties, including thermal conductivity and convective heat transfer coefficients.

Corrosion

Corrosion due to hydrolysis of acid gases is a common phenomenon for carbon and low alloy steels. Pure, dry CO_2 is essentially noncorrosive at ambient to moderate temperatures. However, CO_2 in the presence of water will form carbonic acid, which will corrode pipeline steel. When H_2O is present, the main factor affecting corrosion rate can be related to pH. For iron or steel, the corrosion rate increases significantly below pH 3.5. SO_2 and NO_2 have high solubility in water/moisture. It is not clear to what degree they will lower the pH of any aqueous phase in pipeline condensate; however, this is an area that should be studied to further understand its effect on internal corrosion rates.

There is a need to understand the interaction of SO_x and NO_x with CO_2 under pressurized wet conditions. A fundamental approach is suggested to qualitatively evaluate the effect of small amounts of SO_x and NO_x in CO_2 on corrosion properties of pipeline material. Among the necessary data needed are

- The effect of pH on the rate of CO_2 corrosion.
- Dissolution behavior of SO_2 and NO_2 in moisture, which generates sulfurous and nitrous acids.
- The effect of O_2 on these acids, which can further oxidize them to H_2SO_4 and HNO_3 (strong acids).
- The combined effect of these additional acids with carbonic acid, which will change the pH of the corroding media.

REFERENCES

1. T.P. Shaw, P.W. Hughes, Optimize CO_2 removal, *Hydrocarbon Process.* **80** (5) (2001).
2. Compressed Gas Association, Commodity Specification for Carbon Dioxide, fourth ed., CGA G-6.2-2000, 2000, 20 pp.
3. R. Pierantozzi, Carbon dioxide, in: J.I. Kroschwitz, M. Howe-Grant (Eds.), Kirk-Othmer Encyclopedia of Chemical Technology, vol. 5, Wiley, New York, 1993, pp. 35–53, .
4. J.C. Polasek, J.A. Bullin, Selecting amines for sweetening units. *Proceedings of the Processors Association Regional Meeting Tulsa,* OK, 1994. www.bre.com/papers/selecting_amines_2.htm.
5. J.C. Polasek, G.A. Iglesias-Silva, J.A. Bullin, Using mixed amine solutions for gas sweetening, *Proceedings of the Seventy-First Gas Processors Association*, Gas Processors Association, Tulsa, OK, 1992, pp. 53–63, www.bre.com/papers/mixed_amines_2/mixed_amines_2.htm.
6. T. Suda, M. Fujii, K. Yoshida, M. Iijima, T. Seto, S. Mitsuoka, Development of flue gas carbon dioxide recovery technology, *Energy Convers. Manage.* **33** (5–8) (1992) 317–324.
7. R. Barchas, R. Davis, The Kerr-McGee/ABB Lummus crest technology for the recovery of CO_2 from stack gases, *Energy Convers. Manage.* **33** (5–8) (1992) 333–340.
8. H. Herzog, E. Drake, J. Tester, A Research Needs Assessment for the Capture, Utilization, and Disposal of Carbon Dioxide from Fuel-Fired Power Plant, vol. 2, Topical Reports, 1993.
9. M. Simmonds, P. Hurst, M.B. Wilkinson, C. Watt, C.A. Roberts, *A Study of Very Large Scale Postcombustion CO_2 Capture at a Refining and Petrochemical Complex*, Paper, 6 pp, J. Gale and Y. Kala (eds), 2003, Proceedings of the 6th Intl. Conference on Greenhouse Gas Control Technologies, 1–4 Oct. 2002, Kyoto, Japan.
10. United States Department of Energy, Carbon Sequestration: State of the Science, 1999.
11. H.B. Thomsen, S. Geleff, Recovery of CO_2 from exhaust gasses—another way to improve the economical and environmental aspects of CHP-Plants, Union Engineering, paper, 9 pp, 2002.
12. A. Levy, R.E. Barrett, R.D. Giammar, H.R. Hazard, Coal combustion, in: R. Meyers (Ed.), Coal Handbook, Marcel Dekker, New York, 1981, The composition of coal combustion gases is discussed. Also the impact of sulfur content on SO_3 and SO_2 production as well as NO_X production as a function of the fuel nitrogen level I described. The effect of several coal flue-gas cleaning techniques on flue gas composition is also discussed, Chapter 8.
13. H.D. Weiss, Guide to Organic Reactions, Burgess Publishers, Minneapolis, MN, 1969.
14. S. Wong, W.D. Gunter, D. Law, M.J. Mavor, Economics of Flue Gas Injection and CO_2 Sequestration in Coalbed Methane Reservoirs. Alberta Research Council, Edmonton, Alberta, Canada, 2000.
15. D. Wendt, M. McKellar, Absorption of CO_2 by aqueous diethanolamine solutions in a bortex tube gas–liquid contactor and separator, *Proceedings, Second Annual Conference on Carbon Sequestration*, May 5–8, Alexandria, VA, 2003.
16. J.T. Cullinane, G.T. Rochelle, Properties of concentrated aqueous potassium carbonate/piperazine for CO_2 capture, *Proceedings, Second Annual Conference on Carbon Sequestration*, May 5–8, Alexandria, VA, 2003.
17. D.A. Green, R.P. Gupta, B.S. Turk, W.J. McMichael, J.W. Portzer, Capture of carbon dioxide from flue gas using a cyclic alkali carbonate-based process, *Proceedings, Second Annual Conference on Carbon Sequestration*, May 5–8, Alexandria, VA, 2003.
18. S. Reddy, C.A. Roberts, Econamine FGSM plus: an enhanced amine based CO_2 capture process, *Proceedings, Second Annual Conference on Carbon Sequestration*, May 5–8, Alexandria, VA, 2003.
19. M. Simmonds, P. Hurst, M.B. Wilkinson, S. Reddy, S. Khambaty, Amine based CO_2 capture from gas turbines, *Proceedings, Second Annual Conference on Carbon Sequestration*, May 5–8, Alexandria, VA, 2003.
20. Prosint Av., Benfica/AGA, Reduced CO_2 Emissions to Atmosphere and Production of Food Grade CO_2, CO_2 Capture Demonstration Plant, Rio de Janeiro, Brazil, 1997.
21. Wittemann Carbon Dioxide Equipment, Web Site (Carbon-dioxide-equipment.com), Stack Gas CO_2 Recovery Systems.
22. Carbon Dioxide Capture from Power Stations.
23. Union Engineering, Plant Worldwide 2, CO_2 Recovery Plants, Hydrogas, Malaysia, Kuala Lumpur, Malaysia, 168 MTPD, 1998.

24. N.C. Dave, G.J. Duffy, J.H. Edwards, A. Lowe, Economic Evaluation of Capture and Sequestration of CO_2 from Australian Black Coal-Fired Power Stations, Greenhouse Gas Control Technologies, vol. I, Elsevier, Amsterdam, 2003, pp. 173–178.
25. T. Supap, A. Aroonwilas, A. Veawab, A. Chakma, P. Tontiwachwuthikul, B.D. Kybett, An investigation of solvent stability in amine based systems used to capture CO_2, Greenhouse Gas Control Technologies, vol. I, Elsevier, Amsterdam, 2003, University of Regina, Saskatchewan, pp. 211–216.
26. T. Mimura, K. Matsumoto, M. Iijima, S. Mitsuoka, Development and application of flue gas carbon dioxide recovery technology, Greenhouse Gas Control Technologies, vol. I, Elsevier, Amsterdam, 2003, Kansai Electric Power Co. & Mitsubishi Heavy Industries, Japan, pp. 138–141.
27. G.S. Goff, G.T. Rochelle, Oxidative degradation of aqueous monoethanolamine in CO_2 capture systems under absorber conditions, *Greenhouse Gas Control Technologies*, vol. I, Elsevier, Amsterdam, 2003, pp. 115–119.
28. G. Clerici, E.D. Addario, M. Musicanti, G. Pulvirenti, S. Serenellini, M.G. Valdiserri, Environmental analysis of different options of CO_2 capture in power generation from natural gas, *Greenhouse Gas Control Technologies*, vol. I, Elsevier, Amsterdam, 2003, pp. 63–68.
29. M. Simmonds, P. Hurst, M.B. Willkinson, C. Watt, C.A. Roberts, A study of very large scale postcombustion CO_2 capture at a refining and petrochemical complex, *Greenhouse Gas Control Technologies*, vol. I, Elsevier, Amsterdam, 2003, pp. 39–44.
30. M. Wilson, P. Tontiwachwuthikul, A. Chakma, R. Idem, A. Veawab, A. Aroonwilas, D. Gelowitz, J. Barrie, C. Mariz, Test results from a CO_2 extraction pilot plant at boundary dam coal-fired power station, *Greenhouse Gas Control Technologies*, vol. I, Elsevier, Amsterdam, 2003, pp. 31–36.
31. M.A. Abdi, M.M. Golkar, Improve contaminant control in amine systems, *Hydrocarbon Process.* (2001) 102C–102J.
32. B.P. Callahan, R. Wolfenden, Migration of methyl groups between aliphatic amines in water, *J. Am. Chem. Soc.* **125** (2003) 310–311.
33. S. Bishnoi, G.T. Rochelle, Thermodynamics of piperazine/methyldiethanolamine/water/carbon dioxide, *Ind. Eng. Chem. Res.* **41** (2002) 604–612.
34. C.L. Yaws, J.R. Hopper, S.R. Mishra, Solubility and Henry's Law constants for amines in water, *Chem. Eng.* (2001) 84–88. www.che.com.
35. C. Hanisch, Exploring options for CO_2 capture and management, *Environ. Sci. Tech./News* (1999) 66A–70A.
36. N.P. Chopey, C. Cooper, C. Crable, G. Ondrey, Technology to cool down global warming, Chem. Eng. (1999) 37–41.
37. M.A. Pacheco, G.T. Rochelle, Rate-based modeling of reactive absorption of CO_2 and H_2S into aqueous methyldiethanolamine, *Ind. Eng. Chem. Res.* **37** (1998) 4107–4117.
38. J.C. Polasek, G.A. Iglnlesias-Silva, Using Mixed Amine Solutions for Gas Sweetening, www.bre.com/papers/mixed_amines_2/mixed_amines_2.htm.
39. IEA Greenhouse Gas R&D Programme, Carbon Dioxide Capture from Power Stations, 1993, www.ieagreen.org.uk/sr2p.htm.
40. M. Capone, Sulfurization and Sulfurchlorination, Sulfur removal and recovery, Kirk-Othmer Encyclopedia of Chemical Technology, vol. 23, 1997, pp. 432–452.
41. A. Meisen, X. Shuai, Research and development issues in CO_2 capture, *Energy Convers. Manage.* **38** (Suppl.) (1997) S37–S42.
42. Gas processes 2000: A special report, Hydrocarbon Process. Hydrocarbon Processing, International Edition, **79**(4) (2000), 56–89.
43. H.K. Shethna, Thermodynamic Analysis of Chemisorption Processes, Thesis Submitted to the University of Manchester Institute of Science and Technology, for PhD, March 1996.
44. S. Freguia, G.T. Rochelle, Modeling of CO_2 capture by aqueous monoethanolamine, *AIChE J.* **49** (7) (2003) 1676–1686.
45. J. Kiepe, S. Horstmann, K. Fischer, J. Gmehling, Experimental determination and prediction of gas solubility data for $CO_2 + H_2O$ mixtures containing $NaNO_3$ or KNO_3, *Ind. Eng. Chem. Res.* **42** (2003) 3851–3856.
46. M.S. Ahmed, Y.A. Attia, Multi-metal oxide aerogel for capture of pollution gases from air, *Appl. Therm. Eng.* **18** (9–10) (1998) 787–797.

47. G. Curanov, B. Rumpf, A. Smirnova, G. Maurer, Solubility of single gases carbon dioxide and hydrogen sulfide in aqueous solutions of *N*-methyldiethanolamine in the temperature range 313–413 K at pressures up to 5 MPa, *Am. Chem. Soc.* (1996) 1959–1966.
48. F.-Y. Jou, F.D. Otto, A.E. Mather, Vapor liquid equilibrium of carbon dioxide in aqueous mixtures of monoethanolamine and methyldiethanolamine, *Ind. Eng. Chem. Res.* **33** (1994) 2002–2005.
49. Y.-G. Li, A.E. Mathers, Correlation and prediction of the solubility of carbon dioxide in a mixed alkanolamine solution, *Ind. Eng. Chem. Res.* **33** (1994) 2006–2015.
50. R.H. Weiland, J.E. Dingman, D.B. Cronin, Heat capacity of aqueous monoethanolamine diethanolamine, *N*-methyldiethanolamine, and *N*-methyldiethanolamine-based blends with carbon dioxide, *J. Chem. Eng. Data* **42** (1997) 1004–1006.
51. L.-F. Chiu, H.-F. Liu, M.-H. Li, Heat capacity of alkanolamines by differential scanning calorimetry, *J. Chem. Eng. Data* **44** (1999) 631–636.
52. D.J. Seo, W.H. Hong, Effect of piperazine on the kinetics of carbon dioxide with aqueous solutions of 2-amino-2-methyl-1-propanol, *Ind. Eng. Chem. Res.* **39** (2000) 2062–2067.
53. M. Wilson, P. Tontiwachwuthikull, A. Chakma, R. Idem, A. Veawab, A. Aroonwilas, D. Geowitz, J. Barrier, C. Mariz, *Test Results from a CO_2 Extraction Pilot Plant at Boundary Dam Coal-Fired Power Station*, Paper A4-3, J. Gale and Y. Kala (eds), 2003, *Proceedings of the 6th Intl. Conference on Greenhouse Gas Control Technologies*, 1–4 Oct. 2002, Kyoto, Japan.
54. M. Slater, E. West, C.L. Mariz, Carbon Dioxide Capture from Multiple Flue Gas Sources, Paper C4-1, J. Gale and Y. Kala (eds), 2003, *Proceedings of the 6th Intl. Conference on Greenhouse Gas Control Technologies*, 1–4 Oct. 2002, Kyoto, Japan.
55. See Ref. [9].
56. Gas Processing, Fluor Web Page.
57. Carbon Dioxide Capture from Power Stations, Absorption Technologies web page.
58. W. Breckenridge, A. Holiday, J.O.Y. Ong, C. Sharp, Use of Selexol process in coke gasification to ammonia project, Laurance Reid Gas Conditioning Conference, February 27–March 1, University of Oklahoma, Norman, OK, Selexol™ Process, Gas Processing, 2000, www.uop.com.
59. See Ref. [11].
60. L. Smith, N. Gupta, B. Sass, T. Bubenik, Carbon Dioxide Sequestration I Saline Formations—Engineering and Economic Assessment, Final Technical Report under DOE Contract DE-RAC26-98FT35008, Battelle, July 9, 2001.
61. D.G. Chapel, C.L. Mariz, J. Ernest, Recovery of CO_2 from flue gases: commercial trends, Presented at the Canadian Society of Chemical Engineers Annual Meeting, October 4–6, Saskatoon, Saskatchewan, Canada, 1999.
62. Luzhou Natural Gas Chemicals Group, Manufacture Ammonia and Urea and Capture CO_2, CO_2 Capture Demonstration Project, Luzhou City, China.
63. U.S. Department of Energy, Working Paper on Carbon Sequestration Science and Technology, Office of Science, Office of Fossil Energy, U.S. Department of Energy, 1999.
64. C.B. Farris, Unusual design factors for supercritical CO_2 pipelines, *Energy Prog.* **3** (3) (1983) 150–158.
65. Berkestad H., Separation and compression of CO_2 in an O_2/CO_2-fired power plant, Masters Thesis, Chalmers University of Technology, Goteborg, Sweden, 2002.
66. G.G. King, Design Considerations for Carbon Dioxide Pipelines, Pipe Line Industry, 1981, pp. 125–132.
67. J.E. Powers, Prediction of Thermodynamic Properties and their Use in the Design of Compressors, Conference Proceedings—Natural Gas Resource Technology, 1971.
68. B.C. Price, Processing high CO_2 gas, *Energy Prog.* **4** (3) (1984) 169–174.
69. A.S. Fayed, CO_2 injection for enhanced oil recovery benefits from improved technology, *Oil Gas J.* (1983) 92–96.
70. Compressed Gas Association, Handbook of Compressed Gases, Reinhold Publishing Corporation, New York, 1966.
71. C.W. Gibbs, Compressed Air and Gas Data, Ingersoll-Rand Company, New York, 1969.
72. K.E. Starling, Thermo data refined for LPG—Part 1: equation of state and computer prediction, *Hydrocarbon Process.* **50** (3) (1971) 101–104.
73. K.E. Starling, P.N. Batford, Y.C. Kwok, Thermo data refined for LPG—Part 12: carbon dioxide, *Hydrocarbon Process.* **51** (2) (1972) 101–104.

74. N.B. Vargaftik, Y.K. Vinogradov, V.S. Yargin, Handbook of Physical Properties of Liquids and Gases—Pure Substances and Mixtures, Begell House, New York, 1996.
75. C.J. Fesmire, R. Simon, R.M. Dicharry, F.H. Vorhis, Application of Compressibility Factors for High Ratio Carbon dioxide–Methane Gas Mixtures, Society of Petroleum Engineers, Paper Number SPE 5052, 1974.
76. E.T. Zana, G.W. Thomas, Some Effects of Contaminants on Real Gas Flow, Society of Petroleum Engineers, Paper Number SPE 2577, 1969.
77. NTSB report number PAR-03/01. Natural Gas Pipeline Rupture and Fire Near Carlsbad, New Mexico, August 19, 2000.
78. NACE Standard Recommended Practice RP0102-2002, In-Line Inspection of Pipelines.
79. See Ref. [60].
80. L.G.H. Van der Meer, The conditions limiting CO_2 storage in aquifers, *Energy Convers. Manage.* **34** (9–11) (1993) 959–966.
81. S. Farthing, Internal Corrosion Risk Reduced with Methodic Attention, Training, Pipe-Line.com, 81(8) August 1998.
82. K.J. MacIntyre, Design consideration for carbon dioxide injection facilities, *J. Can. Petrol. Technol.* **25** (2) (1986) 90–94.
83. M. Mohitpour, H. Golshan, A. Murray, Pipeline *Design and Construction: A Practical Approach*, American Society of Mechanical Engineers Press, New York, 2000.
84. R. Razgaitis, J.H. Payer, G.H. Stickford, S.G. Talbert, E.L. White, R.A. Cudnik, D.W. Locklin, C.A. Farnsworth, Research on Heat-Exchanger Corrosion, Gas Research Institute report GRI-84/0157, 1984.
85. Corrosion Doctor's Website, http://www.corrosion-doctors.org/Localized/iron-corrosionX.htm.
86. M.B. Kermani, Carbon dioxide corrosion in oil and gas production—a compendium, *Corrosion* **59** (8) (2003) 659–683.
87. B. Mishra, Development of a predictive model for activation controlled corrosion of steel in solutions containing carbon dioxide, *Corrosion* **53** (11) (1997) 852.
88. N. Sridhar, Effects of water and gas compositions on the internal corrosion of gas pipelines—modeling and experimental studies, *Corrosion* **57** (3) (2001) 221–235.
89. C. De Waard, U. Lotz, A. Dugstad, Influence of Liquid Flow Velocity on CO_2 Corrosion: A Semi-Empirical Model, Corrosion 95, Paper No. 128, National Association of Corrosion Engineers, Houston, TX, 1995.

Carbon Dioxide Capture for Storage in Deep Geologic Formations, Volume 2
D.C. Thomas and S.M. Benson (Eds.)

Chapter 18

EFFECT OF IMPURITIES ON SUBSURFACE CO_2 STORAGE PROCESSES

Steven Bryant and Larry W. Lake

Department of Petroleum and Geosystems Engineering,
The University of Texas at Austin, Austin, TX, USA

ABSTRACT

We examine the potential effect of highly reactive impurities (SO_x, NO_x) on two important aspects of large-scale geological storage of CO_2: well injectivity and enhanced oil recovery processes. The primary influence on well injectivity is expected to be geochemical alteration of the near-well formation. Our simulations of a "worst-case" scenario indicate that the net change in mineral volume is likely to be small, even though extensive changes in the type of minerals may occur. Thus, the effect on injectivity is likely to be insignificant. The presence of impurities in their likely concentrations of less than 1 mole% may speed up the reactions, but otherwise should have little incremental effect on the injectivity. The effectiveness of enhanced recovery processes using CO_2 depends on factors such as minimum miscibility pressure (MMP), mobility ratio, and gravity number. Correlations for these factors developed over several decades of field experience in CO_2 flooding indicate that impurities at the levels typical of flue gases are unlikely to affect recovery adversely.

INTRODUCTION

The CO_2 Capture Project objectives include maximizing safe geologic storage, assessing and mitigating storage risks while minimizing costs. The costs of storing carbon dioxide (CO_2) captured from typical combustion sources—boilers, turbines, heaters—can be significantly reduced if stored without purification. The effect of impurities, specifically commonly occurring concentrations of sulfur and nitrogen oxides (SO_x/NO_x) as found in flue gases is a concern to the injection of CO_2. Likewise, if the costs of storage are to be offset through enhanced oil recovery (EOR), it could result in a detrimental effect of the SO_x/NO_x on the oil-recovery ability of the CO_2. Our goal is to survey the effect of impurities on such issues.

The work summarized is based on the following two assumptions about impurities in CO_2:

1. At reservoir/aquifer conditions, the CO_2 to be stored is a super-critical fluid that is as much like a liquid as a gas, compared to CO_2 at standard conditions or even to SO_x/NO_x at aquifer/reservoir conditions. Thus, diluting CO_2 with SO_x/NO_x will result in a more gas-like mixture than is CO_2 alone. As we will see, such dilution has the effect of making the CO_2 more immiscible with reservoir crude, though this effect on recovery may be offset by other factors.
2. The concentration of SO_x/NO_x to be encountered in the stored fluid will be modest, at most around 5 mole%, but usually less. At these concentrations the effect of impurities on density, viscosity and interfacial tension is likely to be small.

Abbreviations: DOE, Department of Energy; EOR, enhanced oil recovery; MME, minimum miscible enrichment; MMP, minimum miscibility pressure; MOC, method of characteristics; SPE, Society of Petroleum Engineers; WAG, water-alternating-gas.

Injectivity and Subsurface Storage
Many methods proposed for storing CO_2 in subsurface formations (saline formations and hydrocarbon reservoirs) take well injectivity for granted. That is, the injection pressure required to maintain CO_2 flow at a prescribed rate is assumed to remain constant over the course of the injection operation. The geochemical perturbation arising from the dissolution of CO_2 into subsurface brines is a mechanism for altering rock properties. Those alterations in turn have the potential for changing rock permeability and thus injectivity. Even small changes in injectivity could have substantial impact on the economics of storage, including direct costs associated with drilling and completing new wells, well stimulation treatments, and consequent interruptions in the flue gas pipeline operation, and indirect costs, including longer times or higher pressures required to inject the design CO_2 volume.

The rock properties within a few meters of a wellbore typically control injectivity. Because the CO_2 to be stored enters the formation through this small near-wellbore region, the region will be flushed by millions of (local) pore volumes of gas. Thus, even slow incremental geochemical alterations may have a significant cumulative effect over the months or years of the storage operation. The presence of impurities such as NO_x and SO_x in the CO_2 stream increases the possibility of rock alteration, since these components are even more reactive than CO_2 when dissolved in brine.

Yet another feedback loop arises when the injected CO_2 is undersaturated with water vapor. Thus, after displacing water to irreducible saturation in the near-wellbore pore-space, the injected CO_2 will gradually dehydrate or dry the near wellbore region. This process will concentrate the dissolved species and increase their reactivity. At the same time, dehydration will reduce the grain surface area in contact with the irreducible water saturation, reducing the reaction rates and hence limiting the extent of the alteration.

The geochemical reactions are likely also to change the surface energies of the water/mineral interface. Feedback, therefore, exists at the grain scale as well, since the contact between irreducible water and the rock grains is controlled by capillary forces. Investigation of this phenomenon is beyond the scope of the project proposed here. For the purpose of our study, we will assume geochemical reactions are unaffected by capillary forces.

Injectivity losses have been reported in EOR field projects in which CO_2 and water are injected alternately (Patel [1]; also see Grigg, this volume). The most likely explanation for these observations involved multiphase flow behavior, rather than geochemical alteration. There is little information currently available to answer the question of how injecting only CO_2 would affect injectivity. In the latter case, the extent of mineral dissolution and the extent of precipitation of new minerals are the most important factors. These will be the focus of this study.

CO_2 Storage and Enhanced Oil Recovery
The revenue from oil produced miscibly or near-miscibly by injected CO_2 has the potential offset to the cost of CO_2 storage. Determining whether impurities would adversely or positively affect the MMP of CO_2/crude mixtures and consequently, oil recovery is of significance.

MMP is defined as the pressure at which CO_2 develops miscibility with oil. Miscibility means that there is no interface (zero interfacial tension) between oil and an enriched CO_2 mixture. The miscibility suppresses capillary forces allowing excellent oil recovery efficiency on the pore scale. CO_2 EOR is complex because miscibility often develops when CO_2 extracts intermediate components from the crude into the CO_2-rich phase even under circumstances where CO_2 alone is not miscible with the oil. The rate of extraction depends on the purity of the solvent. CO_2-N_2 mixtures are poorer extractors than CO_2 alone, whereas CO_2-H_2S mixtures are better extractors. MMP decreases as the solvent becomes more "oil-like", supporting the conclusion that impurities in a CO_2 stream that are more like the oil than CO_2 will decrease the MMP while impurities that are less like the oil will increase the MMP.

CO_2 has been used as an EOR agent since the early 1970s, and research on this technology predates this by nearly 20 years. More than 50 CO_2 field floods have been conducted, primarily in West Texas, with a corresponding accumulation of field experience and production data out of which has grown a large technology base for the process. The dependency of MMP upon oil composition was addressed most definitively by Holm [2]. Most of the data on MMP has been captured as statistical correlations. Correlations dealing with CO_2 stream impurity also exist. In Ref. [3], the authors account for the influence of any component in the CO_2 on MMP by means of the critical temperature for that component. The Holm [2] and Sebastian [3] correlations account for the major influences on MMP.

EXPERIMENTAL/APPROACH

Many numerical simulators of flow accompanied by transport and chemical reactions have been developed, with varying capabilities and sophistication [4]. In this study, we have chosen to use a relatively simple simulator. The concept of *fronts* (also referred to as *waves*) propagating through a porous medium is a powerful method for understanding behavior. A front separates regions of different chemical (and in this context mineral) composition and moves at some fraction of the fluid velocity. When the reactions are fast relative to the flow rate, the fronts can be determined analytically.

Our study considers the following idealizations aimed at producing a "worst-case" scenario where the maximum conceivable mineralogical alteration occurs. Specifically, we consider 1D flow with intra-aqueous reactions (e.g. dissociation of dissolved CO_2 into bicarbonate anion and hydronium cations) in local equilibrium with a few simple minerals.

RESULTS AND DISCUSSION

Enhanced Oil Recovery

In most projects, the CO_2 injected for EOR is nearly pure. A notable exception was the Slaughter Estate Unit (SAU) in which a mixture of 21 mole% H_2S and 79% CO_2 was injected, to apparent beneficial effect [6]. The ultimate recovery of oil by CO_2 injection depends on many things including: operating strategy, reservoir pressure, reservoir temperature, solvent purity, well spacing, well conditions, and reservoir heterogeneity. Many factors actually operate in opposing fashion [7]. Thus, it is impossible to estimate ultimate recovery except through an involved calculation that accounts for all these factors. Numerical simulators to do this are in common use. It is possible, however, to qualitatively assess the effect of impurities through the use of dimensionless groups (scale factors) that generally influence ultimate EOR recovery. These groups are the capillary number, the mobility ratio, the gravity number, and the continuity number. Injectivity is also important, and this will be addressed as a separate matter in the second half of this report. The following is a summary of effects and tendencies taken from Ref. [5] that are specific to the issue of CO_2 storage in oil reservoirs. The CO_2-rich injectant here will be called a "solvent".

Capillary Number: The capillary number expresses the ratio of capillary forces to viscous forces. It is usually defined as [5]

$$N_c = \frac{\text{viscous}}{\text{capillary}} = \frac{\mu u}{\sigma} \tag{1}$$

where μ is the solvent viscosity [=] F − t/L^2; u, superficial velocity of the flood [=] L/t and σ is the interfacial tension between solvent and oil [=] F/L. (In this equation and those following [=] means "has units of". F is force, L length, m mass, and t time.) Statistical correlations have shown that oil recovery increases as the capillary number (N_c) increases [8–11]. Thus, if impurities increase N_c, the ultimate oil recovery on a small scale (displacement efficiency) will increase. Conversely, if the N_c is lowered by impurities, recovery will decrease.

Though there are three quantities in Eq. (1), in practice only the interfacial tension can be practically changed sufficiently to be considered significant (the relationship between ultimate recovery and N_c is logarithmic). If N_c is very large, the displacement approaches miscibility.

Minimum miscibility pressure

MMP is regarded as one of the primary determinants to the success of an oil-recovering CO_2 flood [12]. If the average reservoir pressure is not greater than the MMP, the flood will be immiscible and expected oil recovery will be much lower.

CO_2 recovers oil by mass transfer and if mass transfer is strong enough, the CO_2 will become miscible with resident crude. The term "mass transfer" means that components transfer from the oil to the solvent or vice versa. For CO_2 injection transfer of intermediates from the oil to the solvent is most important. The mass transfer induces miscibility. Miscibility results in the total suppression of capillary forces (infinite N_c) and leads to large displacement efficiency. In most cases the CO_2 will cause intermediate components from the crude to pass into the CO_2-rich phase creating, after a few contacts, an enriched mixture that is miscible with the crude. In this vaporizing gas drive process, the CO_2, though initially immiscible with the crude is said to have *developed* miscibility through extraction. The extraction depends strongly on the purity of the solvent; CO_2–nitrogen mixtures are poorer extractors than CO_2 alone and CO_2–H_2S mixtures are better extractors.

A common way to gauge the approach to miscibility is through slim tube experiments. In these experiments a permeable medium is saturated with a crude, and then, at constant temperature, subjected to a succession of CO_2 displacements. Before each displacement the tube is re-saturated with crude and the displacement performed at successively higher pressures. The ultimate oil recovery in these experiments tends to increase with pressure and the recovery tends to level off at a specific threshold pressure. This threshold pressure is defined as the MMP. If attaining miscibility is difficult, the MMP will be large; if easy it will be small. MMP is an analogue quantity for minimum miscible enrichment (MME) previously proposed for miscible gas injection [13].

Interestingly, though MMP is defined in terms of oil recovery in a flowing medium, it is also associated with the phase behavior of CO_2–crude mixtures. This means that MMP has a thermodynamic basis in that the mass transfer is that which occurs after repeated contacts in which equilibrium is successively established.

There are three ways to estimate MMP:

1. *Experimental*. Within this category are
   (a) The slim tube experiment discussed above. This is the most common way.
   (b) The rising bubble apparatus [14]. This technique estimates miscibility as the height in a column of crude at which a bubble of CO_2 disappears under gravity rise. The height can be translated into a pressure and thence to MMP. This technique is offered as a commercial service but is much less used than the slim tube experiment.
   (c) Vanishing interfacial tension. This technique is based on the direct correspondence between miscibility and non-zero interfacial tension [15]. The extrapolation of IFT to zero with increasing pressure determines the MMP. This technique, being much newer than the others, is the least used of all [16].
2. *Theoretical*. Because of the thermodynamic basis of MMP, it can be estimated by calculation. There are two categories for this
   (a) Mixing cell models. These are calculations in which solvent is mixed with crude in a succession of well-mixed cells. The solvent–crude mixtures come to equilibrium after each mixing and the equilibrated gaseous phase (solvent plus extracted intermediate components) pass to another cell for further equilibration. There are specialized simulators for this, but the calculation can be done with any commercial compositional simulator.
   (b) Method of characteristics (MOC) models. In the limit of zero dispersion—which all mixing cell models possess to some degree—the MOC provides a direct measure of the MMP through quasi-analytic calculation. This procedure has been developed by the Stanford group [17–19].
3. *Correlation*. Because of the expense of making site-specific MMP measurements, much of the slim tube data has been captured through correlation. There are two types:
   (a) Neural networks. The complexity of the dependencies of MMP has led some to exploit artificial neural networks for correlation [20]. The success of this has been good in matching data but a neural network is not easy to use for prediction.

(b) Statistical correlation. By far the most common type of correlation has been statistical correlation and this, along with the MOC approach above, would be the recommended approach for subsequent estimation. Before reviewing these, we review some of the qualitative trends of MMP with other quantities.

That the composition of crude oils is highly variable accounts for the complexity of MMP correlations. General trends are

(i) MMP should increase with temperature. That is, it should be more difficult to develop miscibility at high temperatures than at low.
(ii) MMP should decrease with increases in the intermediate content in the crude. This is a logical consequence of the development of miscibility depending on the extraction of intermediates. There is, unfortunately, no simple way of simply characterizing the intermediate content in crude.
(iii) MMP should decrease as the solvent becomes more "oil-like". This means that impurities in a CO_2 stream that are more like the oil than CO_2 will decrease the MMP. Impurities that are less like the oil will increase the MMP. Figure 16 in Ref. [20] gives a good summary of this effect. See also Refs. [21–23]. The "oil-like" nature of the solvent will figure into other effects as they are covered.

We now turn, in conclusion, to the more significant MMP correlations. Yellig [24] gave one of the earliest correlations. This correlation was only for West Texas crude and did not contain a composition dependence. The compositional dependency was addressed by Johnson [21] and, most significantly, by Holm [2, 25–27]. Correlations dealing with CO_2 stream impurity also exist. Perhaps the best of these is by Sebastian [3] who attempted to correlate deviations away from the pure CO_2 critical temperature.

$$\frac{P_{MM}}{(P_{MM})_{CO_2}} = 1.0 - (2.13\times10^{-2})(T_{pc} - T_c) + (2.51\times10^{-4})(T_{pc} - T_c)^2 - (2.35\times10^{-7})(T_{pc} - T_c)^3 \quad (2)$$

where P_{MM} is the MMP of a solvent against a particular crude [=] F/L^2; $(P_{MM})_{CO_2}$, the MMP of a pure CO_2 solvent as from, for example, the Holm [2] correlation [=] F/L^2; T_c, critical temperature of CO_2 [=] degrees Rankine and T_{pc} is the pseudocritical temperature of a multicomponent solvent [=] R, where

$$T_{pc} = \sum_{i=1}^{i=N} y_i T_{ci}$$

and y_i is the mole fraction of component i in the solvent.

The combination of Eq. (2) and Sebastian [3] correlations gives the combined effects of temperature, oil composition, and the characteristics of the solvent. The accuracy is within about 50 psia, which should be sufficient for screening.

We apply this correlation with SO_2 and NO as representative impurities. The critical temperatures of SO_2 and NO are 430.5 and 180 K, respectively. The critical temperature of CO_2 is 304 K. Thus, the pseudo-critical temperature can be written as

$$T_{pc} = 304y_{CO_2} + 430.5y_{SO_2} + 180y_{NO}$$

Inspection of the MMP correlation shows that SO_2 will decrease the MMP relative to the pure CO_2 value, while NO will increase it. Because the coefficients in the correlation are small and because the expected values of SO_2 and NO mole fractions are small, the effect on MMP is also small, as illustrated in Figure 1.

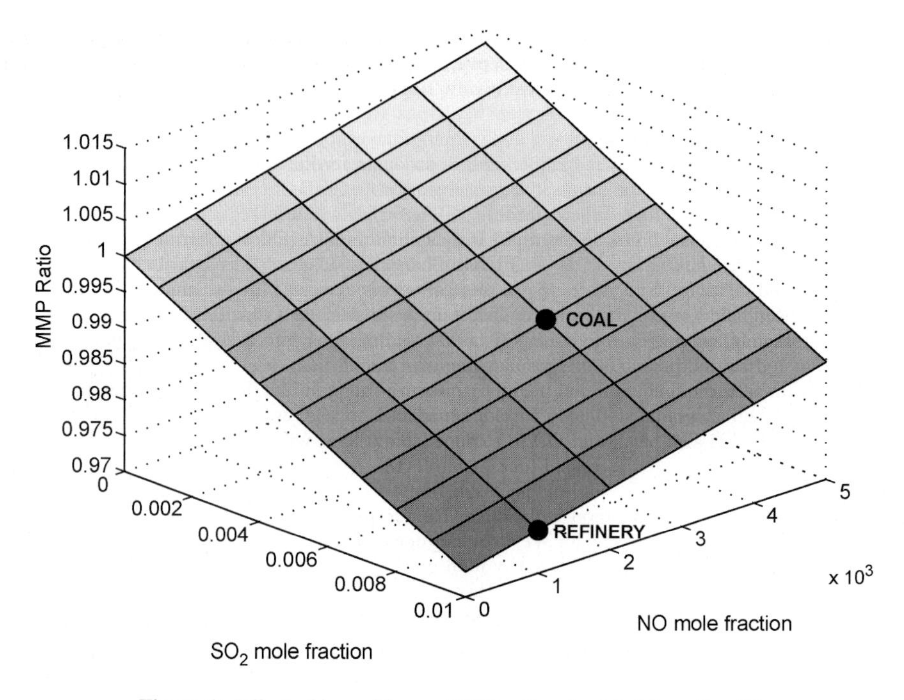

Figure 1: Effect of impurities on MMP ratio calculated from Eq. (2).

Concentrations typical of coal combustion flue gas and a refinery fuel flue gas are plotted; the basis for determining the composition neglects water vapor and nitrogen. The changes in MMP over this range of impurities (less than 1% for SO_2, less than 0.5% for NO) are less than 2%. Thus, we expect little overall effect on MMP. In other words, a candidate flood that would nominally be miscible with pure CO_2 would also be miscible with a diluted solvent.

Factors influencing immiscible and miscible displacements
Several factors affect the efficiency of displacement processes, whether the process is carried out above the MMP or below. Factors that can be affected by the presence of impurities include mobility ratio and gravity number. The subsequent discussion applies to immiscible floods and to miscible floods alike; for these purposes both types can be regarded as EOR.

Mobility ratio (M). While the capillary number and MMP are indicators of ultimate oil recovery on a small scale, the mobility ratio indicates the ability of a solvent to contact volumes of a reservoir, a quantity usually known as volumetric sweep efficiency. M is the ratio of the mobility of the displacing fluid to that of the displaced fluid.

$$M = \frac{(k_r/\mu)_{\text{displacing}}}{(k_r/\mu)_{\text{displaced}}} \approx \frac{\mu_{\text{displaced}}}{\mu_{\text{displacing}}} \tag{3}$$

where k_r is the relative permeability.

As Eq. (3) indicates, M is strongly affected by the viscosities of the fluids. The connection between ultimate recovery and M is complicated but there is a universal tendency that applies to both miscible and immiscible floods. As M increases—displacing fluid become more mobile—volumetric sweep decreases because the displacing fluid now has a greater tendency to bypass the resident or displaced fluid. The more SO_x/NO_x

there is in a CO_2-rich solvent (the more gas-like it becomes), the smaller its viscosity, the larger the M and the smaller the ultimate oil recovery. Unlike the MMP issue, impurities here decrease ultimate oil recovery. However, the concentration of SO_x/NO_x being considered is small enough to make this effect small.

Gravity number (N_g). The gravity number is a dimensionless ratio of gravity (buoyancy) forces to viscous forces. The usual definition is

$$N_g = \frac{\text{buoyancy}}{\text{viscous}} = \frac{kk_r}{\mu u}\Delta\rho g \tag{4}$$

where k is the permeability [=] L^2; g, gravitational acceleration constant [=] L/t^2 and $\Delta\rho = \rho_{\text{displacing}} - \rho_{\text{displaced}}$ is the density difference [=] m/L^3.

This number was derived in Ref. [28]. N_g expresses the propensity for a solvent to segregate within a reservoir. For $\Delta\rho > 0$ the solvent would fall to the bottom of a reservoir; for $\Delta\rho < 0$, the normal case in solvent flooding, the solvent would move to the top of the reservoir causing overriding. Either type of segregation would result in poor volumetric sweep and small ultimate oil recovery.

As Eq. (4) shows, five quantities normally determine the magnitude of N_g; however, the permeability seems to have the most influence. Large k reservoirs inevitably show extensive segregation. The principal effect of solvent impurities would be to decrease $\rho_{\text{displaced}}$ (increase $\Delta\rho$) thereby increasing segregation to the result that ultimate oil recovery would be decreased. However, as in the case of the mobility ratio, the decreases are not likely to be significant at the concentrations envisioned.

Vertical communication. The relationships between ultimate recovery and the quantities discussed above are complicated. Some of this complication can be conveyed through an effective aspect ratio R_L that, which of itself being of little importance to ultimate recovery, is a determinant of the strength of the relationships we have been discussing. The effective aspect ratio is

$$R_L = \frac{L}{H}\sqrt{\frac{k_v}{k_h}} \tag{5}$$

where L is the length of the reservoir (usually the well spacing) [=] L; H, the thickness of the reservoir [=] L; k_v, the harmonic average vertical permeability [=] L^2 and k_h is the arithmetic average horizontal (lateral) permeability [=] L^2.

The relationship between volumetric sweep efficiency E_v and the above three quantities can be written as

$$E_v = f(M, N_g, R_L). \tag{6}$$

When R_L is small the relationship between E_v and N_g is weak; when it is large, the relationship is strong. Sometimes the two effects—buoyancy and vertical communication—are combined into a single group as

$$N_g' = \frac{k_v k_r L}{\mu u H}\Delta\rho g. \tag{7}$$

Finally, comments about the effect of buoyancy on solvent floods must be tempered by a consideration of the dip of the reservoir. In a steeply dipping reservoir, the buoyancy effects can be used to increase the ultimate recovery of oil by injecting the light fluid at the top of a structure. Indeed, some of the largest ultimate recoveries reported in injection process have been by displacements that are made to be gravity stable even sometimes at the expenses of miscibility. In these cases, a large N_g would increase ultimate oil recovery, meaning that impurities would be beneficial.

Factors particular to immiscible displacement include changes in oil properties (swelling, viscosity alteration, etc.). These arise because of mass transfer between the CO_2 and oil. SO_x/NO_x impurities are more "gas-like" than "oil-like". The mass transfer that occurs in immiscible floods is not large enough to cause miscibility. Thus, the incremental changes in oil properties caused by the presence of these impurities are anticipated to be small. It is unlikely that the behavior of an immiscible displacement would be noticeably different than that of a pure CO_2 flood.

Injection Rate

The fluid velocity appeared in the above groups involving viscous forces. The superficial velocity u is directly proportional to the ability to inject fluid into a reservoir, a large injection rate means a large u. But the effect on N_c and N_g on injection rate is secondary given the sensitivities of the other quantities in the groups. Injection rate is far more important, however, than these groups indicate because it is the major determinant of how many wells are needed for a storage project. Because most of these remarks would apply equally well to storage in a saline formation as to an oil reservoir, we refer to the CO_2-rich injectant as simply the "injectant".

The injectivity of a well is defined as

$$J = \frac{q}{\Delta p} [=] \frac{L^5}{t - F} \tag{8}$$

The usual oilfield units for J are barrel/(day-psi). Other terms in Eq. (8) are q is the injection rate [=] L^3/t and $\Delta p = p_{wf} - \bar{p}$ is the pressure difference [=] F/L^2, where p_{wf} is the bottom hole flowing well pressure [=] F/L^2; $\bar{p}$ is the volumetrically average reservoir pressure [=] F/L^2.

Other definitions for injectivity are possible [29] but all lead to the same conclusion that the larger the J, the greater the injection rate for a given pressure difference.

We now turn to a discussion of factors that could cause injectivity changes. An expression for J that is consistent with the injectivity definition is

$$J = \frac{4\pi k k_r H}{\mu\left(\ln\left(\frac{4A}{1.781 C_A r_w^2}\right) + 2s\right)} \tag{9}$$

where A is the area being flooded by the well [=] L^2; r_w, well radius [=] L; C_A, a shape factor and s is the skin factor.

Eq. (9) is the injectivity of a vertical well in a region having flooded area A. The shape factor can be picked to express the geometry of the flooded area, the tilt of the well, the existence of partially perforated intervals, or even, with suitable adjustment, vertical wells. The skin factor s represents "damage" to the well caused by impairment in the well itself, or in the immediate vicinity of the well. One immediately apparent effect of CO_2 impurities would be to increase injectivity through a lessened viscosity, the injectant now being more gas-like.

Injectivity in CO_2 oil recovery floods has long been recognized as a factor that is important to economic success [30,31]. The seminal reference on this subject is by Patel [1] who noted that CO_2 injectivity was unexpectedly smaller than what was estimated. Patel [1] attributed this to one of three causes:

1. Interactions between the CO_2 and crude oil that resulted in solids precipitation. This would increase s in Eq. (9).
2. Mineralization induced by reactions between CO_2-saturated water and host minerals. This would lower k in Eq. (9).
3. Relative permeability hysteresis caused by decreasing the oil saturation (in a preceding waterflood) and then increasing it in a banked up oil zone. This would cause a small k_r in Eq. (9).

Patel [1] concluded that the third effect was the most significant, even though all effects could be present. The first effect would only be present for storage in oil reservoirs, but the other two effects could be important. The following discusses the second and third effects. Interestingly, the two effects combined seem to shed the most insight into predicted behavior.

Mineralization

Water saturated with CO_2 is slightly acidic with pHs in the range of 3–6 depending on the partial pressure of the CO_2-rich phase with which it is in equilibrium. Such lowered pHs can increase the solubility of resident minerals, especially carbonates and some clay minerals. This would increase J by increasing k through enlargements of the pore space as a result of mineral dissolution.

However, the possibility exists that ions liberated from the dissolution of the primary mineral can combine with ions in the water to form a secondary mineral. The secondary precipitation can decrease J by decreasing k. Indeed, calculations based on sea water injection show that it is possible for the volume of secondary minerals and yet-undissolved primary minerals can be greater than the volume of the initial mineral assemblage [32]. Secondary mineralization can also be beneficial because the secondary mineral can be a carbon-storing mineral. Indeed co-injection of the CO_2 with water may be a way to induce such storage.

To investigate these effects, we performed calculations of mineralization in one-dimensional media undergoing inorganic reactions in local equilibrium with a flowing solution, as described in Ref. [4]. Assuming local equilibrium, the specific velocity of a component in single-phase flow is

$$v_{\Delta C_i} = \frac{1}{1 + D_i} \tag{10}$$

where D_i is the dimensionless retardation factor for component i and $v_{\Delta C_i}$ is the specific velocity relative to water velocity.

Here and elsewhere retardation factors are always positive. Our calculations were based on three base components, carbon, calcium and sulfur, for which the primary mineral was $CaSO_4$, which dissolution could result in precipitation of the secondary mineral calcite $CaCO_3$. The aqueous phase was everywhere in local equilibrium with a gaseous injectant that was mainly CO_2, which also could contain varying amounts of SO_2. The reason for this is discussed below. The SO_2 is a surrogate for the SO_x/NO_x impurities.

Ideal chemistry (speciation and dissociation reactions involving aqueous phase components) was simulated. Repeating simulations with corrections for non-ideal solutions (activity coefficients) made little difference. The continuous flow of the saturated brine provides an overestimate of the extent of dissolution and precipitation. In practice brine will flow for a relatively short time in the near-well region.

The model formation contains a mineral that dissolves when contacted by the acidic brine. Both equilibrium and rate limited dissolution were modeled. Though unlikely to be attained under realistic field conditions, the equilibrium assumption establishes a worst case (maximum possible mineral dissolution and precipitation). The dissolved mineral releases divalent cation into the CO_2-saturated brine, causing the precipitation of a carbonate mineral. This is the simplest model that captures two key features: mineral precipitation, which is the only geochemical mechanism likely to cause injectivity loss, and supplying divalent cations for precipitation by dissolving in situ minerals. A more complicated suite of precipitation/dissolution reactions can occur in the field, but their unfavorable influence on injectivity depends only on the extent of precipitation. Thus for the limited purposes of this study, this simplified model was used.

Results show a SO_2-free injectant will indeed dissolve the primary mineral and precipitate a secondary mineral. Typical retardation factors for the first dissolution are 2–3 for $CaSO_4$ and 5–6 for $CaCO_3$. The replacement of $CaSO_4$ by $CaCO_3$ was on an equal molar basis or less. This means that total mineral concentration, the sum of $CaSO_4$ and $CaCO_3$, nowhere exceeded the initial $CaSO_4$ concentration. Results with an injectant that contained SO_2 showed that even modest concentrations of SO_2 reduce the pH of the water even further, down to 1–2 compared to 5–6 for the SO_2-free injectant. This decrease has two causes: SO_2 is more soluble in water than is CO_2, and the acid formed by SO_2 in water, sulfurous acid or H_2SO_3, is a far stronger acid than the carbonic acid H_2CO_3 formed by CO_2. The lowered pH has two effects on the flow: the primary mineral is dissolved much faster, and a secondary mineral never precipitates. The presence of impurities in this regard is entirely beneficial to good injectivity.

Other simulations indicate that the *net change* in mineral volume is likely to be small, even when extensive changes in mineral state may occur. The physical reason is simple: the injected fluids contain no cations, and thus any mineral that precipitates *must* use cations released from dissolution of existing minerals. For example, the carbonic acid-driven attack of a calcium feldspar, anorthite, leads to precipitation of calcite and kaolinite.

$$\text{Anorthite} + 2\text{H}^+ + \text{CO}_3^{2-} + \text{H}_2\text{O} \rightarrow \text{calcite} + \text{kaolinite}$$

The increase in mineral volume associated with this reaction going to completion is 20%. If anorthite occupies 10% of the bulk volume of a formation, this reaction would reduce the porosity by 2%. Even with a strong dependence of permeability on porosity the effect on injectivity is likely to be relatively small regardless of the initial mineral state. This assumes that the precipitated minerals are distributed uniformly within the pore space. Current understanding offers little insight into the likelihood of non-uniform precipitation, but it should be noted that preferential precipitation in pore throats would cause a much larger reduction in permeability.

Relative permeability effects

We consider only the injection of a CO_2-rich injectant into a one-dimensional aquifer here. Initially the gas saturation in the aquifer is zero and the injectant displaces water immiscibly. If the initial water is free of CO_2, a portion of the leading edge of the CO_2 front dissolves into the water resulting in a shock. If the injectant is free of water (dry) eventually, all of the water dissolves in the CO_2, leaving a completely dry region in the vicinity of the injection well. Schematically the sequence is

$$\left\{\begin{array}{c}\text{Inlet, } J \\ f_g = S_g = 1\end{array}\right\} \xrightarrow[\text{slow(trailing)shock}]{} \left\{\begin{array}{c}f_g \text{ varies} \\ C \text{ constant}\end{array}\right\} \xrightarrow[\text{fast(leading)shock}]{} \left\{\begin{array}{c}\text{Initial, } I \\ f_g = S_g = 0\end{array}\right\}$$

Here I and J refer to initial and inlet (injected) fluids. The specific velocity of the fast (leading) shock, the CO_2 dissolution shock, is

$$v_{\text{leading}} = \frac{f_g^- + \dfrac{C_{CO_2}^a}{(C_{CO_2}^g - C_{CO_2}^a)}}{S_g^- + \dfrac{C_{CO_2}^a}{(C_{CO_2}^g - C_{CO_2}^a)}} \tag{11}$$

where f_g^- is the fractional flow of gas just upstream (toward J) of the leading shock; S_g^-, the gas saturation just upstream of the leading shock; $C_{CO_2}^a$ and $C_{CO_2}^g$, concentrations of CO_2 in the aqueous and gas phases in the region between the leading and trailing shocks.

The latter concentrations are set by the partial pressure in the injectant. The specific velocity of the trailing or drying shock is

$$v_{\text{trailing}} = \frac{f_g^+ - \dfrac{C_J^g - C_{CO_2}^a}{C_{CO_2}^g - C_{CO_2}^a}}{S_g^+ - \dfrac{C_J^g - C_{CO_2}^a}{C_{CO_2}^g - C_{CO_2}^a}} \tag{12}$$

where f_g^+ is the fractional flow of gas just downstream (toward I) of the trailing shock; S_g^+, the gas saturation just downstream of the trailing shock and C_J^g, the inlet or injected CO_2 concentration.

Between these two limits the specific velocity of a gas saturation is the well-known Buckley–Leverett velocity

$$v_{S_g} = \frac{df_g}{dS_g} \tag{13}$$

Eq. (13) assumes that the phase concentrations do not change in the region between the leading and trailing shocks. Setting Eq. (11) equal to Eq. (13) defines the downstream shock saturation, the relationship $f_g = f_g(S_g)$ being known and independent of phase concentrations. Setting Eq. (12) equal to Eq. (13) defines the upstream shock saturation. These equations admit to a graphical solution [5]. Eqs. (11) and (12) are of the form

$$v = \frac{f_g + D}{S_g + D}. \tag{14}$$

The retardation factors for both the leading and trailing waves are small; for the leading shock it is about 0.05 and substantially less for the trailing shock.

For a reacting component in the aqueous phase the specific velocity is

$$v_{\Delta C_i} = \frac{f_w}{S_w + D_i} = \frac{f_g - 1}{S_g - (1 + D_i)} \tag{15}$$

where f_w is the fractional flow of aqueous phase and S_w is the saturation of aqueous phase.

Eq. (15) is the generalization of Eq. (10) to two-phase flow.

Now for mineralization in two-phase flow to occur we must have

$$v_{\Delta C_i} > v_{\text{trailing}}.$$

This inequality means that mineralization can only occur when water is present. If the inequality is satisfied, the mineralization occurs where there is remaining water. Substituting from Eqs. (10) and (12) gives

$$\frac{f_g - 1}{S_g - (1 + D_i)} > \frac{f_g - \dfrac{C_J^g - C_{CO_2}^a}{C_{CO_2}^g - C_{CO_2}^a}}{S_g - \dfrac{C_J^g - C_{CO_2}^a}{C_{CO_2}^g - C_{CO_2}^a}}. \tag{16}$$

The ratios of concentration differences on the right side of Eq. (16) are close to one; hence, the only way for this inequality to hold is if $D_i < 0$, which is a physical impossibility. Mineralization cannot occur during the injectant cycle because the CO_2 dries out the medium before this can happen [34]. Stated differently, the trailing (drying) shock is the mineralization wave. This wave is very slow and, in any event, precipitates only the minerals that are dissolved in the water when dried. This amount is too small to cause damage because, once again, there is insufficient supply of minerals.

Injectivity may actually increase because of the drying in the near-wellbore region. If the residual water saturation is large, the end-point relative permeability to the injectant can be small. When the residual water is removed, this injectant relative permeability increases, possibly by a large amount.

After injection, the CO_2 slug may be driven with additional water injection, or natural water flow may displace the CO_2.

$$\left\{\begin{matrix}\text{Inlet, } K \\ f_g = S_g = 0\end{matrix}\right\} \xrightarrow[\text{slow(trailing)shock}]{} \left\{\begin{matrix} f_g \text{ varies} \\ C \text{ constant}\end{matrix}\right\} \xrightarrow[\text{fast(leading)shock}]{} \left\{\begin{matrix}\text{Initial, } J \\ f_g = S_g = 1\end{matrix}\right\}.$$

Here J and K refer to initial and inlet (injected) fluids after CO_2 injection has ended. Water displacing a CO_2-rich injectant is described by the same equations as above, Eqs. (10)–(12), except that the fractional flow function, being hysteretic, now follows a different path. The principal effect of hysteresis is that there is now a residual gas phase that is removed by dissolution in water. The specific velocity of this wave is given by Eq. (11) now written as

$$v_{\text{trailing}} = \frac{0 + \dfrac{C^{a}_{CO_2}}{(C^{g}_{CO_2} - C^{a}_{CO_2})}}{S_{gr} + \dfrac{C^{a}_{CO_2}}{(C^{g}_{CO_2} - C^{a}_{CO_2})}}. \tag{17}$$

The formerly leading wave now becomes the trailing wave. Mineralization would now occur when

$$v_{\text{trailing}} < v_{\Delta C_i}. \tag{18}$$

Combining Eqs. (10) and (17) shows that it is possible for mineralization to occur according to inequality equation (18) if the water displacing the CO_2 is saturated with CO_2. It therefore appears that mineralization (and well damage if water is being injected) most likely occurs during the water-displacing injection where water is in single-phase flow in the presence of a trapped gas saturation. If the displacing water is free of CO_2 the trailing wave will be fast and mineralization suppressed. There is anecdotal evidence of diminished water injectivity following a CO_2 oil recovery flood, though this effect is more likely caused by the trapped gas saturation than mineralization.

The gradual dissolution from the trapped gas saturation provides a CO_2 source for the water that causes the mineralization. That an excess of CO_2 was needed for mineralization was why we modeled the mineralization as a single-phase injection of water in the presence of an excess of injectant. As noted before, the effect of the impurities is to suppress secondary precipitates and advance primary mineral dissolution. So the effect of water displacing an impure CO_2 stream seems unlikely to result in injectivity reduction.

CONCLUSIONS

The several decades of literature on field and laboratory CO_2 flooding provide a reasonably sound foundation for estimating the effects of impurities in the CO_2 phase on EOR behavior. The MMP is a widely used measure of enhanced recovery performance for CO_2 flooding. The most common and most convenient technique for determining MMP is statistical correlations. We have identified correlations of MMP with solvent composition that allow reasonable estimates of the tradeoff between oil recovery vs. the cost of impurity removal. These indicate that the likely range of impurity concentrations (SO_x, NO_x mole fractions of order 1%) will change the MMP only a few percentage points from its pure-CO_2 value.

Several other mechanisms influence the performance of an EOR process. These are applicable to immiscible as well as miscible displacements. The most important effects are captured in two terms, the mobility ratio and the gravity number. Impurities are likely to reduce the viscosity of the solvent, thereby increasing the mobility ratio (ratio of solvent mobility to oil mobility; mobility is phase relative permeability divided by phase viscosity). Impurities are likely to reduce the density of the solvent, thereby decreasing the gravity number (ratio of buoyancy forces to viscous forces). Both these effects will tend to reduce the effectiveness of the process. Factors particular to immiscible displacement include changes in oil properties (swelling, viscosity alteration, etc.). The incremental changes in oil properties due to the presence of these impurities in the displacing gas are anticipated to be small. For the small concentrations of impurities likely to be encountered, all these effects are quite small and should not significantly influence the process.

The injectivity of the disposal well is of primary importance in the logistics and economics of subsurface CO_2 storage. Geochemical alteration of the formation mineralogy is the main potential influence on well injectivity. In the absence of impurities, CO_2 injection is unlikely to have a significant deleterious effect on injectivity. The presence of SO_x and NO_x in the injected stream will significantly increase the acidity of the aqueous phase in the formation. This speeds up dissolution of native minerals, but does not significantly change the ultimate extent of the changes in mineralogy. Thus, these impurities are likely to have little incremental effect on injectivity.

In summary, this study anticipates no substantial disadvantages to injecting an impure CO_2 stream, relative to a purified CO_2 stream. The presence of impurities may even offer slight advantages. Thus, the additional costs required to remove SO_x or NO_x impurities, motivated by concerns about storage reservoir performance, are unlikely to be justified.

RECOMMENDATIONS

1. From the subsurface perspective, impurities can be safely disregarded when developing a storage scheme. But the possible deleterious effects of impurities on surface handling facilities and on the wellbore itself should also be taken into account (cf. Gupta [33] in this volume).
2. Certain waste gases such as hydrogen sulfide may even improve enhanced recovery processes, and the feasibility of co-storage of multiple gas streams should be studied.

ACKNOWLEDGEMENTS

This research was supported by the CO_2 Capture Project. Larry W. Lake holds the W. A. (Monty) Moncrief Centennial Chair at The University of Texas.

REFERENCES

1. P.D. Patel, R.G. Christman, J.W. Gardner, Investigation of unexpectedly low field-observed fluid mobilities during some CO_2 tertiary floods, *SPE Reservoir Eng.* (1987) 507–513. November.
2. L.W. Holm, V.A. Josendal, Effect of oil composition on miscible-type displacement by carbon dioxide, *Society of Petroleum Engineers Journal* **22** (1982) 87–98.
3. H.M. Sebastian, R.S. Winger, T.A. Renner, Correlation of minimum miscibility pressure for impure CO_2 streams, SPE/DOE 12648, *Presented at the Fourth Joint Society of Petroleum Engineers/ Department of Energy Symposium on Enhanced Oil Recovery*, Tulsa, Oklahoma, 1984.
4. L.W. Lake, S.L. Bryant, A.N. Araque-Martinez, Geochemistry and Fluid Flow, Elsevier, New York, 2002.
5. L.W. Lake, Enhanced Oil Recovery, Prentice-Hall, Englewood Cliffs, NJ, 1989.
6. J.C. Ader, M.H. Stein, Slaughter estate unit CO_2 pilot reservoir description via a black oil model waterflood history match, SPE/DOE 10727, *Presented at the 1982 SPE/DOE Third Joint Symposium on Enhanced Oil Recovery of the Society of Petroleum Engineers*, April 4–7, Tulsa, OK.
7. M.K. Roper, Simulation of tertiary carbon dioxide injectivity, PhD Dissertation at UT, December 1994.
8. M. Delshad, G.A. Pope, K. Sepehrnoori, A compositional simulator for modeling surfactant enhanced aquifer remediation, 1. Formulation, *J. Contam. Hydrol.* **23** (1996) 303–327.
9. K.D. Pennell, G.A. Pope, L.A. Abriola, Influence of viscous and buoyancy forces on the mobilization of residual tetrachloroethylene during surfactant flushing, *Environ. Sci. Technol.* **30** (4) (1996) 1328–1335.
10. G.A. Pope, W. Wu, G. Narayanaswamy, M. Delshad, M.M. Sharma, P. Wang, Modeling relative permeability effects in gas-condensate reservoirs with a new trapping model, *SPE Reservoir Eval. Eng.* **3** (2) (2000).
11. G.W. Paul, L.W. Lake, G.A. Pope, G.B. Young, A simplified predictive model for micellar/polymer flooding, SPE 10733, *Presented at the California Regional Meeting of the Society of Petroleum Engineering*, San Francisco, California, 1982..
12. H. Yuan, Application of miscibility calculations to gas floods, PhD Dissertation, The University of Texas at Austin, August 2003.
13. A.L. Benham, W.E. Dowden, W.J. Kunzman, Miscible fluid displacement—prediction of miscibility, *Trans. Soc. Petrol. Eng. AIME* **219** (1961) 229–237.
14. A.M. Elsharkawy, C.U. Suez, F.H. Poettmann, R.L. Christiansen, Measuring minimum miscibility pressure: slim-tube or rising-bubble method? SPE 24114, *Proceedings of the Society of Petroleum Engineers/Department of Energy Eighth Symposium on Enhanced Oil Recovery*, Tulsa, OK, April 22–24, 1992.
15. D. Rao, A new technique of vanishing interfacial tension for miscibility determination, *Fluid Phase Equilib.* **139** (1997) 311–324.

16. D. Rao, J.I. Lee, Determination of gas–oil miscibility conditions for interfacial tension measurements, *J. Colloid Interface Sci.* **262** (2003) 474–482.
17. Y. Wang, F.M. Orr Jr., Analytical calculation of minimum miscibility pressure, *Fluid Phase Equilib.* **139** (1997) 101–124.
18. Y. Wang, F.M. Orr Jr., Calculation of minimum miscibility pressure, *J. Petrol. Sci. Eng.* **27** (2000) 151–164.
19. B. Dindoruk, F.M. Orr, R.T. Johns, Theory of multicontact miscible displacement with nitrogen, *SPE J.* **2** (1997) 268–279. .
20. Y.F. Huang, G.H. Huang, M.Z. Dong, G.M. Feng, Development of an artificial neural network model for predicting minimum miscibility pressure in CO_2 flooding, *J. Petrol. Sci. Eng.* **37** (2003) 83–95.
21. J.P. Johnson, J.S. Pollin, Measurement and correlation of CO_2 miscibility pressures, SPE/DOE 9790, *Presented at the 1981 Joint SPE/DOE Symposium on Enhanced Oil Recovery*, Tulsa, Oklahoma, 1981.
22. W.R. Whitehead, O.K. Kimbler, R.M. Hoshman, J.R. Hervey, Investigations of Enhance Oil Recovery Through Use of Carbon Dioxide, U.S. Department of Energy Progress Report, DOE/MC/03103-2, U.S. Department of Energy, 1980.
23. R.S. Metcalfe, Effects of impurities on minimum miscibility pressure and minimum enrichment levels for CO_2 and rich-gas displacements, *Soc. Petrol. Eng. J.* **22** (1981) 219–225.
24. W.F. Yellig, R.S. Metcalfe, Determination and predication of CO_2 minimum miscibility pressures, *J. Petrol. Technol.* (1980) 160–168. January.
25. L.W. Holm, A.K. Csaszar, Oil recovery by solvents mutually soluble in oil and water, Society of Petroleum Engineers Reprint Series No. 8, *Miscible Processes*, 1965, pp. 31–38..
26. L.W. Holm, V.A. Josendal, Mechanisms of oil displacement by carbon dioxide, *J. Petrol. Technol.* **26** (1974) 1427–1438.
27. L.W. Holm, A comparison of propane and CO_2 solvent flooding processes, *Am. Inst. Chem. Eng. J.* (1961) 179–184. June.
28. M. Shook, D. Li, L.W. Lake, Scaling immiscible flow through permeable media by inspectional analysis, *In Situ* **16** (4) (1992) 311–349.
29. M.P. Walsh, L.W. Lake, A Generalized Approach to Primary Hydrocarbon Recovery, Elsevier, New York, 2003.
30. M.K. Roper Jr., G.A. Pope, K. Sepehrnoori, Analysis of tertiary injectivity of carbon dioxide, SPE 23974, *Presented at the Society of Petroleum Engineers Permian Basin Oil and Gas Recovery Conference*, Midland, TX, March 18–20, 1992a.
31. M.K. Roper Jr., C.T. Cheng, J.E. Varnon, G.A. Pope, K. Sepehrnoori, Interpretation of a CO_2 WAG injectivity test in the San Andres Formation using a compositional simulator, SPE/DOE 24163, *Presented at the Society of Petroleum Engineers/Department of Energy Eighth Symposium on Enhanced Oil Recovery*, Tulsa, OK, April 22–24, 1992b.
32. A. Araque-Martinez, L.W. Lake, A simplified approach to geochemical modeling and its effect on mineral precipitation, *Soc. Petrol. Eng. J.* (2001) 98–107. March.
33. B. Sass, B. Monzyk, S. Ricci, A. Gupta, B. Hindin, N. Gupta, CO_2 purity tradeoff-surface facilities, this volume.
34. M.-Y. Noh, L. Lake, S. Bryant, A. Araque-Martinez, Implications of coupling fractional flow and geochemistry for CO_2 injection in aquifers, SPE89341, *Presented at SPE/DOE 14th Symp. Improved Oil Recovery*, Tulsa, April 17–21, 2004.

SECTION 4: MONITORING AND VERIFICATION

MONITORING AND VERIFICATION PREFACE

Mike Hoversten,

Lawrence Berkeley National Laboratory Berkeley, CA, USA

When considering the enormous challenges of dealing with the global need to reduce CO_2 in the atmosphere one of many possible partial solutions is the storage of CO_2 underground. Among the many questions that must be answered before any implementation of geologic storage can be considered are questions that center on our ability to monitor the placement and movement of CO_2 in underground systems. Seen in a temporal sequence our monitoring needs require that we be able to know where we are placing the CO_2 in the short run, how the host reservoir responds to the injection, where the CO_2 moves within the reservoir during the injection phase, and ultimately where the CO_2 moves over long time periods after the injection phase is concluded.

The monitoring of the CO_2 location and movement is important in the short term to maximize the operational efficiency of any injection project, maximize the amount of CO_2 stored, minimize the effected volume, minimize the risk of any un-planned leakage, and minimize the cost of the operation. In the long term, monitoring is required to verify the site integrity and to demonstrate that the storage is performing according to long range predictions. In the event of unforeseen leakage, the ability to detect such a leak early while remediation can take place prior to significant quantities of CO_2 reaching the atmosphere is important.

The focus of this section is on the remote monitoring of CO_2. The approaches considered here can be placed in two groups, first the monitoring of CO_2 which reaches the surface (Chapters 19, 20, 21 and 24) and secondly the monitoring of CO_2 within the reservoir and potential leakage pathways (Chapters 22 and 23). Taken together they represent the current state-of-the-art in remote monitoring of CO_2 and point the way forward to areas of further research that are required to meet the challenges ahead.

Carbon Dioxide Capture for Storage in Deep Geologic Formations, Volume 2
D.C. Thomas and S.M. Benson (Eds.)

Chapter 19

MONITORING OPTIONS FOR CO_2 STORAGE

Rob Arts and Pascal Winthaegen

Netherlands Institute of Applied Geoscience TNO-National Geological Survey,
PO Box 80015, 3508 TA Utrecht, The Netherlands

ABSTRACT

In this paper an overview of various monitoring techniques for CO_2 storage has been given, structured into three categories: instrumentation in a well (monitoring well); instrumentation at the (near) surface (surface geophysical methods); and sampling at the (near) surface measuring CO_2 concentrations (geochemical sampling techniques).

An overview of what these techniques can monitor has been provided in terms of features, events and processes (FEPs). The main categories of FEPs identified in this report are: cap rock integrity (leakage); ground movements (uplift, earthquakes); lateral spreading of the CO_2 plume; and verification of mass balance.

For the geophysical methods the physically measurable parameters have been provided and the effects of CO_2 on these parameters are discussed and partially quantified.

INTRODUCTION

The objectives of monitoring underground CO_2 storage are to ensure:

- the integrity of CO_2 storage;
- the safety requirements for subsurface activities during and after the operational phase; and
- the injection process takes place as planned in the intended formation.

The first objective is focused on providing information relevant to tariffs and legislation, i.e. whether the agreed quota as originally planned for CO_2 storage are met and maintained.

The second objective focuses on safety at the storage site. The main safety risks can be categorized as follows.

- Leakage to the atmosphere or other geological formations, including possible groundwater contamination. A number of more specific features, events and processes (FEPs) have been identified influencing the future integrity of the seal. A summary is given in Table 1.
- Uplift of the subsurface (overburden) due to injection of CO_2 or subsidence due to production or to a lesser extent migration of CO_2 may cause damage to structures in the vicinity of the storage project.

Monitoring efforts should be focused on these issues.

A secondary goal of monitoring is research and development regarding underground CO_2 storage. Gaining a greater understanding of the physical and chemical processes occurring in the reservoir is important for the optimization of storage sites in the future.

This study is directed to the improvement of long-term monitoring and verification for storage of CO_2 in various geological media [1]. The experience from other projects (SACS I&II, RECOPOL, Coal and gas

TABLE 1
FEPS IDENTIFIED INFLUENCING THE SEAL OF A RESERVOIR DESIGNED FOR CO_2 STORAGE

Fracturing or fault activation due to increased CO_2 pressure
Dissolution or dehydration of seal due to the presence of CO_2
Casing or cementation defects due to improper design or construction
Deterioration of cement plug after abandonment due to CO_2
Corrosion of casing due to CO_2
Formation damage due to drilling of well
Operational failure of well
Unrecognized features in seal like faults, joints or fractures

Thermie B, NASCENT, Dutch NOVEM study) has been used to set up guidelines for an optimum monitoring strategy for the different scenarios in different geological settings.

To monitor CO_2 storage it is important to have baseline measurements available prior to CO_2 injection, so that storage-induced changes can be measured. This implies that a monitoring technique actually has to be selected at the earliest stage of each storage project in order to have a "baseline". This study provides a "best practice" guideline for selecting monitoring techniques by defining the key geological parameters and an estimation of the accuracy of the available monitoring approaches.

CO_2 STORAGE MONITORING TECHNIQUES

A number of different monitoring techniques are available. Basically the systems are classified into three categories:

- instrumentation in a monitoring well;
- instrumentation at the (near) surface (surface geophysical methods); and
- sampling at the (near) surface measuring CO_2 concentrations (geochemical sampling techniques).

Monitoring in a well within the reservoir can be of great value for determining the CO_2 distribution within the reservoir, monitoring the solution of CO_2 in water and calibration of other monitoring techniques. However, penetrating the seal of the storage formation should be avoided as much as possible because these penetrations might affect the seal integrity. Monitoring of wells in aquifers above the reservoir can provide information regarding seal integrity and leakage. Pressure measurements, water analysis and saturation can all be monitored above the storage formation if wells are available.

From surface geophysical monitoring methods, time-lapse seismics have grown over the last decade to a mature technique with wide applications and with a number of recent successes. Depending on the type of reservoir, changes in fluid composition and reservoir pressure have been observed as any change over time. Within the European SACS project, seismic monitoring has been applied for the first time over CO_2 injected into a saline formation at depths of approximately 800–1000 m. The major success of the SACS project has been the demonstration that conventional, time-lapse, p-wave seismic data can be a successful monitoring tool. Even with the CO_2 in a supercritical, rather than a gaseous state, it has been shown that CO_2 accumulations with a thickness as low as about a meter can be detected at these depths, about seven times below the conventional seismic resolution. Even such thin accumulations cause significant, observable and measurable changes in the seismic signal, in both amplitude and traveltime. Of course the sensitivity of these seismic observables depends heavily on the type of reservoir and its overburden and a sensitivity study must be done for each situation.

Figure 1 shows an example of the time-lapse seismic data acquired at Sleipner.

In general it can be stated, that seismic monitoring potentially provides an image of the spatial distribution injected CO_2.

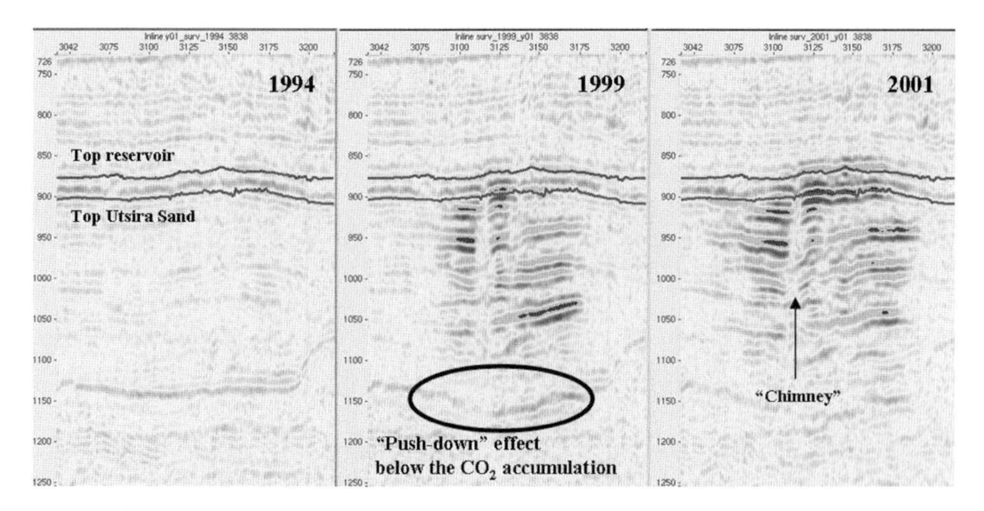

Figure 1: An inline through the injection area for the 1994, 1999 and the 2001 surveys (from Ref. [2]).

In Tables 2–4 an overview of the different monitoring techniques is given indicating what features, events or processes can be monitored. Of course, the sensitivity and accuracy of all these monitoring techniques depends on the geology of the storage site, the size of the storage project and a number of other factors. By combining monitoring methods, the sensitivity and accuracy can be improved (see Hoversten, this volume).

From Table 3b it may be obvious, that the FEPs causing leakage are very difficult to monitor from the surface at an early stage. From Table 3a it is clear that it is more likely that migration of the CO_2 plume can be detected. Fault activation or well bore failure (casing, cement plug) are difficult to detect with surface monitoring methods.

Physical Parameters

The geophysical methods mentioned in the previous section are based on changes in physical parameters. The main parameters responsible for detecting leaking CO_2 are enumerated hereafter. After each parameter the monitoring techniques that are sensitive to the parameter are mentioned.

Bulk density (seismic methods, gravity)

With the $P–T$ conditions known in the reservoir, the density of CO_2 can be determined quite accurately. Under supercritical conditions values for the density are typical in the range of 600–700 kg/m^3 [3]. This implies an important contrast with both densities of water and gas favorable to seismic and gravity methods. Seismic methods are sensitive to contrasts in bulk density. As an example, the change in bulk density of a 100% water saturated (purely quartz constituted) sandstone with a porosity of 20% would change from 2340 to 2260 kg/m^3 when 90% saturated with CO_2.

Compressibility (seismic methods)

The compressibility of CO_2 can be determined quite accurately based on the $P–T$ conditions in the reservoir. The compressibility of the CO_2 directly affects the seismic velocity in the reservoir. For the Utsira Formation, the compressibility of CO_2 is close to the compressibility of a gas ($K \sim 0.1$ GPa) causing very low seismic velocities in the reservoir. In saline formations and depleted oil fields these compressibilities give rise to large impedance contrasts. However, in a depleted gas reservoir with residual gas present, seismic methods might not be able to detect impedance contrasts due to compressibility effects. The small amount of residual gas has already lowered the overall compressibility.

TABLE 2
SUITABILITY OF MONITORING WELL TECHNOLOGY WITH RESPECT TO DIFFERENT FEPS

	Pressure–temperature sensors	Resistivity	TDT	Micro-seismic	VSP	Crosswell	Fluid from reservoir	Fluid from aquifer above reservoir
Cap rock integrity (leakage)	Good	Monitor above the reservoir	Monitor above the reservoir	Good	Good in area of investigation	Good in area of investigation	x	Lab tests
Ground movements	x	x	x	Detection of (small) earthquakes	x	x	x	x
Lateral spreading	Presence monitoring well	Presence monitoring well	Presence monitoring well	Possible	Limited area, calibration for seismics	Limited area, calibration for seismics	Presence monitoring well	Samples around reservoir
Verification or mass balance	x	x	x	x	Calibration for seismics	Calibration for seismics	x	x

TABLE 3A
SUITABILITY OF SURFACE GEOPHYSICAL MONITORING TECHNIQUES WITH RESPECT TO DIFFERENT FEPS

	Time-lapse seismic	**Subbottom profiling**	**Sonar**	**Gravity**	**EM**	**Geodetic**	**InSAR**	**Tilt meters**
Cap rock integrity (leakage)	Good	In case of leakage to the sea	In case of leakage to the sea	Low resolution	Low resolution	x	x	x
Ground movements	x	x	x	x	x	Good	Good	Good
Lateral spreading	Good	x	x	Low resolution	Low resolution	x	x	x
Verification or mass balance	Fair	x	x	Too low resolution	Too low resolution	x	x	x

TABLE 3B
SPECIFICATION OF THE SUITABILITY OF SURFACE GEOPHYSICAL MONITORING TECHNIQUES WITH RESPECT TO SPECIFIC FEPS RELATED TO CAP ROCK INTEGRITY

	Time-lapse seismic	**Subbottom profiling**	**Sonar**	**Gravity**	**EM**	**Geodetic**	**InSAR**	**Tilt meters**
Fault activation (high pressure)	Not likely	x	x	x	x	Not likely	Not likely	When downhole
Dissolution or dehydration of seal	Not likely	x	x	x	x	x	x	x
Casing/cementation failure	x	x	x	x	x	x	x	x
Deterioration cement plug	x	x	x	x	x	x	x	x
Corrosion of casing	x	x	x	x	x	x	x	x
Formation damage due to drilling	Not likely	x	x	x	x	x	x	x
Operational well failure	x	x	x	x	x	x	x	x
Fractures seal	Possible	x	x	x	x	x	x	x

TABLE 4
SUITABILITY OF "GEOCHEMICAL SAMPLING" MONITORING TECHNIQUES WITH RESPECT TO DIFFERENT FEPS

	Groundwater sampling	**(Isotopic) tracers**	**Atmospheric monitoring network**	**Geobotanical monitoring**
Cap rock integrity (leakage)	In case of leakage to the surface	Injected CO_2 discrimination	In case of leakage to the surface	In case of leakage to the surface
Ground movements	x	x	x	x
Lateral spreading	x	x	x	x
Verification or mass balance	x	x	x	x

For the RECOPOL project [4,5] monitoring of the ECBM process is carried out through crosswell seismics. The basic idea is that CO_2 molecules are adsorbed by the coals freeing CH_4 gas. The expectation is that not all the CO_2 can be adsorbed immediately by the coals, leaving free CO_2 in the system. The free CO_2 lowers the overall compressibility of the coal layer leading to a seismic contrast. Crosswell models have been run simulating the free CO_2 front mixed with freed CH_4. As an example some of the results are presented here. In a coal seam of 5 m thickness at a depth of about 1000 m, two vertical wells are drilled with a spacing of 400 m. CO_2 is injected in well 2, while well 1 produces CH_4.

Figure 2 shows a modeled shot gather (before injection, after injection and the difference) obtained with a crosswell geometry.

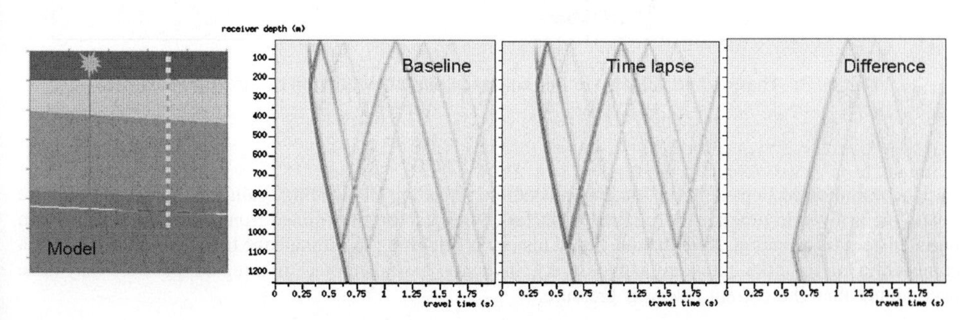

Figure 2: Shot gather of a time-lapse crosswell acquisition geometry with a source at $x = 300$ m and $z = 10$ m and receivers in well 2 at $x = 700$ m from 0 to 1250 m.

The first dipping event indicates the direct arrival or the first p-wave. Around a depth of 1100 m the first arrival reaches the coal bed layer. From that point a strong dipping event going in the opposite direction can be observed. This event is a result of energy reflected on the coal bed layer and then reaching the geophones as an upward traveling wavefield. This reflection is clearly visible on the difference plot since the CO_2/CH_4 has altered the reflection coefficient of the layer. The small part of the direct wave visible on the difference plot at depths larger than 1100 m is caused by the difference in traveltime of the energy going through the coal bed layer.

Effective pressure (seismic methods)
The velocity of sediments freshly deposited on the seafloor approximate the velocity of sound in water. Due to the growing overburden in time (sedimentation) an increasing pressure is applied on these sediments and they compact. The effect of this compaction is a reduction in porosity and an increase in the velocity related to the increasing stiffness of the material. The maximum velocity is determined by the velocities of the constituent grains with a porosity approaching zero. The velocity-effective stress relation for non-decreasing effective stress states is generally referred to as the virgin compaction curve (Figure 3). Note that this curve will flatten at a certain pressure [6].

Most of the porosity loss and velocity gain occurring during compaction is permanent. This means, that the velocity in the rock will actually not decrease along the virgin compaction curve when the effective pressure is released. Instead the so-called unloading curve will be followed (see Figure 3), showing higher velocities than on the virgin compaction curve.

If the effective stress is subsequently increased again, the velocity will go back up the unloading curve until the virgin compaction curve is reached. Beyond this point the velocity will once again follow the virgin compaction curve.

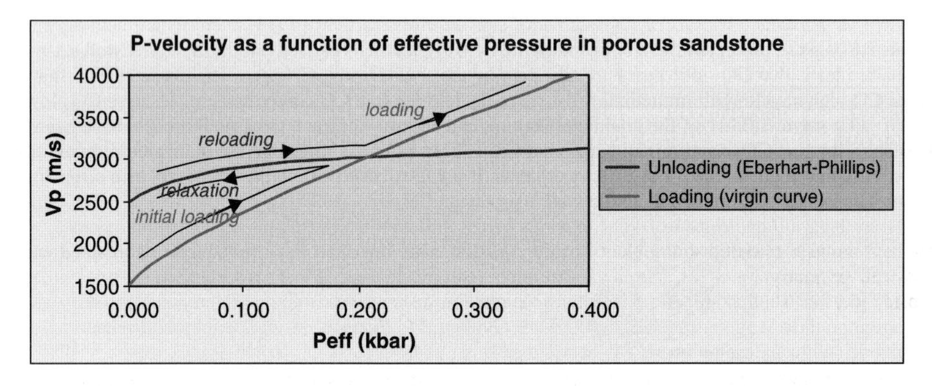

Figure 3: The p-wave velocity in porous sandstone as a function of effective pressure.

The above-indicated steps of initial loading, relaxation, reloading and loading again, are illustrated in Figure 3 with the arrows. In practice the virgin compaction curve can be determined from log measurements and burial history information. Note that the burial history is important to estimate the transition from the virgin compaction curve to the unloading curve. If, e.g. inversion has taken place in a region, pressure may have been higher than the current reservoir pressure.

The unloading curve is determined for various rocks by (numerous) laboratory experiments using ultrasonic measurements [7–9]. Different models (generally empirically determined) describing the unloading curves are available. The disadvantage of these models is that they are only valid for certain rock types under specific conditions, such as the Eberhart-Phillips relation [10] for porous sandstones as used in this study.

In this section only the effective pressure has been mentioned. Effective pressure is the pressure that balances the overburden pressure due to the weight of rock (which forms a matrix) and fluid (which fills the matrix) overlying this point, leading to the following equilibrium relation:

$$P_{\text{effective}} = P_{\text{overburden}} - nP_{\text{pore}}$$

where n is known as the Biot effective stress coefficient equal to 1 for soft sediments and < 1 for cemented rocks. In the next example n has been chosen to be 1. A more elaborate study on the behavior of n can be found in a recent publication of Siggins and Dewhurst [11]. The process of injection causes an increase in the pore pressure. The overburden pressure can be considered constant. As a consequence, the effective pressure will decrease. Note that a decrease in the effective pressure will always follow the unloading (relaxation) curve (Figure 4).

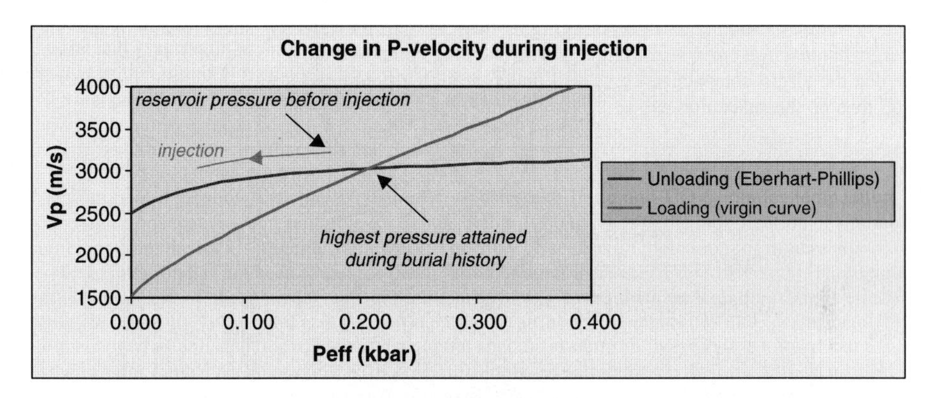

Figure 4: Example of the effect caused by the process of CO_2 injection on the effective pressure and the p-velocity.

The process of production decreases the pore pressure (Figure 5). The overburden pressure remains constant, since nothing changes in the overburden. (Note that this is not necessarily true, e.g. in the case of subsidence, but whether the effect is noticeable remains to be seen.)

As a consequence, the effective pressure will increase, leading to an increase in the velocity as well. In the case of production, it is not obvious which curve (the virgin compaction curve or the unloading curve) the velocity increase will follow. This depends much on the burial history of the reservoir determining the maximum effective pressure ever reached. Reconstruction of this history is recommended.

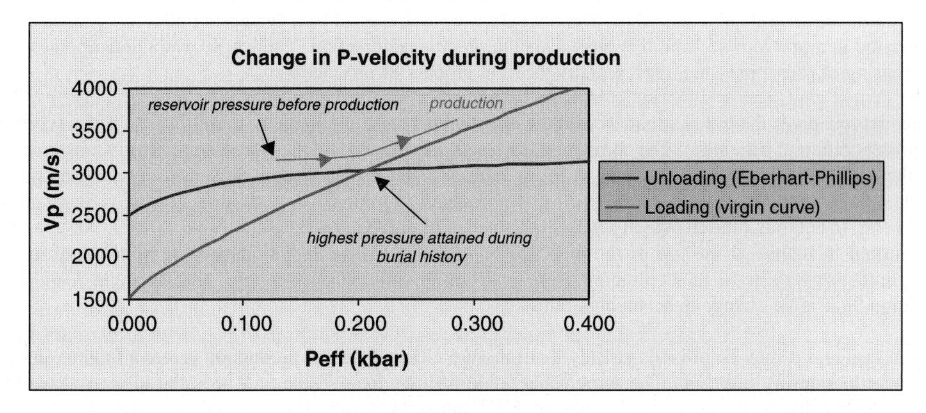

Figure 5: Example of the effect caused by the process of CO_2 production on the effective pressure and the p-velocity.

As an example a sensitivity analysis in terms of seismic measurable parameters such as amplitude and two-way traveltime (TWT) of p-waves has been carried out. Note that the analysis in this section is restricted to p-waves, though s-waves are probably more sensitive to pressure changes [12]. However, the use of shear waves for monitoring purposes is a less mature technology. The reservoir is assumed to be representative of a Rotliegend sandstone gas reservoir in the Dutch subsurface. The reservoir is at a depth of 2500 m and under normal hydrostatic pressure. The thickness of the reservoir is 100 m. The velocity in the overlying seal (anhydrites) is 5700 m/s. The bulk density is 2850 kg/m^3 in the seal and 2300 kg/m^3 in the reservoir. The velocity–pressure relation is determined by the unloading curve (see Figure 3) in the range of effective pressures from 0 to 300 bar. For pressures higher than 300 bar, the velocity–stress relation is governed by the virgin compaction curve (see Figure 3). The resulting composed curve is shown in Figure 6. Note once more that the virgin compaction curve will flatten at higher pressures as well, however, the effect is less drastic than on the relaxation curve.

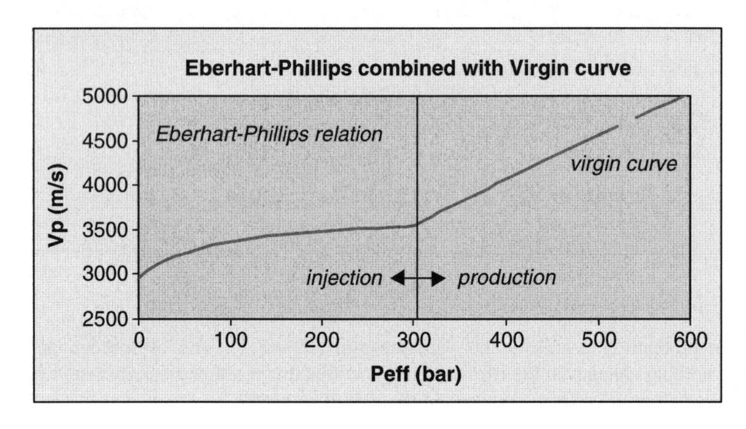

Figure 6: Estimated stress–velocity relation for a Rotliegend sandstone reservoir at an effective pressure of 300 bar.

Table 5 shows the results of the sensitivity analysis. The first row indicates the initial state of the reservoir, approximately at a depth of 2500 m. The next six rows correspond to a decreasing effective pressure. As indicated in a previous section, this represents the process of injection. The last six rows correspond to an increasing effective pressure, representative for the process of production.

The first row gives the initial situation with the effective pressure at 300 bar (column 2) at t_0. The next rows represent different time steps. The corresponding reservoir velocities at these time steps are given in column 3, the TWT in the reservoir in column 4. From column 5 and further, the actual sensitivity analysis starts. Column 5 gives the increase/decrease in effective pressure. Column 6 shows the effect on the velocity in the reservoir. In column 7 the change in TWT is indicated. Column 8 gives the reflection coefficient for p-waves at normal incidence at the top of the reservoir at t_0 and t_1. The relative change (in percentage) in the amplitude is given in the final column 9. Note that the key columns are column 7 (difference in TWT) and column 9 (relative change in seismic amplitude).

From Figure 6 it may be obvious already that pressure changes during production create a larger velocity change than during injection. The same observation follows from Table 5. If, e.g. the effective pressure drops 50 bar due to injection, the change in TWT amounts to only 0.47 ms and the relative change in amplitude 1%. On the other hand, an increase of 50 bar due to production results in a change of TWT of -3.93 ms and a relative change in amplitude of -10%.

TABLE 5
SENSITIVITY ANALYSIS OF THE EFFECT OF STRESS CHANGES ON THE SEISMIC MEASUREMENTS, TWT AND AMPLITUDE

$P_{eff}(t_1) < P_{eff}(t_0)$ means injection (unloading); $P_{eff}(t_1) > P_{eff}(t_0)$ means production (loading)

State	P_{eff} (bar)	V_p (m/s)	TWT (ms)	dP_{eff} (bar)	dV (m/s)	Increase TWT (ms)	Refl. coeff.	Relative change in amp (%)
Initial	300	3541	56.48				0.332	
Injection	240	3506	57.05	−60	−35	0.57	0.337	1
Injection	250	3512	56.96	−50	−29	0.47	0.336	1
Injection	260	3517	56.87	−40	−24	0.38	0.335	1
Injection	270	3522	56.78	−30	−19	0.30	0.334	1
Injection	280	3528	56.70	−20	−13	0.21	0.334	1
Injection	290	3533	56.61	−10	−8	0.13	0.333	0
Production	310	3595	55.64	10	54	−0.84	0.325	−2
Production	320	3648	54.83	20	107	−1.66	0.319	−4
Production	330	3701	54.04	30	160	−2.44	0.312	−6
Production	340	3753	53.28	40	212	−3.20	0.306	−8
Production	350	3806	52.55	50	265	−3.93	0.300	−10
Production	360	3858	51.84	60	317	−4.64	0.293	−12

Conductivity (EM)
The difference in electrical conductivity between CO_2 and brine is the basis for this monitoring technique. See Hoversten, this volume, for a discussion of electromagnetic methods for measuring electrical conductivity.

Fracturing (seismic methods, EM)
(Micro-)fractures can be a migration pathway for CO_2. In principle fractures can be detected by seismic or EM methods especially in the case of aligned systems of micro-cracks. In these cases anisotropy measurements (on seismic velocities or on EM) could provide insight in the preferential orientation of the system and hence the preferred migration pathway [8]. Most suitable are probably azimuthal VSP or crosswell measurements. The quantification of these systems in terms of an effective permeability, however, is highly speculative.

Porosity reduction/increase (seismic methods)
Chemical reactions might cause an increase or a decrease in porosity in the order of 3–4% (see Bryant et al., this volume). For the cap rock, an increase in porosity or permeability would be the most important parameter to monitor. In theory, seismic methods or even gravity methods should be able to detect these changes. However, in most cases such measurements are at the limit of resolution and are only useful when supported by other measurements.

RESULTS AND DISCUSSION

In this project monitoring of CO_2 storage has been approached in a systematic manner.

First a short inventory has been made of why CO_2 storage should be monitored. The answer to this question should determine what parameters should be monitored and the resolution needed. For example, is it sufficient to know that the CO_2 is not leaking to the surface (or overburden), or is it important to know where CO_2 migrates to within the reservoir. In this report, a broad approach has been chosen taking into account as many monitoring techniques as possible.

Globally three areas of investigation for monitoring have been identified:

- the reservoir containing the CO_2 (pressure, temperature, spreading and long-term behavior of the CO_2);
- the integrity of the seal (fractures, faults, wells, heterogeneous permeability); and
- the overburden and the atmosphere with possibly CO_2 leaking (migration pathways of CO_2).

The first and especially the second are probably the most important for monitoring. They provide an early warning system for possible leakage. In the ideal case, when leakage does not occur, no changes would be expected in the properties of the overburden.

CONCLUSIONS

In this paper a short description of various monitoring techniques has been given. To structure the discussion, monitoring techniques have been divided into three categories, namely:

- instrumentation in a well (monitoring well);
- instrumentation at the (near) surface (surface geophysical methods); and
- sampling at the (near) surface measuring CO_2 concentrations (geochemical sampling techniques).

An overview of what these techniques actually can monitor has been provided in terms of FEPs. The main categories of FEPs identified in this report are:

- cap rock integrity (leakage);
- ground movements (uplift, earthquakes);
- lateral spreading; and
- verification or mass balance.

For the seismic methods the physical measurable parameters have been provided and the effect of CO_2 on these parameters are discussed and partially quantified.

RECOMMENDATIONS

As a follow-up to this project the following recommendations are made.

1. The modeling should be extended to different migration pathway scenarios. Especially storage in a depleted gas-field requires more modeling. For most methods it is very difficult to separate effects of residual gas and stored CO_2. A more detailed analysis on a specific case (e.g. a Rotliegend gas-field) is recommended.
2. The FEP matrices showing which monitoring techniques can be applied should be updated. For example, the FEP analysis in the SAMCARDS Project will provide more insight in the most likely leakage scenarios and, more importantly, to the mechanisms causing the leakage. Monitoring techniques and strategies must be focused on these mechanisms at the earliest stage possible.

ACKNOWLEDGEMENTS

The authors would like to thank the CCP for funding this study. Furthermore we would like to thank Mike Hoversten, Jos Maas and Dan Ebrom for fruitful discussions and comments.

REFERENCES

1. T. Holt, J.I. Jensen, E. Lindeberg, Underground storage of CO_2 in acquifers and oil reservoirs, *Energy Convers. Manage.* **36** (6–9) (1995) 535–538.
2. R. Arts, O. Eiken, A. Chadwick, P. Zweigel, L. van der Meer, B. Zinszner, Monitoring of CO_2 injected at Sleipner using time lapse seismic data, in: J. Gale, Y. Kaya (Eds.), Greenhouse Gas Control Technologies, Elsevier, Oxford, 2003, pp. 347–352.

3. E. Lindeberg, P. Zweigel, P. Bergmo, A. Ghaderi, A. Lothe, Prediction of CO_2 dispersal pattern improved by geology and reservoir simulation and verified by time lapse seismic, in: R.A. Durie, D.J. Williams, A.Y. Smith, D. McMullan, C.A.J. Paulson (Eds.), Greenhouse Gas Control Technologies, CSIRO Publishing, Collingwood, Australia, 2000, pp. 372–377.
4. F. van Bergen, H.J.M. Pagnier, L.G.H. van der Meer, F.J.G. van den Belt, P.L.A. Winthaegen, R.S. Westerhoff, The RECOPOL project: developing a field experiment of CO_2 storage in coal seams in Poland, *Proceedings of the International Workshop Present Status and Perspective of CO_2 Sequestration in Coal Seam* in Tokyo, Japan, 2002.
5. P.L.A. Winthaegen, R.S. Westerhoff, Seismic CO_2 monitoring feasibility study, *Proceedings of the International Workshop Present Status and Perspective of CO_2 Sequestration in Coal Seam* in Tokyo, Japan, 2002.
6. G.L. Bowers, Pore pressure estimation from velocity data: Accounting for overpressure mechanisms besides undercompaction, *SPE Drilling Conference*, No. 27488, 1994, pp. 515–530.
7. T. Bourbie, O. Coussy, B. Zinszner, *Acoustics of Porous Media*, Gulf Publishing, Houston, 1987.
8. R.J. Arts, A study of general anisotropic elasticity in rocks by wave propagation—theoretical and experimental aspects, *Ph.D. Thesis*, University Pierre et Marie Curie (Paris VI), Editions Technip, Paris, 1994.
9. M.W.P. Dillen, H.M.A. Cruts, J. Groenenboom, J.T. Fokkema, A.J.W. Duijndam, Ultrasonic velocity and shear-wave splitting behavior of a Colton sandstone under a changing triaxial stress, *Geophysics* **64** (5) (1999) 1603–1607.
10. D.M. Eberhart-Phillips, D.H. Han, M.D. Zoback, Empirical relationships among seismic velocity, effective pressure, porosity, and clay content in sandstone, *Geophys. Soc. Expl. Geophys.* **54** (1) (1989) 82–89.
11. A.F. Siggins, D.N. Dewhurst, Saturation, pore pressure and effective stress from sandstone acoustic properties, *Geophys. Res. Lett.* **30** (2) (2003) 1089.
12. D. Ebrom, P. Heppard, L. Thomsen, Numerical modelling of PS moveout as a function of pore pressure, *SEG Expanded Abstracts*, 2002, pp. 1634–1637.

Carbon Dioxide Capture for Storage in Deep Geologic Formations, Volume 2
D.C. Thomas and S.M. Benson (Eds.)

Chapter 20

ATMOSPHERIC CO_2 MONITORING SYSTEMS

Patrick Shuler and Yongchun Tang

Tang Associates, Covina, CA, USA

ABSTRACT

Monitoring for atmospheric CO_2 concentrations may be an integral part of any subsurface storage project. Several CO_2 measurement methods may be used to meet the monitoring objectives of (1) assuring there are no large leaks at the surface that might pose a health risk and (2) verifying that the injected CO_2 remains trapped below the Earth's surface.

Options include (1) remote sensing from satellites or aircraft, (2) open path instruments that can sample over significant distances and (3) a network of conventional fixed-point detectors. NASA indicates satellite surveys might be useful for a "global" view of CO_2. Aircraft surveys may be a fast means to collect data near ground level, but this is only practical in an infrequent basis. Instruments located near ground level that are based on open path sampling may offer the most efficient means to monitor long term over a large surface area. They could have the capability to detect increases of just a few percent of CO_2 above normal background, over a sample path of tens of meters, and continuously with unattended operation. Many different commercial fixed-point units based on infrared (IR) spectroscopy are available. These detectors may be better suited to monitor sensitive, high-risk points of leakage rather than be deployed in a network to monitor large surface areas.

Besides reviewing atmospheric monitoring options, this chapter also quantifies the capability of ground-level instruments to identify leakages of carbon dioxide from the subsurface. In particular, the objective is to successfully detect the uniform leakage of as little as 1% of the total carbon dioxide injected into the subsurface over 100 years. This analysis suggests the local increased concentration of carbon dioxide into the atmosphere due to such a leak depends greatly on the leakage area, time duration, atmospheric conditions and proximity of the detector to the leak. In some scenarios such a leak would cause an increase of at least tens of ppmv of carbon dioxide in the near-surface atmosphere and likely would be detected by commercially available instruments as being above the natural background variations of carbon dioxide.

INTRODUCTION

Desirable attributes for such monitoring tools include: (1) low cost, (2) accurate measurements of CO_2, (3) measurement over a small as well a large surface area, (4) remote, automated, long-term operation, and (5) reliable and safe to use. One motivation for a monitoring program is to assure the public and the project employees that there are no very large gas releases that pose a risk to human health. A second concern is to locate quickly any smaller leaks that may compromise the permanent capture of the injected CO_2.

Carbon dioxide is a relatively benign chemical, but at very high concentrations it does pose a risk. Atmospheric concentration of two percent carbon dioxide will cause a 50% increase in breathing rate; concentrations exceeding 1000 ppm (0.1%) cause noticeable symptoms in some people (drowsiness, headaches). The OSHA (Occupational Safety and Health Administration) maximum acceptable level is

---

Abbreviations: CCP, carbon capture project; EOR, enhanced oil recovery; MCT, mercury cadmium tellurium (HgCdTe); TEC, thermal electric cooling.

5000 ppm. Because the natural background concentration of CO_2 is 300–400 ppmv in ambient air, one needs only to determine if there is a *very* significant increase before there are any human health concerns.

A complete surface monitoring program would consider several geographic scales:

- large areas—even beyond project boundaries—perhaps tens of square kilometers
- within project boundaries and at the "fence line"—cover several or more square kilometers
- at higher risk points of leakage at the field site such as wellheads and compressors, etc.
- inside or near control rooms where workers are located
- personal monitors for workers who travel to any higher risk areas.

Different monitoring "tools" will be required to fulfill all the measurement requirements. For very large areas, instrumentation mounted in satellite or low-flying aircraft could be a practical approach. Within a project area, sensors that can measure CO_2 over open path lengths of hundreds of meters may be attractive. Fixed CO_2 sensors could play a role at critical points in the facilities such as near compressors in control rooms. Finally, there are a number of portable CO_2 detectors that should be suitable for individuals to use when entering higher risk areas.

After reviewing different monitoring options, the latter portion of this chapter assesses the capability of ground-level instruments to successfully detect leakages of carbon dioxide. In particular, the focus is on whether ground-level instruments can detect leakage to the atmosphere of as little as 1% of the cumulative total of the carbon dioxide injected into the subsurface over a 100-year period (leakage of 0.01%/year for 100 years). The calculation methodology and results are presented below. Different anticipated scenarios for CO_2 leaks are considered (1) uniform CO_2 leakage over an area of multiple square kilometers, and (2) leakage from a point source.

EXPERIMENTAL/STUDY METHODOLOGY

Literature Review

This effort uses information from the open literature, plus contacting NASA and industrial sources.

Capability of Ground-Based Instruments to Measure Carbon Dioxide

Based on instrument specifications and making key assumptions concerning the storage project operation (such as the source of leak, detector location, weather conditions, variations in background concentrations, etc.), one can calculate whether a particular instrument package would successfully identify that leak of carbon dioxide.

Leakage over a wide surface area

From the following sequence of calculations one can estimate the increase in CO_2 concentration to the atmosphere (near ground level) from leakages over a relatively large surface area. These include (1) compute the total mass of CO_2 injected, (2) assume some percent of this gas leaks to the surface (default is 1% of total injected), (3) input the surface area and duration of time of the leak (determines a flux of CO_2), (4) calculate the volume of an imaginary "box" near ground level where the sensor is located—use a height of 3.3 m (10 ft)—and calculate the mass of CO_2 added to this volume daily, and (5) include a dilution factor to account for atmospheric conditions that would deplete the added carbon dioxide to this "box" volume.

The next step is to compare the calculated increase of CO_2 concentration to the two main uncertainties in the measurement: (1) the uncertainty of the instrument measurement, and (2) the natural variation in the background CO_2 concentration. For the former uncertainty, one may consult the detector performance for the selected instrument, plus one must pay attention to proper calibration of any instrument. For the latter uncertainty factor, the natural variation of CO_2 would depend on the sampling location and time of year (or even time of day). The atmospheric CO_2 concentration near the ground can be significantly affected by fluxes of CO_2 with terrestrial vegetation, types of soils, subsurface moisture, and water bodies. Diurnal and seasonal variations of several ppm or more are typical above a vegetated land surface. For example, Conway [1] reports a monthly variation of approximately 5 ppmv CO_2 at one fixed location. Even without local fluxes, the background CO_2 concentration varies significantly and on a range of timescales, as a result of natural and industrial fluxes. The annual average carbon dioxide concentration also has been increasing about 0.5 ± 0.3%, adding to the background levels. If the actual increase in CO_2 concentration is substantially greater than all these uncertainties, the instrument measurement will recognize that there is a leak.

Leakage from a point source

The second type of calculation considers if instead the leakage is described better as a point source (e.g. leaks from around a wellbore) instead of a uniform leakage over a significant surface area. We use a simplified approach of a Gaussian distribution analysis to illustrate the general procedure to analyze the situation where the gas is venting to the surface at a single spot [2,3]. Figure 1 shows a sketch of the problem we are considering, the dispersion of this contaminant plume.

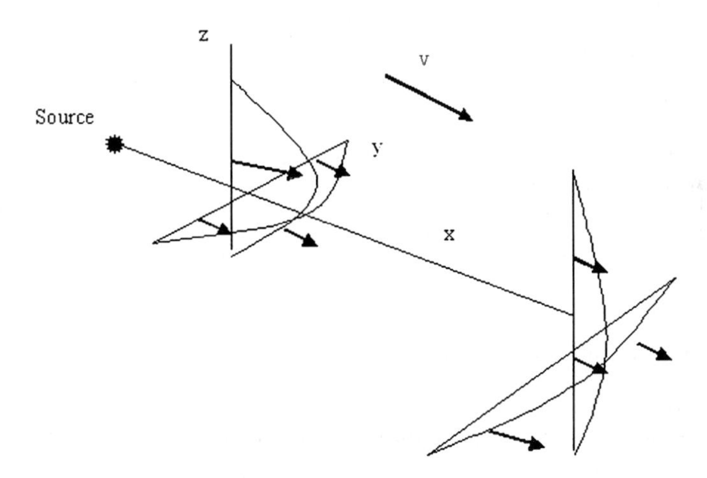

Figure 1: Schematic of movement of a plume of carbon dioxide coming from a point source and depleting in concentration as it moves downwind.

$$\mathrm{Conc}(x) = Q/(\pi\sigma_y\sigma_z u)$$

where Conc(x) is the concentration of added CO_2 at ground level, center of the plume; Q the uniform emission rate of carbon dioxide (g/s); σ_y the standard deviation of plume concentration distribution, horizontal direction (m); σ_z the standard deviation of plume concentration distribution, vertical direction (m) and u the mean wind speed affecting the plume (m/s).

The plume spread has a Gaussian distribution in the horizontal and vertical planes. The calculated concentration is for the additional CO_2 concentration (in excess of the local background level) that is in the center line, downwind of the source. To use this equation (1) input the total mass of CO_2 injected and the percent assumed to leak, (2) choose a time duration for the leak, (3) specify a wind speed, and (4) estimate the horizontal and vertical dispersion coefficients. This last step becomes somewhat involved as it requires referring to tables and graphs. First, one needs to select the atmospheric conditions, done via the so-called Pasquill Stability Class, as described in Table 1. The stability classification ranges from A through F, based on wind speed, time of day, and the degree of overcast. Next, one refers to graphs to determine the horizontal and vertical dispersion coefficients (Figures 2 and 3).

More recent models of pollution dispersion have advanced beyond the Gaussian model and Pasquill Stability Classes utilized here. These more sophisticated models would be appropriate to forecast and analyze leakage behavior for specific storage projects where one would want to account for the local

TABLE 1
PASQUILL STABILITY CLASSIFICATIONS

Pasquill Stability Classes
A: Extremely unstable conditions
B: Moderately unstable conditions
C: Slightly unstable conditions
D: Neutral conditions
E: Slightly stable conditions
F: Moderately stable conditions
G: Extremely stable

Meteorological conditions defining Pasquill Stability Classes

Surface wind speed (m/s)	**Daytime insolation**			**Night-time conditions**	
	Strong	**Moderate**	**Slight**	**Thin overcast or >4/8 low cloud**	**≦ 4/8 cloudiness**
<2	A	A–B	B		
2–3	A–B	B	C	E	F
3–5	B	B–C	C	D	E
5–6	C	C–D	D	D	D
>6	C	D	D	D	D

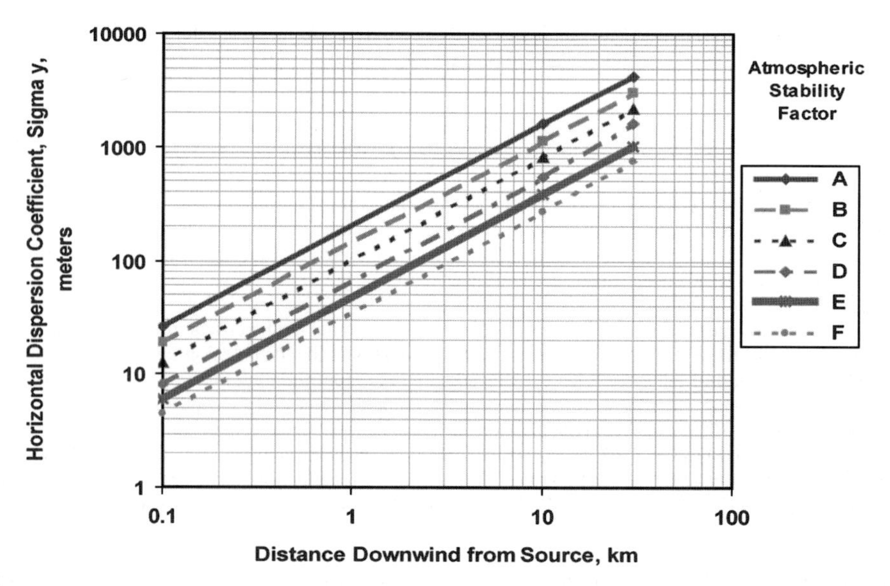

Figure 2: Horizontal standard deviation, dispersion factor, versus the distance downwind from the point source for different atmospheric conditions.

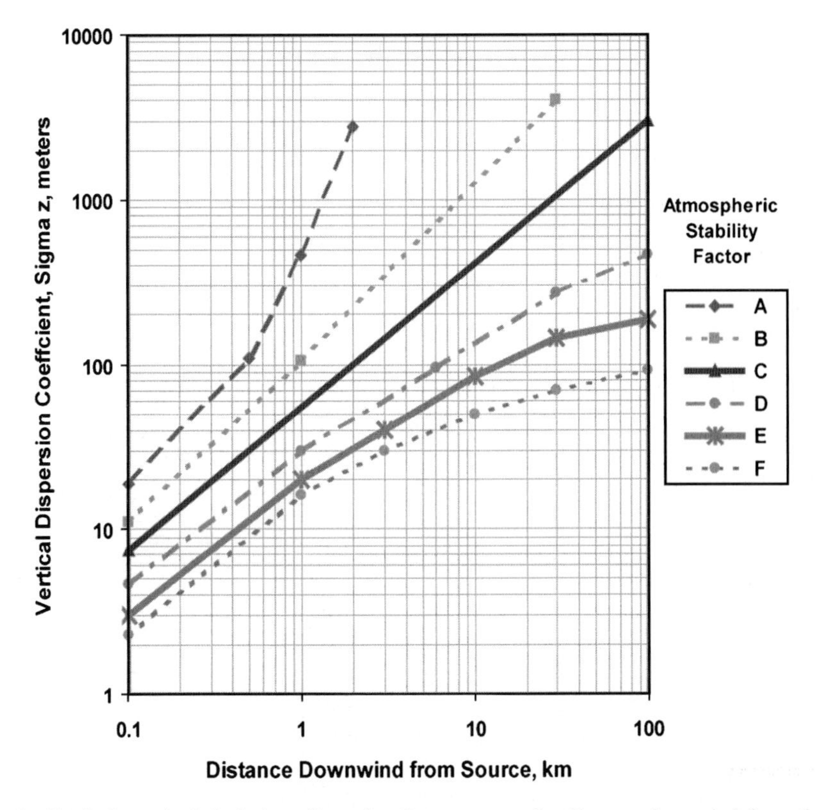

Figure 3: Vertical standard deviation, dispersion factor, versus the distance downwind from the point source, for different atmospheric conditions.

topography and other site-specific details. Newer analytical techniques take into account more fully the physical processes and structure of the atmosphere, while even more sophisticated models incorporate actual topography and dynamic meteorology. Some atmospheric models are offered as packages with license fees, others as shareware. Some of these models are endorsed by regulatory bodies such as environment protection authorities.

RESULTS AND DISCUSSION

Review of Schemes for Detecting CO_2 Concentrations in Ambient Air

See Table 2.

Infrared analysis—general background

Although there are number of different approaches for CO_2 measurement in the ambient air, variations of infrared (IR) detection is the most common technique. CO_2 has unique absorption bands in the IR. Table 3 lists the absorption strength at different CO_2 bands. IR analysis in the open air can measure directly the bulk CO_2 concentration.

The band chosen for CO_2 analysis is based on its absorption strength and the potential interferences from other gases. High absorption bands (such as at 4.25 μ) can detect very low concentrations of CO_2 over even a short sample path length. Most fixed and portable commercial CO_2 monitoring systems are based on IR

TABLE 2
COMPARISON OF METHODS FOR MEASUREMENT OF ATMOSPHERIC CARBON DIOXIDE

Measurement type	Description (application)	Sensitivity/cost for ambient air sampling	Advantages	Disadvantages
Satellite	Remote sensing (potential to cover hundreds of square miles/survey; for infrequent large area sampling)	Costs can be of the order of $\$10^4$–$\$10^5$ per survey. NASA claims can resolve to 100 ft^2	Covers very large area	Only a "2D view", not sample at ground level for direct CO_2 measurement
		Hyperspectral survey can resolve to a few meters	Technology development sponsored at least in part by the government	Available satellites might not cover project area
Airborne	Remote sensing (potential to cover tens to hundreds of square miles; for infrequent sampling)	Estimated at \$100 per survey Single measurement to 3% ±	Cover large area Fairly fast over tens of square miles	Only practical for occasional "snapshot" surveys
Open path laser spectrometer	Ground level (potential to cover several square miles with one device; can be main instrument for long-term monitoring)	Estimated \$1000 per unit Instrument needs development, but estimate can be 3% ± or better	Potential for one fixed instrument to cover large area Measurement could be automated, continuous	Technology for long, open light path detection is still under development
Fixed-point detectors	Ground level (sample at single fixed points of high risk of leaks)	Fairly cheap (circa \$1000) Routinely better than 3% ± . Less than 1% ± available	Fairly cheap and proven technology Best used as points of higher risk	Only measure CO_2 at the detector location Require multiple sensors to sample even a small area
Portable detectors	Personal protection and scan for equipment leaks	Very cheap; units can cost <\$500 Better than 5% ±	Very cheap; can move to suspect "hot spots"	Only suitable for spot checking CO_2 concentrations

TABLE 3
WAVELENGTHS FOR CARBON DIOXIDE ABSORPTION

Wavelength (μm)	Relative absorption strength
1.432	1
1.570	3.7
2.004	243
2.779	6800
4.255	69,000

absorption at 4.25 μ and use a very short optical path at this band along with a filament light source. Usually, the absorption of IR light passing through a confined gas cell is measured. The light is normally generated with a metal filament, giving out radiation from 3 to 10 μ and a power of several microwatts and filtered to 4.25 μ. The detector often is an MCT (Mercury Cadmium Tellerium, HgCdTe) detector with TEC (Thermal Electric Cooling). Such highly absorbing bands, however, are then limited in their maximum concentration detection limit or path length before over-saturating the detector. Bands with low absorption of CO_2 are more suitable for measuring high concentrations of CO_2 or to measure its concentration over a long path length.

The second strongest absorption band by CO_2 is around 2.7 μ; the relative absorption strength here is about 1/10th of the absorption strength at 4.25 μ. This band is also very sensitive and relatively free of interference from other gases. It has been used, e.g. to measure CO_2 levels by the Mars Explorer by NASA. However, there are no commercial diode lasers for this band and NASA had to develop a custom laser. Its relatively strong absorption also does not allow this band to be used for CO_2 detection over a long beam path either.

One other band is the 4.41 ~ 4.45 μ band, which is the absorption band for $^{13}CO_2$. Because ^{13}C occurs at a much lower level than ^{12}C (about 1/100th as much), this band allows detection of much higher level of total CO_2. This method allows detection of much higher concentrations of CO_2, up to 0.27% with a path length of 200 m. However, because the isotope ratio of ^{13}C and ^{12}C varies from site to site, this approach is not generally reliable unless one has an independent measurement of that ratio.

Another potential band is the 2 μ band, with the absorption strength for CO_2 being at least 250 times weaker than at 4.25 μ. The interferences of other gases are also much weaker than CO_2 if a narrow light source is used as the probe. This weak absorption band has already been used for detection of CO_2 in combustion environment [4]. One advantage of this wavelength is the availability of lower cost diode lasers with very narrow (0.01 cm^{-1}) bandwidth at this band. Another advantage is the availability of InGaAs detectors with much better signal-to-noise ratio compared to MCT (HgCdTe) detectors used for 4.25 μ. Based on the absorption strength of CO_2 at this band, in theory, one can measure CO_2 concentration as high as 0.5% over a path length distance of 200 m.

There is a third band at 1.57 μ for the adsorption of CO_2's overtone. The absorption by CO_2 at this band is much weaker (close to 1/100) than the band at 2.01 μ, and is only 1/20,000th compared to the absorption at 4.25 μ. This band is almost completely free of interference by other gases. This band has been investigated as a means for CO_2 detection in a combustion chamber [5]. They found the band to be free of interference from other gases, but it is too weak for short path detection of CO_2. However, this wavelength should be well suited for long path CO_2 detection at concentrations typical of ambient air. Based on the absorption strength of CO_2 at this band, we would expect to be able to detect 1% of CO_2 over a kilometer light path. The band at 1.43 μ is even weaker. Because water absorption is significant at this wavelength, this is not appropriate for detection of CO_2 over a long path.

Infrared analysis—long open path measurement as a newer technology
One attractive concept is to measure absorption loss (and hence CO_2 concentration) across a long, open air, optical path. This has the distinct advantage of having an individual instrument collecting carbon dioxide

concentration data over an extended distance. If there is further modification that the instrument can rotate and reflect a signal from multiple retro-reflectors, then a single laser could sample out several direction and distances and thereby sample an area of the order of a square kilometer. The disadvantage to this scheme is that the absorption (measured concentration) represents a cumulative effect over the entire light path. Thus, one cannot distinguish whether an elevated reading for carbon dioxide might represent a modest, uniform increase over the entire sample path, or could as well be from a larger jump in CO_2 concentrations over a small portion of the light path. Hence, if this device measures a significant increase in CO_2 concentrations, one would have to sample further in the suspicious area with perhaps a portable unit to pinpoint the source(s) of the elevated CO_2 in that sampling path.

After reaching the laser, the retro-reflected beam will be focused onto a detector and recorded. The signal is the ratio between the detector after the collection lens and the reference signal. One instrument's specification (from Air Instrument and Measurements), using a 15 cm cell and such a light source, is able to measure ambient (around 360 ppm) CO_2 levels with a precision of ca. 100 ppb. Beyond the 360 ppm level, the signal registered on the detector falls to the same level as noise. Based on this result, one should be able to probe CO_2 concentration up to 360 ppm over a distance of 15 m range by concentration product of 5400 m ppm.

One could select other wavelengths for performing the measurement where the CO_2 absorption is even weaker (Table 1). With that approach, the light path for the sampling can be much longer, and still provide good measurement of carbon dioxide in the range of interest of ca. 360 ppm. For example, for the band at 1.57 μ where the absorption by CO_2 is much weaker (only 1/20,000th compared to that for 4.25 μ), one could in theory detect up to 1% of CO_2 over a light path as long as 1 km.

The cost for an open path instrument is about $50,000, and a whole detector system with multiple retro-reflectors could be as much as $150,000 [6]. Another vendor provides a cost estimate of about $50,000/ month to conduct a full-service detailed study of a point source problem (e.g. fumes from a dump site, see Ref. [7]). The design of a similar open-path instrument, but specifically designed only for duty as a carbon dioxide detector at storage sites may result in a less costly version of the technology. For carbon dioxide, the common approach for single point detectors is to use the very strong IR absorption band at 4.2 μm. For application as an open path detector, it is recommended that the wavelength of 1.57 ~ 1.60 μm be used. At this wavelength the absorption for carbon dioxide is quite low and largely free of interferences such as water vapor. With recent technology advances in the telecom and other electronic industries, it is conceivable that off-the-shelf parts could be assembled to build such a lower cost open-path instrument.

Solid-state chemical sensors

Based on the ionic reaction of $A^+ + OH^- + CO_2 = AHCO_3$ (A: Na or Li) in phosphate electrolyte, such sensors detect CO_2 level by measuring the potential between the chemical sensors' electrodes. Because of the specific chemical reaction, this type of sensor is very selective. Such detectors could have linear voltage response to the log of CO_2 concentration when the value changes from 100 ppmv to 5 vol.%. But, it is subject to water condensation and therefore not reliable [8–10]. For example, the reading of potential changed by as much as 25% when the water concentration goes up from 0.7 to 30 vol.% [8].

Based on semiconductor oxides' (e.g. $BaTiO_3$ and SnO_2) response to CO_2 it is shown that the sensors can exhibit very good linear response to the log of CO_2 concentration when the sensor is made of nanocrystalline materials [9]. But the long-term stability and signal drift of such sensors are still a problem for such detectors to become commercially available. For example, the nanocrystalline material changed its structure after several days, degrading sensitivity [9]. Micromechanical detectors sense the change of mass of a polymer, which in turn responds to CO_2. Such sensors are still in the developmental stage, as they also have water condensation and selectivity problems [10]. All the above chemical solid-state sensors could be made into very small inexpensive packages, but each sensor could only measure CO_2 at a single point.

Gas chromatography

Carbon dioxide may be measured easily to within a few ppm by standard gas chromatography methods. This is not used very much currently for atmospheric analysis, but it is a standard method for indoor air quality. OSHA uses this as a benchmark to compare against other proposed measurement techniques. Their concern of course is to determine worker exposure to CO_2. For more details, see, e.g. http://www.osha-slc.gov/dts/sltc/methods/inorganic/id172/id172.html.

Chemical reaction/visual indication
Another method to measure carbon dioxide in the ambient air is the so-called "Draeger tubes". The method of detection here is based on drawing in a fixed volume of air with a hand pump through a glass tube containing a granular packed material. The material inside reacts with the CO_2 brought in to create a color change. The concentration of CO_2 may be read from the length of the stain. These tubes come in a variety of concentration ranges in order to improve the accuracy of the measurement. The cost of each disposable tube is a few dollars.

CO_2 MONITORING PROGRAMS IN CURRENT SUBSURFACE (EOR) GAS INJECTION PROJECTS

Several operators of ongoing CO_2 injection projects were contacted for comments concerning current practices to monitor for atmospheric carbon dioxide concentrations. For ongoing industry projects where carbon dioxide is injected for enhanced oil recovery (EOR), monitoring for CO_2 seems to be a fairly low priority. In particular, for projects where H_2S is present in the gas streams along with the carbon dioxide, emphasis is placed on monitoring and preventing human exposure to leakages of the much more dangerous hydrogen sulfide. For example, at Chevron's Rangely Field in Colorado and Kinder Morgan's EOR project in Snyder, Texas, the operators are aggressive in guaranteeing that no person is exposed to even small releases of H_2S gas. Engineers we contacted at these companies said that state-of-the-art (a sensitive gas detector, remote data acquisition, and alarm system) H_2S detection schemes have been placed at selected critical points, with each unit costing of the order of $3000/installation.

These operators said there were minimal legal requirements for monitoring of CO_2 gas as it is considered a non-toxic substance. One engineer contacted at Kinder Morgan, a major producer of CO_2, said detectors typically are placed only at the highest risk points such as near compressors and perhaps in control rooms. Minimal steps are taken to monitor leaks by *chemical detection methods* from carbon dioxide pipelines transporting the gas to various oil industry EOR locations. Pipeline operators rely more on indirect indicators of pipeline leaks such as changes in flow and pressure readings. New CO_2 subsurface injection projects where the main motivation is for storage have paid more attention to monitoring issues, particularly measurements to detect the subsurface migration of injected carbon dioxide.

Summary of Remote Sensing Technology (i.e. NASA) for CO_2 Measurement
Key NASA projects concerning carbon dioxide monitoring have been focused at three NASA sites (JPL—Jet Propulsion Laboratory/Pasadena, CA; Langely/Hampton, VA; and Goddard Space Flight Center/Greenbelt, MD). NASA has an active research program to study the Earth's weather and atmosphere, and global warming and carbon cycle issues in particular. These and other related NASA research areas are of potential interest to the goals of monitoring CO_2 concentrations at carbon storage projects. These other projects include advanced laser and instrumentation methods, and also the study of carbon dioxide and the other components in the atmospheres of other planets. Indirect measurement techniques offer an interesting alternative approach, such as monitoring remotely for subtle changes in the flora at ground level. In fact, one project sponsored by the CCP investigated this concept [11]. Another indirect approach is remote surveys for detecting subtle changes in the surface deformation. These changes reflect movement of pressure changes subsurface associated with CO_2 injection [12].

One common opinion from NASA experts is that satellite monitoring (or that using very high altitude aircraft like a modified U-2) using spectrometers can scan for carbon dioxide over large areas [13,14]. One can resolve carbon dioxide concentrations in blocks perhaps as small as 100 m^2. If one averages over a larger area (such as a square mile) then the total measured concentration of carbon dioxide has improved accuracy. The disadvantage of these measurements is that they sample the entire air column. That is, typically these surveys provide carbon dioxide concentrations only in "two-dimensions". That is, they are not yet able to sample selectively in the third, vertical dimension, and focus their detection to just near ground level, which is of primary interest to this application. Increases in near-surface CO_2 levels due to leakage of injected gas might be detected, but increases in CO_2 in the upper atmosphere for other reasons also would be detected. That is, this approach might be subject to “false positives”. Thus satellites might be a good tool as a “screening” method to spot unusual changes in CO_2 levels; but those changes may not necessarily be at ground level nor related to activity at the storage project.

If there are any satellites that have the correct sensors and fly over the project area, then there may be the opportunity to have data on a quite frequent basis. One might be able to take advantage of already planned and funded NASA projects to collect data of interest. Low-level aircraft surveys are an alternative movable platform for more detailed remote measurement over a near ground-level carbon dioxide concentration. One can choose the exact area to perform such a survey, but the cost and logistics may make this impractical for frequent sampling.

As expected, NASA has its focus on interplanetary space exploration and high atmospheric research for Earth [14–16]. While perhaps not directly applicable for storage monitoring goals, improved laser detection and associated measurement research at NASA could prove useful. For example, projected research at JPL includes development of a superior InGaAsSb/GaSb laser that can detect spectra 2–5 μ. Intersubband Quantum Cascade lasers are being developed for Mars exploration. The wavelengths are in the range of 4–11 μ, typically with a power of 10–20 mW.

Vendor Products/Commercial Carbon Dioxide Detector

There are a number of commercially manufactured carbon dioxide detectors. Typically the detector itself is an NDIR (non-dispersive IR) type. The cost of just the detector can be less than $1000. Adding a visual readout or rudimentary data acquisition capability can increase the price to as much as $2000. A full gas sensor system rated as explosion proof can approach $4000 per installation. Most of these devices are intended as room gas monitors.

Advantages of these instruments are that they are relatively low cost and can indicate at least any large shift in the atmospheric CO_2 concentration. Their responses to changes in the CO_2 concentration are no more than a few seconds, and they have the capability to provide a continuous read out of results. A major limitation of these devices is that typically they will sample the atmospheric gases at one fixed point. Thus a great many commercial sensors would be required in order to cover a substantial area. This means the associated cost to collect the data in this network, plus process all the data, increases rapidly as the number of fixed sensors and the area covered increases. These associated installation costs likely would exceed the cost of the individual detectors.

The claimed accuracy of these instruments varies significantly. Some of the low-cost devices (around $500) are accurate to only $\pm 5\%$ of full scale. Other vendors claim their instruments can achieve an accuracy of $\pm 2\%$, or better. More expensive ones are accurate to $\pm 1\%$, and one vendor claims an accuracy to 1 ppm or better.

Portable (hand-held) detectors are appropriate for personal protection as there is some mild health concern with people being exposed to high levels of carbon dioxide gas. Workers who are in the project area on a regular basis should have access to devices before entering any higher risk area. The resolution of these devices is typically no better than 100 ppm. This is sufficient accuracy if the main purpose is just to verify that the local CO_2 concentration does not pose a health concern. These portable meters commonly are less than $1000 each, and most use IR detection.

Quantitative Analysis of Capabilities for Detecting CO_2 in Ambient Air

First, consider the scenario where the carbon dioxide leak occurs uniformly over a substantial area (say over a square kilometer or more). The graphs below illustrate changes that would occur in the added concentration of carbon dioxide to the background levels, under different assumptions. These calculations presume gas injection for 20 years at a rate of 10 million cubic meters/day (basis of 1 atm and 15 °C), and that 1% of that total injection gas then does leak. Figures 4 and 5 illustrate that either decreasing the surface area or the time duration over which the leakage of carbon dioxide occurs increases its concentration near ground level.

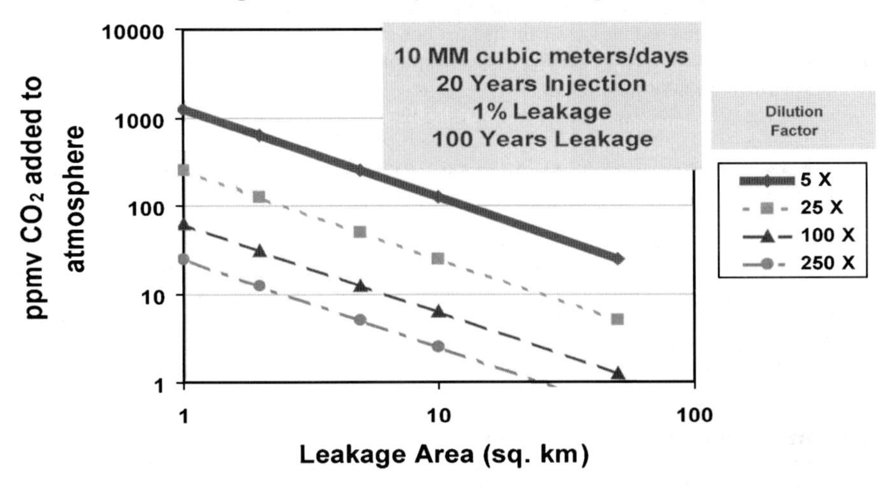

Figure 4: Example calculation results illustrating that the added carbon dioxide concentration to the near ground-level atmosphere increases with a decrease in the leakage area.

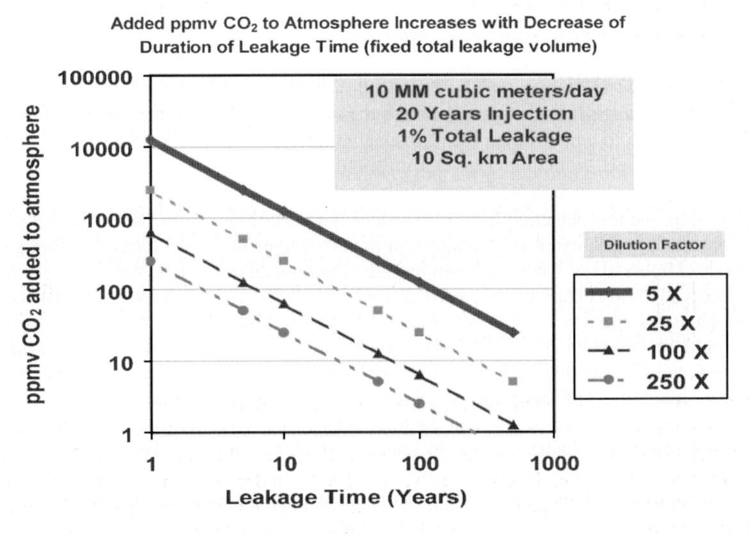

Figure 5: Example calculation results illustrating that the added carbon dioxide concentration to the near ground level atmosphere increases with a decrease in the leakage time.

It is not surprising from the above graphs that there are scenarios where the local addition of CO_2 either may or may not be easily detectable. For example, Figure 6 illustrates a scenario where the local additional CO_2 concentration approaches 25 ppmv. This is substantially greater than the uncertainty associated with variations in the background and measurement fluctuations in this illustration. The next example (Figure 7) is a contrary case where the increase in added CO_2 concentration would be difficult to detect as it is no more than the measurement uncertainties.

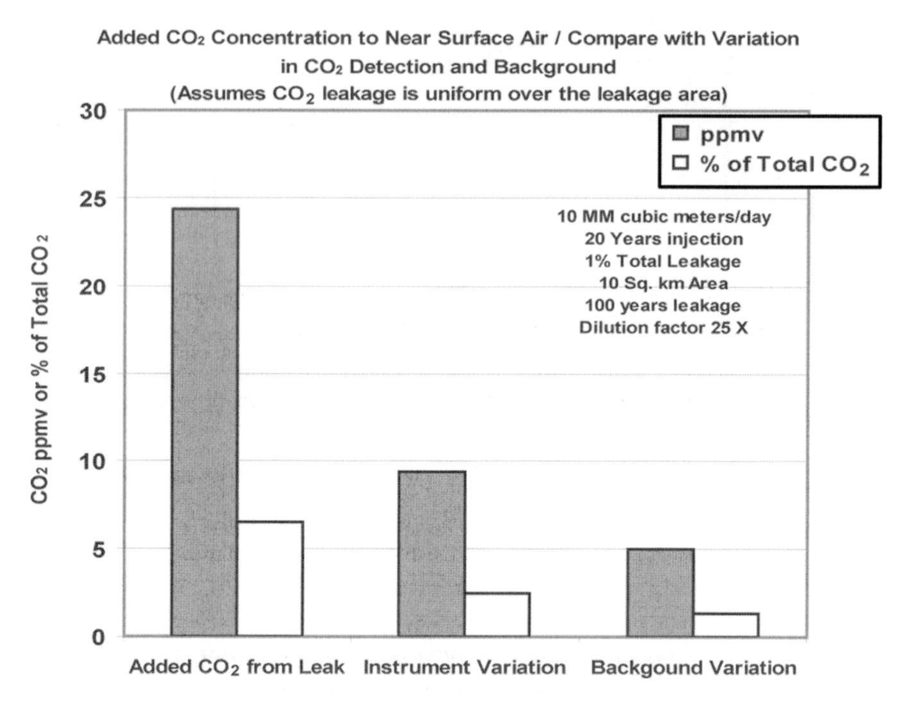

Figure 6: Example calculation result illustrating a scenario where the increase in the carbon dioxide concentration from a leak is substantially above the uncertainties of the measurement and the background concentrations.

One way to improve this detection limit is to reduce the uncertainty in the background noise. This could be accomplished by careful, extensive background measurements of carbon dioxide before gas injection over time periods of hours, days, weeks, or even months (interannual differences can be large). Incorporating the measurements methods detailed here, vertical profiles of CO_2 from towers combined with micrometeorological techniques could be used to determine CO_2 fluxes. Some of the established measurement network stations could be of useful (e.g. the Global Atmospheric Watch network of the World Meteorological Organization, the Fluxnet flux stations). From such careful background data one could quantify better and account for this source of uncertainty. In any case, it is good engineering practice to establish the background responses of the instrument package under field conditions selected before the initiation of carbon dioxide injection. Another tact is to select a detector to improve the accuracy of the measurement.

The other scenario considered is when the leak occurs at a point. Examples include a localized leak with gas coming up a wellbore or leaks from a piece of faulty surface equipment. Figures 8 and 9 are calculated results of the profile of CO_2 concentration from point source leaks versus the distance away, directly downwind. These examples show a very wide range of responses. Note that Figure 8 considers the case

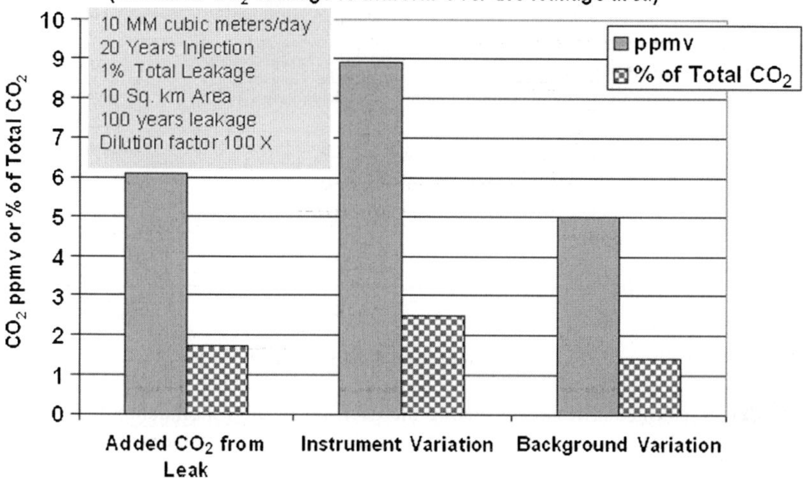

Figure 7: Example calculation result illustrating a scenario where the increase in the carbon dioxide concentration from a leak is substantially below the uncertainties of the measurement and the background concentrations.

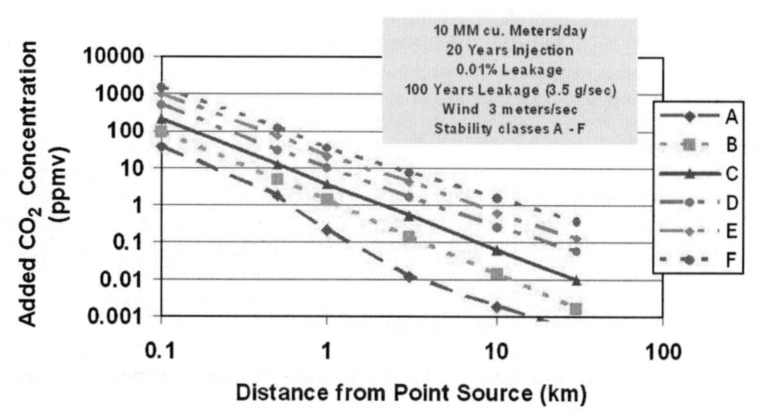

Figure 8: Example calculation result illustrating the decrease in the concentration of CO_2 versus the distance downwind. Results are shown for different atmospheric stability conditions and for a leakage of 0.01% of injected carbon dioxide over a 100 year period.

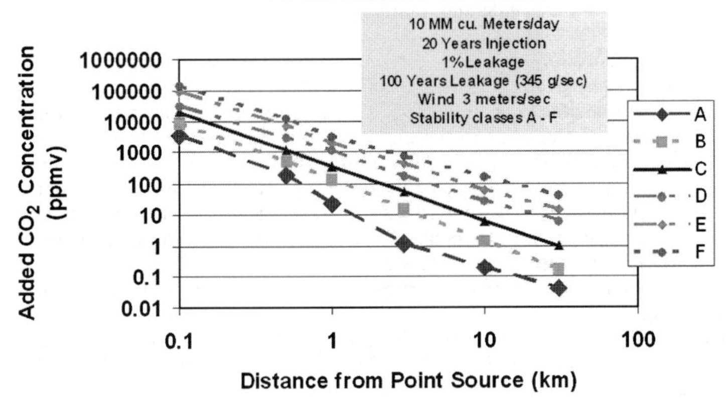

Figure 9: Example calculation result illustrating the decrease in the concentration of CO_2 versus the distance downwind. Results are shown for different atmospheric stability conditions and a leakage of 1% of injected carbon dioxide over 100 years.

where the CO_2 leakage is 0.01% of the total injected CO_2. In contrast, Figure 9 presents similar calculations, but now for the scenario of 1% of the total injected gas escapes at a single point leak over 100 years. Not surprisingly, the concentration of *added* CO_2 (above the background level) versus distance from the point source is much greater for the latter case with the 100-fold greater leakage rate. In both figures we show that the atmospheric conditions can have a substantial effect. As expected, as one increases the stability of the atmospheric conditions, the increase in CO_2 centerline concentration increases.

Table 4 below compares the distance from the point source of the leakage where the concentration of the added carbon dioxide falls to 10 ppmv. At these distances and closer, CO_2 concentrations are high enough so it is likely that many commercial detectors located downwind would determine there is a leak. These results emphasize changing the atmospheric conditions causes a wide variation in the calculated results. At one extreme of a relatively small leak and unstable atmospheric conditions, the distance is significantly less than a kilometer. At the other extreme of a large point leak scenario and stable atmospheric conditions, the concentration of added carbon dioxide can persist above 10 ppmv for several kilometers.

TABLE 4
DISTANCE (KM) DOWNWIND OF A POINT SOURCE OF CARBON DIOXIDE LEAKAGE WHERE ADDED CONCENTRATION FALLS TO 10 PPMV FOR DIFFERENT DISCHARGE RATES

		Pasquill stability class					
Percent of injected	**Leakage rate (g/s)**	**A (km)**	**B (km)**	**C (km)**	**D (km)**	**E (km)**	**F (km)**
0.01	3.5	0.2	0.35	0.5	1	1.7	2.3
0.1	34.5	0.7	1.3	2.5	5	7.5	15
1	345	1.2	3	8	22	37	80

CONCLUSIONS

From this literature review we conclude the following.

1. A suite of different types of CO_2 detection methods are available to acquire atmospheric data to verify the integrity of a subsurface injection project from small to large area.
2. A newer concept for ground-level measurement of carbon dioxide measurement is to use an open-path instrument. These detectors respond to the CO_2 concentration, averaged over the entire sample path length. With this arrangement, a single laser instrument could sample several directions, thereby covering a wide area. This concept could be more efficient than using a large network of commercial, single-point detectors to measure carbon dioxide concentration at ground level over a large storage project area. Existing open path instruments are relatively expensive, but costs might be reduced if a customized device is constructed that only need measure carbon dioxide.
3. Regarding the calculations of required performance of ground-level instruments to identify leakages from the subsurface:
   - Key factors that determine the increase in the ground-level carbon dioxide concentrations include (1) the total mass amount of CO_2 leakage, (2) the leakage surface area and duration of the event, and (3) atmospheric effects that dilute the influx of added CO_2.
   - Calculations suggest a leakage of just 1% of the total carbon dioxide injected could add tens of ppmv of this gas to the local air environment if the leak occurs uniformly over a few square kilometers or smaller area and/or in a time period of several months or shorter. Such leaks would be identified if the detector is in close proximity.
   - For leakages emanating from a point source, the CO_2 concentration downwind of the leak increases with an increase in the (1) mass rate of discharge, (2) stability of atmospheric conditions, and (3) proximity of the sensor to the leak. Example calculations illustrate the atmospheric conditions alone can change by an order of magnitude the distance from which a sensor can recognize a leak is occurring.

RECOMMENDATIONS

Development should be encouraged for less expensive long, open-path instruments to measure CO_2 in the ambient air. Potentially, a single such laser device could sample a radius of several square kilometers. Such a device would have the distinct advantages of (1) continuous monitoring, (2) accuracy to within a couple of percent and (3) remote and unattended operation.

Further discussions are encouraged with NASA with regards to their research activities and plans for monitoring greenhouse gases. NASA has several separate research efforts that bear directly or indirectly on the CO_2 monitoring requirements for geologic storage.

Track future developments in laser/detection technology because improvements in this hardware can aid in creating more cost-effective CO_2 measurement devices.

Use ongoing CO_2 storage project sites and oil field injecting CO_2 for EOR as test beds to evaluate and further develop these CO_2 monitoring concepts. Also natural sites where there are elevated CO_2concentrations (e.g. volcanic activity) are candidate field evaluation sites.

Track further developments in laser spectroscopy technology that can measure in real time carbon and oxygen isotopes; such data could serve as tracers for the fate of transported or injected CO_2. This approach would complement the ongoing CCP supported project that is evaluating isotopic analysis of noble gases as a tracer for gas migration in storage projects.

ACKNOWLEDGEMENTS

The authors wish to acknowledge the support of the CO_2 Capture Project for this work.

REFERENCES

1. T.J. Conway, P.P. Tans, L.S. Waterman, K.W. Thoning, D.R. Kitzis, K.A. Masarie, and N. Zhang, *J. Geophys. Res.* **99D** (1994) 22831–22855.
2. Workbook of Atmospheric Dispersion Estimates, from EPA, web site: http://www.waketech.edu/math/projects/Weston/Dispersion.htm.
3. Air Pollution Workbook, Atmospheric rate modeling, web site: http://homepages.utoledo.edu/aprg/courses/iap/TEXT/workbook/chap4.html.
4. M.E. Webber, S. Kim, S.T. Sanders, D.S. Baer, R.K. Hanson, Y. Ikeda, *Appl. Optics* **40** (6) (2001) 821–828.
5. D.M. Sonnenfroh, M.G. Allen, *Appl. Optics* **36** (15) (1997) 3298–3300.
6. E. Thorton, N. Bowmar, The application of a laser based open-path spectrometer for the measurement of fugitive emissions and process control, presented at A&WM Association Conference, Raleigh, NC, 28 October, 1999.
7. Minnich and Scotto, web site: use of open-path FTIR spectroscopy during site remediation, www.msiair.net/openpathspec.html.
8. N. Imanaka, *Sensor. Actuat. B-Chem.* **24** (1995) 380–382.
9. P. Kellera, H. Ferkelb, K. Zweiackera, J. Naserb, J.-U. Meyera, and W. Richemannb, *Sensor Actuat B-Chem.* **57** (1–3) (1999) 39–46.
10. Q.Y. Cai, A. Cammers-Goodwin, C.A. Grimes, *J. Environ. Monitoring* **2** (6) (2000) 556–560.
11. W.L. Pickles, Geobotanical hyperspectral remote sensing, presented at the *CCP Carbon Sequestration Conference*, Potsdam, Germany, 30 October–1 November, 2001.
12. H.A. Zebker, J. Harris, Monitoring the injection and storage of CO_2 using satellite radar interferometry, presented at the *CCP Carbon Sequestration Conference*, Potsdam, Germany, 30 October–1 November, 2001.
13. Abshire, James, Personal Communication, NASA, Goddard Space Center, Space Born Laser Development.
14. R.W. Carlson, A tenuous carbon dioxide atmosphere on Jupiter's moon Callisto, JPL Report, 3 November, 1998.
15. G. Vane, Airborne visible/infrared imaging spectrometer (AVIRIS), JPL Publication 87-38, 15 November, 1987.
16. M.A. Vincent, S.S. Saatchi, Comparison of remote sensing techniques for measuring carbon sequestration, JPL Publication, February, 1999.

Carbon Dioxide Capture for Storage in Deep Geologic Formations, Volume 2
D.C. Thomas and S.M. Benson (Eds.)

Chapter 21

DETECTING LEAKS FROM BELOWGROUND CO_2 RESERVOIRS USING EDDY COVARIANCE

Natasha L. Miles[1], Kenneth J. Davis[1] and John C. Wyngaard[2]

[1]Department of Meteorology, The Pennsylvania State University, University Park, PA, USA
[2]Departments of Meteorology, Mechanical Engineering, and GeoEnvironmental Engineering, The Pennsylvania State University, University Park, PA, USA

ABSTRACT

We describe the eddy covariance method of measuring earth–atmosphere CO_2 exchange, including past applications to measurements of volcanic venting of CO_2. The technique involves continuous atmospheric measurements of both CO_2 mixing ratio and atmospheric winds from a tower platform. Equipment is robust and commercially available, and the methodology is well established.

The surface area covered by the measurement is described. The upwind coverage is typically $(10-100)z_m$, where z_m is the measurement height, and the cross-wind extent of this area is of the order of the upwind distance. Thus, a 10-m high tower detects fluxes from an upwind distance of 100–1000 m, and an area of order 10^4-10^6 m^2. The eddy covariance method yields continuous measurements of earth–atmosphere exchange over such areas, typically expressed as averages over hourly or half-hourly time periods. The area measured depends on wind speed, wind direction, surface roughness, and stability of the atmospheric surface layer. The measurement works best under well-mixed atmospheric conditions which frequently occur on a daily basis, often for a majority of the day.

We assess the ability to detect leaks from geologic CO_2 reservoirs by comparing expected leakage rates to typical ecological flux rates. While the character and magnitude of ecological fluxes are well established, reservoir leakage rates and areas are uncertain. Fairly conservative estimates based on ensuring the economic viability of CO_2 storage are constructed. Our estimates of leakage rate and area yield leakage fluxes that range from 1 to 10^4 times the magnitude of typical ecological fluxes. The flux measurement areas readily encompass the assumed leakage areas ($10-10^5$ m^2). We conclude that this approach shows promise for the monitoring of belowground CO_2 storage. Leak detection is shown to be a simpler problem than leak quantification, but both can in principle be accomplished using eddy covariance under conditions favorable for the measurement.

INTRODUCTION

Eddy covariance is a possible method to monitor for economically undesirable and potentially dangerous CO_2 leaks from CO_2 storage reservoirs. Although eddy covariance is relatively new to the geologic community, it has been used extensively in the meteorology and ecology communities to study CO_2 exchange between vegetation and the atmosphere [1–4]. The technique has recently been applied successfully to volcanic regions [5–8]. In this chapter, we describe the eddy-covariance method and evaluate its ability to detect leaks from deep aquifers.

EXPERIMENTAL/STUDY METHODOLOGY

Basic Principles of the Eddy-Covariance Technique

The derivation presented here follows work previously published by Yi et al. [9]. The conservation equation for CO_2 in the atmospheric boundary layer can be written as

$$\frac{\partial c}{\partial t} + u\frac{\partial c}{\partial x} + v\frac{\partial c}{\partial y} + w\frac{\partial c}{\partial z} = S_C, \tag{1}$$

where c is the CO_2 mass density (kg CO_2 m^{-3} air), S_C is a source or sink of CO_2 in the atmosphere (kg CO_2 m^{-3} s^{-1}), u and v are wind speeds (m s^{-1}) in the horizontal (x, y) plane, w is the wind speed in the vertical z direction, and t represents time. Molecular diffusion, insignificant for atmospheric transport at spatial scales greater than ~1 mm [10] has been neglected. While oxidation of hydrocarbons and CO does lead to production of CO_2 in the atmosphere [11] this has a characteristic time scale of weeks to months and can be ignored over the time scales of turbulent eddies in the atmosphere (seconds to minutes, Ref. [10]); thus, we set $S_C = 0$. Further, we apply Reynolds decomposition and averaging in combination with the turbulent continuity equation and align the x-coordinate along the mean horizontal wind to obtain

$$\frac{\partial \bar{c}}{\partial t} + \bar{u}\frac{\partial \bar{c}}{\partial x} + \bar{w}\frac{\partial \bar{c}}{\partial z} + \frac{\partial \overline{u'c'}}{\partial x} + \frac{\partial \overline{v'c'}}{\partial y} + \frac{\partial \overline{w'c'}}{\partial z} = 0, \tag{2}$$

where the overbar represents the ensemble-averaged mean and the prime terms represent fluctuations about the mean. In practice, time-averages of point time-series data are used in place of ensemble averages. We integrate from the surface $(z = z_0)$ to the altitude of a sensor $(z = z_m)$ and obtain

$$\int_{z_0}^{z_m} \left(\frac{\partial \bar{c}}{\partial t} + \bar{u}\frac{\partial \bar{c}}{\partial x} + \bar{w}\frac{\partial \bar{c}}{\partial z} + \frac{\partial \overline{u'c'}}{\partial x} + \frac{\partial \overline{v'c'}}{\partial y} \right) dz + \overline{w'c'}_{z_m} = \overline{w'c'}_{z_0}. \tag{3}$$

The term on the right-hand side of Eq. (3) is the flux of CO_2 at the Earth's surface, F_0. The last term on the left-hand side, the covariance of turbulent fluctuations in the vertical wind and the CO_2 density, is the turbulent flux of CO_2 measured at some height above the surface. With negligible net longitudinal and lateral (mean and turbulent) transport, and negligible mean vertical velocity, Eq. (3) simplifies to

$$\int_{z_0}^{z_m} \frac{\partial \bar{c}}{\partial t} dz + \overline{w'c'}_{z_m} = F_0. \tag{4}$$

F_0 (the surface flux of CO_2) is known as the net ecosystem–atmosphere exchange (NEE) in the ecological literature. It is the sum of the turbulent flux of CO_2 across a horizontal plane above the plant canopy and the rate of accumulation of CO_2 below the plane. This is illustrated in Figure 1. The assumption of zero net lateral transfer is generally satisfied when atmospheric turbulence is moderate to vigorous (e.g. sunny and/or windy days), but is often violated in very calm conditions (e.g. cold, clear, calm nights). Extensive evaluation of these assumptions exists in the micrometeorological literature [9, 12–17]. When the atmospheric surface layer is unstable, the accumulation of CO_2 near the Earth's surface is negligible, and the surface–atmosphere exchange rate from Eq. (4) is

$$\overline{w'c'}_{z_m} = F_m = F_0, \tag{5}$$

where F_m is the flux of CO_2 at the measurement height. This flux measurement method is commonly referred to as eddy covariance (hereafter EC).

An example of data used to compute the vertical flux of CO_2 over the averaging time (typically 30 min or 1 h (e.g. Refs. [18,19]) is shown in Figure 2. The data were collected over a forest at midday in the summer. Both positive deviations in CO_2 during downdrafts (e.g. at 17–18 min) and negative deviations in CO_2 during updrafts (e.g. at 13–14 min) contribute to negative flux values. The hourly mean turbulent flux in the example is -0.21 ppm m s^{-1} (-7.2 μmole C m^{-2} s^{-1} or 3.1×10^{-7} kg CO_2 m^{-2} s^{-1}). The negative sign means that, on average, turbulent eddies transport CO_2 towards the Earth's surface, where CO_2 is consumed by photosynthesis. This example also illustrates the variety of units used to describe CO_2 fluxes. We present units of both mass and molar flux when possible, since both are fairly common.

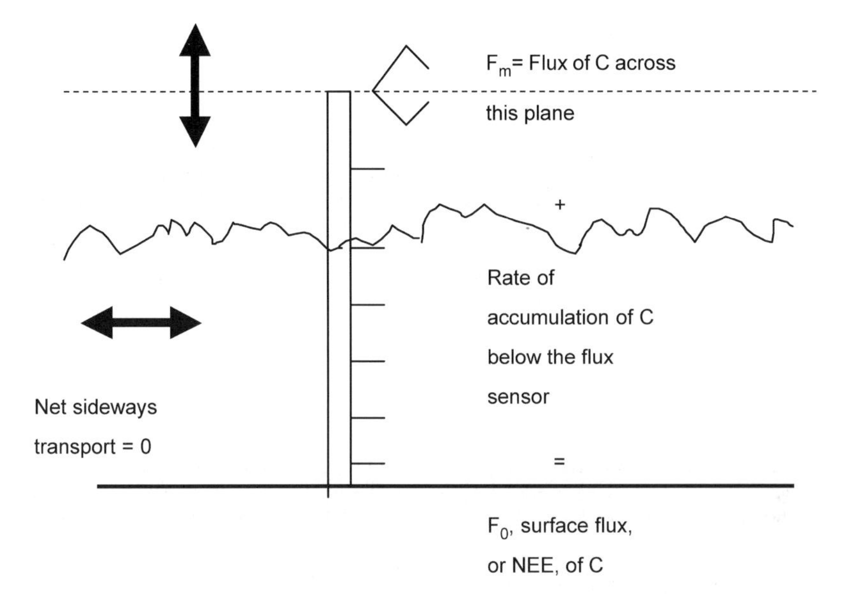

Figure 1: Schematic diagram of the eddy-covariance method of measuring the surface flux F_0 or net ecosystem–atmosphere exchange (NEE) of a scalar such as CO_2. An idealized instrumented tower and flux measurement sensor that rises a height z_m above the Earth's surface is shown.

Molar flux units (μmole C m^{-2} s^{-1}) are most common in the ecological literature. Molar mixing ratios (moles CO_2 per million moles dry air, ppm) are common units in studies of atmospheric composition and transport. Further eddy covariance examples can be found in Ref. [10].

Area Represented by EC Flux Measurements: The Flux Footprint

An EC measurement captures fluxes corresponding to surface areas upwind of a tower, with areas closer to the tower being weighted more heavily. The per unit contribution to surface flux (either a positive or negative flux) and area of each element of upwind surface to the flux at a given point downwind is called the "flux footprint" [20–22].

The mass conservation equation of a diffusing material in the atmosphere is linear, which gives it the attractive mathematical property of superposable solutions. This allows multiple sources of CO_2 to be treated by the superposition of the solutions for individual sources. It also enables a spatially distributed source on the surface to be treated as the superposition of a number of individual point sources. Horst and Weil [20] used this superposition property to rigorously define a flux footprint function f that through a convolution integral relates $F_m(x_m, y_m, z_m)$, the vertical turbulent flux of the diffusing material measured at position (x_m, y_m, z_m), to $F_0(x, y, 0)$, the upwind spatial distribution of its surface flux:

$$F_m(x_m, y_m, z_m) = \int_{-\infty}^{\infty} \int_{-\infty}^{x_m} F_0(x', y', 0) f(x_m - x', y_m - y', z_m) \mathrm{d}x' \, \mathrm{d}y', \tag{6}$$

where x' and y' are dummy variables. This equation indicates that the measured flux at height z_m is the integral of contributions from all upwind surface elements; the flux footprint f gives the weighting of each elemental surface flux. In addition to environmental factors such as wind speed, wind direction, surface roughness, and stability, f depends on both the height z_m at which the downwind flux is measured and the upwind position on the surface. It is conventional to assume that the turbulent flow is horizontally homogeneous, so that the footprint function depends only on the separation between the measurement point

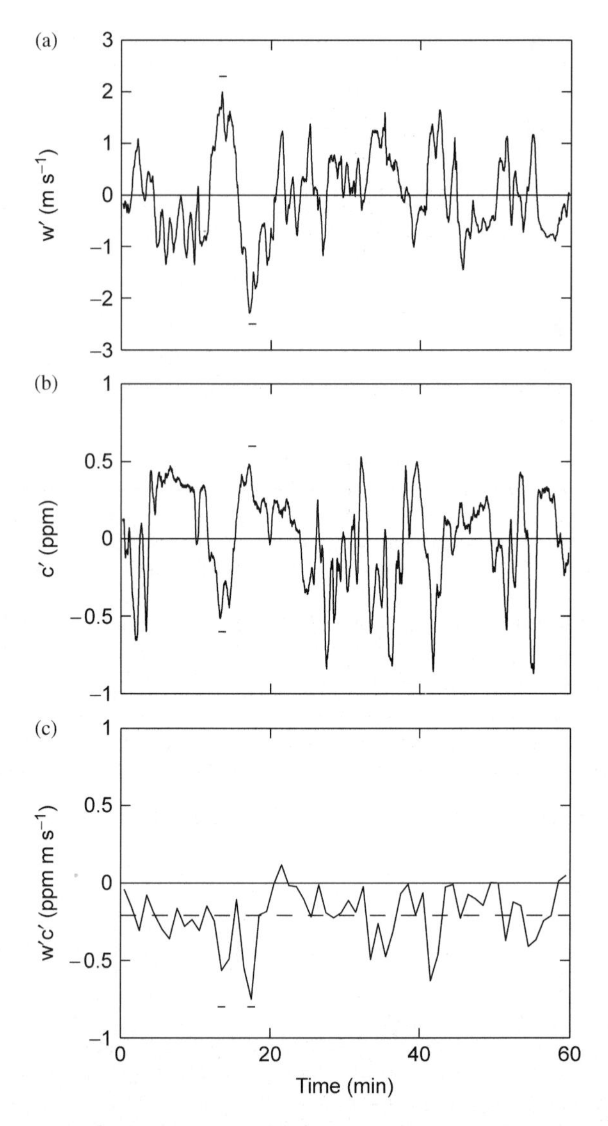

Figure 2: Example of 1 h of data measured at 122 m on a tower in northern Wisconsin during the afternoon on June 15, 1999. 30-s averages of (a) deviations from the mean vertical velocity and (b) deviations from the mean CO_2 concentration. (c) 1-min averages of the eddy covariance. In each panel, small horizontal lines indicate the times corresponding to the examples of an updraft and downdraft described in the text. The mean EC for the hour in this example is -0.21 ppm m s^{-1} (shown as a dashed line in (c)). 1 ppm $CO_2 = 1.5 \times 10^{-6}$ kg CO_2 at a typical air density for the Earth's surface (1 kg air m^{-3}).

and each elemental piece of upwind surface. With the mean wind in the x-direction, the streamwise separation is $x_m - x'$ and cross-wind separation is $y_m - y'$, as indicated in Eq. (6). When z_m is in the surface or "constant-flux" layer, the integral of f over all upwind surface area is 1. For cases where the surface flux F_0 is uniform in space, Eq. (6) simplifies to $F_m = F_0$.

Horst and Weil [20] showed that f can be interpreted as the solution to a point-source problem. If the upwind surface flux is produced by a point source of emission rate Q (mass/time) at position $(x_s, y_s, 0)$, so that

$$F_0(x', y', 0) = Q\delta(x' - x_s)\delta(y' - y_s), \quad (7)$$

then Eq. (6) becomes

$$F_m(x_m, y_m, z_m) = f(x_m - x_s, y_m - y_s, z_m)Q, \quad (8)$$

and

$$f(x_m - x_s, y_m - y_s, z_m) = \frac{F_m(x_m, y_m, z_m)}{Q}. \quad (9)$$

Thus, the footprint function f at a point (x_s, y_s) on the surface upwind can be interpreted as $F_m(x_m, y_m, z_m)$, and the flux at the downwind measurement point, divided by Q, the strength of the point source on the surface at the upwind point (x_s, y_s).

There is no known way to find solutions for statistical properties such as f from the equations governing turbulent flow; any such calculations require that the equations can be approximated in some way before they are solved [20]. Horst and Weil [20] have done such approximate calculations for the footprint function over a range of meteorological conditions in the surface layer. The evidence to date [7,23] suggests these calculations are reliable to at least within a factor of two in typical field conditions.

The lateral extent of the flux footprint f, the area monitored by an EC measurement, is approximately $4\sigma_v x_u/U$, where σ_v is the root mean square lateral wind velocity, x_u is the upwind extent of the footprint, and U is the mean wind speed [8,23]. This width is typically roughly equal to x_u, the upwind extent. The upwind extent is affected by both the measurement height above the surface z_m and the atmospheric stability; typical values of the upwind extent of the flux footprint range from $(10-100)z_m$, depending strongly upon atmospheric stability. A maximum upwind extent is of order 10 km for a very tall tower [4, 24]. Airborne EC can be used to estimate flux from very large regions [25,26], but only for a short time.

Examples of the upwind extent of the footprint for a 20-m tower as a function of atmospheric stability are shown in Figure 3. Unstable atmospheric conditions correspond to very convective conditions, i.e. strong sunlight and a large rate of buoyant production of atmospheric turbulence, and in general a well-mixed atmosphere. In unstable conditions the footprint function has a smaller spatial extent, meaning that fluctuations in mixing ratio are rapidly homogenized and the flux measured at the tower is influenced by areas closer to the tower. Neutral atmospheric stability corresponds to conditions when wind shear is a dominant source of atmospheric turbulence (e.g. an overcast day). Stable atmospheric conditions represent conditions governed by air near the Earth's surface that is colder than air aloft, as can occur through net radiative cooling of the Earth's surface. In neutral and stable conditions, vertical mixing is weak, and mixing-ratio fluctuations are transported long distances before becoming homogenized by turbulence. When cooling is very strong and winds weak, lateral flows can become strong and traditional application of the EC method becomes problematic. Atmospheric stability in the surface layer is quantified via a parameter known as the Monin–Obukhov length [10] and is readily estimated operationally by basic observations such as incoming solar radiation and wind speed [27].

Application of Eddy Covariance to Volcanic Regions

Although EC has been used extensively in meteorological and ecological applications, recent work applying it to volcanic regions (e.g. Refs. [5–8]) is more relevant to detection of stored CO_2 leaks. In most ecological applications, the source or sink of CO_2 is assumed to be homogeneous across the Earth's surface. In volcanic applications, CO_2 fluxes are often spatially heterogeneous. The use of the method can be further complicated by significant topography and large surface heat fluxes. Nevertheless, EC measurements have compared well with chamber measurements under a broad range of atmospheric conditions (e.g. Refs. [5–8]). Emission of CO_2 from a volcanic area is analogous to a distribution of leaks from a belowground CO_2 reservoir. The measured EC flux is the convolution of the surface fluxes and the footprint function, Eq. (6), as shown in Figure 4 for a specific example [6]. Werner et al. [8] calculated that EC could be used

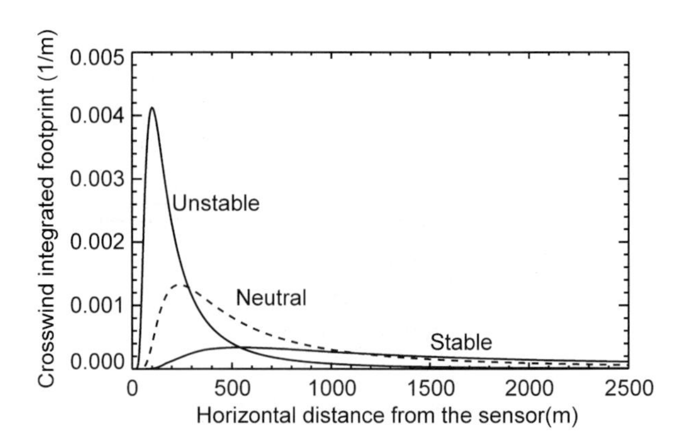

Figure 3: Cross-wind (y-direction) integrated footprint function f for a 20-m tower as a function of upwind distance (x) for different atmospheric stabilities. The calculation is based on Horst and Weil, [20], and assumes a surface roughness of 0.1 m and a displacement height of 0 m. The upwind distance plotted here scales roughly linearly with the measurement height.

to detect even a small volcanic eruption or a slow volcanic leak. We shall extend this approach to detection of leaks from geologic storage after describing the instrumentation.

Typical Instrument Setup

EC flux measurements in the atmospheric surface layer require instruments with fast time-response (10-Hz measurement frequency is typically sufficient) and good precision. The most common sensors used for EC measurements of CO_2 fluxes are infrared gas analyzers, such as those made by LI-COR, Inc., Lincoln, NE, USA. Both open-path (e.g. LI-COR 7500) and closed-path (e.g. LI-COR 7000) instruments have been used for EC measurements. The instruments are robust and stable for long periods of time (months to years) and relatively easy to deploy. Periodic calibration with gases of known CO_2 mixing ratio is required, though this can be done quite infrequently (e.g. monthly) as absolute accuracy in the mixing ratio measurements is not required. For closed-path systems, a reference gas is required for leak detection. This can be either a gas with a known CO_2 mixing ratio (differential mode) or a non-absorbing gas such as N_2 (absolute mode). It is also necessary that air be pumped relatively rapidly through the cell to ensure sufficient time-response at the desired measurement frequency. Long-term application of closed-path infrared gas analyzers for flux measurements is described by several authors (e.g. Refs. [19,28]). Open-path measurements are also common in the CO_2 flux literature (e.g. Ref. [8]).

Also required for EC flux measurements is a sonic anemometer (e.g. Campbell Scientific Inc., Model CSAT3, Logan, Utah) to measure the vertical velocity. This instrument measures orthogonal (component) wind speeds and sonic temperature which can be converted to virtual temperature by determining the time of flight of sound between pairs of ultrasonic signal transducers. Since the CO_2 and wind sensors are not perfectly co-located, there is often a small lag in time between the two data streams. By maximizing the correlation coefficient of w' and c', the lag between the signals can be determined [7,19] and EC fluxes can be computed.

A typical data recovery rate for a flux tower in the AmeriFlux network is 70%, including losses due to instrument failure and exclusion of data during periods in which vertical mixing is very weak (e.g. Ref. [29]). Data exclusion is more frequent at night when the atmosphere is typically stable (as a result of radiational cooling from the Earth's surface) and thus mixing is weak. Long data gaps can be avoided with periodic instrument maintenance.

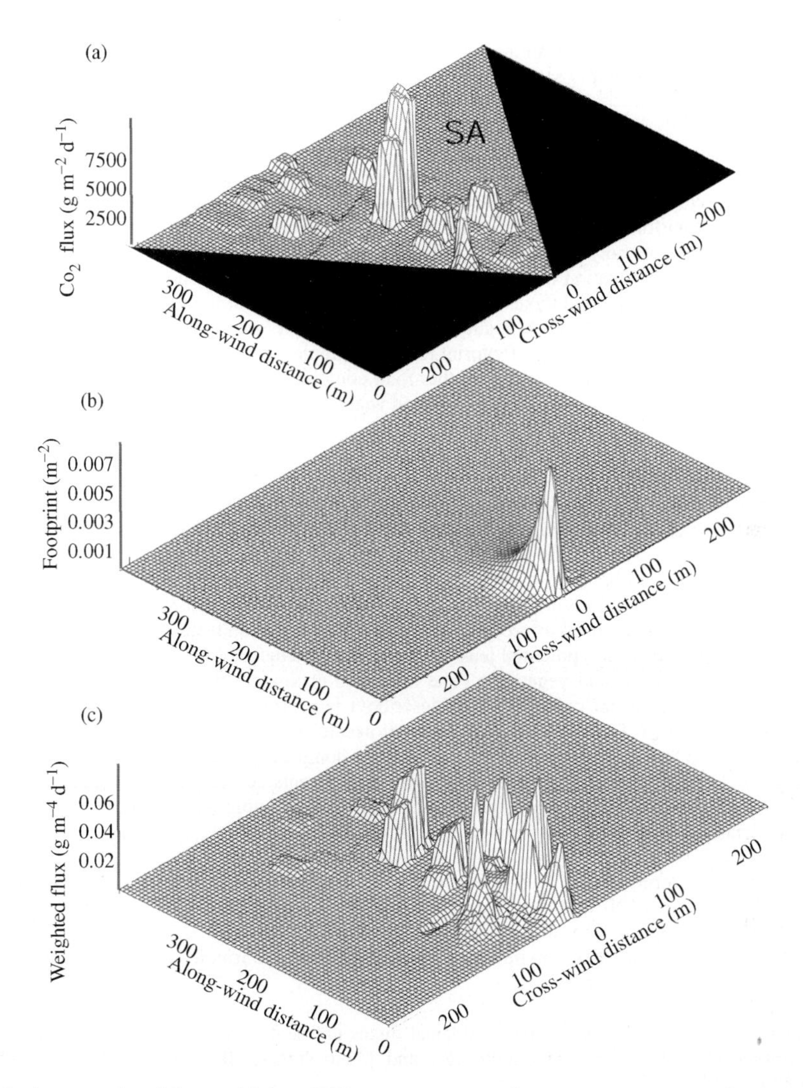

Figure 4: An example of the sensitivity of EC measurements to heterogeneous sources distributed within the EC flux footprint, reprinted from Ref. [6] with the permission of the author. (a) CO_2 flux distribution as measured by chambers in volcanic area in Yellowstone National Park. The source area (SA) contributing to the flux measured at a 2-m tower located at $x = 0$, $y = 0$ is also shown. 1 g CO_2 m^{-2} d^{-1} = 0.26 μmole C m^{-2} s^{-1}. (b) Flux footprint for the 2-m tower for a moderately unstable atmosphere. (c) Weighted flux, a convolution of the flux footprint and the flux distribution. An integral of the weighted flux over the surface yields the observed EC flux at the 2-m tower (Eq. (6)).

LEAK DETECTION

Leak detection can be accomplished by establishing background fluxes for a site, then continuously monitoring the site for significant deviations from these background fluxes. This is a significantly different problem than that of measuring long-term NEE of CO_2, leading to differences in site selection criteria and treatment of missing data. Quantification of the leak can be attempted with multiple flux measurements using one or more measurement systems. Details of this overall approach follow.

Site Selection

Site selection for belowground CO_2 storage depends primarily on geology since a deep storage formation is required. Also, a remote location is preferable, allowing for time to react to a leak, as well as avoiding anthropogenic sources of CO_2 (such as those from power plants, nearby roads, etc.). While flat terrain and an extensive fetch of uniform vegetation are important in order to precisely measure the magnitude of fluxes [30], for leak detection we only need to detect changes and thus do not have such terrain and vegetation requirements. The characterization of background fluxes described below assumes a uniform fetch where most of the variance in background fluxes is described by parameters that influence ecological metabolism, such as temperature and sunlight. A highly heterogeneous site may require further segregation of background fluxes according to the flux footprint (e.g. in a simple case, dividing background flux data into a small number of distinct wind directions). Other than complicating the characterization of background fluxes, however, a non-ideal site in terms of terrain and vegetation cover does not prevent the application of EC to the problem of leak detection.

Background (Ecological) Fluxes

A first step in leak detection is the establishment of background (ecological) CO_2 fluxes for the area near a CO_2 belowground reservoir. Predicting the range of variability in ecological NEE of CO_2 is necessary if the area-integrated flux (Eq. (6), Figure 4) from a hypothesized leak is not significantly larger than the ecological background flux. In order to obtain continuous datasets of NEE for ecological studies, methods based on environmental conditions are currently employed to "gap fill" the missing data (e.g. Refs. [4,29]), and similar techniques can be used to predict environmental (background) fluxes for the purposes of leak detection. EC measurements at a potential leak site must be made before CO_2 injection, or at a second site with a similar flux footprint and vegetation. The measured ecological fluxes can then be characterized as a function of environmental conditions that describe a large fraction of the variance in ecosystem–atmosphere CO_2 exchange. The resulting parameterization can be used with measurements of radiation and temperature to create "modeled" fluxes which can then be compared to ongoing EC measurements at the site where leak detection is required. Measured fluxes that lie outside the range of natural variability, as described by the "gap-filling" functions, can be established as possible leaks (Figure 5). We shall now describe the details of establishing a parameterization for NEE.

In systems without underground sources of CO_2 (i.e. lacking both volcanic activity and leaky underground storage), the surface flux, or NEE of CO_2, depends primarily on temperature, light, and the amount of green vegetation. Hourly ecological CO_2 fluxes are typically within the range of ± 20 μmole C m^{-2} s^{-1} during the growing season, with winter-season fluxes being much smaller (± 2 μmole C m^{-2} s^{-1}) in regions where snow and ice are common [4,30,31].

An established method [4] for predicting ecological fluxes is based on well-documented [32,33] soil and plant responses to soil or air temperature (T) and photosynthetically active radiation (PAR). The equation

$$\mathrm{NEE} = a_0 \mathrm{e}^{a_1(T-a_2)} + b_2 - \frac{b_0 \mathrm{PAR}}{(\mathrm{PAR} + b_1)} \tag{10}$$

can be fitted to measurements of T, PAR, and NEE of CO_2 on data obtained without the possibility of leaks. a_0, a_1, a_2, b_0, b_1, and b_2 are parameters describing characteristics of the ecosystem. Parameters include photosynthetic light response (b_0), base respiration rate (a_0), temperature sensitivity of respiration (a_1), and photosynthetic light saturation level (b_1). Another method, similar to the parameterization described above, is to produce a look-up table based on measurements of PAR and air temperature at a site [29]. Both parameterizations and look-up tables produce small errors when the amount of missing data is small [29]. The parameters obtained from fitting Eq. (10) to tower flux data are similar for similar ecosystems [34–36].

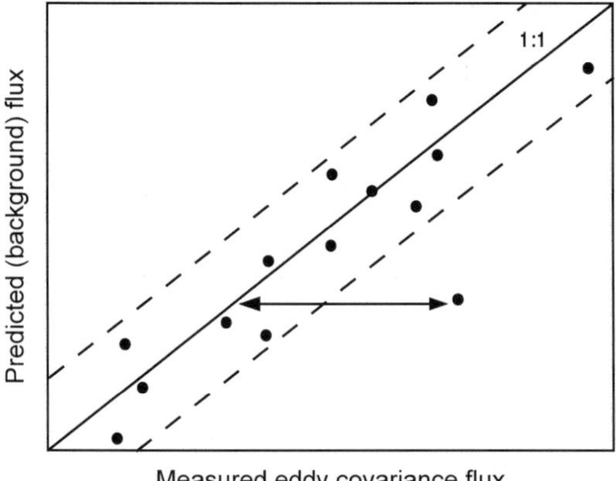

Figure 5: Schematic diagram of leak detection methodology. The x-axis represents ecological background fluxes predicted by characterizing the ecosystem fluxes as a function of environmental conditions (Eq. (10)). The y-axis represents hypothesized hourly EC flux observations from a measurement system over a geologic sequestration site. The dashed line represents random variability in EC fluxes caused by limited sampling of a turbulent atmosphere [39]. Measurement fluxes that lie outside the range of neutral variability can be established as leaks or other anomalous fluxes.

The parameters (or lookup table values) vary slowly as a function of season, in concert with ecological processes such as leaf-out and leaf-fall [37]. A large fraction of the hour-to-hour variability in tower-based EC flux measurements over ecosystems can be explained by variations in environmental conditions, particularly PAR [29,38]. Most of the remaining variability can be explained by limited sampling of a turbulent field [39]. Hour-to-hour variability in ecological CO_2 fluxes is typically similar to or less than the mean flux magnitude [19].

Leak Detection Sensitivity

A leak from an underground CO_2 storage reservoir can be detected only if its flux increases the total flux significantly beyond the range of background (ecological) fluxes normally observed in the specific environment of the measurement. This concept is illustrated in Figure 5. A detection limit of about 10 μmole C $m^{-2}\,s^{-1}$ (4.4×10^{-7} kg CO_2 $m^{-2}\,s^{-1}$) is realistic for hourly measurements in a biologically active area. Longer sampling times reduce the detection limit since the variability due to turbulence is random and decreases with increased time-averaging [39], thus narrowing the range of random flux variability shown in Figure 5.

We turn to volcanic emissions of CO_2 as an analogue to leaks from belowground CO_2 reservoirs. Emissions from volcanic activity can be quite large: a flux of 5×10^{-5} kg $m^{-2}\,s^{-1}$ ($\sim 1 \times 10^3$ μmole C $m^{-2}\,s^{-1}$) was measured in an area with significant tree kill [13] fluxes between 10^{-6} and 10^{-4} kg $m^{-2}\,s^{-1}$ ($20–2 \times 10^3$ μmole C $m^{-2}\,s^{-1}$) were measured in Yellowstone [7], and the Lake Nyos 1986 disaster was associated with fluxes near 10^{-2} kg $m^{-2}\,s^{-1}$ or 2×10^5 μmole C $m^{-2}\,s^{-1}$ [40]. Emissions of this magnitude are readily detectable using EC [7, 8]. Saturation of the CO_2 sensors (i.e. CO_2 mixing ratios that exceed the range of sensitivity of the chosen gas analyzer) is possible, but did not occur in the geologic measurements by Werner [6–8], even though there were large variations in the CO_2 flux. If sensor saturation does occur, this would provide the necessary leak detection, though the EC flux measurement would be rendered invalid.

We shall estimate leaks from geologic storage based on estimated reservoir size, and consider both slow and catastrophic reservoir failure. According to Herzog et al. [41], CO_2 storage is economical as long as the leak rate does not exceed 2.5% of the total CO_2 stored in 100 years. Assuming 100 million tonnes of stored CO_2 as a typical reservoir size [13], leakage of 1% of this total reservoir amount over an area of $10\,m^2$ spread evenly over a 100-year time period would result in a flux on the order of $10^{-2}\,kg\,m^{-2}\,s^{-1}$ ($2 \times 10^5\,\mu mole\,C\,m^{-2}\,s^{-1}$), four orders of magnitude larger than typical growing season ecological fluxes. A diffuse leak of 1% of the entire reservoir (e.g. through faults) over an area of $(300\,m)^2 = 10^5\,m^2$, and distributed over 100 years, would lead to a flux of order $10^{-6}\,kg\,m^{-2}\,s^{-1}$ ($20\,\mu mole\,C\,m^{-2}\,s^{-1}$), which is of the same magnitude as vigorous ecological fluxes. This would be detectable if background ecological fluxes are characterized in advance via typical sunlight and temperature relations (Eq. (10)). This crude analysis implies that eddy covariance is a promising technology for monitoring CO_2 reservoirs both for hazardous leaks and for leaks that would damage the economic viability of belowground storage.

Leak detection could fail in the case of a sudden, catastrophic leak during low-turbulence or unfavorable wind conditions. EC measurements might not promptly detect a sudden event if, at the time of the event, turbulent mixing was very weak or the flux footprint did not encompass the leak area. Relatively short gaps in the data are not necessarily problematic for detecting and quantifying slow leaks, but could prevent timely detection of such a catastrophic event. It seems prudent, therefore, to combine EC with chamber and/or mass balance [42] measurements near sites where leaks are possible.

LEAK QUANTIFICATION

Leak detection differs from leak quantification. This is evident in Figure 4c, which shows the weighted contributions of spatially distributed surface fluxes to the EC flux measured at a tower site, the result of a convolution of the flux footprint f and the surface flux field $F_0(x, y, 0)$. As is clear from this figure, the location of a source of CO_2 has a strong impact on the flux measured at the sensor. The precise location and magnitude of a CO_2 source from geologic storage are not likely to be known. In this case a single observation, while likely to detect the leak if it falls within the main region of the flux footprint, will not allow the position and magnitude of the leak to be quantified. Therefore, (1) it is important that the potential source region for a leak be located within the tower flux footprint and further (2) quantification of leaks will either require independent verification of their location (e.g. chamber measurements once a leak has been detected via EC), or the application of multiple EC measurements with different flux footprints. Some combination of these approaches is possible. Multiple EC measurements with different flux footprints can be used to identify the magnitude and location of a leak because a large number of independent observations (F_m, Eq. (6)) can be satisfied by only a limited number of possible source distributions (F_0). This could be accomplished with multiple flux towers. Alternatively, if the source is relatively steady over time, a single tower will provide measurements that are mathematically equivalent to multiple towers since the flux footprint changes over time because of changes in wind direction and stability. It is likely, therefore, that a small number of flux towers clustered around potential leak locations can provide accurate leak detection as well as leak quantification, though the latter will be more challenging.

CONCLUSIONS

The eddy-covariance method for monitoring earth–atmosphere CO_2 exchange can be used to measure fluxes with hourly temporal resolution over areas of order 10^4–$10^6\,m^2$. Instrumentation is robust and can be deployed in remote locations to collect data continuously. Suitable meteorological conditions exist on a daily basis at most locations and on average for roughly three-quarters of any given day. The method has been shown to be able to retrieve volcanic emissions in field tests [8].

We judge the method also as promising for the monitoring of leakage from geologic storage reservoirs. Our estimates are based on assumptions regarding reservoir size, area of leakage, and the total amount of CO_2 that escapes over an assumed time. These parameters are quite uncertain, but we have chosen what we believe characterize two important limits of the issue—catastrophic leakage and economically undesirable leakage. These leakage rates are compared to ecological fluxes which serve essentially as background noise for this application. We conclude that, using EC, CO_2 storage could be verified to be within the limits set by the economic viability of the storage. More careful assessment of likely leakage rates and useful detection limits, and field test of this approach are warranted based on our findings.

RECOMMENDATIONS

This work should be followed by a more accurate analysis of the likely magnitude, area and duration of potential leaks from geologic storage of CO_2. This initial study has shown promise in utilizing EC to monitor storage sites, but is dependent upon rough estimates of leak rates and areas of emission.

Second, a discussion of the need for leak detection only, or leak quantification is needed. Both are possible using EC methods, but leak quantification is more technically demanding. The methodology should be evaluated in light of the cost of monitoring and the economic and environmental benefits of CO_2 storage. These discussions will guide future system design and testing.

Finally, field testing is warranted, particularly for the topic of leak quantification. Leak detection should be possible within the ranges of emissions and footprint areas described in this chapter. Leak detection experiments should be focused primarily on determining the operational costs and benefits of EC methods. Leak quantification, a more challenging technical problem, should be demonstrated in the field, followed by evaluation of the operational costs and benefits.

NOMENCLATURE

$a_0, a_1, a_2, b_0, b_1, b_2$	NEE fit parameters
c	CO_2 mass density in air
$\bar{c}$	mean CO_2 mass density in air
c'	fluctuations in the CO_2 mass density in air about the mean
EC	eddy correlation
f	footprint function
F_0, F_m	vertical turbulent flux of CO_2 at heights z_0, z_m
L	Monin–Obukhov length
NEE	net ecosystem–atmosphere exchange of CO_2
PAR	photosynthetically active radiation
S_C	source or sink of CO_2 in the atmosphere
T_s	soil temperature
Q	point source emission rate
u, v, w	wind speeds: along-wind, cross-wind and vertical
$\bar{u}$, $\bar{v}$, $\bar{w}$	mean wind speeds
u', v', w'	fluctuations in wind speed about the mean
$\overline{u'c'}$, $\overline{v'c'}$, $\overline{w'c'}$	turbulent fluxes of CO_2 in the along-wind, cross-wind and vertical directions
$\overline{w'c'}_{z_m}$, $\overline{w'c'}_{z_0}$	vertical turbulent flux of CO_2 at heights z_m, z_0
x, y, z, t	along-wind, cross-wind and vertical directions, and time
x_m, y_m, z_m	position of a measurement
x_s, y_s	position at the Earth's surface
x_u	upwind distance
z_0	surface of the Earth
δ	Kronecker delta function
σ_v	root mean square lateral wind velocity
x', y'	dummy variables

ACKNOWLEDGEMENTS

This study was prepared with the support of the U.S. Department of Energy, under Award No. DE-FC26-01NT41145, and any opinions, findings, conclusions, or recommendations expressed herein are those of the authors and do not necessarily reflect the views of the DOE. We thank Daniel Ricciuto and Weiguo Wang for assistance with data processing and figures.

REFERENCES

1. R.L. Desjardins, E.R. Lemon, Limitations of an eddy correlation technique for the determination of the carbon dioxide and sensible heat fluxes, *Boundary-Layer Meteorol.* **5** (1974) 475–488.
2. R.L. Desjardins, A technique to measure CO_2 exchange under field conditions, *Int. J. Biometeorol.* **18** (1974) 76–83.
3. S.W. Running, D.D. Baldocchi, D. Turner, S.T. Gower, P.S. Bakwin, K.A. Hibbard, A global terrestrial monitoring network integrating tower fluxes, flask sampling, ecosystem modeling and EOS satellite data, *Remote Sens. Environ.* **70** (1999) 108–127.
4. K.J. Davis, P.S. Bakwin, C. Yi, B.W. Berger, C. Zhaos, R.M. Teclaw, J.G. Isebrands, The annual cycles of CO_2 and H_2O exchange over a northern mixed forest as observed from a very tall tower, *Global Change Biol.* **9** (2003) 1278–1293.
5. D.E. Anderson, C.D. Farrar, Eddy correlation measurement of CO_2 flux to the atmosphere from an area of high volcanogenic emissions, Mammoth Mountain, California, *Chem. Geol.* **177** (2001) 31–42.
6. C. Werner, CO_2 emissions in Yellowstone, USA, and Solfatara Volcano, Italy: use of eddy correlation and mass flux modeling. *PhD Dissertation*, The Pennsylvania State University, 2002, 126 pp.
7. C. Werner, J.C. Wyngaard, S.L. Brantley, Eddy-correlation measurement of hydrothermal gases, *Geophys. Res. Lett.* **27** (2000) 2925–2928.
8. C. Werner, G. Chiodini, D. Voight, S. Caliro, R. Avino, M. Russo, T. Brombach, J. Wyngaard, S. Brantley, Monitoring volcanic hazard using eddy correlation at Solfatara volcano, Naples, Italy, *Earth Planet. Sci. Lett.* **210** (2003) 561–577.
9. C. Yi, K.J. Davis, P.S. Bakwin, B.W. Berger, L.C. Marr, The influence of advection on measurements of the net ecosystem–atmosphere exchange of CO_2 observed from a very tall tower, *J. Geophys. Res.* **105** (2000) 9991–9999.
10. R.B. Stull, An Introduction to Boundary Layer Meteorology, Kluwer Academic Publishers, Dordrecht, 1988.
11. J.H. Seinfeld, S.N. Pandis, Atmospheric Chemistry and Physics: From Air Pollution to Climate Change, Wiley, New York, 1999.
12. D.D. Baldocchi, B.B. Hicks, T.P. Meyers, Measuring biosphere–atmosphere exchanges of biologically related gases with micrometeorological methods, *Ecology* **69** (1988) 1331–1340.
13. D.H. Lenschow, Micrometeorological techniques for measuring biosphere–atmosphere trace gas exchange, in: P.A. Matson, R.C. Harriss (Eds.), *Biogenic Trace Gases: Measuring Emissions from Soil and Water*, Blackwell Science, Cambridge, MA, 1995, pp. 126–163, Chapter 5.
14. M.L. Goulden, J.W. Munger, S.-M. Fan, B.C. Daube, S.C. Wofsy, Measurements of carbon sequestration by long-term eddy covariance: methods and a critical evaluation of accuracy, *Global Change Biol.* **2** (1996) 169–182.
15. J.B. Moncrieff, Y. Malhi, R. Leuning, The propagation of errors in long-term measurements of land–atmosphere fluxes of carbon and water, *Global Change Biol.* **2** (1996) 231–240.
16. J. Finnigan, A comment on the paper by Lee (1998): on micrometeorological observations of surface-air exchange over tall vegetation, *Agric. For. Meteorol.* **97** (1999) 55–64.
17. W.J. Massman, X. Lee, Eddy covariance flux corrections and uncertainties in long term studies of carbon and energy exchanges, *Agric. For. Meteorol.* **113** (2002) 121–144.
18. H.P. Schmid, C.S. Grimmond, F. Cropley, B. Offerle, H.-B. Su, Measurements of CO_2 and energy fluxes over a mixed hardwood forest in the mid-western United States, *Agric. For. Meteorol.* **103** (2000) 357–374.
19. B.W. Berger, K.J. Davis, C. Yi, P.S. Bakwin, C.L. Zhao, Long-term carbon dioxide fluxes from a very tall tower in a northern forest: flux measurement methodology, *J. Atmos. Ocean. Technol.* **18** (2001) 529–542.
20. T.W. Horst, J.C. Weil, Footprint estimation for scalar flux measurements in the atmospheric surface layer, *Boundary-Layer Meteorol.* **59** (1992) 279–296.
21. P.H. Schuepp, M.Y. Leclerc, J.I. Macpherson, R.L. Desjardins, Footprint prediction of scalar fluxes from analytical solutions of the diffusion equation, *Boundary-Layer Meteorol.* **50** (1990) 353–373.

22. H.P. Schmid, Footprint modeling for vegetation atmosphere exchange studies: a review and perspective, *Agric. For. Meteorol.* **113** (2002) 159–183.
23. T.W. Horst, J.C. Weil, How far is far enough? The fetch requirements for micrometeorological measurements of surface fluxes, *J. Atmos. Ocean. Technol.* **11** (1994) 1018–1025.
24. J.C. Weil, T.W. Horst, Footprint estimates for atmospheric flux measurements in the convective boundary layer, in: S.E. Schwartz, W.G.N. Slinn (Eds.), *Precipitation Scavenging and Atmosphere–Surface Exchange*, vol. 2, Hemisphere Publishing, Washington, DC, 1992, pp. 717–728.
25. S.P. Oncley, D.H. Lenschow, K.J. Davis, T.L. Campos, J. Mann, Regional-scale surface flux observations across the boreal forest during BOREAS, *J. Geophys. Res.* **102** (1997) 29147–29154.
26. S.K. Kaharabata, M.Y. Leclerc, R.L. Desjardins, J.I. MacPherson, P.H. Schuepp, S. Ogunjemiyo, S. Shen, Footprint considerations in BOREAS, *J. Geophys. Res.* **102** (1997) 29113–29124.
27. S.P. Arya, Air Pollution Meteorology and Dispersion, Oxford University Press, New York, 1999.
28. A. Grelle, A. Lindroth, Eddy-correlation system for long-term monitoring of fluxes of heat, water vapour, and CO_2, *Global Change Biol.* **2** (1996) 297–307.
29. E. Falge, D. Baldocchi, R. Olson, P. Anthoni, M. Aubinet, C. Bernhofer, G. Burba, R. Ceulemans, R. Clement, H. Dolman, A. Granier, P. Gross, T. Grunwald, D. Hollinger, N.O. Jensen, G. Katul, P. Keronen, A. Kowalski, C.T. Lai, B.E. Law, T. Meyers, H. Moncrieff, E. Moors, J.W. Munger, K. Pilegaard, U. Rannik, C. Rebmann, A. Suyker, J. Tenhunen, K. Ju, S. Verma, T. Vesala, K. Wilson, W. Wofsy. Gap filling strategies for defensible annual sums of net ecosystem exchange, *Agric. For. Meteorol.* **107** (2001) 43–69.
30. D. Baldocchi, E. Falge, L.H. Gu, R. Olson, D. Hollinger, S. Running, P. Anthoni, C. Bernhofer, K. Davis, R. Evans, J. Fuentes, A. Goldstein, G. Katul, B. Law, X.H. Lee, Y. Malhi, T. Meyers, W. Munger, W. Oechel, K.T. Paw, K. Pilegaard, H.P. Schmid, R. Valentini, S. Verma, T. Vesala, K. Wilson, S. Wofsy, FLUXNET: a new tool to study the temporal and spatial variability of ecosystem-scale carbon dioxide, water vapor and energy flux densities, *Bull. Am. Meteorol. Soc.* **82** (2001) 2415–2435.
31. T.P. Meyers, A comparison of summertime water and CO_2 fluxes over rangeland for well watered and drought conditions, *Agric. For. Meteorol.* **106** (2001) 205–214.
32. J. Lloyd, J.A. Taylor, On the temperature dependence of soil respiration, *Funct. Ecol.* **8** (1994) 315–323.
33. G.J. Collatz, J.T. Ball, C. Grivet, J.A. Berry, Physiological and environmental-regulation of stomatal conductance, photosynthesis and transpiration—a model that includes a laminar boundary-layer, *Agric. For. Meteorol.* **54** (1991) 107–136.
34. B.E. Law, E. Falge, L. Gu, D.D. Baldocchi, P. Bakwin, P. Berbigier, K. Davis, A.J. Dolman, M. Falk, J.D. Fuentes, A. Gold Stein, A. Granier, A. Grelle, D. Hollinger, I.A. Janssens, P. Jarvis, N.O. Jensen, G. Katul, Y. Mahli, G. Matteucci, T. Meyers, R. Monson, W. Munger, W. Oechel, R. Olson, K. Pilegaard, K.T. Paw, H. Thorgeirsson, R. Valentini, S. Verma, T. Vesala, K. Wilson, S. Wofsy, Environmental controls over carbon dioxide and water vapor exchange of terrestrial vegetation, *Agric. For. Meteorol.* **113** (2002) 97–120.
35. E. Falge, D. Baldocchi, J. Tenhunen, M. Aubinet, P. Bakwin, P. Berbigier, C. Bernhofer, G. Burba, R. Clement, K.J. Davis, J.A. Elbers, A.H. Goldstein, A. Grelle, A. Granier, J. Guomundsson, D. Hollinger, A.S. Kowalski, G. Katul, B.E. Law, Y. Malhi, T. Meyers, R.K. Monson, J.W. Munger, W. Oechel, K.T. Paw, K. Pilegaard, U. Rannik, C. Rebmann, A. Suyker, R. Valentini, K. Wilson, S. Wofsy, Seasonality of ecosystem respiration and gross primary production as derived from FLUXNET measurements, *Agric. For. Meteorol.* **113** (2002) 53–74.
36. E. Falge, J. Tenhunen, D. Baldocchi, M. Aubinet, P. Bakwin, P. Berbigier, C. Bernhofer, J.M. Bonnefond, G. Burba, R. Clement, K.J. Davis, J.A. Elbers, M. Falk, A.H. Goldstein, A. Grelle, A. Granie, T. Grunwald, J. Gudmundsson, D. Hollinger, I.A. Janssens, P. Keroner, A.S. Kowalski, G. Katul, B.E. Law, Y. Malhi, T. Meyers, R.K. Monson, E. Moors, J.W. Munger, W. Oechel, K.T. Paw, K. Pilegaard, U. Rannik, C. Rebmann, A. Suyker, H. Thorgeirsson, G. Tirone, A. Turnipseed, K. Wilson, S. Wolfsy, Phase and amplitude of ecosystem carbon release and uptake potentials as derived from FLUXNET measurements, *Agric. For. Meteorol.* **113** (2002) 75–95.
37. D. Baldocchi, E. Falge, K. Wilson, A spectral analysis of biosphere–atmosphere trace gas flux densities and meteorological variables across hour to multi-year time scales, *Agric. For. Meteorol.* **107** (2001) 1–27.

38. I. Baker, A.S. Denning, N. Hanan, L. Prihodko, M. Uliasz, P.-L. Vidale, K.J. Davis, P.S. Bakwin, Simulated and observed fluxes of sensible and latent heat and CO_2 at the WLEF-TV tower using SiB2.5, *Global Change Biol.* **9** (2003) 1262–1277.
39. D.H. Lenschow, J. Mann, L. Kristensen, How long is long enough when measuring fluxes and other turbulence statistics?, *J. Atmos. Ocean. Technol.* **11** (1994) 661–673.
40. S. Holloway, Safety of the underground disposal of carbon dioxide, *Energy Convers. Mgmt.* **38** 241–245.
41. H. Herzog, K. Caldeira, J. Reilly, An issue of performance: assessing the effectiveness of temporary carbon storage, *Clim. Change* **59** (2003) 293–310.
42. O.T. Denmead, L.A. Harper, J.R. Freney, D.W.T. Griffith, R. Leuning, R.R. Sharpe, A mass balance method for non-intrusive measurements of surface-air trace gas exchange, *Atmos. Environ.* **32** (1998) 3679–3688.

Carbon Dioxide Capture for Storage in Deep Geologic Formations, Volume 2
D.C. Thomas and S.M. Benson (Eds.)

Chapter 22

HYPERSPECTRAL GEOBOTANICAL REMOTE SENSING FOR CO_2 STORAGE MONITORING

William L. Pickles[1] and Wendy A. Cover[2]

[1]Lawrence Livermore National Laboratory, Livermore, CA, U.S.A.
[2]University of California at Santa Cruz, Santa Cruz, CA, U.S.A.

ABSTRACT

This project has developed an airborne remote sensing method for detection and mapping of CO_2 that might be leaking up from an underground storage formation. The method uses high-resolution hyperspectral imagery to detect and map the effects of elevated CO_2 soil concentrations on the roots of the local plants. The method also detects subtle or hidden faulting systems which localize the CO_2 pathways to the surface. Elevated CO_2 soil concentrations deprive the plant root systems of oxygen which is essential for a healthy plant. Excessive soil CO_2 concentrations are observed to significantly affect local plant health, and hence plant species distributions. These effects were studied in a previous remote sensing research program at Mammoth Mountain, CA, USA. This earlier research showed that subtle hidden faults can be mapped using the spectral signatures of altered minerals and of plant species and health distributions. Mapping hidden faults is important because these highly localized pathways are the conduits for potentially significant CO_2 leaks from deep underground formations.

The detection and discrimination methods we are developing use advanced airborne reflected light hyperspectral imagery. The spatial resolutions are 1–3 m and 128 band to 225 wavelength resolution in the visible and near infrared. We are also using the newly available "Quickbird" satellite imagery that has spatial resolutions of 0.6 m for panchromatic images and 2.4 m for multispectral. These are two commercial providers of the hyperspectral imagery acquisitions, so that eventually the ongoing surveillance of CO_2 storage fields can be contracted for commercially. In this project we had a commercial provider acquire airborne hyperspectral visible and near infrared reflected light imagery of the Rangely, CO enhanced oil recovery field and the surrounding areas in August 2002. The images were analyzed using several of the methods available in the suite of tools in the "ENVI" commercial hyperspectral image processing software to create highly detailed maps of soil types, plant coverages, plant health, local ecologies or habitats, water conditions, and man-made objects throughout the entire Rangely oil field and surrounding areas. The results were verified during a field trip to Rangely, CO in August 2003. These maps establish an environmental and ecological baseline against which any future CO_2 leakage effects on the plants, plant habitats, soils and water conditions can be detected and verified. We have also seen signatures that may be subtle hidden faults. If confirmed these faults might provide pathways for upward CO_2 migration if that occurred at any time during the future.

INTRODUCTION

The purpose of this research program has been to further develop remote sensing methods that can detect and discriminate the effects of elevated soil CO_2 concentrations on the local plants, their local habitats or ecologies, and to map possible hidden faulting systems at the surface above underground geological CO_2 storage formations. These effects were studied in a previous remote sensing research project at Mammoth Mountain, CA, USA (Figures 1–5). This earlier research mapped areas of tree kills and surrounding regions of tree plant stress, created by elevated CO_2 soil concentration levels. These elevated soil concentrations reach as high as 98% and are caused by CO_2 effluents from the magma interactions with formations below

Figure 1: At Mammoth Mountain CO_2 emission levels burst on short time scales to hazardous levels in small areas.

Figure 2: Trees killed at Mammoth Mountain, CA by highly elevated CO_2 soil concentrations. This area is near Horseshow Lake.

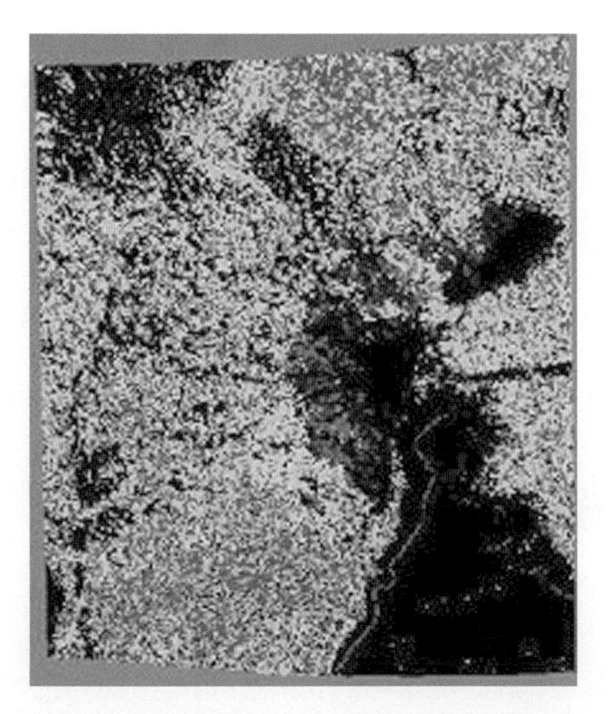

Figure 3: Tree health from hyperspectral imagery: dark, unhealthy; light and speckled; healthy.

the mountain. The mapping was produced by analysis of advanced airborne reflected light hyperspectral imagery acquired by a commercial provider. The spatial resolution was 5 m and wavelength resolution was 128 bands in the visible and near infrared reflected light spectrum between 420 and 2500 nm. The bands are of equal width and are contiguous covering the entire wavelength region.

Hidden faults were also located using the hyperspectral imagery at Mammoth Mountain, CA and similar hyperspectral imagery of the geothermal region near Dixie Valley and Dixie Meadows, NV, USA. Please see the first six references for discussion of the methods and results of the earlier research projects [1–6].

Subtle hidden faults have been detected by mapping mineralization and plant signature shifts in the hyperspectral images (Figure 5). This is potentially important for CO_2 sequestration. CO_2 escaping from an underground storage formation would probably convect along cracks, joints, and faults if there were any. Mapping all the subtle faulting in area above a CO_2 underground storage formation will help focus locations for leak monitoring efforts. The CO_2 escaping into the air at Mammoth Mountain is highly localized spatially and has large variations in emission rates (Figure 4).

The localization of CO_2 effluent was measured by the USGS Menlo Park personnel using hand-held CO_2 instruments at Mammoth in the air, just above the ground. They found very high spatial variability on the order of a few feet and very large changes in effluent rates [7,8].

In this study we are extending these techniques and experience to an enhanced oil recovery (EOR) field at Rangely, CO, USA. The field has been injected with CO_2 for 15 years for EOR. We acquired airborne hyperspectral imagery of the Rangely oil field, the surrounding areas including the town of Rangely, CO in August 2002. Two extensive field trips have been conducted to Rangely, one in August 2002 and a second in August 2003.

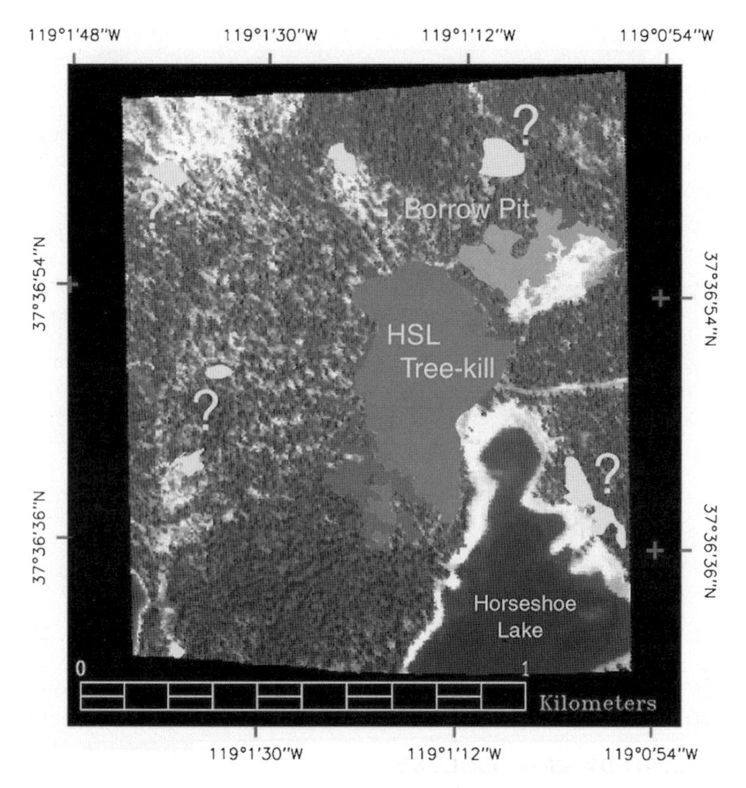

Figure 4: Plant mapping from hyperspectral imagery [3].

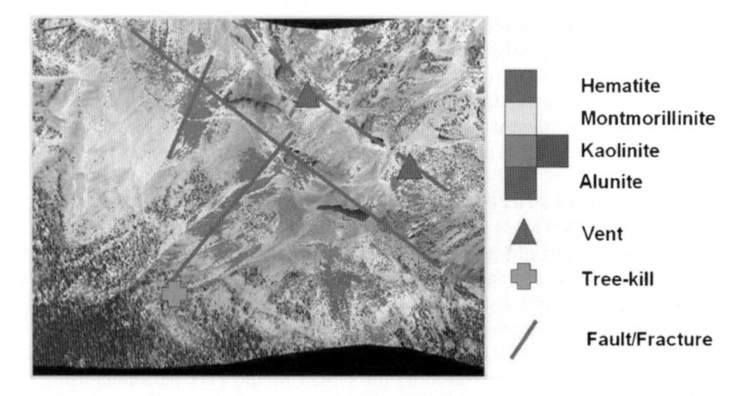

Figure 5: Hidden faults, mineralization and tree kill mapping from the hyperspectral imagery at Mammoth.

In our study of the plant life at Rangely, CO oil field and surrounding areas on the second field trip in August 2003, we did not observe any plant life effects that might be due to CO_2 effluents. There are some CO_2 and methane soil concentrations at about a dozen sampling locations in the area that are elevated above normal levels. Ron Klusman, (Colorado School of Mines in Golden, CO, USA) made measurements as part of a DOE funded project to study CO_2 and CH_4 concentrations in soils and in the air at the Rangely oil field. The elevated readings he observed were as great as 100 times the natural CO_2 concentrations caused by the activities of the microorganisms in the soil. He found winter to summer variations. He also did isotopic analysis on the CO_2 and CH_4 at many locations. These measurements were made over several years. The soil CO_2 concentration levels that were measured do not affect the "plant health" noticeably as represented by any measurements we made, in the imagery and on the ground using the field portable spectroradiometer in August 2000. What we did discover during our fieldwork at Rangely this summer is that the hyperspectral imagery of this relative dry high desert area maps the complex spatial distributions of a number of subtlety different "habitats". We speculate that if CO_2 soil concentrations were to start rising "significantly" above normal levels in well-defined spatial zones that the habitats in that area might change their boundaries or the spacing of plants within the habitats. If the average concentrations of soil CO_2 were high enough for several months it would be possible to species populations changing. The hyperspectral imagery analysis does map the habitat boundaries and does allow some species type differentiation so it could provide a mapping of some of these changes over time if the average CO_2 soil concentrations start rising to levels like those observed at Mammoth Mountain. However, the lowest levels of CO_2 soil concentration that would begin to affect plant health, or the shape and types of local ecologies or habitats are not known. The effects of time dependent and spatially dependent CO_2 effluent variations on the plants are also not known.

The hyperspectral geobotanical remote sensing techniques that we are developing use advanced commercial airborne imaging spectrometer systems now available in the USA and worldwide. The sensor system we normally contract for in our overhead imaging missions produces visible and near IR reflected light images with spatial resolution of 1–3 m in 128 wavelength bands (see http://www.hyvista.com/).

The HyVista sensor spatial and wavelength imagery resolution and signal-to-noise ratio of over 1000 to 1 allows us to detect and discriminate individual species of plants as well as the complexities of the geological and man-made objects in the images. The imagery is of sufficient quality that subtle local plant ecologies can be discriminate in verifiable detail.

EXPERIMENTAL/STUDY METHODOLOGY

The experimental method we have used is summarized in the following series of sequentially executed steps:

(1) Determine the area above the formation to be monitored including some surrounding areas that are thought to be outside the influence of any CO_2 that might migrate to the surface.
(2) Work with the airborne hyperspectral image acquisition contractor to develop a set of flightlines along which the images will be acquired (Figure 6).
(3) If possible make a group trip to the area to start to become familiar with the special characteristics of the region at the time planned for the airborne overhead image acquisition. Using hand-held DGPS and digital cameras visually record the soils, plants, minerals, waters, and man-made objects in the area.
(4) Remain on location to be available to the pilots and image sensor operator while the imagery is being acquired. This is usually 1 day between 10:00 and 14:00. But it can be 2 days. We normally all meet at the aircraft before and after the day's flight to go over the plans and check results.
(5) After the acquisition is completed our team reviews the imagery and georectification at the plane or in a motel. Our acquisition contractor then sends it to their main facility for final post-processing.

The set of all flightline images is returned on DVDs as three products; raw, corrected to reflectance including atmospheric absorptions, and georectification control files. This usually takes less than a few weeks. The imagery on the DVDs is analyzed using the ENVI commercial computer software on Windows and/or Unix platforms, by our researchers at UCSC, LLNL, and HyVista Corp., working as a team. ENVI is considered as the standard for the hyperspectral image analysis community worldwide.

The algorithms in the ENVI program are used to produce classification regions in the imagery that correspond to plant species types, plant health within species types, soil types, soil conditions, water bodies,

Figure 6: Morning preparations of the B 300 Twin Otter aircraft rented by HyVista. The LLNL UCSC team meets with the HyVista team while they are preparing the HyMap sensor, the georectification system, and computer system for the hyperspectral image acquisition. Note the clear skies, which is ideal.

water contents such as algae or sediments, mineralogy of exposed formations, and man-made objects such as roads, buildings, playgrounds, golf courses, etc. Some of the classification regions are distinct "ecologies". The classification regions derived from the imagery analysis are then studied to look for species of plants where they would normally not be found, relative plant health patterns, altered mineral distributions, soil type distributions, soil moisture distributions, water and water contents, and other categories.

We then return to the field with our analysis to verify and further understand the complex classification regions produced. Based on the verification results the analysis can be "fine-tuned" in the field to produce more accurate results. Since the imagery is georectified and the pixel size is 3 m individual objects such as trees, outcropping minerals, jeep trails, well heads, and pads can all be located using the maps and a hand-held GPS. The verified maps are extremely accurate. We also use a backpackable field spectroradiometer to measure reflectance spectra from a large number of plants, soils, and minerals. Figure 7 shows the backpackable ASD spectroradiometer being used at Rangely during our second field trip in August 2003. The spectrometer itself is in the backpack. The computer that controls the spectrometer and displays the acquired spectra is in a front pack sling. The pistol grip lens is connected to the spectrometer by a fiber optic cable.

The classification regions derived from the analysis of the imagery are a snapshot of the conditions at the time of acquisition. They show any areas of existing anomalous conditions such as plant kills and linear species modifications caused by hidden faults. They are also the "baseline" that is used to chart any future changes that are not due simply to normal seasonal and weather variations. This is accomplished by reimaging the area routinely over the years to monitor and document any effects that would be caused by significant CO_2 leakage reaching the surface and near subsurface defined by the root depth of the local plants.

The sensor used for the image acquisition at Rangely, CO is the HyMap™ hyperspectral scanner manufactured by Integrated Spectronics Pty Ltd (Figures 8–9). The HyMap sensor provides 126 bands

Figure 7: Using the ASD field spectroradiometer to measure reflectance spectra at Rangely in August 2003. The spectrometer is in the backpack. The computer that controls the spectrometer and displays the acquired spectra is in a front pack sling. The pistol grip lens is connected to the spectrometer by a fiber optic cable.

across the reflective solar wavelength region of 0.45–2.5 nm with contiguous spectral coverage (except in the atmospheric water vapor bands) and bandwidths between 15 and 20 nm.

The sensor operates on a 3-axis gyro-stabilized "IMU" platform to minimize image distortion due to aircraft motion. The HyMap sensor provides a signal-to-noise ratio ($>$500:1). Laboratory calibration and then daily operational system monitoring is done by HyVista to ensure that the calibration of the imagery is stable which is required for our very demanding spectral mapping tasks. Geolocation and image geocoding is achieved with an-on board differential GPS (DGPS) and an integrated IMU (inertial monitoring unit). Typically the HyMap sensor is operated with an angular field of view (IFOV) of 2.5 mr along track, 2.0 mr across track, FOV of 61.3° (512 pixels), DGPS and an integrated IMU, GIFOV—3–10 m (typical operational range). The DGPS and IMU is fully integrated with the image acquisition. The latitude and longitude of each pixel is recorded along with the hyperspectral image. Experiments to date indicate that the accuracy of this method is about one or two pixels over flat ground. The fact that the imagery and/or analysis results can be georectified at any time allows analysis of the data in any series of complex steps, without having the georectification process to influence the results. Then the georectification process can be applied to convert the final product to a highly accurate georectified form that can be verified and studied in the field.

Figure 8: The hyperspectral sensor is shown in the aircraft used by our acquisition contractor for image acquisitions.

The georectification subroutine is built into the ENVI software. It uses the georectification data files (IGM files) recorded in the airplane.

The sensor characteristics and a discussion of overhead hyperspectral imagery acquisition can be found at the web site of our imagery acquisition contractor http://www.hyvista.com/ and at http://www.intspec.com/.

RESULTS AND DISCUSSION

Many analysis of the August 2002 Rangely hyperspectral imagery have been done using ENVI. The analysis results have been combined with the photos, topographic maps, and digital elevation models of the Rangely oil field, town and surrounding areas. All of the products are georectifiable using the GLT files provided by HyVista as part of the imagery, as explained in the experimental section of this report. The accuracy of the precision of the georectification is about two pixels or 6 m. The accuracy is also about 6 m on flat ground. We made a second field trip to Rangely in August 2003 with all our analysis results. The complex and highly detailed classification region patterns that emerged from the analysis were easily verified because of the accuracy of the georectification. The ENVI "SAM" analysis picked out classification regions that were verified to be "habitats" or local ecologies. The perimeter of these habitats was mapped using a hand-held DGPS and recording a "waypoint" every few meters. The downloaded waypoint list from the hand-held DGPS was then compared to the pattern of the perimeter of the classification region identified as the habitat area. The two agreed in all the details, to within about 3 m.

The Field Site

The White River Basin is shown running from the center right to the lower left corner (Figure 10). The Rangely oil field basin is in the center of the figure. The 18 flightlines that were flown to acquire the 18 strip images are shown as dark lines. They are exactly due north and south by design. The folded formations whose motion created the oil field are easily seen running from southeast to northwest on either side of the basin and east–west across the top. Mellen Hill and the Mellen Hill fault can be easily seen at the northwest end of the oil field basin.

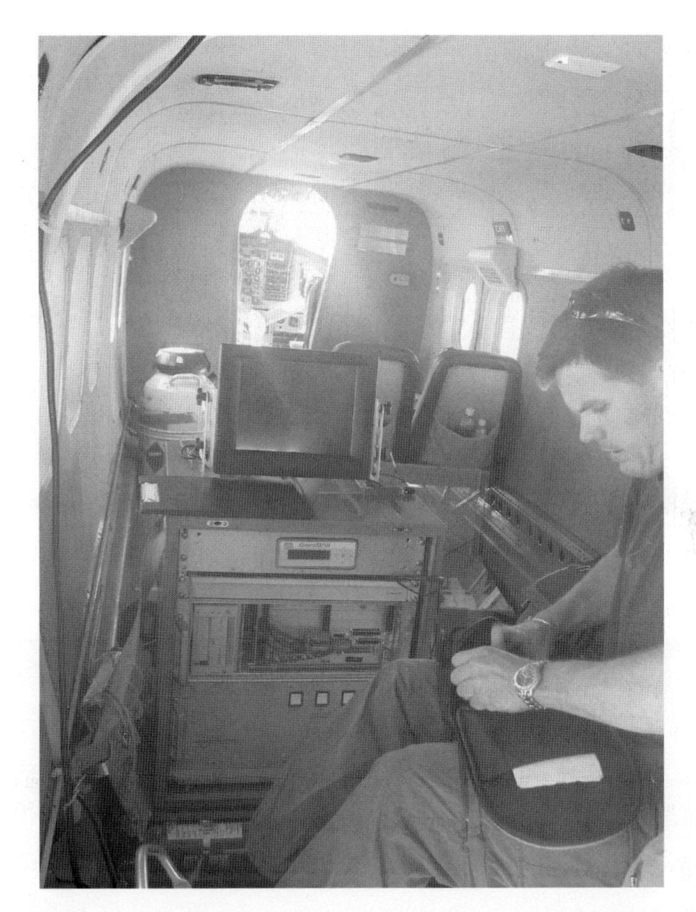

Figure 9: The acquisition contractor's sensor operator and flight commander. He is shown with the onboard computer system that controls the sensor systems and records the image, and the exact geolocation of each pixel in the image as it is acquired.

The individual flightlines are all georectified and are mosiaced together to produce an image of the whole region (Figure 11). All of the imagery shown in this report is georectified and true north is straight up on the page.

We have analyzed all the flightlines for the level of "plantness" with an "NDVI" ratio formula that uses individual bands in the hyperspectral image. The result is a computed ratio formula image that is normally called the "NDVI" or "Vegetation Index" image. The NDVI numerical value is high for lush green plant life and low for soils or man-made objects.

The NDVI images of each flightline were mosiaced together to produce the composite NDVI image of the whole region. The resulting NDVI image is shown in Figure 12 as a grayscale image. It can be presented as a pseudocolor image just as easily, but the grayscale image shows the results in a more understandable way. The brighter, or whiter, that a pixel appears, the more "healthy plant like" the contents of the pixel are. This means that the brighter the pixel in the image is, the bigger percentage of the pixel area is healthy vegetation and the healthier the plants. So the brightness of the pixel is a combination of the percentage of plant area coverage and the distribution of levels of health of the plants within the pixel. Understanding the causes

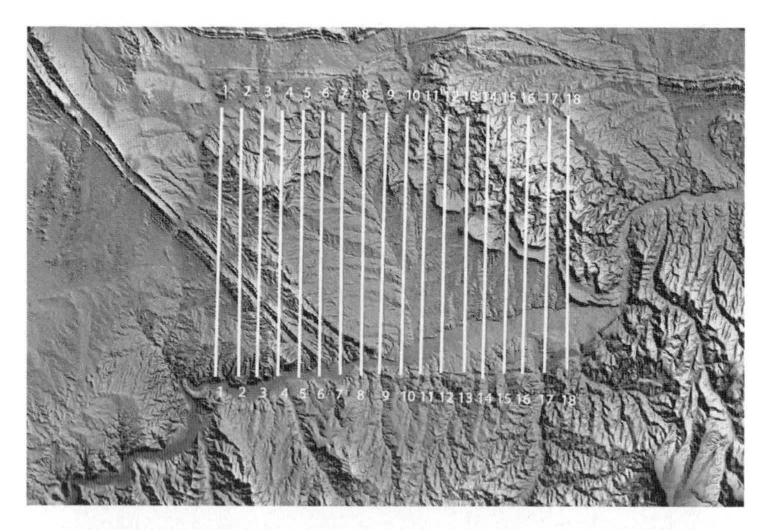

Figure 10: This is a digital elevation model of the Rangely oil field basin and surrounding formations with flightlines.

Figure 11: All the flightlines used to make a "true color" RGB image of the whole Rangely oil field basin, the surrounding formations and the town of Rangely.

of variation in the NDVI image is critical in this arid high desert environment, with its sparse vegetation. In general most plants are somewhat smaller than the 3 m pixel size of the Rangely imagery. In between the sparse plants the spectra from the low grasses that may be seasonally dry or bare soil will mix with the plant spectra. A very healthy plant smaller than 3 m surrounded by dirt will have some intermediate NDVI value.

Figure 12: All the flightlines images converted to NDVI images and mosiaced. This is a gray scale image. The whiteness or brightness of each image pixel means both that there is a higher percentage of plant coverage in the pixel area and/or that the plants are exposing more chlorophyll in their leaves and stems.

It is a mistake to use an intermediate NDVI value to say that there are no healthy plants in that pixel. Users interpreting geobotanical features from an NDVI remote sensing analysis unfortunately often make this kind of mistake. An additional mistake can easily be made using the NDVI for mapping plant health in high desert regions even when there is 100% of a pixel covered by vegetation. Some plants in these regions can look desiccated and brown on the outside while they are perfectly healthy on the inside. They are just waiting for water, or they are in their seasonal cycles. With all those cautions we still can make important use of this NDVI mapping of the Rangely region for CO_2 leak detection and baselining.

The NDVI map shows the exact location of almost every live plant in the entire region on the acquisition day. We can infer that there is less percentage of plant coverage in the darker pixels. Darker pixels may also, however, have reasonable percentage of plant coverage, but the plants may appear highly desiccated or woody on the outside. Or both may be true. This can be resolved, for a particular location, by additional hyperspectral image analysis or by simply driving to the exact location using a DGPS and observing what the plant coverage and health are.

The NDVI analysis then does provide an important map for locating exact places for further study. In addition, it does show the very complex patterns of plant coverage/health. It provides a very good baseline snapshot of the geobotanical conditions in the field. If a CO_2 leak of significant magnitude did develop, it is likely to darken the NDVI pixels in the area where the soil CO_2 concentration has risen significantly, either because of decreasing plant percentage coverage or decreasing plant health or both. Judging from the tree kill areas at Mammoth Mountain, both would occur.

By inspecting the NDVI composite image in Figure 12, we see that the oil field basin appears generally dark. The basin does have relatively low percentage plant coverage in most open flat areas, and the plants also appear woody or desiccated. Most of the plants in the basin are in fact perfectly healthy; they are just well adapted to their environment. In the arroyos, streambeds and drainage patterns there is a significantly higher percentage of vegetation coverage, and many different types of species that appear less woody and have more green showing. These areas in the basin are brighter in the NDVI image. The surrounding areas that are higher in altitude have much higher vegetation coverage and species that are very much greener. These areas are also much brighter in the NDVI. The White River Basin that runs across the bottom of the image has very high plant coverage and it is lush green. These areas are the brightest of all in the NDVI. The town of Rangely that is just south of the White River has many lush green areas such as the golf course which is the very bright object in the lower right corner.

The NDVI mosaic shown in Figure 12 clearly shows the difference in the plant health before and after the first rainstorm of the season. The flightlines on the left, numbers 1–12 were acquired on Tuesday. Then it rained for the first time in the summer on Wednesday and Thursday. Then the flightlines on the right, numbers 12–18, were acquired on Friday, which was a clear day. The flightline number 12 was reacquired to provide a direct comparison of before and after rain on the plants in the same flightline. The two flightlines 12A and 12B are shown in Figure 13.

The NDVI produced from flightline 12A acquired before the rain storms is shown in Figure 13. The NDVI produced from flightline 12B is shown on the right. In studying this carefully we find that most pixels are brighter, or whiter, in 12B. This is what we expect if the rain did trigger the high desert plants into increased photosynthetic activity. There are some pixels in 12B that remain as dark as the same pixel in 12A. These are probably pixels with only soils or man-made objects in them. When we were on our second field trip at Rangely we noted that most areas had vegetation cover. The rain probably activated many of these high desert species. This hypothesis is in agreement with the widespread elevation of the NDVI values between 12A and 12B throughout most of the area imaged.

The arroyos and washes in the basin and the junipers on top of the surrounding higher elevations seemed to have brightened the most, as you would expect. Also it appears that the complex pattern of vegetation distribution has not changed, only the relative observable photosynthetic activity. This is a very important observation. The plants have not had time in 2 days of rain to change their habitat distributions. This comparative NDVI mapping shows that clearly. This shows that we have a reasonably correct view of the overall response of plants and habitats to natural environmental variations and understand the time scales involved. That increases our chances of being able to detect and discriminate CO_2 soil concentration induced anomalous habitat distribution changes at an early stage. It also shows the power of having repeat imaging of the area to be monitored. Imaging seasonal and weather-induced variations may be very powerful for the early detection of elevated soil CO_2 concentration effects on the plants and habitats.

This area shown in the photo in Figure 14 is located at about the top three quarters mark in flightline 3 image shown in Figure 15. Examples of detailed mapping in the oil field using the hyperspectral image analysis are presented in the following sections and include the area in the photograph. The well pads, roads, vegetation, vegetation patterns or habitats and various soils can be seen in the photo and the classified regions in the flightline image.

Figure 15 is an unsupervised classification of flightline 3 done using ENVI. In this analysis, an algorithm separates the pixels in the image into some number of groups, in this case 35 groups. Individual pixels are grouped so that the complete spectral signatures and brightness are most similar for the pixels in each group. This process is what you would do if asked to sort a pile of multicolored "Indian corn" kernels into smaller piles where the kernels in the new piles were most like each other in size, colors, and patterns. As you went through the process you would create more and more separate piles. Then if you wanted to limit the number of piles you would be forced to recombine piles. In recombining two piles you might decide to resort the piles along with several other piles to reduce the overall number of piles. This could iterate for a long time so at some point we would call halt to the process unless you had decided that the piles were not changing much between iterations. This is exactly what the unsupervised classification algorithm does using the brightness and spectral shapes in each pixel. This process is "blind" in that there is absolutely no information about what is causing the brightness or spectral shape in the pixels involved. This is always a wise step to take at the beginning of analyzing imagery of a new area, because it alerts you to the complexities of the region. This result is georectified as are all the analysis products after they are created. Therefore, it can be used as a very effective tool to sort out different types of soils, plants, etc. during

Figure 13: NDVI line 12A left, 12B right.

a subsequent field trip to the site. Unfortunately, this method of classification is very sensitive to the overall brightness of the pixel. We used this type of analysis initially to tell us the quality of image data that had been acquired. As can be seen in the figure shown on the left the large number of categories and the complexity of the patterns indicate that the imagery is extremely information rich.

The two images on the right are the analysis of flightline 3 using the ENVI minimum noise fraction (MNF) procedure. In this procedure the original 128 bands of spectral information in each pixel are transformed

Figure 14: View of the Rangely oil field south of Mellen Hill.

into a new smaller set of objects that are referred to as MNF "pseudo-bands". They are not spectra any more, but they are "pseudo-bands" that contain the most noise free information that was in the original hyperspectral image. MNF "pseudo-bands" 1,2,3 are presented as RGB in the center image. MNF pseudo-bands 4,5,6 are presented on the far right. This step in the ENVI "hourglass" procedure shows that there is a very large and detailed amount of information contained in the hyperspectral image. The MNF results were not the final step.

We continued on to do a Spectral Angle Mapper (SAM) analysis that is shown in Figure 16.

The ENIV "hourglass" set of algorithms that ends with a SAM analysis is the method that is the most able to detect and discriminate plant habitats or ecology types, soils, mineralization, water conditions, and man-made objects. Figure 16 provides an example of the result of using this process on the flightline 11 hyperspectral imagery. Fifty-eight different categories were found. Many categories were tentatively identified using the USGS mineral spectral library, which we found to be a reasonable guide to what we actually found when we visited the locations of the categories during the second field trip taken in August 2003. The unknown categories were found to be mixed vegetation "habitats" upon on-site inspection in the field with these analysis products.

There is a "True" color image made from flightline 11 bands 15,9,2 as RGB on left of Figure 16. Note the difference between a color "picture" on the left and the SAM analysis for mixed vegetation "habitats", soils, and water on the right. This process was repeated for all the flightlines.

The ENVI SAM analysis has picked out classification regions that are a finely detailed mapping of local ecologies or "habitats". Some of these habitats are found to extend across the entire Rangely oil field and into the surrounding areas. These ecologies are made up of a narrow range of percentage admixtures of two or three very specific plant types and soil types.

The products are all georectified and so we were able to drive to the exact locations that were being picked out by the analysis and visually inspect them. We were able to walk back and forth between adjacent areas

Figure 15: Unsupervised minimum noise fraction (MNF).

that were being picked out by the analysis. The computer analysis was sorting out regions based on the relative mixture of plant types, plant sizes, plant spacings, intervening ground cover such as grass types and soils. Within these regions things like roads, paths, animal trails, man-made objects such as oil well pad areas, tanks, buildings are all also apparent and are picked out separately by the algorithm.

The area where we discovered that our analysis was picking out habitats is in the flightline 3 image. The SAM analysis of flightline 3 is shown on the right-hand side of Figure 17. The SAM analysis categories shown in light blue and in green were found to be two distinct mixed vegetation "habitats" surrounded

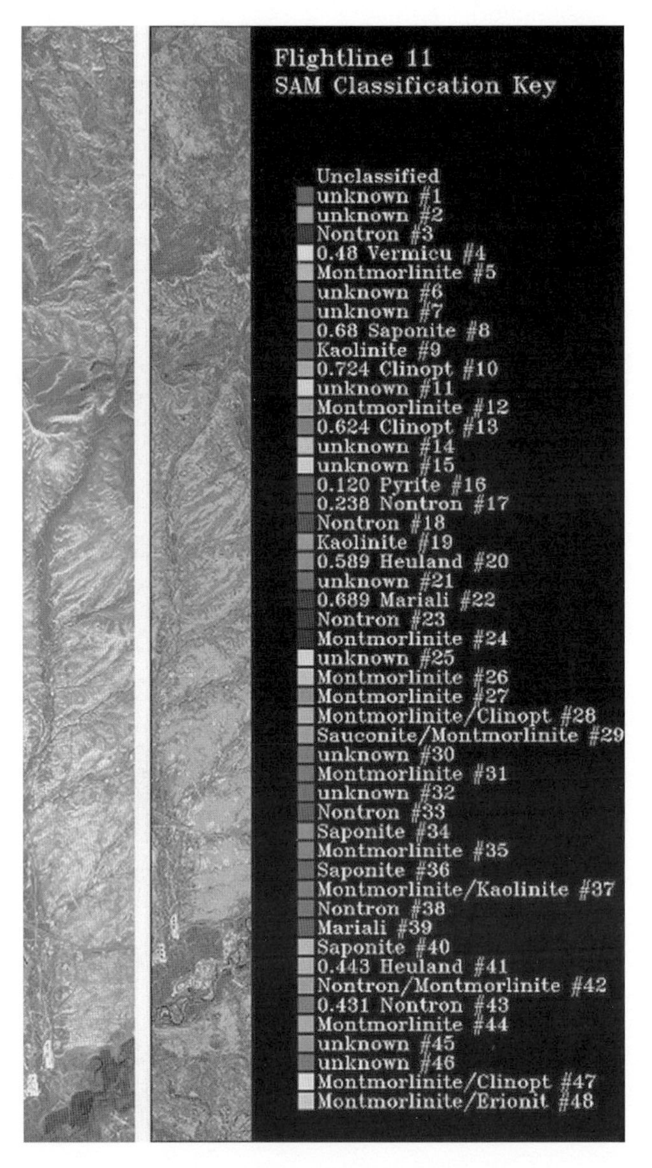

Figure 16: Detailed habitat mapping that results from an ENVI "hourglass" process. This process ended with a Spectral Angle Mapper (SAM) analysis of flightline 11 that is shown in the center. On the left is a "true color" RGB image of flightline 11.

by a third distinct habitat when we went back into the field at Rangely with these analysis products. The habitats consist of healthy sagebrush, mixed with golden dry cheek grass, and a percentage of dry soils. We found these two habitats were all over the Rangely region once we learned to recognize them from the SAM analysis. This result was unexpected and is a powerful means of mapping subtle meso-ecologies or "habitats" with mixed vegetation and soils.

Figure 17: Habitat mapping using flightline 3 hyperspectral image. True color (15,9,2 RGB) on right, infrared is in center, SAM habitats on the right.

The other images shown in Figure 16 are a "true color" of flightline 3 made with bands 15,2,9 as RGB on the left. In the center is a different 3-band image used to produce the equivalent of an infrared photograph, called a "color infrared". Then an analysis that maps the mineral Montmorillonite and Kaolinite in soil mixtures, shown in yellow–orange, has been laid on top. The lush green vegetation that is found primarily near the White River appears in red near the bottom of the image. Healthy vegetation reflects very strongly in the near infrared and so it appears very bright.

The "color infrared" image is made from the hyperspectral image using the bands 27, 16, and 7 as red, green, and blue, respectively.

Figure 18 is an enlarged view of the SAM analysis for the top of flightline 3 showing two of the habitats discovered by using the imagery analysis to guide us in the field. Light blue was found to be healthy sagebrush, mixed with golden dry cheek grass, and almost zero percentage of dry soils. Dark green was found to be smaller sagebrush plants mixed with cheek grass but with dry soil showing over about 50% of the area between the sagebrush plants. We walked the edges of some of these areas with DGPS and found the mapping to be accurate to 1 or 2 pixels. We speculate that any CO_2 leakage would begin to affect the

Figure 18: Close-up of line 3 SAM analysis showing the newly discovered habitats.

shape of these habitats and hence be easily seen in subsequent reimaging as a change. Of course that would only target the changed area and map it so that a team could go to that spot and check for excessive CO_2. This would also target a place to put CO_2 sensors.

Photos of these two habitats are shown in Figures 21 and 22. The top two photos in Figure 21 are of the habitat mapped by the light blue SAM category in Figure 18. The bottom photo is of the habitat mapped by the green category in the SAM analysis. The location of this habitat area is shown in Figures 19 and 20 by the DGPS waypoints shown as dots. The DGPS was left on all day each day in the SUV and recorded the "trip" (Figure 19). Figure 22 shows taking reflectance spectra of a plant using the ASD field spectrometer at the southern boundary between these two habitats (Figure 23).

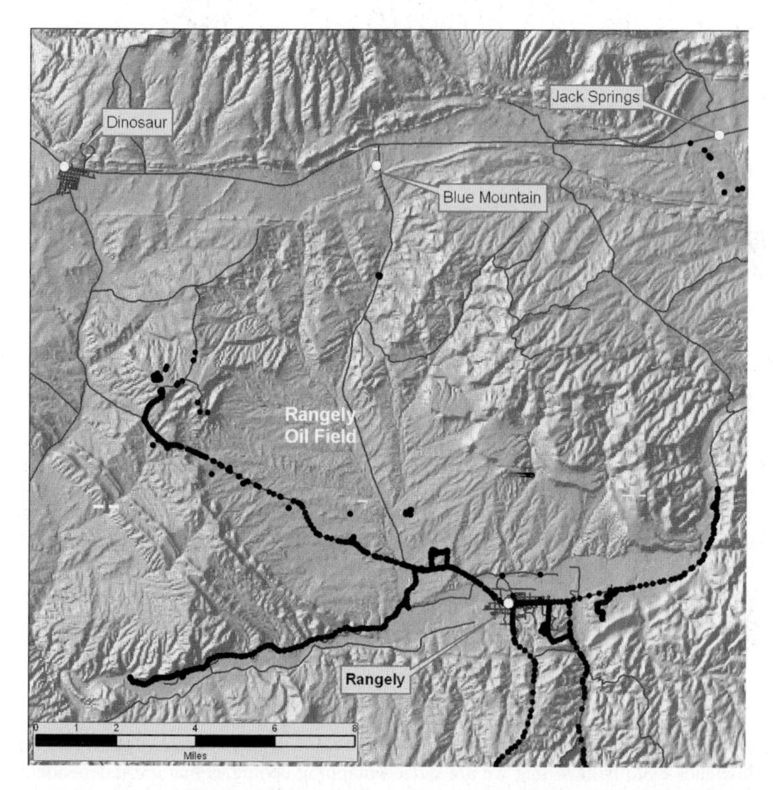

Figure 19: Continuous recording of our hand-held DGPS "track" during 1 day of the July 2003 field trip showing the sites visited that lead to the creation of the "habitat" mapping concept.

The match between the DGPS waypoints (dots) collected by walking the perimeter of the habitat shown in the top half of Figure 20 and the 1990 airphoto is remarkably accurate just as it is with the SAM analysis of the hyperspectral image. The exception is the area shown in the upper left. Apparently the habitat has been stable over the 23 years, except that junipers have intruded into the area along the two roads to the northwest just outside the boundary observed in July 2003. The area outside the red dot DGPS boundary to the lower left is the other habitat shown in the lower half of Figure 20. The circular pattern in the lower left is a well pad. Photos of the habitat outside or at the edge between the habitats are shown in Figures 24 and 25.

By carefully selecting ENVI SAM "endmembers", four of the most obvious "habitats" or ecologies, were mapped in all the flightlines imaged at Rangely. These ecologies are discernable and mappable even though

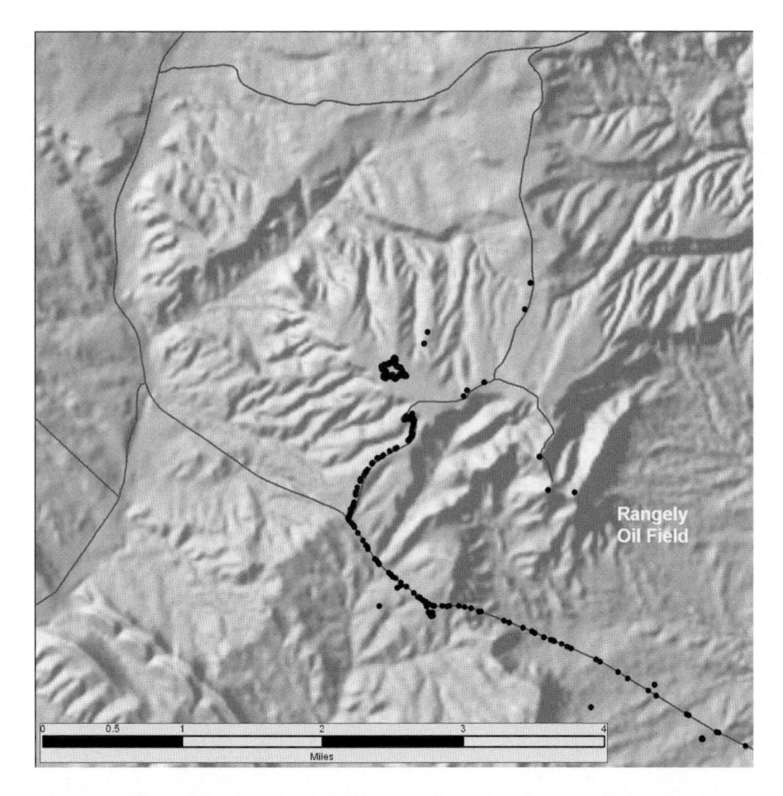

Figure 20: Close-up view of the location of the area where we discovered that the SAM analysis was mapping "habitats". The DGPS waypoints acquired while walking the habitat picked by the SAM analysis of the line 3 hyperspectral image are shown as the group of dots at the top center. The other dots on the roads to the northeast (upper right) were places mapped by the SAM analysis as the same habitat types. We drove to these sites guided by our georectified SAM imagery analysis and found they were indeed the same type of habitats. Once we became educated about we began to recognize these habitats.

the eastern 1/3 of the flightlines that were acquired on Friday after the heavy rains on Wednesday and Thursday. This leads us to believe that we are indeed mapping ecologies that are independent of detailed weather conditions by using this SAM analysis. Figure 26 shows the four ecologies in four different colors. The black pixels are all other categories. The mapping is accurate to the individual pixel (3 m in size) level throughout all 18 flightlines of the entire Rangely region, including the imaging on the western side before and imaging on the eastern side after the first seasonal rainstorm.

This research has demonstrated ability to do regional scale meso-ecology mapping from the hyperspectral imagery. It also establishes the basis for further progress in the refinement of these methods. The serendipitous rainfall that occurred during the data collection period provided the opportunity to observe short-term changes in the vegetation hyperspectral images. Observing deviations from these normal patterns will allow targeting of specific locations for making on-the-ground measures to detect CO_2 releases.

A new very high-resolution commercial satellite called QuickBird is now being operated by the Digital Globe Corp. The panchromatic imagery has a resolution of 0.6 m, and for the band multispectral

Figure 21: The two adjacent habitats discovered by using the line 3 SAM analysis categories.

Figure 22: Measuring reflectance spectra of a plant at the border between two habitats. Notice the well pad.

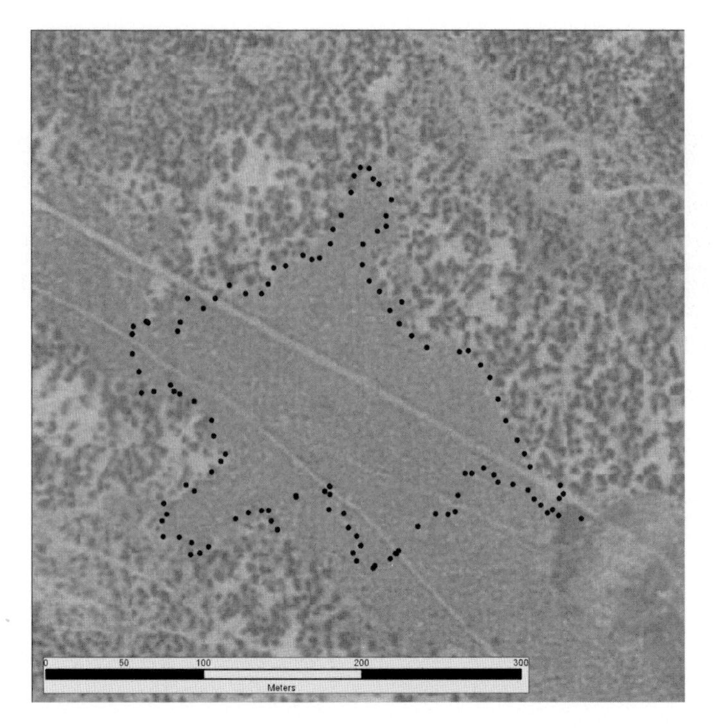

Figure 23: The dots are DGPS waypoints collected by walking the perimeter of the habitat shown in the top half of Figure 20. The DGPS points are shown overlaid on the 1990 airphoto.

Figure 24: Northern boundary of habitat and the next habitat of Junipers. The edge is accurately mapped by SAM analysis and the DGPS waypoints (dots).

Figure 25: Close-up of the northern boundary between habitats. This is the top if the "finger" in the top of the DGPS waypoints (dots) shown in Figure 23.

Figure 26: Three distinct habitats (yellow, green, and brown) and a soil type (white) mapped across the entire Rangely region.

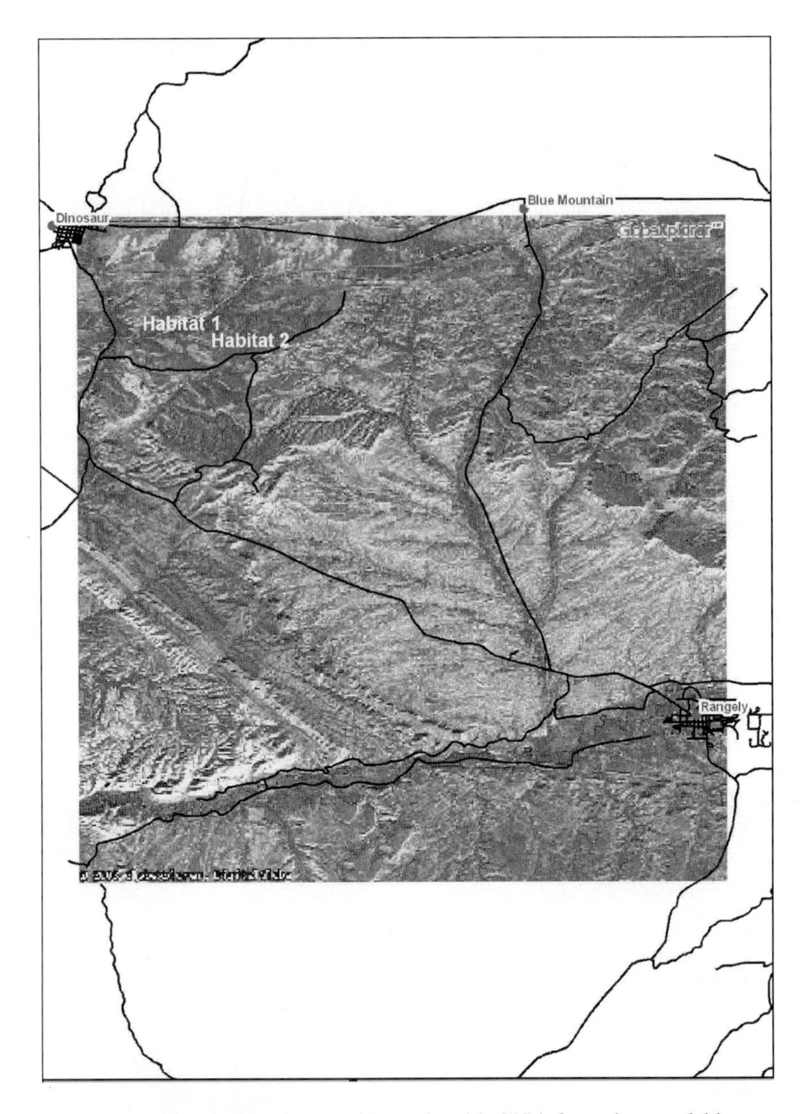

Figure 27: Airphoto image of Rangely with GIS information overlaid.

imagery spatial resolution is 2.4 m. Quick Bird has not yet imaged Rangely, CO. Estimated costs for this would be $4000. Figure 27 shows what a QuickBird multispectral image of the Rangely oil field and surrounding areas would look like. This imagery is an orthorectified airphoto. The town and roads are shown. The areas that established the habitat mapping results are labeled Habitat 1 and Habitat 2 in the upper left.

CONCLUSIONS/RECOMMENDATIONS

In conclusion, we have found an unexpected result that is potentially very important to the task of monitoring for CO_2 that has leaked to within the plant root depths near the surface. The discovery is that one of our analysis techniques has picked out finely detailed mapping of local ecologies that extend across the entire Rangely oil field and surrounding areas. These ecologies appear to be made up of a fairly narrow range of percentage admixtures of two or three very specific plant types and soil types. The products are all georectified and so we were able to drive to the exact locations that were being picked out by the analysis and visually inspect them. We were able to examine adjacent areas in detail that were being picked out by the analysis. The computer analysis was sorting out regions based on the relative mixture of plant types, plant sizes, plant spacings, intervening ground cover such as grass types and soils. Within these regions, things like roads, paths, animal trails, man-made objects such as oil well pad areas, tanks, buildings, town, golf courses, are all also apparent and are picked out separately by the algorithm.

The results show georectified, detail and complexity in the mapping of ecologies, soil types, plant types, plant health, water conditions, and human use features. This work does provide a "snap shot" of the ecological complexity of the entire area as of August 2002. Interestingly, we found that the August 2002 imagery analysis seemed to be completely valid in August 2003 during our return field trip. We have not found any evidence in the imagery analysis or anywhere on the ground, during our field trips in August 2002 and 2003 of any plant life responses that might indicate CO_2 leakage from the formation below. In fact, detailed analysis and observations of areas where elevated CO_2 and methane fluxes had been observed provided no indication of any effect on the soils or vegetation.

We strongly recommend a long-term research effort that will establish what CO_2 soil concentration levels produce observable changes in the biosphere and the corresponding subtle and complex ecological distributions in various environments (including terrestrial and marine). This is an extremely important and highly relevant task for CCP SMV to pursue. The biosphere is always integrating, and responding to and creating changes. We are well advised to learn to read and understand all the subtle signs it is providing to us, continuously.

We also recommend trying to measure directly CO_2 and CH_4 gas concentrations in the air using airborne hyperspectral imagers. We recommend using infrared hyperspectral imaging spectrometer sensors that have enough wavelength resolution to measure the CO_2 and CH_4 absorption resonances in the infrared that are due to rotational, stretch, and vibrational molecular absorption mechanisms. The sensors will have to be able to distinguish the CO_2 and CH_4 resonances from other resonances caused by molecules that are likely to be present. The sensitivity will have to be great enough to measure normal background concentrations in air, so that anomalies can be detected.

We also recommend starting a program to develop inexpensive, nanotechnology sensor to detect CO_2 and CH_4 concentrations in soils and at the surface. New sensors that are unpowered autonomous and read out by pulsed rf or optical interrogation during a fly over could be placed along faults, near well head, and at any features thought to possible venting paths. These would be permanently installed and readout at whatever interval was required.

NOMENCLATURE

FOV	field of view
GPS	global positioning system
DGPS	differential GPS using the two GPS designated satellites for reference
IMU	inertial monitoring unit
GIFOV	ground field of view nominal
mr	milliradians of angular view
LLNL	Lawrence Livermore National Laboratory
UCSC	University of California Santa Cruz
EOR	enhanced oil recovery (field)
CO_2	carbon dioxide

REFERENCES

1. W.L. Pickles, G.D. Nash, W.M. Calvin, B.A. Martini, P.A. Cocks, T. Kenedy-Bowdoin, R.B. Mac Knight IV, E.A. Silver, D.C. Potts, W. Foxall, P. Kasameyer, Geobotanical Remote Sensing Applied to Targeting New Geothermal Resource Locations in the US Basin and Range with a focus on Dixie Meadows NV, Geothermal Resources Council Transactions, vol. 27, October 12–15, 2003, Geothermal Resources Council, Davis, California, USA, UCRL-JC-153443.
2. B.A. Martini, E.A. Silver, W.L. Pickles, P.A. Cocks, Hyperspectral Mineral Mapping in Support of Geothermal Exploration: Examples from Long Caldera, CA and Dixie Valley, NV, USA, Geothermal Resources Council Transactions, vol. 27, October 12–15, 2003, Geothermal Resources Council, Davis, California, USA.
3. B.A. Martini, New insights into the structural, hydrothermal, and biological systems of long valley Caldera using hyperspectral imaging, a dissertation submitted in partial satisfaction of the requirements for the Degree of Doctor of Philosophy in earth sciences, UC Santa Cruz, December 2002, http://eed.llnl.gov/other/martini/.
4. B.A. Martini, Thesis Chapter 3 and 4 are found at: Hyperspectral imaging in long valley Caldera: volcano-associated biological communities, http://eed.llnl.gov/other/martini/chapter3.pdf, http://eed.llnl.gov/other/martini/Ch3_Figures.pdf.
5. Hyperspectral imaging in Long Valley Caldera: tracking volcanogenic CO_2 and its lethal effects, http://eed.llnl.gov/other/martini/chapter4.pdf, http://eed.llnl.gov/other/martini/Ch4_Figures.pdf.
6. K.A. McGee, T.M. Gerlach, Annual cycle of magmatic CO_2 at Mammoth Mountain, California: implications for soil acidification, *Geology* **26** (1998) 463–466.
7. J.D. Rogie, D.M. Kerrick, M.L. Sovey, G. Chiodini, Measurement and Analysis of Diffuse Degassing of Magmatic CO_2 at Mammoth Mountain, California, 1997–2000, Eos Trans. AGU, 81 (48), Fall Meet. Suppl., Abstract V72D-12, 2000.
8. J.D. Rogie, D.M. Kerrick, M.L. Sovey, G. Chiodini, D.L. Galloway, Dynamics of carbon dioxide emission at Mammoth Mountain, California, *Earth Planet. Sci. Lett.* **188** (2001) 535–541.

Carbon Dioxide Capture for Storage in Deep Geologic Formations, Volume 2
D.C. Thomas and S.M. Benson (Eds.)

Chapter 23

NON-SEISMIC GEOPHYSICAL APPROACHES TO MONITORING

G.M. Hoversten and Erika Gasperikova

Lawrence Berkeley National Laboratory, Berkeley, CA, U.S.A.

ABSTRACT

This chapter considers the application of a number of different geophysical techniques for monitoring geologic storage of CO_2. The relative merits of the seismic, gravity, electromagnetic (EM) and streaming potential (SP) geophysical techniques as monitoring tools are examined. An example of tilt measurements illustrates another potential monitoring technique, although it has not been studied to the extent of other techniques in this chapter. This work does not represent an exhaustive study, but rather demonstrates the capabilities of a number of geophysical techniques on two synthetic modeling scenarios. The first scenario represents combined CO_2 enhance oil recovery (EOR) and storage in a producing oil field, the Schrader Bluff field on the north slope of Alaska, USA. The second scenario is of a pilot DOE CO_2 storage experiment scheduled for summer 2004 in the Frio Brine Formation in South Texas, USA. Numerical flow simulations of the CO_2 injection process for each case were converted to geophysical models using petrophysical models developed from well log data. These coupled flow simulation–geophysical models allow comparison of the performance of monitoring techniques over time on realistic 3D models by generating simulated responses at different times during the CO_2 injection process. These time-lapse measurements are used to produce time-lapse changes in geophysical measurements that can be related to the movement of CO_2 within the injection interval.

The time-lapse performance of seismic, gravity, and EM techniques are considered for the Schrader Bluff model. Surface gravity, surface tilt and SP measurements are considered for the Frio brine formation model. These two models represent end members of a complex spectrum of possible storage scenarios. EOR/storage projects in general and Schrader Bluff in particular represent relatively thin injection intervals with multiple fluid components (oil, hydrocarbon gas, brine, and CO_2) while brine formations such as the Frio will usually have much thicker injection intervals and only two component (brine and CO_2) systems.

INTRODUCTION

Cost effective monitoring of reservoir fluid movement during CO_2 storage is a necessary part of a practical geologic storage strategy.

In this chapter, we evaluate seismic and alternative approaches for long-term monitoring. In order to evaluate alternative geophysical monitoring techniques two numerical simulations of CO_2 storage scenarios are considered. The time-lapse performance of seismic, gravity, and EM techniques are examined using models derived from reservoir flow simulation of the CO_2 EOR/storage process for the Schrader Bluff reservoir on the North Slope of Alaska, USA. Surface gravity, surface tilt, and SP measurements are considered for the Frio brine formation test in south Texas, USA. These two models represent end members of a complex spectrum of possible storage scenarios. EOR/storage projects in general and Schrader Bluff in particular represent relatively thin injection intervals with multiple fluid components (oil, hydrocarbon gas, brine, and CO_2) while brine formations will usually have much thicker potential injection intervals and only two component (brine and CO_2) systems.

Petroleum reservoirs and brine formations offer the two most obvious storage targets. Petroleum reservoirs have the natural advantages that they are already well characterized, have a demonstrated seal, have an existing infrastructure, and offer cost offsets in the form of enhanced petroleum production as CO_2 is injected. From a monitoring standpoint, petroleum reservoirs offer more challenges than brine formations because they typically have less vertical extent ($\sim$25 m for oil vs. hundreds of meter for brine formations) and have multiple in situ fluids. Not withstanding their inherent monitoring challenges, petroleum reservoir will undoubtedly provide many of the early storage examples.

We have chosen to include seismic modeling of the Schrader Bluff scenario for comparison with the non-seismic techniques. While the work presented here is all forward modeling of responses, future work will concentrate on inversion of data to produce quantitative estimates of reservoir properties from the various techniques. Simulation of the seismic response for the same models considered for non-seismic techniques will allow a side-by-side quantitative comparison. Within the seismic modeling section for Schrader Bluff we have included some models appropriate for brine formations using a published rock-properties model.

The Schrader Bluff model used here for analysis began with a three dimensional (3D) flow simulation model provided by BP Alaska. In addition, we developed a detailed rock-properties model from log data that provides the link between the reservoir parameters (porosity, pressure, saturations, etc.) and the geophysical parameters (velocity, density, electrical resistivity). The rock-properties model was used to produce geophysical models from the flow simulations. The same procedures were used to produce numerical models for the Frio brine formation pilot test.

On-shore EOR project—Schrader Bluff, Alaska

One site being considered for geologic storage is the Schrader Bluff reservoir on Alaska's North Slope (Figure 1). Preliminary evaluations show that a CO_2-based enhanced oil recovery could increase oil recovery by up to 50% over water-flooding [1]. Furthermore, the studies concluded that up to 60% of the CO_2 injected as part of the EOR scheme would remain in the reservoir. A schematic geological cross-section through the Schrader Bluff Formation is shown in Figure 2.

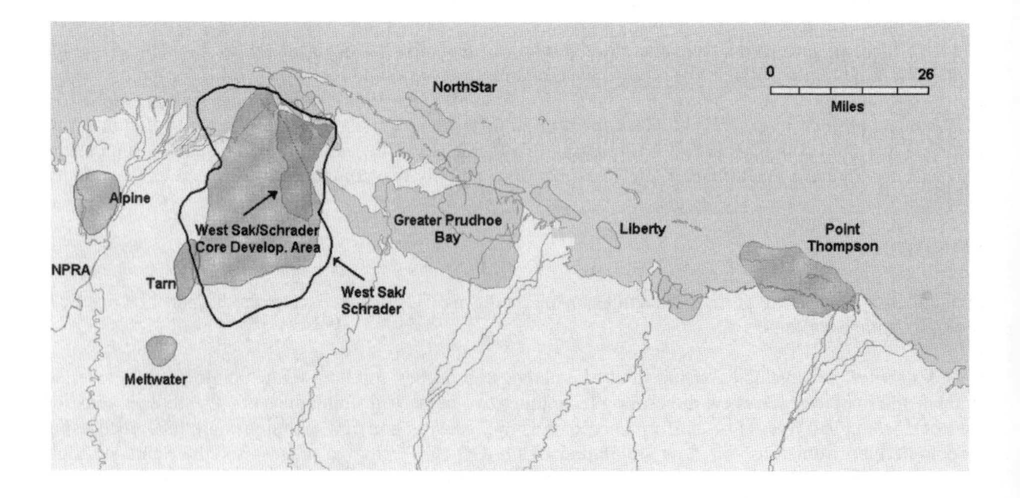

Figure 1: Location of Schrader Bluff reservoir on Alaska's North Slope.

In order to compare the spatial resolution and sensitivity of various geophysical techniques being considered for CO_2 storage monitoring, a 3D flow simulation model of the reservoir provided by BP was used in conjunction with rock-properties relations developed from log data to produce geophysical models from the flow simulations. The Schrader Bluff reservoir is a sandstone unit, between 25 and 30 m

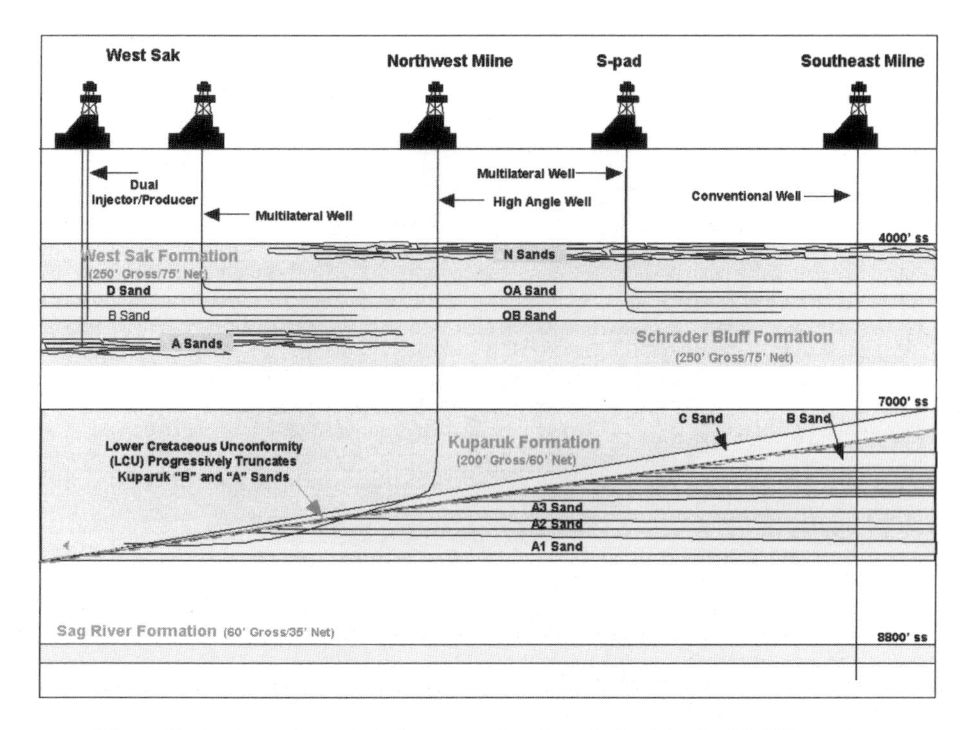

Figure 2: A schematic geological cross-section through the Schrader Bluff Formation.

thick, at a depth of 1100–1400 m. Figure 3 shows a 3D view of the portion of the reservoir under consideration for a CO_2 storage test. The reservoir unit gently dips to the east with major faulting running mainly north–south. Two faults with offsets in excess of 75 m cut the reservoir with several smaller sub-parallel faults present. Time-lapse snap shots of the reservoir at initial conditions and 5-year increments out to 2035 were used. A water after gas (WAG) injection strategy is considered which produces complicated spatial variations in fluid (CO_2, brine, oil and gas) saturation within the reservoir over time.

Rock-properties model

A rock-properties model was developed from log data for the reservoir. This model relates reservoir parameters to geophysical parameters, and is used to convert the flow simulation model parameters to geophysical parameters (acoustic velocity V_p, shear velocity V_s, density and electrical resistivity). We have assumed the unconsolidated sand model where the effective pressure is equal to lithostatic pressure minus the pore pressure. As noted by Brandt [2] as cementation of the sand grains increases the effective pressure would be the lithostatic minus some fraction of the pore pressure. Pressure effects are included through the effective pressure on the dry frame and through the effects of pore pressure on the fluids used in the Gassmann fluid substitutions. A description of the rock-properties modeling process is given by Hoversten et al. [3]. Archie's law is used for electrical resistivity as a function of porosity and water saturation. Figure 4 shows the rock-properties parameters along with the predicted values of V_p, V_s and density compared to the log values from the MPS-15 well. The model V_p, derived from the flow simulation, at initial conditions is shown in Figure 4 on a east–west cross-section through two injection wells.

A critical porosity [4] appropriate for sandstone of 35% was assumed. Oil API gravity and brine salinity are taken from measured values. The regression-determined values of the grain shear modulus and Poisson ratio are appropriate for quartz grains. The model parameters are determined for the reservoir interval in the logs. The full geophysical models are built by interpolating available well logs in 3D using the seismic reservoir

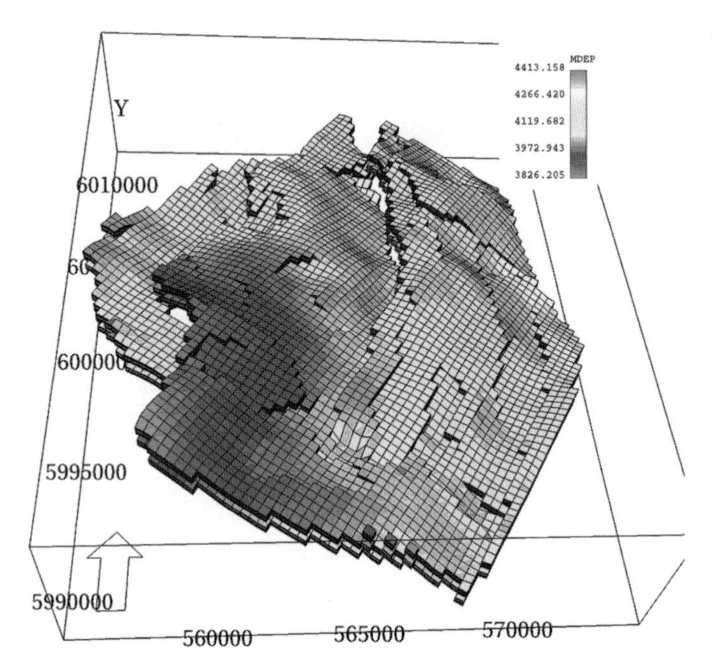

Figure 3: Three-dimensional view of the portion of the reservoir under consideration for CO_2 sequestration test at Schrader Bluff. Depths range between 3800 and 4400 ft (1158 and 1341 m) true vertical depth.

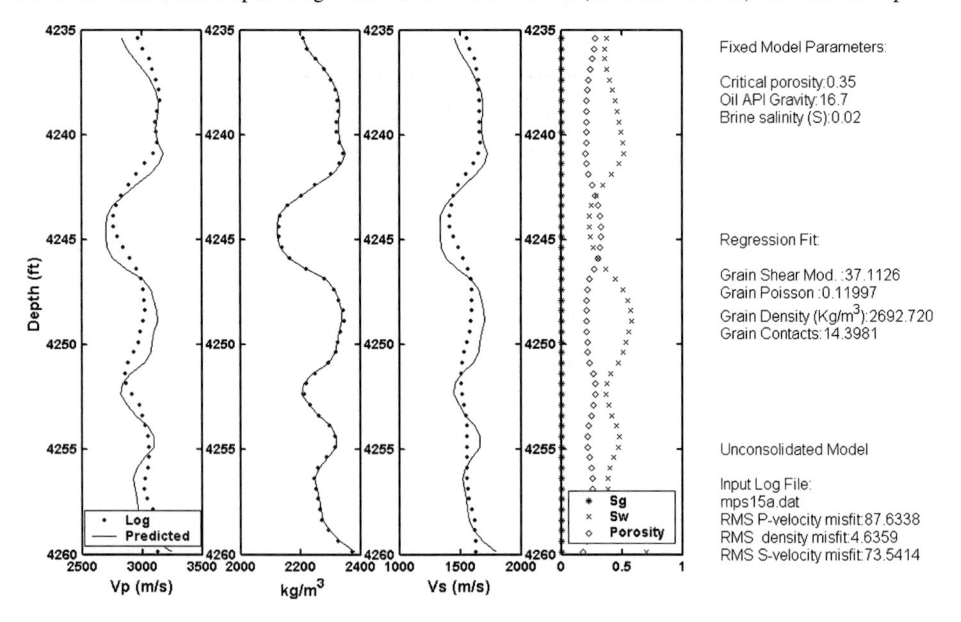

Figure 4: Rock-properties model based on un-consolidated sandstone model [4]. Measured log values shown as dots. Parameters (right side) are derived from a simplex minimization of the misfit between observed and calculated V_p, V_s and density logs. Predicted V_p, V_s and density are shown as solid lines.

surfaces as a spatial guide. This produces a background model of V_p, V_s, density, and resistivity. The reservoir flow simulations, which only cover the reservoir interval, are then filled in at the time intervals where flow simulations were done. The model shown in Figure 4, along with Archie's law, is used to convert the porosity, water saturation, oil saturation, gas saturation, CO_2 saturation, pressure, and temperature from the flow simulation to V_p, V_s, density, and electrical resistivity.

Gravity modeling

A snapshot of the model at initial conditions, before CO_2 injection begins, is shown in Figure 5. Figure 5a is an east–west cross-section of bulk density as a function of depth and horizontal distance between a pair of injection wells. In this figure, gravimeters are located in two wells roughly 8 km apart. The reservoir interval is outlined in white; the positions of the gravimeters are indicated by black squares. Since they are very closely spaced they overlap and show as an image of a well. Figure 5b is a plan view of the density at initial conditions at a depth of 1200 m with positions of 23 injecting wells taken from the reservoir simulation. The circled well location in the upper part of Figure 5b indicates a well for which borehole gravity responses are shown in Figures 11 and 12.

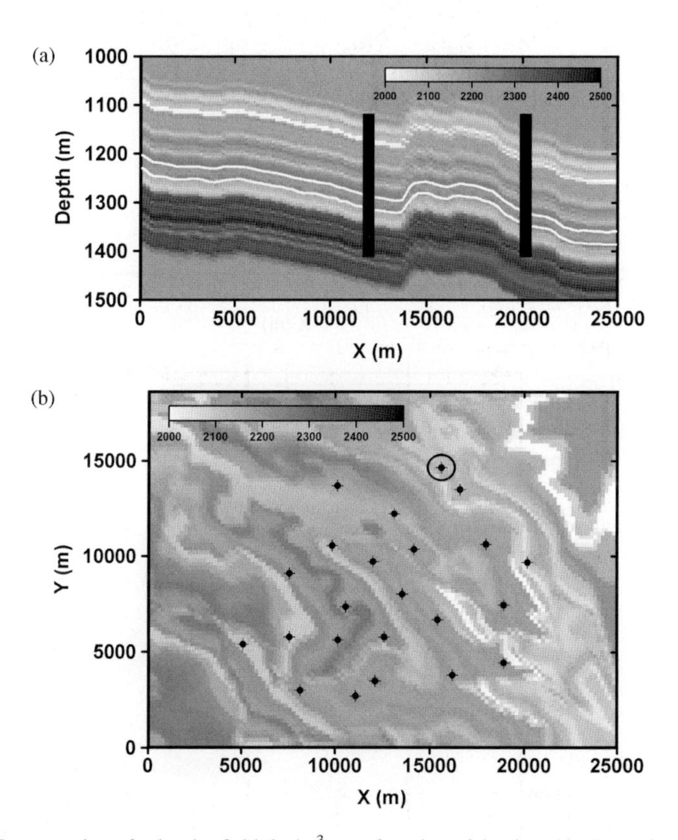

Figure 5: (a) Cross-section of a density field (kg/m^3) as a function of depth and horizontal position. (b) Plan view of a density (kg/m^3) field at a depth $z = 1200$ m. The circled well indicates the well location used for borehole gravity calculations shown in Figures 11 and 12.

The surface gravity response was calculated on a grid of stations with 1 km spacing from 2000 to 22,000 m in the x direction, and from 2000 to 16,000 m in the y direction. In general, since CO_2 is less dense (at reservoir conditions) than either oil or water, addition of CO_2 to the reservoir causes a reduction in the measured gravitational attraction either at the surface or in a borehole. (We assumed that porosity does not vary as gas is injected.)

The change in the vertical attraction of gravity (G_z) at the ground surface between 2020 and initial conditions is overlaid as black contours in Figure 6a on the net density changes within the reservoir.

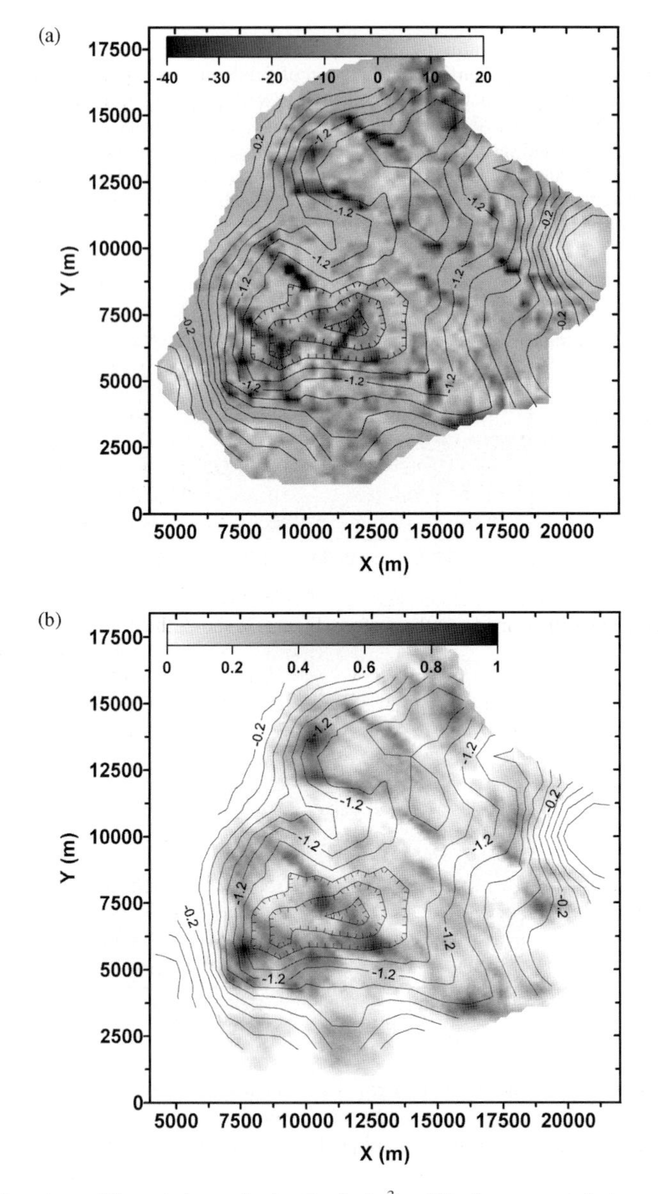

Figure 6: (a) Plan view of the net change in density (kg/m^3) within the reservoir between 2020 and initial conditions. (b) Plan view of the net changes in CO_2 saturation within the reservoir. The change in G_z at the surface for the same time interval is shown as black contours with hatch marks indicating decreasing G_z values [33].

The peak-to-peak change in G_z is on the order of 3 μgal, which is right at the level of repeatability of a field survey using current technology. The changes in the vertical gradient of gravity (dG_z/dz) between 20 years into CO_2 injection and initial conditions (not shown) are approximately 0.02 Eötvös units (EU), below the noise level of current instruments. The high spatial variations of the net density changes within the reservoir are expressed as a smoothed response at the surface (due to the depth of the reservoir) and only show the average changes on a larger scale.

It should be noted that petroleum reservoirs in general, and this reservoir in particular, are thinner (30 m) than many brine formations considered for CO_2 storage (100–200 m). This difference means that while the calculated response for Schrader Bluff at the surface are below current technology repeatability, brine formations at the same depths would produce measurable responses. This is the experience at the Sleipner CO_2 project [5] for a gravity survey conducted in 2002. These results suggest that future analysis with maximum sensitivity of G_z and dG_z/dz (that could be obtained by permanent emplacement of sensors with continuous monitoring coupled with surface deformation measurements to reduce noise levels) would be required.

Figure 6b shows the change in surface gravity G_z as black contours overlaid on the net change in CO_2 saturation within the reservoir. Because the density changes within the reservoir are caused by a combination of CO_2, water, and oil saturation changes as the WAG injection proceeds, there is not a one-to-one correlation in space between the net change in CO_2 saturation (S_{CO_2}) and the change in surface G_z. There is, however, a correlation between the change in surface G_z and the net change in S_{CO_2}, averaged on a large scale. For example, the largest changes in S_{CO_2} occur in the south–west quadrant of the image (Figure 6b) where the largest change in G_z occurs. This scenario, injecting CO_2 into an oil reservoir with multiple fluid components, is a worst case for the use of gravity to directly map changes in S_{CO_2}. In a case of CO_2 injection into a brine formation, there would only be water and CO_2, and the net changes in density within the reservoir would directly correlate with the net changes in S_{CO_2} as would the change in G_z at the surface.

Access to boreholes allows gravity measurement to be made closer to the reservoir, thus strengthening the signal compared to observations made on the surface. Figure 7a shows the change in G_z (2020–initial) at a depth of 1200 m (just above the reservoir in this section of the field), while Figure 7b is a change in dG_z/dz at the same depth. In both the figures, the data were calculated on the same grid of 1 km by 1 km site locations as on the surface. The shaded images in Figure 7a and b are the net density changes in the reservoir from Figure 6a. The changes in G_z and dG_z/dz, respectively, correlate directly with the maximum density changes. The magnitude of the changes in both G_z and dG_z/dz is larger than for surface measurements, although only the change in G_z would be measurable in the boreholes with current commercial technology. It should be noted, however, that work on more sensitive borehole G_z and dG_z/dz meters is ongoing and has the potential to significantly lower the sensitivity of such devices in the near future [6].

While Figure 7 illustrated the potential resolution by measuring close to the reservoir, access through only the existing injection wells would substantially reduce the data coverage. Figure 8a shows a map of contoured changes in G_z measured only in the 23 injection wells at a depth of 1200 m. Figure 8b is a net change of CO_2 saturation for comparison. Figure 8a was generated using a minimum curvature algorithm for data interpolation; however, it is representative of the general features present in all of the other types of interpolation tested. In general, interpretation of the interpolated G_z changes from the existing 23 boreholes would lead to an overestimate of the CO_2 saturation changes in the reservoir. This problem is particularly evident at the north end of the reservoir where increased CO_2 saturation at two isolated wells produces an interpolated image that would be interpreted as increased CO_2 between the wells where none exists. Borehole measurements need to be used in conjunction with some form of surface measurement to guide the interpolation between wells. Alternatively, pressure testing between wells could provide estimates of spatial variations in permeability that could be used to condition, in a statistical sense, interpolation of the borehole gravity data. Many possibilities exist for combining the borehole data with other information in order to produce more accurate maps of change within the reservoir. This is an area where further work could be done.

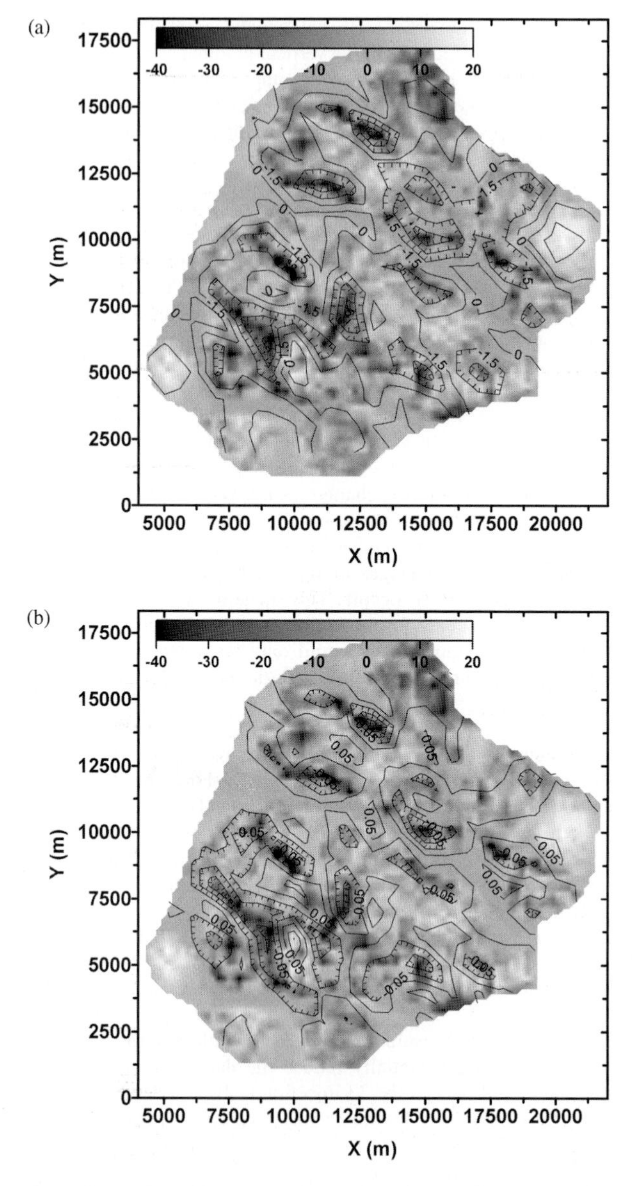

Figure 7: (a) Plan view of the net change in density (gray scale) within the reservoir (2020-initial). The change in G_z (μGal) at a depth of 1200 m is overlaid as black contours. The peak-to-peak change in G_z is approximately 10 μGal. (b) The change in dG_z/dz (EU) at a depth of 1200 m overlaid on the net change in density. The peak-to-peak change in dG_z/dz is approximately 0.3 EU.

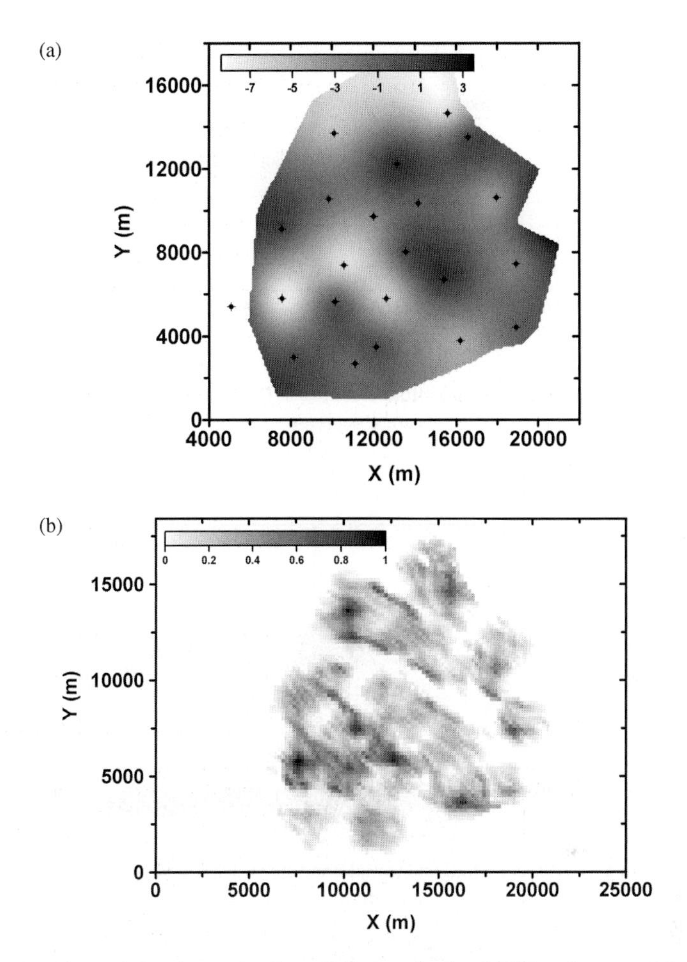

Figure 8: (a) Plan view of the change in G_z (μGal) at a depth of 1200 m between 20 years into CO_2 injection and initial conditions using 23 wells indicated by black symbols. (b) Plan view of the net change in S_{CO_2} within the reservoir between 20 years into CO_2 injection and initial condition.

In addition to considering spatial variations in G_z and dG_z/dz both on the surface and at a single depth within boreholes, the response of G_z and dG_z/dz in vertical *profiles* down boreholes was calculated. Figure 9 is the change in S_w between 2020 and initial conditions along a vertical slice through the reservoir at an injection well indicated by a circle in Figure 5b. Figure 10 shows the change in S_{CO_2} between 2020 and initial conditions. At the top of the reservoir near the injection well, S_w decreases while S_{CO_2} increases. At the bottom of the reservoir, both S_{CO_2} and S_w increase slightly. G_z measured in the borehole, shown in Figure 11a, reflects this change by a decrease in the response at the top of the reservoir, and an increase in the response at the bottom. The change in G_z is ± 8 μGal. The reservoir interval is between 1325 and 1350 m at this location. The change in G_z between 2020 and initial conditions (Figure 11b) clearly identifies the position of fluid saturation changes within the reservoir. The sign of the change reflects the changes in the local densities caused by the combined changes in all fluids (oil, brine and CO_2). The reservoir is outlined by the shaded gray area. The vertical gradient response (dG_z/dz) is shown in Figure 12a, and the change between 2020 and initial conditions is shown in Figure 12b. The change in the response is about 10 EU.

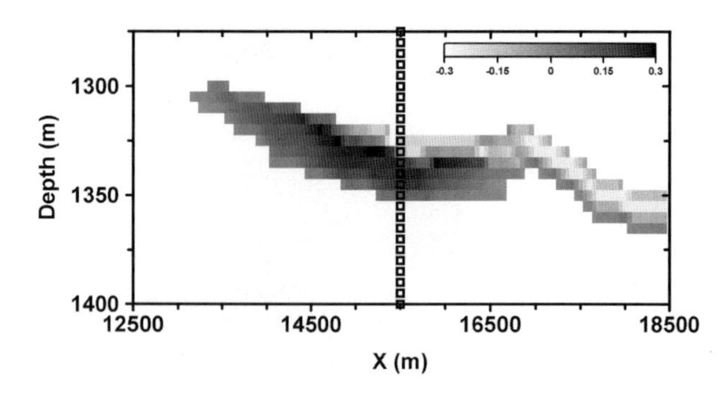

Figure 9: Change in S_w between 2020 and initial conditions. Dark colors are an increase in S_w, light colors are a decrease.

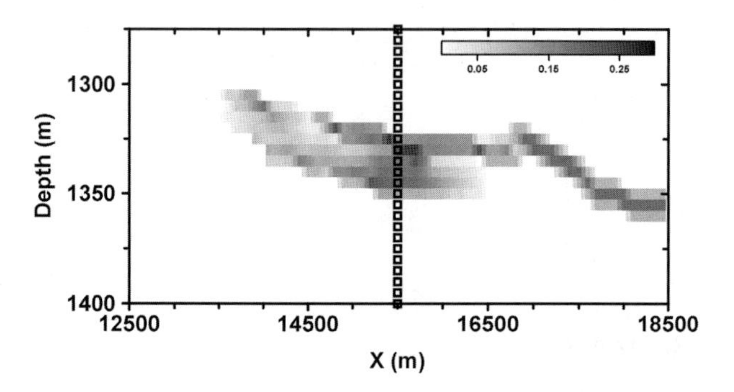

Figure 10: Change in S_{CO_2} between 2020 and initial conditions. Dark colors are an increase in S_{CO_2}, light colors are a decrease.

Popta et al. [7] showed that a geological structure with a sufficient density contrast can be detected by borehole gravity measurements if the observation well is not further away than one or two times the thickness of the zone of density contrast. Figure 13 shows a CO_2 wedge of 250 m radius and density of 2260 kg/m^3 (representing 20% CO_2 saturation in 20% porosity) inside of 100 m thick sand layer with a density of 2285 kg/m^3 at the depth of 1 km. The background density is 2160 kg/m^3. The borehole gravity response as a function of distance from the edge of the wedge is shown in Figure 14a. The maximum response at the edge of the CO_2 wedge is 10 μGal (due to 1% change in density). The responses decrease with distance away from the wedge: 50 m away from the wedge the response is 6 μGal, 100 m away response decreases to 4.4 μGal, and 200 m away it is down to 2.5 μGal. The borehole vertical gradient response for the same model is shown in Figure 14b. The response changes from 7 EU at the edge of the CO_2 wedge to 1 EU 50 m away from the edge. Current borehole gravimeter technology has a repeatability of around 5 μGal for G_z, this means that with current technology borehole measurements are sensitive to changes in a zone up to distances equal to the zone thickness away from the zone edge.

Seismic modeling

The flow simulation models for Schrader Bluff have been converted to acoustic velocity (V_p), shear velocity (V_s) and density, as previously described. A simulated seismic line (isotropic finite-difference algorithm with uniform overburden) has been calculated, running approximately N45°E across the reservoir.

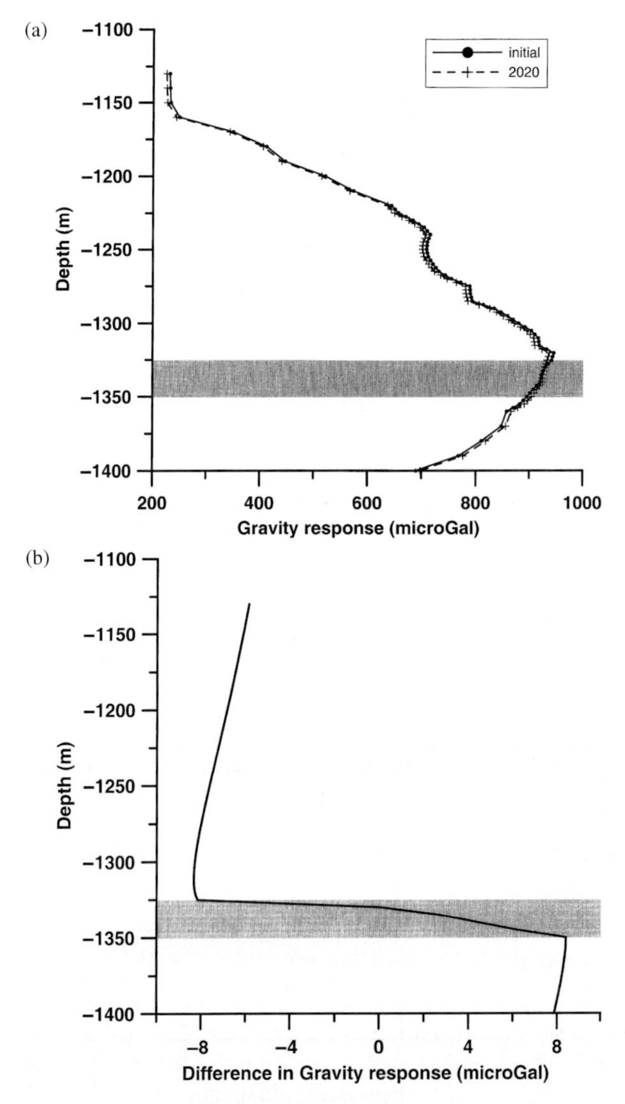

Figure 11: (a) Borehole G_z for initial conditions (circle) and 2020 (plus), (b) Change in G_z between 2020 and initial conditions. The reservoir interval is indicated by the light gray area.

The elastic response to a 50 Hz Ricker wavelet was calculated. The general increase in S_{CO_2} in portions of the reservoir near injection wells produces an approximately 20% decrease (between 2020 and 2005) in seismic velocity V_p as shown in Figure 15. The S_{CO_2} and S_w changes are shown in Figures 16 and 17, respectively. The seismic P-wave responses, for a single shot located at 7500 m (covering the area of the reservoir with maximum change in S_{CO_2}) on the 2D profile, for 2005 and 2020 are shown in Figure 18 with the difference shown in Figure 19. As discussed later, there is a significant Class 3 [8] type AVO effect as S_{CO_2} increases in the reservoir.

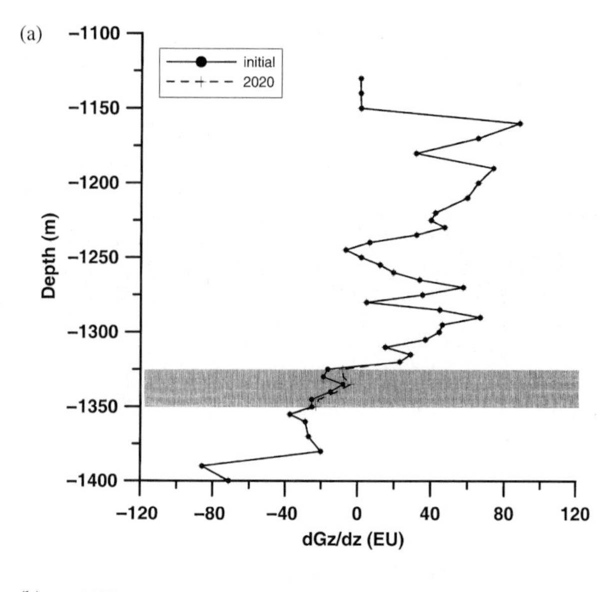

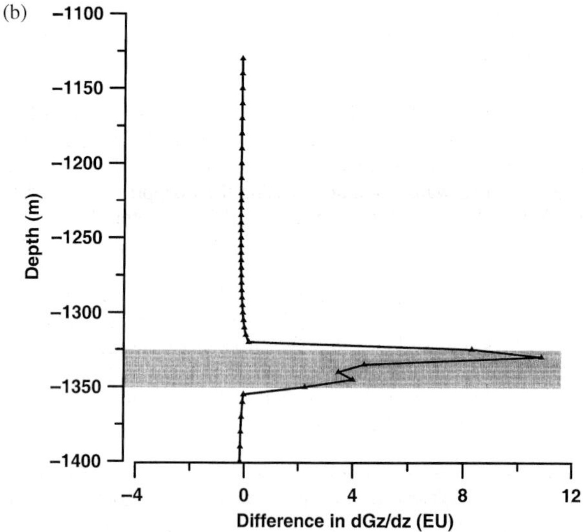

Figure 12: (a) Borehole vertical gradient response (dG_z/dz) for initial conditions (circle) and 2020 (plus), (b) Change in dG_z/dz between 2020 and initial conditions. The reservoir interval is indicated by the light gray area.

The P-wave response was sorted to CDP gathers, NMO corrected and stacked to produce the sections for 2005 and 2020 shown in Figure 20. The gray line is a constant time horizon within the reservoir for reference. The 30 m reservoir interval is not uniform and is comprised of 5 m thick substrata, each of which has reflection coefficients at their top and base that vary with S_{CO_2}. These sub-strata are all below the seismic tuning thickness. This produces a seismic response without a clear top and base reflector. There is a significant increase in S_{CO_2} to the right of CDP 8412.5 producing the large change in the stacked sections shown in Figure 20.

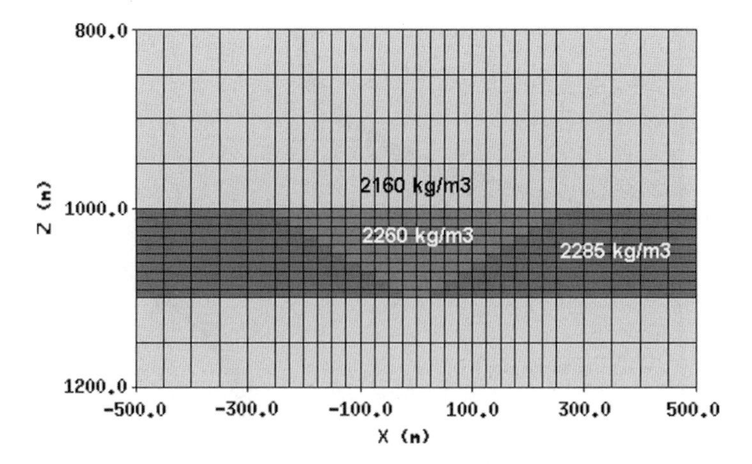

Figure 13: CO_2 wedge model.

The difference between the stacked sections between 2020 and 2005 is shown in Figure 21. Below the areas of major change in the reservoir (to the right of CDP 8412.5) the decrease in the velocity of the reservoir produces a time shift in the 2020 seismic responses below the reservoir, resulting in the events around 1100 ms that do not reflect CO_2 saturation changes at this depth, only the time shift from CO_2 above.

There is a large, and easily measurable, change in the stacked trace amplitude associated with the reservoir caused by the changes in S_w and S_{CO_2}. In addition, there is a change in the AVO effects as seen in Figure 19. Both amplitude and AVO can be exploited to make quantitative estimates of saturation changes under certain conditions. Convolutional forward calculations using the Zoeppritz equation for both the 2005 and 2020 models provide insight into the AVO dependence on model parameters. The forward modeling creates a synthetic seismic gather from a given set of elastic parameters V_p, V_s and density as a function of depth. The full Zoeppritz equation is used to compute the acoustic to acoustic (pp) reflection coefficient $R_{pp}(\theta)$ for each angle and at each layer boundary. Synthetic seismic CDP gathers are calculated by convolving the angle-dependent reflection coefficients with a 50 Hz Ricker wavelet. The convolution model assumes plane-wave propagation across the boundaries of horizontally homogeneous layers, and takes no account of the effects of geometrical divergence, inelastic absorption, wavelet dispersion, transmission losses, mode conversions and multiple reflections. Hence, it is easier to understand intuitively than the finite-difference modeling of Figures 18–21, but demonstrates similar features.

The change in V_p, V_s, and density within the reservoir (depth between 1250 and 1275 m) is shown in Figure 22. The synthetic CDP gathers as a function of angle are shown in Figure 23a and b for 2005 and 2020, respectively. The change in reflection amplitude between 2020 and initial conditions is shown in Figure 24. The AVO response of the composite reflections from the reservoir interval shows increasing negative amplitude with offset, a typical Class 3 gas response. The negative trough (associated with the top of the reservoir) increases its magnitude with offset and is followed by a peak, also increasing with offset.

Use of AVO in fluid saturation prediction. The AVO attributes of reflections from the reservoir can be used to estimate fluid saturations under certain circumstances. AVO data can be used to estimate the acoustic and shear impedance of the reservoir [9]. When used in a time-lapse sense, these data can provide estimates of the change in water saturation and pressure within the reservoir [10].

The ability to predict changes in water saturation and pressure within a reservoir is illustrated in Figure 25. Here, the methods referenced above, and the rock-properties model derived for the North Sea sands

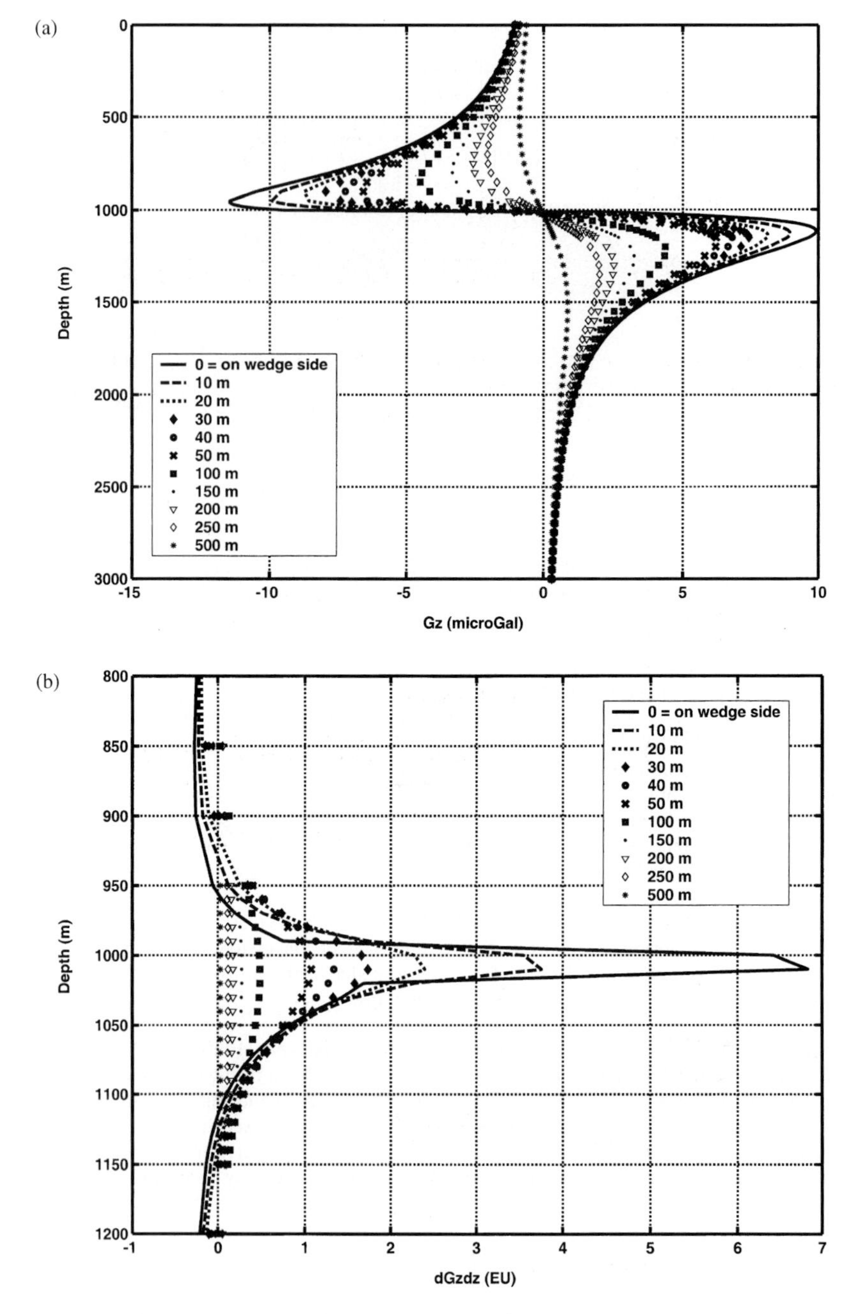

Figure 14: (a) Borehole gravity response of the model in Figure 13 as a function of distance from the wedge edge. (b) Borehole vertical gradient gravity response of the model in Figure 13 as a function of distance from the wedge edge.

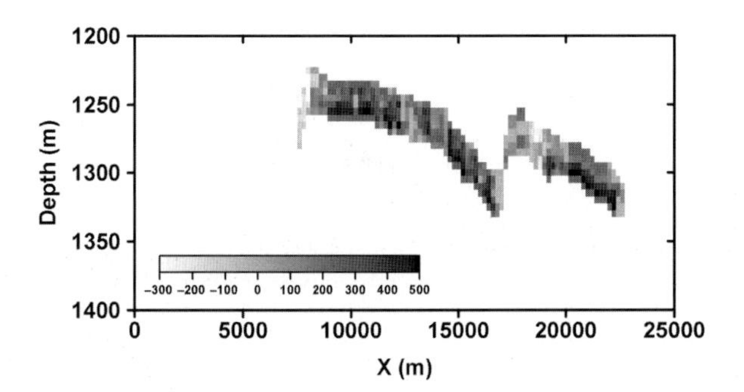

Figure 15: Change in the acoustic velocity (V_p) between 2020 and 2005 along a 2D profile extracted form the 3D model volume. The profile runs N45°E across the 3D model. Note the significant decrease in V_p associated with the increase in S_{CO_2} (Figure 16).

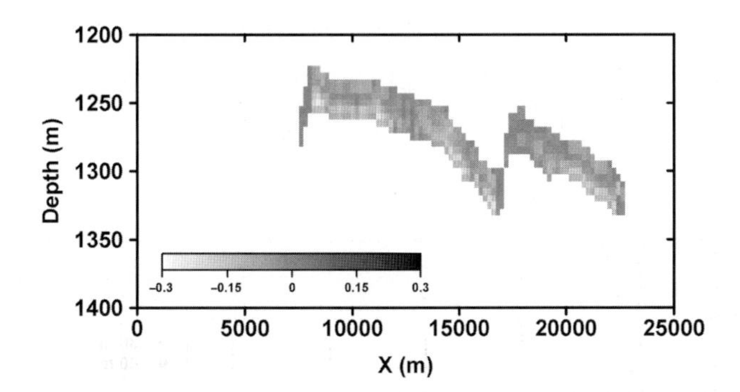

Figure 16: Change in the S_{CO_2} between 2020 and 2005.

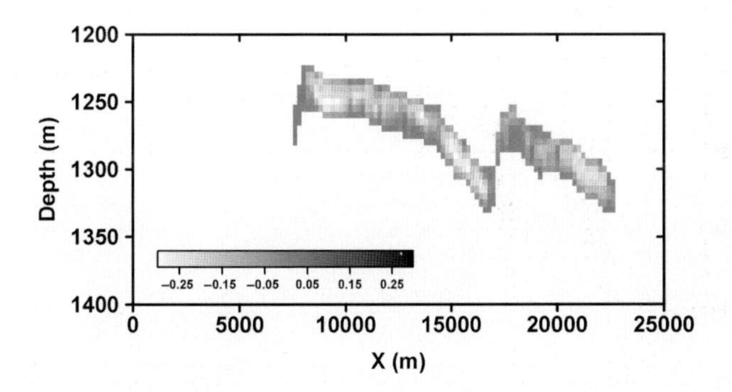

Figure 17: Change in S_w between 2020 and 2005.

of the Troll reservoir [4] is used to calculate the changes in shear and acoustic impedance of the reservoir as the water saturation and pore pressure for two cases of oil saturation as CO_2 is introduced. The first case (open circles) has initial oil and water saturation of 50%, as CO_2 is introduced it replaces water. The second case (closed circles) has an initial oil saturation of 60 and 40% water, with CO_2 replacing water. In both

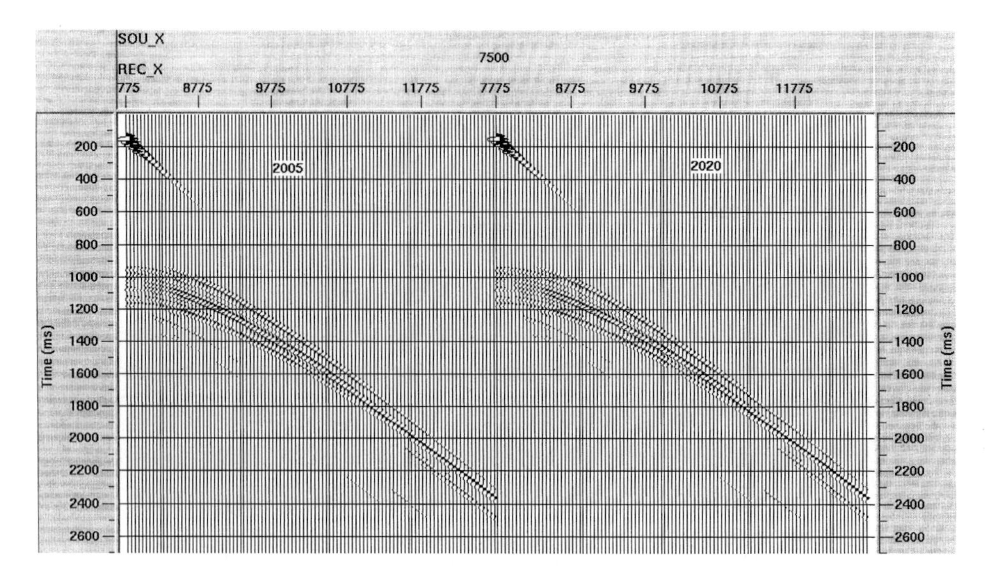

Figure 18: Seismic pressure response (shot gather) for 2005 and 2020.

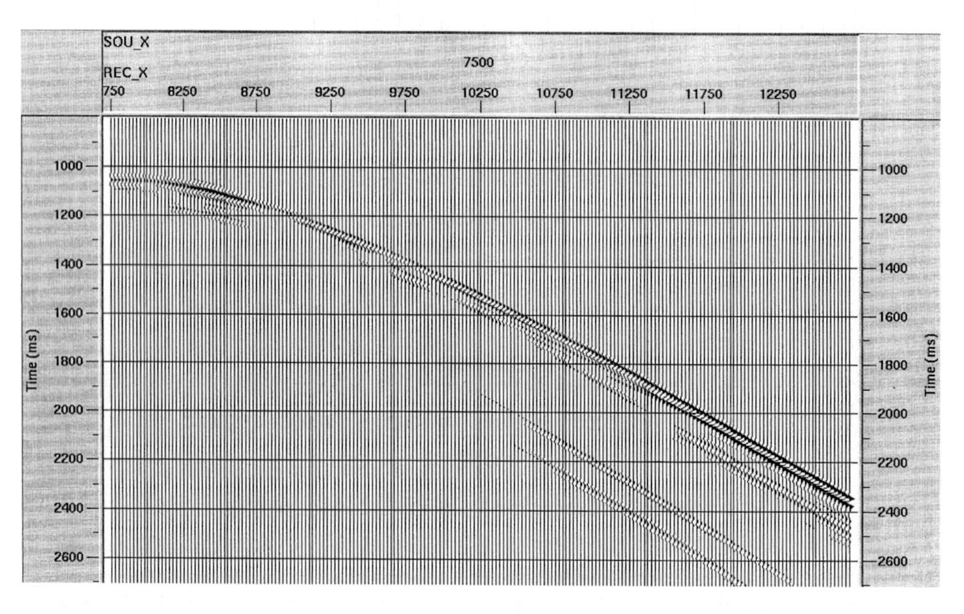

Figure 19: Change in pressure response (shot gather) between 2020 and 2005. Note amplitude change and AVO effects associated with S_w and S_{CO_2} changes in the reservoir.

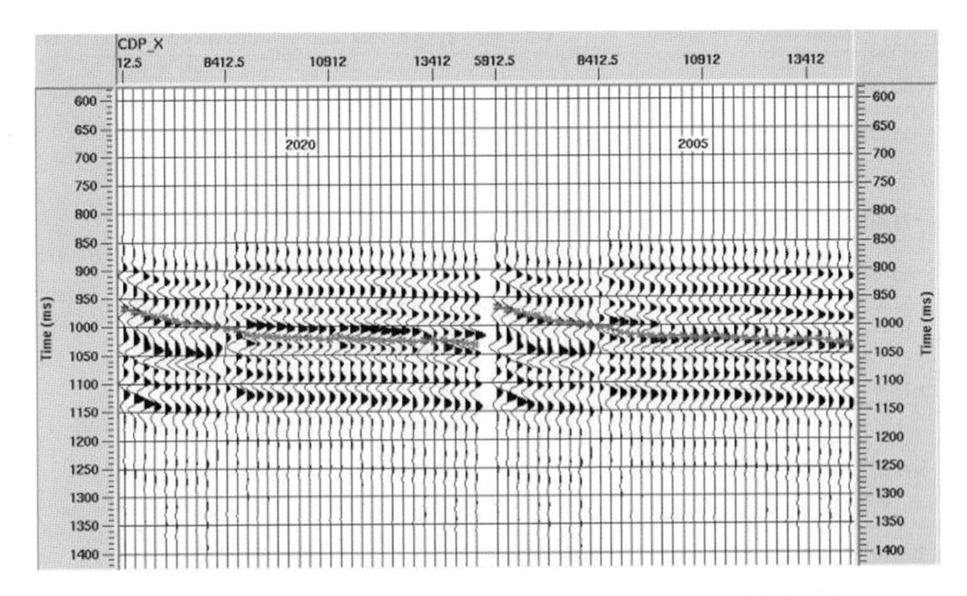

Figure 20: Stacked section for 2005 and 2020, gray line is constant time pick for reference.

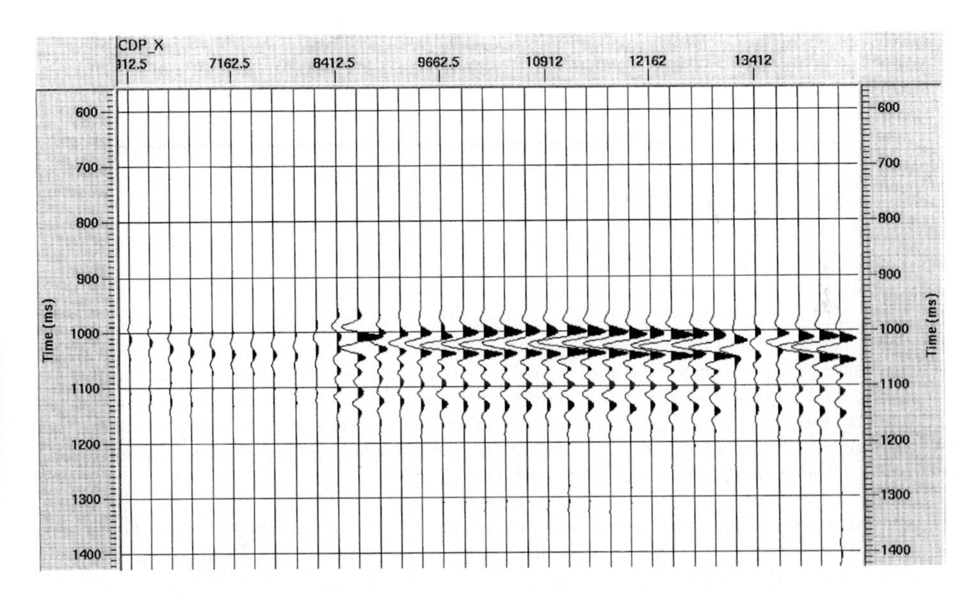

Figure 21: Change in the stacked sections between 2020 and 2005 (2020–2005).

cases S_{CO_2} ranges from 0 to 30%. Each point in the figure represents a unique value of S_w and S_{CO_2} with the oil saturation held fixed at either 50 or 60%. S_{CO_2} values increase in increments of 0.015% from right to left on the figure, and pore pressure increases and decreases (indicated by arrows) from the reference pressure of 24.24 MPa by increments of 0.7 MPa.

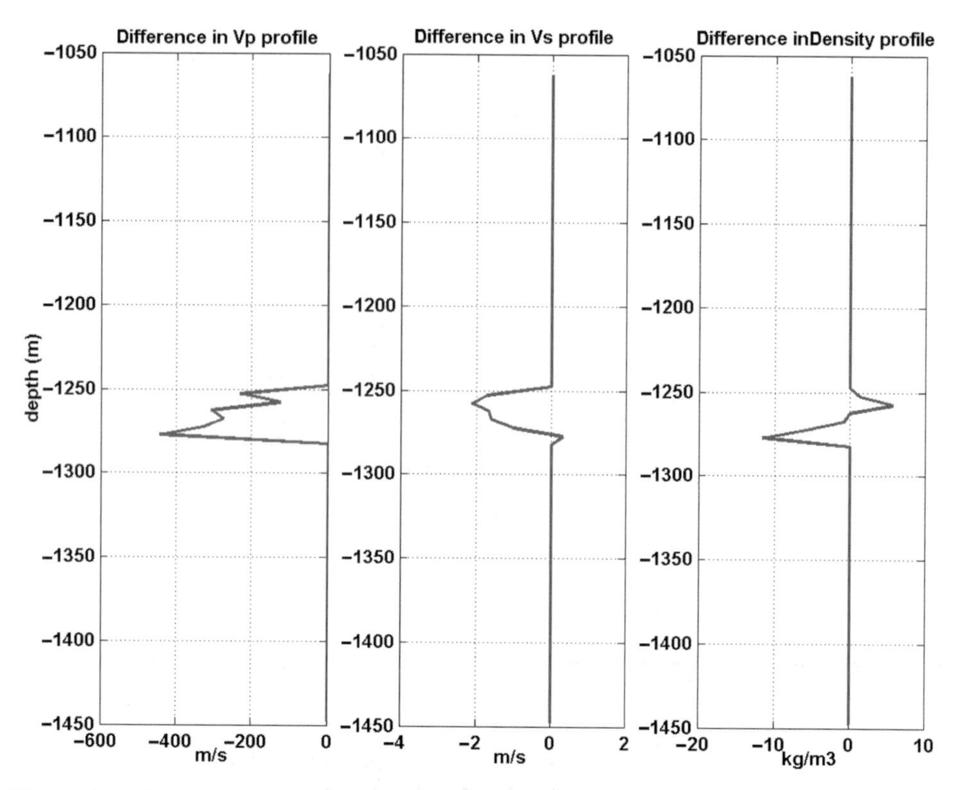

Figure 22: Difference in V_p, V_s, and density profiles between 2020 and 2005 for the Schrader Bluff model at the center of maximum CO_2 saturation increase.

Figure 25 illustrates four important points: (1) if the oil saturation is known, the changes in shear and acoustic impedance of the reservoir can determine the change in pressure and CO_2 saturation, (2) the changes in the shear impedance required to make the estimates is quite small and would require extremely good shear data, (3) an uncertainty in the oil saturation level of 10% in this example has only a small effect on the estimated values of changes in S_{CO_2} and almost no effect on the estimates of pressure change, (4) in this model, fluid saturation changes affect mostly the acoustic impedance, while fluid pressure changes affect mostly the shear impedance. In this example the change in the acoustic impedance alone could provide estimates of the change of S_{CO_2} even if the pressure changes could not be estimated due to insufficient accuracy on the shear impedance estimates.

An uncertainty on the value of oil saturation has limited effects in these calculations because of the relative similarity of the bulk modulus and density of oil, compared to water, when either is compared to CO_2. The situation is significantly different if there is hydrocarbon gas (such as methane) in the reservoir. In this case (due to the extreme differences between the properties of methane and water) even a small uncertainty in the hydrocarbon gas saturation leads to very large uncertainties in the estimated values of pressure and CO_2 saturation changes, making this technique essentially unusable unless an independent estimate of water saturation or gas saturation can be obtained from other methods [3].

While estimation of changes in fluid saturation using AVO is complicated by the multiple fluid components in oil or gas reservoir, the situation is simpler in a brine reservoir. For cases where CO_2 is injected into

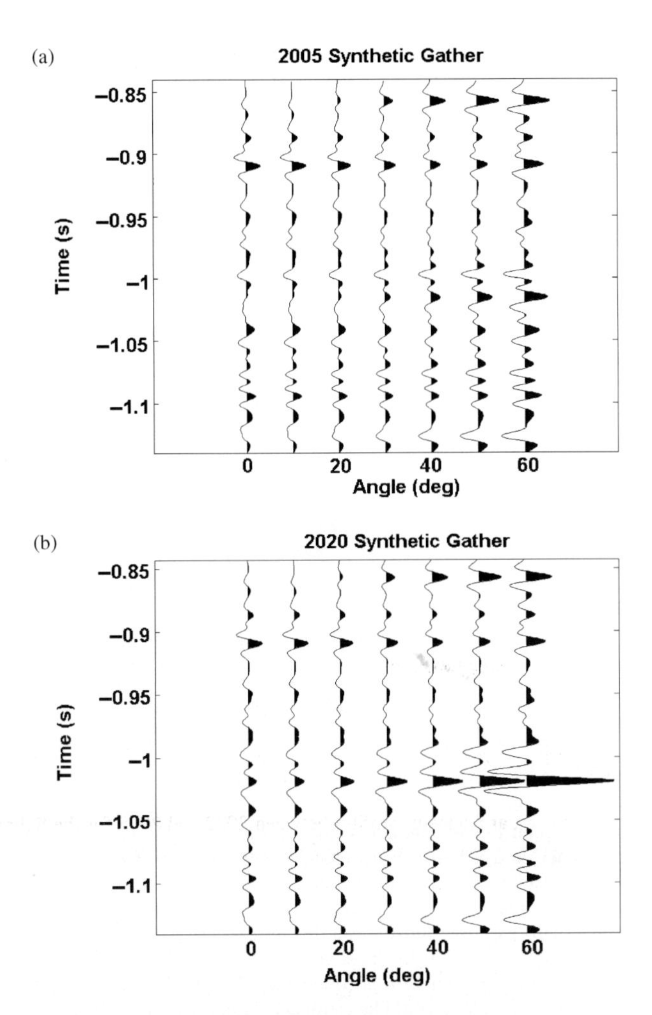

Figure 23: Synthetic gather for (a) 2005 and (b) 2020.

a brine reservoir, there are only two fluid components (brine and CO_2) and the added constraint that their saturations levels sum to one. In this case, AVO information can more easily be used to estimate the level of CO_2 in the reservoir. The following example illustrates this process. An unconsolidated North Sea sand of the Troll reservoir [4] encased in shale is assumed to contain 50% brine and 50% CO_2 as the reference point for these calculations. Pressure and temperature are such that the CO_2 is in the liquid state. The values of CO_2 (and hence water) saturation and pore pressure are varied about this starting point and the acoustic and shear velocities as well as density are calculated.

The reflection coefficient at the top of the reservoir can be approximated [11] by

$$R(\theta) \approx A + B\sin^2(\theta) + C\sin^2(\theta)\tan^2(\theta) \tag{1}$$

where θ is the average of the reflection and transmission angle for a plane wave hitting the interface. The constants A and B are referred to as the intercept and slope, respectively, in the AVO literature.

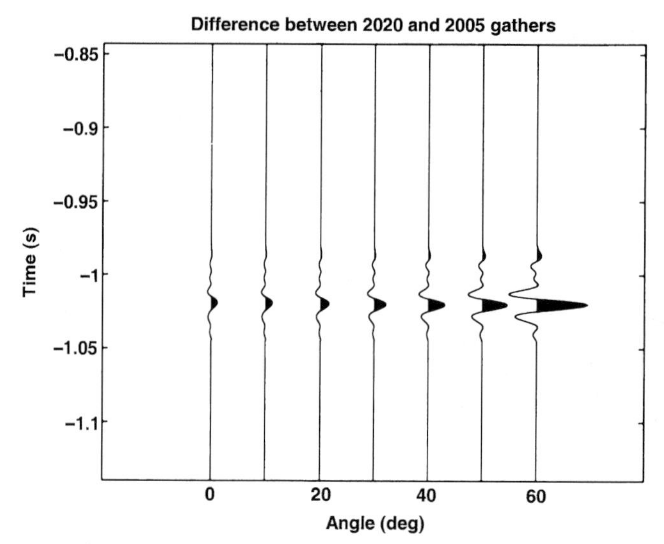

Figure 24: Difference between 2020 and 2005 gathers.

The constants A, B and C are functions of the velocity and density of the media on either side of the reflecting interface and are given by

$$A = 1/2(\Delta V_p/\langle V_p\rangle + \Delta\rho/\langle\rho\rangle) \tag{2}$$

$$B = 1/2(\Delta V_p/\langle V_p\rangle - 2(\langle V_s\rangle/\langle V_p\rangle)^2(2\Delta V_s/\langle V_s\rangle + \Delta\rho/\langle\rho\rangle)) \tag{3}$$

$$C = 1/2(\Delta V_p/\langle V_p\rangle) \tag{4}$$

where ΔV_p is the change in acoustic velocity across the interface and $\langle V_p\rangle$ is the average acoustic velocity across the interface, ΔV_s, $\langle V_s\rangle$, $\Delta\rho$, and $\langle\rho\rangle$ are changes and averages for shear velocity and density, respectively. If time-lapse seismic data is acquired, and A and B are estimated from the AVO data and used to calculate ΔA and ΔB, the associated ΔS_{CO_2} and ΔP_p can be estimated from model-based calculations such as illustrated in Figure 26. This example illustrates a theoretical case without noise in the seismic data; in practice estimation of the "curvature", C, is the most difficult. Extremely high signal-to-noise (S/N) seismic data would be required even for estimates of B accurate enough to make pressure change estimates. Even with poor estimates of B changes, in S_{CO_2} could be estimated from the changes in the zero offset impedance (A) because the contours in Figure 26a are nearly orthogonal to the ΔA axis.

Electromagnetic modeling

The electrical resistivity of reservoir rocks is highly sensitive to changes in water saturation. This can be seen from Archie's Law [12], which is commonly used to describe the electrical resistivity of sedimentary rocks as a function of water saturation, porosity, and pore fluid resistivity. Figure 27 shows the rock bulk resistivity (in Ωm) as a function of gas saturation ($S_g = 1 - S_w$) for a reservoir with brine resistivity equivalent to sea water ($\rho_{brine} = 0.33\,\Omega$m) with 25% porosity. All petroleum fluids (oil, condensate, and hydrocarbon gas) as well as CO_2 are electrically resistive, hence the relation shown in Figure 27 is appropriate for any combination of oil, hydrocarbon gas, condensate, or CO_2.

The bulk resistivity in Figure 27 is plotted on a log scale to span the large range of resistivity values as a function of the gas saturation (S_g). This high sensitivity to water saturation in a reservoir can be exploited by electromagnetic (EM) techniques, where the response is a function of the rock bulk electrical resistivity. Of all the possible combination of EM sources and measured EM fields, one system combines both relative ease of deployment with high sensitivity to reservoirs of petroleum scale and depth. This technique uses a grounded electric dipole that is energized with an alternating current at a given frequency to produce time

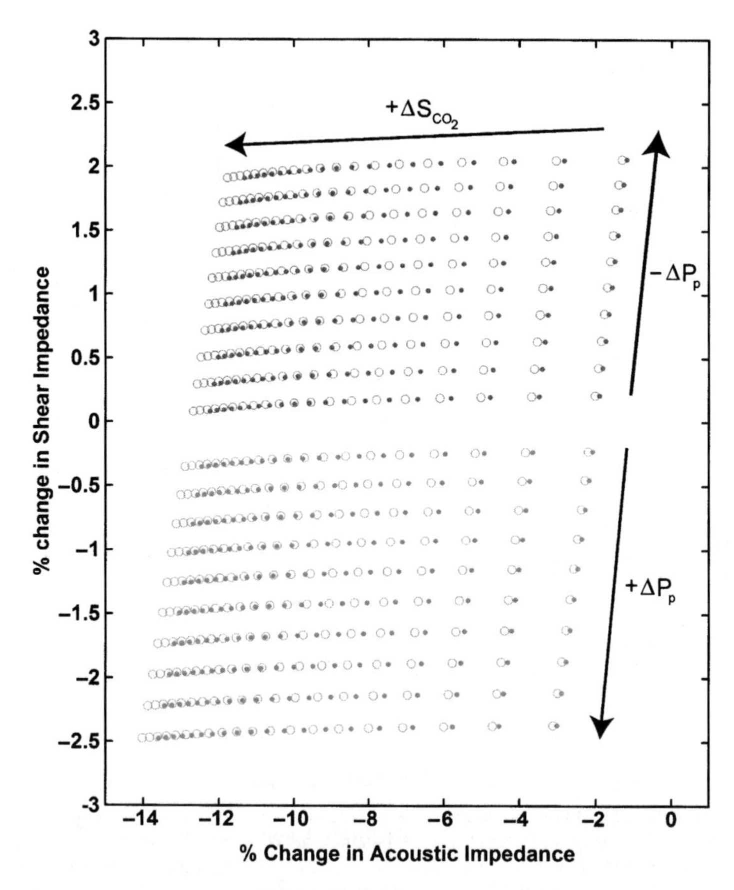

Figure 25: Changes in pore pressure (ΔP_p) and CO_2 saturation (ΔS_{CO_2}) as a function of changes in the shear and acoustic impedance of the reservoir. Open circles represent oil saturation of 50% with CO_2 replacing water. Filled dots represent oil saturation of 60% with CO_2 replacing water. Initial pore pressure is 25.24 MPa, initial S_{CO_2} is 0%. S_{CO_2} increments are 0.015 and pressure increments are 0.7 MPa.

varying electric and magnetic fields that can be measured on the earth's surface. The electric dipole can consist of two steel electrodes (1 m^2 plates or sections of drill pipe) buried at a shallow depth (1–10 m) separated by 100 m and connected by cable to a low-power generator (a portable 5000 W generator is sufficient). The measured data would consist of the electric field at a given separation from the transmitter acquired on the surface or within the near surface.

To simulate such an EM system we have calculated the electric field on the surface of the Schrader Bluff model using 100 m electric dipoles operating at 1 Hz with measurements of the resulting electric field at a separation of 2 km in-line with the transmitting dipole. Figure 28 shows the amplitude of the generated EM field at 2 km separation and 1 Hz together with the natural background electric field generated from worldwide thunderstorms and pulsations in the earth's ionosphere. Figure 28 shows that the generated electric field for the Schrader Bluff model, using only a small portable generator (producing a 10 A current in the source dipole) is an order of magnitude above the background electric field (noise) at the operating frequency of 1 Hz. This means that synchronous detection of the signal combined with stacking can recover signal variations to better than 1%.

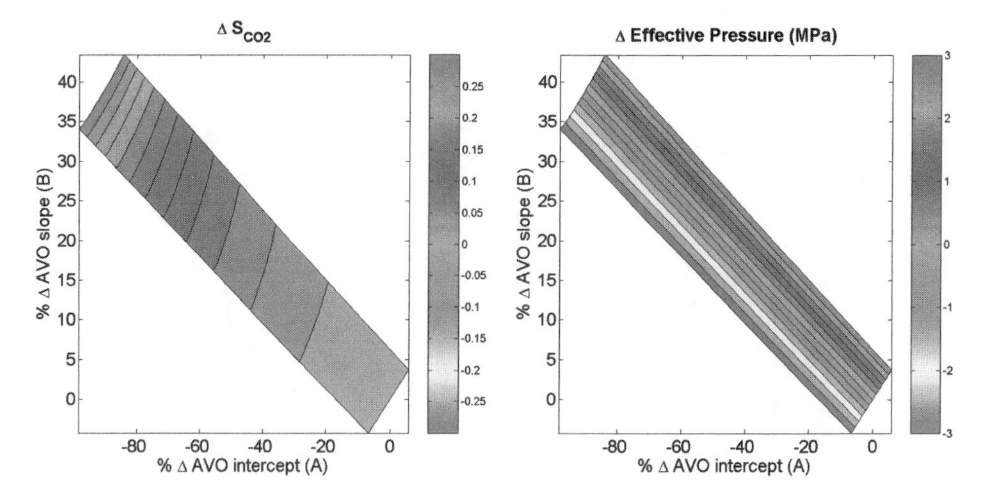

Figure 26: Contours of the change in CO_2 saturation (left panel) and effective pressure (lithostatic − pore pressure) (right panel) as function of the change in the AVO intercept (A) and slope (B) for an unconsolidated sand surrounded by shale.

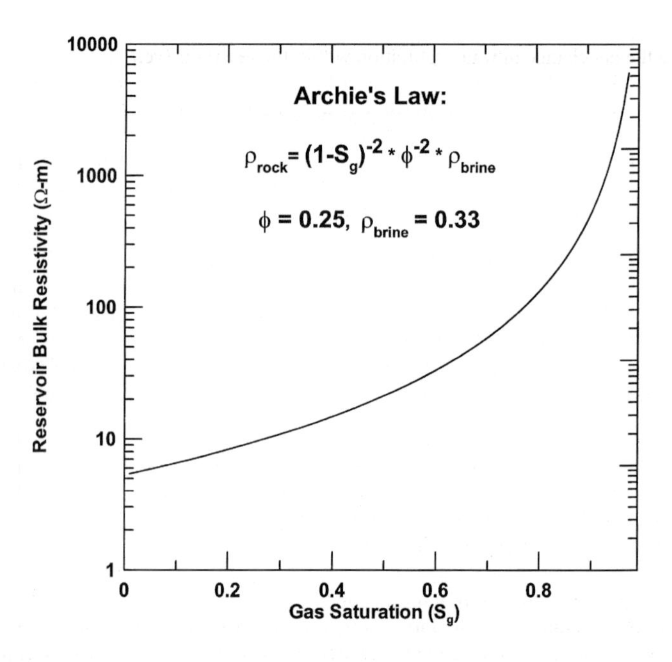

Figure 27: Reservoir bulk resistivity as a function of gas saturation (S_g). Porosity = 25%.

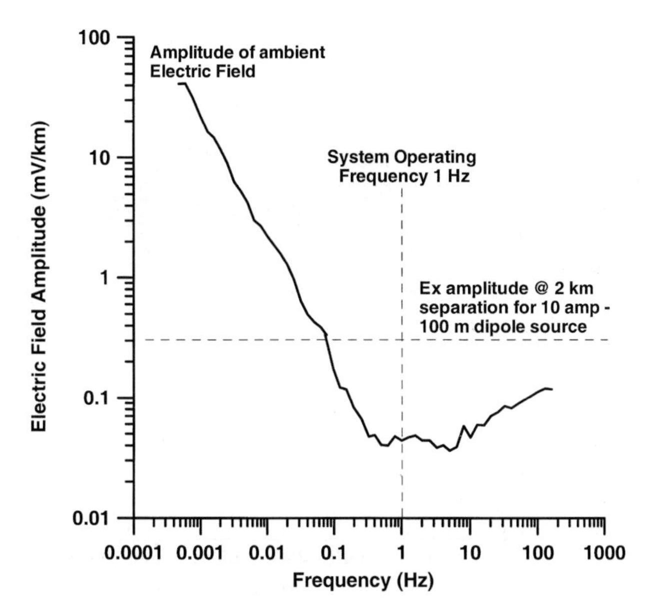

Figure 28: Amplitude of naturally occurring electric field as a function of frequency [34] that would be considered noise to that EM system considered here for monitoring, shown as solid curve. The horizontal dotted line represents the signal amplitude at a source–receiver separation of 2 km at an operating frequency of 1 Hz for a 100 m electric dipole energized with 10 A of current.

Figure 29 shows the net change in water saturation within the reservoir (vertically integrated ΔS_w) between 2020 and initial conditions. The change in the electric field amplitude for the same interval is overlaid as black contour lines, with peak-to-peak amplitude of 1.2%. There is a direct one-to-one correspondence with the change in S_w and the change in the electric field amplitude. While this signal level is low, it can be measured given the S/N ratio of the data (Figure 28). Although this represents a potential low-cost monitoring technique it is best suited for CO_2–brine systems where there is a one-to-one correlation between the change in S_w and the change in S_{CO_2} (since $S_w + S_{CO_2} = 1$).

In a petroleum reservoir such as Schrader Bluff, the presence of hydrocarbons as additional fluids eliminates the one-to-one correlation between changes in S_w and changes in S_{CO_2}. This is illustrated in Figure 30 where the same changes in electric field amplitude are overlaid on the net change in the CO_2 saturation within the reservoir between 2020 and initial conditions. In this case, we see that the correlation between changes in S_{CO_2} and changes in the electric field amplitude are not as good as seen between changes in S_w and the electric field data.

This type of EM technique has not yet been employed as a monitoring tool within the petroleum industry. However, EM technology is currently the subject of a significant upsurge in industry interest. Several commercial contractors are now offering this technique as a survey tool, most notably, in the offshore environment where marine EM is used as an exploration tool [13]. The equipment and service providers exist to apply this technique for monitoring in the future.

On-Shore Saline Aquifer—Frio Formation, Texas
Brine-bearing formations that are below and hydrologically separated from potable water reservoirs above have been widely recognized as having high potential for CO_2 storage. One of the most promising sites is the Frio Formation in Texas, which has been chosen as a field demonstration site as part of the US DOE and

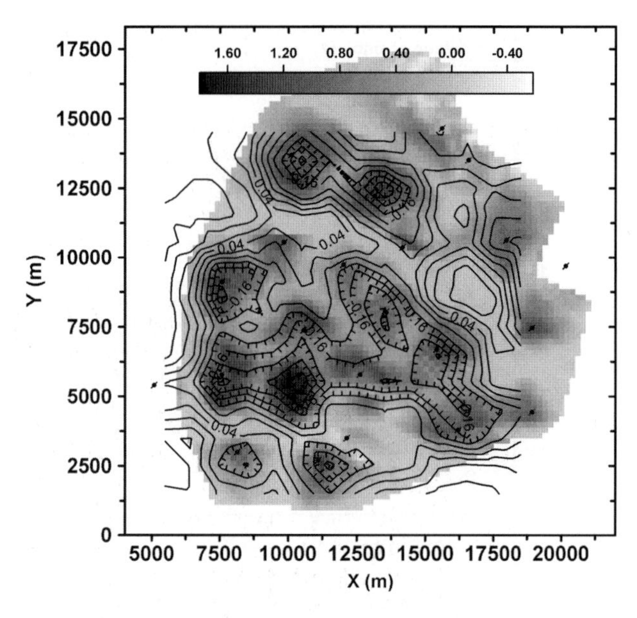

Figure 29: Shaded color map of the net change in water saturation over the vertical interval of the reservoir between 2020 and initial conditions. The change in the amplitude of the electric field from an electric dipole source at a separation of 2 km is overlaid as black contours. The peak-to-peak change in electric field amplitude is 1.2%. Note the direct correlation between decreases in the electric field amplitude and increases in water saturation (decreased electric resistivity of the reservoir). Locations of injection wells are shown by black circles with arrows through them.

National Energy Technology Laboratory (NETL) sponsored GeoSeq project. The test demonstration project has four main goals: (1) demonstrate that CO_2 can be injected into a saline formation without adverse health, safety, or environmental effects, (2) determine the subsurface location and distribution of the injected CO_2 plume, (3) demonstrate an understanding of the conceptual models, and (4) develop experience necessary for the success of future large-scale CO_2 injection experiments [14].

The South Liberty pilot test site lies on the south side of a salt dome (Figures 31 and 32). The injection target is the Frio Formation; strongly compartmentalized by a pattern of high-angle faults radiating from the salt dome and associated cross faults. The structure and fault boundaries used for modeling are based on structure and fault patterns mapped from 3D seismic data. This structural interpretation has a 440 m-wide compartment with fault boundaries on the northwest, northeast, and southeast. A fault boundary in the southwest side of the compartment was not imaged within the seismic volume, so the closure on this side is unknown and is considered as a variable in the modeling experiment. Within the compartment, strata are tilted off the salt dome. At the injection well, the top of the Frio Formation is at about 1500 m depth, strikes N70°W, and dips 15° toward the southwest. Stratigraphy employed for the flow modeling focuses on the selected injection interval, a 12-m thick high-porosity, high-permeability sandstone referred to as the C sand, which is separated into upper and lower halves by a thin (0.3 m) shale layer. The section below the thin shale, an upward-coarsening sand, is the actual injection target. Locally extensive shale deposited within the Frio during cycle-bounding flooding events form sealed boundaries at the top and bottom of the C sand. The thick regionally extensive shale of the Anahuac Formation overlies the Frio Formation and provides an additional impermeable boundary isolating CO_2 from the land surface. The regional geothermal gradient is taken to be 32.6 °C/1000 m [15]. For Frio water chemistry at these depths, reasonable values are

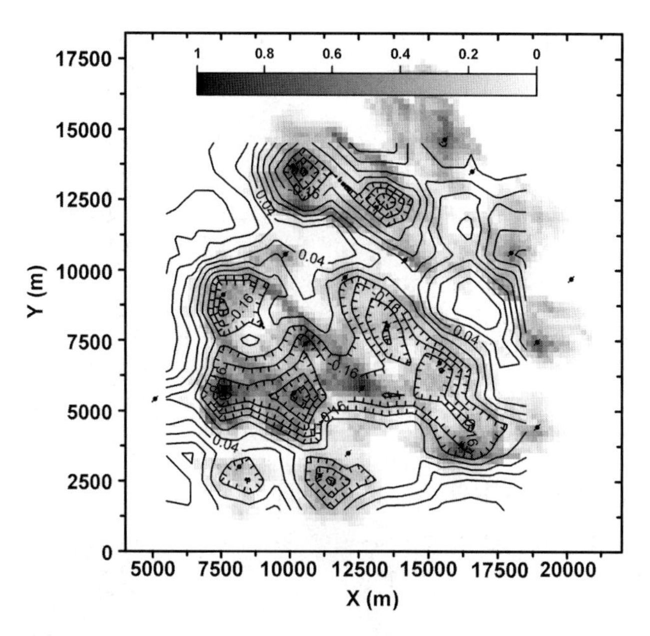

Figure 30: Shaded color map of the net change in CO_2 saturation (ΔS_{CO_2}) over the vertical interval of the reservoir between 2020 and initial conditions. The change in the amplitude of the electric field from an electric dipole source at a separation of 2 km is overlaid as black contours. The peak-to-peak change in electric field amplitude is 1.2%. Location of injection wells are shown by black circles with arrows through them.

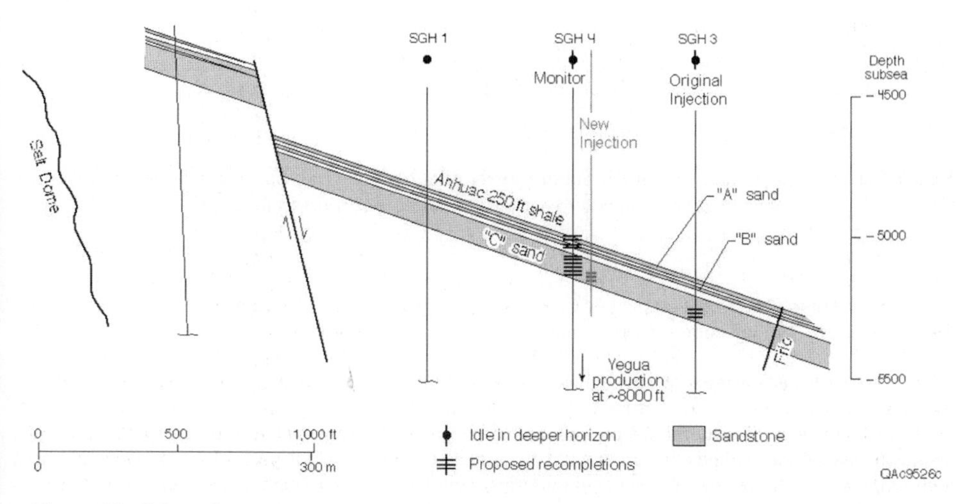

Figure 31: Schematic cross-section of the Frio Formation at the South Liberty pilot test site, Texas.

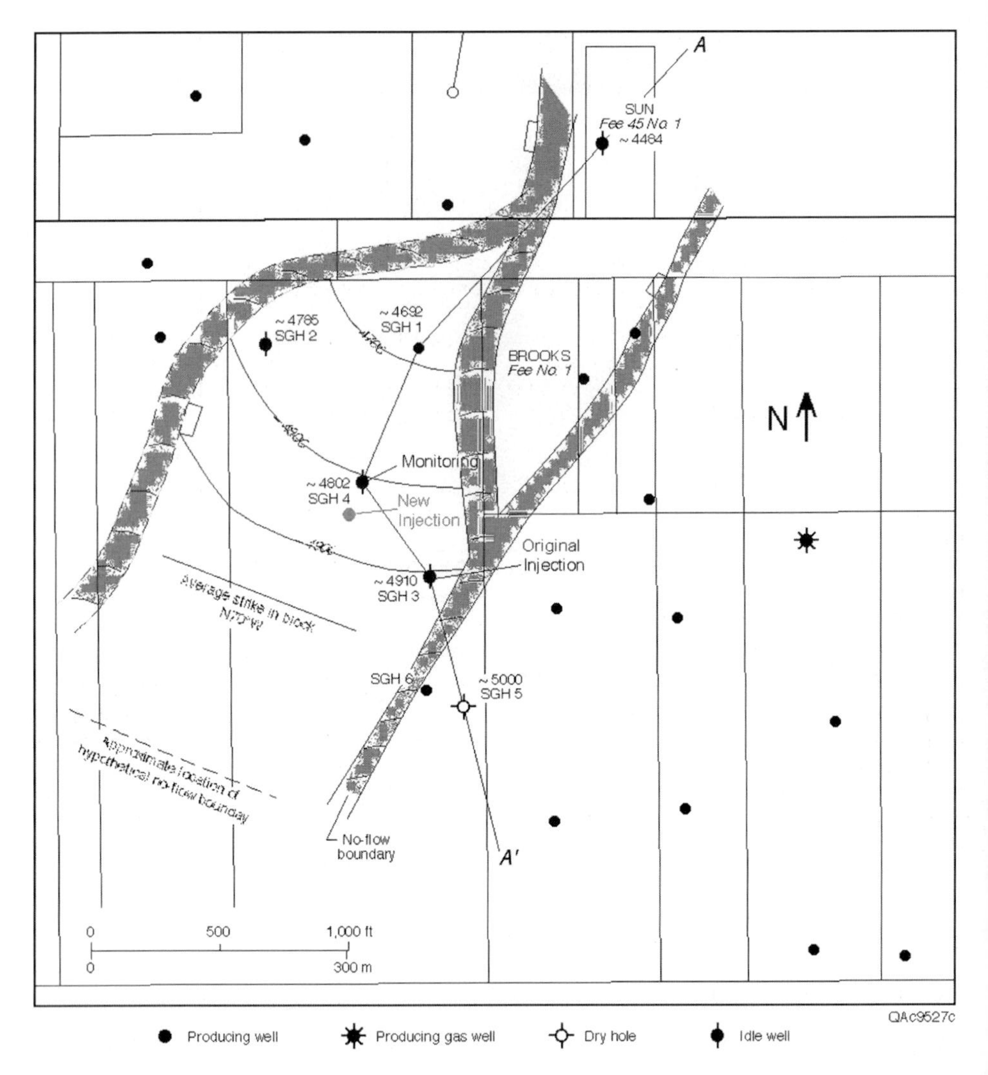

Figure 32: Schematic plan view of the South Liberty pilot test site. The shaded bands show sub-vertical faults that are assumed to act as impermeable barriers to fluid flow.

TDS 100,000 ppm, Na 35,000 ppm and Cl 45,000 ppm [16]; these values affect the fluid resistivities discussed below. The injection interval is non-productive of hydrocarbons.

CO_2 will be trucked to the site and injected into the high-permeability C sand within the upper Frio formation. There will be a series of field monitoring experiments before, during, and after CO_2 injection. These experiments will test effectiveness of a spectrum of CO_2 monitoring techniques and compare the results to validate the methods. Injection will be completed within 15–20 days, followed by up to a year of monitoring and assessment. There is one monitoring well, located about 30 m up-dip of the injection well (Figures 31 and 32).

Based on the geological setting of fluvial/deltaic Frio Formation, a 3D stochastic model of the C sand was created for fluid flow and transport modeling using a two-phase (liquid, gas), three-component (water, salt, and CO_2) system in the pressure/temperature regime above the critical point of CO_2 ($P = 73.8$ bars, $T = 31\,°C$) [18]. When CO_2 is injected in a supercritical state it has a much lower density and viscosity than the liquid brine it replaces, making buoyancy flow a potentially important effect. The model is bounded above and below by closed boundaries, which represent continuous shale. Three of the four lateral boundaries are closed to represent the edges of the fault block. CO_2 is injected at a rate of 250 metric tons per day (2.9 kg/s) for a period of 20 days, and then the system is monitored for an additional year. Initial formation conditions are $P = 150$ bars, $T = 64\,°C$ and TDS = 100,000 ppm. Under these conditions, supercritical CO_2 has a density of 565 kg/m^3 and a viscosity of 4.3×10^{-5} Pa s. In the reservoir, about 15% of the CO_2 dissolves in the brine, with the remainder forming an immiscible gas-like phase.

During the 20-day injection period, flow simulations show the distribution of CO_2 is nearly radially symmetric around the injection well (Figure 33). The plume arrives at the monitoring well in 2–3 days. After injection ends, the modeled plume begins to spread and it does not take long (approximately 30 days) for gas saturation to decrease to the residual value, making the plume essentially immobile.

During this test, less than 5000 tons of CO_2 will be injected into a 6 m thick sand unit at a depth of 1500 m. As such, it is a good limiting case for detection and resolving capabilities of geophysical monitoring techniques. A flow simulation model of the injection target was created using geo-statistical realizations of the sand shale distributions based on log data. Log data were used to construct rock-properties models that relate the reservoir parameters to geophysical parameters. These relations were used to convert the flow simulation model to geophysical models.

Streaming potential measurements
Fluid flow within a porous media can produce an electrical potential due to the separation of ions across flow boundaries. This phenomenon is the basis of the Streaming Potential (SP) method. SP has been used in geothermal exploration [19], in earthquake studies [20,21]), and in engineering applications [22–24]. Early model studies were based on polarized spheres or line dipole current sources. These techniques provided very little information about the nature of the primary sources. Marshall and Madden [25] discussed source mechanisms in detail and provided a technique for the solution of coupled flows that incorporated the primary driving potential. Sill [26] presented an alternative method for the solution of coupled flow problems that explicitly models both the primary flow and the induced secondary electric potential.

The measurement of the SP generated electric fields is a relatively simple and low cost measurement. The ease of the measurement coupled with the fact that the data is generated directly by the flow phenomena suggests a potential technique for low-cost, low-resolution monitoring.

The gradient of the electric potential (electric field) produced at a flow boundary by the SP is given by:

$$\nabla\phi = L\frac{\Gamma\mu}{k\sigma}$$

where L is the so-called "coupling coefficient", Γ the primary fluid flux, related to the pressure gradient by Darcy's Law, k the solution dielectric constant, σ the bulk conductivity of the rock, and μ the fluid viscosity.

A review of the literature showed that there was very little data on the coupling coefficient, L, for flow of CO_2 within sedimentary rocks. This led to a program of laboratory studies to measure this parameter. In the following sections we describe the laboratory and the numerical modeling studies.

SP laboratory studies. Laboratory studies were done for the SP due to CO_2 injection in Berea sandstone (Lang Stone, Columbus, OH). These are the first such measurements for CO_2 to our knowledge. The testing device held a 127 mm long core of 25 mm diameter (Figure 34). Tests were run on two different rock samples. Each sample was saturated prior to testing under vacuum for a period no less than 1 day. The pore

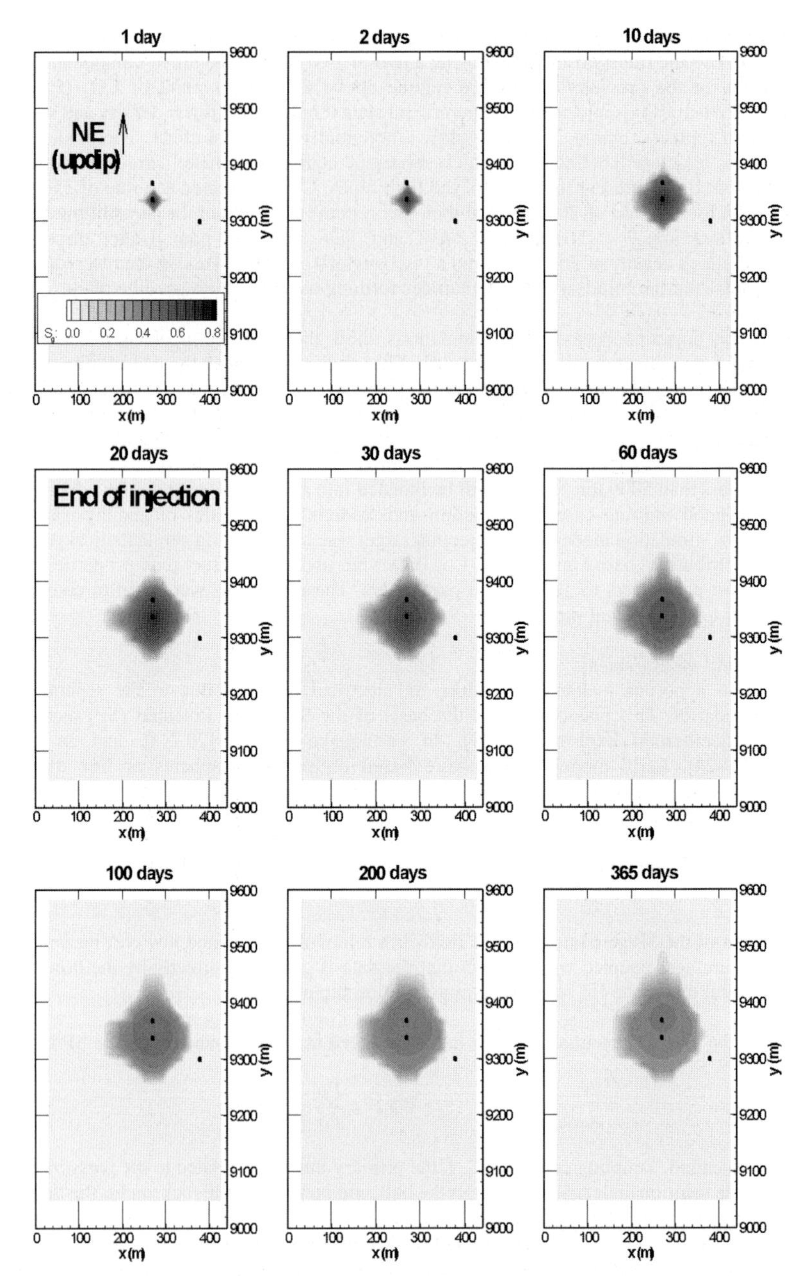

Figure 33: Plan view of gas saturation (S_{CO_2}) distribution at the top of the injection interval within the C sand, for a series of times during and after CO_2 injection. The three black dots show the locations of well SGH-3, well SGH-4, and the new injection well (see Figures 31 and 32).

Figure 34: Testing device containing Berea sandstone core. Sample is 127 mm long and 25 mm diameter.

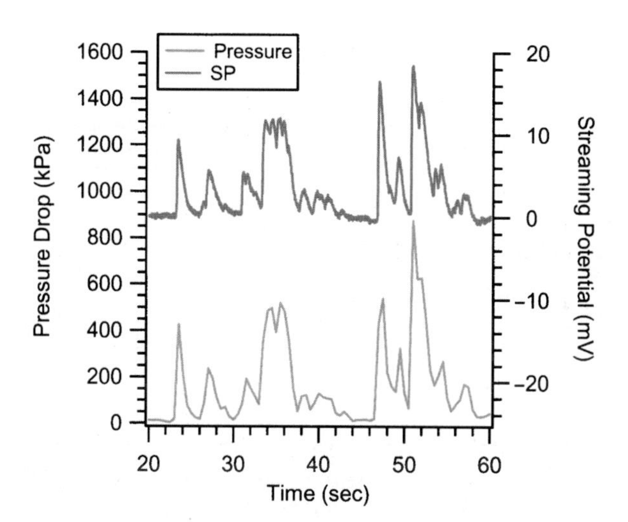

Figure 35: Streaming potential and pressure changes as a function of time as CO_2 is injected into the core sample.

fluid for initial saturation was Berkeley tap water, tested to have a resistivity of 125 Ω m. The coupling coefficient for the rock/water case was determined both before and after each CO_2 flood of two samples using a low-pressure static head method. Between these tests, liquid CO_2 was flowed over each sample. Test 1 allowed liquid CO_2 to flow through the sample for $1\frac{1}{2}$ h, while test 2 lasted 1 h. Figure 35 illustrates that the observed potentials and applied pressure changes correlated well throughout the testing. For these low-pressure tests, results indicate linear correlation of applied pressure and observed potential, as illustrated in Figure 36. When liquid CO_2 was applied to the sample, the water in the sample pore space was displaced, while reacting with the CO_2 to form carbonic acid. The coupling coefficient evolved over time in response to the mixing and displacing of the pore water. Figure 37 shows the coupling coefficient evolution of both tests

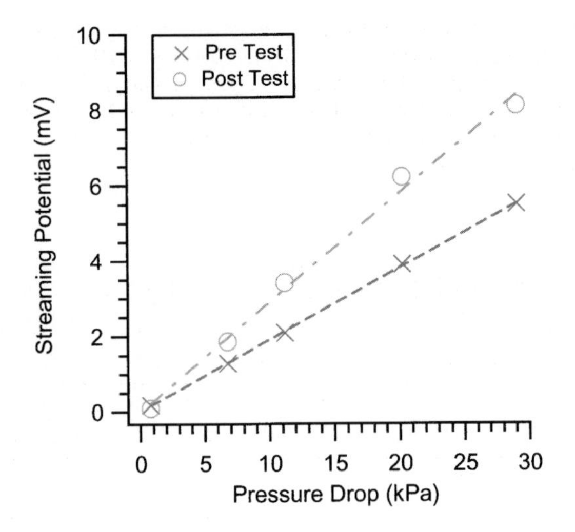

Figure 36: Results for static head testing to determine water-only coupling coefficient both prior to and following CO_2 injection test 2. Resistivity of pore fluid was 125 Ω m. Slope of line indicates coupling coefficients of 20 mV/0.1 MPa (pre) and 30 mV/0.1 MPa (post).

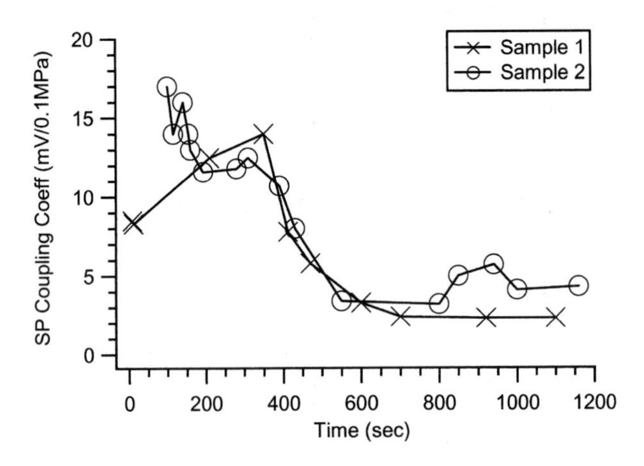

Figure 37: Coupling coefficients as a function of time for the first 20 min of CO_2 injection for samples 1 and 2. Coupling coefficient values were steady for times greater than 700 s, and remained steady throughout the remaining testing time.

for the 20 min following CO_2 injection. The results of the test are summarized in Table 1. As the CO_2 displaced the water the coupling coefficient decreased. On average, the coupling coefficients observed for steady CO_2 flow is about 10 times lower than for water flow in the same sample. Since the liquid CO_2 coupling coefficient is smaller than that of water, the most effective way to monitor spatial variation in injected CO_2 flow is to monitor the progressing CO_2/water front, where the coupling coefficient is largest.

SP modeling. In order to determine the magnitude of the SP response a 2D numerical model based on the geology and configuration of the Liberty Field CO_2 injection test was used. The model consists of a 10 m thick sand unit at a depth of 1,500 m embedded in shale. The resistivity of the sand unit is 2 Ω m, while the resistivity of surrounding shale is 1 Ω m. The flow rate of CO_2 is 350 kg/s; the viscosity of CO_2 is 0.073×10^{-3} Pa s and the density of CO_2 is 788 kg/m^3 at a temperature of 70 °C and a pressure of 30 MPa. The model is shown in Figure 38a. The 2D algorithm developed by Sill [26] was used. This algorithm assumes the fluid sources to be a line perpendicular to the geologic variation at steady state conditions (constant flow of a single-phase fluid).

TABLE 1
SUMMARY OF COUPLING COEFFICIENT RESULTS

	Pre-test (water)	**During (CO_2)**	**Post-test (water)**
Sample 1	45	2.5	15
Sample 2	20	3.5	30

All units are in mV/0.1 MPa.

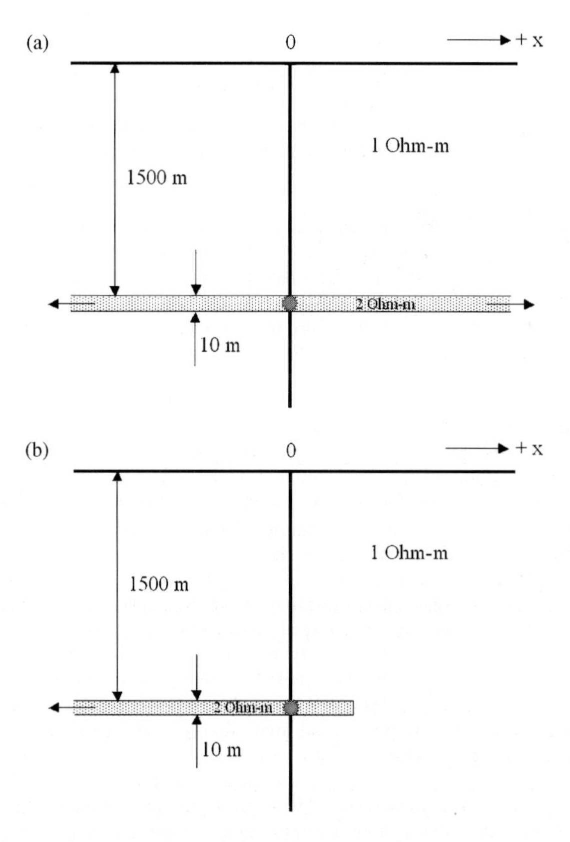

Figure 38: (a) Continuous layer model simulating the Liberty Field geology: 10 m thick sand layer at a depth of 1500 m. (b) Layer truncated at +300 m in x.

Figure 39a shows the pressure distribution for the model in Figure 38a with the associated electric potential shown in Figure 39b. In general, SP noise sources are on the order of a few to 10 s of mV although this number is highly site specific. SP signals over 10 mV are considered large.

The model shown in Figure 38b has the same parameters as the model in Figure 38a, except that the sand layer is terminated at +300 m. Comparison of results from these two models give an indication of the ability of SP surface measurements to resolve lateral variations in the subsurface flow of CO_2. The largest effect of the layer truncation is to concentrate the flow onto the left side of the model, increasing the flux and the pressure gradient there, thus increasing the magnitude of the SP observed at the surface. The truncation of the layer also introduces an asymmetry in the surface SP response (solid curve in Figure 40). The response is 10 mV higher on the truncated side than on the continuous side. The ability to differentiate this spatial variation in the signal will depend on the background noise level in the electric fields on the surface.

The effects of layer depth on the SP response are shown in Figure 41. A 100 m thick sand layer (properties taken from the Liberty test site) is placed at depths of 500, 1000, 1500, and 2000 m, respectively. The deeper the sand layer is the smaller is the signal amplitude on the surface.

Another aspect of interest is the effects of CO_2 flow rate on the SP response. Figure 42 illustrates that the SP response increases with CO_2 flow rate. The model used in this figure has a 100 m thick layer at a depth of 1000 m; all other parameters were the same as previous models. The flow rates used were 440, 293, and 40 L/s m, respectively.

To study the relationship between the thickness of the layer and the SP response models with 10, 30, 100, and 200 m thick sand layer at the depth of 1000 m were run; all other parameters were unchanged. Figure 43 shows that the amplitude of the SP response is inversely proportional to the thickness of the layer. The 10 m thick layer produces the largest response. The thinnest layers produce the largest response because the SP response is linearly proportional to the fluid flux, so that for a given injection rate, the thinner layers have a higher fluid flux.

Figure 44 illustrates how the SP response depends on the coupling coefficient *L*. The Liberty Field injection target is a 10 m thick layer at 1500 m depth with a lateral extent of 500–600 m. Its permeability is 150 milliDarcies, the flow rate is 4 L/s, and the viscosity of CO_2 is 73 μPa s. The model was run for three different values: 15 mV/atm (0.148 V/MPa), 57 mV/atm (0.5625 V/MPa), and 100 mV/atm (0.9869 V/MPa) representing a linear progression from potable water ($L = 15$) to resistive benzene ($L = 100$). Figure 44 shows linear dependence between the cross-coupling coefficient and the SP response.

Gravity modeling

In order to set some limits on the size and depths of CO_2 plumes that can be detected and resolved by surface gravity measurements, a wedge model of 240 m radius at the depth of 1000 and 2000 m was considered. The rock parameters were taken as general onshore Texas values of density. The surrounding shale was modeled having a density of 2240 kg/m^3 with the sand layer having 20% porosity and being brine saturated with a density of 2280 kg/m^3. The 3D wedge of CO_2 saturated sand was considered to be 100% saturated with CO_2, which resulted in a density of 2200 kg/m^3 for the wedge.

Figure 45 shows three surface response curves of the vertical component of the gravity field for the top of the wedge at 2000 m depth. The radius of the wedge is 240 m. The simulation was run for 100, 50, and 30 m thick wedges. A reasonable number for land gravity sensitivity levels is 2 μGal. For this depth, even the response of the 100 m thick wedge is below this level. This wedge (with thickness of 100 m) contains the equivalent amount of CO_2 produced by a 1000 MW US coal fired power plant in 41 days. Since the response of the 100 m thick wedge is just below the 2 μGal level, this indicates that amounts larger than 41 days production could be detected but not resolved.

A second set of models with the wedge at 1000 m depth were run; their responses are shown in Figure 46. With the CO_2 plume at 1000 m, both the 50 and 100 m thick volumes are detectable. The observed gravity response for the 100 m wedge is large enough to be resolved to some degree. Our conclusions to date are that gravity will most likely only be a useful monitoring technique for accumulations of CO_2 with depths on

(a)

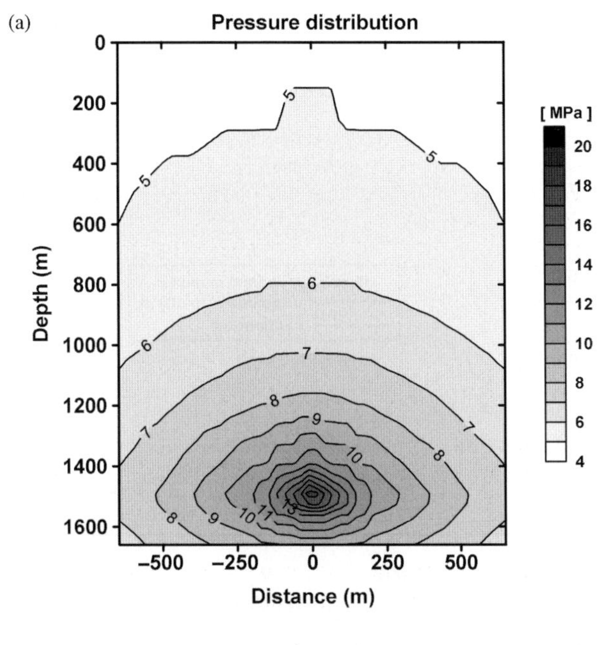

(b)

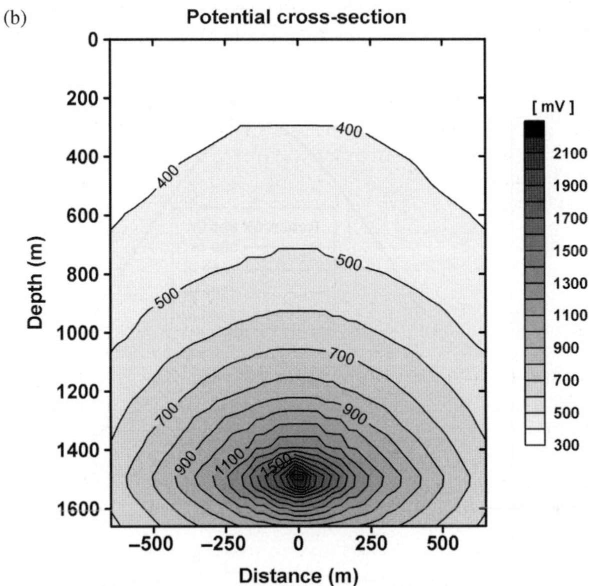

Figure 39: (a) Pressure distribution for the model from Figure 38a. (b) Electric potential cross-section for model in Figure 38a with coupling coefficient, $L = -15\,\mathrm{mV/atm}$.

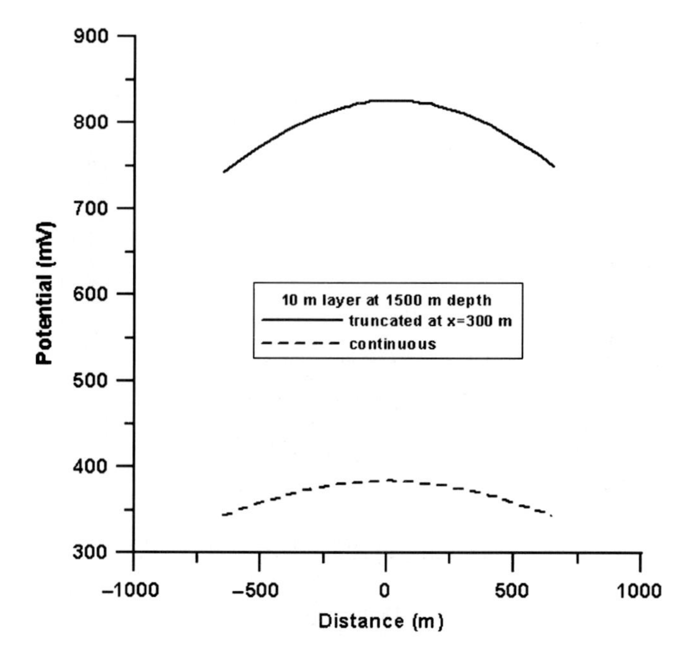

Figure 40: Surface SP response for models shown in Figure 38. Dash curve is for continuous layer; solid curve is for the truncated layer.

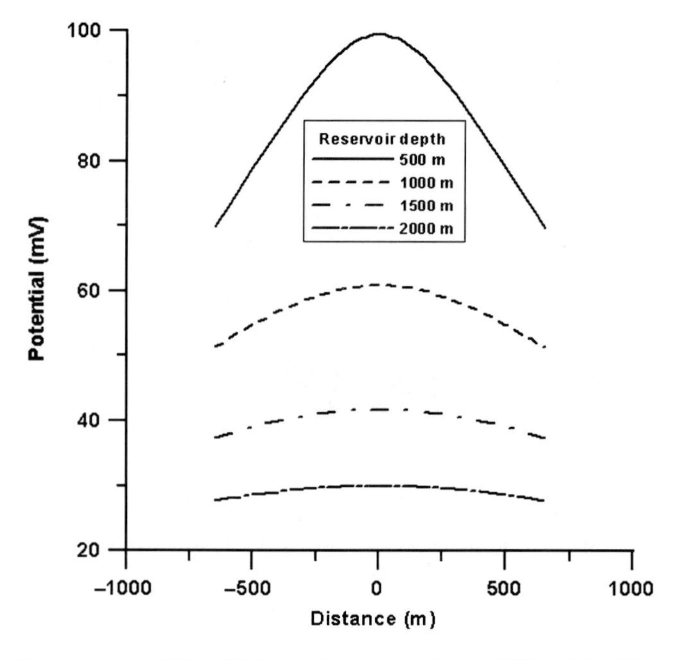

Figure 41: SP response for 100 m thick sand layer at the depth of 500, 1000, 1500, and 2000 m.

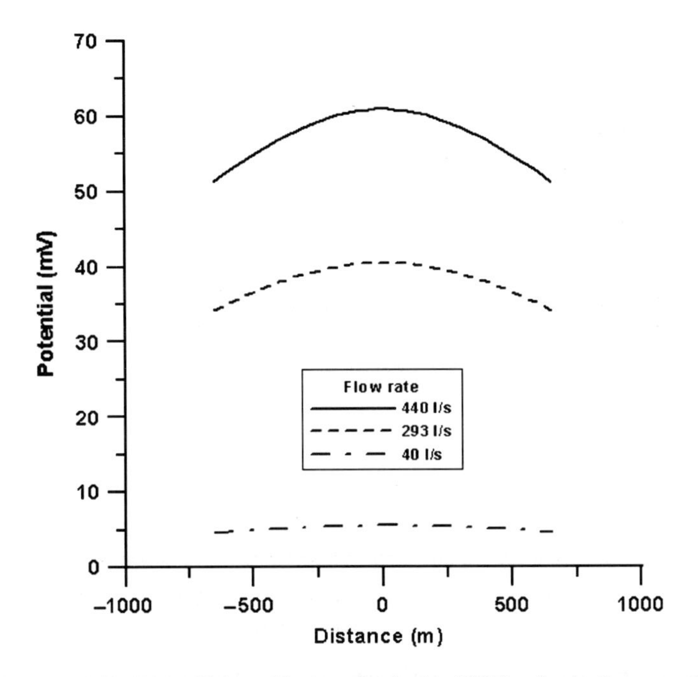

Figure 42: SP response for 100 m thick sand layer at the depth of 1000 m for the flow rate of 440, 293, and 40 L/s m.

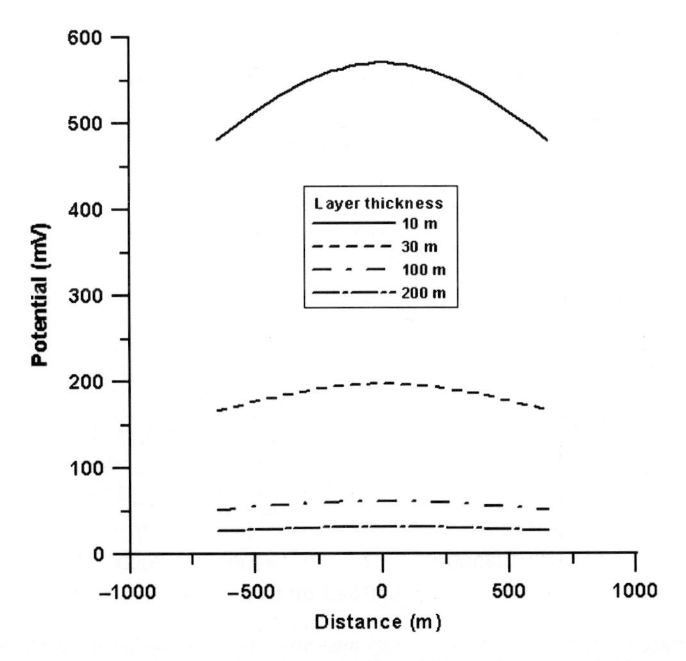

Figure 43: SP response of the 10, 30, 100, and 200 m thick sand layer at the depth of 1000 m.

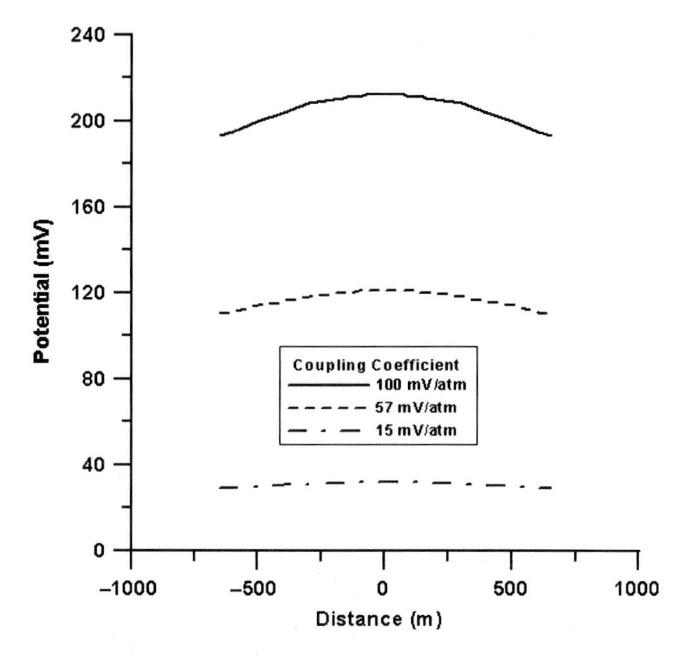

Figure 44: SP response of the Liberty Field reservoir for the coupling coefficient of 15, 57, and 100 mV/atm.

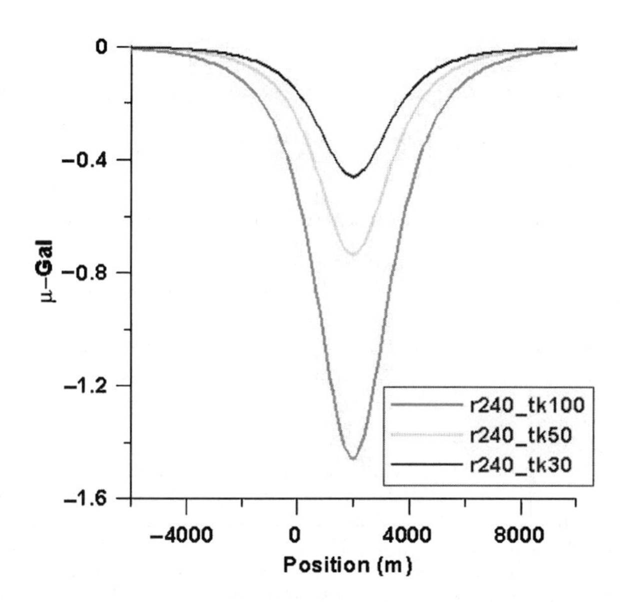

Figure 45: Surface vertical component of gravity measured over a 3D wedge at a depth of 2000 m. The wedge radius is 240 m with thickness of 100, 50, and 30 m.

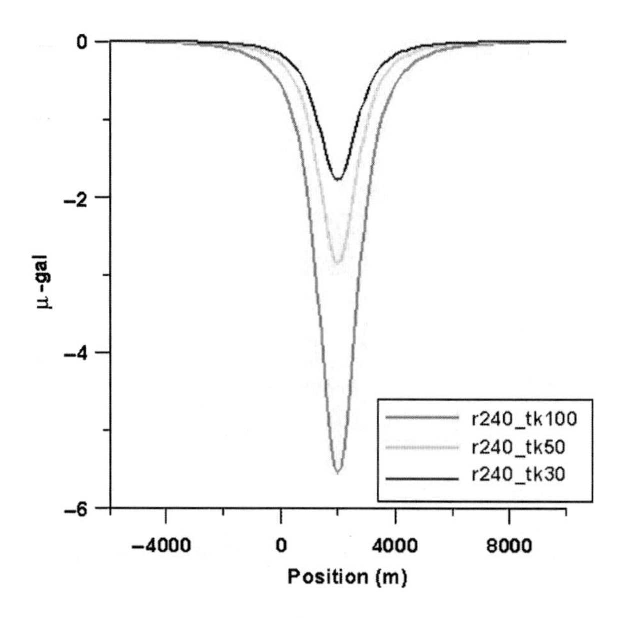

Figure 46: Surface vertical component of gravity measured over a 3D wedge at a depth of 1000 m. The wedge radius is 240 m with thickness of 100, 50, and 30 m.

the order of 1 km. The volumes affected for deeper targets will have to be much larger. These results are model-specific to the Texas gulf coast.

Tilt calculations

Recent advances in satellite imaging provide new opportunities for using land surface deformation and spectral images to indirectly map migration of CO_2. Ground surface deformation can be measured by satellite and airborne interferometric synthetic aperture radar (InSAR) systems [27,28]. Tilt meters placed on the ground surface can measure changes in tilt of a few nano-radians [29]. Taken separately or together these measurements can be inverted to provide a low-resolution image of subsurface pressure changes. While these technologies are new and have not yet been applied for monitoring CO_2 storage projects, they have been used in a variety of other applications, including reservoir monitoring [30] and groundwater investigations [30,31].

Numerical modeling work done in preparation for the DOE GeoSeq CO_2 field test in the Liberty Field, Texas provides an illustration of the application of surface deformation as a monitoring tool. The presence of the sealing faults acts to confine pressure build-up to the fault block, thus increasing the magnitude of the surface deformation.

As CO_2 injection proceeds, there is an associated pressure build up in the storage unit. This pressure increase translates into strain changes that propagate to the surface and manifest themselves as surface deformation. Figure 47 shows the change in pressure (left panel) within a 15 m thick sand unit at a depth of 1500 m from the flow simulation model of the Liberty field project as well as the inversion (right panel) of the resulting surface tilt data [30,32]. The surface tilt is shown in Figure 48. The response is dominated by the fact that the injection occurs in a bounded fault block, thus amplifying the surface tilt above the injection point. The inverted pressure distribution has captured the large-scale pressure increase trending from southwest to northeast across the center of the section. The calculated tilt values are easily observable in the field, since it is possible to achieve an accuracy of 1 nano-radian in field tilt measurements. While the limited spatial extent of this model with the presence of bounding faults (increasing the pressure buildup) dominate the response, it is clear that these measurements can be made in the field over very small quantities of injected CO_2.

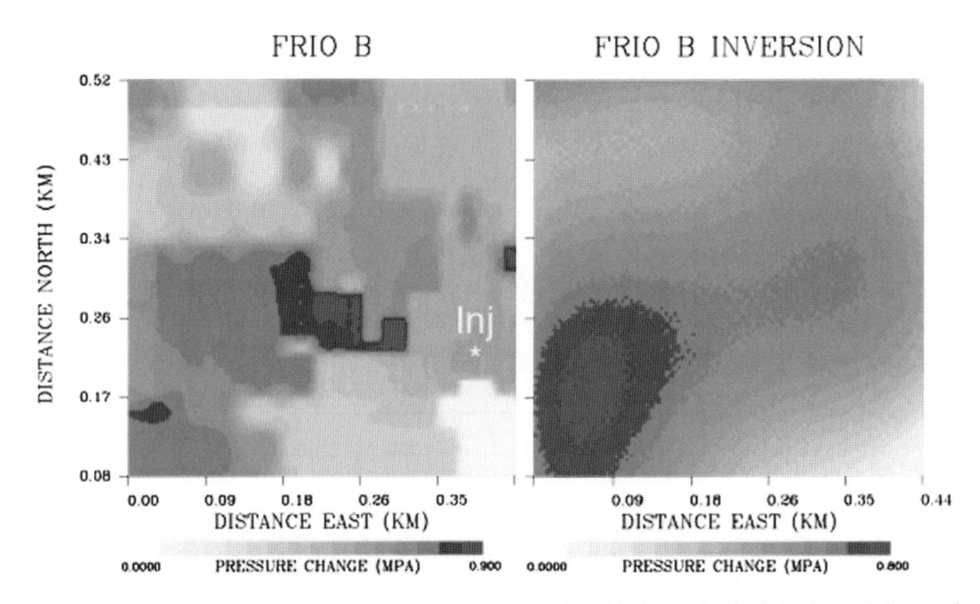

Figure 47: Left panel: pressure buildup in Frio B sand after 30 days of CO_2 injection. Right panel: inversion for pressure change from surface tilt measurements. The section shown is bounded by faults on left, right and top and is open to the bottom. CO_2 concentration is centered on the injector well but permeability variations within the unit cause the maximum pressure increase to be offset from the injection well.

The tilt measurements sensitivity to pressure changes provides an ability to map vertically integrated permeability within the injection unit. In this model the injection well is in the lower right corner of the figures. The permeability model was generated as a geostatistical realization. The model has a zone of increased permeability in the lower portions of the model below the main injection sand unit. When this unit is pressured up, the pressure front moves ahead of the injected CO_2 and pressurizes the zones with higher permeability. This causes the vertically integrated pressure change to have a maximum toward the center of the model away from the injection well. The tilt responses to this pressure increase, therefore, maps the high net permeability regions of the injection interval, ahead of the arrival of the CO_2 itself, providing a means of mapping future migration pathways.

CONCLUSIONS

Both surface and borehole gravity measurements have been modeled for Schrader Bluff. The injection of CO_2 produces a bulk density decrease in the reservoir that in turn produces a reduction in the gravitation attraction from the reservoir. The spatial pattern of the change in the vertical component of gravity (G_z) as well as the vertical gradient of gravity (dG_z/dz) is directly correlated with the net change in density of the reservoir. The difference in the vertical component of gravity on the surface caused by CO_2 injection over a 20-year period is on the order of 2 μGal, which is below the level of repeatability of current field surveys [33]. However, measurements made in boreholes just above the reservoir interval (1200 m depth) are sensitive enough to observe measurable changes in G_z as CO_2 injection proceeds. Such measurements made in numerous wells could map the areas of net density changes caused by injected CO_2 and water within

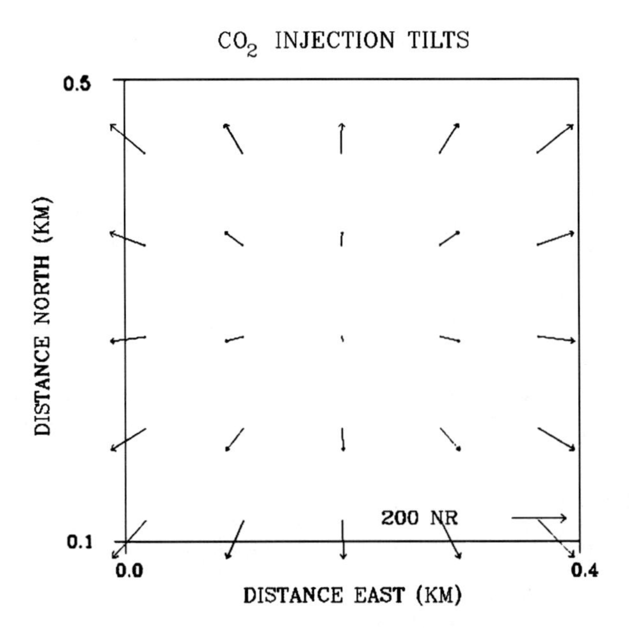

Figure 48: Surface tilt calculated for the pressure change shown in Figure 47 and rock properties representative of the Liberty Field geology. Vectors show the orientation and magnitude of the tilt. The center of the bulge over the maximum pressure is flat and has little tilt. The bounding faults truncate the pressure field and produce locations of maximum tilt.

the reservoir. The time-lapse changes in the borehole G_z and dG_z/dz clearly identify the vertical section of the reservoir where fluid saturations are changing.

There is a clear change in seismic amplitude associated with the reservoir caused by the changes in water and CO_2 saturation. In addition, there is a change in the seismic AVO effects. Both seismic amplitude and AVO can be exploited to make quantitative estimates of saturation changes, subject to modeling assumptions. Forward calculations using the isotropic Zoeppritz equation for both 2005 and 2020 models support this argument. The applications of seismic data for monitoring are covered further in Chapter 22.

The electrical resistivity of rocks is primarily a function of porosity and water saturation (S_w). When the porosity is known, or can reasonably be assumed to have small spatial variation, the changes in electrical resistivity are directly related to the changes in water saturation. EM techniques can be used to map such spatial variations in electrical resistivity. Of all the possible EM field systems, one combines both relative ease of deployment with high sensitivity to reservoirs of petroleum scale and depth. This technique uses a grounded electric dipole energized with an alternating current at a given frequency to produce time varying electric and magnetic fields that are measured on the earth's surface. This EM configuration was simulated for the Schrader Bluff model using 100 m electric dipoles operating at 1 Hz and measuring the resulting electric field at a separation of 2 km in-line with the transmitting dipole. The generated electric field for the Schrader Bluff model, using only a small portable generator is an order of magnitude above the background electric field (noise) at the operating frequency of 1 Hz. This means that synchronous detection of the signal combined with stacking can recover signal variations to better than 1%. There is a direct one-to-one correspondence with the change in S_w and the change in the electric field amplitude. While this signal level is low, it can be measured given the S/N ratio of the data. Although this represents a potential low-cost monitoring technique it is best suited for CO_2–brine systems where there is a one-to-one correlation between the change in water saturation and the change in CO_2 saturation (since $S_w + S_{CO_2} = 1$). In petroleum reservoirs such as Schrader Bluff, the presence of hydrocarbons as additional fluids eliminates the one-to-one correlation between changes in S_w and changes in S_{CO_2}.

Electric potentials are generated when fluid flows through a porous media. Measurement of these SPs is easily done at low cost. The technique is used routinely to locate leaks in fluid containment structures such as waste pits and dams. Laboratory studies coupled with numerical simulations show that the SP coupling coefficients for CO_2 flow are large enough to cause a measurable SP signal in the field. As the CO_2 displaces water in a formation, the coupling coefficient decreases. On average, the coupling coefficients observed for CO_2 flow is about 10 times lower than for fresh water flow in the same sample. Two-dimensional steady-state calculations based on the Frio brine pilot case, using laboratory-derived coupling coefficients, indicate that the technique is a potential low-cost, low-resolution monitoring technique.

Surface and borehole tilt measurements can be used to monitor the stain changes in the reservoir and overburden associated with CO_2 injection. Inversion of the data can produce estimates of the pressure changes within the reservoir as well as estimates of permeability. While this technique has not been tested in the field over CO_2 injection sites, it offers the potential for predicting permeability pathways within the reservoir ahead of injected fluids.

The non-seismic techniques presented here show enough promises as low-cost supplements to seismic monitoring that we believe further work needs to be done to assess their spatial resolution under a wider range of conditions. A number of areas should be considered further. Borehole gravity measurements should be used in conjunction with pressure test data and/or surface seismic data to do statistical interpolation of predicted changes in S_{CO_2}. This may provide a low-cost way of monitoring changes within the reservoir with only the initial 3D seismic survey being relatively expensive. A field demonstration of the EM technique should be considered to demonstrate its potential. Surface tilt measurements coupled with pressure and injection data should be jointly tested following the work of Vasco et al. [30]. SP modeling codes that can model 3D transient multi-phase flow should be developed to more realistically address the potential of SP as a monitoring tool. SP modeling developments should be done in conjunction with field SP measurements over an injection test site. A future study of resolution that can be achieved by inversion of gravity, electrical, and SP data should be done and compared to seismic resolution.

ACKNOWLEDGEMENTS

This work was supported in part by a Cooperative Research and Development Agreement (CRADA) between BP Corporation North America, as part of the CO_2 Capture Project (CCP) of the Joint Industry Program (JIP), and the US Department of Energy (DOE) through the National Energy Technologies Laboratory (NETL), and by the Ernest Orlando Lawrence Berkeley National Laboratory, managed by the University of California for the US Department of Energy under contract DE-AC03-76SF00098.

REFERENCES

1. G. Hill, B. Moore, M. Weggeland, The CO_2 Capture Joint Industry Project: GHGT-5, Australia, 2000, pp. 248–253.
2. B.S. Brandt, A study of the speed of sound in porous granular materials, *J. Appl. Mech.* **22** (1955) 479–486.
3. G.M. Hoversten, R. Gritto, J. Washbourne, T.M. Daley, Pressure and fluid saturation prediction in a multicomponent reservoir using combined seismic and electromagnetic imaging, *Geophysics* **68** (2003) 1580–1591.
4. J. Dvorkin, A. Nur, Elasticity of high-porosity sandstones: theory of two North Sea data sets, *Geophysics* **61** (1996) 1363–1370.
5. S.L. Nooner, M.A. Zumberge, O. Eiken, T. Stenvold, G.S. Sasagawa, Seafloor micro-gravity survey of the Sleipner CO_2 sequestration site, *EOS Trans. AGU* **84** (46) (2003) Fall Meet. Suppl. Abstract GC31A-01.

6. L.A. Thomsen, J.L. Brady, E. Biegert, K.M. Strack, A Novel Approach to 4D Full Field Density Monitoring, SEG Workshop, 2003.
7. J.V. Popta, J.M.T. Heywood, S.J. Adams, D.R. Bostock, Use of Borehole Gravimetry for Reservoir Characterization and Fluid Saturation Monitoring, SPE 20896, 1990, pp. 151–160.
8. S.R. Rutherford, R.H. Williams, Amplitude-versus-offset variations in gas sands, *Geophysics* **54** (1989) 680–688.
9. J.P. Castagna, H.W. Swan, J.F. Forster, Framework for AVO gradient and intercept interpretation, *Geophysics* **63** (1998) 948–956.
10. M. Landro, Discrimination between pressure and fluid saturation changes from time-lapse seismic data, *Geophysics* **66** (2001) 836–844.
11. Shuey, A simplification of the Zoeppritz equations, *Geophysics* **50** (1985) 609–614.
12. G.E. Archie, The electrical resistivity log as an aid in determining some reservoir characteristics, *Trans. AIME* **146** (1942) 54–62.
13. S. Ellingsrud, T. Eidesmo, S. Johansen, M.C. Sinha, L.M. MacGregor, S. Constable, Remote sensing of hydrocarbon layers by seabed logging (SBL): results from a cruise offshore Angola, *The Leading Edge* **21** (2002) 972–982.
14. S.D. Hovorka, P.R. Knox, Frio brine sequestration pilot in the Texas Gulf Coast, *Sixth International Conference on Greenhouse Gas Control Technologies (GHGT-6)*, Kyoto, Japan, 1–4 October, 2002.
15. R.G. Loucks, M.M. Dodge, W.E. Galloway, Regional controls on diagenesis and reservoir quality in lower Tertiary sandstones along the lower Texas Gulf Coast, in: D.A. McDonald, R.C. Surdam (Eds.), *Clastic Diagenesis: American Association of Petroleum Geologists Memoir*, vol. 37, 1984, pp. 15–46.
16. C.W. Kreitler, M.S. Akhter, A.C.A. Donnelly, W.T. Wood, Hydrology of formations for deep-well injection, The University of Texas at Austin, Bureau of Economic Geology, Texas Gulf Coast, 1988, unpublished contract report, 204pp.
17. G.L. Macpherson, Regional variation in formation water chemistry; major and minor elements, Frio Formation fluids, Texas, *American Association of Petroleum Geologists Bulletin* **76** (5) (1992) 740–757.
18. C. Doughty, K. Pruess, Modeling supercritical CO_2 injection in heterogeneous porous media, *Proceedings*, TOUGH Symposium, May 12–14, 2003.
19. R.F. Corwin, D.B. Hoower, The self-potential method in geothermal exploration, *Geophysics* **44** (1979) 226–245.
20. D.V. Fitterman, Electrokinetic and magnetic anomalies associated with dilatant regions in a layered earth, *J. Geophys. Res.* **83** (B12) (1978) 5923–5928.
21. R.F. Corwin, H.F. Morrison, Self-potential variations preceding earthquakes in central California, *Geophys. Rev. Lett.* **4** (1977) 171–174.
22. A.A. Ogilvy, M.A. Ayed, V.A. Bogoslovsky, Geophysical studies of water leakages from reservoir, *Geophys. Prospect.* **17** (1969) 36–62.
23. V.A. Bogoslovsky, A.A. Ogilvy, Deformations of natural electric fields near drainage structures, *Geophys. Prospect.* **21** (1973) 716–723.
24. D.V. Fitterman, Self-potential surveys near several Denver Water Department dams: US Geol. Surv. Open File Report, 1983, pp. 82–470.
25. D.J. Marshall, T.R. Madden, Induced polarization, A study of its causes, *Geophysics* **24** (1959) 790–816.
26. W.B. Sill, Self-potential modeling form primary flows, *Geophysics* **48** (1983) 76–86.
27. H. Zebker, Studying the Earth with Interferometric Radar, *Comput. Sci. Engng* **2** (2000) 52–60.
28. Y. Fialko, M. Simons, Deformation and seismicity in the Coso Geothermal Area, Inyo County, California: observations and modeling using satellite radar interferometry, *J. Geophys. Res.* **21** (2000) 781-21, 793.
29. C. Wright, E. Davis, W. Minner, J. Ward, L. Weijers, E. Schell, S. Hunter, Surface tiltmeter fracture mapping reaches new depths-10,000 feet and beyond?, *Soc. Petrol. Engng* (1998) 39919.
30. D.W. Vasco, K. Karasaki, K. Kiyoshi, Coupled Inversion of Pressure and Surface Deformation Data, *Water Resour. Res.* (2001) 3071–3089.
31. J. Hoffmann, H.A. Zebker, D.L. Galloway, F. Amelung, Seasonal subsidence and rebound in Las Vegas Valley, Nevada observed by synthetic aperture radar interferometry, *Water Resour. Res.* **37** (2001) 1551.

32. D.W. Vasco, K. Karasaki, L.R. Myer, Monitoring of fluid injection and soil consolidation using surface tilt measurements, *J. Geotechn. Geoenviron. Engng* **124** (1998) 29–37.
33. J.L. Hare, J.F. Ferguson, C.L.V. Aiken, The 4-D microgravity method for waterflood surveillance: a model study from the Prudhoe Bay reservoir, Alaska, *Geophysics* **64** (1999) 78–87.
34. E. Gasperikova, G.M. Hoversten, M.P. Ryan, J.P. Kauahikaua, G.A. Newman, N. Cuevas, Magnetotelluric investigations of Kilauea volcano, Hawaii. Part I: experiment design and data processing, *J. Geophys. Res.* (2004) in review.

Carbon Dioxide Capture for Storage in Deep Geologic Formations, Volume 2
D.C. Thomas and S.M. Benson (Eds.)

Chapter 24

THE USE OF NOBLE GAS ISOTOPES FOR MONITORING LEAKAGE OF GEOLOGICALLY STORED CO_2

Gregory J. Nimz and G. Bryant Hudson
Lawrence Livermore National Laboratory, Livermore, CA, USA

ABSTRACT

One of the primary concerns in CO_2 storage is monitoring the storage site on a long-term basis for possible leakage of CO_2. Concentrations of CO_2 vary widely in the Earth's crust, making detection of very small releases difficult. Small amounts of noble gas isotopes can be dissolved into the CO_2 being injected for storage and used as tracers to monitor CO_2 movement. Noble gases are chemically inert, environmentally safe, and are persistent and stable in the environment. The unique isotopic compositions that can be imparted to the CO_2 can be unambiguously identified during monitoring. Among the noble gases, xenon isotopes have commercial costs and availability suitable for use in large CO_2 storage operations. Required xenon volumes are low, simplifying handling and injection. Multiple batches of injected CO_2 at the same site could be imparted with different xenon isotopic compositions, making each of them identifiable with only a single xenon analysis. These characteristics are believed to make xenon a superior tracer to other option, SF_6 and $^{14}CO_2$. A case study in noble gas tracing at the Mabee Enhanced Oil Recovery field in West Texas indicates that unique noble gas isotopic compositions within a CO_2 injection stream can be detected and readily identified in outlying wells, and that noble gas behavior in a CO_2 storage setting will be systematic and predictable.

INTRODUCTION

Noble gases (helium, neon, argon, krypton, xenon; Table 1) can be dissolved into CO_2 injected into geological formations for long-term storage and used as tracers when monitoring for CO_2 leakage or subsurface migration. Injected CO_2 is in a supercritical state and the noble gases will remain dissolved in that liquid. Using noble gases for subsurface tracing in this form is similar to using any common type of chemical tracer. However, leaking CO_2 will become a gas as it migrates to the Earth's surface. It is here that the noble gases become unique and highly valuable tracers. At the pressure and temperature conditions in which the supercritical CO_2 becomes a gas, the noble gases will also be released as gases. The noble gases will thereby track CO_2 gas migration toward the surface.

Noble gases become distinctive tracers when non-natural isotopic compositions are used. In the natural environment all of the noble gases have multiple isotopes, atoms of the same element with different numbers of neutrons. While the atomic ratios of the noble gas isotopes are generally very similar throughout the planet and atmosphere, commercial isotope separation makes available significant volumes of noble gases with certain isotopes enhanced over their natural abundances. Xenon, for example, occurs naturally in nine different isotopic states: ^{124}Xe, ^{126}Xe, ^{128}Xe, ^{129}Xe, ^{130}Xe, ^{131}Xe, ^{132}Xe, ^{134}Xe, and ^{136}Xe. The fractional abundance of each of these varies only slightly in nature; for instance, ^{136}Xe comprises about 8.9% of all natural xenon. However, pure xenon gas can be purchased that contains about 60% ^{136}Xe. Adding this to injected CO_2 would create a distinctive tracer with non-natural xenon isotopic ratios that later could be unambiguously identified when monitoring for leakage.

Since they are chemically inert and non-radioactive, noble gas tracers are persistent and stable in the environment. They are non-toxic and environmentally safe. After injecting CO_2 spiked with noble gas isotopes into a reservoir, the region surrounding and above the storage site could be monitored to detect the distinctive noble gas isotopic signatures for decades to millennia. Because CO_2 will always be detected within the Earth's crust, the question for monitoring purposes will be whether its origin is natural or from injection. The isotopic signature of the noble gases measured with the CO_2 would indicate whether or not the gases originated within the storage site.

TABLE 1
NOBLE GAS PROPERTIES AND CONCENTRATIONS RELEVANT TO THEIR USE IN CO_2 STORAGE

Noble gas	Atomic number	Atomic radius[a] (Å)	Natural stable isotopes	Atmospheric volume fraction of dry air[a]	Average seawater[a] (cm^3 STP/g)	Dakota aquifer[b] (cm^3 STP/g)	Dogger aquifer[c] (cm^3 STP/g)	Suggested primary tracer isotope (s)
He	2	1.8	2	5.2×10^{-6}	4.0×10^{-8}	1.0×10^{-5}	6.2×10^{-4}	^{3}He
Ne	10	1.6	3	1.8×10^{-5}	1.7×10^{-7}	2.0×10^{-7}	3.2×10^{-7}	^{22}Ne
Ar	18	1.9	3	9.3×10^{-3}	3.5×10^{-4}	3.5×10^{-4}	3.4×10^{-4}	^{36}Ar
Kr	38	2.0	6	1.1×10^{-6}	8.5×10^{-8}	8.2×10^{-8}	8.3×10^{-8}	–
Xe	54	2.2	9	8.7×10^{-8}	1.1×10^{-8}	1.1×10^{-8}	–	$^{124,129,136}Xe$

The Dakota and Dogger aquifer concentrations given can be considered typical for deep continental groundwaters, although significant variations can occur.
[a] Source: Ref. [1].
[b] Source: Ref. [2]. Average of Group 3 waters, central Kansas.
[c] Sources: Ref. [3] for He, Ne, Ar and Ref. [4] for Kr. Values are averages of listed wells.

For the purposes of our initial calculations and assessments, we are assuming that the noble gas tracers move conservatively with the stored CO_2. In reality, there will be partitioning of the noble gases between the phases present in the system (water, hydrocarbons, and a gas phase). Partitioning would not only affect noble gas tracers, but any potential tracer except for $^{14}CO_2$ ($^{14}CO_2$ is discussed below). Noble gas solubilities in waters and brines, and their temperature dependence, are fairly well known [1,5–9]. Data on noble gas solubilities in some hydrocarbons (hexane, decane, benzene) at 25 °C and 1 atm are available [10]. Kharaka and Specht [11] determined noble gas solubilities in two crude oils (API gravity 25 and 34) over the temperature range 25–100 °C. However, very little is known about the partitioning/solubility behavior of noble gases or other possible tracers in the $P-T-x$ conditions of the CO_2 storage environment. Much more information will be needed before a complete assessment of noble gas or other tracers can be made. One of our recommendations provided at the end of this chapter is for research to obtain this information.

This chapter discusses a methodology for using noble gases in CO_2 storage. It covers injection methods, costs, detection and monitoring scenarios, and compares noble gases with other potential tracers (SF_6 and $^{14}CO_2$). We also present the results of a noble gas "tracer" study we performed in an enhanced oil recovery (EOR) field in the Permian Basin of West Texas.

EXPERIMENTAL/STUDY METHODOLOGY

Noble gas tracing has been used successfully for large-volume groundwater tracing in several different locations [12]. The initial investigation, therefore, was to determine whether this technique was also suitable for CO_2 tracing. This involved an analysis of noble gas tracer costs, availability, detection limits, and a comparison with other potential tracers. It required an assessment of the amounts of tracer needed per mass of stored CO_2, as well as an assessment of probable monitoring strategies. The amounts of any tracer required is a function of the type of system to be monitored; possibilities include groundwater, deep soil or formation gases, and ground surface emissions. An analysis of injection methods must also be made: is it feasible to label the entire CO_2 injection stream with noble gas tracers? If only portions of the injected stream can be labeled, monitoring will be compromised.

The investigation then turned to a "field demonstration" at an active EOR location. During the course of this study, no CO_2 storage operations were being conducted in which a noble gas tracer could be added and traced in the subsurface. The EOR setting was the closest available analogue. Fortuitously, the CO_2 being injected into the subsurface in the Permian Basin of West Texas contains noble gases that have very unique and recognizable isotopic characteristics [13]. The CO_2 originates from extensive CO_2 deposits ("domes") in northern New Mexico and southern Colorado (Figure 1), and is piped to West Texas.

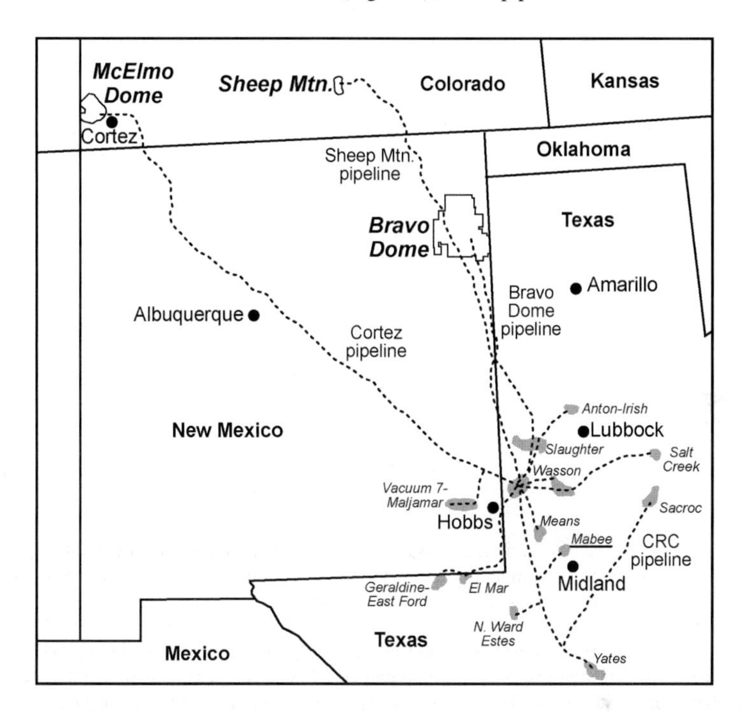

Figure 1: CO_2 distribution via pipelines (dotted lines) from the McElmo Dome, Sheep Mountain, and Bravo Dome CO_2 deposits to the Enhanced Oil Recovery fields in the Permian Basin. Only major or pipeline-termination EOR fields are shown, and many more exist. The Mabee EOR field is shown located just north of Midland, Texas.

Isotopic compositions of the noble gases in the dome CO_2 are unlike those of the noble gases naturally present within the Permian Basin, which will be typical crustal values (Figure 2). They are also distinct from atmospheric values. Large-scale CO_2 injection to enhance oil recovery has been going on since the 1970s. This permits an assessment of whether noble gas isotopic compositions would serve as a tracer of CO_2. Rather than having to artificially add a noble gas tracer to a CO_2 stream and wait months to years for that CO_2 to migrate to a monitoring well or to the ground surface, the "experiment" had already begun many years ago.

We were given access to the Mabee EOR field north of Midland, Texas by ChevronTexaco, its owners and operators (Figure 1). Samples were collected for noble gas analysis of the dome CO_2 prior to injection, and of gases being extracted in 13 outlying oil production wells within the Mabee field. The extracted gases from many wells covering large sections of the Mabee field is typically combined into a single return pipeline and added to incoming (new) dome CO_2 for reinjection. A sample of this "blend" CO_2 was also collected and analyzed for noble gas isotopic compositions. Thus we were able to simulate a field demonstration of noble gas isotopic tracing in which we could compare the isotopic compositions of injected noble gases and those of CO_2-related gases outlying from the point of injection. This mimics a situation in which CO_2 would be injected for storage and then monitored through outlying wells in order to

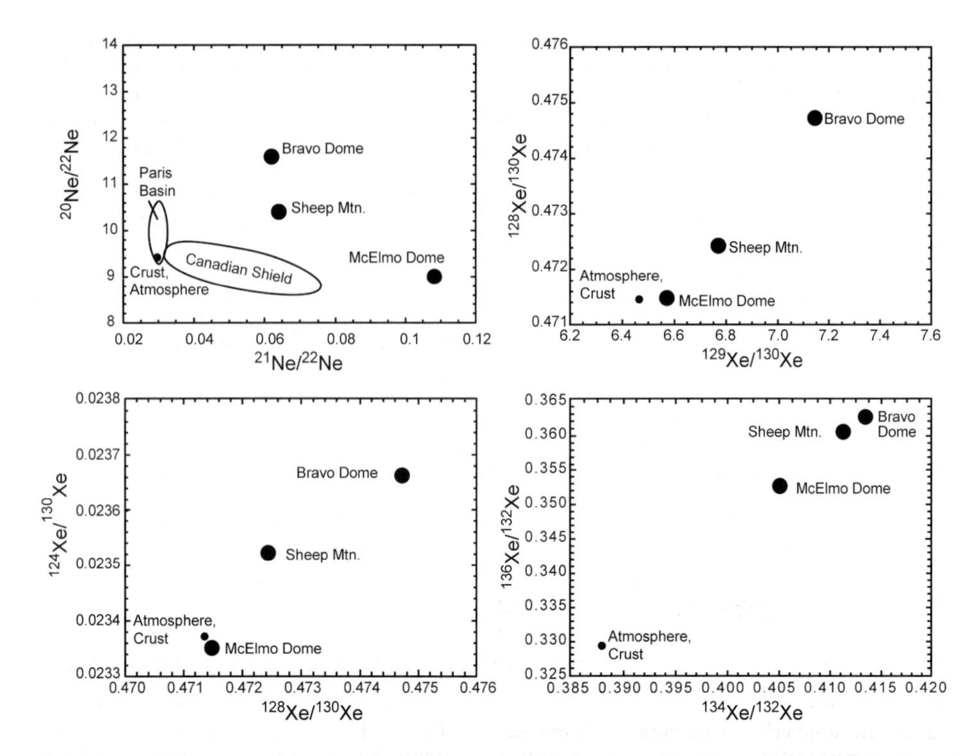

Figure 2: Isotopic compositions of neon and xenon in CO_2 from the McElmo Dome, Sheep Mountain, and Bravo Dome CO_2 deposits. Atmospheric values given for comparison. Typical crustal values will be closer to atmospheric values than to CO_2 deposit values, as shown by the neon isotope fields from the Canadian Shield and Paris Basin. McElmo Dome, Sheep Mountain, and Bravo Dome data are from Caffee et al. [13]; Canadian Shield data from Bottomley et al. [14]; Paris Basin data from Castro et al. [3].

understand its subsurface distribution. It is important to note that the noble gas concentrations in the natural dome CO_2 are far less than those that could be imparted to CO_2 being injected for storage. In this respect, the Mabee analogue represents the most difficult monitoring situation we would expect to encounter.

Gas samples were collected in double-ended stainless steel high pressure bottles (<2000 psi) and shipped to Lawrence Livermore National Laboratory for noble gas isotopic analysis. Analytical methods were similar to those presented in Caffee et al. [13]. The noble gas lab consists of two VG5400 noble gas mass spectrometers. The first spectrometer is set up to analyze xenon isotope ratios with very high precision. Major isotope ratios (^{129}Xe, ^{131}Xe, ^{132}Xe, ^{134}Xe and ^{136}Xe) are reproducible in air and water standards at 0.05%. This instrument is also used for precise $^{3}He/^{4}He$ measurements and the determination of tritium by the ^{3}He in-growth method. The second spectrometer is dedicated to measuring large variations in noble gas isotope ratios and samples with high $^{3}He/^{4}He$ and high tritium. This instrument also performs isotope dilution measurements of He, Ne, Ar, Kr and Xe abundances. Both spectrometer systems have automated, multi-port, sample processing systems able to handle gas and water samples.

RESULTS AND DISCUSSION

Comparison of Tracers: Noble Gas Isotopes, SF_6, and $^{14}CO_2$

Theoretical aspects of the utility of potential tracers for CO_2 storage monitoring can be assessed on the basis of availability, costs, ease-of-use, detectability, and environmental safety. A generalized assessment of each of these is impossible without making assumptions concerning the amounts of CO_2 to be stored, the rate at which it is put into storage, the purpose and location of the monitoring, and other specific aspects of the storage operation. For example, the cost of any tracer is likely to decrease with increasing amounts purchased. Likewise, detectability is a function of the purpose and location of the monitoring—monitoring for verification of subsurface location is different from ground surface monitoring for leakage.

For the purposes of an assessment of detectability, we considered the most difficult monitoring setting, ground surface emissions. For this we also considered a very conservative monitoring strategy in which the concentration of CO_2 in the soil gas, normally at about 1% in the environment, is raised an additional 1% (i.e. grows to 2% total). This is within the natural range of variations in soil gas CO_2, such that CO_2 monitoring alone would not detect an abnormal variation. It is also far below the value at which vegetation stress would occur, giving an obvious signal of CO_2 leakage. For the assessment, we also assumed that all potential tracers move conservatively with the leaking CO_2. The concentration of tracer required within the injected CO_2 will then be a function of the detection limits for the tracer and the level at which we desire to detect leaking CO_2. For the assumption that we desire to detect a 1% CO_2 increase in soil gas, known natural background levels in the atmosphere (Table 1) and instrumental detection limits determined the amount of tracer required per unit mass of CO_2 (e.g. tracer/ton CO_2). This then permitted the assessment of costs per unit mass CO_2 and potential availability. An assessment of the amounts of tracer required per year can be made by assuming an amount of CO_2 to be stored per year. For the purpose of discussion, we have used as an example the amount of CO_2 being injected at the Mabee EOR field. For the entire injection manifold, CO_2 storage would occur at the rate of 3.8×10^3 metric tons/day, or about 1.4×10^6 metric tons/year at full performance. Table 2 compares such criteria with respect to several noble gas tracers and two other potential tracers, SF_6 and $^{14}CO_2$.

TABLE 2
COMPARISON OF TRACERS FOR MONITORING SHALLOW SOILS FOR LEAKING CO_2

Tracer	Atmospheric concentration (cm^3/cm^3 air)	Minimum detectable variation (%)	Required tracer concentration in stored CO_2 (cm^3/cm^3 CO_2, STP)	Required tracer per $10^{-6} m^3$ CO_2 (L, STP)	Tracer cost ($US/l)	Tracer cost/metric ton CO_2 ($US)	Required tracer per year (L, STP)
^3He	7.2×10^{-12}	300	2.17×10^{-9}	2.2	100	0.11	1532
^{22}Ne	1.7×10^{-6}	0.1	1.68×10^{-7}	168	50	4.27	118629
^{36}Ar	3.2×10^{-5}	0.1	3.16×10^{-6}	3161	1000	1610	2234515
^{124}Xe	8.7×10^{-11}	0.2	1.75×10^{-11}	0.02	20000	0.18	12
^{129}Xe	2.5×10^{-8}	0.05	1.25×10^{-9}	1.2	1000	0.64	883
^{136}Xe	8.7×10^{-9}	0.10	8.70×10^{-10}	0.9	300	0.13	615
SF_6	1.0×10^{-11}	1000	1.00×10^{-8}	10	1	0.005	7070
$^{14}CO_2$	1.0×10^{-14}	300	3.00×10^{-12}	0.003	5000	0.008	2

Minimum detectable variation values are based on observed natural atmospheric variations ("background") and available analytical precision. They represent the minimum recognizable non-natural shift in isotopic ratios (or change in SF_6 concentration) that would provide a clear signal of the presence of the tracer in soil gas samples. Calculations assume soil gas is 1% natural CO_2 by volume. Calculated amounts of required tracer in the stored CO_2 are for detection of an additional 1% contribution from leaking CO_2 (i.e. total $CO_2 = 2\%$). Minimum variation for $^{14}CO_2$ (300%) is relative to ambient soil CO_2 (assumed 1%); with the additional 1% stored contribution it is equivalent to an isotopic shift of 150% (1.5 times modern atmospheric $^{14}C/^{12}C$). Tracer required per year is for the Mabee storage analogue discussed in the text (storage of 1.937×10^{-6} m^3 CO_2/day (STP) for the entire injection manifold).

Clearly the best tracer is $^{14}CO_2$. Its cost is low ($0.008 per ton CO_2) and only 2 L (STP) per year for the Mabee storage analogue would be required. Of all potential tracers, $^{14}CO_2$ is the most likely to migrate conservatively with leaking CO_2, since the leaking CO_2 will itself be partially $^{14}CO_2$. However, ^{14}C is radioactive. Two liters (STP) of $^{14}CO_2$ would contain about 5.6 Ci of activity. This is actually a small amount of radioactivity on the scale of CO_2 storage. It would impart an activity to the stored CO_2 equivalent to about half that attained in the CO_2 in the northern hemisphere atmosphere in the mid-1960s from uncontained nuclear testing in the Pacific. For a leakage that raises soil gas CO_2 concentrations by 1%, the increased radiation would be difficult or impossible to detect in the ambient air ($\sim$80 pCi per m^3 air). Measurement would require accumulation of soil gas CO_2. Handling the $^{14}CO_2$ 2-L canisters during the injection process could be done safely with very simple protocols. The potential commercial availability of ^{14}C is very high, since it is abundantly produced in power plant nuclear reactors. By scientific and technical standards, $^{14}CO_2$ might be a desirable tracer and warrants further investigation. However, public perception is another matter. It is possible that public reaction to injection of radioactivity, no matter how small, would doom attempts to use $^{14}CO_2$ as a CO_2 tracer.

SF_6 is a well-understood synthetic tracer that is widely used in groundwater and other applications [15]. It is inexpensive and would be cost effective for CO_2 storage monitoring (US$1/L to $0.005/ton CO_2 stored). It has become increasingly prevalent in the atmosphere, and would therefore require fairly large amounts to be injected with the stored CO_2 ($\sim$7000 L (STP) per year for the Mabee storage analogue). Atmospheric concentrations of SF_6 are rapidly rising [16] and it is difficult to determine present injection requirement for detection in future decades or centuries (Table 2 lists a conservative estimate for long-term monitoring). Although SF_6 is a relatively stable molecule, it can be expected to decompose over time in the subsurface. Its long-term, centuries to millennia, reliability as a CO_2 tracer is unclear.

As Table 2 indicates, not all noble gases will be cost effective tracers of CO_2. The high atmospheric abundance of ^{36}Ar necessitates a high concentration of ^{36}Ar in the injected CO_2, resulting in a cost of nearly $1600 (US) per metric ton CO_2. The required yearly volume for the Mabee analogue, over 2 million liters of ^{36}Ar, is also prohibitive—for both availability and ease-of-use. A high atmospheric concentration also makes ^{22}Ne an expensive choice for ground surface monitoring. However, the low commercial cost of neon (Table 2) and its high availability suggest that ^{22}Ne could be considered in special circumstances. For example, it would be useful for subsurface tracing of supercritical CO_2 where the natural background is insignificant, and therefore injected concentrations would be substantially lower. Although costs for ^{3}He are low ($0.11/ton CO_2) and the necessary quantities required for the Mabee analogue are within obtainable amounts, ^{3}He is very rare (only 0.00014% of natural helium is ^{3}He). For large-scale CO_2 storage, availability would be problematic.

The three xenon isotopes listed in Table 2 appear to be efficient to use and inexpensive relative to CO_2 storage costs. For the Mabee storage analogue, only 12 L of ^{124}Xe (at STP) would be required per year, at a cost of $0.18 per metric ton CO_2. This compares to current CO_2 industrial separation costs in the range of $50–$100 per metric ton. The requirement for ^{129}Xe is 883 L/year, equivalent to about four 55-gallon drums (at STP), and about 12% of the required volume of SF_6. The small volumes required greatly simplify injection logistics.

The fact that multiple xenon isotopes are available, inexpensive, simple to use, and highly detectable allows the possibility that there may be occasions where injecting batches of CO_2 at the same location containing distinctly different xenon tracer isotopic compositions (i.e. different proportions of ^{124}Xe, ^{129}Xe, and ^{136}Xe). This would be useful in tracing subsurface migration of CO_2 batches injected at different geographical locations within a field. It would also be useful for monitoring of locations where several different batches of CO_2 were stored, perhaps by different corporations, in different geologic formations, or at different times. The unique isotopic signal would indicate which batch or batches were leaking to ground surface. Since the analysis would be only for xenon, only one measurement would be needed. Multiple tracers, or even multiple noble gases, would each require a separate analysis.

Availability of Noble Gas Tracers

The commercial source for all of the noble gases is the Earth's atmosphere. Separation of noble gases from air, and from one another, can be accomplished by liquefaction/cryogenic methods. Separation of the individual noble gas isotopes is accomplished by gas centrifugation such as the technique that separates $^{235}UF_6$ from $^{238}UF_6$ for nuclear fuel (uranium enrichment). The commercial availability of xenon is currently adequate for the short term, and can be expected to increase in the future. Xenon is commonly used as an anesthetic and recently has begun to be used to enhance Magnetic Resonance Imaging (MRI) resolution. The MRI application uses exclusively ^{129}Xe. A driving force for an increase in xenon availability is its use as a propellant in spacecraft ion engines, such as the NASA High Power Electric Propulsion (HiPEP) engine. A commercial demand for xenon for CO_2 storage would result in a rapidly increased production.

A basic calculation indicates the total scope of potential xenon availability. The Earth's atmosphere contains an inventory of about 7×10^{13} L of ^{124}Xe [1]. Total US required CO_2 storage is expected to be about 1×10^9 metric tons/year [17]. If injection for this storage amount were to occur for a century, a total of 1×10^{11} metric tons of CO_2 would be stored (about 5×10^{13} m^3 CO_2, STP). The requirement for ^{124}Xe tracer is 0.02 L per 1×10^6 m^3 CO_2, STP (Table 2). The total 100 year requirement for ^{124}Xe would therefore be about 1×10^6 L, or about 1.5×10^{-6}% of the atmospheric xenon. Clearly, the xenon inventory is sufficient.

Gas centrifuge technology required to separate the individual noble gas isotopes is abundant worldwide and currently underutilized. An example of the volumes of isotope separates that could be made available is the double-beta neutrino mass experiment being conducted as a US–European collaboration [18]. For the experiment, 10 metric tons (1.65×10^6 L, STP) of ^{136}Xe is being used. This volume of ^{136}Xe would be sufficient for ~2700 years of CO_2 injection at the Mabee storage analogue (Table 2). For the total US required CO_2 storage of 1×10^9 metric tons/year, this would be a 4-year supply. This would be sufficient time to "replenish" the supply using the currently available gas centrifuge technology.

Addition of Noble Gases into the CO_2 Injection Stream

The low annual volumes of xenon tracer required for CO_2 storage monitoring simplify the methods required for injection. Figure 3 shows a generalized schematic of an injection system. To insure uniform solution of the xenon (or other noble gas) into the CO_2 stream, a side-track make up flow is partitioned from the main CO_2 delivery manifold by pressure regulation. Noble gas tracer is bled into this stream at a rate of perhaps a few cm^3 per minute (STP). A compressor in the side-track make up flow line insures solution of the tracer into the liquid CO_2. A tracer flow regulator coupled to the main CO_2 manifold flow meter would insure constant concentration of the tracer in the injected CO_2.

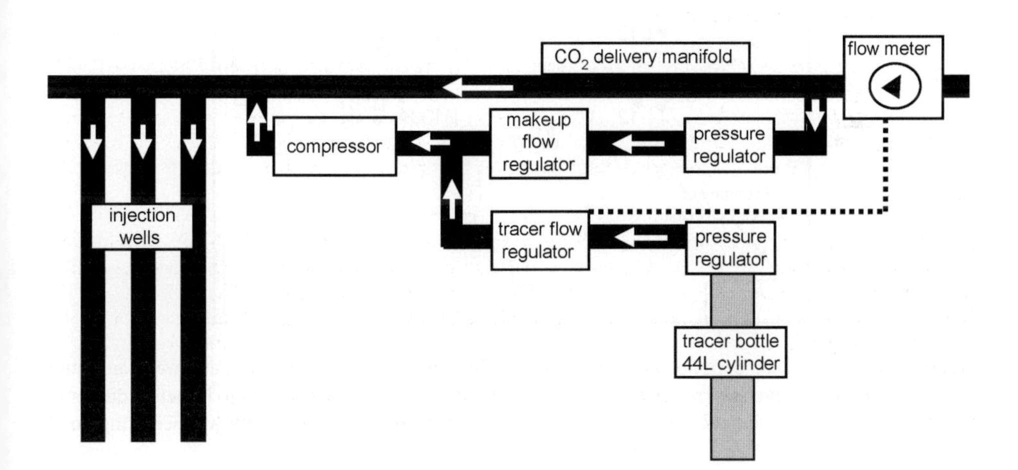

Figure 3: Conceptual schematic for the addition of noble gas tracers to CO_2 during injection for long-term storage. The noble gases are mixed into a side-track make up CO_2 flow and compressed for solution. A CO_2 flow meter and the tracer flow regulator can be coupled (dotted line) to insure a constant concentration of noble gases in the injected CO_2. Tracer bottle cylinder size can be selected for optimal desired lifetime of injection (e.g. 1 year).

The size of the tracer bottle, depicted as a standard 44 L cylinder in Figure 3, could be chosen for ease of use and handling. If only 12 L of tracer are used per year, as in the Mabee storage analogue using ^{124}Xe, one cylinder per year would be sufficient. However, in this case the tracer flow would be extremely small, about 0.02 cm^3/min. This suggests that it may be preferable for the tracer bottle to contain addition gas, probably CO_2, such that total flow would be more easily regulated.

Field Demonstration: Noble Gas Tracing at the Mabee EOR Field, West Texas

The fact that CO_2 containing unusual noble gas isotopic compositions had been injected into Permian Basin oil fields for EOR since the 1970s permitted an analysis of the utility of using noble gases as CO_2 storage tracers. A staged field demonstration, in which a noble gas tracer is artificially added to a CO_2 stream, will be a valuable trial in the future. However, no new field CO_2 demonstrations were being initiated during the time of our study, and unless an aggressive subsurface sampling strategy was to be employed, the demonstration could take months to years to complete. Ground surface monitoring in a field demonstration would only be useful if CO_2 leakage to the surface were induced during the demonstration. Therefore, the use of the Permian Basin was advantageous to the assessment of noble gas behavior during injection of CO_2.

Figure 4 is a map of the Mabee EOR field in the Permian Basin. The locations of the 13 production wells at which gas samples were collected are shown, as are the locations of CO_2 injection wells known to be operative during the sampling period. Because injection has been occurring at Mabee for many years, the entire subsurface has been affected by CO_2 flooding. It is unclear whether the currently active injection had direct effect on the composition of the samples collected. However, the data clearly indicate that noble gas compositions in the collected gas are directly affected by compositions closely similar or identical to those in the CO_2 being injected at the time of sampling.

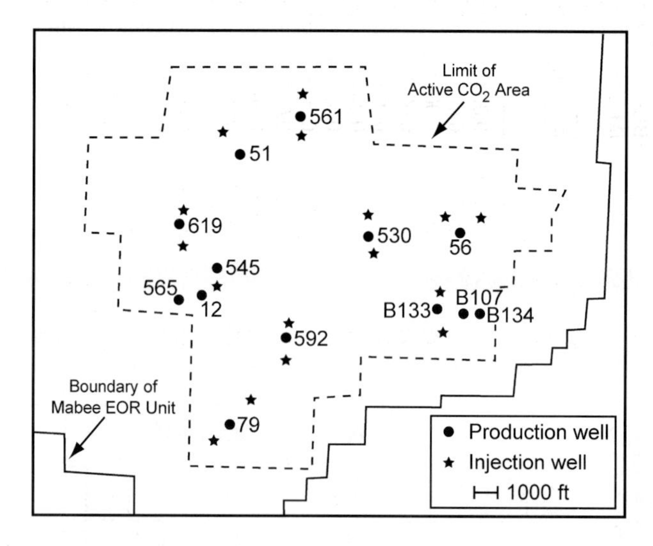

Figure 4: Map of the southern portion of the Mabee EOR field. The sampled production wells are shown with adjacent sample numbers (see Tables 3 and 4). Injection wells are those known to be active during the sampling interval. The field contains many other production wells than those shown, occurring on approximate 1000 ft spacings.

Helium, neon, argon, and xenon concentration and isotopic data are presented in Tables 3 and 4, and shown in Figures 5 and 6. Sample "KMCO2" is CO_2 directly from the supply pipeline feeding the Mabee field (Figure 1). Sample "Blend CO2" is a composite gas collected from a CO_2-return pipeline that mixes produced gases from many production wells in the field, including some of those sampled individually by us. The produced CO_2 from this pipeline is mixed with the incoming KMCO2 pipeline

TABLE 3

CONCENTRATIONS AND ISOTOPIC COMPOSITIONS OF HELIUM, NEON, AND ARGON IN GASES FROM THE MABEE EOR FIELD

Sample	^{4}He (ppmv)	^{3}He (ppmv)	^{3}He/^{4}He (÷ ^{3}He/^{4}He air)	^{20}Ne (ppmv)	^{21}Ne/^{20}Ne	^{22}Ne/^{20}Ne	^{36}Ar (ppmv)	^{40}Ar/^{36}Ar	Excess ^{21}Ne (ppmv)	Excess ^{40}Ar (ppmv)
KM CO2	1086	2.42×10^{-4}	0.161	0.0055	0.00996	0.1086	0.0185	12432	3.82×10^{-5}	225
Blend CO2	703	2.39×10^{-4}	0.246	0.0138	0.00504	0.1040	0.0556	3255	2.83×10^{-5}	165
Mabee 545	826	3.61×10^{-4}	0.316	0.0211	0.00449	0.1035	0.1018	1984	3.16×10^{-5}	172
Mabee 12	838	3.14×10^{-4}	0.217	0.0148	0.00480	0.1036	0.0614	3114	2.66×10^{-5}	173
Mabee 565	550	3.25×10^{-4}	0.427	0.0177	0.00398	0.1026	0.0797	1757	1.76×10^{-5}	116
Mabee 619	881	3.20×10^{-4}	0.262	0.0193	0.00480	0.1046	0.0705	2979	3.50×10^{-5}	189
Mabee 561	821	2.92×10^{-4}	0.257	0.0130	0.00509	0.1043	0.0558	3333	2.73×10^{-5}	169
Mabee B133	–	–	–	0.0182	0.00432	0.1033	0.0705	2636	2.42×10^{-5}	165
Mabee 79	708	3.24×10^{-4}	0.331	0.0179	0.00422	0.1032	0.0759	2148	2.20×10^{-5}	141
Mabee 51	691	3.42×10^{-4}	0.358	0.0205	0.00397	0.1028	0.0878	1879	2.01×10^{-5}	139
Mabee 530	693	2.87×10^{-4}	0.300	0.0143	0.00458	0.1041	0.0578	2751	2.28×10^{-5}	142
Mabee 56	449	2.68×10^{-4}	0.431	0.0204	0.00387	0.1026	0.0759	1792	1.79×10^{-5}	114
Mabee 592	773	3.68×10^{-4}	0.344	0.0194	0.00426	0.1032	0.0962	1850	2.46×10^{-5}	150
Mabee 107	799	3.55×10^{-4}	0.321	0.0154	0.00455	0.1037	0.0642	2633	2.40×10^{-5}	150
Mabee B134	508	2.91×10^{-6}	0.414	0.0133	0.00400	0.1031	0.0543	2247	1.35×10^{-5}	106
Air	5.22	7.22×10^{-4}	1.000	16.45	0.00299	0.1020	31.6	295.5	0	0

"Excess ^{21}Ne" and "Excess ^{40}Ar" refer to excess abundances of ^{21}Ne and ^{40}Ar, relative to ^{20}Ne and ^{36}Ar, compared to atmospheric values; the excess abundances are due to subsurface contributions.

TABLE 4
CONCENTRATIONS AND ISOTOPIC COMPOSITIONS OF XENON IN GASES FROM THE MABEE EOR FIELD

Sample	^{132}Xe (ppmv)	$^{124}Xe/^{132}Xe$	$^{126}Xe/^{132}Xe$	$^{128}Xe/^{132}Xe$	$^{129}Xe/^{132}Xe$	$^{130}Xe/^{132}Xe$	$^{131}Xe/^{132}Xe$	$^{134}Xe/^{132}Xe$	$^{136}Xe/^{132}Xe$
KMCO2	4.52×10^{-5}	0.00358	0.00345	0.07094	0.97643	0.14976	0.77812	0.40606	0.35329
		0.00009	0.00009	0.00007	0.00049	0.00015	0.00039	0.00020	0.00018
Mabee 561	2.35×10^{-4}	0.00351	0.00333	0.07127	0.98316	0.15102	0.78747	0.39057	0.33316
		0.00009	0.00008	0.00007	0.00049	0.00015	0.00039	0.00020	0.00017
Mabee 565	4.02×10^{-4}	0.00358	0.00336	0.07145	0.98214	0.15139	0.78834	0.38956	0.33132
		0.00009	0.00008	0.00007	0.00049	0.00015	0.00039	0.00019	0.00017
Mabee 619	2.84×10^{-4}	0.00367	0.00350	0.07132	0.98009	0.15076	0.78724	0.39109	0.33361
		0.00009	0.00009	0.00007	0.00049	0.00015	0.00039	0.00020	0.00017
Mabee B134	3.16×10^{-4}	0.00358	0.00336	0.07133	0.98294	0.15115	0.78855	0.38969	0.33142
		0.00009	0.00008	0.00007	0.00049	0.00015	0.00039	0.00019	0.00017
Air	2.34×10^{-2}	0.00354	0.00330	0.07136	0.9832	0.1514	0.7890	0.3879	0.03294
		0.00001	0.00002	0.00009	0.0012	0.00012	0.0011	0.0006	0.0004
^{238}U sf							0.1549	1.458	1.761
							0.0059	0.033	0.031
McElmo Dome		0.00348	0.00323	0.07017	0.97745	0.14883	0.77892	0.40514	0.35276
		0.00002	0.00001	0.00007	0.0049	0.00010	0.00039	0.00020	0.00018

The values for "^{238}U sf" refer to isotopic ratios derived during the spontaneous fission of uranium.

CO_2 for reinjection. Thus Blend CO2 can be considered an overall average of the CO_2 from the Mabee active CO_2 area.

The incoming pipeline CO_2 (KMCO2) is isotopically similar to CO_2 from the McElmo Dome. Although there are opportunities for mixing of the CO_2 from the various CO_2 deposits during transport to West Texas, it appears that the injected Mabee CO_2 is mostly from the McElmo Dome. It had been thought by the operators of the Mabee field that the CO_2 originated from the Bravo Dome, and we had begun this study anticipating Bravo isotopic compositions.

Relative to KMCO2, the Mabee production samples have all acquired a noble gas isotopic component similar to that in the Earth's atmosphere ("Air"; Figure 5a). The origin of this component could be either native to the oil producing geologic formations (and therefore present prior to any oil production), water that is injected in large amounts alternating with CO_2 injection (a process called "wagging", or "WAG", water-alternating-gas/CO_2), or from atmospheric contamination during sampling. The latter can be seen not to be the case by the $^{20}Ne-^{36}Ar$ trend shown in Figure 5c. Mixing KMCO2 directly with atmosphere ("Air") during sampling would produce a trend toward higher ^{20}Ne relative to ^{36}Ar. Instead, the samples follow a trend similar to that involving mixing of KMCO2 with water. The difference between the two trends is derived from the variable solubility of neon and argon in water, which results in a different Ne/Ar ratio in

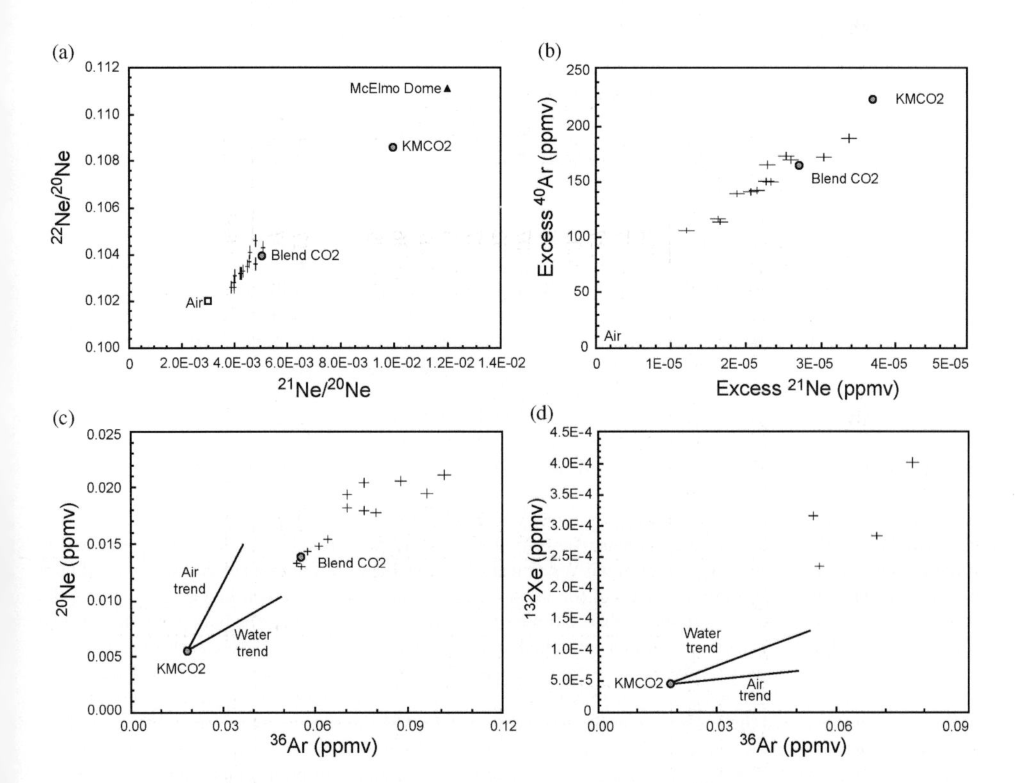

Figure 5: Isotopic compositions and concentrations for neon, argon, and ^{132}Xe from gases sampled at the Mabee EOR field. Data from production wells are shown as small crosses indicating analytical precision. Sample KMCO2 is taken directly from the CO_2 supply pipeline at the Mabee field prior to injection. Blend CO_2 is taken from the return pipeline, and represents an overall mixing of gases captured at active wellheads throughout the active CO_2 area (see Figure 4). The significance of the water and air trends is discussed in the text.

water than in air. This might indicate that noble gases dissolved in the WAG water have a strong effect on the isotopic compositions of the produced gases.

The water trends on Figure 5 depict trajectories for KMCO2 mixing with water in equilibrium with the atmosphere. Figure 5b indicates that there must be a component present in addition to, or perhaps other than, a component that would be provided by atmosphere-equilibrated water. The ^{40}Ar and ^{21}Ne depicted in Figure 5b represent the excess amounts of these isotopes, relative to ^{36}Ar and ^{20}Ne, beyond those present in the atmosphere (i.e. by definition air contains no excess ^{40}Ar or ^{21}Ne). The excess amounts are from subsurface production of ^{40}Ar and ^{21}Ne, through radioactive decay of ^{40}K and the nucleogenic reaction $^{18}O(\alpha,n)^{21}Ne$. If the KMCO2 component mixed only with air, or with water in equilibrium with air, the Mabee samples should trend toward zero ^{40}Ar and ^{21}Ne (the origin), which they do not. They trend toward a position with elevated Excess ^{40}Ar relative to Excess ^{21}Ne. Thus the component mixing with KMCO2 must originate within the subsurface.

Xenon isotopic systematics also show the presence of a subsurface component in addition to KMCO2 (Table 4). The production well gases appear to be enriched in ^{134}Xe relative to ^{136}Xe in a manner consistent with the presence of xenon from spontaneous fission of uranium in the subsurface (Figure 6). Spontaneous fission would also increase ^{132}Xe concentrations, and Figure 5d shows ^{132}Xe enhanced relative to both water and air trends originating at the KMCO2 composition.

We were not able to collect samples of the WAG water and do not know if it contained the subsurface component indicated in Figures 5 and 6. If it did, the WAG water must be fairly old and derived from a deep groundwater aquifer. Instead, however, we tend to believe the subsurface component originated within the oil producing reservoir itself. The distributions of the isotopic compositions shown in Figures 5 and 6 could be derived from a simple two-component mixture of KMCO2 and a native subsurface component with air-like neon isotopic *ratios*, slightly enhanced ^{20}Ne over ^{36}Ar, and enhanced ^{132}Xe, ^{134}Xe relative to ^{136}Xe, and Excess ^{40}Ar.

How does all of this relate to noble gas tracers in CO_2 storage? The Mabee field "demonstration" resulted in several important observations. First, and most important, the noble gas tracer (KMCO2) is not completely "lost" in the subsurface. What was injected can be detected. The production well data require the presence of the KMCO2 component. Tracer tests, particularly those in the deep subsurface, are known to be problematic in that the tracer apparently can be fractionated away from the traced fluid and lost to the surroundings. This did not occur, as indicated by the KMCO2 component present in all samples. Second, even though a subsurface component and possibly a WAG water component, is detected in the produced gases, the KMCO2 component is always clearly identifiable. If today the injected CO_2 migrates away from the Mabee active CO_2 injection area (Figure 4) and is collected from external wells, the KMCO2 noble gas fingerprint will be clearly identifiable. Third, the McElmo Dome CO_2 sampled as KMCO2 has a ^{124}Xe concentration of about 1.62×10^{-7} ppmv (Table 4). The required concentration for ground surface monitoring (Table 2) is about two orders of magnitude higher than this. Therefore, the tracer amount calculated for surface monitoring will clearly be sufficient for subsurface monitoring. This is especially true if the subsurface component originates in the WAG water, since CO_2 stored without WAG water would not be affected. This was an important check of the verity of the calculations presented in Table 2. Fourth, the isotopic compositions observed in the Mabee sample suite were very systematic. The data generally formed linear trends (Figure 5) with deviations explainable by known subsurface components (e.g. uranium spontaneous fission). Alternatively, had the data set been chaotic and unexplainable, it would have raised serious questions about the behavior and thus utility of noble gases in a CO_2 storage setting. Finally, the Mabee test demonstrated the analytical ability to detect in a CO_2 storage setting, even very subtle variations in isotopic compositions. Compositions imparted by xenon isotopic tracers would not be nearly so subtle. Thus the Mabee test demonstrated that the noble gas tracer technique can be a robust and reliable tracer method in CO_2 injection setting.

Monitoring for Noble Gas Isotopes

Strategies for noble gas monitoring at CO_2 storage sites will depend heavily on site-specific geological and hydrological characteristics, and on risk assessment parameters such as leak location probabilities

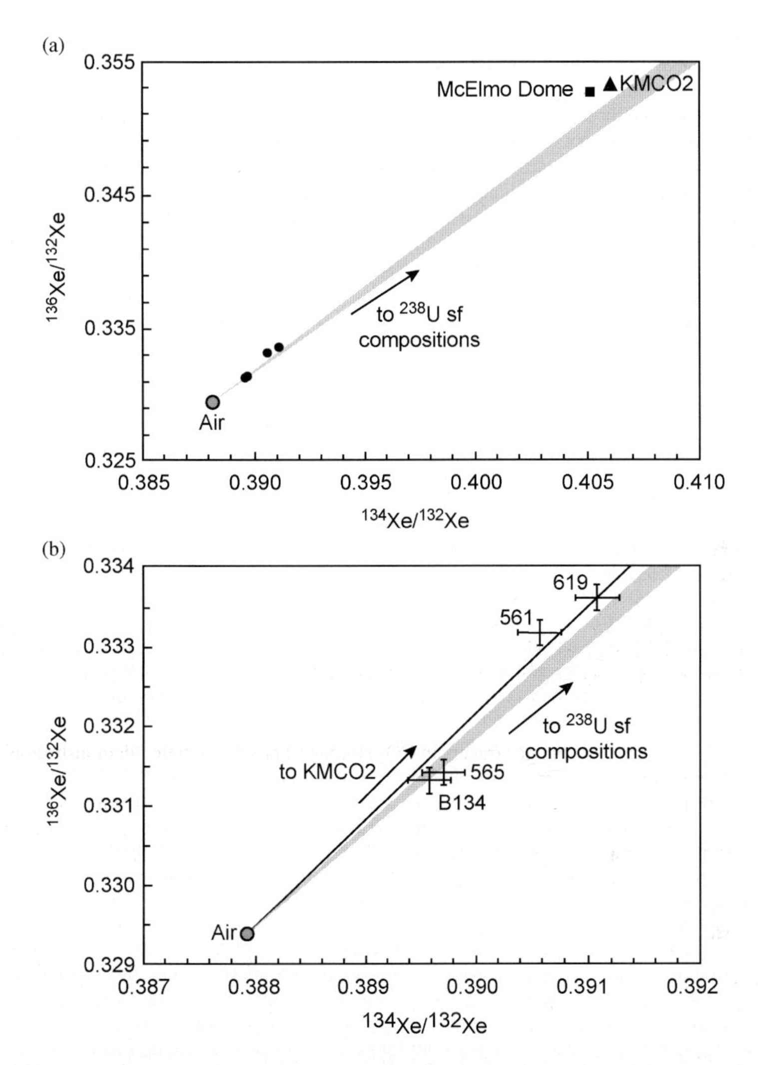

Figure 6: Isotopic compositions for xenon from gases sampled at the Mabee EOR field. Sample KMCO2 is taken directly from the CO_2 supply pipeline at the Mabee field prior to injection, and can be seen to have the approximate xenon isotopic composition of the McElmo Dome. Gray shaded field from Air to ^{238}U spontaneous fission (sf) values represents potential compositions derived by subsurface addition of fission-derived ^{136}Xe and ^{134}Xe to air. Bottom diagram is a close up showing analytical precision of the Mabee samples, and their relation to the spontaneous fission field and KMCO2-Air mixing line.

(e.g. borehole leakage) and degree of early warning required. A significant amount of work is yet required concerning both of these topics before definitive monitoring strategies can be developed.

In general, monitoring could be of two types. Noble gases and CO_2 escaping from deep formations may dissolve into waters and brines of overlying hydrostratigraphic units. In such cases, monitoring deep groundwaters for the distinctive noble gas isotopic tracer may be preferable. A considerable amount of work has been done concerning natural noble gas signatures in both deep and shallow groundwater [2,4,19–23]. As indicated in Table 1, noble gas concentration in continental aquifers are typically about two orders of magnitude lower per gram water (i.e. 1 cm^3), than atmospheric concentrations per cm^3 air. For a given gas flux derived from stored CO_2, noble gas tracers could be easier to detect in these aquifers than in soil gases, provided that the noble gas flux dissolves into the aquifer water. Sampling would also be somewhat more certain for aquifers than for soil gases. As soil gas is collected, air from above ground surface may be pulled into the sampler; such atmospheric "contamination" is much easier to avoid in ground water sampling. The desirability of deep aquifer monitoring is dependant on the expected behavior of leaking CO_2 and noble gases in the strata overlying the storage formation. Much more work, including expansion of numerical models of CO_2 and noble gas migration, is required.

Ultimately, even though aquifers may be monitored, ground surface monitoring will be desirable. Table 2 was constructed on this premise. Any of the various methodologies available for collection of soil gases could be employed. The amount of gas required for analysis by noble gas mass spectrometry is very small, and the problem will not be in the volume of gas to be collected, but in insuring that the gas collected is representative of the soil gas being emitted. However, multi-liter sampling and repetitive sampling will likely be sufficient.

A study was conducted at our laboratory several years ago that would appear significant for soil gas monitoring for CO_2 leakage [24]. In the study, SF_6 and ^3He tracers were released together by a chemical detonation at about 350 m depth. The ground surface was monitored at several locations above the release, including along an adjacent geologic fault. SF_6 was detected in the fault zone 50 days after detonation (nothing was ever detected outside the fault zone). The ^3He was not detected for an additional 325 days. It had been expected, following conventional wisdom, that ^3He would be detected first. The small atomic size of helium would cause it to be less impeded in its migration than the larger SF_6. Numerical modeling using the NUFT unsaturated transport code was successful in reproducing the tracer breakthrough sequence and timing. The helium apparently had access to micropathways that exclude SF_6 due to its size. The SF_6 migration path had less tortuosity and was more directly upward within the fault zone. This study raises the possibility that a xenon isotopic tracer (atomic radius 2.2 Å; Table 1) could migrate to ground surface faster than leaking CO_2 (1.4 Å C–O linear bond distance). If so, it could provide an early-warning system for leaking CO_2.

An additional monitoring consideration is that geologic CO_2 storage may be sited in oceanic shelf/basin regions such as the North Atlantic oil fields (e.g. Sleipner). Monitoring in these locations would involve direct monitoring of seawater. This is perhaps the easiest scenario for noble gas tracer detection due to the low noble gas concentrations in seawater (Table 1). Oceans have a very significant capacity for buffering CO_2 concentration. Initial releases of leaking CO_2 would not be detectable by direct CO_2 monitoring nor by monitoring biological effects associated with increased aqueous HCO_3. Noble gas isotopic compositions within bottom waters could be significantly altered by noble gas emissions from leaking CO_2 storage sites. However, assessment of oceanic monitoring will require an examination of the effects of ocean bottom currents, ocean sediment pore water behavior, and pore water diffusion, among other processes.

CONCLUSIONS

Noble gas isotopes, particularly xenon isotopes, may provide a mechanism for leakage monitoring and subsurface tracing of stored CO_2. They are chemically inert, environmentally safe, persistent, and stable in all environments. While ^3He and ^{36}Ar may be suitable only for special uses due to availability and high costs, ^{124}Xe, ^{129}Xe, and ^{136}Xe are inexpensive and readily available. Only small volumes of these xenon isotopes would be needed (<900 L per year for the Mabee storage analogue), simplifying handling and injection. The Mabee field test conducted as part of this study demonstrated that unique noble gas isotopic compositions injected with CO_2 can be readily detected in production wells. Even though other isotopic components are present in the subsurface, the unique isotopic fingerprint can always be identified. The noble gases behave in a systematic and predictable manner in the CO_2 injection setting, indicating that the noble gas tracing technique would be a robust and reliable method. Many aspects of potential monitoring methods must yet be formulated, but monitoring in deep aquifers, ocean waters, and the ground surface, for which the calculations presented here are intended, all appear achievable.

RECOMMENDATIONS

Very important questions remain concerning both the partitioning behavior of tracers in the CO_2 storage environment and the techniques for the monitoring process itself. The lack of data on partitioning was discussed above. Monitoring strategies that need to be addressed include deep versus shallow (Earth surface) monitoring, the value of focusing monitoring along geologic structures, and the possibility that some tracers may serve as an early warning of CO_2 migration to the surface. These issues lead to the following recommendations for future work:

1. Laboratory-scale studies should be conducted to document partition coefficients and solubilities of noble gases and other tracers for the phases, and the $P-T$ conditions, to be encountered in the CO_2 storage environment.
2. Based on this partitioning and solubility data, numerical modeling should be conducted to simulate the behavior within the storage reservoir and overlying strata of noble gases and other tracers dissolved in injected CO_2. This will provide a mechanism for assessing a variety of monitoring strategies.
3. Field tests should be conducted examining the upward migration of CO_2 and large- and small-atomic radii noble gases for the purpose of assessing the possibility that certain tracers may be capable of providing an early warning of CO_2 migration to the surface. These tests could be conducted at shallow depths (perhaps <300 m).
4. A field demonstration at a potential storage site is needed in which only CO_2 and noble gas tracers are injected (no EOR, no WAG water), and for which leakage is induced and aggressive monitoring is conducted. It is only in this more realistic setting that a true evaluation can be made of the potential for noble gas tracing of CO_2.

NOMENCLATURE

EOR enhance oil recovery
ppmv parts per million by volume
STP standard temperature and pressure (= 25 °C and 1 atm)
WAG water-alternating-gas (CO_2)

ACKNOWLEDGEMENTS

We would like to thank Jeff Woliver and Craig Goodyear of ChevronTexaco Corporation in Midland, Texas for giving us access to the Mabee EOR field, for providing us with information concerning the production and injection wells, and for helping with field logistics for sampling. We also thank two anonymous reviewers for their helpful comments which improved the content and presentation of this chapter. This work was performed under the auspices of the US Department of Energy by the Lawrence Livermore National Laboratory under contract W-7405-Eng-48.

REFERENCES

1. M. Ozima, F.A. Podosek, Noble Gas Geochemistry, second ed., Cambridge University Press, Cambridge, UK, 2001.
2. J.F. Clark, M.L. Davisson, G.B. Hudson, P.A. Macfarlane, *J. Hydrol.* **211** (1998) 151.
3. M.C. Castro, A. Jambon, G. de Marsily, P. Schlosser, *Water Res. Res.* **34** (1998) 2443.
4. D.L. Pinti, B. Marty, *Geochim. Cosmochim. Acta* **59** (1995) 3389.
5. R.W. Potter, M.A. Clynne, *J. Solution Chem.* **6** (1978) 837.
6. S.P. Smith, B.M. Kennedy, *Geochim. Cosmochim. Acta* **49** (1985) 893.
7. H.L. Clever (Ed.), Solubility Data Series, v1. International Union of Pure and Applied Chemistry, Pergamon Press, Oxford, New York, 1979.
8. H.L. Clever (Ed.), Solubility Data Series, v2. International Union of Pure and Applied Chemistry, Pergamon Press, Oxford, New York, 1979.

9. H.L. Clever (Ed.), Solubility Data Series, v4. International Union of Pure and Applied Chemistry, Pergamon Press, Oxford, New York, 1979.
10. R.J. Wilcock, W.F. Danforth, E. Wilhelm, *J. Chem. Thermodyn.* **10** (1978) 817.
11. Y.K. Kharaka, D.J. Specht, *Appl. Geochem.* **3** (1988) 137.
12. J.F. Clark, G.B. Hudson, M.L. Davisson, G. Woodside, R. Herndon, *Groundwater* **42** (2004) 167.
13. M.W. Caffee, G.B. Hudson, C. Velsko, G.R. Huss, E.C. Alexander, A.R. Chivas, *Science* **285** (1999) 2115.
14. D.E. Bottomley, J.D. Ross, W.B. Clarke, *Geochim. Cosmochim. Acta* **48** (1984) 1973.
15. T.F. Kraemer, D.P. Generaux, in: C. Kendall, J.J. McDonnell (Eds.), *Isotope Tracers in Catchment Hydrology*, Elsevier, Amsterdam, 1998.
16. M. Maiss, C.A.M. Brenninkmeijer, *Environ. Sci. Technol.* **32** (1998) 3077.
17. US Climate Change Technology Program, Technology Options for the Near and Long Term, November, 2003.
18. M. Danilov, R. DeVoe, A. Dolgolenko, G. Giannini, G. Gratta, P. Picchi, A. Piepke, F. Pietropaolo, P. Vogel, J.-L. Vuilleumier, Y.-F. Wang, O. Zeldovich, *Phys. Lett.* **B480** (2000) 12.
19. A. Zaikowski, B.J. Kosanke, N. Hubbard, *Geochim. Cosmochim. Acta* **51** (1987) 73.
20. K. Osenbruck, J. Lippman, C. Sonntag, *Geochim. Cosmochim. Acta* **62** (1998) 3041.
21. A. Battani, P. Sarda, A. Prinzhofer, *Earth Planet. Sci. Lett.* **181** (2000) 229.
22. C.J. Ballentine, M. Schoell, D. Coleman, B.A. Cain, *Nature* **409** (2001) 327.
23. R. Kipfer, W. Aeschbach-Hertig, F. Peeters, M. Stute, in: D. Porcelli, C.J. Ballentine, R. Wieler (Eds.), *Noble Gases in Geochemistry and Cosmochemistry*, Reviews in Mineralogy and Geochemistry, vol. 47, Mineralogical Society of America, Washington, DC, 2002.
24. C.R. Carrigan, R.A. Heinle, G.B. Hudson, J.J. Nitao, J.J. Zucca, *Nature* **382** (1996) 528.

SECTION 5:
RISK ASSESSMENT

RISK ASSESSMENT PREFACE

Sally M. Benson

Lawrence Berkeley National Laboratory, Berkeley, CA, USA

Assessing the risk of CO_2 leaking out of the storage structure and the consequences thereof is one of the most important and demanding tasks for assuring that a geologic storage project is safe and effective. Thoroughness and scientific rigor are needed at every stage in the process to ensure that good decisions are made about siting storage projects, operating parameters, designing and implementing monitoring networks, and developing mitigation and intervention strategies. Transparency in all aspects of the risk assessment process will also be required to gain regulatory and public acceptance for a storage project.

The risk assessment portion of this book is divided into two parts. The first part reviews literature about risk assessment, regulation, and remediation for analogues to CO_2 storage projects. The second part presents original research on the development and application of risk assessment methodologies.

The first part begins by reviewing risk assessment and mitigation practices used in industrial analogues for storage (Benson, Chapter 25). Hepple (Chapter 26) then set geologic storage of CO_2 in the context of the global carbon cycle, industrial uses of CO_2 and human health effects from exposure to elevated concentrations of CO_2. Chapter 27 (Apps) reviews the regulatory approach used in the United States governing the widespread practice of disposing liquid wastes in deep geologic formations, drawing parallels and highlighting differences with geologic storage of CO_2. Potential intervention and remediation methods for storage projects that do leak are then presented by Benson and Hepple (Chapter 28).

The remainder of this section is devoted to original research on potential risks of CO_2 storage and risk assessment methodologies. Oldenburg and Unger (Chapter 29) present a new model for predicting the consequences of CO_2 leakage between the vadose zone and the atmosphere. Onstott (Chapter 30) presents the first analysis of the impacts of subsurface storage on microbial communities in the deep subsurface and how microorganisms will modify abiotic geochemical reactions. Stenhouse et al. (Chapter 31), Wo et al. (Chapter 32) and Wildenborg et al. (Chapter 33) describe formal procedures and methods for performing probabilistic risk assessment.

Together, this collection of studies provides a snapshot of the rapidly evolving status of risk assessment for geologic storage of CO_2. Much progress has been made, but more is expected in the future as these ideas and methods are tested on real-world problems.

Carbon Dioxide Capture for Storage in Deep Geologic Formations, Volume 2
D.C. Thomas and S.M. Benson (Eds.)

Chapter 25

LESSONS LEARNED FROM INDUSTRIAL AND NATURAL ANALOGS FOR HEALTH, SAFETY AND ENVIRONMENTAL RISK ASSESSMENT FOR GEOLOGIC STORAGE OF CARBON DIOXIDE

Sally M. Benson

Lawrence Berkeley National Laboratory, Berkeley, CA, USA

ABSTRACT

This literature survey was conducted to gather and interpret information regarding potential approaches for assessing, managing and mitigating risks associated with the deep geologic storage of CO_2. Information was gathered from three principle sources: (1) industrial analogs such as natural gas storage, deep injection of hazardous wastes and nuclear waste storage and (2) natural analogs, especially those with CO_2 leaks at the surface and (3) industrial uses of CO_2 for a variety of applications. A set of lessons learned from these analogs was compiled and forms the basis for recommendations in the areas of risk assessment framework and methodology, risk management approaches and risk mitigation and remediation methods. Lessons learned include:

1. There is an abundant base of experience to draw on that is relevant and suggests that CO_2 can be stored safely if geologic storage sites are carefully selected and monitored.
2. The human health effects of exposure to elevated concentrations of CO_2 have been extensively studied and occupational safety regulations are in place for safe use. Ecosystem impacts from elevated soil gas concentrations are less well characterized and may require additional research.
3. The hazard created by CO_2 releases depends more on the nature of the release rather than the size of the release. In particular, since CO_2 is denser than air, hazardous situations arise when large amounts of CO_2 accumulate in low-lying, confined or poorly ventilated spaces. Releases, even large ones, do not pose a hazard if they are quickly dissipated in the atmosphere, such as from tall industrial stacks or explosive volcanic events.
4. Many of the risks of CO_2 storage are well understood based on experience from natural gas storage and deep injection of hazardous waste. Experience from these analogs suggest that the biggest risks from CO_2 storage will be due to: leakage through poor quality or aging injection well completions; leakage up abandoned wells; leakage due to inadequate cap rock characterization; and inconsistent or inadequate monitoring of injection wells, groundwater in overlying formations and leakage from abandoned wells.
5. Regulatory paradigms and approaches for the industrial analogs vary and none address all the issues that are important for CO_2 storage.

This chapter reviews the lessons learned and also provides recommendations for additional research to address gaps in knowledge and risk management approaches.

INTRODUCTION

Three operations are currently underway in the US that provide useful insights for geologic storage of CO_2, namely: (1) deep well injection of industrial wastes; (2) natural gas storage; and (3) industrial use of CO_2 for a variety of applications. An assessment of these activities and the lessons they provide was performed to assist the CO_2 Capture Project in selecting a portfolio of R&D projects that could improve health, safety and environmental (HSE) risk assessment for geologic storage of CO_2. In particular, the following were reviewed:

- history, status and scope of the activity;
- risk assessment framework and methods, including key issues, performance specifications and performance assessment methods;
- risk management approaches, including regulatory oversight and permitting, site characterization methods, monitoring and performance confirmation;
- risk mitigation and remediation methods employed or planned in the event that performance specifications are not met or other unintended consequences arise; and
- case studies documenting responses to historical accidents.

HSE risk assessment for geologic storage will be driven by the hazards associated with exposure to elevated concentrations of CO_2; therefore, in addition to reviewing the history and status of these activities, we also reviewed information about human and ecological health risks from exposure to elevated levels of CO_2, information from natural analogs for CO_2 storage, industrial uses of CO_2 and monitoring technology for CO_2 detection. A complete version of the extended literature review is available in Benson et al. [1].

REVIEW OF INDUSTRIAL ANALOGS RELEVANT TO CO_2 STORAGE

Natural Gas Storage

Underground natural gas storage projects have been operated successfully in the US for almost 90 years and today, 450 projects store approximately 139 million metric tonnes (MMT) of natural gas in 30 states (see also Perry [2]). The majority of storage facilities are in depleted oil and gas reservoirs, but 51 are in brine-filled aquifers and 40 in salt caverns. Experience has shown that there are a number of factors critical to the success of these projects:

- first, it is important to have a site that is adequately characterized (i.e. permeability, thickness and extent of storage reservoir, cap rock integrity, geologic structure, lithology, etc.);
- second, the storage formation should be deep enough to allow sufficiently high gas pressures for the economic success of the operation;
- third, injection/withdrawal wells must be properly designed, constructed, monitored and maintained;
- fourth, overpressuring the storage reservoir should be avoided;
- finally, abandoned wells in and near the project must be located and plugged.

While underground natural gas storage has been used safely and effectively, there have been a number of documented cases where leakage has occurred [1,2]. In over one-half of these cases, leakage was caused by defective wells (poorly constructed or improperly plugged abandoned wells). Over time, as engineering practices have improved and regulatory oversight has grown more stringent, fewer accidents have occurred, and modern procedures have made underground natural gas storage a safe and effective operation.

One of USEPA's primary regulatory responsibilities is to protect drinking water aquifers from detrimental effects caused by underground gas storage. The USEPA has delegated authority to most of the states, which have effective regulations for permitting, operating and monitoring underground gas storage fields [3]. Regulations differ from state to state and are tailored to local concerns, such as in Pennsylvania where extra measures are taken to avoid leakage of gas into underground coalmines. In several states with abundant oil, gas, coal and/or mineral resources, a protection (or buffer) zone is established to avoid or reduce the risk of accidents caused by human intrusion.

Monitoring is an important part of the regulatory oversight of these projects [1–3]. While regulations on monitoring and reporting vary among states, almost all monitoring requirements focus on assuring that the wells are not leaking (e.g. pressure measurements and down hole logs such as temperature, pressure, noise/sonic and casing inspection logs). Observation wells installed and monitored for the purpose of verifying that gas has not leaked into shallower strata are rarely required; however, a few storage projects have over 100 wells for this purpose. Geophysical techniques to monitor the operation are not required.

Depleted oil and gas reservoirs are easier to develop than aquifer storage projects because the geologic structure and cap rock are usually well characterized from existing wells. Moreover, since the structure is known to have trapped and stored hydrocarbons over geologic time periods, it is likely to be effective for natural gas storage. Standard natural gas reservoir engineering practices are used during the permitting process and storage operations. For aquifer gas storage projects, extensive site characterization is required and well testing methods specifically for evaluating the permeability and continuity of the cap rock have been developed.

In the event that leakage occurs, remediation is possible by producing or venting the gas accumulated in shallower layers and/or reducing reservoir pressure [2,4]. In most cases, leakage is caused by the presence of leaking or abandoned wells, which should be identified and plugged as soon as possible. Some projects, such as the Herscher storage project in Illinois, continue to operate even though leakage continues. Here shallow extraction wells are used to capture the gas that leaked from the storage interval.

When a natural gas storage site is shut down, as much of the gas as is practical is removed from the formation. The injection wells are then plugged and abandoned using prescribed procedures. No long-term monitoring is required after a project has been shut down.

Deep Injection of Liquid and Hazardous Waste

The USEPA's Underground Injection Control Program recognizes five classes of injection wells, including [3,5]:

- *Class I*: wells used to inject hazardous, industrial or municipal waste beneath the lowermost formation containing an underground source of drinking water;
- *Class II*: wells used to inject fluids related to the production of oil or natural gas;
- *Class III*: wells used to inject for the extraction of minerals such as sulfur, salt, potash, or metals such as uranium by solution mining;
- *Class IV*: wells used to dispose off hazardous or radioactive waste into or above a formation that contains a USDW or an exempted aquifer. These wells are now effectively prohibited;
- *Class V*: injection wells not included in Classes I, II, III or IV.

Class I and Class II wells are most relevant to geologic storage of CO_2, particularly with regard to the potential for contaminating drinking water aquifers. However, it is important to recognize that regulations regarding the HSE effects of surface facilities and leakage of CO_2 back into the atmosphere are likely to be regulated through other programs. Confusion and inefficiencies from overlapping jurisdictions and requirements may create a regulatory morass. Early attention to this issue may prevent decades of frustration with an overly complicated and inefficient set of regulations.

Industrial liquid waste disposal by deep-well injection was initiated in 1939. Since that time the practice has expanded so that now, 9 billion gallons per year of hazardous, industrial and municipal wastes are injected into 485 Class I wells. In 1987, the cost of liquid hazardous waste disposal ranged from $49 to $207/ton. Early performance was mixed, with many examples of well failures and contamination of drinking water aquifers. Failures were attributed to: (1) poor characterization of the confining units; (2) improper well completion techniques; (3) use of well construction materials that were incompatible with the waste streams and, consequently, corroded; (4) inconsistent or inadequate monitoring; and (5) leakage through abandoned wells. Because of these problems and the inconsistent approach to oversight, progressively more stringent regulations were put in place to make the practice of industrial waste disposal by liquid injection safer. By 1988, the current set of regulations was put in place and since that time there have been no incidents where drinking water contamination has been reported.

EPA has adopted the approach of stringent regulation of deep-well-injection operations, with the goal of ensuring that contamination does not occur in the first place [5]. To obtain a permit for hazardous waste disposal by deep-well injection, the operator must demonstrate that "No Migration" of the waste will occur outside the formation into which it is injected. The formation must contain over 10,000 ppm of dissolved solids, be overlain by a suitable cap rock and be separated from a drinking water aquifer by at least one other impermeable formation.

The regulations mandate stringent controls for the siting, operation, reporting and abandoning of injection wells. Experience has shown that leaks from injection and abandoned wells were the most frequent short-term failure mechanisms. Consequently, much of the current regulatory approach focuses on minimizing the possibility of such failures. Current well completion and rehabilitation techniques appear to be adequate to prevent leakage, although finding abandoned wells remains a significant challenge. As for contamination of drinking water aquifers distant from the wellbore, some efforts to detect transmissive faults between the injection zone and overlying aquifers are mandated, and if monitoring wells are already in place in overlying protective aquifers, EPA also requires that these must be monitored for contamination. The permitting process for hazardous waste injection wells is extensive, time consuming and expensive. According to USEPA [5], factoring in the costs for geologic testing and modeling, a "No Migration" petition can cost in excess of $2,000,000.

When a facility is shut down, the EPA is particularly concerned that deep injection wells, especially those that have injected hazardous waste, are properly plugged and abandoned. Upon closure, a Class I hazardous waste well must be plugged with cement in a manner that will not allow the movement of fluids into or between drinking water aquifers. Class I hazardous waste well operators must also prepare and comply with a plan for post-closure care. The plan must include the predicted position of the waste front at closure, the status of any cleanups required and the estimated cost of proposed post-closure care. In addition, the owner or operator must continue to conduct any required groundwater monitoring until pressure in the injection zone decays to the point that the well's cone of influence no longer intersects the base of the lowermost drinking water aquifer. The owner or operator must demonstrate and maintain financial responsibility for post-closure care. This obligation survives the termination of a permit or the cessation of injection and is enforceable regardless of whether the requirement is a condition of the permit.

For deep-well injection of liquid wastes, the density of the injected fluid is usually within $\pm 5\%$ of the surrounding formation fluids [6]. In this case, the injected wastes tend to migrate away from the injection well with little buoyant force driving it up or down. For CO_2 storage in oil or water-filled geological formations, this will not be the case. Buoyancy forces will tend to drive CO_2 upward. A case study of municipal waste disposal in Florida demonstrates that under these conditions, containment can be more difficult and there is evidence that the less dense effluent is migrating in the opposite direction than originally anticipated based on regional hydrologic gradients. This is an important lesson for geologic storage of CO_2 and highlights the unique requirements for characterizing sites where the injected fluid will migrate under the action of gravity and not necessarily follow the migration path of or move at the same rate as regional groundwater [6].

Industrial Use of CO_2 and Human Health Effects from CO_2 Exposure

Carbon dioxide is generally regarded as a safe and non-toxic, inert gas. It is an essential part of the fundamental biological processes of all living things. It does not cause cancer, affect development or suppress the immune system in humans. Carbon dioxide is a physiologically active gas that is integral to both respiration and acid–base balance in all life. Exposure to elevated concentrations of CO_2 can lead to adverse consequences, including death. The effects of elevated CO_2 depend on the concentration and duration of exposure.

Ambient atmospheric concentrations of CO_2 are currently about 370 ppm. Humans can tolerate increased concentrations with no physiological effects for exposures up to 1% CO_2 (10,000 ppm) [7]. For concentrations of up to 3%, physiological adaptation occurs without adverse consequences. A significant effect on respiratory rate and some discomfort occurs at concentrations between 3 and 5%. Above 5%, physical and mental ability is impaired and loss of consciousness can occur. Severe symptoms, including rapid loss of consciousness, possible coma or death, result from prolonged exposure above 10%. Experiments conducted on a group exposed to up to 3% CO_2 for many weeks and short-term exposures to even higher concentrations have shown that all effects are reversible except for prolonged coma, the consequences of prolonged hypoxia (lack of oxygen) and death. These experiments, however, have been conducted on healthy adults and these conclusions may not be applicable to other more sensitive populations. Loss of consciousness occurs within several breaths and death is imminent at concentrations above 25–30%. Deaths from catastrophic releases of CO_2 are known from industrial accidents and natural disasters.

Carbon dioxide is used in a wide variety of industries: from chemical manufacture to beverage carbonation and brewing, from enhanced oil recovery to refrigeration and from fire suppression to inert-atmosphere food preservation [7]. Sources of CO_2 include natural reservoirs, separation from crude oil and natural gas and as a waste product of industrial processes (chemical manufacture), combustion processes (energy production) and biological respiration (brewing). Because of its extensive use and production, the hazards of CO_2 are well known and routinely managed. Engineering and procedural controls are well established for dealing with the hazards of compressed and cryogenic CO_2. Nevertheless, the hazards of CO_2 are significant as fatalities from fire-suppression system malfunctions and confined-space accidents attest.

Carbon dioxide is regulated by Federal and State authorities for many different purposes, including occupational safety and health, ventilation and indoor air quality, confined-space hazard and fire suppression, as a respiratory gas and food additive, for animal anesthesia and the humane slaughter of livestock, transportation and most recently as a greenhouse gas (UNFCCC). Federal occupational safety and health regulations set three limits:

- 0.5% or 5000 ppm for an average 8 h day or 40 h week;
- 3% or 30,000 ppm for an average short-term 15 min exposure limit;
- 4% or 40,000 ppm for the maximum instantaneous exposure limit above which is considered immediately dangerous to life and health.

Most industrial and safety regulations for CO_2 focus on engineering controls and specifications for transportation, storage containers and pipelines.

Monitoring is a routine part of industrial use and production of CO_2. Both real-time monitors and air sampling are used to ensure that levels remain within the regulatory guidelines. In addition, CO_2 concentrations are routinely measured and used as a proxy for air quality in buildings.

LESSONS LEARNED FROM NATURAL AND INDUSTRIAL ANALOGS

A large amount of valuable information was obtained from investigating natural and industrial analogs for geologic storage of carbon dioxide. What follows here is a synthesis of this information into five lessons that can be used to identify issues that must be addressed, identify best practices that could be adopted for geologic storage, find technologies that may be applicable for risk management or mitigation, and avoid the pitfalls encountered in the industrial analogs.

Lesson 1. There is an abundant base of experience to draw on that is relevant and suggests that CO_2 can be stored safely if geologic storage sites are carefully selected and monitored.

This includes relevant experience from the following.

- Natural gas storage projects, depleted oil and gas reservoirs, as well as aquifers have been successfully used for the purpose of providing local storage to meet fluctuating daily and seasonal demand for natural gas. Today, in the US 139 MMT of natural gas is stored annually in over 450 projects in 35 states. Over 90 years of relevant experience is available that buoyant fluids such as CO_2 can be safely stored.
- Deep injection of liquid and hazardous wastes—deep geologic formations, far below the depth of the deepest drinking water aquifer, are used to dispose off hazardous and other liquid wastes. Today, approximately 9 billion gallons of hazardous waste are disposed off this way and nearly 300 billion gallons of oil field brines are also injected into deep geologic formations. Deep injection of liquid and hazardous waste has been implemented safely since an adequate set of regulations and rigorous enforcement have been established.
- CO_2 enhanced oil recovery—59 projects are currently underway in the US that uses CO_2 to enhance oil recovery from depleted reservoirs. Most of these projects use CO_2 produced from natural reservoirs, specifically, the McElmo and Bravo Domes on the Colorado Plateau. Extensive experience with production, injection and transportation of CO_2 in long pipelines has been obtained from these projects. In addition, natural CO_2 reservoirs demonstrate that CO_2 can be stored underground for geologic time

periods (thousands to millions of years) and can be used to help understand the chemical interactions that take place between the stored CO_2 and the rock formations.
- Food preservation, beverage carbonation, and fire suppression—CO_2 is safely used in a wide range of industrial applications, from food preservation to fire suppression. This experience and the regulations for safe work practices provide further evidence that CO_2 can be managed safely and the risks are well understood.

Lesson 2. The human health effects of exposure to elevated concentrations of CO_2 have been extensively studied and occupational safety regulations are in place for safe use. Ecosystem impacts from elevated soil gas concentrations are less well characterized and may require additional research.

Lesson 3. The hazard created by CO_2 releases depends more on the nature of the release rather than the size of the release. In particular, since CO_2 is denser than air, hazardous situations arise when large amounts of CO_2 accumulate in low-lying, confined or poorly ventilated spaces. Releases, even large ones, do not pose a hazard if they are quickly dissipated in the atmosphere, such as from tall industrial stacks or explosive volcanic events. This conclusion is based on the lack of correlation between the size and consequences of releases from examples such as large volcanic eruptions, natural ecosystem fluxes, refinery emissions, small but fatal confined space releases (fire suppression) and the hazardous limnic releases that recently occurred in Cameroon (e.g. Lakes Nyos and Manoon in Cameroon).

Lesson 4. Many of the risks of CO_2 storage are well understood based on experience from natural gas storage and deep injection of hazardous waste. Experience from these analogs suggest that the biggest risks from CO_2 storage will be due to: leakage through poor quality or aging injection well completions; leakage up abandoned wells; leakage due to inadequate cap rock characterization; and inconsistent or inadequate monitoring of injection wells, groundwater in overlying formations and leakage from abandoned wells.

Lesson 5. Regulatory paradigms and approaches for the industrial analogs vary and none address all the issues that are important for CO2 storage. For example, (1) some regulations rely on performance-based requirements while others use practice-based requirements, (2) some activities are regulated by the states while others have federal regulatory oversight, and (3) there is not a consistent approach or requirement for requirements for short and long-term monitoring. Perhaps more important than these differences, none of the regulations fully address several issues that are important for CO_2 storage in geologic formations.

- Storage is needed over a comparatively long time frame to ensure that geologic storage is an effective method for decreasing greenhouse gas concentrations in the atmosphere (hundreds to thousands of years), therefore, performance requirements, regulations and liability issues over this extended time period need to be addressed [8–10].
- Additional storage security is provided by dissolution of CO_2 in pore fluids, residual gas trapping and mineral trapping; consequently, a regulatory framework that includes the storage benefits of these geochemical trapping processes is needed [6,11].
- Migration is strongly controlled by the density contrast between CO_2 and native pore fluids, leading to buoyancy-driven flow and subsequent trapping beneath low permeability and capillary barriers. Regulations specific to CO_2 should be considered that fully account for migration driven by buoyancy forces [6].
- The fact is that CO_2, unlike most other substances that are regulated for environment, health and safety purposes, is not only non-hazardous at low concentrations but also an essential part of all living systems. Alternatives to regulatory approaches used to protect groundwater quality, which are based on avoiding exposure to very low concentrations of contaminants, may be needed.

RECOMMENDATIONS FOR FUTURE RESEARCH AND EVALUATION

From the lessons learned that are described above and an evaluation of the gaps between future needs and the current knowledge, we have identified a number of recommendations for future research and evaluation. These recommendations are divided into three categories, namely, those related to: (1) risk assessment methodology; (2) risk management approaches; and (3) risk mitigation and remediation. These recommendations are summarized in Table 1. As described in this volume, many of these issues are being addressed through research and development projects sponsored by the CCP.

TABLE 1
SUMMARY OF RECOMMENDATIONS FOR RISK ASSESSMENT, RISK MANAGEMENT, AND REMEDIATION OF GEOLOGIC STORAGE PROJECTS

Topic	Recommendations
Risk assessment methodology, including key issues, performance specifications and performance assessment methods	(1) Develop a common health, safety and environmental (HSE) risk assessment framework, including treatment of uncertainty, for geological sequestration f CO_2 (2) Develop performance requirements for each of the critical components of the sequestration system, namely, at a minimum, specifications for: – Injection well completion and monitoring; – Acceptable leakage rates from the primary containment structure; and – Surface concentrations of CO_2 that could effect human health or ecosystems (3) Identify and quantify risks to ecosystems and natural resources in the vicinity of surface leaks (4) Develop and test coupled atmosphere, land-surface and subsurface models that predict atmospheric dispersion of CO_2 from leaks (5) Adapt the *F*eatures–*E*vents–*P*rocesses (FEP) procedure used in nuclear waste storage for identifying and ranking importance of critical performance parameters for geologic storage of CO_2
Risk management approaches, including regulatory oversight and permitting, site characterization methods, monitoring and performance confirmation	(1) Develop a single, consistent regulatory approach that addresses HSE issues, especially those issues dealing with surface leakage – Local safety concerns – Effectiveness for greenhouse gas control (2) Identify and investigate the effectiveness of multi-containment concepts (e.g. solubility and mineral trapping) (3) Develop well completion methods, well abandonment procedures and methods for sealing abandoned wells that are compatible with long-term containment of CO_2 (4) Develop a risk management strategy that couples monitoring requirements to performance confirmation

(*continued*)

TABLE 1
CONTINUED

Topic	Recommendations
Risk mitigation and remediation methods employed or planned in the event that performance specifications are not met or other unintended consequences arise	Methods for mitigating and remediating risks caused by leakage of CO_2 from the primary storage reservoir should be developed. Examples of potential mitigation methods include (1) controlling the pressure within the storage reservoir to prevent leakage or damage to the cap rock, (2) in the event that leakage has occurred, gas that has accumulated in shallow traps can be pumped out to prevent further migration and surface releases, (3) injection well monitoring to detect damage or leakage and (4) repair of leaking injection, production or abandoned wells

In addition, other general recommendations that will expedite development of safe and effective methods for CO_2 storage in deep geologic formations include: (1) considering the implications of others gases on HSE risk assessment (H_2S, SO_x, NO_x, hydrocarbons); (2) investigating natural analogs for HSE risks from surface leakage; (3) supporting the development and systematic evaluation of computational models that include the full set of physical, geochemical and geomechanical processes that influences the safety and effectiveness of geologic storage; and (4) supporting pilot tests in a number of geologic setting. This last recommendation is particularly important as experience has shown that most knowledge is gained from real-world experience and invariably surprises arise that were not anticipated during planning studies.

CONCLUSIONS

In conclusion, the key poorly understood HSE concerns surrounding geologic storage of CO_2 relate to the potential for unanticipated leakage. Such releases could be associated with surface facilities, injection wells or natural geological "containers" and may range from small-scale diffuse leaks to large catastrophic incidents. Extensive industrial experience with CO_2 and gases in general shows that the risks from industrial storage facilities are manageable using standard engineering controls and procedures. Serious accidents have occurred, but the incidents were preventable and experience teaches us how to operate these facilities even more safely. On the other hand, our understanding of and ability to predict CO_2 releases and their characteristics in any given geologic and geographic setting is far more challenging. Certainly there are many sites, such as oil and gas reservoirs where the probability of leakage is very low. However, brine formations, which generally are not well characterized and do not have cap rocks or seals that have stood the test of time, will require significant effort to evaluate potential risks, and these risks must be taken seriously.

ACKNOWLEDGEMENTS

This work was sponsored and funded by the CO_2 Capture Project, a Joint Industry Project supported by BP, ChevronTexaco, ENI, Norsk Hydro, PanCanadian, Statoil, Shell International, and Suncor. The author thanks the two peer reviewers who provided valuable comments that improved this manuscript.

REFERENCES

1. S.M. Benson, R. Hepple, J. Apps, C.-F. Tsang, M. Lippmann, Lessons learned from natural and industrial analogues for storage of carbon dioxide in deep geologic formations, Lawrence Berkeley National Laboratory Report, LBL-51170, Berkeley, CA, 2002.

2. K. Perry, Natural gas storage industry experience: analogue to CO_2 storage, 2005, Lessons Learned from Industrial and Natural Analogs for Health, Safety and Environmental Risk Assessment for Geologic Storage of Carbon Dioxide, Vol. 2. Geologic Storage of Carbon Dioxide with Monitoring and Verification, Elsevier Publishing.
3. E.J. Wilson, T.L. Johnson, D.W. Keith, Regulating the ultimate sink: managing the risks of geologic CO_2 storage, *Environ. Sci. Technol.* **37** (2003) 3476–3483.
4. S.M. Benson, R. Hepple, Prospects for early detection and options for remediation of leakage from geologic storage projects, 2005, Lessons Learned from Industrial and Natural Analogs for Health, Safety and Environmental Risk Assessment for Geologic Storage of Carbon Dioxide, Vol. 2. Geologic Storage of Carbon Dioxide with Monitoring and Verification, Elsevier Publishing.
5. J. Apps, The regulatory climate governing the disposal of liquid wastes in deep geologic formations: a paradigm for regulations for subsurface disposal of CO_2, 2005, Lessons Learned from Industrial and Natural Analogs for Health, Safety and Environmental Risk Assessment for Geologic Storage of Carbon Dioxide, Vol. 2. Geologic Storage of Carbon Dioxide with Monitoring and Verification, Elsevier Publishing.
6. C.-F. Tsang, S.M. Benson, B. Kobelski, R. Smith, Scientific considerations related to regulation development for CO_2 storage in brine formations, *Environ. Geol.* **42** (2–3) (2002) 275–281.
7. R. Hepple, S.M. Benson, Health, safety and environmental risk assessment for CO_2 exposure, 2005, Lessons Learned from Industrial and Natural Analogs for Health, Safety and Environmental Risk Assessment for Geologic Storage of Carbon Dioxide, Vol. 2. Geologic Storage of Carbon Dioxide with Monitoring and Verification, Elsevier Publishing.
8. R.P. Hepple, S.M. Benson, Implications of surface seepage on the effectiveness of geologic storage of carbon dioxide as a climate change mitigation strategy, *Environ. Geol.* (2004) in press.
9. E. Lindeberg, The quality of a CO_2 repository: what is the sufficient retention time of CO_2 stored underground, in: J. Gale, Y. Kaya (Eds.), *Proceedings of Greenhouse Gas Control Technologies Sixth International Conference (GHGT-6)*, vol. 1, Elsevier, Amsterdam, 2003, pp. 255–260.
10. S. Pacala, Global constraints on reservoir leakage, in: J. Gale, Y. Kaya (Eds.), *Proceedings of Greenhouse Gas Control Technologies Sixth International Conference (GHGT-6)*, vol. 1, Elsevier, Amsterdam, 2003, pp. 267–272.
11. W.D. Gunter, S. Bachu, S. Benson, The role of hydrogeological and geochemical trapping in sedimentary basins for secure geological storage of carbon dioxide, in: S.J. Baines, J. Gale, R.H. Worden (Eds.), *Geological Storage of Carbon Dioxide for Emissions Reduction: Technology*, 2003, Geologic Society Special Publication, Bath, U.K.

Carbon Dioxide Capture for Storage in Deep Geologic Formations, Volume 2
D.C. Thomas and S.M. Benson (Eds.)

Chapter 26

HUMAN HEALTH AND ECOLOGICAL EFFECTS OF CARBON DIOXIDE EXPOSURE

Robert P. Hepple

University of California, Davis, Davis, CA, USA

ABSTRACT

Understanding of human health and ecosystem impacts from exposure to elevated concentrations of CO_2 in air, soils and water is needed to assess the consequences of leakage from geologic storage projects. This chapter places CO_2 storage in the context of the global carbon cycle, reviews information on human health effects and ecosystem impacts from exposure to high concentrations of CO_2, and reviews industrial uses of CO_2 and describes the regulations put in place to protect workers and the public. This information provides the foundation for understanding and assessing risks of leakage from geological storage projects.

INTRODUCTION

To begin a risk assessment of geologic storage, we must first understand both the context for evaluating CO_2 exposures as well as the human health and environmental impacts of exposure to elevated concentrations of CO_2. Fortunately, there is a large amount of information to draw on in this regard. Carbon dioxide was one of the first gases identified, and it remains widely used in industry. Regulations are well developed for using CO_2 in occupational and industrial settings and for storing and transporting it. Moreover, the central role that CO_2 plays in living systems and ecosystem processes has motivated the development of an enormous knowledge base from which to begin this assessment.

We begin this chapter by placing CO_2 storage in the context of the global carbon cycle. We then summarize what is known about the basic physiology of CO_2 and how exposure to elevated concentrations leads to human and ecological risks. A review of industrial sources, uses, and accidents follows, and finally, we summarize current regulations and monitoring approaches for occupational and industrial exposures to CO_2.

CO_2 STORAGE IN THE CONTEXT OF THE GLOBAL CARBON CYCLE

Carbon dioxide is ubiquitous in the natural world. It undergoes an endless cycle of exchange among the atmosphere, living systems, soil, rocks, and water. Volcanic outgassing, the respiration of living things from humans to microbes, mineral weathering, and the combustion or decomposition of organic materials all release CO_2 into the atmosphere. Atmospheric CO_2 is then cycled back into plants, the oceans, and minerals through photosynthesis, dissolution, precipitation, and other chemical processes. Biotic and abiotic processes of the carbon cycle on land, in the atmosphere, and in the sea are connected through the atmospheric reservoir of CO_2.

Figure 1 illustrates the primary compartments of the global carbon cycle and the fluxes between them. The atmosphere contains approximately 755,000 Mt of carbon (Mt C), the terrestrial biosphere 1,960,000 Mt C and the oceans 38,100,000 Mt C. Carbon dioxide from the atmosphere is converted into biological matter by photosynthesis. The process, called primary production, converts approximately 60,000 Mt C (225,000 MMT CO_2) into biomass each year. This flux is balanced by a nearly equal flux of CO_2 back into the atmosphere, resulting from the respiration of living organisms and the decomposition of organic matter. Differences between these two competing fluxes determine whether the terrestrial biosphere is a net source or sink of CO_2. Carbon dioxide also dissolves in surface, ground, and ocean water, mostly as bicarbonate (HCO_3^-; $CO_2 + H_2O \rightarrow H_2CO_3 \rightarrow HCO_3^- + H^+$), and in shallow tropical waters, it precipitates out as carbonate rocks such as limestone ($CaCO_3$; $2HCO_3^- + Ca^{2+} \rightarrow CaCO_3 + H_2O + CO_2$). Annually, approximately 90,000 Mt C are exchanged between the ocean and the atmosphere. These quantities provide a context for evaluating the 6000 Mt C that is currently generated from fossil fuel combustion and the fraction of that amount that may be stored in geological formations in the future.

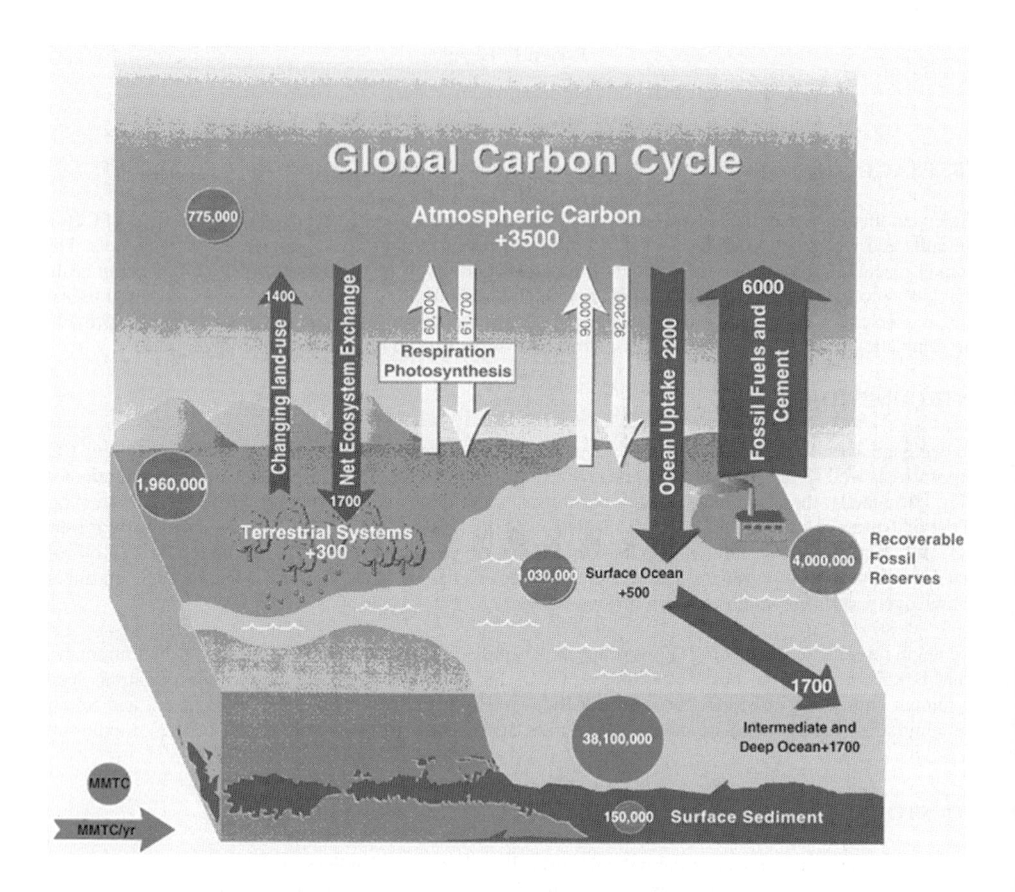

Figure 1: Global biogeochemical carbon cycle. Includes human influence from fossil fuel combustion and changing land-use patterns. Black arrows indicate net fluxes and white arrows indicate gross fluxes. Annual net additions are shown as + numbers, and pool sizes (circles) are shown in gray. All quantities are in million metric tonnes Carbon, MMTC, and all fluxes are in MMTC/yr (modified from US DOE, 1999).

The quantities of CO_2 that might be stored may also be put in the context of other known carbon reservoirs, as well as industrial and natural emissions. Figure 2a and b tabulate many of the known carbon reservoirs and fluxes, and compare them to the carbon storage goals identified by the US DOE. The current US DOE target for global annual storage capacity by 2025 is 1000 Mt C/yr [1]. As shown in Figure 2b, this is nearly equal to (for example) the US annual petroleum consumption or global annual natural gas consumption in 1998. The US DOE goal for global carbon storage capacity in 2050 is 4000 Mt C/yr, which is comparable to the total US natural gas reserves as assessed in 1998. While these comparisons point to the very large

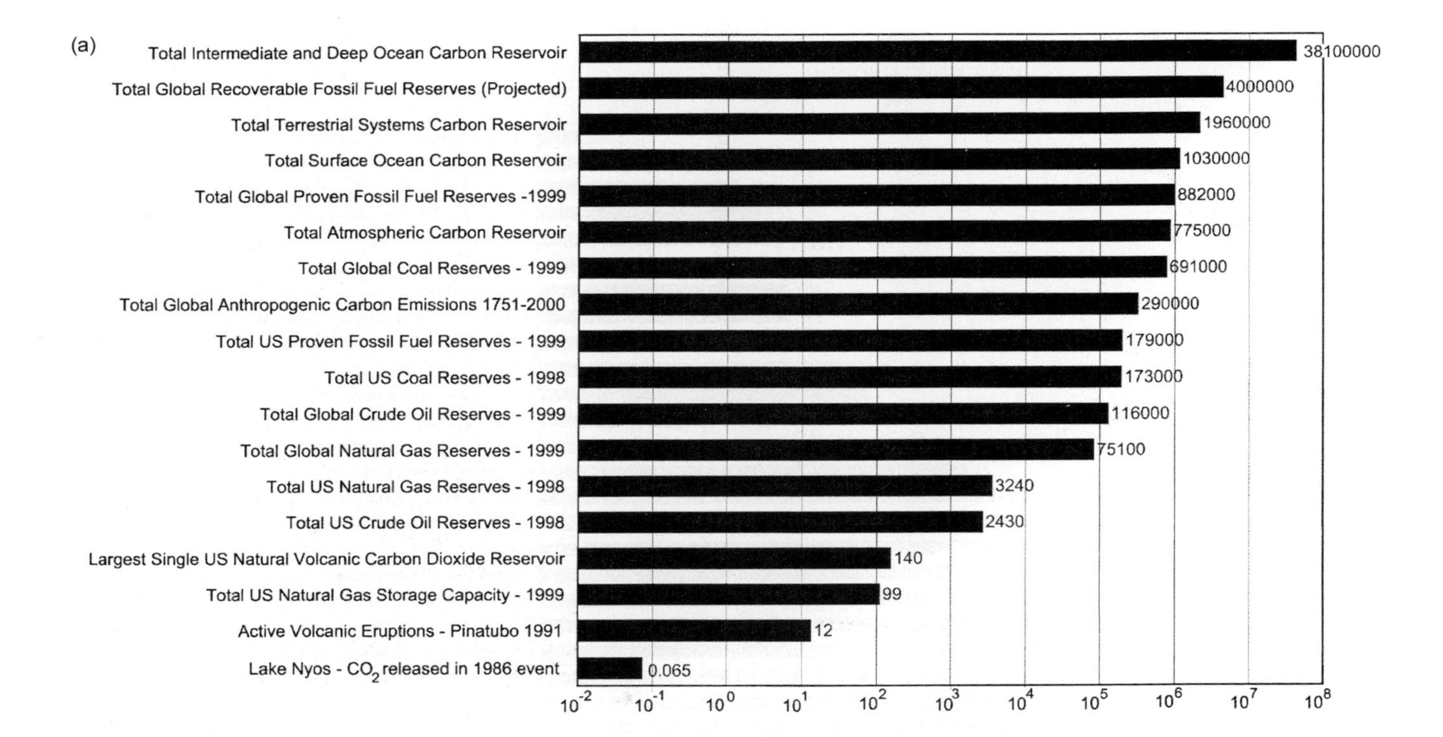

Figure 2: (a) Comparison of carbon reservoirs and one-time events. Data tables, references, and conversions can be found in Ref. [77]. Estimates are order of magnitude only and may include small conversion discrepancies due to independent rounding. (b) Comparison of carbon fluxes and target sequestration rates. Estimates are order of magnitude only and may include small conversion discrepancies due to independent rounding.

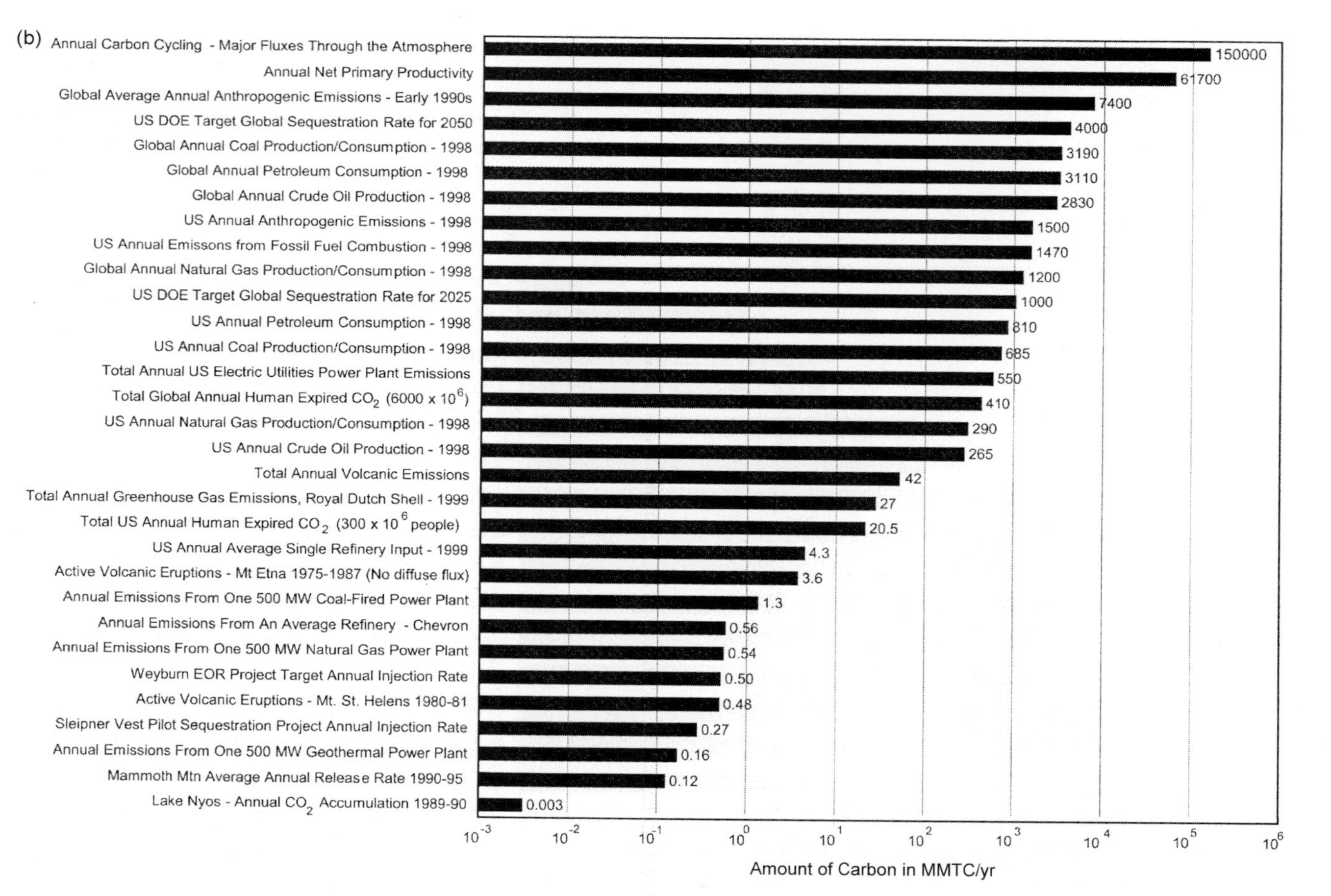

Figure 2: Continued.

quantities of CO_2 that may be stored, they are still small in comparison to the 90,000 MMTC/yr exchanged annually between the atmosphere and ocean and the 60,000 MMTC/yr exchanged between the atmosphere and the terrestrial biosphere. On the other hand, Figure 2b illustrates that the storage target is large compared to the global volcanic emissions of 42 Mt C/yr.

From a risk-assessment perspective, a more useful comparison may be the quantity of CO_2 associated with an individual storage project. Each facility is anticipated to store 0.25–10 Mt C/yr. For example, the Sleipner storage project in the North Sea currently injects 1 Mt CO_2 or 0.27 MMTC (1 tonne carbon = 3.667 tonnes CO_2) into the Utsira Formation beneath the sea floor. While this is small in comparison to the reservoirs and fluxes mentioned thus far, it is twice the annual release of CO_2 at Mammoth Mountain in California between 1990 and 1995, where over 100 acres of trees were killed by the natural release of magmatic CO_2. The 1986 Lake Nyos event in Cameroon released 0.24 MMT CO_2 (0.07 MMTC), approximately one quarter of the annual amount stored annual at Sleipner. This natural CO_2 release led to 1746 people and many animals being killed, up to 14 km away and 24 h after the initial event. However, a significantly larger release in 1991 from an eruption at Mt Pinatubo ejected 11.5 Mt C in one massive event, but the gas dispersed high in the atmosphere and did not pose a direct hazard.

In addition to providing a context for evaluating the magnitude of CO_2 that may be stored, these comparisons illustrate the important point that the risk associated with CO_2 storage depends much more on effective dispersion than the total quantity of CO_2 released. A small leak may pose significant risk to exposed humans, animals, or ecosystems if it becomes concentrated. Conversely, a very large release, even over a short period of time can have little effect if it is discharged high above the ground surface and dispersed by wind. It also points out the fact that large releases from a storage project could result in a significant hazard if confirmed to a small area.

GENERAL PHYSIOLOGY OF CARBON DIOXIDE

Understanding the general physiology of CO_2 provides a context for evaluating the environmental health impacts of CO_2 releases. Carbon dioxide is an important biological compound because it is the ultimate source of carbon for all life. Organic chemistry, the chemistry of biological compounds, is the study of carbon chemistry. Also, the biological cycling of carbon between photosynthesis and cellular respiration is a major portion of the global carbon cycle and is mediated through atmospheric CO_2 [2].

Primary producers, such as plants and photosynthetic microbes, use energy from sunlight, water, and CO_2 absorbed from the atmosphere to generate all of their organic constituents. The primary product of carbon fixation or photosynthesis is the carbohydrate glucose. A simple empirical formula for carbohydrates is CH_2O. Photosynthesis uses energy (E), CO_2, and water (H_2O) to make carbohydrates (CH_2O) and oxygen (O_2). In the evolution of the biosphere, this process generated virtually all of the oxygen in the atmosphere and remains central to the world around us.

$$\text{Photosynthesis} = CO_2 + H_2O + E \rightarrow CH_2O + O_2$$
$$\text{Respiration} = CH_2O + O_2 \rightarrow CO_2 + H_2O + E; \quad E \rightarrow \text{ATP}$$

Cellular respiration is the controlled reverse of photosynthesis, and the two together are integral to the flow of energy and carbon through the biosphere, as shown in Figure 3. Respiration, as depicted in Figure 4, is the combustion or oxidation of carbohydrates coupled to gas exchange and to reactions that produce ATP (adenosine tri-phosphate), the chemical energy currency of life. ATP is the primary form of energy used by most life for biosynthesis, metabolism, and movement. Some plants and microorganisms can produce every organic compound they need from glucose as a carbon skeleton starting material and energy source. The biosynthetic pathways retained in animals are relatively limited, so animals must consume organic material to obtain energy in the form of glucose and diverse raw starting materials. The processes of photosynthesis in primary producers and respiration, which is nearly universal among all forms of life, are of ancient origin and highly conserved through evolution. The few exceptions are chemoautotrophic bacteria. They survive on alternative abiotic energy sources and are typically thermophiles (heat-lovers), thiophiles (sulfur-lovers), or obligate anaerobes (oxygen-haters, e.g. methanogens) [3].

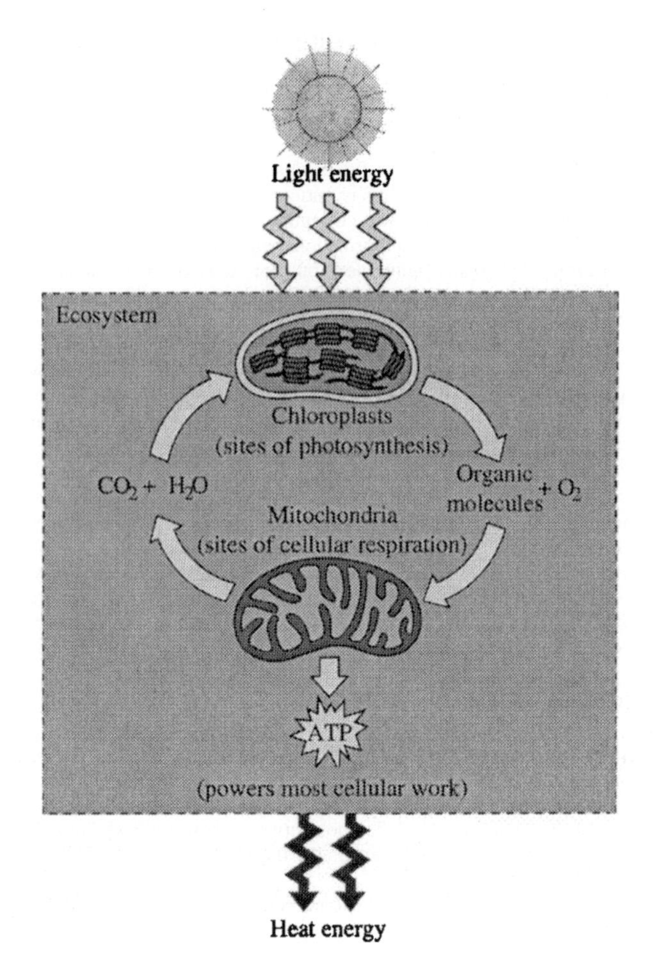

Figure 3: Fundamental biological carbon and energy cycles involving photosynthesis and respiration (Campbell et al. [4]).

HUMAN PHYSIOLOGY OF CARBON DIOXIDE: NORMAL AND HAZARDOUS EXPOSURE

Human Physiology of Carbon Dioxide

In humans, like the vast majority of organisms, cellular respiration consumes O_2 and generates CO_2. Breathing is the process by which we obtain oxygen from air and remove CO_2 from our bodies. Figure 5 illustrates how the coupling between the circulatory system and the respiratory systems transports O_2 to cells throughout our bodies and removes respired CO_2. Air breathed into the lungs contains 21% O_2 and 0.04% CO_2, and exhaled air is 16% O_2 and 3.5% CO_2 on average, though it can exceed 5% CO_2 during strenuous exercise.

Blood returns from the periphery through the right side of the heart to the lungs and contains 5% O_2 and 6% CO_2. Carbon dioxide diffuses out of the blood and into the lungs, and O_2 diffuses in the opposite direction, from the lungs to the blood. Blood leaving the lungs has 5% CO_2 and 14% O_2 and travels through the left

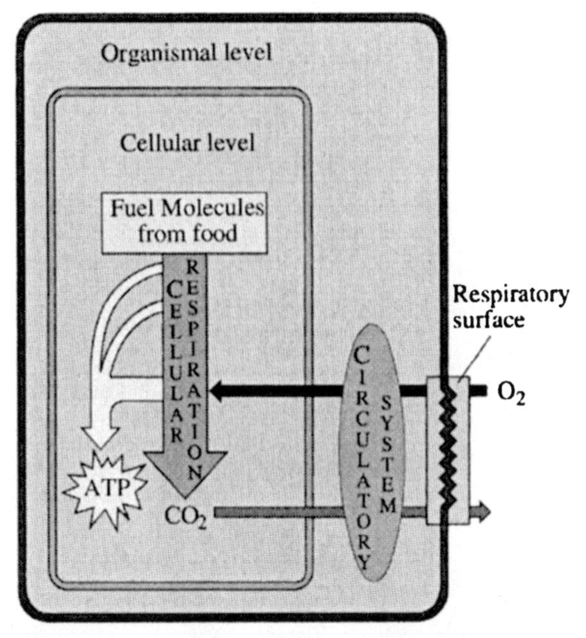

Figure 4: The role of gas exchange and respiration in bioenergetics. Illustration depicts the coupling of gas exchange and cellular respiration via the respiratory and circulatory systems (Campbell et al. [4]).

side of the heart, then on to the periphery. Oxygenated blood flows through capillaries surrounded by extracellular fluid. Oxygen is pulled out of the blood and into the cells because of its constant consumption by cellular respiration which maintains the low concentration within the cells, generally less than 5% O_2. The concentration of CO_2 in the blood, 5%, is lower than in the cells, where respiration produces CO_2, so CO_2 is absorbed into the blood and transported to the lungs [4].

Carbon dioxide is involved in several physiological functions aside from cellular respiration and bioenergetics. It is the primary regulator of breathing in coordination with two regions of the brain, the pons and the medulla. Most CO_2 is transported in red blood cells in its dissolved, hydrated form of bicarbonate (HCO_3^-). When CO_2 dissolves in the blood, it increases H^+ concentration or decreases pH, and humans are very sensitive to changes in pH. The concentrations of CO_2, electrolytes, and total weak acids determine blood pH [5], which is normally 7.4, and Van de Ven et al. [6] considered a pH drop of -0.04 to indicate acute metabolic acidosis. The relationship between CO_2 and pH is the most likely basis for CO_2 toxicity. The medulla monitors CO_2 levels in the blood by measuring subtle changes in pH, and lowered pH stimulates the need to breathe. Sensors in the aorta and carotid bodies detect blood oxygen, but oxygen levels only affect breathing when dangerously low, as at altitude. Via its role in acid–base and electrolyte balance, bicarbonate is involved in other processes including bone buffering and renal regulation [7–10].

Hyperventilation leads to hypocapnea or alkalosis, which is low blood CO_2 and high pH. Extreme stress and anxiety causes rapid breathing, which quickly lowers blood CO_2 levels and increases blood pH. The initial symptoms of hypoxia and feeling out-of-breath are indistinguishable from alkalosis. The breathing control center does not tell the body to breathe when blood pH is elevated, as long as there is sufficient oxygen. In a paradoxical, reinforcing reaction, one panics even more and continues taking rapid, deep breaths that keep

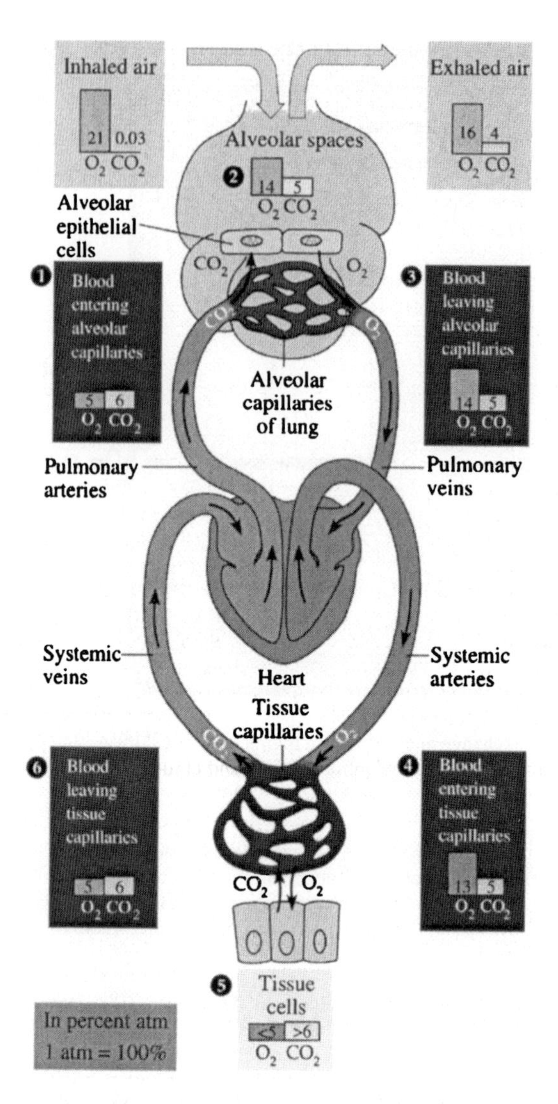

Figure 5: Gas exchange and variation in respiratory gas concentration along the coupled respiratory and circulatory systems (Campbell et al. [4]).

blood pH elevated. One feels dizzy and out-of-breath, which further exacerbates the feelings of anxiety and panic. Taking slow, normal breaths or rebreathing exhaled air from a paper bag allows blood CO_2 levels to increase and return the control of breathing to the normal mechanism. This happens in any case after the person loses consciousness and the body's autonomic systems take over. At high altitude, hyperventilation is more serious because low CO_2 reduces the drive to breathe, while reduced partial pressures of oxygen require more vigorous breathing. For pilots to avoid losing consciousness, supplied oxygen is necessary and often contains added CO_2 to augment the physiological drive to breathe [11,12].

Human exposure to elevated levels of CO_2 can be hazardous either by (1) reducing the oxygen content of the ambient air and causing hypoxia or (2) through direct carbon dioxide toxicity. For example, the National

Institute of Occupational Safety and Health (NIOSH) confined-space-hazard classification system defines CO_2 as a nontoxic, inert gas that displaces oxygen. In most cases of hazardous CO_2 exposure, it is presumed to act as a simple asphyxiant, even though extensive research indicates that exposure to elevated concentrations of CO_2 has significant effects before oxygen dilution could be physiologically significant. Typically, the ambient oxygen concentration is 21%, and the normal range is from 19.5 to 23.5%. Below 17% O_2, hypoxia leads to weakened night vision, increased breathing rate and volume per breath, and increased heart rate. Declining muscle coordination, rapid fatigue, and intermittent respiration are observed between 14 and 16% O_2, in addition to increased volume per breath and accelerated heart rate. Nausea, vomiting, and unconsciousness occur between 6 and 10%. Below 6%, loss of consciousness is rapid, and death takes place within minutes [13].

Effects of Low-level and Chronic Exposure to CO_2

At exposure to slightly elevated concentrations of CO_2, such as in rebreathing masks on airplanes at high altitude, the effects of elevated CO_2 can be beneficial, but that changes rapidly when concentrations exceed a few percent. In the year 2000, the average concentration of CO_2 in the atmosphere was 370 ppm. Studies show the threshold for perceiving stale air is 800 ppm. Carbon dioxide is used to assess adequate ventilation in buildings, and standards are set to ensure indoor odor control and comfort. Sick building syndrome (SBS) is a broad suite of health problems and complaints associated with inadequately ventilated buildings. Research shows that CO_2 is a good proxy for SBS and sufficient ventilation. Carbon dioxide builds up in enclosed spaces where occupants respire it, but no causal connection between SBS and CO_2 is known at this time. No physiological compensation or adverse health effects have been noted at or below 1% CO_2, though no controlled studies of exposure to such low levels have been done yet for longer than 6 weeks. Most studies involved healthy young male subjects, especially in controlled atmospheres such as submarines. Carbon dioxide tolerance in highly susceptible subgroups such as children, the elderly, or people with respiratory deficiency has not been studied—except for some work on chronic acidosis resulting from respiratory impairments and the observation of decreased ventilatory response to CO_2 in infants who were developmentally exposed to cocaine [8,14–20].

Carbon dioxide acts as a respiratory stimulant above 1%, and chronic exposure to 1.5–3% CO_2 results in physiological adaptation without adverse consequences. The only lingering effects are increased alveolar dead space (alveoli are the microscopic air sacs in the lungs where gas exchange takes place) and decreased sensitivity to increased concentrations of CO_2 as measured by respiratory stimulation. Exposure to 1.5–3.0% CO_2 leads to hypercapnea (elevated levels of blood CO_2). Because of the direct relationship between dissolved blood CO_2 and pH, hypercapnea is synonymous with decreased blood pH or acidosis. The immediate reaction is increased breathing depth and rate (respiratory compensation). In response to chronic acidosis, the body compensates by bicarbonate reabsorbtion in the blood and through bone buffering and renal regulation. Increased urine production aids in excreting excess hydrogen ions (H^+) and bicarbonate. Calcium deposition may increase transiently, but the body eventually attains homeostatic compensation as long as the chronic level of CO_2 exposure does not exceed 3%.

Elevated CO_2 levels in the air or blood limit the capacity for exercise and require increased respiration and long-term metabolic compensation. Below 3%, no adverse effects appear aside from the awareness of increased breathing rate and effort, mild headache, and sweating. No deleterious long-term consequences have been observed for chronic exposure to 3% CO_2 or less, and all symptoms of short-term exposure to such levels have proven to be short lived and reversible [8,21,22].

Acute Exposure to Elevated Concentrations of CO_2

The most striking effect of CO_2 levels over 3% is the exponential increase in minute volume, the average volume breathed during 1 min. Minute volume increases from 7 L/min at 0.03% CO_2 to 8 L/min at 1%, 9 L/min at 2%, 11 L/min at 3%, 26 L/min at 5%, and 77 L/min at 10.4%. Volume per breath increases from 440 to 2500 ml during exposure to 10.4% CO_2.

Hearing loss and visual disturbances occur above 3% CO_2. Carbon dioxide also acts as a local vasodilator and a potent cerebral vasodilator. This may explain many of the symptoms associated with CO_2 toxicity, including narcosis, headache, and dizziness. Healthy young adults exposed to more than 3% CO_2 during exercise experience adverse symptoms, including labored breathing, headache, impaired vision, and mental confusion.

Exposure to 4–5% CO_2 for a few minutes leads to headache, dizziness, increased blood pressure, and uncomfortable dyspnea (difficulty breathing). Suppressed shivering is observed at 7.5% CO_2 for 15 min in 5 °C. Seven to 10% CO_2 for several minutes to an hour results in headache, increased heart rate, shortness of breath, dizziness, sweating, rapid breathing, and near or full unconsciousness. (The lowest published lethal concentration for humans, reported in 1933, was 9% CO_2 for 5 min—Vermont SIRI 2001.) Eye flickering, psychomotor excitation, myoclonic twitching, headache, dizziness, dyspnea, sweating, restlessness, and "fullness in head" were observed at 10% carbon dioxide. Dizziness, drowsiness, severe muscle twitching, and unconsciousness occur after one to several minutes' exposure to 10–15% CO_2. Above 15%, loss of consciousness occurs in less than 1 min. Narcosis, respiratory arrest, convulsions, coma and death due to depression of the central nervous system can take place rapidly with continued exposure. Death occurs within minutes at 30% CO_2 [8,21,22]. An interesting aside to the discussion of hazardous CO_2 exposure is the routine use of a single breath of 20–35% CO_2 to diagnose and treat panic disorder [23–26].

Figure 6 summarizes information about natural occurrences of carbon dioxide compared with physiologically relevant concentrations and thresholds at which human health effects become noticeable or significant (see a discussion of regulatory limits for occupational CO_2 exposure later in this chapter). Clearly, CO_2 is not toxic at parts per million or even low percentage levels, but someone enveloped in a cloud of highly concentrated CO_2 is in imminent danger. The risk of exposure to dangerous levels whenever CO_2 is concentrated in large amounts or under pressure must be considered in the context of geologic CO_2 storage for both surface facilities and leakage from geological formations. Fortunately, industry has long experience with CO_2 and routinely controls this hazard in settings such as breweries, beverage carbonation facilities, and enhanced oil recovery (EOR) operations—through engineering and procedural controls and monitoring. Although individual susceptibility to CO_2 is variable, general guidelines are straightforward and useful, especially in light of the precautionary principle used in setting occupational exposure limits.

ECOLOGICAL AND ENVIRONMENTAL IMPACTS OF CO_2 RELEASES

The environmental impacts of CO_2 releases are not well understood despite numerous natural and man-made examples and extensive physiological research. Nevertheless, a summary of the existing literature that is qualitatively relevant to the potential risks of geologic CO_2 storage is helpful. Respiratory physiology and pH control are the primary physiological mechanisms controlling responses of different forms of life to hazardous CO_2 exposures. Information on the response of animals and vegetation to elevated CO_2 and low levels of O_2 can be found in diverse locations, including physiology, respiratory physiology, comparative physiology, plant physiology, botany, food preservation, and aerospace literature. Human responses are useful models for other mammals, and for all air breathers and large terrestrial animals, because of the universal nature of respiration. The death of animals and people in similar areas from the plume of natural CO_2 released from Lake Nyos, Cameroon in 1986 supports this observation. Plants usually have a higher tolerance for CO_2 than mammals, as evidenced by the lack of broad vegetation die-off at Lake Nyos. A standard amount used to preserve food from insects, microbes, and fungi is 40% CO_2; at this amount, insects are incapacitated or killed and microbes and fungi either die or experience severely retarded growth rates. Comparative physiology reveals that gas exchange mechanisms and organs, respiratory medium, and pH and osmotic homeostatic regulation vary among organisms and according to the ecological niche inhabited. These factors determine tolerance to elevated CO_2. The physiological basis of CO_2 tolerance and ecosystem response will be reviewed by looking at respiration and gas exchange in simple organisms, animals, and plants.

Simple Organisms: Cellular Respiration, Homeostasis, Diffusion, and Increasing Complexity

Cellular respiration, especially the preliminary step called glycolysis, which does not require oxygen, is almost universal among organisms from the most simple to the most complex. Photosynthetic organisms store solar energy as the chemical energy of organic compounds by converting water and CO_2 to simple carbohydrates and oxygen. This provides the organic material consumed by all heterotrophic organisms. Oxidative respiration is the reverse—the harnessing of chemical energy via the breakdown of carbohydrates back to CO_2 and water. As a result, CO_2 is the primary metabolic waste product of all oxygen-consuming

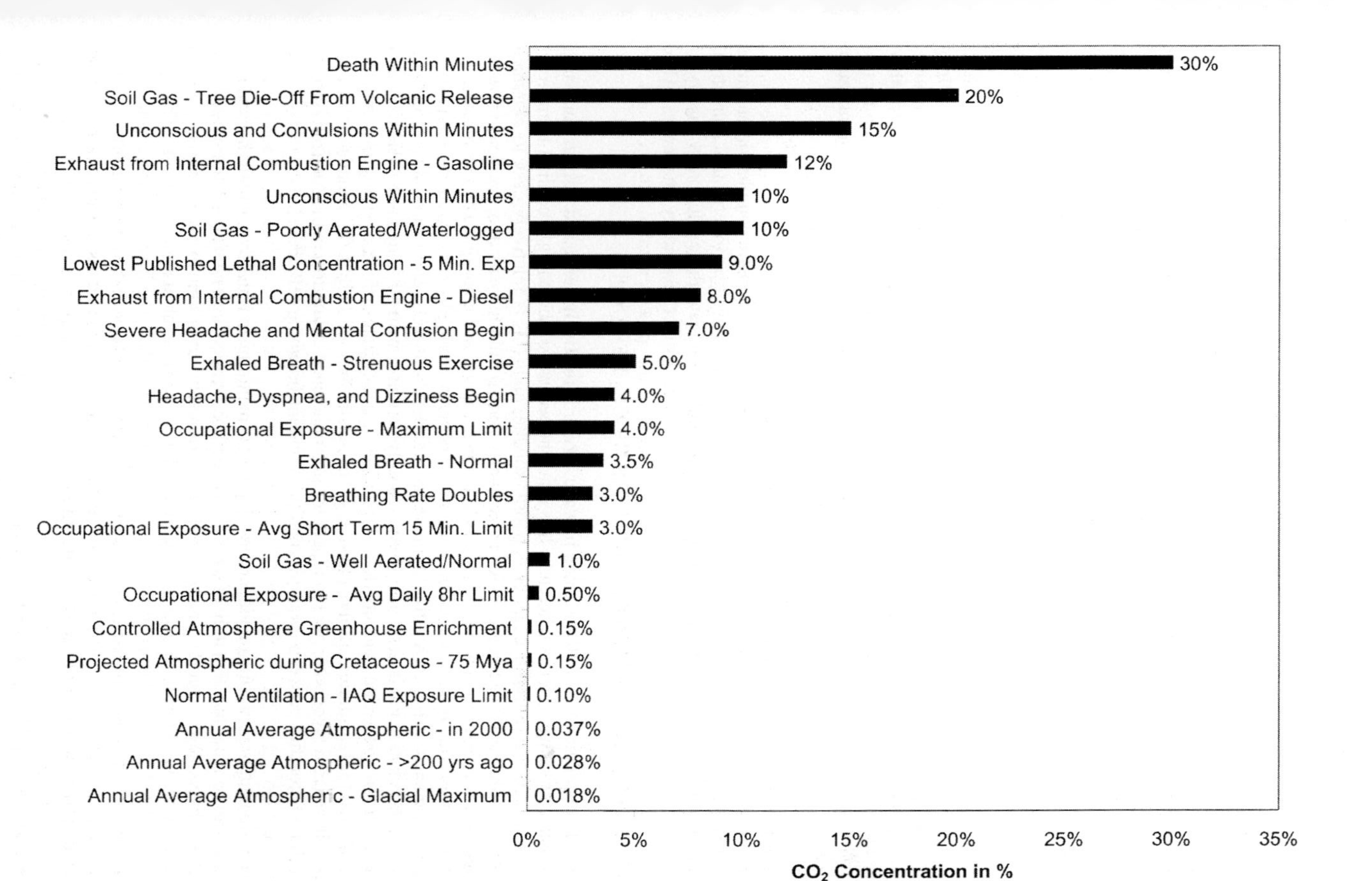

Figure 6: Comparison of ambient concentrations of CO_2 and risks of exposure. Data tables with references in Ref. [77].

organisms. Respiration requires regular gas exchange between an organism and its environment. Oxygen is required as the terminal electron acceptor in oxidative respiration and CO_2 must be eliminated. All single-celled organisms maintain some level of homeostasis, especially osmotic and pH, by controlling what passes through their cell membranes; but in the case of respiratory gases, diffusion is usually the operative, uncontrolled process. Microbes have adapted to virtually every environment that exists on Earth, and they use biochemical mechanisms to adjust to variable environmental conditions [29]. Some microbes can survive in virtually 100% CO_2 as long as trace amounts of O_2 are available. The suspension of metabolism under extreme desiccating conditions is an analogous survival mechanism. As a result of variation in environmental conditions and such adaptive capabilities, only a qualitative generalization is useful about the level of CO_2 that is toxic to microbes or bacteria. Although the range is broad and CO_2 concentrations below 10% kill some simple organisms, in general 50% CO_2 has a significant inhibitory if not lethal effect [27,28].

Fungi are not simple organisms from an evolutionary perspective, but their respiratory gas exchange is controlled by diffusion as opposed to more complex, specialized systems. As with insects and microbes, the majority of information on the tolerance of fungus to elevated CO_2 comes from food-preservation literature. This source biases our understanding toward the amount and duration of exposure at which virtually nothing survives as opposed to defining the minimum level at which the most sensitive are harmed. Temperature, relative humidity, oxygen concentration, and CO_2 concentration all have significant effects on the growth of fungi. Significant inhibition of growth and the germination of spores were observed at 15–25% CO_2 for two types of fungi. At 30% CO_2, no measurable growth was observed, and 50% CO_2 prevented the germination of spores [30,31].

Animals: Comparative Physiology and Active Bulk Gas Exchange in Water and Air

To summarize the effects of elevated CO_2 on animals, a brief review of respiration in complex organisms is required. The transition from water to air as the respiratory medium, the transition to terrestrial habitation, and increasing metabolic rates that could ultimately support flight dominate the evolutionary history of respiration. Aqueous, soil, terrestrial, and aerial environments offer distinct respiratory challenges. Also, respiration encompasses two interrelated yet distinct processes, the procurement of oxygen and the elimination of carbon dioxide.

The oldest form of respiration is simple diffusion in water. Water is relatively dense and viscous compared to air, so gas exchange using water is inherently less efficient because of the energy required to move it. Oxygen is more difficult to obtain because it is poorly soluble, and CO_2 is relatively easy to eliminate because it is highly soluble. Paleoclimatic evidence suggests that 70 million years ago, previous climates were typically much warmer, biological productivity was much higher, and consequently tropical swamps were widespread. These conditions are thought to be one of the main driving forces behind the evolution of air breathing because such waters have little O_2 and very high levels of CO_2. The end product of organic matter decay in such reducing environments is predominantly methane, with 60% methane, 30% CO_2, and 10% hydrogen, carbon monoxide and ethylene typical of carbohydrate-rich decomposition (refer to references in Maina [32]). Studies of the physiology of inhabitants of high CO_2 aquatic or marine environments such as Lake Nyos were not found, but the preponderance of lungfish in the Amazon Basin is suggestive of the relationship between swampy conditions and air breathing [10,32,33].

Terrestrial habitation requires two major adaptations. The first is to minimize water loss, and the second is to increase metabolism to meet the demands of terrestrial mobility. Supporting and moving the entire body weight without the buoyancy of water is energy intensive. Fortunately, air is a much more efficient medium for delivering oxygen, in contrast to aqueous breathing, but CO_2 disposal becomes difficult. The CO_2 carrying capacity of water is high, especially when bicarbonate is included. The limiting factor, then, is the exchange rate at the respiratory interface. Developing or co-opting the enzyme carbonic anhydrase solves this problem, because it catalyzes the bi-directional conversion of CO_2 to bicarbonate and back again, and accelerates the reaction rate in the range of a millionfold [4,32].

Ambient O_2, CO_2, and pH are determining factors for what types of organisms inhabit a given environment. A rapid or significant change in any of these conditions would cause biological stress, and the type of respiratory organ tells us much about an organism's normal environment and its ability to adapt or survive.

Aquatic and marine animals use skin diffusive respiration, the gill, the water lung, or the placenta. The transport of respiratory gases in the circulatory system, the diffusion of gases between the blood and the cells, and the placenta in utero are the connection of humans with water as a respiratory medium. Gills are considered either simple or complex, and they are involved in many different processes, including respiration, feeding, ammonia excretion, locomotion, and the regulation of osmotic pressure, acid–base balance, and some hormones. Because of the extensive buffering capacity of the oceans, pH varies little in marine environments, but the concentrations of O_2 and CO_2 can be dynamic. This is especially true in enclosed or stagnant bodies of water where mixing is not thorough. Lakes that are stably stratified by salinity or temperature contrasts are particularly susceptible to variations in respiratory gas concentrations. In contrast, freshwater does not have much buffering capacity, so CO_2 released into freshwater could change pH significantly. The dearth of macrofauna in or near geothermal efflux or soda springs suggests that CO_2 leakage may have significant localized impacts. The effects of such change would depend upon the natural variability of pH in that specific environment and an organism's physiological ability to adapt.

Typically, a change in pH of a few tenths would be a significant stressor, if not fatal. However, the impact of CO_2 released into a body of water depends upon the amount and rate of release, the water body's buffering capacity, and its mixing dynamics. Studies of the natural CO_2 release at Mammoth Mountain, California, indicated that large amounts of CO_2 were dispersed through the groundwater system and released quickly upon exposure to the atmosphere. In fact, no evidence of a high CO_2 flux remained in water even a few hundred meters downstream of the source (Kennedy, 2001, personal communication). Evidence from fish kills and swamps suggests that O_2 is the key respiratory gas among aquatic and marine organisms instead of CO_2, except to the extent that CO_2 could affect environmental pH [32,33].

The transition from water to air breathing and from aquatic to terrestrial habitat involved bimodal breathing, the combination of an air-breathing lung with remnant gills or skin diffusive respiration. Some current bimodal breathers are exclusively aquatic (e.g. lungfish), some are primarily terrestrial (e.g. land crabs), and some live in both worlds (e.g. amphibians). Most bimodal breathers are amphibians, gastropod molluscs, crustaceans, or lungfish. Such animals obtain most of their oxygen via their lungs and eliminate most CO_2 through the skin or gills, effectively separating these processes; but they often have multiple modes of breathing available, depending upon the respiratory medium and medium of immersion. No studies were found that specifically addressed the tolerance of bimodal breathers to elevated CO_2 concentrations.

Food preservation research has shown that insects have much higher tolerance to CO_2 than vertebrates. Mortality data for the rusty grain beetle compiled by Mann et al. [34] varies from 15% CO_2 for 42 days to 100% CO_2 for 2 days. Table 1 shows a subset of recommended CO_2 concentrations and exposure time. Even after hours to days of exposure to high CO_2, many insects can recover. Other insects' tolerances have been measured, and 35% CO_2 is the minimum concentration needed for effective control of all but a few unusually CO_2 tolerant species [35]. Another common trend is decreasing mortality with increasing CO_2 concentrations above 85% [36,37]. Zhou et al. [38,39] are investigating the precise mechanisms of elevated CO_2 effects on insects.

TABLE 1
CONCENTRATIONS OF CO_2 AND THE DURATION REQUIRED FOR EFFECTIVE CONTROL OF THE RUSTY GRAIN BEETLE (VIRTUALLY 100% KILLED)

CO_2 concentration (%)	**Exposure time (days)**
15+	42
40	8–13
60	3–4
80	3
100	2

Mann et al. [34].

Information on CO_2 tolerance was not compiled for each type of animal, but some further studies on burrowing animals and soil invertebrates were found.

Fossorial animals, more commonly called burrowers, live within soils, where environmental conditions are extremely variable. Even though soils are extremely heterogeneous, CO_2 in the soil atmosphere normally increases with depth along a diffusion gradient. Diffusion is the primary transport mechanism for O_2 into the soil air and for CO_2 out to the atmosphere, and diffusion limits respiration in the soil environment because respiration itself is a constant source of CO_2. In well-aerated soils, the CO_2 concentration can remain below 1% at 1 m depth; but in poorly aerated, waterlogged soils, CO_2 levels can exceed 10%. In fact, the major controls on O_2 and CO_2 levels in soils are the amount of respiration, the moisture level, and the specific soil chemistry [32,40–42].

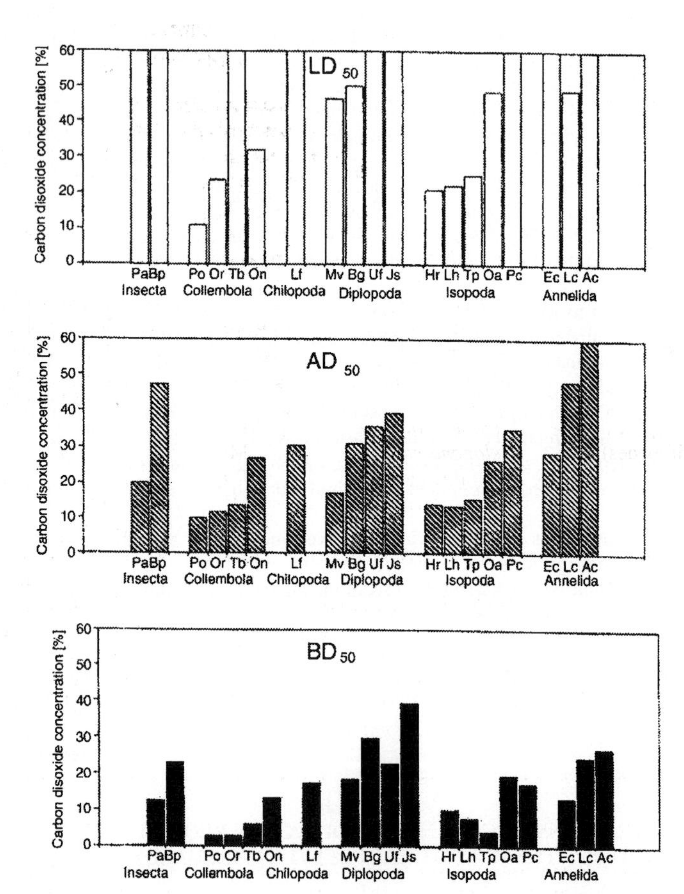

Figure 7: Response of 19 species of soil invertebrate to elevated levels of CO_2. The concentrations of CO_2 at which behavioral reactions (BD50), constant paralysis (AD50), or mortality (LD50) appeared in 50% of animals. Species abbreviations are shown in Table 2. LD50 was higher than 60% CO_2 for some species, but the range in this figure is limited to 60% (reproduced from Sustr and Simek [43]).

The response of soil invertebrates to CO_2 shows inter- and intra-species variation and depends upon their ecomorphological niche. The results of the study by Sustr and Simek [43] are shown in Figure 7. Behavioral changes in half of the observed individuals for a given species occurred between 2 and 39% CO_2 in the ambient soil air, with the majority of species affected by 20%. Paralysis in half of observed individuals was apparent between 10 and 59% CO_2. Half of the species were paralyzed by 30%, and all but one were paralyzed by 50%. Carbon dioxide levels from 11 to 50% were lethal for half of the species investigated. The effects of CO_2 depend upon temperature, humidity, and oxygen concentration, but (according to previous research) the dilution of oxygen in soil environments at high CO_2 levels is not significant. Breathing mechanism had an obvious effect, with gills and skin-diffusive respiration being more sensitive to CO_2 levels than tracheal and pseudotracheal respiratory mechanisms or skin breathing earthworms with closed circulatory systems. Biochemical mechanisms also appeared significant based upon the range in sensitivity of springtails (Collembola), a group using skin diffusive respiration [43] (Table 2).

TABLE 2
SPECIES ABBREVIATIONS AND DESCRIPTIONS MODIFIED FROM SUSTR AND SIMEK [43]

Species		**Abbreviation**	**Notes on age, habitat, and ecomorphological type**
Insecta (Insects)	*Pyrrhocoris apterus*	Pa	Adults, soil surface
	Bibio pomonae	Bp	Larvae, litter
Collembola (Springtails)	*Pogonognatellus flavescens*		Epigeic (surface) species
	Orchesella flavescens	Or	Atmobiotic species
	Tetrodontophora bielanensis	Tb	Hemiedaphic species (part-time soil dweller)
	Onychiurus cf. ambulans	On	Euedaphic species (soil dweller)
Chilopoda (Centipedes)	*Lithobius forficatus*	Lf	Litter
Diplopoda (Millipedes)	*Melogona voigti*	Mv	Litter
	Blaniulus guttulatus	Bg	Litter
	Unciger foetidus	Uf	Litter
	Julus scandinavius	Js	Litter
Isopoda (Pill bugs)	*Hyloniscus riparius*	Hr	Litter, hygrophilous (water/moisture loving)
	Trichoniscus pussillus	Tp	Litter, hygrophilous
	Oniscus asellus	Oa	Litter, hygrophilous
	Porcellium collicolla	Pc	Litter
Enchytraeidae (Potworms)	*Enchytraeus crypticus*	Ec	Litter
Lumbricidae (Earthworms)	*Lumbricus castaneus*	Lc	Epigeic species
	Aporrectodea caliginosa	Ac	Endogeic (subsurface) species

Another group of insects analogous to soil dwellers are beetles and fly larvae that are specially adapted to living in dung pats. Microbial activity in fresh pats is substantial, so the dung air may have O_2 concentrations below 1%, CO_2 concentrations from 20 to 30%, and methane concentrations from 30 to 50%. The dung insects generally could adapt to 20% CO_2, and some larvae remained visibly unaffected up to 43% CO_2. Yet tolerance varies greatly, and some surface-dwelling insects are paralyzed by 8% CO_2 [41].

Burrowers have the highest CO_2 tolerance among vertebrates because soil air often contains high levels of CO_2. Most burrowers inhabit open tunnels and spend only part of their time underground. Such animals include gophers, many rodents, and some birds. The CO_2 content of their respective burrows have been measured as high as 4, 2, and 9%. Concentrations of CO_2 as high as 13.5% have been found in the dens of hibernating mammals (see references in Maina [32]).

Among the major classes of terrestrial vertebrates—reptilia, mammalia, and aves—the lung takes on several forms. Reptilian lungs are morphologically the most diverse, but they are also the least efficient. Aerobic capacity in reptiles is a fraction of that in mammals. In contrast, anaerobic capacity and tolerance for hypoxia is greater. As a result, reptiles are probably more tolerant to elevated CO_2 than mammals. Turtles are specially adapted to hypoxia via depressed metabolism. They can withstand complete anoxia for days or even months and a decrease in brain pH to 6.4. The diaphragm affects the complete functional separation of the thoracic and abdominal cavities in mammals and is a unique characteristic. The homogeneity of form and function of the mammalian lung is another striking feature: all are tidally (rhythmically) ventilated, dead-end sacs. The lungs of bats are proportionally much larger in order to sustain flight and increased aerobic capacity, yet bats are aerobically inefficient relative to birds. The lung airsac system of birds is closest to the multicameral reptilian lung. There is remarkable morphological and functional homogeneity among bird lungs, as with mammals, but the lung airsac is a highly efficient gas exchanger. The airsac changes volume by only 1–2% per breath, but it allows a constant unidirectional flow of air through the lungs. Along with several other structural and functional characteristics (including countercurrent exchange), the avian lung is unquestionably the most efficient vertebrate gas exchange system known. Birds can sustain increases in aerobic capacity by a factor of 20–30, while elite human athletes can manage similar increases for a few minutes at most. Birds are uniquely tolerant to low-pressure oxygen deficiency and low CO_2 from flying at altitude and sustaining high-energy output. Specific references to the tolerance of birds to elevated CO_2 were not found.

Effects of Elevated CO_2 Concentrations on Plants

At slightly enriched levels (500–800 ppm) over atmospheric background (370 ppm), carbon dioxide usually stimulates growth in plants, depending on the mechanism of introducing CO_2 into the photosynthetic or Calvin cycle—C3, C4, or CAM (crassulacean acid metabolism). The majority of plants are C3, like trees, and first make a three-carbon acid when fixing CO_2. As part of adapting to arid conditions, the need to minimize water loss during hot, dry days, and perhaps low levels of CO_2, C4 and CAM plants such as grasses and succulents first make four-carbon acids. C4 plants separate CO_2 uptake and fixation spatially by segregating the processes in different cell types, and CAM plants separate uptake and fixation temporally by absorbing CO_2 at night and fixing it during the day [44]. The experimental increase of ambient CO_2, called free air CO_2 enrichment (FACE), initially causes proportional increased growth in C3 plants, followed by a tapering down to slightly elevated growth rates above unenriched levels. It also increases water-use efficiency and changes carbon allocation among tissue types. The growth rate of C4 and CAM plants is not limited by CO_2 availability as it is for C3 plants. As a result, the response of C4 and CAM plants is usually more complex and of lesser magnitude, so no simple generalization can be made about the effects of FACE [45]. Enhanced growth of plants in controlled-atmosphere greenhouses with enriched CO_2, optimally between 1000 and 2000 ppm, is the result of elevated CO_2 in conjunction with elevated temperature, plentiful water, and intensive fertilization [46]. Individual plants adapt easily to small changes in ambient CO_2, perhaps changing the allocation of biomass among roots, stems, and leaves, but over decades to centuries, plant species composition may change at the ecosystem level, generally in favor of C3 plants, with a consequent change in ecosystem composition and type. Research is underway to investigate the response of all aspects of ecology and the environment to elevated CO_2, projected temperature increases, and alterations to the hydrologic cycle (see articles in Refs. [45,47–52]).

The range and effects of high levels of CO_2 on plants, between FACE and lethal levels, are not clearly delineated. The precise mechanisms of tree kill in events like the outgassing from Mammoth Mountain, California, are poorly understood. The most likely cause is suppression of root-zone respiration via hypoxia, hypercapnia, or acidification of the soil environment. Long-term exposure over weeks or months to 20% or more CO_2 in soil gas led to dead zones where no macroscopic flora survived. The distribution of effects relative to CO_2 concentration suggests that 20–30% is a critical threshold for plants and ecosystems in general. Although some plants will die quickly from severe hypoxia, the lack of vegetation killed by the natural release of CO_2 at Lake Nyos indicates that plants generally have a much higher tolerance than animals to extremely high, short-lived exposures.

Concluding Remarks Regarding Ecosystem Exposure to Elevated Levels of CO_2

In future, a more thorough review of comparative, plant, and ecosystem physiology relating to hypoxia, hypercapnea, pH tolerance, and biochemical mechanisms of homeostasis may prove fruitful. In fact, modeling of ecosystem response to various scenarios of CO_2 release will require rigorous, quantitatively defined thresholds or probability distributions correlating CO_2 concentrations with specific impacts. Human tolerance provides a convenient rule of thumb for environmental CO_2 exposure limits, based on this review of physiology.

With regard to geologic storage, subsurface CO_2 storage and leakage may lead to the dissolution of minerals, the mobilization of metals in the aqueous phase, and the potential concentration of organic compounds in supercritical CO_2 due to its solvent properties [53–55]. The rates, likelihood, and potential significance of these processes (if any) are not well known. The risks associated with these types of processes are not addressed here.

Carbon dioxide outgassing near geothermal vents, fumaroles, and soda springs may provide an opportunity for research on the environmental effects of a range of concentrations and duration of exposure. The scarcity of macrofauna in such high CO_2 environments may be indicative of the physiological stress caused by elevated CO_2. However, the presence of other gases such as H_2S in some cases may make it difficult to draw quantitative conclusions about the effects of elevated CO_2 concentrations from such an evaluation.

More importantly, while there has been a great deal of research about the ecological effects of slightly elevated concentration of CO_2 and on the high concentrations that are known to create a lethal response, there is little research about short or long-term exposure to intermediate concentrations of CO_2. (For example, what would be the ecological consequence of prolonged exposure to 3–5% CO_2?) Soil gas and biological surveys near natural CO_2 surface releases and above underground reservoirs would be useful to address issues such as these.

INDUSTRIAL SOURCES AND USES OF CO_2

The risk-assessment process for CO_2 can also be informed by a review of industrial sources and uses, safety issues and procedures, and accidents. Carbon dioxide was one of the first chemicals identified, and it has diverse uses. Table 3 is a summary of US emissions and sinks of CO_2 in 1998. Fossil fuel combustion generated 1468.2 million metric tonnes of carbon equivalents (Mt CE). All other industrial processes utilized and ultimately emitted 18.4 Mt CE. In addition to being a by-product of fossil fuel combustion, CO_2 is a by-product of pH control/acid neutralization, cement manufacture, and the chemical production of lime, ammonia, ethyl alcohol, hydrogen, ethylene oxide, and synthetic natural gas.

Yet CO_2 is also a commodity with diverse applications. The main sources of CO_2 for industrial use are natural reservoirs, the by-product of chemical manufacture, and separation from crude oil or natural gas. Manufacturing carbonates, urea, and methanol use CO_2 as a reactant. Carbon dioxide is used to carbonate beverages, and when produced by yeast, it is the leavening agent in baking and the by-product of fermentation in the manufacture of alcoholic beverages. We use it to preserve food and extinguish fires. Under regulations for the humane slaughter of livestock, we anesthetize and kill animals with it. The oil industry, in a process called EOR, pumps CO_2 into hydrocarbon reservoirs to aid in the secondary and tertiary recovery of oil and gas. In EOR, carbon dioxide can form an immiscible mixture with the reservoir oil, thus making it easier to extract it from the reservoir. When it dissolves in the crude oil, it causes a decrease in the fluid viscosity and density. This drive to increase volume maintains reservoir pressure and increases the proportion of original-oil-in-place recovered. Enriched levels of CO_2 in greenhouses enhance the growth of plants, and dry ice and liquid CO_2 are used as refrigerants. Carbon dioxide is also used as a pressurizing agent and a supercritical solvent. A list of applications provided by Airgas is listed in Table 4 [8,56–59].

For 1999, an industry research group called Freedonia reported the shipment of 1.81 Mt C of liquid carbon dioxide (1 Mt C = 3.67 million metric tonnes of CO_2) through the merchant market and 1.87 Mt C (6.86 Mt CO_2) total production, including on-site captive consumption. Another Freedonia CO_2 industry study from 1991 estimated that 20% of CO_2 sold on the merchant market came from natural reservoirs and 80% from captured emissions. The EPA's emissions estimate of 0.4 MMTC comes from assigning 80% of the 1.87 Mt C to emissions accounted for elsewhere and only the 20% derived from natural reservoirs uniquely to carbon dioxide consumption. Total EOR demand in 1999 according to Freedonia was 1.79 Mt C

TABLE 3
U.S. SOURCES OF CO_2 IN 1998

Source	Amount in 1998
Fossil fuel combustion	1468.2
Industrial processes	18.4
Cement manufacture	10.7
Lime manufacture	3.7
Limestone and dolomite use	2.4
Soda ash manufacture and consumption	1.2
Carbon dioxide consumption[a]	0.4
Iron and steel production[b]	23.9
Ammonia manufacture[b]	6.3
Ferroalloy production[b]	0.5
Aluminum production[b]	1.6
Natural gas flaring	3.9
Waste combustion	3.5
Land use change and forestry (sink)	(210.8)
International bunker fuels	31.3
Total emissions	1494.0
Net emissions (sources and sinks)	1283.2

Emissions and sinks in MMTC (US EPA [59]).
[a] Includes food processing, chemical production, carbonating beverages, and EOR. Primary sources include natural reservoirs, chemical manufacture, and separation from crude oil and natural gas.
[b] Emissions from these processes are primarily due to energy consumption and are included in the total for fossil fuel combustion.

TABLE 4
COMMON INDUSTRIAL USES OF CO_2

Beverage carbonation	The characteristic tingle and fizz of carbonated beverages results from the interaction of CO_2 and H_2O molecules in beverage solutions
Fire protection	CO_2 is widely used in fire extinguishers for both hand held and fixed systems. It is also used in "blanketing" to displace oxygen to prevent combustion. A major advantage of CO_2 is its cleanliness
Enhanced recovery of petroleum products	CO_2 is used in various processes of oil and natural gas well stimulation to enhance productivity
Molded product deflashing	Molded products, especially rubber compounds, often have undesirable flashings where mold sections were joined. CO_2 is used to cool and embrittle the flashings in preparation of mechanical removal, saving the high cost of hand trimming and buffing
pH control of waste water	One of the critical aspects of effluent disposal is its degree of alkalinity. CO_2 is one of the safest, cleanest and most economical means of reducing the pH of waste water
Foam expansion	The use of CO_2 as an expanding agent in polyurethane foams eliminates the use of volatile organic compounds and chlorofluorocarbons to provide a safe, low cost alternative to these harmful chemicals

(*continued*)

TABLE 4
CONTINUED

Shielded arc welding	CO_2 vapor is used to displace oxygen at the point of contact in arc welding. Speed, efficiency, quality and cost factors have stimulated wide use of this application in the Welding Industry
Low temperature grinding	CO_2 is added to heat sensitive materials in grinding operations for heat removal to prevent product softening or melting
Aerosol propellant	CO_2 is a cost effective alternative pressure medium in many non-water based aerosol products, eliminating the use of hazardous solvents and chlorofluorocarbons
Recarbonation of potable water	As a result of typical municipal potable water softening operations, the pH level of the water is raised which results in a chemically unstable water condition. The application of CO_2 (recarbonation) establishes a chemical balance and minimizes mineral deposits in the water distribution system
Purging and inerting	Fuel tanks, pipelines and other containers with explosive or combustible vapors must be purged prior to some types of maintenance and/or change in usage. CO_2 is an effective method for purging vessels of many unwanted vapors
Foundry core hardening	As an alternative to the conventional process of baking foundry cores, CO_2 is used in conjunction with a treated (silica) sand to form high quality cores resulting in time and energy savings
Chemical reactant	CO_2 is used in the production of various carbonate compounds, in controlling pH and in many other processes involving chemical reactions
IQf freezing	Cooling and freezing operations are integrated into high-speed production lines with CO_2 tunnel and spiral freezers. Advantages include reduction of cold storage space, bacteria retardation, greater refrigeration efficiency, enhanced product quality and more efficient space utilization
Shrink fitting	Machined metal products such as bushings, collars and seats which require a "tight fit" can be easily assembled by cooling with dry ice
Refrigeration in mixing and blending	CO_2 injection reduces heat buildup induced by blade friction in mixing and blending of meat products and firms it in preparation of the forming process. Semi-automatic operation reduces manpower and minimizes space requirements
Low-temperature testing	CO_2 is used as a refrigerant for testing products by simulating ambient temperatures down to -109.8F. CO_2 is easily stored, readily available and can be piped for automatic operation
Pest control in stored grain	Fumigating coffee, tea, tobacco and grains with CO_2 has been successful in controlling insects in storage. CO_2 provides a safe, clean alternative to environmentally hazardous fumigants
Greenhouse atmosphere enrichment	CO_2 is an essential raw material used by green plants in photosynthesis. Increasing the amount of CO_2 available to plants can greatly increase plant growth and yields
In-transit refrigeration of processed foods	Perishable processed foods can be refrigerated with CO_2 during processing, enabling direct loading onto trucks and sustained safe temperatures during mechanical refrigeration temperature pull-down. Valuable freezer space is conserved and risk of spoilage is minimized. CO_2 can also be used in limited applications as the sole refrigerant
In-flight food refrigeration	CO_2 in its solid form (dry ice) is used to refrigerate In-flight Modules by the Airline Industry. Alternative methods are not as reliable or cost effective
Non-destructive cleaning	CO_2 cleaning utilizes dry ice to remove contaminants from most surfaces, greatly reducing waste products and the need for chemical solvents, sand, water and other media
Modified atmosphere packaging	Packaging perishable food products with CO_2 greatly extends the product shelf life by limiting the growth of aerobic microorganisms. Other benefits include reduced development of rancidity and odors, and better color retention

(6.56 Mt CO_2), of which 5.3% or 0.09 Mt C (0.35 Mt CO_2) was supplied by the merchant market. The remaining 1.70 Mt C (6.21 Mt CO_2) was supplied via pipeline from natural reservoirs, separated from crude oil or natural gas, or recycled in existing EOR projects. Combining Freedonia's estimates for liquid CO_2 sold on the merchant market with the EOR estimates yields a total of 3.56 MMTC (13.1 MMT CO_2) utilized in 1999. The summary report on industrial gases for 1999 from the Department of Commerce (DOC) estimates total production of 3.25 MMTC (11.9 MMT CO_2). The difference of 0.31 MMTC (1.2 MMT CO_2) between the numbers from Freedonia and the DOC can be ascribed to uncertainty in the estimates of CO_2 separated from crude oil and natural gas or recycled within existing EOR projects. The proportion of CO_2 that comes from natural reservoirs versus recycled or separated from crude oil and natural gas has not been evaluated [59,60–62].

Industry experience with CO_2 also provides insights into the safety concerns of any geologic storage project. Refineries process large quantities of hydrocarbons, on average 4.3 Mt CE per refinery in the US during 1999. According to the Texas Natural Resources Conservation Commission, one large refinery in Harris County, TX, vented 5.4 Mt CO_2 in 1998 [63], yet that CO_2 posed no immediate human health or environmental hazard because it dispersed from high smokestacks. Chevron Research and Technology Corporation's Health, Environment and Safety Group estimates that an average refinery emits 5600 tonnes CO_2/day or 2 Mt CO_2/yr (Chevron, 2001, personal communication). Shell, one of the largest international energy companies, estimated their total annual global emissions in 1999 to be 90 Mt of CO_2 alone or 27 Mt CE including other greenhouse gases [64]. CO_2 pipelines are mostly associated with EOR, and their accident record is available through the Office of Pipeline Safety (OPS) in the Department of Transportation (DOT). Eight accidents are on record for CO_2 pipelines from 1968 to 2000: three in 1994, one in 1995, three in 1996, and one in 1997. There were no injuries and no fatalities. A failed weld caused one and corrosion caused another. Three were failures of control or relief equipment, and two more resulted from other failed components. Outside force caused the other one [65].

Catastrophic pipeline failures are considered unlikely and the environmental consequences of a massive CO_2 pipeline rupture are expected to be minimal because of engineering controls. One attempt to model the impacts of rupture was reported by Kruse and Tekiela [66]. Typically, the main procedural controls are maintenance routines and visual inspections via plane, truck, or walking the line. If odorants and colorants are used, they render small leaks easier to detect. Vegetation that has been killed or that is visibly under stress is used to locate leaks in natural gas pipelines, especially where they are underground. In addition to manufacturing standards for the pipeline materials, automatic pressure control valves are placed regularly along the length of pipelines in the case of catastrophic ruptures. The safety control valves shut down the flow of gas if the pressure in the pipeline exits a preset range. The pressure drop of a large leak or rupture would trip the shut-off valves, so only the gas between two safety control valves could vent to the atmosphere. The spacing of such control devices is set according to regulations and safety considerations depending upon proximity to human residences.

The EPA published a review of the risks of CO_2 as a fire suppressant. Carbon dioxide is used in 20% of fire protection applications, and it is common in large industrial systems. The EPA report included a summary of 51 incidents that occurred between 1975 and 2000. These involved a total of 72 deaths and 145 injuries. In a characteristic incident at Idaho National Engineering and Environmental Laboratory, an accidental release during the maintenance of an electrical system resulted in one fatality and 12 injuries. The sudden discharge filled an enclosed space with 2.5 tonnes of CO_2 and created an atmosphere with approximately 50% CO_2 and 10.5% O_2.

Engineering controls and procedures set by the DOT, National Fire Protection Association (NFPA), American National Standards Institute (ANSI), Compressed Gas Association (CGA), American Society of Mechanical Engineers (ASME), and Occupational Safety and Health Administration (OSHA) are used to ensure the safety of refineries, pipelines, fire suppression systems, and any transportation of CO_2 whether it is pressurized, cryogenic, solid, liquid, or gas [67–69].

The cumulative experience of industry suggests two conclusions. First, carbon dioxide is a familiar and integral part of our everyday lives that is generally regarded as safe; and second, concentrated CO_2 in confined spaces poses a significant but well-known hazard that falls within standard industry practice, engineering controls, and safety procedures. The environmental consequences of CO_2 separation facilities and pipelines are the same as such facilities used for other purposes. Geologic carbon storage does not pose any new or uncertain hazards in its surface facilities.

CARBON DIOXIDE REGULATIONS

Regulations for CO_2 have been promulgated by a number of organizations for a variety of purposes. These guidelines roughly reflect our collective experience with, understanding of, and attitude toward CO_2 and provide further context for risk assessment. We discuss the regulations briefly and summarize them in Tables 5 and 6.

Occupational Health Standards for Carbon Dioxide

The OSHA in the Department of Labor (DOL) sets the most directly relevant regulations regarding CO_2. The NIOSH in the Center for Disease Control and Prevention (CDC) is part of the Public Health Service (PHS) in the Department of Health and Human Services (DHHS) and recommends exposure limits. These agencies regulate CO_2 as an occupational air contaminant. The general permissible exposure limit (PEL) set by OSHA is a time-weighted average (TWA) of 5000 parts per million by volume (ppm) (0.5%) for an 8-h work day and a 40-h workweek. The NIOSH recommended exposure limit (REL) is a 10-h/day and 40-h/week TWA of 5000 ppm, a 15-min TWA short-term exposure limit (STEL) of 30,000 ppm (3%), and 40,000 ppm (4%) as the level immediately dangerous to life and health (IDLH). All IDLH atmospheres require the use of respiratory protection equipment. The *Occupational Health Guideline for Carbon Dioxide*, published jointly by OSHA and NIOSH, is included as Appendix 7, and the NIOSH *Pocket Guide to Chemical Hazards* entry for CO_2 is attached as Appendix 8 [8,21,69–71].

The American Conference of Governmental Industrial Hygienists (ACGIH) is the source of OSHA standards for construction and recommends a 5000 ppm TWA threshold limit value (TLV) and a 30,000 ppm TWA-STEL. ACGIH and NIOSH criteria documents are the core sources of occupational exposure limits through their own research and references to primary literature. The limit for CO_2 in surface and underground metal and nonmetal mines set by the Mine Safety and Health Administration (MSHA) in DOL is referenced to the ACGIH as well [8,22,69].

The DOT regulates carbon dioxide through the Federal Aviation Administration (FAA) as an air contaminant and as a surrogate for adequate ventilation in cabin air (5000 ppm CO_2) [14,69].

Ventilation and Indoor Air Quality

The FAA uses the criteria for occupational exposure to CO_2 to set its ventilation rates, but that is unusual. General building ventilation and indoor air-quality requirements are set by American National Standards Institute (ANSI)/ASHRAE (American Society of Heating, Refrigerating and Air-Conditioning Engineers) Standard 62-1999 at 700 ppm above ambient outdoor CO_2 levels, or about 1000 ppm CO_2 for HVAC (Heating, Ventilating and Air-Conditioning) industrial ventilation systems. This standard was set using comfort and odor control criteria. HVAC systems often monitor CO_2 concentration as a general proxy for indoor air quality because it is the primary contaminant produced by occupants. Ventilation rates that keep CO_2 levels below 1000 ppm are proven to reduce SBS, complaints such as irritated eyes, nose, and throat; headache, coughing, nausea, and dizziness [15,18–20].

Confined Space Hazard and Fire Suppressant

NIOSH is the single best source of information on confined space hazards, and OSHA is the regulatory body with oversight responsibility. OSHA establishes labeling, warning, and training requirements for confined space hazards like CO_2. In occupational settings such as silos, manure pits, breweries, and ship holds, CO_2 is recognized as a serious inert gas danger that creates oxygen-deficient atmospheres. Other OSHA regulations control the use of CO_2 as a fire suppressant and require a discharge alarm, time to exit before discharge, and employee training about the hazards associated with the use of CO_2 to fight fires. Many of the OSHA rules regarding fire protection come from the National Fire Prevention Association (NFPA) [13,68,69,71–75]. The Emergency Management Institute of the Federal Emergency Management Agency (FEMA), professional and academic emergency management programs, and underground utilities organizations are additional sources of regulations, information, and training regarding confined space hazards.

Breathing Gas, Respiratory Protection, and Controlled, Self-Contained Atmospheres
Academic medical researchers and governmental aviation and aerospace organizations such as NASA and the US Naval Medical Research Institute have investigated the physiology of CO_2 and the engineering controls needed to sustain humans in controlled and self-contained environments. The compressed-breathing-gas CO_2 limit for OSHA/CGA Grade D breathing air used in respiratory protection and Self Contained Underwater Breathing Apparatus (SCUBA) equipment is 1000 ppm. Through the Coast Guard, the DOT establishes a limit for CO_2 at 1000 ppm in SCUBA breathing gas for commercial diving.

NIOSH and PHS also regulate the CO_2 content of breathing gas for self-contained breathing apparatus (SCBA) and supplied air respirators. These limits are the same CGA standard of 1000 ppm, but also mandate maximum inspired CO_2 content for rebreathed air while using an SCBA (as shown in Table 5). Because humans at rest exhale 3.5% CO_2 on average, some exhaled air in the mask of an SCBA is rebreathed. The equipment design must ensure that the average CO_2 content of inhaled air does not exceed the tolerances listed in Table 5 [58,69,76].

TABLE 5
MAXIMUM ALLOWED PERCENTAGE OF CO_2 IN MIXED SUPPLIED/ REBREATHED AIR FROM SCBA APPARATUS

Service Time (h)	**Maximum allowed CO_2 content in %**
$<1/2$	2.5
1	2.0
2	1.5
3	1.0
4	1.0

US GPO, 2000-42 CFR 84.97.

Food Additive and Medical Gas
The DHHS sets rules for or defines uses of CO_2 as a general food additive, a leavening agent, a diagnostic indicator of severe disorders associated with changes in body acid–base balance, and as a medical gas. As long as CO_2 is manufactured in accordance with current good manufacturing practices (CGMP) as defined in 21 CFR sections 210–211, it is generally recognized as safe (GRAS) as a food additive. The CO_2 limit for medical gas is 500 ppm, as set by the CGA, United States Pharmacopeia (USP), and the National Formulary (NF).

Chemical Safety, Hazard Communication, and Hazard Response
Information on the hazards of CO_2 and recommended responses to its release are available through the OSHA-mandated Material Safety Data Sheets (MSDS) produced by manufacturers. Other sources about hazards include FEMA's Hazardous Material Guide, the DOT's Emergency Response Guide, toxicological information from the Registry of Toxic Effects of Chemical Substances (RTECS), and the International Chemical Safety Card (ICSC). The International Programme on Chemical Safety is a joint project of the United Nations Environmental Programme (UNEP), the World Health Organization (WHO), and the International Labour Office (ILO) that produces the ICSCs.

The response to CO_2 releases or hazards is the same as for any IDLH atmosphere. First, rescuers must wear respiratory protection. Victims are removed to a well-ventilated area and provided with supplementary oxygen if available. Aggressive ventilation and release to the atmosphere disperse the CO_2.

Transportation
Most regulations regarding CO_2 by the DOT refer to engineering controls on equipment used to transport CO_2 (such as tanks and pipelines) and include the OPS. The CGA, ANSI, ASME, and NFPA are other good sources of information and regulations pertaining to the transport of carbon dioxide by various means.

Toxic and Hazardous Substances: Where CO_2 is Not Regulated

The regulations that do not include carbon dioxide are equally interesting. As with any substance, the dose makes the poison. Even oxygen is toxic at high concentrations, so while CO_2 is a physiologically active gas and lethal above 15– 30%, it is not regarded as a toxic substance for regulatory purposes because it has no known toxicological effects (such as causing cancer, impairing the immune system, or causing birth defects). The EPA enforces the Clean Air Act by regulating ambient outdoor air-quality contaminants, and carbon dioxide is not included. The EPA does not set a limit for the amount of CO_2 allowed in food, as it does for other pesticides. Carbon dioxide is not suspected of any harmful effects in small concentrations (ppm), so the National Toxicology Program (NTP) has not studied it yet. None of the following organizations lists or studies CO_2 as a toxic substance: the Agency for Toxic Substances and Disease Registry (ATSDR) or NIOSH in the CDC, the National Institute of Environmental Health Science (NIEHS) in the National Institutes of Health (NIH), the National Center for Toxicological Research (NCTR) in the FDA, or the EPA. Nor do the following regulations identify or regulate CO_2 as a toxic or hazardous material: the Federal Insecticide, Fungicide, and Rodenticide Act of 1972 (FIFRA), the Resource Conservation and Recovery Act of 1976 (RCRA), the Comprehensive Environmental Response, Compensation, and Liability Act of 1980 (CERCLA or Superfund), and the Superfund Amendments and Reauthorization Act of 1986 (SARA). Only the inventory list for the Toxic Substances Control Act of 1976 (TSCA), the NIOSH confined-space hazard classification system, and FEMA's hazardous materials guide treat CO_2 as a hazardous substance to the extent that any concentrated or pressurized gas poses a danger. In all cases, it is included in the least hazardous category.

Summary of Regulations Related to CO_2

Table 6 is a summary of established exposure limits, and Table 7 is a list of the majority of regulations from the Code of Federal Regulations that pertain to CO_2.

TABLE 6
SUMMARY OF INFORMATION REGULATORY LIMITS FOR EXPOSURE TO CO_2

Organization	Regulation type	Regulation limit
OSHA	Occupational	5000 ppm TWA PEL; 30,000 ppm TWA STEL
NIOSH	Occupational	5000 ppm TWA REL; 30,000 ppm TWA STEL; 40,000 ppm IDLH
ACGIH	Occupational	5000 ppm TWA TLV; 30,000 ppm TWA STEL
ASHRAE	Ventilation	1000 ppm
OSHA/NIOSH/CGA/ USP/NF	Compressed breathing gas for respiratory protection SCBA and SBA	1000 ppm
OSHA/CGA/Coast Guard	SCUBA breathing gas	1000 ppm
FDA/CGA/USP/NF	Medical gas	500 ppm

CONCLUSIONS

Carbon dioxide is generally regarded as safe and non-toxic, inert gas. It is an essential part of the fundamental biological processes of all living things. It does not cause cancer, affect development, or suppress the immune system in humans. CO_2 is a physiologically active gas that is integral to both respiration and acid–base balance in all life. However, exposure to elevated concentrations of CO_2 can lead

TABLE 7
CODE OF FEDERAL REGULATIONS (CFRS) RELATING TO CARBON DIOXIDE

CFR	Government branch	Regulated as	Description	Regulation (limit/max)
9 CFR 313.5	FSIS, DOA	Anaesthetic and asphyxiant	Humane slaughter of livestock	XX
14 CFR 25.831	FAA, DOT	Ventilation air contaminant	In airplane cabins	5000 ppm (0.5%) by volume
21 CFR 137.180, 137.185, 137.270	FDA, DHHS	Leavening agent	In self-rising cereal flours	Must exceed 5000 (0.5%)
21 CFR 184.1240	FDA, DHHS	Direct food substance	GRAS—generally recognized as safe	GRAS
21 CFR 201.161	FDA, DHHS	Medical drug	Exempt from labeling requirements of 21 CFR 201.100	Exempt from labeling
21 CFR 210-211	FDA, DHHS	Medical gas	Current good manufacturing practices (CGMP)	CGMP
21 CFR 582.1240	FDA, DHHS	General purpose food additive	GRAS—generally recognized as safe	GRAS
21 CFR 862.1160	FDA, DHHS	Clinical chemistry test system	Diagnostic of blood acid–base imbalance	XX
29 CFR 1910.134	OSHA, DOL	Compressed breathing gas	In respiratory protection equipment CGA and USP	CGA breathing air Grade D—1000 ppm (0.1%)
29 CFR 1910.146	OSHA, DOL	Confined space hazard	General environmental controls	Permit required to enter
29 CFR 1910.155-1910.165 Subpart L	OSHA, DOL	Fire suppressant and confined space hazard	Required engineering controls on fire-fighting systems and equipment, employee training, and respiratory protection—NFPA	XX

29 CFR 1910.430	OSHA, DOL	Compressed breathing gas	Commercial diving operations—SCUBA	1000 ppm (0.1%)
29 CFR 1910.1000 Table Z-1	OSHA, DOL	Air contaminant	General occupational exposure limits	5000 ppm (0.5%) TWA PEL
29 CFR 1915.1000 Table Z	OSHA, DOL	Air contaminant	Exposure limits for shipyard employment	5000 ppm (0.5%) TWA PEL
29 CFR 1926.55	OSHA, DOL	Air contaminant	Exposure limits for construction	ACGIH: 5000 ppm (0.5%) TWA TLV
30 CFR 56.5001	MSHA, DOL	Air contaminant	Exposure limits for surface mines	ACGIH: 5000 ppm (0.5%) TWA TLV
30 CFR 57.5001	MSHA, DOL	Air contaminant	Exposure limits for underground mines	ACGIH: 5000 ppm (0.5%) TWA TLV
40 CFR 180.1049	EPA	Pesticide, insecticide	Tolerance for pesticide chemical in food	Exempt from tolerance
42 CFR 84.79	NIOSH, PHS, DHHS	Compressed breathing gas	SCBA	USP/NF, CGA: 1000 ppm (0.1%)
42 CFR 84.97	NIOSH, PHS, DHHS	Inspired air from SCBA	Test of inspired air in SCBA—control of rebreathing	$>$30 min./2.5%; 1 h/2.0%; 2 h/1.5%; 3 h/1.0%; 4 h/1.0%
42 CFR 84.141	NIOSH, PHS, DHHS	Compressed breathing gas	Supplied air respirators	CGA: 1000 ppm (0.1%)
46 CFR 197.340	Coast Guard, DOT	Compressed breathing gas	Commercial diving operations—SCUBA	1000 ppm (0.1%)
49 CFR 100-180	DOT	Transportation material	General transportation requirements	
49 CFR 190-199	OPS, DOT	Gas or hazardous liquid	Engineering safety controls on pipelines	

to adverse consequences, including death. The effects of exposure to CO_2 depend on the concentration and duration of exposure.

Ambient atmospheric concentrations of CO_2 are currently about 370 ppm. Humans can tolerate increased concentrations with virtually no physiological effects for exposures that are up to 1% CO_2 (10,000 ppm). For concentrations of up to 3%, physiological adaptation occurs without adverse consequences. A significant effect on respiratory rate and some discomfort occurs at concentrations between 3 and 5%. Above 5%, physical and mental ability is impaired and loss of consciousness can occur. Severe symptoms, including rapid loss of consciousness, and possible coma or death result from prolonged exposure above 10%. Experiments conducted on a submarine crew exposed to up to 3% CO_2 for many weeks and short-term exposures to even higher concentrations have shown that all effects except for prolonged coma, consequences of prolonged hypoxia (lack of oxygen), and death are reversible. Loss of consciousness occurs within several breaths and death is imminent at concentrations above 25–30%. Deaths from catastrophic releases of CO_2 are known from industrial accidents and natural disasters.

The potential for lethal or otherwise harmful exposure depends on the nature of the release rather than on the concentration of CO_2 or the size of the release. In particular, since CO_2 is denser than air, hazardous situations arise when large amounts of CO_2 accumulate in low-lying, confined, or poorly ventilated spaces. Releases, even large ones, that are quickly dissipated in the atmosphere, such as those that occur during explosive volcanic releases or from tall industrial stacks, do not pose a hazard.

Evidence for the effects of exposure to elevated concentrations of CO_2 on natural resources and ecosystems comes from many sources, including volcanic releases, soda springs, comparative, respiratory and fundamental physiology, free-air CO_2 enrichment studies, food preservation literature, and space science research. Among the major classes of terrestrial vertebrates, respiratory physiology and mechanisms for acid–base balance (pH regulation) vary widely, so tolerance to CO_2 exposure varies as well. Tolerance for CO_2 also correlates to ecological niche suggesting evolutionary adaptation to environmental conditions. Plants, insects, and soil-dwelling organisms have higher tolerance to CO_2 than most other forms of life. In spite of these differences, all air-breathing animals including humans have similar respiratory physiology and therefore broadly similar tolerance to CO_2, and prolonged exposure to high CO_2 levels, above 20–30%, will kill virtually all forms of life except some microbes, invertebrates, fungi, and insects. Some microbes can survive in a pure CO_2 atmosphere as long as trace amounts of oxygen are available. However, the identity and physiology of microorganisms dwelling in deep geologic formations is largely unknown, so the effects of CO_2 on them are uncertain.

Ecosystem impacts from exposure to elevated concentrations of CO_2 are poorly understood. Plants in general are even more tolerant than invertebrates to elevated CO_2, so any small-scale, short-term gas leaks would have minimal impacts. Persistent leaks, however, could suppress respiration in the root zone or result in soil acidification, and catastrophic releases could certainly kill vegetation as well as animals. Most of the controlled experiments have focused on the moderate increases in CO_2 concentrations that are expected to occur due to atmospheric buildup of CO_2 from the continued use of fossil fuels or that stimulate plant productivity in greenhouses. The studies have shown that moderate increases in CO_2 concentrations stimulate plant growth, while decreasing the loss of water through transpiration. At the other end of the scale, tree kills associated with soil gas concentrations in the range of 20–30% CO_2 have been observed at Mammoth Mountain, California, where volcanic outgassing of CO_2 has been occurring since at least 1990. Little information is available in the intermediate range of 2–30%. In addition, information on the tolerance of aquatic ecosystems to short-term, catastrophic releases was not found during this literature search and may need to be researched.

Carbon dioxide is used in a wide variety of industries: from chemical manufacture to beverage carbonation and brewing, from EOR to refrigeration, and from fire suppression to inert-atmosphere food preservation. Sources of CO_2 include natural reservoirs, separation from crude oil and natural gas, and as a waste product of industrial processes (chemical manufacture), combustion processes (energy production), and biological respiration (brewing). Because of its extensive use and production, the hazards of CO_2 are well known and routinely managed. Engineering and procedural controls are well established for dealing with the hazards of compressed and cryogenic CO_2. Nevertheless, the hazards of CO_2 are significant as fatalities from fire-suppression system malfunctions and confined-space accidents attest.

CO_2 is regulated by Federal and State authorities for many different purposes, including occupational safety and health, ventilation and indoor air quality, confined-space hazard and fire suppression, as a respiratory gas and food additive, for animal anesthesia and the humane slaughter of livestock, transportation and most recently, as a greenhouse gas (UNFCCC). Federal occupational safety and health regulations set three limits:

- 0.5% or 5000 ppm for an average 8-h day or 40-h week,
- 3% or 30,000 ppm for an average short-term 15-min exposure limit,
- 4% or 40,000 ppm for the maximum instantaneous exposure limit above which is considered IDLH.

Most industrial and safety regulations for CO_2 focus on engineering controls and specifications for transportation, storage containers, and pipelines.

In addition to understanding when and how CO_2 is regulated for industrial and occupational settings, it is important also to know that CO_2 is not regulated, studied, or suspected as a toxic substance by the following federal agencies or regulations, including: Clean Air Act 1970, 1990, Federal Insecticide, Fungicide, and Rodenticide Act (FIFRA) 1972, Resource Conservation and Recovery Act (RCRA) 1976, Comprehensive Environmental Response, Compensation, and Liability Act (CERCLA or Superfund) 1980, Superfund Amendments and Reauthorization Act (SARA) 1986, NTP, ATSDR or the NIOSH within the CDC, NIEHS in the NIH, and the NCTR in the FDA. Only the inventory list for the TSCA of 1976, the NIOSH confined space hazard classification system, and FEMA's hazardous materials guide treat CO_2 as a hazardous substance to the extent that any concentrated or pressurized gas poses a danger.

In conclusion, the key poorly understood health, safety, and environmental concerns surrounding geologic storage of CO_2 relate to the potential for unanticipated leakage. Such releases could be associated with surface facilities, injection wells, or natural, geological "containers" and may be small-scale diffuse leaks or large catastrophic incidents. Long industrial experience with CO_2 and gases in general shows that the risks from industrial storage facilities are manageable using standard engineering controls and procedures. Serious accidents have occurred but the incidents described were preventable and experience teaches us how to operate these facilities even more safely. On the other hand, our understanding of and ability to predict CO_2 releases and their characteristics in any given geologic and geographic setting is far more challenging. Certainly there are many sites, such as oil and gas reservoirs where the probability of leakage is very low. However, brine formations, which generally are not well characterized and do not have cap rocks or seals that have stood the test of time, will require significant effort to evaluate potential risks, and these risks must be taken seriously.

To date, the majority of the thought process regarding the risks of CO_2 geologic storage has revolved around human health risks. This study raises the issue that, if leakage occurs, ecosystem risks may also be significant, particularly for soil dwelling or ground hugging organisms. In addition, acidification of soils in the vicinity of surface leaks may also harm plants. Similarly, persistent low-level leakage could affect aquatic ecosystems by lowering the pH, especially in stagnant or stably stratified waters.

REFERENCES

1. USDOE, Carbon Sequestration Research and Development, National Technical Information Service, Springfield, VA, 1999, www.ornl.gov/carbon_sequestration/.
2. N.E. Tolbert, J. Preiss (Eds.), Regulation of Atmospheric CO_2 and O_2 by Photosynthetic Carbon Metabolism, Oxford University Press, New York, NY, 1994.
3. S.E. Humphris, T.A. Zierenberg, L.S. Mullioneaux, R.E. Thompson, Geophysical Monograph 91: Seafloor Hydrothermal Systems: Physical, Chemical, Biological, and Geological Interactions, American Geophysical Union, Washington, DC, 1995.

4. N.A. Campbell, L.G. Mitchell, J.B. Reece, Biology, fifth ed., Benjamin/Cummings, Menlo Park, CA, 1999, plus earlier Publishing editions.
5. J.A. Kellum, Determinants of blood pH in health and disease, *Crit. Care* **4** (1) (2000) 6–14.
6. M.J.T. Van de Ven, W.N.J.M. Colier, M.C. van der Sluijs, B. Oeseburg, H. Folgering, Ventilatory response in metabolic acidosis and cerebral blood volume in humans, *Respir. Physiol.* **124** (2001) 105–115.
7. J.D. Jones, Comparative Physiology of Respiration, *Special Topics in Biology Series*, Edward Arnold Publishers, London, 1972.
8. NIOSH (National Institute of Occupational Safety and Health), Criteria for a Recommended Standard. Occupational Exposure to Carbon Dioxide, NIOSH Publication No. 76-194, US GPO, Washington, DC, 1976, www.cdc.gov/niosh/, www.gpo.gov.
9. N. Heisler, Interactions between gas exchange, metabolism, and ion transport in animals: an overview, *Can. J. Zool.* **67** (1989) 2923–2935.
10. F.N. White, Carbon dioxide homeostasis, in: S.C. Wood (Ed.), *Lung Biology in Health and Disease, Comparative Pulmonary Physiology—Current Concepts*, Vol. 39, Marcel Dekker, New York, NY, 1989, pp. 439–466.
11. Naval Aerospace Medical Institute (NAMI), United States Naval Flight Surgeon's Manual, third ed., 1991, http://www.vnh.org/FSManual/fsm91.html.
12. USAF (United States Air Force), USAF Flight Surgeon's Guide, online ed., 2001, http://wwwsam.brooks.af.mil/ram/fsr-web/fsr-default.htm.
13. NIOSH, Criteria for a Recommended Standard. *Working in Confined Spaces*, NIOSH Publication No. 80-106, US GPO, Washington, DC, 1979, www.cdc.gov/niosh/, www.gpo.gov.
14. NRC (National Research Council) Committee on Airliner Cabin Air Quality, The Airliner Cabin Environment: Air Quality and Safety, National Academy Press, Washington, DC, 1986, www.nap.edu.
15. USEPA and NIOSH, Building Air Quality: A Guide for Building Owners and Facility Managers, US GPO, Washington, DC, 1991.
16. USEPA, Indoor Air Pollution: An Introduction for Health Professionals, US GPO Publication No. 1994-523-217/81322, USEPA, Washington, DC, 1994, www.epa.gov/iaq/pubs/hpguide.html.
17. J.L. Gingras, A. Muelenaer, L.B. Dalley, K.J. O'Donnell, Prenatal cocaine exposure alters postnatal hypoxic arousal responses and hypercarbic ventilatory responses but not pneumocardiograms in prenatally cocaine-exposed term infants, *Pediatric Pulmonol.* **18** (1) (1994) 13–20.
18. OSHA (Occupational Safety and Health Administration), Indoor Air Quality, Federal Register # 59: 15968-16039, US GPO, Washington, DC, 1994, www.cdc.gov/niosh/, www.gpo.gov.
19. NIOSH, NIOSH Facts—Indoor Environmental Quality (IEQ), Document No. 705002, US GPO, Pittsburgh, PA, 1997, www.cdc.gov/niosh/, www.gpo.gov.
20. ASHRAE Standards Committee, ANSI/ASHRAE Standard 62-1999: Ventilation for Acceptable Indoor Air Quality, American Society of Heating, Refrigerating, and Air-Conditioning Engineers, Atlanta, GA, 1999, www.ashrae.org.
21. NIOSH (Occupational Health Guidelines for Chemical Hazards), Occupational Health Guidelines for Chemical Hazards, NIOSH Publication No. 81-123, US GPO, Washington, DC, 1981, www.cdc.gov/niosh/, www.gpo.gov.
22. ACGIH (American Conference of Governmental Industrial Hygienists), Threshold Limit Values for Chemical Substances and Physical Agents and Biological Exposure Indices (1994–1995), ACGIH, Cincinnati, OH, 1994.
23. W. Coryell, Hypersensitivity to carbon dioxide as a disease-specific trait marker, *Biol. Psychiatry* **41** (1997) 259–263.
24. J.P. Forsyth, C.W. Lejuez, C. Finlay, Anxiogenic effects of repeated administrations of 20% CO_2-enriched air: stability within sessions and habituation across time, *J. Behav. Therapy Exp. Psychiatry* **31** (2000) 103–121.
25. W. Coryell, A. Fyer, D. Pine, J. Martinez, S. Arndt, Aberrant respiratory sensitivity to CO_2 as a trait of familial panic disorder, *Biol. Psychiatry* **49** (2001) 582–587.
26. D. Koszycki, J. Bradwejn, Anxiety sensitivity does not predict fearful responding to 35% carbon dioxide in patients with panic disorder, *Psychiatry Res.* **101** (2001) 137–143.
27. M. McIntyre, B. McNeil, Dissolved carbon dioxide effects on morphology, growth, and citrate production in *Aspergillus niger* A60, *Enzyme Microb. Technol.* **20** (1997) 135–142.

28. M.H.J. Bennik, W. Vorstman, E.J. Smid, L.G.M. Gorris, The influence of oxygen and carbon dioxide on the growth of prevalent *Enterobacteriaceae* and *Pseudomonas* species isolated from fresh and controlled-atmosphere-stored vegetables, *Food Microbiol.* **15** (1998) 459–469.
29. J.K. Frederickson, T.C. Onstott, Microbes deep inside the Earth, Earth from the Inside Out, Scientfic American, New York, NY, 2000, pp. 10–15.
30. I. Haasum, P.V. Nielsen, Preincubation of *Penicillium commune* Conidia under modified atmosphere conditions: influence on growth potential as determined by an impedimetric method, *J. Stored Prod. Res.* **32** (4) (1996) 329–337.
31. S. Tian, Q. Fan, Y. Xu, Y. Wang, A. Jiang, Evaluation of the use of high CO_2 concentrations and cold storage to control *Monilinia fructicola* on sweet cherries, *Postharvest Biol. Technol.* **22** (2001) 53–60.
32. J.N. Maina, Zoophysiology Vol. 37: The Gas Exchangers—Structure, Function, and Evolution of the Respiratory Processes, Springer, Berlin, 1998.
33. S.K. Hamilton, S.J. Sippel, J.M. Melack, Oxygen depletion and carbon dioxide and methane production in waters of the Pantanal wetland of Brazil, *Biogeochemistry* **30** (1995) 115–141.
34. D.D. Mann, D.S. Jayas, N.D.G. White, W.E. Muir, Mortality of adult *Cryptolestes ferrugineus* (Stephens) exposed to changing CO_2 concentrations, *J. Stored Prod. Res* **35** (1999) 385–395.
35. P.C. Annis, R. Morton, The acute mortality effects of carbon dioxide on various life stages of *Sitophilus oryzae*, *J. Stored Prod. Res.* **33** (2) (1997) 115–123.
36. E.C.W. Leong, S.H. Ho, Effects of carbon dioxide on the mortality of *Liposcelis bostrychophila* Bad. and *Liposcelis entomophila* (End.) (Psocoptera: Liposcelididae), *J. Stored Prod. Res.* **31** (3) (1995) 185–190.
37. G.N. Mbata, C. Reichmuth, The comparative effectiveness of different modified atmospheres for the disinfestation of Bambarra groundnuts, *Vigna Subterranea* (L.) Verde, infested by *Callosobruchus subinnotatus* (Pic) (Coleoptera: Bruchidae), *J. Stored Prod. Res.* **32** (1) (1996) 45–51.
38. S. Zhou, R.S. Criddle, E.J. Mitcham, Metabolic response of *Platynota stultana* pupae to controlled atmospheres and its relation to insect mortality response, *J. Insect Physiol.* **46** (2000) 1375–1385.
39. S. Zhou, R.S. Criddle, E.J. Mitcham, Metabolic response of *Platynota stultana* pupae during and after extended exposure to elevated CO_2 and reduced O_2 atmospheres, *J. Insect Physiol.* **47** (2001) 401–409.
40. R.G. Amundson, E.A. Davidson, Carbon dioxide and nitrogenous gases in the soil atmosphere, *J. Geochem. Expl.* **38** (1990) 13–41.
41. P. Holter, Tolerance of dung insects to low oxygen and high carbon dioxide concentrations, *Eur. J. Soil Biol.* **30** (4) (1994) 187–193.
42. J. Pinol, J.M. Alcaniz, F. Roda, Carbon dioxide efflux and pCO_2 in soils of three Quercus ilex montane forests, *Biogeochemistry* **30** (1995) 191–215.
43. V. Sustr, M. Simek, Behavioural responses to and lethal effects of elevated carbon dioxide concentration in soil invertebrates, *Eur. J. Soil Biol.* **32** (3) (1996) 149–155.
44. H. Mohr, P. Schopfer, Plant Physiology, Springer, New York, NY, 1995, translated by G. Lawlor and D.W. Lawlor.
45. D.M. Orcutt, E.T. Nilsen, *Physiology of Plants Under Stress*, Wiley, New York, NY, 2000, 683pp.
46. J.W. Mastalerz, The Greenhouse Environment, Wiley, New York, NY, 1977, 629pp.
47. M.C. Press, J.D. Scholes, M.G. Barker (Eds.), *Physiological Plant Ecology*, British Ecological Society/Blackwell, Malden, MA, 1999.
48. IPCC, IPCC Second Assessment Report: Climate Change 1995, Cambridge University Press, Cambridge, 1996.
49. IPCC, IPCC Third Assessment Report: Climate Change 2001, Cambridge University Press, Cambridge, 2001.
50. M.H., Jones, P.S., Curtis, (eds.), Bibliography on CO_2 Effects on Vegetation and Ecosystems: 1990–1999 Literature, ORNL/CDIAC-129, Oak Ridge National Laboratory, US Department of Energy, Oak Ridge, TN, 2000. http://cdiac.esd.ornl.gov/epubs/cdiac/cdiac129/cdiac129.html.
51. CDIAC http://cdiac.esd.ornl.gov/, Oak Ridge National Laboratory, www.ornl.gov/, US Department of Energy, Oak Ridge, TN, 2001a.
52. CDIAC, Free Air CO_2 Enrichment (FACE), Oak Ridge National Laboratory, US Department of Energy, Oak Ridge, TN, 2001, http://cdiac.esd.ornl.gov/programs/FACE/face.html.
53. J.A. Hyatt, Liquid and supercritical carbon dioxide as organic solvents, *J. Org. Chem.* **49** (26) (1984) 5097–5101.

54. J. Pawliszyn, N. Alexandrou, Indirect supercritical fluid extraction of organics from water matrix samples, *Water Pollut. Res. J. Can.* **24** (2) (1989) 207–214.
55. W.D. Spall, K.E. Laintz, A survey on the use of supercritical carbon dioxide as a cleaning solvent, in: J. McHardy, S.P. Sawan (Eds.), *Supercritical Fluid Cleaning*, Noyes Publishing, Westwood, NJ, 1998, pp. 162–194.
56. IEAGHG, Carbon Dioxide Utilization, 1995, www.ieagreen.org.uk/.
57. CGA, G6: Carbon dioxide, fifth ed., CGA, Arlington, VA, 1997, www.cganet.com.
58. Airgas, Inc., Applications of CO_2, www.airgas.com/carbonic/applications.htm, 2001.
59. USEPA, Carbon Dioxide as a Fire Suppressant: Examining the Risks, US GPO, Washington, DC, 2000, www.epa.gov/ozone/title6/snap/co2report.html.
60. Freedonia Group, Carbon Dioxide, Business Research Report B286, Freedonia Group, Inc., Cleveland, OH, 1991, 46pp.
61. Freedonia Group, Report 1345: Industrial Gases to 2004, Freedonia Group, Inc, Cleveland, OH, 2000.
62. USDOC (United States Department of Commerce), Current Industrial Reports—1998, Economics and Statistics Administration, Bureau of the Census, 1999.
63. Texas Natural Resources Conservation Commission (TNRCC), Point Source Emissions database, 2001, www.tnrcc.state.tx.us.
64. Shell, Royal Dutch, web site, http://www.shell.com/royal-en/content/0,5028,25544-51037,00.html, 2001.
65. USDOE, Annual Energy Report 1999, Energy Information Administration, US DOE, Washington, DC, 2000, DOE/EIA-0384(99)www.eia.doe.gov/aer/contents.html, www.eia.doe.gov/pub/pdf/multi.fuel/038499.pdf.
66. H. Kruse, M. Tekiela, Calculating the consequences of a CO_2-pipeline rupture, *Energy Convers. Manage.* **37** (6–8) (1996) 1013–1018.
67. USDOE, Type A Accident Investigation Board Report of the July 28, 1998 Fatality and Multiple Injuries Resulting From Release of Carbon Dioxide at Building 648, Test Reactor Area, Idaho National Engineering and Environmental Laboratory, 1998, www.id.doe.gov/doeid/foia/archive.htm, www.id.doe.gov/doeid/foia/tra-accident.htm.
68. USEPA, An Annotated Summary of Climate Change Related Resources, National Service Center for Environmental Publications, Cincinnati, OH, 2000.
69. US GPO (United States Government Printing Office), Code of Federal Regulations, US GPO, Washington, DC, 2000, www.access.gpo.gov.
70. NIOSH, Documentation for Immediately Dangerous to Life and Health Concentrations, NTIS Publication No. PB-94-195047, National Technical Information Service, Springfield, VA, 1994, www.cdc.gov/niosh/, www.ntis.gov.
71. NIOSH, NIOSH Pocket Guide to Chemical Hazards, NIOSH Publication No. 97-140, National Technical Information Service, Springfield, VA, 1997, www.cdc.gov/niosh/, www.gpo.gov.
72. NIOSH, Worker Deaths in Confined Spaces: A Summary of NIOSH Surveillance and Investigative Findings, NIOSH Publication No. 94-103, Publications Dissemination, DSDTT, NIOSH, Cinncinnati, OH, 1994.
73. CGA, P-14: Accident Prevention in Oxygen-Rich and Oxygen-Deficient Atmospheres, CGA, Arlington, VA, 1992, www.cganet.com.
74. CGA, SB-2: Oxygen Deficient Atmospheres, CGA, Arlington, VA, 1992, www.cganet.com.
75. CGA, SB-15: Safety Bulletin—Avoiding Hazards in Confined Work Spaces During Maintenance, Construction, and Similar Activities, CGA, Arlington, VA, 1998, www.cganet.com.
76. CGA (Compressed Gas Association, Inc.), G7: Compressed Air for Human Respiration, fourth ed., CGA, Arlington, VA, 1990, www.cganet.com.
77. S.M. Benson, R. Hepple, J. Apps, C.-F. Tsang, M. Lippman, Lessons learned from natural and industrial analogues for storage of carbon dioxide in deep geologic formations, Lawrence Berkeley National Laboratory Report, LBL-51170, Berkeley, CA, 2002.

Carbon Dioxide Capture for Storage in Deep Geologic Formations, Volume 2
D.C. Thomas and S.M. Benson (Eds.)

Chapter 27

THE REGULATORY CLIMATE GOVERNING THE DISPOSAL OF LIQUID WASTES IN DEEP GEOLOGIC FORMATIONS: A PARADIGM FOR REGULATIONS FOR THE SUBSURFACE STORAGE OF CO_2?

John A. Apps

Ernest Orlando Lawrence Berkeley National Laboratory, Berkeley, CA, USA

ABSTRACT

Federal and state regulations covering the deep injection disposal of liquid waste have evolved over the last 30 years in response to legislation designed to protect underground sources of drinking water (USDW). These regulations apply to so-called Class I wells, and address issues relating to the confinement of hazardous and nonhazardous wastes below the lowermost USDW. They have been made progressively more stringent with time, and are now quite effective in protecting USDWs. The deep injection disposal of compressed carbon dioxide (CO_2) into similar environments will undoubtedly require similar regulation. Accordingly, the history relating to the development of legislation to protect groundwater supplies, and resulting regulations is reviewed and conclusions drawn regarding the extent to which these regulations might eventually be applied to CO_2 injection.

INTRODUCTION

The technology of deep well injection disposal of liquid wastes has many similarities to that envisioned for the storage of CO_2 in deep saline formations. The issues raised—technical, legislative, regulatory and social—would be similar to those relating to disposal of hazardous liquid wastes in comparable subsurface environments. However, stabilizing or halting the increasing United States inventory of CO_2 in the atmosphere by subsurface storage would require the disposal of volumes of CO_2 approximately two orders of magnitude larger than those required for hazardous liquid waste. Concerns over the consequences of storing such large volumes of CO_2 in deep geologic formations will likely generate public apprehensions similar to those raised over deep well injection disposal of hazardous liquid waste at so-called off-site facilities.

A review of deep well injection technology and the regulatory framework governing the disposal of liquid wastes by this method is particularly valuable in anticipating corresponding issues affecting the subsurface disposal of CO_2. In this chapter, we consider the historical, technical, and regulatory basis for deep injection disposal of liquid industrial and municipal wastes with particular emphasis on regulations governing hazardous waste disposal. We then consider the implications for the future regulatory climate governing CO_2 disposal by similar means.

STUDY METHODOLOGY

History of Underground Disposal of Liquid Wastes

Deep well injection storage of industrial wastes came into prominence in the United States following World War II. But the technology had its roots much earlier during the first part of the 20th century, when substantial quantities of saline brines were co-produced in oil and gas fields. The brines and associated oil field wastes were initially discarded in surface evaporation or infiltration pits. However, this disposal method compromised the integrity of shallow groundwater aquifers, and states banned the practice. The oil and gas industry therefore turned to injection of liquid wastes. Currently, more than 300 billion gallons (1.1 billion m^3) of brine are injected yearly into approximately 175,000 wells [1,2]. The disposal of oilfield brines is still not without attendant risks of groundwater pollution. In fact, according to an earlier report by Gordon and Bloom [3] 17 of the 32 oil and gas producing states had reported groundwater contamination resulting from the storage of these brines.

By the early 1970s, nearly 90% of the US population had become dependent on groundwater for domestic use and for agricultural irrigation. At the same time, industry was increasingly looking for alternatives to surface effluent discharges, which had become an undesired focus of attention under the Clean Water Act of 1972. The disposal of any liquid waste by injection down wells was technically feasible, and particularly attractive, for the chemical and petrochemical industries, which produced large-volume, dilute hazardous waste streams that were difficult to dispose of by other means. Furthermore, although the capital cost of developing an injection facility was high, the operating costs were usually low. The lack of public awareness of this disposal method initially allowed industry to proceed without close scrutiny or adequate regulation.

The number of deep disposal facilities for hazardous liquid waste initially grew rapidly. In 1950, there were only five such facilities. By 1963, there were 30, [4], and between 1973 and 1975 the number peaked at about 270 following passage of the Clean Water Act of 1972. Ten years later, the total number of wells injecting hazardous waste had fallen slightly to 252 at 95 facilities. Since then, with implementation of more stringent regulations in 1988 following passage of the Resource Conservation and Recovery Act (RCRA) in 1984, the number of facilities had fallen to about 50, comprising 163 so-called Class I wells (i.e. wells injecting below the lowest aquifer containing a potential source of drinking water) injecting hazardous waste. Most of these wells are found in Texas (78) and Louisiana (18). Another 221 facilities comprising 366 Class I wells injecting nonhazardous waste are also in operation, the majority of which are found in Florida (112) and Texas (110). Of those operating in Florida, 104 are dedicated to the disposal of municipal waste, the only state operating this class of well [26]. Figure 1 illustrates the variation in the number of deep well injection facilities over time.

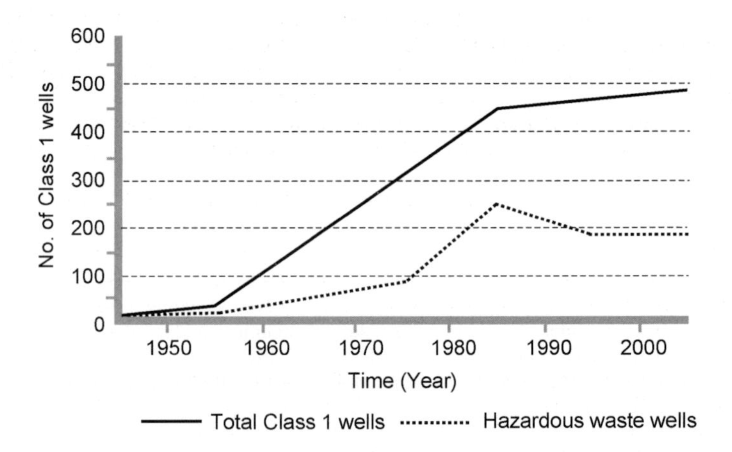

Figure 1: Number of Class I wells operating in the United States on a yearly basis since 1950 (from Ref. [13]).

Sites favorable for deep well disposal commonly overlie sedimentary basins, where deep formation waters are highly saline, and where permeable aquifers are interspersed with relatively impermeable shale "confining beds". The 48 contiguous states are endowed with several such basins, some of which are strategically located with respect to centers of industrial development, especially those in the raw materials sector. Figure 2 illustrates the distribution of these major sedimentary basins.

Figure 2: Map of the contiguous 48 states, showing major stratigraphic features in relation to the location of deep injection disposal wells (after Warner [29]).

Most of the deep injection disposal wells are located on the coastal plain of the Gulf Coast, and in states surrounding the Great Lakes. In the Great Lakes region, deep injection well depths range from 1700 to 6000 ft (520–1830 m), whereas those along the Gulf Coast range from 2200 to 9000 ft (670–2740 m) [5]. About 60% of the wells are located within the EPA jurisdiction of Region VI, which includes Texas and Louisiana. According to Gordon and Bloom [3], manufacturers of organic chemicals account for nearly 65% of the injected volume, while the petroleum refining and petrochemical industries account for a further 25%.

The quantities of hazardous liquid waste injected at deep well disposal facilities are enormous. By 1985, 11.5 billion gallons (43.5 million m^3) of industrial liquid wastes were injected annually [28]. At the time, this was 10 times the amount going to landfills and twice that going to surface impoundments [3]. By 1990, the quantity injected had fallen to 9 billion gallons (34 million m^3) [6] and currently remains at about the same level [2,7]. The injected hazardous liquid waste constitutes approximately 60% of all such waste generated in the United States.

Deep well injection disposal of hazardous waste remains a viable method of disposal, particularly on the basis of cost. In 1987, the cost of liquid hazardous waste disposal ranged from $49 to $207 per ton [8]. This compared with $776–1426 per ton for incineration, $85–394 per ton for chemical treatment, and $131–329 per ton for resource recovery (in the case of organics from the aqueous phase). Brower et al. [9] estimated that the pre-treatment of injected waste to remove the hazardous components could increase the operating costs 3–40 times. Furthermore, alternative disposal methods, involving treatment and surface disposal, increase the potential risks of adverse consequences to the environment and public health. Thus, subsurface injection will remain a preferred method of hazardous liquid waste disposal for the foreseeable future.

Legislative history governing deep well injection disposal

Early deep well disposal practices commonly resulted in poorly engineered or constructed facilities, which were carelessly operated, and resulted in an increasing number of reported occurrences of potable aquifer contamination. Consequently, the Federal Water Quality Administration (FWQA) published policy guidelines governing the deep well injection of hazardous wastes [10], which "opposed the disposal or storage of wastes by subsurface injection without strict controls and a clear demonstration that such [injected] wastes will not interfere with present or potential use of subsurface water supplies" [11]. Furthermore, the policy provided for critical evaluation by the FWQA of all proposals for subsurface injection of wastes to ensure that the fate of the wastes could be predicted, and that the waste would not interfere with the use of water resources or cause environmental hazards. Waste injection had also to be continuously monitored and the injection well properly plugged following cessation of operations. The FWQA emphasized that subsurface disposal of wastes was to be considered a temporary expedient until alternative methods providing better environmental protection were developed.

With the reorganization in 1970 of federal government agencies charged with protection of the environment, the FWQA was absorbed by the United States Environmental Protection Agency (EPA), which subsequently issued a Technical Studies Report concerning deep well injection disposal [27]. This report noted that many problems arising through the use of this technology could be avoided if the fate of the injected wastes could be monitored. The report concluded that deep well injection should be regulated through a system of laws, and that a permitting process should be implemented, based on both injection site and the nature of waste injected (as noted in Ref. [11]).

EPA eventually set forth its own policies regarding deep well injection [12]. EPA was also opposed to the storage or disposal of contaminants by subsurface injection "...without strict control and clear demonstration that such wastes will not interfere with present or potential use of subsurface water supplies" [13]. But EPA also recognized that for some industries, such practice was then the only feasible means of disposal as was clearly the case in the oil and gas and geothermal industries where reinjection of large volumes of liquid wastes had been a standard practice for several decades.

Shortly after the EPA had published its policy on deep well injection, Congress passed the Safe Drinking Water Act (SDWA) of 1974. Part C of the SDWA is the Underground Injection Control (UIC) program, which implemented EPA's policy concerning deep well injection and mandated controls on injection practices. According to Herbert [11], the SDWA was essentially the first federal statute to address deep well injection practices. Furthermore, it provided for a joint system involving both state implementation and federal oversight, in which EPA would implement the policy guidelines set forth by the federal government by setting minimum requirements for state programs. EPA was to be allowed discretion in requiring states to use a permit system, rule making, or both to control underground injection. The reason for this discretion was to allow compatibility with permit provisions already in place under the Federal Water Pollution Control Act (FWPCA) of 1970.

EPA responded with the publication of technical UIC regulations in June 1980. It was also in these regulations that an Underground Source of Drinking Water (USDW), as set forth in 40 CFR Part 144.3, was first defined as containing fewer than 10,000 mg/L of dissolved solids, and was capable of providing a sufficient quantity of groundwater to supply a public water system. The regulations also categorized injection wells into five classes to deal with the multiplicity of waste streams and well functions, as set forth in 40 CFR Part 144.6—Classification of Wells. The classes most relevant to deep injection disposal of CO_2 are Class I, that class of wells injecting waste below the deepest USDW, and Class II, in which fluids relating to oil and gas production can be injected. Under certain circumstances, however, CO_2 could be injected into wells under the Class V designation.

Although the SDWA was promulgated to ensure the protection of the nation's water supplies, it did not specifically address the improper handling of hazardous waste. This omission was rectified through passage of RCRA in 1976. With this act, Congress made it a national policy to eliminate, or at least reduce, hazardous waste generation as expeditiously as possible. The act also designated responsibility to EPA for promulgating regulations governing the treatment, storage, and disposal of hazardous waste, including hazardous waste injection wells.

EPA classified hazardous wastes in 40 CFR Part 261—Identification and Listing of Hazardous Waste. In general, hazardous wastes are either "listed" or identified by their "characteristics". The characteristics are subdivided into six groups and given respective Hazard Codes. Each characteristic causes or significantly contributes to increased mortality or serious illness, or possesses a substantial present or potential threat to human health and the environment. The criteria for listing a hazardous waste are extensive and are provided in Subpart C of Part 162. Waste that is not classified as hazardous is, by default, nonhazardous. However, such waste cannot be allowed to contaminate a USDW unless it meets the criteria set forth in 40 CFR Part 141—National Primary Drinking Water Regulations. These criteria are sufficiently restrictive that most wastes would not be eligible for direct injection into a USDW.

Because of the technical complexity of the issues involved, and an overlap between SDWA and RCRA, EPA decided to coordinate their implementation by regulating aboveground facilities under RCRA, but injection wells under SDWA. However, deficiencies in EPA's coordination of SDWA and RCRA with respect to well disposal of hazardous wastes and protection of USDWs were revealed by the discovery of groundwater contaminated with hazardous chemicals due to malfunctioning and poorly regulated hazardous-waste injection wells [14]. In 1982, therefore, Congress gave EPA specific directives regarding the implementation of its UIC program to ensure that vulnerable subsurface drinking water supplies were adequately protected as specified in the Hazardous and Solid Waste Amendments (HSWA) to RCRA in 1984. Congress further mandated that land disposal of hazardous waste was allowed only if an applicant for a permit exempting restriction on land disposal could demonstrate that no migration of the waste would occur [11].

EPA responded to the 1984 RCRA amendments in 1988 with revised UIC regulations governing hazardous-waste injection, otherwise known as the "Land Ban" regulations. Henceforth, the subsurface injection of hazardous wastes would be prohibited unless EPA was to issue a permit exempting the operator of a deep well injection facility from the prohibition. To obtain a permit, the operator had to petition EPA and provide supporting documentation demonstrating that the injected waste would not migrate outside of a designated injection zone within 10,000 years, or that the waste would become nonhazardous. As noted above, Congress allowed EPA to delegate responsibility to the states to administer their own UIC programs, should a state wish to assume primacy. States also had the option of administering all or part of the UIC program. To date, 34 states have been delegated full authority to regulate Class I wells within their territory, and two share responsibility with the federal government. The remainder is administered under the federal program [15]. Four states have placed an outright ban on Class I wells. Figure 3 illustrates the distribution of states with primacy.

Although states with primacy follow federal regulations quite closely, specific variations do occur. These variations are generally more restrictive than the corresponding parts of federal regulations [16,17]. For example, several states require that a critical area surrounding the injection well must be established, based on logging and testing the injection well and surrounding formation, known as an "Area of Review" (AoR), which is larger than required by federal regulations, e.g. Texas requires 2.5 mile (4 km), Louisiana requires 2 mile (3.2 km), and Florida and Kansas require a 1 mile minimum (1.6 km) [17].

Current regulations have generally proven effective in protecting USDWs, as shown by relatively recent independent investigations, e.g. see Ref. [17]. There has been no evidence of subsurface leakage into USDWs from Class I hazardous waste injection wells since 1988, and no evidence of contamination of USDWs due to the migration of hazardous waste from well injection zones. With respect to the disposal of nonhazardous wastes, only one serious problem has arisen; that concerning the disposal of sewage waste in Class I wells in Florida where leakage outside the confining zone has been observed in several instances.

Difficulties remain, however, in providing a technically convincing demonstration of waste containment in the injection zone. These deficiencies have been exploited by environmental groups in their opposition to the underground disposal of hazardous liquid wastes. Opposition is not limited to technical issues related to the ultimate fate of the injected waste, but to environmental, social and quality of life issues arising from surface facilities. Off-site injection facilities are particularly subject to criticism, because the hazardous waste must be trucked in from various sources and transferred and temporarily stored in surface tanks or impoundments. Wastes from different sources may vary widely in chemical composition and react with undesired consequences when mixed. Off-site facilities are commonly prone to toxic releases to the atmosphere and to contamination of surface waters and shallow ground waters. They also tend to be more frequently in violation of federal and state regulations governing deep well injection disposal. Recently, the General Accounting Office [18] has questioned the adequacy and timing in soliciting public

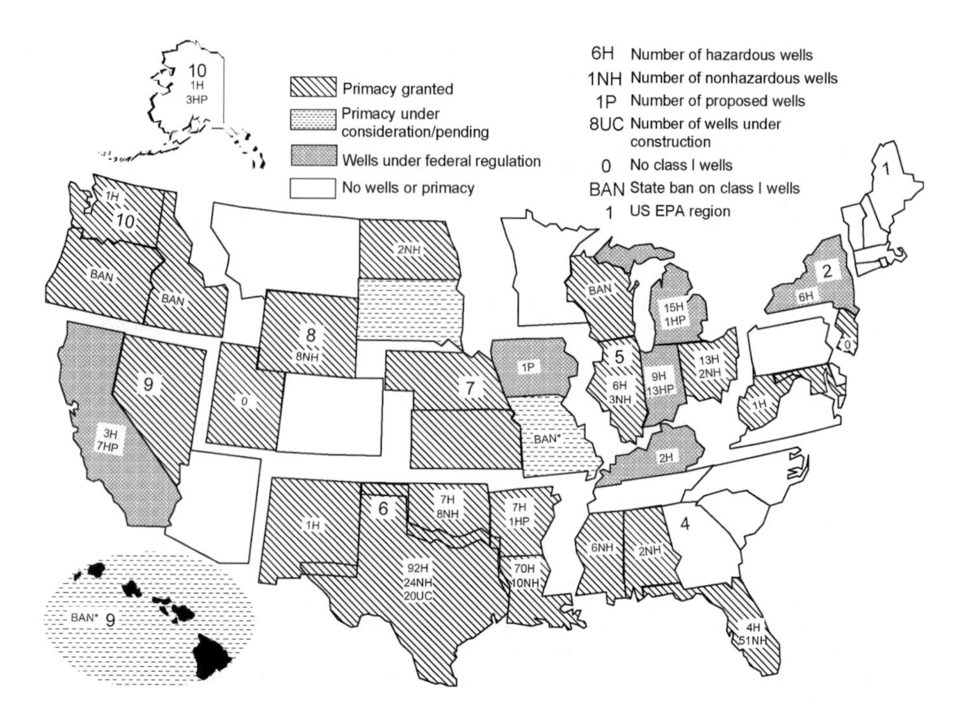

Figure 3: Regulatory status of Class I wells in the U.S. (from Ref. [9]).

comments during the permitting process of off-site Class I hazardous waste wells, and has recommended that the public be involved at an earlier stage in the process (see below). EPA, in response, believes that many of the issues raised by local communities are not relevant to the mission of the UIC program.

For the foreseeable future, deep well injection disposal will continue to be the technology of choice for the elimination of liquid wastes. Despite impediments to its use, regulatory policy, which considers deep injection waste disposal as an interim expedient, owner liability, and opposition from environmental groups, the technology remains one of the cheapest, safest, and most convenient disposal options for many hazardous waste generators. CO_2 storage in deep geologic formations by injection would similarly be the technology of choice, particularly as current estimates suggest a cost of between only $3 and $10 per ton CO_2 injected.

Current Regulations Governing Deep Well Injection Disposal of Liquid Wastes

Current regulations governing the deep well disposal of wastes are found in the Code of Federal Regulations, Chapter 40, Parts 144–148:

- Part 144—Underground Injection Control Program
- Part 145—State UIC Program Requirements
- Part 146—Underground Injection Control Program: Criteria and Standards
- Part 147—State Underground Injection Control Programs
- Part 148—Hazardous Waste Injection Restrictions.

In addition, 40 CFR Part 124—Procedures for Decision-Making, includes public-participation requirements that must be met by UIC programs.

This section focuses on regulations pertaining to Class I wells, because these wells usually penetrate to considerable depths and discharge their waste below aquifers containing potable water. Operating conditions are therefore somewhat similar to those expected of wells injecting supercritical CO_2, where the optimum depth range would fall between 3000 and 6000 ft (915–1830 m).

The following discussion refers only to federal regulations; where states have primacy, the regulations sometimes differ. The criteria and standards applicable to Class I wells are very stringent, and even more so for those Class I wells injecting hazardous waste. Those applicable to wells injecting nonhazardous waste are given in Subpart B of Part 146, and those for wells injecting hazardous waste are given in Subpart G of Part 146. Each subpart is loosely subdivided into seven categories covering the following requirements:

- Information Required for Authorization (Permitting) by the Director of EPA.
- Siting
- Construction
- Operation
- Monitoring
- Ambient monitoring
- Reporting
- Closure and post-closure requirements.

Application for a permit to operate a Class I well

For authorization to operate a Class I hazardous waste well, the owner or operator must first submit a so-called "No-Migration Petition" to EPA describing all aspects of the proposed operation, including well siting, design and operation, and conduct hydrologic modeling and geochemical modeling (if feasible) to demonstrate that migration will not occur beyond a defined injection zone. The need for adequate site characterization, especially for Class I hazardous waste wells, is particularly critical to ensure that no failure occurs for whatever reason, and that hazardous waste will be contained for at least 10,000 years or become nonhazardous during that period.

Because of the 10,000-year period required for post-closure regulatory compliance, experimental verification is not feasible, and therefore much of the justification necessary to demonstrate waste containment must depend on predictive modeling. This modeling usually takes the form of numerical simulations, and conceptual models based on an understanding of the hydrologic and chemical processes occurring in the subsurface environment. The no-migration petition could take one or both of two forms: "A Fluid Flow Petition," or a "Waste Transformation Petition" [17]. Because quantitative information describing the chemical processes that render waste nonhazardous is usually understood only qualitatively, geochemical arguments supporting the fate or attenuation of hazardous wastes are not normally invoked, and therefore waste transformation petitions are rarely submitted. Instead, most modeling invokes hydrologic arguments to demonstrate confinement over the 10,000-year period. Furthermore, because many parameters used in the models are not precisely known, limiting conservative values are usually selected, leading to modeling results that represent worst-case scenarios. If these results show satisfactory containment, then it can be argued that a more realistic assessment would predict an even smaller likelihood of failure. A flowchart illustrating the permitting process is given in Figure 4.

The EPA regional offices are responsible for reviewing all no-migration petitions for Class I hazardous waste wells. The review process takes the best part of a year to accomplish, owing to the vast quantity of information required, the interdisciplinary nature of the technical arguments presented, and the inevitable challenges regarding the adequacy of information presented. Brasier and Kobelski [13], citing an earlier report [6], noted that industry spent $343,000 in preparation and the EPA dedicated over 2000 employee hours to the review of each demonstration. According to USEPA [17], factoring in the costs for geologic testing and modeling, a no-migration petition can cost in excess of $2,000,000.

Each petition is subject to public notice and comment. Notice of the final decision regarding the petition is published in the Federal Register [17]. The duration of the permit for a Class 1 well does not exceed 10 years. The permit may be reissued for a new term, but the entire permit application must be reopened and is subject to revision.

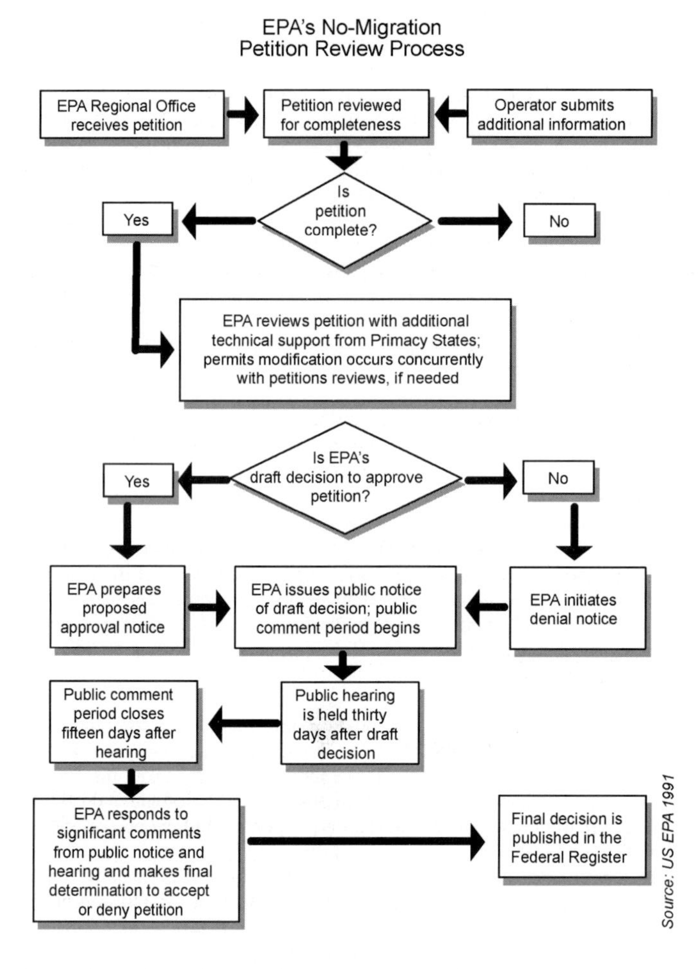

Figure 4: A flowchart illustrating EPA's no-migration review process for Class I deep well injection disposal facilities (from Ref. [16]).

Information required in a permit application. The information needed prior to construction of a disposal well is concerned with the suitability of the subsurface environment to contain the injected waste. Of particular importance is the identification of any potential conduits for the migration of waste, such as other wells in the vicinity, whether operational or abandoned, and geologic faults. To this end, suitable maps and cross sections providing topographic, geologic and hydrologic information are required to show the locations of wells and faults in relation to the proposed injection well and the injection well to USDWs. A plan is required to ensure that any abandoned wells in the vicinity that penetrate the injection zone will be properly plugged prior to waste injection. Details are also required concerning the construction, operation, monitoring, and periodic testing of the well as well as contingency plans in the event of well failure. Finally, EPA requires delivery of a performance bond for its final closure and abandonment.

After construction of the well, EPA requires the submission of the results of a comprehensive test program demonstrating that the well is safe to operate, a description of specific injection procedures, and corrective actions concerning improperly abandoned wells in the vicinity. In addition, an AoR surrounding the injection well must be established.

Siting requirements. All Class I wells must be sited in such fashion that they inject into a formation that is beneath the lowermost formation containing a USDW. Class I hazardous waste injection wells must be restricted to geologically suitable areas. The geology of the area must be described with sufficient confidence that the limits of waste fate and transport can be accurately predicted. The injection zone must have characteristics that prevent migration of fluids into USDWs. The confining zone must also have sufficient structural integrity to prevent the movement of fluids into a USDW, and it must contain at least one formation of sufficient thickness and characteristics to prevent vertical propagation of fractures. Furthermore, the confining zone must be separated from the base of the lowermost USDW by at least one sequence of permeable and less permeable strata, to provide an added layer of protection for the USDW in the event of fluid movement in an undetected transmissive pathway. Finally, within the AoR, the piezometric surface of the fluid in the injection zone must be less than the piezometric surface of the lowermost USDW, or a USDW must be absent.

Construction requirements. The requirements for construction of a Class I well vary somewhat, depending on the nature of the waste, i.e. whether hazardous or nonhazardous, and if nonhazardous, whether or not the waste is treated sewage. The most stringent requirements pertain to wells injecting hazardous waste, and a design incorporating all of the desired features for such wells is illustrated in Figure 5.

All Class I wells must be cased and cemented to prevent the movement of fluids into or between USDWs. Materials used in construction must be designed for the life expectancy of the well and to prevent potential

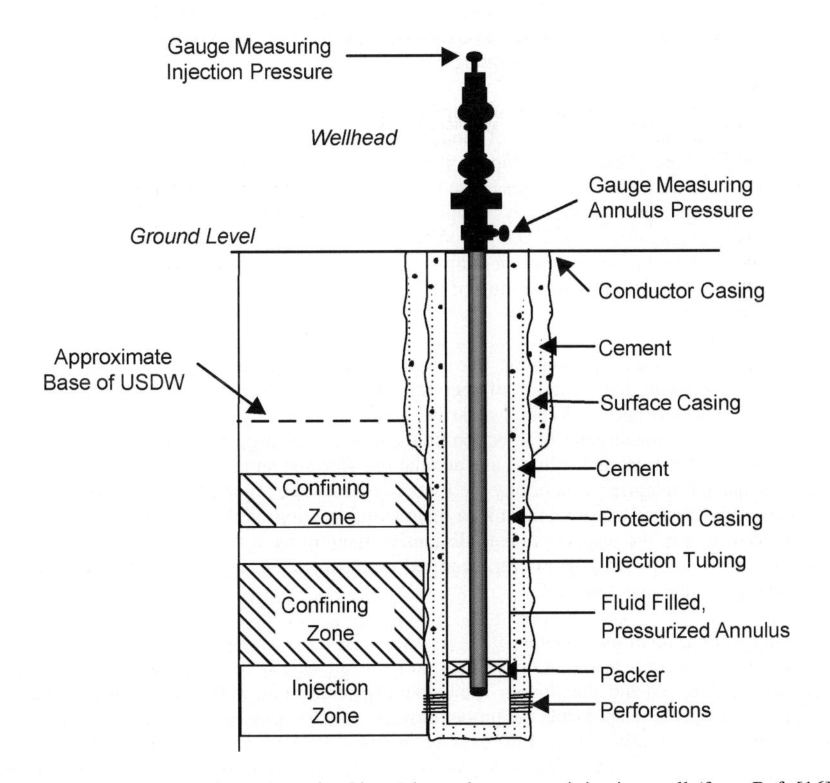

Figure 5: A typical configuration of a Class I hazardous waste injection well (from Ref. [16]).

leaks. Hazardous waste wells must also be constructed and completed to prevent the movement of fluids into any zones other than that selected for injection, and allow continuous monitoring of injection tubing and long-string casing. All Class I injection wells, except those municipal wells injecting noncorrosive wastes, must inject fluids through tubing with a fluid seal or packer set immediately above the injection zone.

Appropriate logs and other tests must be conducted during the drilling and construction of new Class I wells to characterize comprehensively the subsurface environment. Such logs and tests are also needed to establish accurate baseline data against which future measurements may be compared. Numerous logs and tests must also be made to ensure that casing intended to protect USDWs will retain its integrity. Upon completion of a Class I well injecting hazardous waste, pump or injectivity tests must be conducted to verify the hydrogeologic characteristics of the injection zone.

Operating requirements. During operation of a Class I well, the injection pressure at the wellhead must not be so high as to cause hydrofracturing of the injection zone or confining beds, in order to prevent migration of the waste into an overlying USDW. In general, the annulus fluid between the tubing and the casing must be maintained at a suitable overpressure to prevent leakage of waste from the tubing. For hazardous-waste injection, an approved plan for chemical and physical analysis of the waste must be followed. The hazardous waste stream and its anticipated reaction products must not affect the properties of the confining or injection zones.

Monitoring requirements

The operator must monitor both the operation of a Class I well and the volume and composition of the waste stream to ensure compatibility with the well construction materials. He must also monitor injection pressure, and "ambient" conditions in the injection zone, confining beds and adjacent USDWs. For hazardous waste injection, the injection pressure, flow rate and temperature of injected fluids, and the fluid pressure in the annulus, must be monitored and recorded continuously with automatic alarm and shut-off systems in the event of deviation from operating conditions.

The extent of "ambient" monitoring depends upon the potential for fluid movement from the well or injection zone, and on the potential value of monitoring wells to detect such movement. At a minimum, annual monitoring of the pressure buildup in the injection zone is required. EPA may also require continuous monitoring for pressure changes in the first aquifer overlying the confining zone and, if a well is installed, the aquifer must be periodically sampled for chemical analysis. If a well is installed in the lowermost USDW, it must also be analyzed periodically for water quality. If additional wells are available for monitoring contaminant migration or detecting communication with the injection zone through pressure testing, EPA can also mandate the monitoring of these wells.

Reporting requirements

The EPA and state agencies responsible for environmental protection specify minimum reporting requirements for monitoring the operation of deep injection wells. These requirements are specified in 10 CFR Part 146, Subparts B and G. Reports must be submitted quarterly to EPA. They must include the physical and chemical characteristics of injection fluids, monthly average, maximum and minimum values for injection pressure, flow rate and volume, and annular pressure and monitoring results. In addition, results of periodic mechanical-integrity tests or any well workover must be reported to EPA. For wells injecting hazardous waste, the quarterly reports must also include information on the operating conditions and any alarms or shutdowns, and the responses taken. Extensive monitoring is required to ensure the continuing integrity of the well components. However, required ambient monitoring is restricted only to periodic pressure testing using the injection well.

Closure and post-operational monitoring

The EPA is particularly concerned that deep injection wells, especially those that have injected hazardous waste, are properly plugged and abandoned. The owner or operator of a well that injected hazardous waste must, therefore, not only be responsible for proper closure, but also assume responsibility in perpetuity for any contamination to a USDW. At least 60 days beforehand, EPA must be notified of the intended well closure. The owner or operator of a Class I hazardous waste injection well must comply with a closure plan as part of the permit application. Prior to granting approval for the plugging and abandonment, EPA must review details concerning the method of closure and receive information on the casing and any other materials to be left in the well, and any proposed tests or measurements. A Class I hazardous waste well must be plugged with cement in a manner that will not allow the movement of fluids into or between USDWs. Before closing the well, the pressure decay must be recorded and mechanical integrity testing conducted to ensure the integrity of any long-string casing and cement that will be left in the ground after closure.

After closure, the owner must also submit a closure report to EPA. For a Class I hazardous waste well, the owner must prepare and comply with a plan for post-closure care. The plan must also include the predicted position of the waste front at closure, the status of any cleanups required, and the estimated cost of proposed post-closure care. The plan must also assure financial responsibility. The owner or operator must continue groundwater monitoring until pressure in the injection zone decays to the point that waste migration into a USDW would not occur. A number of administrative requirements must be met including submission of a survey plan to the local zoning authority with a copy to EPA, and notification of state and local authorities having cognizance over drilling activities, to enable them to impose appropriate conditions on subsequent drilling activities in the vicinity of the abandoned well.

Reliability of Deep Well Injection Facilities

The purpose of existing federal (and state) regulations and standards governing deep well injection disposal is to ensure protection of USDWs. The effectiveness of these regulations in protecting USDWs serves as a basis for deciding whether corresponding regulations governing CO_2 injection are promulgated, what modifications to these regulations would be necessary to ensure containment of this less dense fluid. This subject can be considered from several perspectives. We can establish from monitoring programs, the frequency and nature of injection facility failures. Such failures can be broadly classified into two categories: those relating to the well failure, and those where the ambient environment has failed to prevent migration of the waste from the injection zone as a result of confining bed failure. The experience thus gained will allow estimates to be made of future system failures. The extent of operator or owner compliance can be assessed, and what enforcement remedies are necessary in the event of noncompliance. Enforcement will in turn ensure the minimization of component failures and increase reliability.

Two studies were conducted prior to 1988 (when more stringent regulations were introduced) to assess the nature and frequency of operational problems [19,20]. According to USEPA [17], the CH2M Hill study identified 26 malfunctions involving 43 wells, suggesting an overall malfunction rate of 9%. Only six wells injecting nonhazardous waste (or 2% of all Class I wells) experienced malfunctions that resulted in contamination of a USDW. The GAO study reported only two cases of USDW contamination and eight cases of contamination of nonpotable aquifers. In all cases, contamination occurred prior to 1980.

Since 1988, when more stringent regulations were introduced, reported substantive incidents involving contamination of formations or USDWs not designated as injection zones have been rare. Of those few known violations since that time, all but one can be traced to actual contamination when less stringent regulations were in force. Contamination of USDWs by Class I nonhazardous waste injection facilities in Florida injecting treated sewage waste is unusual in that violations have occurred more recently. The reasons for the failure of some of the Florida waste wells are discussed elsewhere [21]. These reasons, and the environmental consequences, are relevant to the proposed CO_2 injection into saline aquifers, because the density of the injected sewage, like that of CO_2 is less than the ambient ground waters of the injection zone. Furthermore, the scale of Florida sewage injection would be comparable in size and density to CO_2 injection facilities throughout the United States.

EPA has analyzed mechanical integrity (MI) failures in all Class I wells in selected states between 1988 and 1991 [17]. One hundred and thirty internal MI failures were attributed to leakage from the injection tubing or failure of the long-string casing. Only one external MI failure occurred, involving flow along the outside of the casing. There were four cases of nonhazardous waste migration, three of which were detected by monitoring wells and a fourth during the drilling of a new injection well. A second EPA analysis of MI failures for the time period 1993–1998 [22] showed that the overall rate of failures had declined to half the rate in all states except Texas, where failures increased by 65%.

Cases of noncompliance or violations are handled at a level commensurate with the nature of the violation [28]. However, the extent to which UIC regulations are enforced is not adequately known.

Since 1988, EPA has also pursued, through the US Justice Department, two cases of purported contamination of USDWs through deep well injection. In both cases, the alleged violators were charged with injecting hazardous waste into an aquifer that was claimed by EPA to be a USDW. Continuing operation would, under RCRA, subject each company to substantial fines. In one case, total potential fines could have been several hundred million dollars and would have been, if successfully levied, the largest environmental fine in history. Plans for litigation were abandoned, however, because the evidence was insufficient to make a case that the aquifer was a USDW. In the other case, the company settled with a $3,500,000 fine, which was imposed for both surface and subsurface contamination [23]. In contrast, EPA's response to contamination of potable aquifers in Florida by Class I wells injecting nonhazardous treated sewage was to propose amended regulations rather than mandate cessation of injection and costly remediation.

In summary, the present regulatory climate regarding USDW contamination by waste, whether hazardous or not, is first and foremost to operate a Class I facility in such a manner that the risk of failure is extremely small. Secondly, the operation of the well should be such that failures that can be easily and effectively demonstrated should be tightly regulated. The preferred enforcement approach is through consent decrees and fines, rather than by requiring remediation. The status of deep aquifers in relation to their classification as USDWs is sometimes difficult to establish in the absence of direct evidence from pumping tests. Furthermore, in the event that monitoring wells do suggest contamination, existing remedies can prove to be politically unacceptable.

RESULTS AND DISCUSSION

The experience gained from the regulation of deep well injection of both hazardous and nonhazardous wastes will inevitably be carried over in the regulation of deep injection of supercritical CO_2. There are, however, several major distinctions between CO_2 and hazardous wastes, which will necessitate the promulgation of regulations that recognize these distinctions. Firstly, CO_2 is generally considered to be nonhazardous. Secondly, the volumes of CO_2 being considered for interception prior to atmospheric release exceed by at least two orders of magnitude the quantities of hazardous waste currently being injected. Thirdly, while current regulations require isolation of hazardous waste within the confining zone of an injection well for the lesser of 10,000 years, or that length of time before it is rendered nonhazardous, these restrictions are hardly relevant to supercritical CO_2 injection. Fourthly, regulations governing the diversion of CO_2 production from the atmosphere to the subsurface environment must take into account the extent to which CO_2 can be contained in the subsurface environment over long periods of time and the more stringent monitoring requirements necessary to ensure confinement. Finally, the question arises as to the class under which CO_2 injection wells should be categorized, or whether a special class should be designated. The classification question is complicated by the fact that CO_2 is also used in the tertiary recovery of oil, and its injection for that purpose is regulated under Class II wells.

Compressed carbon dioxide is a relatively benign chemical substance when pure, and is not considered to be hazardous, except under certain environmental conditions. It is limited in its capacity to corrode well tubing and casing, and carbonates casing cements relatively slowly. For these reasons, current regulations specifying the design of nonhazardous injection wells would likely suffice. A critical aspect of well design would be to prevent leakage of CO_2 along the casing and its consequent migration into shallow aquifers or cause surface blowouts. Rigorous monitoring of tubing annulus pressure and proper testing of casing and cement grout integrity would therefore be mandatory. If, however, CO_2 containing significant sulfur dioxide and water vapor were to be injected tubing and casing corrosion could become an issue. Furthermore, the presence of sulfur dioxide could require that the injected gas mixture be subject to regulations similar to those for injection of hazardous waste. Therefore, we can conclude that existing regulations relating to the design and operation of Class I wells could be adapted to CO_2 injection with relatively minor modifications.

If all CO_2 generated by fossil fuel power plants in the United States were to be recovered for sub-surface disposal, approximately 3 Gton/yr would have to be injected [2,24]. Suitable sedimentary formations for CO_2 storage exist under only about half of the land area of the 48 contiguous states, and not all sedimentary formations would be suitable for injection for various geologic, hydrologic and societal reasons. Quantitative estimates of the total volume of CO_2 that could be injected into formations that could contain indefinitely the injected CO_2 have not been made, but rough calculations would suggest that disposal capacity would permit injection for about 500 years. However, although the injection rate would be comparable in tonnage to that of oilfield brines, the latter are essentially refilling a void from which they were previously extracted, whereas CO_2 would eventually require the displacement of a corresponding volume of brine or volumetric compensation by other means.

CO_2 confinement within an injection zone is far more challenging from the regulatory point of view. Supercritical CO_2 is less dense and has substantially lower viscosity than the ambient brine of the injection zone. It will therefore tend to concentrate beneath any confining zone, and spread laterally over a substantial area. In the case of a 1000 MWe plant injecting into a 100 m thick aquifer, the areal extent of the injected plume could attain 120 km^2 [24]. The CO_2 could migrate buoyantly upwards through structural defects into overlying aquifers. The substantial lateral spreading of the CO_2 fluid increases the likelihood that such defects will be encountered. Given the lateral extent to which CO_2 can migrate, it would be important to use whatever means are feasible to monitor potential leakage into formations overlying the injection zone. According to the EPA [28], the installation of monitoring wells is not required for Class I wells, as there was no technology available that would define the siting of these wells. Furthermore, the drilling of multiple monitoring wells into a very deep interval would be prohibitively costly. Available monitoring wells have been limited, mainly to hydrologic testing [25].

Although regulations require containment of hazardous waste within the injection zone of a Class I well for 10,000 years, such a stringent requirement would be unnecessary for the interim storage of CO_2, unless mandated by social, technical and scientific considerations. The argument implicit in CO_2 capture is that this activity would be conducted only until alternative energy sources, conservation, or longer term natural processes would lead to a stabilization of the ambient CO_2 concentration in the atmosphere. Therefore, total confinement of injected CO_2 for very long periods, i.e. over 10,000 years could be unnecessary. Unlike hazardous waste, supercritical CO_2 storage could be considered as a form of interim storage, rather than disposal. Controlled leakage of the stored CO_2 might therefore be acceptable, provided it were sufficiently slow to allow the goals of carbon dioxide capture to be met. However, controlled leakage would invariably result in migration into overlying USDWs where present.

The preceding review is predicated on the assumption that CO_2 injection would be governed by regulations developed and implemented for the protection of USDWs from contamination by injected liquid wastes, particularly those classified as hazardous. However, the cost associated with the petitioning process to obtain a permit for the operation of a Class I hazardous waste injection well is substantial. It can be argued that much of the cost might be avoided through classification of a CO_2 injection well under the Class I nonhazardous category. But in doing so, many of the protections would be lost, which are afforded by the requirements for the injection of hazardous waste, and which would be applicable for the safe operation of a well dedicated to CO_2 injection.

The nonhazardous nature of CO_2 might justify its injection into deep aquifers meeting the definition of a USDW, but which would otherwise be economically unsuitable for exploitation of potable water, and therefore CO_2 injection could be permitted under Class V. Allowing this flexibility would enhance the opportunities for locating injection facilities in regions where USDWs occupy the whole sedimentary sequence above basement rocks, or where aquifers of optimal depth for CO_2 injection, while saline, would still fall within the classification as a USDW. Future regulations could well take into account the potential opportunity to utilize USDWs for injection, provided such sources have no likelihood of being used for potable water. Caution must necessarily be exercised, however, because the consequences of compressed CO_2 penetration into a USDW have not yet been assessed. Although such contamination, whether as a gaseous fluid or dissolved in potable water, may not be particularly hazardous, CO_2 can act as an organic solvent and react with host rock minerals causing some chemical constituents to exceed drinking water standards. Clearly, in formulating future regulations, these issues must be considered.

Given that CO_2 is currently permitted under Class II, a special class designated exclusively for subsurface CO_2 storage might be designated. This class should take into account the current experience relating to Class II injection, and the demonstrated protections afforded by Class I, while permitting restricted operation under what would normally be Class V for the reasons given above. Finally, past experience with off-site hazardous waste facilities indicates that timely input from a concerned public would facilitate acceptance. On-site facilities within the boundaries of existing coal-fired plants would be more likely accepted, particularly if instituted in conjunction with more effective pollution control.

CONCLUSIONS

Regulations designed to protect USDW from contamination by injected industrial and municipal wastes will serve as a basis for regulations governing subsurface disposal of CO_2. Large volumes of CO_2 would require disposal, exceeding by about two orders of magnitude the amount of hazardous liquid waste that is currently injected, only a small portion of which will be diverted for tertiary oil recovery under Class II well regulations. Therefore, injection of the remainder will most likely be subject to federal regulations similar to those promulgated for Class I wells. However, these regulations should not be adopted without modifications to account for the different physical and chemical properties of CO_2 compared with existing wastes disposed by deep well injection.

Pure compressed CO_2 is generally considered in the context of deep well injection disposal to be a nonhazardous noncorrosive chemical compound, and therefore regulations governing the design and operation of nonhazardous Class I wells may be followed. If by-products of the combustion of coal, such as sulfur dioxide and water vapor are not removed special precautions mandated for hazardous waste disposal could be required. The lower density of CO_2 in relation to brines of the injection zone, and its tendency to migrate buoyantly, poses potential problems of containment that are not adequately addressed by Class I well regulations governing the disposal of hazardous waste. Thus, the siting of compressed CO_2 injection wells, like those for the injection of hazardous waste, will be critically dependent on the geology. Furthermore, Class I well hazardous waste ambient monitoring requirements are insufficient to guarantee the protection of USDWs from CO_2 intrusion. Therefore, additional ambient monitoring techniques not currently required for Class I wells may be imposed. The solvent properties and chemical interactions of compressed CO_2 with the subsurface environment and its potential to transport co-contaminants into USDWs are not fully understood. Thus, further study to assess the significance of co-contamination will be necessary before promulgation of regulations could be promulgated.

Opposition to CO_2 injection disposal by environmental groups and local communities could be as vehement as has been the case regarding some off-site Class I well hazardous waste disposal facilities. Prior experience would indicate that on-site disposal at sources of CO_2 generation would mitigate potential opposition and potential litigation. The permitting process for Class I hazardous waste wells under Land-Ban regulations could be adapted to regulate the siting of CO_2 injection wells provided there is timely input to the satisfaction of affected local communities.

RECOMMENDATIONS

If CO_2 deep subsurface storage and storage is to become a reality, the consequence of high pressure CO_2 intruding an overlying USDW must be considered. The buoyant nature of compressed CO_2 could result in preferential migration even to the surface through along faults, fractures and incompletely plugged and abandoned wells. Without a clear understanding of the risks entailed in deep subsurface disposal, the drafting of regulations would necessarily require a degree of conservatism that could inhibit further consideration of this means of managing greenhouse gases. Therefore, first and foremost, comprehensive investigations must be conducted of migration mechanisms and probabilities of USDW contamination by CO_2 and displaced brines from underlying compressed CO_2 reservoirs. Secondly, potential contamination of USDWs by co-contaminants due to the solvent properties and reactivity of compressed CO_2 should be investigated. With such information in hand, appropriate risk analyses should be conducted to assess the probability of barrier failure in the ambient environment. Thirdly, minimum barrier requirements and specifications should be defined for acceptable risk. Finally, minimum ambient monitoring requirements should be formulated to ensure that any potential threat to the integrity of a USDW is identified in a timely manner.

The integrity of natural barriers of the ambient environment and protection of USDWs in the event of failure of these barriers is the most important issue requiring definition. But current practices relating to Class I injection well design construction, operation and monitoring must also be reviewed, and modified to account for the physical and chemical characteristics of compressed CO_2. In particular, the corrosive characteristics of CO_2 containing minor concentrations of sulfur dioxide and water vapor should be investigated, improved casing cement formulations and inclusion of blow-out preventers should be reviewed in the light of current Class I well regulations.

Following the technical definition stage, current regulations governing the injection of both hazardous and nonhazardous waste in Class I wells should be reviewed in relation to CO_2 storage by similar means. Although existing regulations provide a substantial basis for regulating the future disposal of CO_2 will be necessary, and must be debated and evaluated. Consideration should be given to regulations governing CO_2 injection under Class II, and the feasibility of allowing CO_2 injection into USDWs under restricted circumstances. The urgency of such an evaluation will depend on government policy, but in any event, it would be reasonable for EPA either to perform the evaluation in-house, or contract with a suitable company in the private sector.

Finally, due consideration should be given to perceived environmental and social impacts of the subsurface storage of such large anticipated quantities of CO_2. The public comment period for proposed regulations governing CO_2 injection disposal will inevitably become a venue for opponents. However, current regulations should be modified to allow timelier and sufficiently comprehensive input from affected communities. A proactive stance should therefore be taken by evaluating a priori the environmental and social issues arising from CO_2 injection disposal.

ACKNOWLEDGEMENTS

The author also wishes to express his appreciation for fruitful discussions with several of his colleagues in the Earth Sciences Division at the Lawrence Berkeley laboratory, and in particular, Sally Benson, Robert Hepple, Marcelo Lippmann and Chin-Fu Tsang. He would also like to thank staff at USEPA headquarters and at Region VI for discussions concerning UIC regulations. However, the opinions expressed in this chapter are those of the author alone. The preparation of this chapter is supported by the CO_2 Storage Project under the auspices of the US Department of Energy Contract Number DE-AC03-76SF00098.

REFERENCES

1. Ground Water Protection Council, Florida ground water conditions, Ground Water Report to Congress, 2000, pp. 19–20, www.gwpc.site.net/gwreport/states.htm.
2. E.J. Wilson, T.L. Johnson, D.W. Keith, Regulating the ultimate sink: managing the risks of geologic CO_2 storage, *Environ. Sci. Technol.* **37** (2003) 3476–3483.
3. W. Gordon, J. Bloom, Deeper problems: limits to underground injection as a hazardous waste disposal method, Natural Resources Defense Council, 1985, p. 82, Available From NRDC, NY.
4. E.C. Donaldson, Surface disposal of industrial wastes in the United States, US Bureau of Mines Information Circular 8212, 1964.
5. T. Kozlowski, Waste Goes Well Underground, CMA News, November 1997. (Document #2333122), p. 3, 1997, Available from PolicyFax at (312) 377-3000.
6. USEPA, Analysis of the effects of epa restriction on the deep injection of hazardous waste, EPA 570/9-91-031, 1991.

7. Ground Water Protection Council, Underground Injection Fact Sheet, 2001, www.site.net/factshee.htm.
8. H. Sigman, Reforming Hazardous Waste Policy. Essays in Public Policy, Hoover Institution, Stanford, CA, 2001, www-hoover.stanford.edu/publications/epp/93/93a.html.
9. R.D. Brower, I.G. Krapac, B.R. Hensel, A.P. Visocky, G.R. Peyton, J.S. Nealon, M. Guthrie, Evaluation of Current Underground Injection of Industrial Waste in Illinois, Final Draft report HWRIC RR, Hazardous Waste Research and Information Center, State Water Survey Division, Illinois Department of Energy and Natural Resources, 1986.
10. USFWQA, Policy on disposal of wastes by subsurface injection, COM 5040.10, October 15, 1970.
11. E.A. Herbert, The regulation of deep-well injection: a changing environment beneath the surface, *Pace Environ. Law Rev.* **14** (1) (1996) 169–226. www.law.pace.edu/pacelaw/pelr/herbert.htm.
12. USEPA, Subsurface emplacement of fluids by well injection, 39 Fed. Reg. at 12,922, 1974.
13. F.M. Brasier, B.J. Kobelski, Injection of industrial wastes in the United States, in: J.A. Apps, C.F. Tsang (Eds.), Deep injection Disposal of hazardous and Industrial Waste: Scientific and Engineering Aspects, Academic Press, San Diego, CA, 1996, pp. 1–8, Chapter 1.
14. J. Bloom, Resource Conservation and Recovery Act Reauthorization Hearings, supra note 5, at 263 (Statement of Jane Bloom, NRDC) citing General Accounting Office, Hazardous Waste Facilities with Interim Status may be Endangering Public Health and the Environment, 1981. (Cited in Herbert, 1996), 1982.
15. USEPA, State UIC Programs, 2002a, www.epa.gov/safewater/uic/primacy.html..
16. J.E. Clark, An overview of injection well history in the United States, American Institute of Hydrology, *4th USA/CIS Joint Conference*, November 9, Cathedral Hill Hotel, San Francisco, California, 1999.
17. USEPA, Class I underground injection control program: study of risks associated with Class I underground injection wells, Office of Water, EPA 816-R-01-007, 2002b.
18. U.S. General Accounting Office, Deep injection wells. EPA needs to involve communities earlier and ensure that financial assurance requirements are adequate. Report to the Honorable Lynn C. Woolsey, House of Representatives, GAO-03-761, 2003, p. 36.
19. CH2M Hill, A Class I injection well survey, Report Prepared for the Underground Injection Practices Council, 1986.
20. U.S. General Accounting Office, Hazardous waste: controls over injection well disposal operations, Report to Chairman, Environment, Energy, and Natural Resources Subcommittee, Committee on Government Operations, House of Representatives, GAO/RCED-87-170, August 8, 1987.
21. D.F. McNeill, A review of upward migration of effluent related to subsurface injection at Miami-Dade Water and Sewer South District plant, Final Report Prepared for the Sierra Club—Miami Group, 2000, p. 30.
22. ICF Inc., Class I mechanical integrity failure analysis: 1993–1998, Prepared by ICF, Inc., Fairfax, Virginia, for USEPA, Office of Ground Water and Drinking Water, Underground Injection Control Program, 1998.
23. USDOJ, Zeneca Agrees to Pay Civil Penalty, Cease Deep Well Injection, Press release, http://ww.usdoj.gov/opa/pr/1998/388enr.html, 1998, p. 2.
24. C.-F. Tsang, S.M. Benson, B. Kobelski, R.E. Smith, Scientific considerations related to regulation development for CO_2 sequestration in brine formations, *Environ. Geol.* **42** (2002) 275–281.
25. D.L. Warner, Monitoring of Class I injection wells, in: J.A. Apps, C.-F. Tsang (Eds.), Deep Injection Disposal of Hazardous and Industrial Waste: Scientific and Engineering Aspects, Academic Press, San Diego, CA, 1996, pp. 421–432, Chapter 25.
26. USEPA, Deep Wells (Class I), Office of Water, UIC Program, 2001, www.epa.gov/safewater/uic/classi.html.
27. USEPA, Subsurface pollution problems in the united states, technical studies, Report: U.S. EPA Office of Water Programs TS-00-72-02, at 8-9 (May 1972).
28. USEPA, Report to Congress on Injection of Hazardous Waste, EPA 570/9-85-003, With corrections, 2nd printing (May 1985), 1983.
29. D.L. Warner, American Association of Petroleum Geologists, Memoir No. 10, 1968, p. 11.

Carbon Dioxide Capture for Storage in Deep Geologic Formations, Volume 2
D.C. Thomas and S.M. Benson (Eds.)

Chapter 28

PROSPECTS FOR EARLY DETECTION AND OPTIONS FOR REMEDIATION OF LEAKAGE FROM CO_2 STORAGE PROJECTS

Sally Benson[1] and Robert Hepple[2]

[1]Lawrence Berkeley National Laboratory, Berkeley, CA, USA
[2]University of California Davis, Davis, CA, USA

ABSTRACT

Geologic storage projects of CO_2 should be designed to maintain secure storage thousands of years or longer. However, in some cases, leakage may occur and remediation measures, either to stop the leak or to prevent human or ecosystem impacts will be needed. Moreover, the availability of remediation options will reassure the public that geologic storage can be safe and effective and help build confidence in carbon capture and storage.

This study reviews the remediation options available for many of the types of leakage that may occur based on analogous situations in natural gas storage, oil and gas production, groundwater remediation, and soil gas and vadose zone cleanup. Remediation options are discussed for damaged injection wells, leaking abandoned wells, over pressured reservoirs, carbon dioxide accumulations in shallow groundwater, secondary contamination of groundwater by acidification, vadose zone and soil gas accumulations, and surface releases. Examples of remediation options for buildings and surface water are also discussed. This study demonstrates that remediation options are available for many of the leakage scenarios that can be envisioned.

INTRODUCTION

The need for methods of early detection, intervention measures to prevent leakage and remediation of leakage from CO_2 storage projects is a recurrent theme in discussions about the acceptability of geologic storage of CO_2 as an approach to emission reduction. To date, little, if any, research has been done that addresses this issue. The purpose of this study is to identify intervention options to prevent leakage and remediation options that could be used to eliminate or manage risks after leakage has been detected. The approach taken in this study is as follows:

- Identify and develop the leakage scenarios and consequences that are most likely to occur in geologic storage projects (e.g. leakage up abandoned wells, leakage up undetected faults or fractures in the reservoir seal, etc.).
- Calculate a range of hypothetical leakage rates from prototypical storage projects, including those performing effectively and those leaking at unacceptable rates.
- Survey and document remediation practices currently used in natural gas storage, oil and gas production, groundwater and vadose zone remediation.
- Evaluate how and the extent to which existing remediation practices could be employed to remediate leakage in geologic storage projects.
- Identify potential new approaches for remediation of geologic storage projects for scenarios where existing remediation approaches are not sufficient.
- Identify additional knowledge or information needed to develop and build confidence in the effectiveness of new or improved remediation approaches.

LEAKAGE SCENARIOS AND CONSEQUENCES FOR GEOLOGIC STORAGE PROJECTS

To identify options for remediation of CO_2 storage projects it is necessary to first understand potential failure mechanisms and pathways. Potential leakage scenarios are illustrated in Figure 1, which shows the major leakage pathways and potential consequences of leakage. Beginning with the deepest parts of the geologic storage site, we evaluate the principle components of the system and how they may lead to leakage.

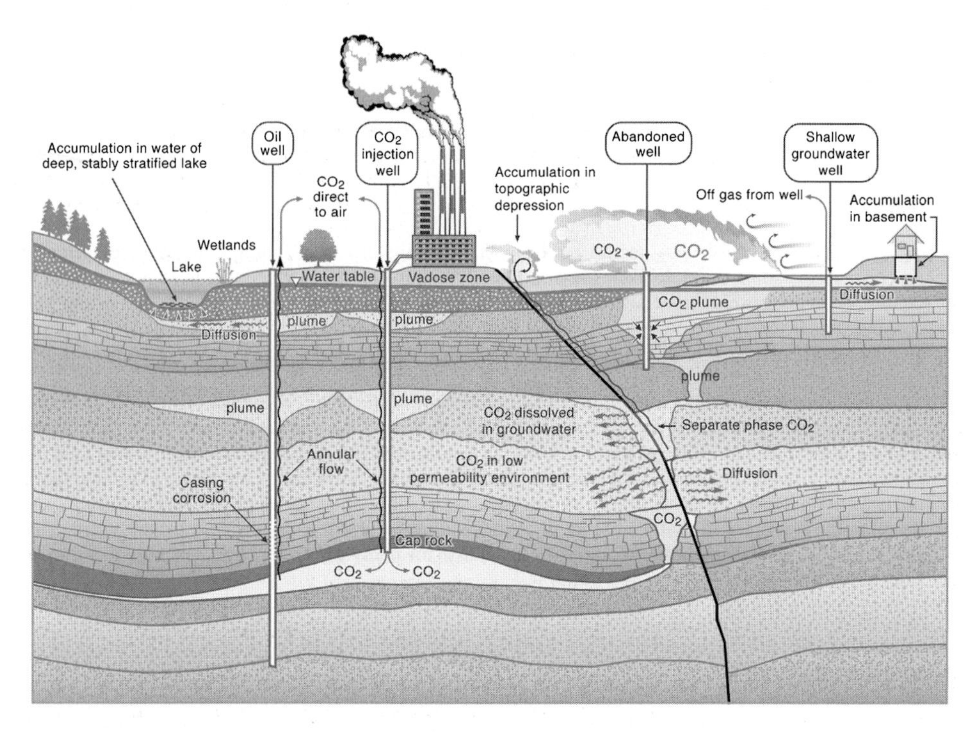

Figure 1: Illustration of potential leakage pathways at poorly selected storage sites and consequences of leakage.

CO_2 will be injected underground using injection wells. Drilling and well completion technology has matured to the point where the wells perform effectively over long periods of time. However, it is possible that poor well construction techniques or inadequate maintenance of those wells can create leakage of CO_2 back to the surface [1–3]. Injection wells may fail in a number of ways: (1) corrosion or mechanical damage to the casing, (2) corrosion or mechanical damage to the injection tubing and packers, and (3) leakage in the annular region between the outside of the casing and the borehole wall. In addition to the injection wells themselves, nearby oil and gas wells, whether used for production, monitoring, injection, or are idle may also provide a leakage pathway towards the surface. One of the greatest uncertainties and potential risk of underground storage is the existence of abandoned wells and the leakage paths they may create [4]. In the past century, millions of wells have been drilled all around the United States, many of which were poorly constructed originally or were never properly plugged and abandoned. Unlike active wells or recently abandoned wells, record keeping on the location and depth of older wells is poor and they will often be difficult to locate. Shallow groundwater wells may also provide a leakage pathway to the surface.

Geologic storage sites will be selected based on careful site investigations which demonstrate the presence of a suitable cap rock or seal for the storage reservoir. The cap rock provides a low permeability barrier that prevents the upward migration of the buoyant CO_2 plume. In some cases, however, it is possible that faults or fractures go undetected or that the CO_2 spreads beyond the intended storage footprint. If the CO_2 plume encounters a permeable fault or fracture it may provide a leakage path of CO_2 towards the surface. CO_2 leakage up faults and fractures is likely to occur as a separate phase. During the process of leakage up faults or fractures, CO_2 may accumulate in secondary, shallower traps such as other subsurface formations or aquifers.

As the CO_2 approaches the land surface, a number of different atmospheric discharge scenarios can be envisioned:

- CO_2 could be directly discharged to the atmosphere through a well;
- CO_2 may accumulate in the vadose zone and be released by advection and diffusion across the land surface;
- in regions with a very shallow water table CO_2 could discharge directly to the atmosphere;
- if overlain by water, CO_2 would dissolve in the ocean or a lake and discharge by diffusion to the atmosphere. If the leakage rate was sufficiently high, CO_2 may bubble through the water column and be discharged to the atmosphere directly. Under unusual circumstances found in deep lakes in the tropics, CO_2 could accumulate in the bottom waters of a deep lake or at sufficient depths in the ocean, possibly leading to eventual eruption following supersaturation or some triggering event—as evidenced at Lake Nyos. If the water was deep and cold enough, the CO_2 may form hydrates that accumulate on the sea floor;
- if overlain by a building, CO_2 could accumulate in the basement or subfloor, leading to the build-up of potentially dangerous concentration of CO_2.

The local health, safety and environmental consequences of each of these discharge scenarios will depend on the size and rate of the release. For example, Oldenburg et al. [5] demonstrate that even large releases from the vadose zone to the atmosphere are unlikely to result in unsafe concentrations of CO_2 in the atmosphere because atmospheric mixing rates are high enough to quickly dilute the CO_2. However, these same scenarios show that even for low release rates, high concentrations of CO_2 can build-up in the soils, exceeding concentrations known to damage vegetation.

POTENTIAL RELEASE RATES FROM LEAKING STORAGE PROJECTS

Estimating the amount of CO_2 that may be released from a failing storage project is an important starting point for assessing detection and remediation options. Two scenarios are examined: (1) a small-scale project storing the emissions from an oil refinery that emits 1 Mt CO_2 per year and (2) a larger scale project storing emissions from a 500 MW coal-fired power plant that emits 3.6 Mt/year. In each case we assume that CO_2 is injected at a constant rate for a 50-year period. For these projects, we calculate how much CO_2 would be released if 0.01, 1 and 10% of the cumulative amount of CO_2 that had been stored was released over a 1-year period. The release rates were selected to cover a broad range of values, not because they are based on actual or calculated release rates from any particular project. Table 1 summarizes the quantity of CO_2 that would be released during a 1-year period after 1, 10 and 50 years of injection.

TABLE 1
SUMMARY OF THE QUANTITY OF CO_2 THAT WOULD BE RELEASED (IN MT/YEAR) FOR LEAKING CO_2 STORAGE PROJECTS FOR LEAKAGE RATES OF 0.01, 1 AND 10% OF THE TOTAL AMOUNT STORED

Scenario	**Refinery (1 Mt CO_2/year)**			**500 MW power plant (3.6 Mt/year)**		
Leakage rate (%stored/year)	0.01	1	10	0.01	1	10
1 year (Mt/year)	0.0001	0.01	0.1	0.00036	0.036	0.36
10 years (Mt/year)	0.001	0.1	1.0	0.0036	0.36	3.6
50 years (Mt/year)	0.005	0.5	5.0	0.0186	1.86	18.6

To put these release rates into perspective it is useful to compare them to the flux of CO_2 that is associated with the natural cycling of CO_2 between the atmosphere and the biosphere. The maximum range of natural ecosystem fluxes between the land surface and the atmosphere is from 10 $\mu mol/m^2/s$ efflux during the peak of night-time respiration to 30 $\mu mol/m^2/s$ drawdown during maximal daytime photosynthesis at the height of the growing season but more typically would be 0.5–2 $\mu mol/m^2/s$ efflux and 2–10 $\mu mol/m^2/s$ drawdown [6,7]. To directly compare the hypothetical releases from a storage project to these natural fluxes, the area over which the release occurs must be specified. The footprint of the underground plume of CO_2 created by a storage project of this magnitude is expected to be on the order of 50–300 km^2 (equivalent to a radius of 5–10 km), based on capacity estimates developed by Doughty et al. [8]. One meaningful comparison is to assume that the release of CO_2 is evenly distributed over the entire footprint of the plume. For this case the fluxes are provided in Figure 2. As shown, for all but the release rate of 10% per year, these fluxes are much lower than or in the range of the natural ecosystem fluxes. This indicates that for release rates of 1% or less that are evenly distributed over the footprint of the plume, detection of the releases would be difficult. For release rates of 10% or greater, the fluxes are significantly greater than the natural fluxes and consequently would be easily detected and likely to have observable ecosystem impacts (see discussion in Ref. [5]).

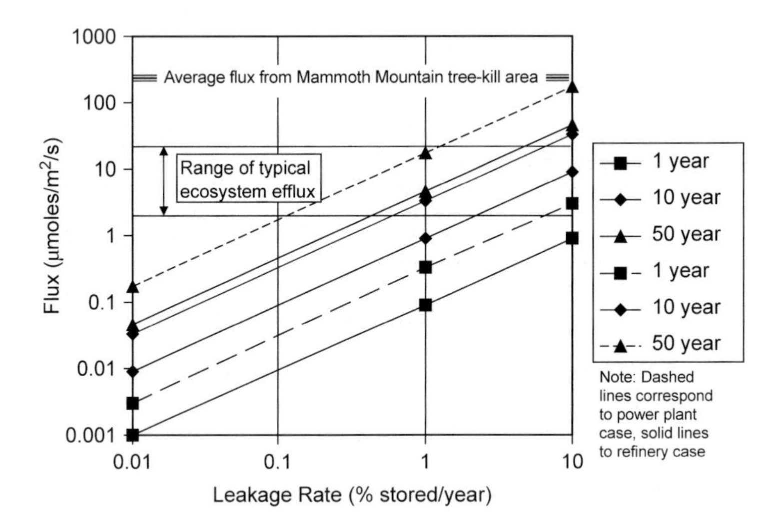

Figure 2: Fluxes of CO_2 for the scenarios listed in Table 1 assuming that the releases are eventually distributed over the footprint of a plume with a radius of 5 km.

In reality, it is very unlikely that a storage site would have leakage that is evenly distributed over the entire footprint of the plume. Instead, as described in the previous section, leakage would be concentrated around aging wells, faults that compromise the integrity of the cap rock, or storage structures with incomplete closure. In each of these cases, leakage would be concentrated within a limited area and therefore, fluxes could be significantly higher than the values shown in Figure 2. To investigate the leakage fluxes for some other scenarios, Figure 3a,b show the estimated flux of CO_2 for each leakage rate at the end of the 50-year project lifetime for four surface release scenarios: radial flux zones with 10 m (area of 300 m^2) and 100 m radius (area of 30,000 m^2) and 1 km linear flux zones either 1, 10, or 100 m wide (areas of 1000, 10,000 and 100,000 m^2, respectively). Note that in all these cases the surface release is concentrated in a very small fraction of the overall dimension of the plume of CO_2 in the storage formation.

For all of the cases provided in Figure 3a,b, the flux is orders of magnitude greater than the natural ecological flux. In fact, at 1% seepage, the fluxes range from 3–6 orders of magnitude greater than background ecological levels and would be greater than those seen at Mammoth Mountain at the height of the flux in the early 1990s—1000–1200 tonnes of CO_2 per day. The area of associated tree-kill is 170 acres (6.9×10^5 m^2) (USGS, http://lvo.wr.usgs.gov/CO2.html, 2003;) leading to an average flux of approximately 400 μmol/m^2/s. Using eddy flux correlation, Anderson and Farrar [9] measured the CO_2 flux as between 180 and 360 μmol CO_2/m^2/s. These calculations suggest that leakage from medium to large size projects, even at rates as low as 0.01%/year could be easily detected if it is confined to relatively small areas because the flux is so much greater than the background rate.

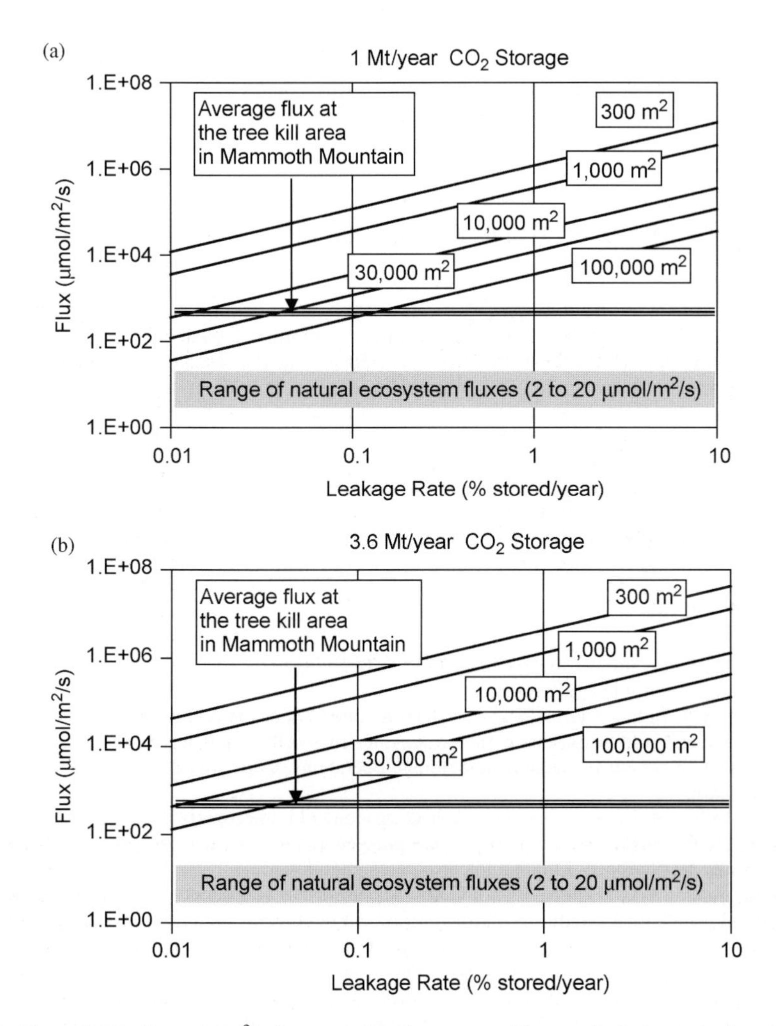

Figure 3: Flux of CO_2 in μmol/m^2/s for several leakage rates. These calculations are for release rates after 50 years of CO_2 injection. (a) Fluxes for a 1 Mt/year storage project. (b) Fluxes for a 3.6 Mt/year storage project.

METHODS FOR EARLY DETECTION OF LEAKAGE

A wide variety of approaches are available for monitoring geologic CO_2 storage operations and potential leakage. Several chapters in this volume summarize these monitoring methods and will not be repeated here [10–16]. Suffice it to say that monitoring approaches are available to monitor the progress of a storage project as the storage reservoir volume fills up and potentially detects leakage long before it approaches the land surface. Early detection of leakage from the storage reservoir will provide the opportunity for intervention and remediation before large amounts of CO_2 are released back into the atmosphere. If CO_2 has migrated to the land surface and is discharged to the atmosphere, methods are available to monitor both the flux and concentration of CO_2 in the atmosphere, soils and surface water [10,11]. As described above, if the release is confined to a limited area, as we expect it to be, fluxes from leaking CO_2 storage projects will be much higher than the natural background flux of CO_2. The high flux, relative to the background rate will make detection relatively easy using commercially available equipment, if the location of the release is known. Subsurface monitoring techniques should be effective in locating the areas where surface leaks are likely to occur. Similarly, understanding of the geologic setting should also provide an indication of the places where leakage is most likely to occur. Remote sensing may also be useful for locating leaks by detecting ecosystem stress from elevated soil-gas CO_2 [12]. In future, remote sensing may also be used to directly monitor CO_2 concentrations if the technology can be improved enough to have sufficient resolution.

REVIEW OF THE RESPONSE TO LEAKAGE IN NATURAL GAS STORAGE PROJECTS

Natural gas storage projects provide a good modern analogue for storage of CO_2 in underground geological formations. The technology for injection, well construction and monitoring is very similar to that which would be used for geological storage of CO_2. It is natural, therefore, to look to these projects for insight into remediation options. In the US alone there are over 450 natural gas storage projects. Of these, a small number have experienced leakage [2]. The projects listed below provide information on how the leakage was managed and what kinds of approaches were used for remediation.

- *Herscher-Galesville, IL.* In mid-1953, several months after natural gas was first pumped into the Galesville formation, bubbles of gas appeared in shallow water wells in the Herscher field. Wells were drilled around the periphery of the field to remove water and thereby minimize the pressure build-up. The water was then reinjected into the Potosi Dolomite (above the Galesville) in order to pressurize the shallower formation. By carefully monitoring the differential pressures and recycling gas from several vent wells in other still shallower formations, the Herscher-Galesville natural gas storage project has been active for almost 50 years. To this day, the cause of the leakage is still not known with certainty [17].
- *Leroy, WY.* At this gas storage site, gas was observed bubbling to the surface; it was reportedly controlled by limiting maximum injection pressures.
- *East Whittier, CA.* In the 1970s, storage gas had migrated out of the original storage footprint and was being produced and sold by another company from a neighboring lease, which according to DOGGR is not an uncommon reason for abandoning storage operations. Currently, this field is in the process of being shut down.
- *West Montebello, CA.* In the 1970s, gas was leaking along old, improperly plugged wells to a shallower zone but not to the surface. Problem wells were plugged and the gas that trapped in the shallower zone may eventually be produced. This field is also being shut down.
- *McDonald Island, CA.* On 17 May 1974, PG&E lost control of a new injection/withdrawal well, Whiskey Slough 14 W, which then caught fire. While pulling out of hole, the well fluid level apparently dropped and was not monitored. The fire was extinguished and the well was controlled after 19 days by drilling a relief well and killing the blowout with heavy mud [18].
- *Indiana.* In the 1960s and 1970s, many water wells in northern Indiana were contaminated with natural gas from a shallow storage aquifer. Under current regulations, such a project would not be allowed.
- *Hutchinson, KS.* Natural gas from the Yaggy gas storage project leaked from an injection/withdrawal well. The storage structure is composed of several mined salt caverns at least 150 m deep. The leaked

gas migrated seven miles to the town of Hutchinson through a 20 ft zone with several dolomite layers interspersed with shale. Within the town, it then flowed up and erupted from old, unplugged wells that no one had known about and that had been used for salt solution mining many decades ago (KGS, www.kgs.ku.edu/PRS/Poster/2002/2002-44/, 2002). Remediation was accomplished by plugging the injection/withdrawal well and the abandoned wells. After this, 12 wells were drilled into the shallow aquifer to intercept gas accumulations and vent the gas into the atmosphere. High-resolution seismic imaging was successfully used to locate gas accumulations in the aquifer and to guide siting the wells used to vent the gas [19]. In addition, electromagnetic and high-resolution magnetic methods were used to locate additional abandoned wells that could provide leakage pathways in the future [20].

These projects demonstrate that each time when a leak was detected, an intervention or remedy was put in place that stabilized the situation. In most cases, the project was able to continue, but in a few others, a decision was made to terminate the project. In the following section, these and other remediation techniques for geologic storage projects are presented.

REMEDIATION OPTIONS FOR LEAKING GEOLOGIC STORAGE PROJECTS

In the process of exploring past practices and potential failure scenarios, seven problems have been identified that require remediation: (1) leaks from the storage reservoir; (2) leakage from active or abandoned wells; (3) contamination of shallow groundwater; (4) vadose zone and soil contamination; (5) localized surface fluxes; (6) leakage of carbon dioxide in indoor spaces, especially basements; and (7) leakage into surface water. Remediation options for each of these types of leaks are described below and summarized in Table 2. In some cases the methods are well established. In others, they are more speculative, but with appropriate research and development, may nevertheless one day become feasible.

Leakage from the Storage Reservoir

There are three basic approaches to stopping leakage from the storage reservoir: (1) the pressure in the storage formation can be reduced; (2) the pressure in the formation into which leakage is occurring can be increased; or (3) the CO_2 plume can be intercepted and extracted from the reservoir before it leaks out of the storage structure. Lowering the pressure in the storage reservoir will help reduce or stop leakage in two ways. First, lowering the pressure will reduce the pressure gradient driving the CO_2 out of the storage reservoir. Second, if faults or fractures have become leakage pathways as a result of the pressure build-up in the storage reservoir, lowering the pressure can mitigate this [21]. Increasing the pressure in the formation into which CO_2 is leaking will decrease the pressure gradient that is causing the storage structure to leak. Finally, extraction of the CO_2 plume before it leaks will directly intercept and prevent leakage. Techniques for accomplishing these three approaches are listed below.

- Lower the reservoir pressure by injecting at a lower rate or through more wells [17].
- Lower the reservoir pressure by removing water or other fluids from the storage reservoir.
- Lower the reservoir pressure by creating a pathway to access new compartments in the storage reservoir, e.g. hydrofracture or a well completion open to two storage zones.
- Increase the upgradient pressure by injecting water or brine ahead of the leak.
- Stop injection in order to lower the reservoir pressure and stabilize the project.
- Stop injection, produce the CO_2 from the storage reservoir and reinject it back into a more suitable storage structure.
- Drill extraction (pumping wells) wells in the vicinity of the leak to stop the leakage and capture the CO_2 before it leaks out of the storage structure. Reinject the CO_2 at a more suitable location.

Leakage from Active or Abandoned Wells

Methods for repairing active and abandoned wells are used on a routine basis in the oil, gas, natural gas storage and waste disposal industries. In addition, for newly drilled wells, federal, state and local regulations have been developed to ensure that wells are drilled and completed safely and will not harm groundwater or other resources. These techniques can be employed to remediate leaking wells in CO_2 storage projects.

Examples of these techniques are provided below.

- Repair leaking injection wells with standard well recompletion techniques such as replacing the injection tubing and packers.
- Repair leaking injection wells by squeezing cement behind the well casing to plug leaks behind the casing.
- Plug and abandon injection wells that cannot be repaired by the methods listed above.
- Stop blowouts from injection or abandoned wells using standard techniques to "kill" a well such as injecting a heavy mud into the well casing. After control of the well is reestablished, the recompletion or abandonment practices described above can be used. If the wellhead is not accessible, a nearby well can be drilled to intercept the casing below the ground surface and "kill" the well by pumping mud down the interception well [18].

Groundwater Remediation

Groundwater remediation methods that may be useful for CO_2 leakage can be categorized as: (1) passive, (2) active, and (3) those meant to deal with contamination caused by dissolution of secondary minerals as a result of groundwater acidification from CO_2. Passive methods utilize natural attenuation of the CO_2 by dissolution in groundwater, dilution and mineralization. Monitoring is used to confirm that the hazard is being remedied at an acceptable rate. Active methods involve injection or extraction of fluids to accelerate removal or stabilization of the CO_2. The most commonly employed method, "pump and treat", removes the groundwater from the aquifer and treats it at the surface to remove the unwanted impurities. For CO_2, this could include both gas phase pumping and groundwater extraction. For gas that remains trapped as a residual and immobile phase, groundwater extraction could also be used to dissolve the plume of CO_2. The greatest need for remediation from the effects of CO_2 leakage may be for removal of elements mobilized by the dissolution of minerals, e.g. arsenic (As) and lead (Pb) caused by acidification of groundwater from CO_2. In addition to the "pump and treat" approach mentioned above, flow-through treatment barriers may be effective for removal of trace elements mobilized by groundwater acidification [22]. Another possible method is to contain the plume of contaminated water by managing hydraulic heads and preventing the flow of contaminated waters [23]. Examples of these approaches are provided below.

- Accumulations of gaseous CO_2 in groundwater can be removed, or at least made immobile, by drilling wells that intersect the accumulations and extract the CO_2. The extracted CO_2 could be vented to the atmosphere or reinjected back into a suitable storage site.
- Residual CO_2 that is trapped as an immobile gas phase can be removed by dissolving it in water and extracting it as a dissolved phase through groundwater extraction wells.
- CO_2 that has dissolved in the shallow groundwater could be removed, if needed, by pumping to the surface and aerating it to remove the CO_2. The groundwater could then either be used directly or reinjected back into the aquifer.
- If metals or other trace contaminants have been mobilized by acidification of the groundwater, "pump-and-treat" methods can be used to remove them. Alternatively, hydraulic barriers can be created to immobilize and contain the contaminants by appropriately placed injection and extraction wells. In addition to these active methods of remediation, passive methods that rely on natural biogeochemical processes may also be used. Treatment walls designed to remove the trace elements could also be used.

Vadose Zone

Vadose zone remediation is a mature field. Similar to groundwater remediation, there is a basic distinction between passive and active methods such as soil vapor extraction (SVE). Passive methods rely on diffusion from the vadose zone to the atmosphere or natural biogeochemical processes to remove the unwanted substance. Passive removal can also be enhanced by using the natural diurnal fluctuations in atmospheric pressure to accelerate diffusive fluxes or accelerated even more by the use of "BaroBalls" [5,24]. Passive methods have the advantage of being less expensive but typically take much longer than active methods.

Active methods of vadose zone remediation that might be applicable to CO_2 removal are generally variations of the industry standard SVE, sometimes with covers or sprinkling/irrigation [25]. The basic mechanism behind SVE is flushing fresh air through the soil and extracting soil gas. SVE systems can be optimized by using vertical wells, horizontal wells, drainage systems or trenches for collection of soil gas, and surface facilities would include a vacuum pump or blower, moisture knockout and treatment facilities. As an alternative, for soluble substances such as CO_2, sprinkling or irrigation can be used to dissolve and move them downward into shallow groundwater which can then be diluted by the groundwater or processed by pump-and-treat. Another alternative method, which is commonly used in landfills, is to cover the surface with an impermeable barrier and install a collection system below the cover. Examples of vadose remediation techniques are provided below.

- CO_2 can be extracted from the vadose zone and soil gas using standard vapor extraction techniques from horizontal or vertical wells.
- Fluxes from the vadose zone to the ground surface could be decreased or stopped using caps or gas vapor barriers. Pumping below the cap or vapor barrier could be used to deplete the accumulation of CO_2 in the vadose zone.
- Since CO_2 is a dense gas it could be collected in subsurface trenches. Accumulated gas could be pumped from the trenches and released to the atmosphere or reinjected back underground.
- Passive remediation techniques that rely only on diffusion and "barometric pumping" could be used to slowly deplete one-time releases of CO_2 into the vadose zone. This method will not be effective for managing ongoing releases because it is relatively slow.
- Acidification of the soils from contact with CO_2 could be remediated by irrigation and drainage. Alternatively, agricultural supplements such as lime could also be used to neutralize the soil.

Large Short Duration Releases of CO_2 to the Atmosphere

Large short duration releases of CO_2 to the atmosphere can be managed by passive or active dilution in the atmosphere. For example

- For releases inside a building or confined space, large fans could be used to rapidly dilute CO_2 to safe levels.
- For large releases spread out over a large area, dispersion from natural atmospheric mixing (wind) will be the only practical method for diluting the CO_2.
- For recurrent or ongoing leakage in confined spaces (e.g. cellar around a wellhead) fans could be used to keep the rate of air circulation high enough to ensure adequate dilution.

Indoor Environments and Basements with Chronic Low Levels of Leakage

The remediation of indoor air contaminants has been studied extensively for volatile organic compounds and radon. Leaking CO_2, like radon, would enter the building from the subsurface, so the remediation techniques for radon should be directly applicable to elevated CO_2 in the indoor environment. The two major techniques used today are subsurface or subslab pressurization and subslab depressurization with venting. Schematics of these two processes are shown in Figure 4. Subslab pressurization pumps outside air into the basement or area beneath the foundation slab; this flushes fresh air through the near-building soil air and disperses contaminants (VOC, Rn, CO_2) away from the building. Subslab depressurization pumps air from beneath the foundation slab out the top of the building; the decreased pressure beneath the slab pulls atmospheric air through the soil and flushes out contaminants. Both methods induce airflow through the near-building soil gas in order to disperse contaminants [26–28].

Venting Systems to Remove CO_2 from Deep Stably Stratified Lakes

The most catastrophic natural disaster known to be directly caused by CO_2 was the 1986 incident at Lake Nyos in Cameroon. Approximately 1700 people and many thousands of cattle were killed in this event in which an enormous amount of CO_2 dissolved in the bottom waters of a crater lake exsolved, flowed down the narrow drainage, and suffocated almost all animals in its path for many miles. The CO_2 had built up slowly through time and was measurably increasing again in the years following the 1986 disaster. An international coalition of scientists studied the problem and one group of them, headed by Michel Halbwachs, designed, built and tested a system for degassing Nyos and a smaller nearby lake called Monoun with a similar CO_2 build-up. Figure 5 depicts the degassing principle, which relies on a controlled natural gas lift through a small diameter pipe (http://perso.wanadoo.fr/mhalb/nyos/project/annexes/safety.PDF). The process is now underway and expected to make Lake Nyos and Lake Manoun safe within the next several years.

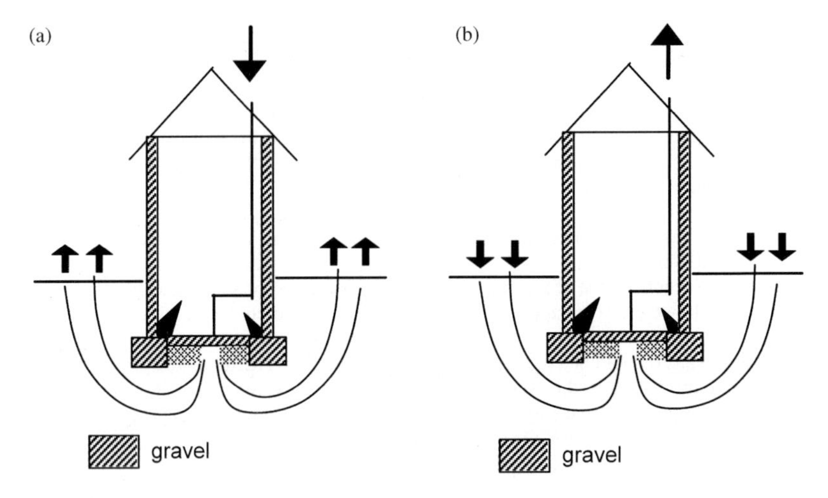

Figure 4: (a) The schematic on the left shows a conceptual representation of a subslab pressurization (ssp) system. (b) The diagram on the right depicts a conceptual representation of a subslab depressurization (ssd) system (after Ref. [26]).

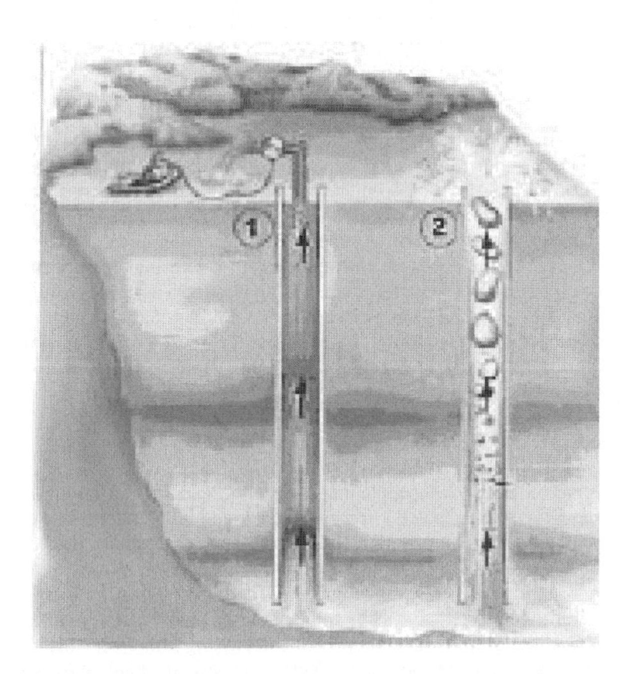

Figure 5: Degassing procedure principle: 1, priming the self-siphon using an electrical pump; 2, autonomous soda fountain (http://perso.wanadoo.fr/mhalb/nyos/).

TABLE 2
OPTIONS FOR REMEDIATION OF LEAKAGE FROM THE STORAGE FORMATION

Scenario	Remediation options
Leakage from the storage reservoir	Lower injection pressure by injecting at a lower rate or through more wells [17]
	Lower reservoir pressure by removing water or other fluids from the storage structure
	Intersect the leakage with extraction wells in the vicinity of the leak
	Create a hydraulic barrier by increasing the reservoir pressure upstream of the leak
	Lower the reservoir pressure by creating a pathway to access new compartments in the storage reservoir
	Stop injection to stabilize the project
	Stop injection, produce the CO_2 from the storage reservoir and reinject it back into a more suitable storage structure
Leakage from active or abandoned wells	Repair leaking injection wells with standard well recompletion techniques such as replacing the injection tubing and packers
	Repair leaking injection wells by squeezing cement behind the well casing to plug leaks behind the casing
	Plug and abandon injection wells that cannot be repaired by the methods listed above
	Stop blowouts from injection or abandoned wells using standard techniques to "kill" a well such as injecting a heavy mud into the well casing. After control of the well is reestablished, the recompletion or abandonment practices described above can be used. If the wellhead is not accessible, a nearby well can be drilled to intercept the casing below the ground surface and "kill" the well by pumping mud down the interception well [18]
Leakage into shallow groundwater	Accumulations of gaseous CO_2 in groundwater can be removed, or at least made immobile, by drilling wells that intersect the accumula tions and extract the CO_2. The extracted CO_2 could be vented to the atmosphere or reinjected back into a suitable storage site
	Residual CO_2 that is trapped as an immobile gas phase can be removed by dissolving it in water and extracting it as a dissolved phase through groundwater extraction wells
	CO_2 that has dissolved in the shallow groundwater could be removed, if needed, by pumping to the surface and aerating it to remove the CO_2. The groundwater could then either be used directly, or reinjected back into the groundwater
	If metals or other trace contaminants have been mobilized by acidification of the groundwater, "pump-and-treat" methods can be used to remove them. Alternatively, hydraulic barriers created to immobilize and contain the contaminants by appropriately placed injection and extraction wells. In addition to these active methods of remediation, passive methods that rely on natural biogeochemical processes may also be used
Leakage into the vadose zone andaccumulation in soil gas [25]	CO_2 can be extracted from the vadose zone and soil gas using standard vapor extraction techniques from horizontal or vertical wells

(*continued*)

TABLE 2
CONTINUED

Scenario	Remediation options
	Fluxes from the vadose zone to the ground surface could be decreased or stopped using caps or gas vapor barriers. Pumping below the cap or vapor barrier could be used to deplete the accumulation of CO_2 in the vadose zone
	Since CO_2 is a dense gas it could be collected in subsurface trenches. Accumulated gas could be pumped from the trenches and released to the atmosphere or reinjected back underground
	Passive remediation techniques that rely only on diffusion and "barometric pumping" could be used to slowly deplete one-time releases of CO_2 into the vadose zone. This method will not be effective for managing ongoing releases because it is relatively slow
	Acidification of the soils from contact with CO_2 could be remediated by irrigation and drainage. Alternatively, agricultural supplements such as lime could also be used to neutralize the soil
Large releases of CO_2 to the atmosphere	For releases inside a building or confined space, large fans could be used to rapidly dilute CO_2 to safe levels
	For large releases spread out over a large area, dilution from natural atmospheric mixing (wind) will be the only practical method for diluting the CO_2
	For ongoing leakage in established areas, risks of exposure to high concentrations of CO_2 in confined spaces (e.g. cellar around a wellhead) or during periods of very low wind, fans could be used to keep the rate of air circulation high enough to ensure adequate dilution
Indoor environments with chronic low level leakage	Slow releases into structures can be eliminated using techniques that have been developed for controlling release of radon and volatile organic compounds into buildings. The two primary methods for managing indoor releases are basement/substructure venting or pressurization. Both would have the effect of diluting the CO_2 before it enters the indoor environment [26]
Accumulation in surface water	Shallow surface water bodies that have significant turnover (shallow lakes) or turbulence (streams) will quickly release dissolved CO_2 back into the atmosphere
	For deep, stably stratified lakes, active systems for venting gas accumulations have been developed and applied at Lake Nyos and Monoun in Cameroon (http://perso.wanadoo.fr/mhalb/nyos/)

CONCLUSIONS

This study has demonstrated that early detection of CO_2 leakage should be possible and that many remediation options are available that could be applied in the event that leakage occurs, specifically;

- the most probable scenarios for leakage of CO_2 have been identified;
- it has been demonstrated that even small leaks should be detectable if they are confined to limited areas around leaking wellbores, faults or fractures; and
- Remediation options are available that could be used to reduce or stop leakage and control the environmental, health and safety impacts of unintended releases.

While this study has identified many promising options, it must be recognized that remediation of subsurface systems is always expensive, fraught with difficulties and success is not always certain. Additional detailed studies are needed to further assess the feasibility of applying these to geologic storage projects—based on more realistic scenarios, simulations and field studies. In particular, we recommend carrying out controlled release experiments and experiments at sites with natural CO_2 seeps to confirm our ability to detect and remedy leakage.

RECOMMENDATIONS

The study presented here focused largely on identifying existing options for remediating geologic storage projects. While these have been employed with considerable success in analogous situations, particularly for gas storage and cleanup of contaminated soils and groundwater, additional options tailored specifically for geologic storage of CO_2 would be helpful. For example, injecting chemical additives that would increase the dissolution of CO_2 into the in situ fluids or accelerate mineral trapping could be used as a remediation measure. Methods for increasing the extent of residual gas trapping, potentially with surfactant-based foams, could also be useful. Similarly, it may be possible to use foam to block weaknesses in the reservoir seal, at least on a temporary basis. These and other methods should be investigated and tested in real-world situations.

With regard to early detection of leaks, experiments should be conducted to test and improve methods for leak detection. All the techniques for leak detection presented in this book should be employed in concert to determine which of them or combination of them is most effective. Moreover, it is expected that some techniques will be more effective in different geologic settings. Therefore, it is important to repeat these tests in the range of geologic settings where they will be employed.

ACKNOWLEDGEMENTS

The authors gratefully acknowledge the support of the CCP for supporting this work. We also thank the two peer reviewers who provided constructive comments that have significantly improved this paper.

REFERENCES

1. J. Apps, The regulatory climate governing the disposal of liquid wastes in deep geologic formations: a paradigm for regulations for subsurface disposal of CO_2, in: S.M. Benson (Ed.), Carbon Dioxide Capture for Storage in Deep Geologic Formations – Results from the CO_2 Capture Project, Volume 2—Geologic Storage of Carbon Dioxide with Monitoring and Verification, Elsevier, London, 2004.
2. K. Perry, Natural gas storage industry experience and technology: potential application to CO_2 geological storage, in: S.M. Benson (Ed.), Carbon Dioxide Capture for Storage in Deep Geologic Formations – Results from the CO_2 Capture Project, Volume 2—Geologic Storage of Carbon Dioxide with Monitoring and Verification, Elsevier, London, 2004.
3. S.M. Benson, R. Hepple, J. Apps, C.-F. Tsang, M. Lippmann, Lessons Learned from Natural and Industrial Analogues for Storage of Carbon Dioxide in Deep Geological Formations, Lawrence Berkeley National Laboratory (LBNL) Report No. 51170, 2002.
4. R.G. Bryant, A. Guswa, M. Celia, C. Peters, Safe storage of CO_2 in deep saline aquifers, *Environ. Sci. Technol.* **36** (11) (2002) 240A–245A.
5. C.M. Oldenburg, A.J. Unger, Modeling of near-surface seepage of CO_2 for risk characterization, in: S.M. Benson, et al. (Eds.), Carbon Dioxide Capture for Storage in Deep Geologic Formations – Results from the CO_2 Capture Project, Volume 2—Geologic Storage of Carbon Dioxide with Monitoring and Verification, Elsevier, London, 2004.
6. G.B. Drewitt, et al., Measuring forest floor CO_2 exchange in a coastal temperate rainforest, *Agric. For. Meteorol.* **110** (2002) 299–317.
7. I.A. Janssens, A.S. Kowalski, R. Ceulemans, Forest floor CO_2 fluxes estimated by Eddy covariance and chamber-based model, *Agric. For. Meteorol.* **106** (2001) 61–69.

8. C. Doughty, S.M. Benson, K. Pruess, Capacity investigation of brine-bearing sands for geologic sequestration of CO_2, in: J. Gale, Y. Kaya (Eds.), *Proceedings of the Sixth International Conference on Greenhouse Gas Control Technologies*, GHGT-6, September 30–October 4, Kyoto, Japan, vol. 2, 2002, pp. 1645–1648.
9. D.E. Anderson, C.D. Farrar, Eddy covariance measurements of CO_2 flux to the atmosphere from an area of high volcanogenic emissions, Mammoth Mountain, California, *Chem. Geol.* **177** (2001) 31–41.
10. R. Arts, P. Winthaegen, Monitoring options for CO_2 storage, in: S.M. Benson, et al. (Eds.), Carbon Dioxide Capture for Storage in Deep Geologic Formations – Results from the CO_2 Capture Project, Volume 2—Geologic Storage of Carbon Dioxide with Monitoring and Verification, Elsevier, London, 2004.
11. S.M. Benson, L. Myer, Monitoring to Ensure Safe and Effective Geologic Storage of Carbon Dioxide, Intergovernmental Panel on Climate Change (IPCC) Workshop on Carbon Sequestration, November 18–22, Regina, Saskatchewan, Canada, 2002.
12. N.L. Miles, K.J. Davis, J.C. Wyngaard, Detecting leaks from belowground CO_2 reservoirs using Eddy covariance, in: S.M. Benson, et al. (Eds.), Carbon Dioxide Capture for Storage in Deep Geologic Formations – Results from the CO_2 Capture Project, Volume 2—Geologic Storage of Carbon Dioxide with Monitoring and Verification, Elsevier, London, 2004.
13. G.M. Hoversten, E. Gasperikova, Non-seismic geophysical approaches to monitoring, in: S.M. Benson, et al. (Eds.), Carbon Dioxide Capture for Storage in Deep Geologic Formations – Results from the CO_2 Capture Project, Volume 2—Geologic Storage of Carbon Dioxide with Monitoring and Verification, Elsevier, London, 2004.
14. W.L. Pickles, W.A. Cover, Hyperspectral geobotanical remote sensing for CO_2 storage monitoring, in: S.M. Benson, et al. (Eds.), Carbon Dioxide Capture for Storage in Deep Geologic Formations – Results from the CO_2 Capture Project, Volume 2—Geologic Storage of Carbon Dioxide with Monitoring and Verification, Elsevier, London, 2004.
15. P.J. Shuler, Y. Tang, Atmospheric CO_2 Monitoring Systems—A Critical Review of Available Techniques and Technology Gaps: Report for SMV Group, CCP Project Report, 2002.
16. P. Shuler, Y. Tang, Atmospheric CO_2 monitoring systems, in: S.M. Benson, et al. (Eds.), Carbon Dioxide Capture for Storage in Deep Geologic Formations – Results from the CO_2 Capture Project, Volume 2—Geologic Storage of Carbon Dioxide with Monitoring and Verification, Elsevier, London, 2004.
17. T.C. Buschbach, D.C. Bond, Underground storage of natural gas in illinois—1973, Illinois Petroleum 101, Illinois State Geological Survey, 1974.
18. DOGGR (California Department of Oil, Gas, and Geothermal Resources), Sixtieth Annual Report of the State Oil and Gas Supervisor, Report No. PR06, pp. 51–55, 1974.
19. E.S. Nissan, J. Xia, W.L. Watney, Seismic detection of shallow natural gas beneath Hutchinson, Kansas, *Kansas Geological Survey Open File Report* 2002-44, 2002.
20. J. Xia, Natural gas explosions in Hutchinson, Kansas: locating abandoned brine wells be high-resolution magnetic and electromagnetic survey, GSA, North-Central Section—*37th Annual Meeting*, March 24–25, Kansas City, Missouri, 2003.
21. S. Wo, J.-T. Liang, CO_2 Storage in coalbeds: CO_2/N_2 injection and outcrop seepage modeling, in: S.M. Benson (Ed.), Carbon Dioxide Capture for Storage in Deep Geologic Formations – Results from the CO_2 Capture Project, *Volume 2—Geologic Storage of Carbon Dioxide with Monitoring and Verification*, Elsevier, London, 2004.
22. R.D. Vidic, F.G. Pohland, Treatment walls, GWRTC Report, TE 96-01, University of Pittsburg, Pittsburg, PA, 1996.
23. I. Javandel, C. Doughty, C.F. Tsang, Groundwater Transport: Handbook of Mathematical Models, Water Resources Monograph 10, American Geophysical Union, 1984, 228pp.
24. C.M. Lewis, R. Van Pelt, Natural Remediation at Savannah River Site, WSRC-MS-2002-00075, Bechtel Savannah River Inc., Aiken, SC, 2002.
25. B. Looney, R. Falta, Vadose Zone Science and Technology Solutions: Volume II, Batelle Press, Columbus, OH, 2000.
26. A.J. Gadgil, Y.C. Bonnefous, W.J. Fisk, Relative effectiveness of sub-slab pressurization and depressurization systems for indoor radon mitigation: studies with an experimentally verified numerical model, *Indoor Air* **4** (1994) 265–275.

27. M.L. Fischer, A.J. Bentley, K.A. Dunkin, A.T. Hodgson, W.W. Nazaroff, R.G. Sextro, J.M. Daisy, Factors affecting indoor air concentrations of volatile organic compounds at a site of subsurface gasoline contamination, *Environ. Sci. Technol.* **30** (10) (1996) 2948–2957.
28. B.H. Turk, R.J. Prill, W.J. Fisk, D.T. Grimsrud, R.G. Sextro, Effectiveness of radon control techniques in fifteen homes, *J. Air Waste Manage Assoc.* **41** (5) (1991) 723–734.

Carbon Dioxide Capture for Storage in Deep Geologic Formations, Volume 2
D.C. Thomas and S.M. Benson (Eds.)

Chapter 29

MODELING OF NEAR-SURFACE LEAKAGE AND SEEPAGE OF CO_2 FOR RISK CHARACTERIZATION

Curtis M. Oldenburg[1] and André A.J. Unger[2]

[1]Lawrence Berkeley National Laboratory, Berkeley, CA, USA
[2]University of Waterloo, Waterloo, Ontario, Canada

ABSTRACT

The injection of carbon dioxide (CO_2) into deep geologic CO_2 storage sites entails risk that CO_2 will leak away from the primary storage formation and migrate upwards to the unsaturated zone from which it can seep out of the ground. We have developed a coupled modeling framework called T2CA for simulating CO_2 leakage and seepage in the subsurface and in the atmospheric surface layer. The results of model simulations can be used to calculate the two key health, safety, and environmental (HSE) risk drivers, namely CO_2 seepage flux and near-surface CO_2 concentrations. Sensitivity studies for a subsurface system with a thick unsaturated zone show limited leakage attenuation resulting in correspondingly large CO_2 concentrations in the shallow subsurface. Large CO_2 concentrations in the shallow subsurface present a risk to plant and tree roots, and to humans and other animals in subsurface structures such as basements or utility vaults. Whereas CO_2 concentrations in the subsurface can be high, surface-layer winds reduce CO_2 concentrations to low levels for the fluxes investigated. We recommend more verification and case studies be carried out with T2CA, along with the development of extensions to handle additional scenarios such as calm conditions, topographic effects, and catastrophic surface-layer discharge events.

INTRODUCTION

The injection of carbon dioxide (CO_2) into deep geologic formations for CO_2 storage involves the risk that CO_2 will unexpectedly leak away from the target formation and migrate generally upward eventually reaching the shallow subsurface where CO_2 could seep out of the ground. In the near-surface environment, defined here roughly as within 10 m of the ground surface either above or below ground, high concentrations of CO_2 can pose significant health, safety, and environmental (HSE) risks. The assessment of HSE risks is an essential part of public acceptance, planning, and permitting of geologic CO_2 storage projects. Risk assessment in general can be divided into three parts: (1) definition of scenarios of what can go wrong; (2) assessment of the likelihood of those scenarios; and (3) assignment of a measure of severity to the consequences arising from a given scenario. When applying this approach to substances that pose a hazard to human health and ecosystems, the risk assessment process includes hazard identification and risk characterization. For geologic CO_2 storage, a recognized HSE hazard is CO_2 leakage and seepage from the storage site leading potentially to exposure by humans, plants, and animals to elevated CO_2 concentrations in air and water. Risk characterization requires the estimation or calculation of elevated CO_2 concentrations to which humans, plants, and animals may be exposed in the given failure scenarios. The research described here focuses on calculating CO_2 concentrations and fluxes using a coupled subsurface and atmospheric surface-layer numerical simulator.

A formal and consistent terminology is needed to describe the different modes of CO_2 migration. We define *leakage* as migration away from the primary storage formation, whereas *seepage* is CO_2 migration through an interface such as the ground surface, a basement floor or wall, or the bottom of a body of surface water. In Figure 1, we present a schematic of some of the important features that may affect HSE risk characterization for CO_2 leakage and seepage in the near-surface environment, a region that we define as within approximately 10 m of the ground surface, either below (i.e. in the subsurface) or above (i.e. in the atmospheric surface layer). These features include a house with a basement and cracked floor through which CO_2 can seep, and a water well through which water with high dissolved CO_2 content could be produced if CO_2 leaked up through the aquifer. Also shown are plants, a tree, and roots that may be sensitive to elevated CO_2 concentrations in the shallow subsurface. We also show animals that live in the ground and therefore may be susceptible to elevated CO_2 concentrations in soil, along with their burrows that may provide fast-flow paths for CO_2 that enhance mixing by barometric pumping of soil gas and ambient air. Snow cover or ice (not shown) can also affect CO_2 flow and transport. In addition, we show in Figure 1 the saturated zone, unsaturated zone, surface water, and wind in the atmospheric surface layer all of which may be capable of diluting and attenuating leaking and seeping CO_2.

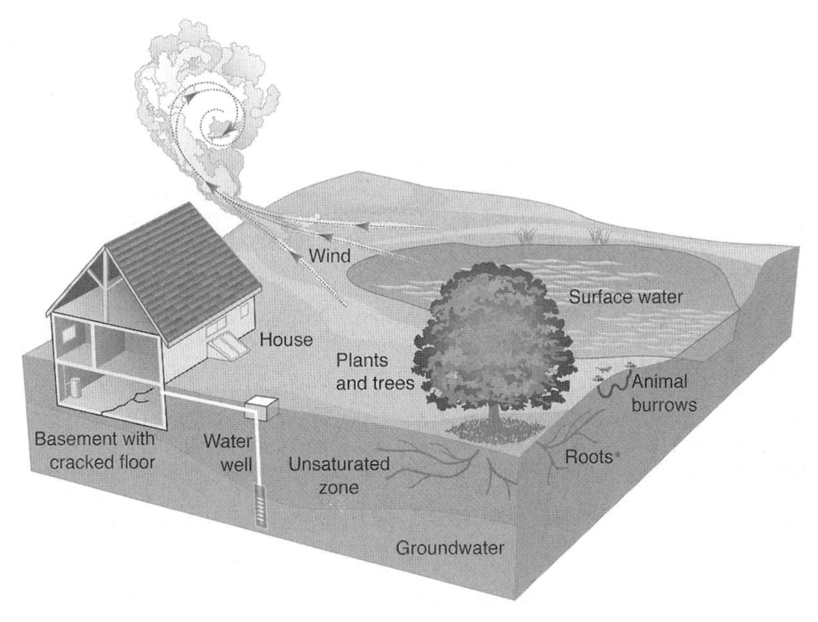

Figure 1: Sketch of near-surface environment with accompanying features relevant to HSE risk associated with CO_2 leakage and seepage.

In this chapter, we summarize our research into the development and demonstration of the coupled modeling framework T2CA applicable to the leakage and seepage of CO_2 from geologic carbon storage sites. The purpose of the coupled model is to calculate CO_2 fluxes and concentrations in the near-surface environment where risk to humans, plants, and animals is highest. The underlying premise of our approach is that the fundamental drivers of the HSE risk are the CO_2 flux and near-surface CO_2 concentrations, and that a capability to calculate these quantities is essential for a defensible HSE risk assessment. A new coupled model is required because to our knowledge there is no existing model that handles both subsurface and atmospheric surface-layer transport and dispersion along with the coupling at the subsurface–surface-layer interface at length scales of order $10^2 – 10^3$ m. The focus of our approach is on diffuse and low-level leakage that could occur through the natural barriers in the subsurface as opposed to catastrophic leakage such as may occur through abandoned wells or well blowouts.

METHODOLOGY

Key Concepts

The methodology and structure of the coupled modeling framework are based on the following key concepts: (1) the human, plant, and animal receptors span the interface between the subsurface and surface layer; (2) the flow processes involved in leakage and seepage are coupled; and (3) the main risk drivers are CO_2 flux and concentration. Before describing the methods and structure, we elaborate on these three key concepts and discuss the time and length scales appropriate to our approach.

First, HSE risk assessment applies to humans, plants, and animals. These environmental receptors live generally near the ground surface but may be entirely below, entirely above, or in both regions at different times. As examples of the importance of the subsurface, surface layer, and in-between environments, consider the house and basement and the burrows of animals shown in Figure 1. Clearly the house and the burrow are open to gas flow from both the subsurface and surface layer and therefore CO_2 in either the subsurface or surface layer has the potential to affect the environment in which people or animals live. The plants, trees and their roots similarly will be affected by CO_2 leakage and seepage in both the subsurface and surface-layer environments. Because exposure to CO_2 in the near-surface environment is the main risk associated with CO_2 leakage and seepage, we have developed a coupled modeling framework that focuses on this region.

Second, CO_2 leakage and seepage are coupled transport processes. Specifically, CO_2 gas in the near-surface environment will flow by advection and diffusion as controlled by pressure, density, and concentration gradients. For example, seeping CO_2 will be strongly advected by surface winds above the ground surface, while atmospheric pressure variations (i.e. barometric pumping) will cause CO_2 to move in and out of the subsurface. However, the low permeability of soils will tend to dampen subsurface advective transport driven by pressure variations and wind in the surface layer. Rainfall infiltration containing dissolved CO_2 can be another mechanism for CO_2 to return from the surface layer to the subsurface. Because these apparent coupled processes occur between the surface layer and subsurface, a coupled modeling framework capable of modeling these interactions is required.

Third, if high CO_2 concentrations are the fundamental adverse condition for HSE risk, then CO_2 seepage flux and near-surface CO_2 concentration are the main risk drivers. Seepage flux in terms of mass has units of kg CO_2 m^{-2} s^{-1} and is a measure of the rate at which CO_2 is passing out of the ground per unit area. If CO_2 is the only component of the gas stream seeping out of the ground, then flux and concentration are directly correlated. However, if the CO_2 is contained within a stream of another component (e.g. with steam in a geothermal vent), then there can be a high CO_2 flux with low CO_2 concentrations. In this sense, flux and CO_2 concentration must be considered independently. In the case where the only component in the seeping gas is CO_2, the seepage flux is a good indicator of whether the given surface-layer winds, surface-water flows, or plant uptake rates are capable of reducing CO_2 concentrations to near-ambient levels. Annual leakage rates given as percentages per year of given CO_2 storage projects should not be used for characterizing risk since they do not provide information on the form or nature of the leakage process. As for CO_2 concentrations, the location of the occurrence of high concentration and nature of the receptor control the attendant risk. For example, high CO_2 concentrations at a depth of 1 m in the ground may cause negligible risk to humans because people live mostly above the ground surface, while such concentrations would pose a serious risk to burrowing animals or to plants through exposure to their roots.

Given these key concepts, it is apparent that a quantitative coupled modeling capability is required to make defensible estimates of CO_2 flux and concentration for various expected leakage and seepage scenarios. Overly simplified models of the subsurface or surface layer alone may not stand up to public and scientific scrutiny. We have used a methodology and structure that is based on multiphase and multicomponent reservoir simulation. The fluxes and concentrations calculated by the coupled framework can be used as inputs to exposure models to calculate defensible HSE risks. The direct output from the present coupled modeling framework is also useful by itself since CO_2 flux and concentration are primary risk drivers. The approach we have taken can be used to model the whole leakage pathway from deep storage site to the surface, but here we focus the model description on the region where the main HSE hazards occur, namely the near-surface environment containing the unsaturated zone and surface layer.

Length and Timescales

With CO_2 storage operations potentially occurring on a large and widespread industrial scale, the length and timescales of interest to CO_2 risk characterization are quite large. Because broad and diffuse CO_2 seepage may occur over large areas for long periods of time, such leakage and seepage may be hard to detect and difficult to mitigate. As such, diffuse seepage is an important focus for risk assessment and risk management. Catastrophic events such as well failures are also relevant, but such events are obviously serious HSE risks and everything possible will be done to stop such events. We have focused on the 10–10^3 m length scale, and the 1 month to 10-year timescale consistent with the diffuse seepage scenario that is our focus. Over these length and timescales, averaging is defensible. For example, constant wind speed, pressure, rainfall infiltration, and other weather-related processes can be used along with appropriate parameterizations since the timescale is relatively long. While the coupled model is capable of nonisothermal simulations, we consider here only isothermal situations and we parameterize turbulence using variable-K theory to model atmospheric dispersion.

Subsurface Flow and Transport

The coupled modeling framework we are using is built on the TOUGH2 code [1], a multiphase and multicomponent integral finite difference reservoir simulator. Briefly, TOUGH2 uses a multiphase version of Darcy's law for fluid flow and the advective–dispersive model for component transport. Readers interested in greater detail and information on the theory or practical implementation of TOUGH2 should consult the user's guide [1] and the website (http://www-esd.lbl.gov/TOUGH2). The coupled model handles five components (H_2O, brine, CO_2, a gas tracer, air) and heat. Air is a pseudocomponent that is approximated as a mixture of 21% oxygen and 79% nitrogen by volume. Real gas mixture properties are calculated so the full range from high-pressure storage-site conditions to low-pressure ambient surface-layer conditions can be modeled. We refer to the coupled model as T2CA, for *T*OUGH2 for *CO*$_2$ and *A*ir. While the discussion below focuses on the CO_2 transport, all the gas-phase components are modeled in the TOUGH2 multicomponent framework, and an analogous treatment can be developed for heat.

Atmospheric Dispersion

The approach we use for atmospheric surface-layer transport is based on gradient transport and variable-K theory [2]. In this approach, the advection and dispersion of CO_2 are modeled with an advective–dispersive transport equation in which advection is unidirectional in the x-direction, velocity varies with height according to the logarithmic velocity profile applicable for neutral stability conditions, and the dispersivities K_y and K_z model eddy diffusion. The advective–dispersive transport equation with x-axis aligned with the unidirectional flow field can be written for CO_2 concentration (c) as

$$\frac{\partial c}{\partial t} + u\frac{\partial c}{\partial x} - \frac{\partial}{\partial y}\left(K_y\frac{\partial c}{\partial y}\right) - \frac{\partial}{\partial z}\left(K_z\frac{\partial c}{\partial z}\right) = 0 \tag{1}$$

where we assume advection dominates transport in the x-direction. The logarithmic velocity profile for neutral stability conditions [2,3] is given by the equation

$$u(z) = \frac{u_*}{k}\ln\left(\frac{z}{z_0}\right) \tag{2}$$

where u_* is the friction velocity, k the von Karman's constant ($k = 0.4$), z_0 the roughness length, and z the height above the ground surface. Turbulent eddies act to disperse gaseous components, and these eddies become larger with elevation above the ground surface. Arya [2] recommends use of an increasing K_z with elevation for neutral stability conditions according to:

$$K_z = ku_*z \tag{3}$$

Lateral dispersion K_y in variable-K theory is less well understood, and we avoid consideration of how to parameterize K_y by adopting a 2D model problem that neglects lateral dispersion of CO_2 and will therefore be conservative in that CO_2 concentrations will be overestimated relative to a case with lateral dispersion.

In summary, for atmospheric dispersion in the surface layer we use variable-K theory and assume neutral stability and a logarithmic velocity profile. The logarithmic velocity profile represents time-averaged surface winds to model advection in the surface layer, with turbulent mixing parameterized by a variable K_z. The velocity field in the surface layer is prescribed as an initial condition and stays constant throughout the simulation. The surface layer is defined simply by setting porosity to unity and layer permeabilities to a range of values, orders of magnitude larger than the subsurface parts of the domain and that specify the desired logarithmic profile for the given boundary conditions. The entire coupled subsurface–surface-layer calculation is carried out using a single grid. Hence, the model regions are implicitly coupled. Full multiphase and multicomponent flow and transport are used throughout the domain.

Field experiments of dense gas dispersion have been used to develop correlations involving the most important parameters controlling atmospheric dispersion such as wind speed, density of released gas, and release flux [4,5]. These correlations were developed based on simple scale and dimensional analyses. One of these correlations relates the seepage flux and average wind speed at an elevation of 10 m to the form of the dispersion process, i.e. whether it is density-dependent or passive (not density-dependent) as appropriate for a gas tracer. In density-dependent dispersion of a dense gas like CO_2, the gas can flow in response to its own density gradient relative to air, and it can resist mixing if contained in a low-lying area such as a valley or other topographic depression. In Figure 2, we have plotted this correlation with values appropriate for CO_2–air mixtures for various source-area length scales along with the typical ecological flux of CO_2 emitted and taken up by plants, soil, and roots known as the net ecosystem exchange (NEE) [6]. As shown in Figure 2, seepage fluxes have to be quite high (note logarithmic scale) for windy situations for the resulting dispersive mixing process to be density-dependent. Note that wind conditions are averages over a period of 10 min.

In prior work [7], we have simulated subsurface migration of leaking CO_2 through the unsaturated zone with rainwater infiltration for various leakage rates specified at the water table. These leakage rates were given as annual mass leakage percentages of the total stored CO_2 of the order of 10^9 kg through a circular region with radius 100 m. Typical seepage fluxes for the 0.1% yr^{-1} leakage rate were of the order of 10^{-5}–10^{-6} kg m^{-2} s^{-1}. As shown in Figure 2, seepage fluxes of this magnitude lead to passive dispersion for all but the calmest wind conditions. It must be emphasized that deriving a leakage rate from annual percent leakage is case-specific in that doing so produces a leakage rate that is dependent on the mass of stored CO_2, i.e. the size of the storage project. For example in this case, if the project were 100 times larger (stored CO_2 of the order of 10^{11} kg), seepage fluxes of the order of 10^{-5}–10^{-6} kg m^{-2} s^{-1} would result from leakage rates of 0.001% yr^{-1} for the same leak geometry. Similarly, 0.1% yr^{-1} leakage from a project 100 times larger would produce fluxes of order 10^{-3}–10^{-4} kg m^{-2} s^{-1} for the same geometry, which could produce density-dependent dispersion at higher wind speeds as shown in Figure 2. In general, the CO_2 leakage and seepage flux are the important quantities governing flow behavior, while percent leakage per year provides information only about mass loss and requires definition of the project size and leakage or seepage area.

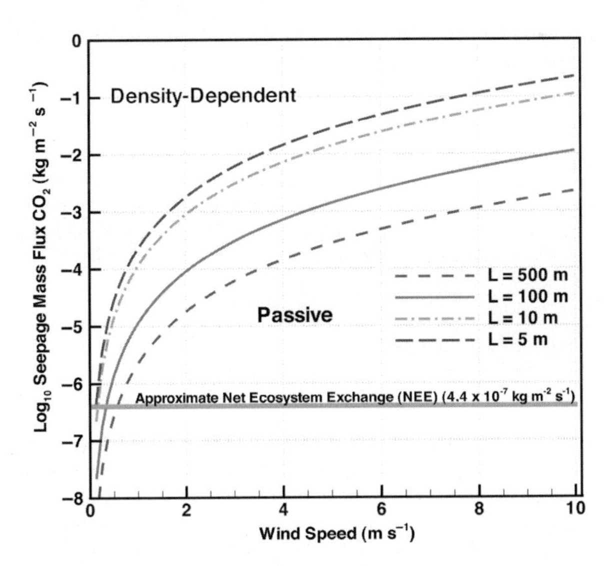

Figure 2: Correlation for density-dependent and passive dispersion in the surface layer as a function of seepage flux and wind speed for four different source length scales.

RESULTS AND DISCUSSION

Verification

The subsurface flow and transport methods in T2CA are well established by virtue of the long history of TOUGH2, the novel part being the equation of state module for the mass components water, brine, CO_2, gas tracer, and air. We have compared physical properties of the gas mixtures in T2CA against independent predictions and observed good agreement [7]. Real gas mixture properties are calculated because they are needed at depth, e.g. below approximately 800 m where CO_2 becomes supercritical, although our focus here is on the unsaturated zone and surface layer where pressures are approximately 1 bar (0.1 MPa).

Here we present verification of the surface-layer methods in T2CA for the special case of uniform velocity and constant eddy diffusivity, in which the approach reduces to the well-known Gaussian plume dispersion model for which there are simple analytical solutions. We present in Figure 3 results of a verification study in which we compared the T2CA result of a 3D Gaussian plume dispersion problem against the analytical solution. In this problem, $u = 1\ \mathrm{m\ s^{-1}}$, $D_{xx} = D_{yy} = D_{zz} = 5\ \mathrm{m\ s^{-2}}$. The point-source strength $Q_{1/4} = 0.0785\ \mathrm{kg\ s^{-1}}$, where $Q_{1/4}$ is the source strength for the one-quarter domain used in the T2CA simulation that takes advantage of the symmetry planes in the horizontal and vertical directions parallel to the flow direction. The main part of Figure 3 shows the 3D plume, while the upper inset shows the y–x plane with comparison of the T2CA result to the analytical solution given by Arya [2]. The agreement is very good and confirms our implementation of surface layer atmospheric dispersion processes in T2CA.

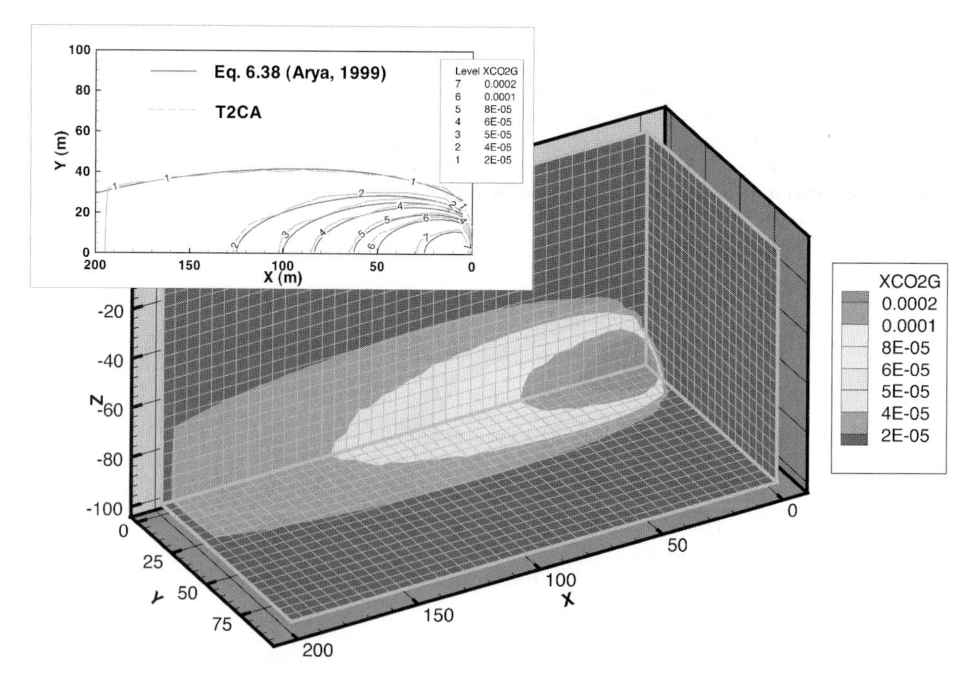

Figure 3: Contours of kg CO_2 $\mathrm{m^{-3}}$ gas from T2CA for the Gaussian plume dispersion verification problem in 3D, and comparison to analytical solution in the x–y plane (inset).

Unsaturated Zone Attenuation

The purpose of this application is to examine the extent to which the unsaturated zone can attenuate CO_2 leakage, full details of which can be found in Ref. [7]. We consider a radial system with a thick (30 m) unsaturated zone into which a CO_2 leakage flux enters from below. The leakage fluxes are arbitrarily set at 4.04×10^{-6}, 4.04×10^{-7}, and 4.04×10^{-8} kg m^{-2} s^{-1}. For reference, a leakage flux of 4.04×10^{-6} kg m^{-2} s^{-1} would correspond to an annual loss through a 100 m radius region of 0.1% yr^{-1} of a 4×10^{9} kg CO_2 storage project, or 0.001% yr^{-1} of a project 100 times bigger (4×10^{11} kg CO_2). We point this out to emphasize again that leakage flux rather than annual percentage loss controls leakage and seepage processes. In some cases leakage rate and leakage flux will be loosely correlated because leakage area may scale with size of project, but in general these quantities represent distinct measures of storage integrity. The leakage area was one of the many properties of the system that was varied as part of the sensitivity analysis discussed below. Rainfall infiltration flows downward through the section and acts to dissolve CO_2 and transport it downward. Additional properties of the system for the base case are provided in Table 1.

Figure 4 shows the steady-state simulation results for the base case at the three different arbitrary leakage rates 4×10^{6}, 4×10^{5} and 4×10^{4} kg yr^{-1}. Steady state is reached after approximately 0.3, 5, and 30 yrs for the three cases, respectively. Carbon dioxide concentrations in the shallow subsurface increase with increasing leakage rate, as diffusion and the specified rainfall infiltration are overwhelmed by larger leakage fluxes. Note further the limited degree to which the CO_2 spreads outward in the unsaturated zone despite the density contrast. Pressure gradients induced by the active leakage flux dominate over gravity effects here and thus lead to predominantly vertical CO_2 flow through the vadose zone to the ground surface [7].

Figure 5 shows seepage flux and near-surface CO_2 concentration (mole fraction) for a large number of simulations carried out as part of a sensitivity analysis [7]. For reference, we have plotted the typical ecological flux or NEE 4.4×10^{-7} kg m^{-2} s^{-1} [6] and the soil–gas CO_2 mole fraction ($x_{gas}^{CO_2} = 0.3$) that appears to have caused tree mortality at Mammoth Mountain, California [8]. As shown, the leakage flux exerts the strongest control on flux and concentration at the ground surface. Permeability and permeability anisotropy are also very important in controlling CO_2 seepage flux and near-surface concentrations. Simulations of barometric pumping presented in prior work [7] show that pressure variations produce local temporal changes in flux and concentration but have little effect on long-term average values for this leakage scenario. The fundamental observation of the simulation results presented here is that subsurface CO_2 concentrations from leakage and seepage can be high in the near-surface environment, even when the fluxes are of the same order of magnitude as the NEE [6].

TABLE 1
HYDROGEOLOGICAL PROPERTIES OF THE UNSATURATED ZONE FOR THE BASE CASE

Property	**Value**
Permeability ($k_r = k_Z$)	1×10^{-12} m^2 (1 Darcy)
Porosity (ϕ)	0.2
Infiltration rate (i)	10.0 cm yr^{-1}
Temperature (T)	15 °C
Residual water saturation (S_{lr})	0.1
Residual gas saturation (S_{gr})	0.01
van Genuchten [10] α	1×10^{-4} Pa^{-1}
van Genuchten [10] m	0.2

Subsurface–Surface-Layer Coupling

We have also applied the new simulation capability to a coupled subsurface–surface-layer cartesian system, properties of which are listed in Table 2. The domain discretization and boundary conditions are shown in Figure 6. The bottom boundary is held at constant pressure, while the top boundary is closed. The side boundaries are closed in the unsaturated zone, and held at constant pressure in the surface layer to prescribe the logarithmic velocity profile. Further details of our modeling approach and this application can be found

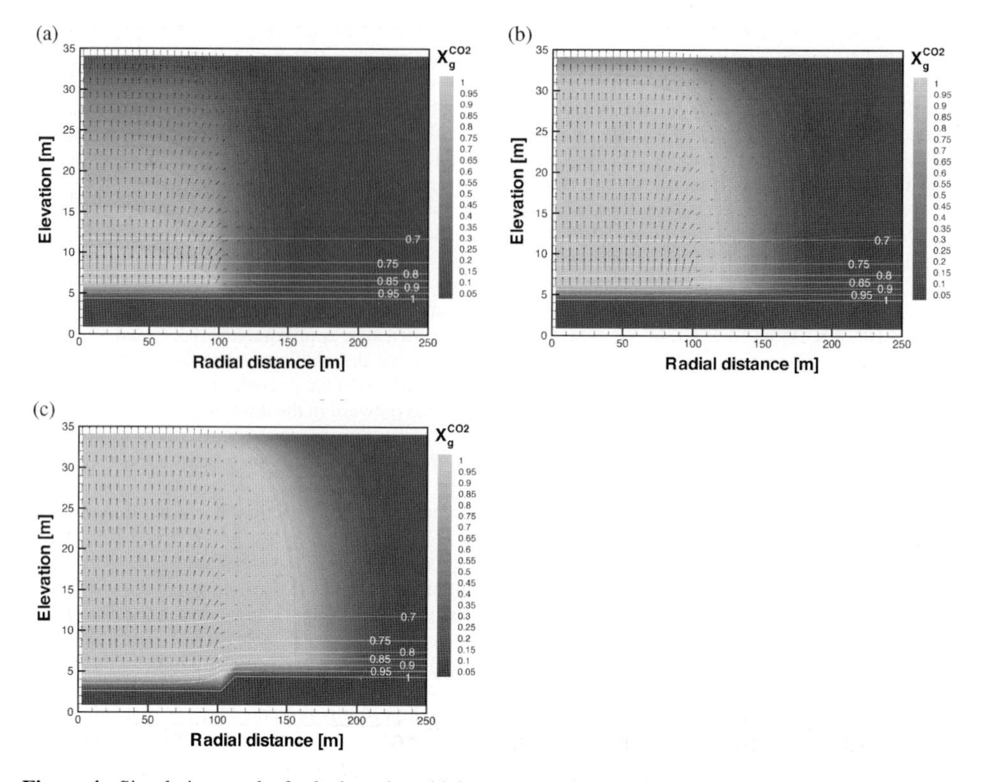

Figure 4: Simulation results for leakage in a thick unsaturated zone where shading indicates mass fraction of CO_2 in the gas phase, and labeled contour lines indicate water saturation, and vectors show gas-phase pore velocity for steady-state leakage rates of 4×10^4, 4×10^5, and 4×10^6 kg yr^{-1}. The maximum vector size represents values of approximately (a) 0.057, (b) 0.53, and (c) 3.6 m d^{-1}.

in Ref. [9]. We present in Figure 7 simulation results after 6 months of leakage showing subsurface and surface-layer CO_2 concentrations (mass fraction) and gas-phase velocity vectors for the cases of winds of 1, 3, and 5 m s^{-1} at a height of 10 m from the ground surface and neutral atmospheric conditions [2,3]. In Figure 7d we show a summary of the temporal evolution of the CO_2 gas mass fraction at $x = 645$ m (approximately 100 m downstream from the edge of the source). Figure 7a–c shows the strong effects of wind and atmospheric dispersion on seeping CO_2. Concentrations downwind from the source are strongly attenuated by turbulent mixing. Note further in Figure 7a–c the downward migration of CO_2 into the subsurface downwind of the source. This process is due to CO_2 dissolution in rainwater that is infiltrating at 10 cm yr^{-1}. It is important to note that in all the simulations we have assumed a zero background CO_2 concentration to emphasize the additional CO_2 that seeps from the ground in the various scenarios. Note that the mass fraction scale in Figure 7 shows that CO_2 concentrations in the surface layer are very low, barely above the background concentration of 370 ppmv which would be 0.00056 by mass fraction. The fundamental conclusion is that surface winds and atmospheric dispersion appear to be very effective at diluting diffuse CO_2 seepage fluxes over flat ground. We note that calm conditions, topographic depressions, and higher CO_2 seepage fluxes not yet analyzed can cause larger CO_2 concentrations to develop.

Given that HSE risks will be calculated based on exposures at certain locations in the flow field, we present in Figure 7d downwind CO_2 concentrations as a function of time for the test problem. Note that

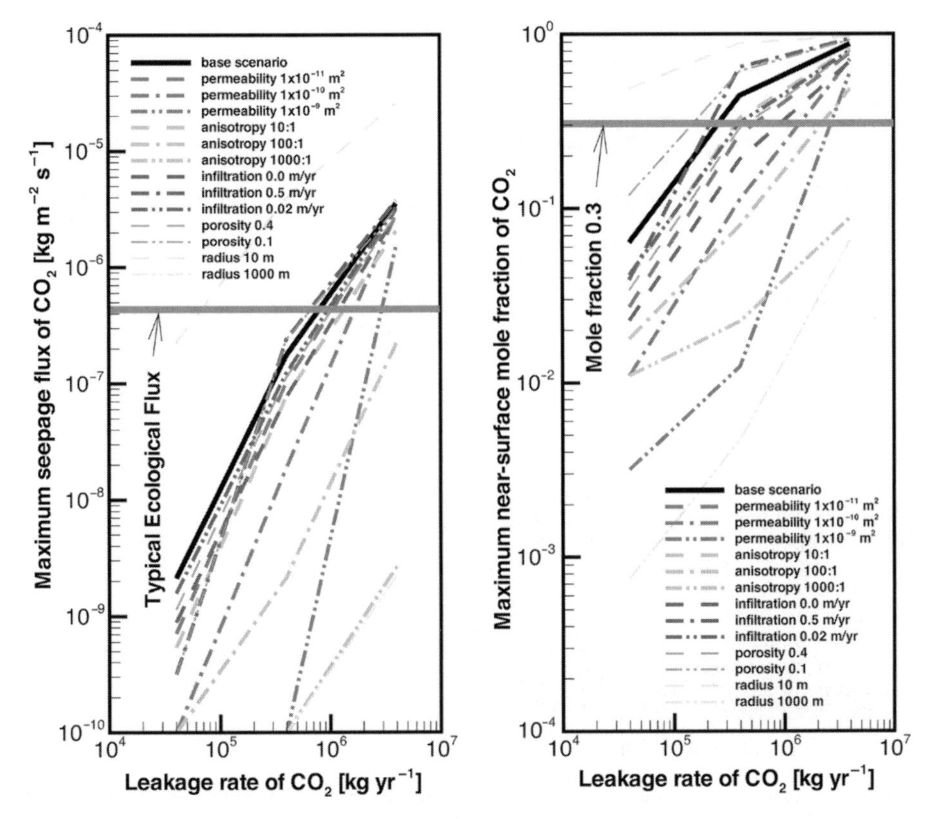

Figure 5: Maximum seepage flux of CO_2 and maximum near-surface gas mole fraction CO_2 as a function of leakage rate at steady-state seepage conditions.

concentrations are conservative because of the assumption of a 2D system and the use of a closed boundary at the top of the surface layer. For this case of diffuse CO_2 seepage, concentrations would be elevated above background by approximately 23 ppmv (3.5×10^{-5} mass fraction) for the 1 m s^{-1} case, and concentrations decrease approximately linearly with reference wind speed. Although this test problem is 2D, the coupled modeling framework is a fully 3D capability.

CONCLUSIONS

We have demonstrated the use of a coupled modeling framework for modeling CO_2 fluxes and concentrations for risk characterization. This work is relevant and important to the development of geologic CO_2 storage because it provides a modeling capability for simulating CO_2 flow and transport from the deep CO_2 storage site all the way to the atmosphere. The approach is built on the assumption that the near-surface environment is the main region in which HSE risks will arise. In this region, CO_2 flux and concentration are the main risk drivers. The coupled model handles subsurface and atmospheric surface-layer flow and transport assuming that dispersion in the surface layer is passive and that the wind is described by a logarithmic velocity profile. Model results show limited unsaturated zone attenuation of leakage flux, with correspondingly large CO_2 concentrations possible in the shallow subsurface. These results suggest that if leakage leads to CO_2 migrating as far as the vadose zone, high CO_2 concentrations can occur in the root zone of the shallow subsurface with potentially harmful effects on plants, as well as

TABLE 2
PROPERTIES OF THE COUPLED SUBSURFACE–SURFACE-LAYER MODEL SYSTEM

Property	Value
Subsurface	
Subsurface region extent (xyz)	1 km × 1 m, 0 m $< z <$ 35 m
Discretization ($N_x N_y N_z$)	100 × 1 × 35
Permeability ($k_X = k_Z$)	1×10^{-12} m^2
Porosity (ϕ)	0.2
Infiltration rate (i)	10.0 cm yr^{-1}
CO_2 flux region	450 m $< x <$ 550 m
CO_2 mass flux (0.1, 0.01, and 0.001% yr^{-1})	4.04×10^{-6}, 10^{-7}, 10^{-8} kg m^{-2} s^{-1}
Residual water sat. (S_{lr})	0.1
Residual gas sat. (S_{gr})	0.01
van Genuchten [10] α	1×10^{-4} Pa^{-1}
van Genuchten [10] m	0.2
Surface layer	
Surface-layer region extent (xyz)	1 km × 1 m, 35 m $< z <$ 45 m
Discretization ($N_x N_y N_z$)	100 × 1 × 20
Pressure in surface layer	1 bar (0.1 MPa)
Temperature (isothermal)	15 °C
Atmospheric stability	Neutral
Velocity profile	Logarithmic
Reference velocity at z = 10 m	1, 3, or 5 m s^{-1}
Friction velocity for u_x = 1, 3, 5 m s^{-1}	0.0868, 0.261, 0.434 m s^{-1}
Roughness length (z_0)	0.10 m

on humans or other animals in poorly ventilated subsurface structures such as basements or burrows. Coupled subsurface–surface-layer demonstration simulations show large degree of dilution that occurs in the surface layer, and the possible reflux of CO_2 to the subsurface that occurs when CO_2 dissolves in infiltrating rainwater.

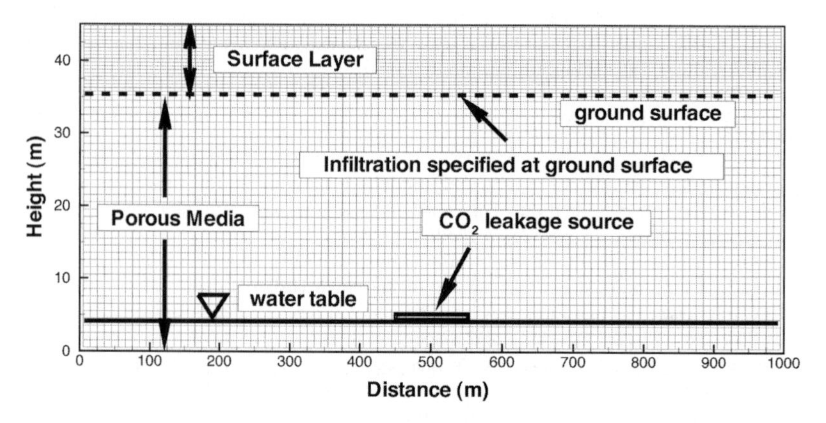

Figure 6: Domain and discretization used in the coupled subsurface–surface-layer test problem.

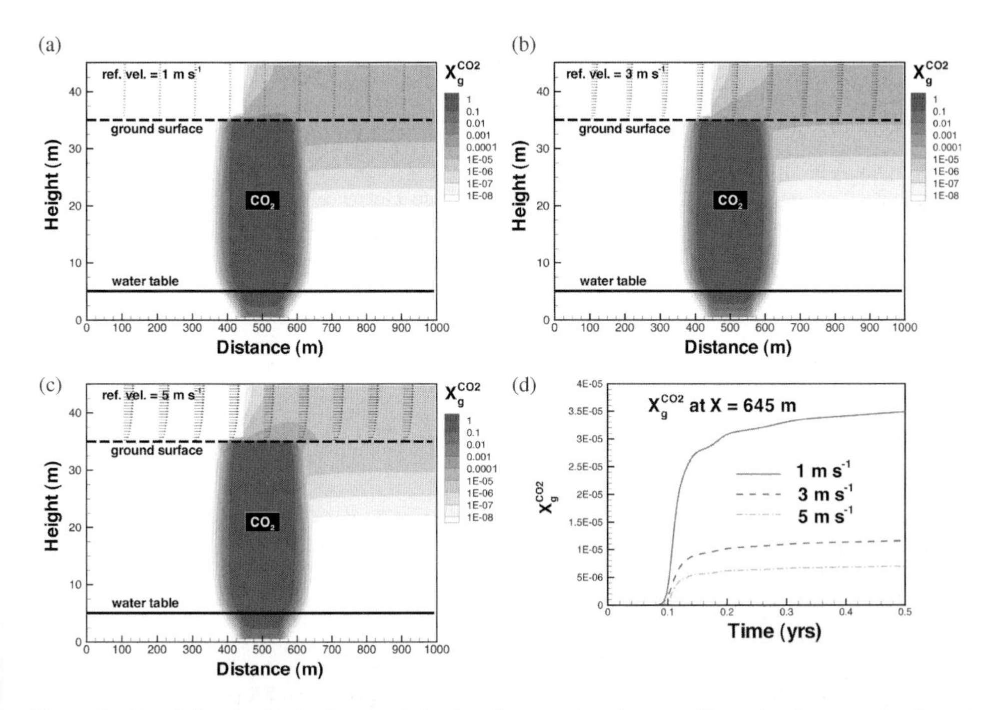

Figure 7: Simulation results for the coupled subsurface–surface-layer problem showing mass fraction of CO_2 in the gas phase and gas velocity vectors: (a) wind speed 1 m s^{-1}; (b) wind speed 3 m s^{-1}; (c) wind speed 5 m s^{-1}; (d) mass fraction CO_2 in the gas vs. time at $x = 645$ m.

RECOMMENDATIONS

We recommend development of additional capabilities for risk characterization related to leakage and seepage, along with further verification and testing of model approaches. Although the coupled modeling framework T2CA is applicable to many important leakage and seepage scenarios, it is neither applicable to absolute calm conditions where dense gas dispersion occurs, nor to very high fluxes such as might occur from an open well or catastrophic tank or pipeline release into the open atmosphere. In addition, buildings are neglected even though it is well established that exposures to people by soil–gas contaminants (e.g. radon) are most likely to occur indoors. We recommend that future research funding be directed toward model development for simulation of the foregoing processes. Finally, the surface-layer methods in T2CA should be compared against other atmospheric dispersion models for verification, and the methods should be refined if necessary.

NOMENCLATURE

2D	two-dimensional
3D	three-dimensional
CO_2	carbon dioxide
HSE	health, safety and environmental (risks)
NEE	net ecosystem exchange (for CO_2)
TOUGH2	reservoir simulator, *T*ransport *O*f *U*nsaturated *G*roundwater and *H*eat 2
T2CA	for *T*OUGH2 for *C*O_2 and *A*ir
x	mole fraction
X	mass fraction

ACKNOWLEDGEMENTS

We thank Chin-Fu Tsang (LBNL) and two anonymous reviewers for constructive comments and reviews of this report. This work was supported in part by a Cooperative Research and Development Agreement (CRADA) between BP Corporation North America, as part of the CO_2 Capture Project (CCP) of the Joint Industry Program (JIP), and the US Department of Energy (DOE) through the National Energy Technologies Laboratory (NETL), and by the Ernest Orlando Lawrence Berkeley National Laboratory, managed by the University of California for the US Department of Energy under contract DE-AC03-76SF00098.

REFERENCES

1. K. Pruess, C. Oldenburg, G. Moridis, TOUGH2 User's Guide Version 2.0, Lawrence Berkeley National Laboratory Report LBNL-43134, November, 1999, 197 pp.
2. S.P. Arya, Air Pollution Meteorology and Dispersion, Oxford University Press, New York, 1999.
3. D.H. Slade (Ed.), Meteorology and Atomic Energy 1968, US Atomic Energy Commission, 1968, Chapter 2.
4. R.E. Britter, *Ann. Rev. Fluid Mech.* **21** (1989) 317–344.
5. R.E. Britter, J. McQuaid, Workbook on the Dispersion of Dense Gases, *Health Saf. Exec. Rep.*, Sheffield, UK, HSE Contract Research Report No. 17/1988, 1988.
6. D.D. Baldocchi, K.B. Wilson, *Ecol. Model.* **142** (2001) 155–184.
7. C.M. Oldenburg, A.J.A Unger, *Vadose Zone J.* **2** (2003) 287–296.
8. C.D. Farrar, M.L. Sorey, W.C. Evans, J.F. Howle, B.D. Kerr, B.M. Kennedy, Y. King, J.R. Southon, *Nature* **376** (1995) 675–678.
9. C.M. Oldenburg, A.J.A. Unger, Coupled vadose zone and atmospheric surface-layer transport of CO_2 from geologic carbon sequestration sites, *Vadose Zone J.* **3** (2004) 848–857.
10. M.T. Van Genuchten, *Soil Sci. Soc. Am. J.* **44** (1980) 892–898.

Carbon Dioxide Capture for Storage in Deep Geologic Formations, Volume 2
D.C. Thomas and S.M. Benson (Eds.)

Chapter 30

IMPACT OF CO_2 INJECTIONS ON DEEP SUBSURFACE MICROBIAL ECOSYSTEMS AND POTENTIAL RAMIFICATIONS FOR THE SURFACE BIOSPHERE

T.C. Onstott

Department of Geosciences, Princeton University, Princeton, NJ, USA

ABSTRACT

Based upon the calculated potential microbial power for microbial redox reactions, the most readily identified impact of CO_2 injections on the subsurface microbial communities was the reduction of one pH unit for the ground water hosted in the siliclastic reservoir. The slightly lower pH is based upon the assumption, yet to be verified, that alteration of detrital feldspars to clay in equilibrium with calcite occurs on the time scale of the injection. The power levels for many of the microbial redox reactions were generally larger than in the original ground water systems but because of this reduction of one pH unit in the ground water, microbial Fe(III) reduction reactions were significantly enhanced over the expected ambient conditions. If sufficient electron donors are available for both biotic and abiotic Fe(III) reducing reactions and sufficient Fe(III) bearing oxides are present in the aquifer (as is usually the case) then these reactions will restore the aquifer's pH to its initial, pre-injection value. CO_2 injection should cause a short-term stimulation of Fe(III) reducing communities. For long-term storage of CO_2 in siliclastic reservoirs the short-term enhancement of Fe(III) reducing microorganisms will increase the pH and most likely lead to the precipitation of various carbonates. As readily available Fe(III) is depleted it can be introduced. If this is not feasible and sulfate is not a major constituent in the ground water, then methanogenic activity will begin to dominate and the proportion of CO_2 converted to CH_4 will depend upon the H_2 and acetate fluxes.

A dolomitic or carbonate aquifer may be more severely impacted by the simulated CO_2 injection because the dissolution of the carbonate failed to restore the pH to a range that is more commensurate with the pH ranges of some of the microorganisms. If mafic igneous rocks host the groundwater and contain Fe bearing clinopyroxene, then the lower pH will automatically stimulate the release of H_2 by the oxidation of this ferrous iron to $Fe(OH)_3$. This, in turn, would lead to stimulation of methanogenic and acetogenic communities and a reduction of the injected CO_2. Fe(III) reducing microbial reactions may also be stimulated by the appearance of $Fe(OH)_3$ leading to Fe(III) reduction and an eventual increase in pH.

For rhizosphere and surface biosphere the most obvious impact would be due to a potential increase in crustal CH_4 flux for carbonate and mafic rock hosted aquifers and a decrease in H_2 flux in all cases. Since the fluxes of both gaseous species from fermentative communities in shallower, organic-rich aquitards are 10–100 times greater than the deep subsurface flux, this probably is not a showstopper.

INTRODUCTION

Liquid CO_2 injection into hydraulically tight, deep permeable formations has been proposed as a means of carbon mitigation and is used to develop oil reservoirs. The extent to which subsurface microbial communities will play a role in the long-term fate of CO_2 is not known and it may depend upon numerous factors including the abundance, diversity and relative proportions of autotrophic to heterotrophic organisms in the community, the abundance of potential electron donors (e.g. H_2, acetate and fermenters), the formation of a separate gas phase in the aquifer, the ambient temperature and pressure.

Deep subsurface microbial communities are dominated by four anaerobic, physiological types, methanogens, sulfate or sulfur reducing bacteria, fermentative anaerobes and Fe(III) reducing bacteria. These encompass the majority of subsurface species encountered to date. Their presence or absence in the 16S rDNA clone libraries can be roughly correlated with the free energy of the redox reactions they utilize for energy maintenance and the availability of the reactants in these redox reactions. The combination of energy and availability, referred to in this report as the potential microbial power, is an important parameter for gauging microbial activity. To ascertain the probable impacts of CO_2 injection upon deep subsurface microbial communities we calculated the potential microbial power for a range of ground water chemistries and temperatures, in a carbonate and siliclastic aquifer subjected to high partial pressures of CO_2.

BACKGROUND

Over the past 15 years scientists have discovered the existence of microbial communities surviving at depths to at least 3.2 km below the terrestrial land surface (kmbls.) [1]. The sessile or rock bound population density declines with depth from $\sim 10^8$ cell g^{-1} at just below the soil zone to 10^2–10^5 cell g^{-1} in solid rock at ~ 3 kmbls., whereas the planktonic cell density ranges between 10^2 and 10^6 cell ml^{-1} and exhibits a slight decline with depth (Figure 1). A majority of these microorganisms represent new species, new genera and perhaps in some cases new phyla on the microbial tree of life. Most of these deep-seated environments include autotrophic methanogens, acetogens and sulfate reducers that utilize CO_2 but struggle in an aqueous environment that is HCO_3^--poor, perhaps even limiting, Ca-rich, alkaline and usually electron-acceptor limited. Ground water dating indicates that these communities can survive for tens to hundreds of millions of years and indirect evidence suggests that they are self-sufficient in terms of nutrient and energy resources [2]. In other words, they do not necessarily rely upon downward transport of growth substrates from the surface photosphere, but are biologically and chemically isolated.

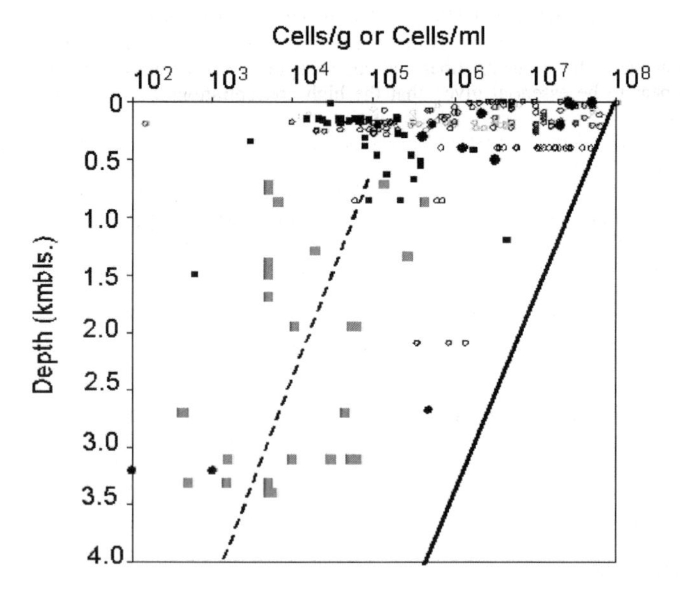

Figure 1: Cellular density as a function of depth for sediments (filled and solid circles) and ground water samples (filled and open squares) from subsurface microbial studies cited in Ref. [4]. Large open squares represent results from the fracture water in the Witwatersrand basin, South Africa (unpublished data) based upon flow cytometry analyses and large filled circles represent results from rock strata in the Witwatersrand Basin, South Africa [5]. The solid line is the least squares best fit from Parkes et al. [6] based upon marine sediment data collected up to a depth of 800 m and the dashed line is an extrapolation of that best fit to greater depths.

The biomass concentration and level of activity are controlled by temperature, water activity, porosity, permeability, substrate and trace metal concentrations, substrate availability and the free energy of the redox reactions utilized for metabolisms. Of the large number of redox reactions that are relevant to H, C, N, O, S and P cycling by microorganisms living at temperatures less than 120 °C [3], we have focused on 47 redox reactions, 42 of which are known to be associated with specific genera and five of which are abiotic (Table 1).

The microbial diversity in deep subsurface environments has begun to be well established with application of 16S rDNA analyses. This data tells us that most of the 47 redox reactions are being utilized by some member of the subsurface microbial community. Biodiversity appears to diminish dramatically with depth with some fractures in the Witwatersrand Basin possessing only one strain of microorganism. Some of the patterns that have emerged include the following:

1. For many aquifers where ambient temperatures are <40 °C and the depths are <500 m, members of the Proteobacteria division are common [7,8]. These include aerobic heterotrophs, methanotrophs, facultative anaerobes bacteria and chemolithotrophs. Members of the Crenarchaeota and Euryarchaeota division of the Archaea are also common in this aquifer. The Proteobacteria also appear in drilling water and mine water and therefore appear to be the dominant type of microbial contaminants.
2. The sulfate and S reducing bacteria of the δ Proteobacteria division, the Firmicutes and occasionally Archaea are common constituents of deep subsurface aquifers, [9] particularly oil reservoirs where sulfate concentrations are >100 μM. The anaerobic hydrocarbon oxidizing sulfate reducing bacteria appear to be confined to the δ Proteobacteria division [10]. Sulfate reducing bacteria that utilize aromatic hydrocarbons, such as benzoate, do occur in injection fluids in North Sea oil fields, but because we have no information regarding the concentration of aromatic hydrocarbons from formation fluids [11] and because these particular organisms have not been found to be indigenous to the subsurface (yet) we have not included this in our modeling.
3. Thiosulfate reducing bacteria that belong to the low G + C Firmicutes group, *Fusibacter paucivorans*, are also found in oil field brines [12]. Other phylogenetically related members of the *Clostridiales* that are fermentative bacteria appear to be subterranean inhabitants capable of not just fermenting organic acids and producing H_2 but of using other electron acceptors, such as thiosulfate, in the process [13]. This would appear to be essential given that the high concentrations of H_2 typically present in the subsurface preclude derivation of ATP from straight fermentation.
4. Thermophilic, Fe(III) reducing members of Firmicutes are found in deep subsurface formations where sulfate concentrations are <1 mM. [14,15] Fe(III) reducing members of the δ Proteobacteria division, e.g. *Geobacter metallireducens* [16] and the γ Proteobacteria domain, e.g. *Shewanella putrefaciens* [17] are found in shallow aquifers and in developed oil reservoirs [18] at mesophilic temperatures.
5. Thermophilic and hyperthermophilic members of Thermus [14] and Archaea [1] are found in deep subsurface aquifers, though rarely, and phylogenetically are closely related to hyperthermophiles and thermophiles found in surface hot springs.
6. In the case of the Witwatersrand Basin, South Africa, approximately one third of the 16S rDNA clone sequences are not closely related to sequences in the ribosomal data base and may represent new species of unknown metabolic function.

These trends have also been seen in the 16S rDNA data of petroleum reservoirs (Hinton, personal communication, 2003).

STUDY METHODOLOGY

The modeling involved the following four steps:

1. *Aquifer prior to CO_2 injection*. Geochemist workbench version 4 (Rockware Inc.) was utilized to calculate the free energy of the 47 reactions in Table 1 for three reservoir temperatures, 20, 45 and 80 °C, and for four ground water compositions (Table 2). We have restricted the maximum temperature to the upper limit for thermophiles, because we have very little evidence to date for the existence of hyperthermophilicity in the deep subsurface with the exception of Stetter et al.'s [19] discovery of hyperthermophilic Archaea associated with oil reservoirs. The four ground water types represent

TABLE 1
MICROBIALLY FACILITATED AND ABIOGENIC NON-PHOTOSYNTHETIC REDOX REACTIONS

Reaction no.	Microbial redox reaction	Classification		Example Phylla
15	$CH_4 + 2O_2 \rightarrow HCO_3^- + H^+ + H_2O$	Aerobic	Methanotrophs	β Proterobacteria: *Methylmonas*
11	Acetate $+ 2O_2 \rightarrow 2HCO_3^- + H^+$			γ Proterobacteria: *Aeromonas hydrophilia*
23	$2H_2 + O_2 \rightarrow 2H_2O$	Chemolithotrophs	Knall Gas Bacteria	β Proterobacteria: *Hydrogenophaga*
19	$2CO + O_2 + 2H_2O \rightarrow 2HCO_3^- + 2H^+$		Carboxydobacteria	α Proterobacteria: *Psuedomonas*
44	$NH_3 + 1.5O_2 \rightarrow NO_2^- + H^+ + H_2O$		Nitrifiers	β Proterobacteria: *Nitrosomonas*
46	$2NO_2^- + O_2 \rightarrow 2NO_3^-$			α Proterobacteria: *Nitrobacter*
17	$HS^- + 2O_2 \rightarrow SO_4^{2-} + H^+$		S-oxidizers	γ Proteobacteria: *Thiothrix*
16	$S_2O_3^{2-} + 2O_2 + H_2O \rightarrow 2SO_4^{2-} + 2H^+$			β Proterobacteria: *T. thioparus*
21	$S + 1.5O_2 + H_2O \rightarrow SO_4^{2-} + 2H^+$			β Proterobacteria: *T. thiooxidans*
18	$2HS^- + 2O_2 \rightarrow S_2O_3^{2-} + H_2O$			β Proterobacteria: *T. thioparus*
25	$2HS^- + O_2 + 2H^+ \rightarrow 2S + 2H_2O$			β Proterobacteria: *T. thioparus*
40	$4Fe^{2+} + O_2 + 10H_2O \rightarrow 4Fe(OH)_3 + 8H^+$		Fe-oxidizers	β Proterobacteria: *T. ferrooxidans*
47	$2Mn^{2+} + O_2 + 2H_2O \rightarrow 2MnO_2 + 4H^+$		Mn-oxidizers	β Proterobacteria: *Leptothrix*
7	$4H_2 + NO_3^- + H + \rightarrow NH_3 + 3H_2O$	Nitrate Reducing		
1	$5H_2 + 2NO_3^- + 2H + \rightarrow N_2 + 6H_2O$			
8	Acetate $+ NO_3^- + H_2O \rightarrow 2HCO_3^- + NH_3$			γ Proterobacteria: *Psuedomonas stutzeri*
2	Acetate $+ 1.6NO_3^- + 0.6H^+ \rightarrow 2HCO_3^- + 0.8H_2O + 0.8N_2$			Firmicutes: *Geobacillus*
4	$4CO + NO_3^- + 5H_2O \rightarrow 4HCO_3^- + NH_3 + 3H^+$			none
6	$2.5CO + NO_3^- + 2H_2O \rightarrow 2.5HCO_3^- + 1.5H^+ + 0.5N_2$			none
9	$S_2O_3^{2-} + NO_3^- + 2H_2O \rightarrow 2SO_4^{2-} + H^+ + NH_3$			β Proterobacteria: *T. denitrificans*
10	$HS^- + NO_3^- + H_2O \rightarrow SO_4^{2-} + NH_3$			β Proterobacteria: *T. denitrificans*
3	$2S + 1.5NO_3^- + 3.5H_2O \rightarrow 2SO_4^{2-} + 2.5H^+ + 1.5NH_3$			β Proterobacteria: *T. denitrificans*
45	$4Mn^{2+} + NO_3^- + 5H_2O \rightarrow 4MnO_2 + 7H^+ + NH_3$			
13	$5Fe^{2+} + NO_3^- + 12H_2O \rightarrow 5Fe(OH)3 + 9H^+ + 0.5N_2$			
14	$NO_2^- + H^+ + NH_3 \rightarrow 2H_2O + N_2$		Anaamox	Planctomycetales
39	Acetate + 4Hematite $+ 15H^+ \rightarrow 8Fe^{2+} + 8H_2O + 2HCO_3^-$	Fe(III) Reducing		Firmicutes
33	CO + Hematite $+ 3H^+ \rightarrow 2Fe^{2+} + H_2O + HCO_3^-$			none
41	H_2 + Hematite $+ 4H^+ \rightarrow 2Fe^{2+} + 3H_2O$			γ Proteobacteria: *S. putrefacians* and Crenarchaeota: *Pyrodictum islandicum*

5	Acetate + $4MnO_2 + 7H^+ \rightarrow 4Mn^{2+} + 4H_2O + 2HCO_3^-$	Mn(IV) Reducing		Firmicutes: *B. Infernus*
28	$4H_2 + H^+ + SO_4^{2-} \rightarrow HS^- + 4H_2O$	Sulfate Reducing		Firmicutes: *Desulfotomaculum*, Euryarchaeota: *Archaeoglobus*
31	Acetate + $SO_4^{2-} \rightarrow 2HCO_3^- + HS^-$			Firmicutes: *Desulfotomaculum*, Euryarchaeota: *Archaeoglobus*
20	$4CO + SO_4^{2-} + 4H_2O \rightarrow 4HCO_3^- + HS^- + 3H^+$			δ Proteobacteria: *Desulfococcus*, Firmicutes: *Desulfotomaculum*
35	$CH_4 + SO_4^{2-} \rightarrow H_2O + HCO_3^- + HS^-$		Anaerobic Methane Oxidation	δ Proteobacteria: *Desulfosarcina* + Euryarcheota: ANME: 1&2
37	$4H_2 + H^+ + 2HCO_3^- \rightarrow$ Acetate + $4H_2O$	CO_2 Reducing	Acetogen	Firmicutes: *Morella thermoacetica*
32	$4H_2 + H^+ + HCO_3^- \rightarrow CH_4 + 3H_2O$		Autotrophic Methanogen	Euryarchaeota: *Methanococcus*
30	4Formate + $H^+ + H_2O \rightarrow CH_4 + 3HCO_3^-$	Fermentation	Methanogen	Euryarchaeota
36	Acetate + $H_2O \rightarrow CH_4 + HCO_3^-$		Aceticlastic Methanogen	Euryarchaeota
38	$S_2O_3^{2-} + H_2O \rightarrow SO_4^{2-} + H^+ + HS^-$		S disproportionation	
24	$S_2O_3^{2-} + 4H_2 \rightarrow 3H_2O + 2HS^-$			δ Proteobacteria: *Desulfocapsa* and Firmicutes: *Desulfotomaculum*
42	Propanoate + $3H_2O \rightarrow$ Acetate + $HCO_3^- + H^+ + 3H_2$		Fermenters	
26	$H_2 + S \rightarrow HS^- + H^+$	S Reducer		Eubacteria: *Thermotogales*: Thermosipho Euryarchaeota: *Thermococcus*
12	Acetate + $4S + 4H_2O \rightarrow 5H^+ + 2HCO_3^- + 4HS^-$			Euryarchaeota: *Thermococcus*
Abiogenic Reactions				
43	HS^- + 4Hematite + $15H^+ \rightarrow SO_4^{2-} + 8Fe^{2+} + 8H_2O$			
22	$4CO + 5H_2O \rightarrow CH_4 + 3HCO_3^- + 3H^+$		Fischer-Tropsch	
27	$3H_2 + CO \rightarrow CH_4 + H_2O$		Fischer-Tropsch	
34	$CO + 2H_2O \rightarrow HCO_3^- + H^+ + H_2$		Gas Shift Reaction	Euryarchaeota: *M. thermoautotrophicum* (reverse reaction)
29	$3H_2 + N_2 \rightarrow 2NH_3$		N_2 fixation	

Other Fe(III) bearing mineral phases modeled included $Fe(OH)_3$, Goethite and Magnetite.

the average of 200 analyses of ground water collected at depths ranging up to 3.2 kmbls. in South Africa and is the only data set with sufficient detailed analyses to permit calculation of the free energy of the reactions in Table 1. The coupling of geochemical analyses of ground water with partial equilibrium calculation of the free energy of redox reactions to determine the principal terminal electron acceptor process has been successfully applied to shallow contaminated aquifers [20]. The four ground water types include dolomitic water, low salinity meteoric water, moderate salinity water and highly saline water. The dolomitic water is typical of carbonate dominated water and in terms of major cation and anion composition is comparable to that reported for the Madison limestone and Floridian aquifers [21,22].

2. *Aquifer during CO_2 injection.* The PCO_2 was set at 200 bars, equivalent to the hydrostatic pressure at 2 kmbls. and the four ground water types were equilibrated with this high PCO_2. We assumed that CO_2 injection occurs as a separate phase. The change in PCO_2, pH and pe will impact the free energy for most of the 47 redox reactions in Table 1. The formation of a separate phase in a H_2O saturated aquifer will result in a reduction of the dissolved gas concentrations and the amount of reduction will depend upon the volume ratio of gas to fluid. We did not attempt to model this effect, because the gas concentrations in Table 1 probably reflect the formation of a separate CH_4 gas phase during depressurization. We also treated the CO_2 injection as a pulse injection so we could examine the processes that could mitigate the reduction in CO_2.
3. *Aquifer following CO_2 injection.* The final parameter variation involved dissolution of aquifer minerals by carbonic acid. For the dolomitic water dolomite and calcite was dissolved until both minerals obtained saturation and the pH and pe stabilized [22]. In the case of low salinity, moderate salinity and high salinity ground water, albite and calcite were chosen as the aquifer minerals. These two mineral phases not only appear to control major cation composition of the ground water types in South Africa but also that of saline water associated with petroleum reservoirs in the Gulf Coast [23]. The dissolution of these two minerals phases proceeded until they both attained saturation at which point pH and pe stabilized. Other minerals were allowed to precipitate during the dissolution reaction. These minerals included sulfide minerals that control trace metal concentrations, clay minerals that mitigate pH and Al concentrations, Chalcedony which controls Si concentrations, and Nahcolite that like Calcite is influenced by the PCO_2. Fe hydroxide surface protonation reactions were not utilized in the simulations at 20 °C, but their effect probably would have been to moderate the acid production of the high PCO_2.
4. To relate the free energy of the microbial redox reactions in Table 1 calculated in the first three steps to microbial activity or ATP production we made three assumptions. The first is that conservation of energy does occur during electron transport processes as first proposed by Thauer et al. [24] for anaerobic reactions occurs for all of the metabolic pathways involved in the 42 microbial redox reactions listed in Table 1. Secondly, we assumed that conversion of this chemical energy to ATP takes place with a maximum efficiency, which is equivalent to saying that a minimum chemical free energy, ΔG, is required for ATP synthesis to occur. For normal bacteria, this minimum energy is 70 kJ mol^{-1} of reactant, but under certain conditions, ATP synthesis has been observed to proceed at 20 J mol^{-1} of reactant [25] and microbial activity has been recorded to occur in the lab at ~12–15 J mol^{-1} of organic reactant with syntrophic microbial consortia [26]. For the purpose of our calculations we have used a value of 20 kJ mol^{-1} as the minimum free energy required for ATP synthesis. Finally, we assumed that the maximum rate at which this energy could be accrued was given by the maximum rate of diffusion of the rate limiting reactant to the microorganism. This rate (mol cell^{-1} s^{-1}) is approximately by $4\pi DrC$, where C is the concentration of the rate limiting reactant (mol kg^{-1}), D the diffusivity of the reactant (cm^2 s^{-1}) and r the radius of the microorganism. We assumed r was 0.5 μm for all simulations. The reactant diffusivity increases with temperature according to the Stokes–Einstein relationship and the values used were from Cussler [27]. This assumption presumes that deep subsurface microorganisms are nonmotile, which is a safe assumption given their extraordinarily slow rates of growth [28,29]. The potential microbial power (J cell^{-1} s^{-1}) for a specific microbial redox reaction is equal to $4\pi DrC\Delta G$. The rate needs to be at least equivalent to the demand by the microorganism as required for its maintenance energy demand in order for the pathway to be viable. In the case of a mesophilic nitrifying bacterium, this maintenance demand is on the order of 1.7×10^{-19} kJ cell^{-1} s^{-1}. In the case of a mesophilic methanogen the maintenance demand is on the order of 1.4×10^{-19} kJ cell^{-1} s^{-1} (Colwell, personal communication, 2004).

TABLE 2
GEOCHEMICAL COMPOSITION OF FOUR DEEP GROUND WATER TYPES USED IN SIMULATIONS

Ground water type	**pH**	**pe**	***T* (°C)**	**TOC (gfw = 12)**	**DOC (gfw = 12)**	**DIC (gfw = 44)**	**Acetate**	**Formate**	**Propionate**	**F**	**Cl**
Anions (ppm)											
Dol	7.62	0.55	26	17.50	5.00	153.33	0.45	1.80	0.01	2.30	26
LowS	8.86	−3.03	41	1.00	1.50	29.99	0.65	0.09	0.05	3.63	102
ModS	8.78	−3.90	43	8.35	5.14	25.05	1.41	0.20	0.12	2.20	1,274
Brine	8.05	−3.89	44	8.83	13.95	8.52	2.98	1.11	0.67	0.93	13,680
	NO_2^-	SO_4^{2-}	HS^-	$S_2O_3^{2-}$	Br	NO_3^-	PO_4^{2-}	Total P as PO_4^{2-}	I		
Dol	0.005	79.08	20.39	0.01	0.59	0.54	0.010	1.07	0.48		
LowS	0.004	14.97	26.17	0.43	1.12	0.04	0.018	0.03			
ModS	0.060	34.42	15.53	0.87	6.50	0.16	0.011	2.08	0.87		
Brine	0.069	120.93	20.72	0.68	76.02	0.14	0.024	6.16	3.22		
Cations and trace metals (ppm)											
	NH_3 (gfw = 14)	Li	Na	Mg	K	Rb	Ca	Sr	Ba	Al	Si
Dol	0.02	0.01	16	29.62	1.41	0.003	57	0.169	0.131	0.468	7.38
LowS	0.18	0.07	78	0.07	2.11	0.181	11	0.156	0.017	0.061	15.45
ModS	0.42	0.49	555	2.93	6.67	0.071	197	4.202	0.649	0.267	9.28
Brine	0.45	2.85	3,876	134.41	46.99	0.826	3,619	96.828	15.172	0.261	8.12
	Mn	Fe	Mo	Cr	Co	Ni	Cu	Zn	As	W	U
Dol	0.042	0.309	0.020	0.037	0.007	0.020	0.008	0.030	0.0250	0.0296	0.0220
LowS	0.004	0.201	0.182	0.004	0.002	0.015	0.038	0.008	0.0115	0.0779	0.0239
ModS	0.746	0.328	0.050	0.019	0.002	0.011	0.007	0.068	0.0293	0.0726	0.0349
Brine	2.527	8.063		0.020	0.004	0.052	0.024	0.037	0.0574	0.0868	0.1027
Dissolved gases (μM)											
	H_2	He	Ar	N_2	CH_4	C_2H_6	C_3H_8	iso-C_4	*n*-C_4	CO	
Dol	0.97	1.30	7.00	350.00	104.91	0.00	0.00	0.00	0.00	0.37	
LowS	0.13	162.49	9.64	1907.54	1150.95	62.84	12.98	0.25	0.88	0.34	
ModS	181.47	441.44	62.63	4019.32	6223.95	109.68	13.14	0.29	1.74	11.44	
Brine	515.20	1082.57	148.86	3625.57	10,740.75	603.77	55.36	0.46	5.25	69.94	

For example, 200 mmol of ATP are required to produce 1 g of anaerobic bacteria (wet weight) if acetate or CO_2 is the carbon substrate. For a cell mass of 10^{-12} g dry weight, this means that 2×10^{-13} mol of ATP or 1.4×10^{-11} kJ are required to produce one cell. If the potential microbial power for aceticlastic, sulfate reduction was $\sim 10^{-11}$ J s^{-1} cell^{-1}, then sufficient energy would be accrued after $\sim 2 \times 10^3$ s for a single cell. If the biomass concentration of the ground water and aquifer was 4×10^9 cells kg^{-1} like that reported for the Middendorf Aquifer by Phelps et al. [28] and all the cells in the ground water were utilizing this one reaction (reaction (31) in Table 1) at a rate of 10^{-16} mol s^{-1}cell^{-1}, then the steady state rate of HCO_3^- production in the aquifer would be 4×10^{-7} mol kg^{-1} s^{-1} or 35 mmol kg^{-1} day^{-1}. This estimated rate is far greater than the ~ 10 and 0.001 μmol kg^{-1} day^{-1} sulfate reduction rates determined by Phelps et al. [28] which were based upon $^{35}SO_4$ measurements for the former and geochemical reaction rates for the Middendorf aquifer calibrated by ^{14}C ages for the latter.

Part of the discrepancy may be explained if just a small portion of the biomass is active, sulfate reducing bacteria. Regardless, the expression, $4\pi DrC\Delta G$, is considered a maximum potential for microbial power, as it ignores enzyme inhibition by competitive species or reactions and the transport rate across the cellular membrane. In the case of solid reactants, such as S and Fe(III) and Mn(IV) bearing oxides, we have assumed that the limiting reactant is the aqueous phase, not the solid phase, which clearly cannot be true all the time. We have also not corrected for the minimum concentration required for an enzyme to function or to be expressed. Nevertheless, we feel that the potential microbial power values for the different redox reactions can be used to assess the relative importance of one type of metabolism versus another.

RESULTS AND DISCUSSION

Initial Conditions

The four ground water types exhibit the following trends with increasing salinity and temperature (deeper ground water tends to be hotter and more saline):

1. The pe becomes more negative. As reliable dissolved O_2 measurements are difficult to make when concentrations are close to the detection limit of 0.03 mM, we have utilized the fO_2 predicted by the pe as our estimate of the dissolved O_2 for the model simulations.
2. With the exception of the dolomite water the pH decreases.
3. Sulfate concentrations increase whereas sulfide concentrations are relatively uniform with the possible exception of the dolomite water.
4. The Fe and Mn concentrations increase.
5. Dissolved reduced gases and hydrocarbons increase.
6. Trace levels of nitrate and nitrite are present throughout with the highest nitrate concentration associated with the dolomite water.
7. The ammonia concentration increases.
8. The concentration of organic acids increases, whereas the concentration of inorganic carbon decreases.

These trends, particularly the increasing dissolved organic acids and reducing potential with depth or temperature is consistent with observations of pore water and ground water from basins where organic matter is far more abundant than our South African aquifers. With the dolomite, moderate salinity and highly saline ground water, calcite is saturated and the concentration of the decreasing DIC is a direct reflection of the increasing Ca concentrations and elevated pH. If the pH were to remain constant during CO_2 injection, the Ca concentrations would remove a large fraction of the CO_2 and precipitate it as calcite. The degree to which the microorganisms would facilitate such a process is a subject for the next stage of investigation (see Recommendations). The first and most important question to be answered by the modeling is as follows. Do any of the microbial redox reactions that yield negative free energies for our subsurface ground water types become positive under the conditions anticipated to occur with CO_2 injection? This would be considered a detrimental impact on those subsurface microorganisms relying upon those specific redox reactions.

The free energy and potential microbial power calculations for the four types of ground water (Tables 3 and 4) provide a baseline against which to compare the community structure inferred from the 16S rDNA results and the simulated geochemical changes associated with CO_2 injection. The free energy calculations revealed the following:

1. The two microbial redox reactions that are the most obviously relevant to CO_2 injection are the CO_2 reducing methanogenesis and acetogenesis reactions (reactions (32) and (37) in Table 3 and Figure 2). The free energy for both the reactions decrease with increasing temperature and are marginally exothermic in the dolomitic and low salinity ground water type where dissolved H_2 concentrations are <1 μM (Table 2), but are exothermic in the moderate and high salinity water where H_2 concentrations are >100 μM.
2. For all four ground water types the ammonia oxidizing and the Mn oxidizing reactions were all positive regardless of temperature (reactions (44)–(47) in Table 3 and Figure 2). The implication is that in order for these reactions to proceed the O_2 concentrations must be much higher than is typical even for microaerophilic ground water. This is also consistent with the absence of nitrifying and Mn oxidizing organisms from the 16S rDNA results for aquifers. Nitrifying bacteria have been found associated with the more oxygenated drilling water. Anaerobic ammonia oxidation by reduction of nitrite, the anammox reaction (reaction (14) in Table 3 and Figure 2) is energetically favorable in all four ground water types, but the microorganisms associated with this reaction belong to the order of the *Planctomycetales* [30] and the 16S rDNA signatures of this order have yet to be identified in the deep subsurface, although the anammox reaction has been detected in shallow marine sediments [31,32].
3. Conversely Mn reduction (reaction (5) in Table 3 and Figure 2) and nitrate reduction (reactions (1)–(4), (6)–(10) and (13)) possessed highly negative free energies. Of the nitrate reducing reactions, those yielding N_2 as the product (reactions (1), (2), (6) and (13) in Table 3 and Figure 1) were more exothermic than those yielding ammonia (reactions (3), (4) and (7)–(10) in Table 3 and Figure 2).
4. Despite the extremely low concentrations of dissolved O_2 predicted by measured pe for the four ground water types, aerobic reactions (reactions (11), (15)–(19), (21), (23) and (25) in Table 3 and Figure 2) still retained highly negative free energies. These free energies increase with temperature as fO_2 increases. The most energetic reaction is acetate oxidation (reaction (11) in Table 3 and Figure 2) followed by CH_4, thiosulfate, CO, HS^-, S, H_2 and the least exothermic aerobic reaction is oxidation of HS^- to S (reaction (25) in Table 3 and Figure 2).
5. Reduction of S compounds to HS^- (reactions (12), (24), (26), (28), (31) and (35) in Table 3 and Figure 2) was energetically favorable. The most exothermic reaction was the reduction of S to HS^- by acetate (reaction (12) in Table 3 and Figure 2). The free energies of S reduction reactions with acetate increased with increasing temperature, whereas those with H_2 decreased with increasing temperature.
6. The anaerobic oxidation of CH_4 coupled to the reduction of SO_4^{2-} to HS^- (reaction (35) in Table 3 and Figure 2) is the least exothermic S reducing reaction. This reaction has been detected in shallow, marine sediments and methane clathrates where it appears to require the syntrophic activity of two microorganisms, one of which is a H_2 utilizing sulfate reducing bacteria and one of which is a CO_2 reducing methanogen. The methanogen is believed to be reversibly oxidizing CH_4 by reaction (32) in Tables 1 and 3. This can only occur if the H_2 concentration is low enough for the free energy of reaction (32) to exceed $+20$ kJ mol^{-1} [33]. The free energy yields for reaction (35), therefore, are not germane even though they would appear to be favorable. Accordingly, Table 3 indicates that only the low salinity ground water at high temperature would be energetically favorable for anaerobic CH_4 oxidation.
7. Although anaerobic oxidation of hydrocarbon by sulfate reduction was not specifically modeled, the free energy for these reactions is slightly greater than that of the anaerobic CH_4 oxidation.
8. The free energy for reduction of hematite to Fe^{2+} is pH and temperature dependent with the reaction favored for low pH and low temperature (reactions (39) and (41) in Table 3 and Figure 2). This holds true for the other Fe(III) oxides as well. In this report, we have restricted the analysis to hematite under the presumption that amorphous $Fe(OH)_3$ and goethite would be the first phases to be reduced leaving hematite as the sole, remaining, Fe(III) oxide for deep, anaerobic environments. Microbial reduction of magnetite would require even lower pH values than is typical of these environments. In the absence of a reaction that would regenerate $Fe(OH)_3$ our model would suggest that microbial Fe(III) reduction would be restricted to mesophilic environments for ground water with pH ~ 7.5. Because microbial Fe(III) reduction by either acetate or H_2 raises the pH of the environment this represents a severe

TABLE 3
FREE ENERGY (KJ MOL^{-1}) FOR REDOX/MICROBIAL REACTIONS IN DOLOMITIC, LOW SALINITY, MODERATE SALINITY AND HIGHLY SALINE GROUND WATER AT 20, 45 AND 80 °C

Microbial redox reactions	1. Do 20	2. Do 45	3. Do 80	4. LS 20	5. LS 45	6. LS 80	7. MS 20	8. MS 45	9. MS 80	10. Br 20	11. Br 45	12. Br 80
(1) $5H_2 + 2NO_3^- + 2H^+ \rightarrow N_2 + 6H_2O$	−983	−957	−920	−923	−892	−848	−1225	−993	−961	−1249	−1015	−984
(2) Acetate + $1.6NO_3^- + 0.6H^+ \rightarrow 2HCO_3^- + 0.8H_2O + 0.8N_2$	−749	−744	−739	−736	−731	−725	−896	−740	−735	−907	−750	−747
(3) $2S + 1.5NO_3^- + 3.5H_2O \rightarrow 2SO_4^{2-} + 2.5H^+ + 1.5NH_3$	−700	−700	−699	−697	−699	−702	−840	−698	−702	−833	−691	−690
(4) $4CO + NO_3^- + 5H_2O \rightarrow 4HCO_3^- + NH_3 + 3H^+$	−655	−645	−627	−673	−665	−654	−855	−707	−703	−884	−732	−728
(5) Acetate + $4MnO_2 + 7H^+ \rightarrow 4Mn^{2+} + 4H_2O + 2HCO_3^-$	−648	−631	−609	−627	−609	−584	−708	−567	−540	−790	−642	−626
(6) $2.5CO + NO_3^- + 2H_2O \rightarrow 2.5HCO_3^- + 1.5H^+ + 0.5N_2$	−567	−559	−547	−576	−569	−559	−724	−597	−591	−740	−612	−608
(7) $4H_2 + NO_3^- + H^+ \rightarrow NH_3 + 3H_2O$	−535	−516	−488	−490	−468	−439	−678	−547	−526	−700	−565	−543
(8) Acetate + $NO_3^- + H_2O \rightarrow 2HCO_3^- + NH_3$	−497	−494	−490	−487	−485	−484	−594	−491	−492	−608	−503	−502
(9) $S_2O_3^{2-} + NO_3^- + 2H_2O \rightarrow 2SO_4^{2-} + H^+ + NH_3$	−463	−458	−444	−468	−465	−462	−565	−466	−463	−560	−460	−455
(10) $HS^- + NO_3^- + H_2O \rightarrow SO_4^{2-} + NH_3$	−448	−441	−432	−437	−430	−423	−525	−429	−422	−529	−432	−422
(11) Acetate + $2O_2 \rightarrow 2HCO_3^- + H^+$	−264	−295	−336	−179	−203	−234	−166	−158	−186	−130	−126	−150
(12) Acetate + $4S + 4H_2O \rightarrow 5H^+ + 2HCO_3^- + 4HS^-$	−264	−295	−336	−179	−203	−234	−166	−158	−186	−130	−126	−150
(13) $5Fe^{2+} + NO_3^- + 12H_2O \rightarrow 5Fe(OH)_3 + 9H^+ + 0.5N_2$	−272	−294	−323	−324	−349	−385	−388	−347	−382	−372	−333	−365
(14) $NO_2^- + H^+ + NH_3 \rightarrow 2H_2O + N_2$	−297	−293	−287	−291	−285	−275	−358	−292	−284	−358	−293	−287
(15) $CH_4 + 2O_2 \rightarrow HCO_3^- + H^+ + H_2O$	−242	−271	−310	−158	−180	−209	−142	−136	−161	−102	−101	−122
(16) $S_2O_3^{2-} + 2O_2 + H_2O \rightarrow 2SO_4^{2-} + 2H^+$	−230	−259	−290	−160	−182	−212	−137	−132	−156	−82	−83	−102
(17) $HS^- + 2O_2 \rightarrow SO_4^{2-} + H^+$	−215	−242	−278	−128	−147	−172	−98	−96	−115	−52	−55	−70

(18) $2HS^- + 2O_2 \rightarrow S_2O_3^{2-} + H_2O$	-200	-225	-266	-97	-112	-132	-58	-60	-74	-21	-26	-37
(19) $2CO + O_2 + 2H_2O \rightarrow 2HCO_3^- + 2H^+$	-211	-223	-236	-182	-191	-202	-214	-187	-198	-203	-177	-188
(20) $4CO + SO_4^{2-} + 4H_2O \rightarrow 4HCO_3^- + HS^- + 3H^+$	-207	-203	-194	-236	-236	-231	-330	-278	-281	-355	-300	-306
(21) $S + 1.5O_2 + H_2O \rightarrow SO_4^{2-} + 2H^+$	-175	-200	-234	-118	-137	-163	-99	-99	-121	-58	-63	-81
(22) $4CO + 5H_2O \rightarrow CH_4 + 3HCO_3^- + 3H^+$	-180	-174	-163	-206	-203	-195	-286	-237	-235	-304	-254	-254
(23) $2H_2 + O_2 \rightarrow 2H_2O$	-151	-158	-167	-91	-93	-94	-125	-107	-110	-111	-94	-95
(24) $S_2O_3^{2-} + 4H_2 \rightarrow 3H_2O + 2HS^-$	-102	-91	-67	-85	-73	-56	-192	-154	-145	-201	-162	-154
(25) $2HS^- + O_2 + 2H^+ \rightarrow 2S + 2H_2O$	-80	-83	-89	-22	-20	*-18*	*3*	*7*	*12*	*14*	*16*	*22*
(26) $H_2 + S \rightarrow HS^- + H^+$	-76	-79	-83	-45	-46	-47	-63	-53	-55	-55	-47	-48
(27) $3H_2 + CO \rightarrow CH_4 + H_2O$	-90	-78	-58	-69	-55	-33	-153	-117	-102	-166	-129	-115
(28) $4H_2 + H^+ + SO_4^{2} \rightarrow HS^- + 4H_2O$	-87	-74	-55	-53	-38	-16	-153	-117	-104	-170	-133	-121
(29) $3H_2 + N_2 \rightarrow 2NH_3$	-87	-74	-55	-57	-44	-29	-131	-100	-90	-150	-115	-102
(30) 4Formate + $H^+ + H_2O \rightarrow CH_4 + 3HCO_3^-$	-77	-69	-60	-47	-37	-24	-63	-44	-33	-91	-69	-61
(31) Acetate + $SO_4^{2} \rightarrow 2HCO_3^- + HS^-$	-49	-53	-58	-51	-55	-62	-68	-62	-70	-79	-71	-80
(32) $4H_2 + H^+ + HCO_3^- \rightarrow CH_4 + 3H_2O$	-61	-45	-24	-24	*-5*	*21*	-108	-77	-58	-120	-87	-69
(33) CO + Hematite + $3H^+ \rightarrow 2Fe^{2+} + H_2O + HCO_3^-$	-54	-41	-20	-37	-23	*-1*	-58	-34	-15	-70	-46	-28
(34) $CO + 2H_2O \rightarrow HCO_3^- + H^+ + H_2$	-30	-32	-35	-46	-49	-54	-44	-40	-44	-46	-42	-46
(35) $CH_4 + SO_4^{2} \rightarrow H_2O + HCO_3^- + HS^-$	-27	-29	-32	-30	-33	-37	-44	-41	-46	-51	-46	-52
(36) Acetate + $H_2O \rightarrow CH_4 + HCO_3^-$	-22	-24	-26	-21	-23	-25	-24	-21	-24	-28	-25	-28
(37) $4H_2 + H^+ + 2HCO_3^- \rightarrow$ Acetate + $4H_2O$	-38	-22	*2*	*-3*	*17*	*46*	-84	-55	-33	-92	-62	-41
(38) $S_2O_3^{2-} + H_2O \rightarrow SO_4^{2-} + H^+ + HS^-$	*-15*	*-17*	*-12*	-32	-35	-40	-39	-36	-41	-31	-29	-33
(39) Acetate + 4Hematite + $15H^+ \rightarrow 8Fe^{2+} + 8H_2O + 2HCO_3^-$	-59	*-12*	*55*	*35*	*90*	*167*	*30*	*78*	*152*	*-5*	*46*	*114*
(40) $4Fe^{2+} + O_2 + 10H_2O \rightarrow 4Fe(OH)_3 + 8H^+$	*24*	*-10*	-57	*19*	*-15*	-63	*55*	*13*	-31	*91*	*46*	*6*
(41) H_2 + Hematite + $4H^+ \rightarrow 2Fe^{2+} + 3H_2O$	-24	*-8*	*14*	*8*	*27*	*53*	*-13*	*6*	*30*	-24	*-4*	*18*
(42) Propanoate + $3H_2O \rightarrow$ Acetate + $HCO_3^- + H^+ + 3H_2$	*23*	*10*	*-9*	*-6*	-22	-44	*56*	*35*	*17*	*63*	*41*	*24*

(continued)

TABLE 3
CONTINUED

Microbial redox reactions	1. Do 20	2. Do 45	3. Do 80	4. LS 20	5. LS 45	6. LS 80	7. MS 20	8. MS 45	9. MS 80	10. Br 20	11. Br 45	12. Br 80
(43) $HS^- + 4\text{Hematite} + 15H^+ \rightarrow SO_4^{2-} + 8Fe^{2+} + 8H_2O$	*− 10*	*41*	*112*	*86*	*145*	*229*	*99*	*140*	*222*	*74*	*117*	*195*
(44) $NH_3 + 1.5O_2 \rightarrow NO_2^- + H^+ + H_2O$	*157*	*130*	*99*	*211*	*190*	*163*	*302*	*232*	*210*	*342*	*267*	*246*
(45) $4Mn^{2+} + NO_3^- + 5H_2O \rightarrow 4MnO_2 + 7H^+ + NH_3$	*151*	*138*	*118*	*139*	*123*	*99*	*114*	*76*	*48*	*182*	*139*	*124*
(46) $2NO_2^- + O_2 \rightarrow 2NO_3^-$	*151*	*140*	*124*	*194*	*186*	*174*	*253*	*203*	*194*	*271*	*220*	*212*
(47) $2Mn^{2+} + O_2 + 2H_2O \rightarrow 2MnO_2 + 4H^+$	*192*	*168*	*136*	*224*	*203*	*175*	*271*	*205*	*177*	*330*	*258*	*238*

The reactions are ordered from most negative to positive with respect to the free energy for the dolomite ground water at 20 °C. The microbial reaction numbers and column heading numbers refer to Figure 2. Values in italics are > -20 kJ mol^{-1} and therefore are not considered to be viable for microbial metabolism.

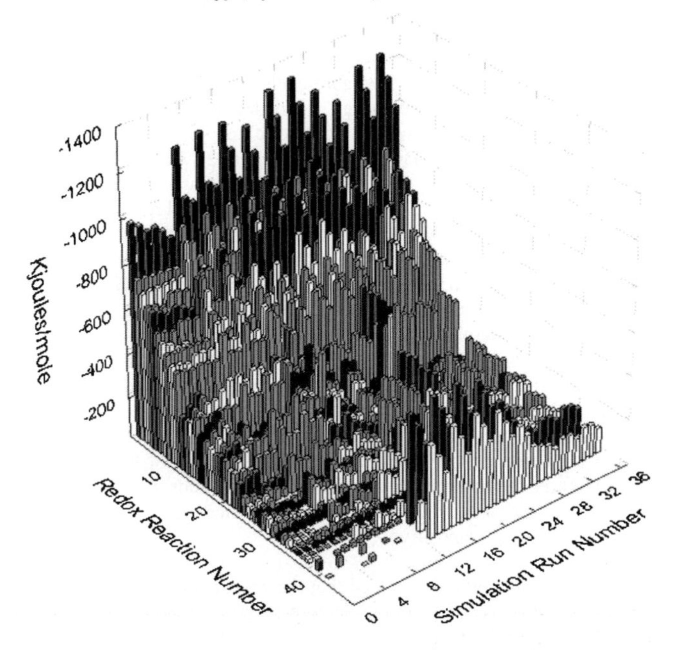

Figure 2: Free energy in kJ mol^{-1} for microbial reactions for four different types of ground water and aquifers under ambient conditions at 25, 45 and 80 °C (Simulation Run Numbers 1–12), during simulated CO_2 injection (Simulation Run Numbers 13–24), and during post-injection equilibration with aquifer minerals (Simulation Run Numbers 24–36), based upon data in Tables 3, 5 and 7.

restriction unless the aquifer is buffered to lower pH or an acid producing reaction occurs at a comparable rate.

9. The free energy for oxidation of Fe^{2+} to $Fe(OH)_3$ is also pH and temperature dependent with the reaction favored for high temperatures and high pH (reactions (13) and (40) in Table 3 and Figure 2). Fe^{2+} oxidation is weakly favored even for the minute amounts of dissolved O_2 present in our simulations. Fe^{2+} oxidation by nitrate, however, is strongly exothermic even for trace amounts of nitrate.
10. Of the organic fermentation reactions formate fermentation to CH_4 and CO_2 (reaction (30) in Table 3 and Figure 2) was the most exothermic, followed by acetate fermentation to CH_4 and CO_2 (reaction (36) in Table 3 and Figure 2) and propionate fermentation to acetate, CO_2 and H_2 (reaction (42) in Table 3 and Figure 2). The free energy for the formate reaction decreased with increasing temperature. The free energy for the acetate fermentation was remarkably constant at ~ -25 kJ mol^{-1} regardless of temperature or ground water type. The free energy for propionate fermentation increased with decreasing temperature and was only microbially favored for the high temperature, low salinity water where the dissolved H_2 concentrations were <0.1 μM. Inorganic fermentation of thiosulfate, or thiosulfate disproportionation, is marginally favorable and sensitive to the pH of the ground water with the free energy increasing as the pH increases. The free energy for formation of acetate from CO_2 and H_2 (reaction (37) in Table 3 and Figure 2) was slightly greater than that of acetate fermentation to CH_4 and CO_2 with the exception of the low salinity water with H_2 concentrations ~ 0.1 μM. This suggests that if μM concentrations of H_2 are maintained then the conversion of CO_2 to CH_4 via acetogenesis and aceticlastic methanogenesis is viable. The production of H_2 by fermentation of propionate (or for that matter benzoate or butyrate) is not energetically

favorable unless the H_2 concentrations ~0.1 μM. In other words, organic fermentative production of H_2 will not sustain the conversion of CO_2 to CH_4 for the conditions encountered in the deep subsurface and alternative abiotic reactions are required to do so.

11. Abiotic conversion of CO to H_2 and CO_2 or the Gas shift reaction (reaction (34) in Table 3 and Figure 2) represents a possible source of H_2, is favorable even for the highly saline water with the highest dissolved H_2 concentrations and its free energy is greater than that of acetogenesis. A competing reaction for consumption of CO, however, is the abiotic conversion of CO to CH_4 and CO_2 or the Fischer–Tropsch reaction (reaction (22) in Table 3 and Figure 2). This is energetically favorable for all ground water types and has a greater free energy than the Gas shift reaction, but its free energy decreases with increasing temperature. This reaction is most favored for the highly saline ground water probably because of the lower HCO_3^- and pH of this ground water. This suggests that if the CO_2 produced by this reaction is converted to carbonate, then the Fischer–Tropsch reaction may compete with microbial conversion of CO_2 to methane. The abiotic conversion of H_2 and N_2 to NH_3 is favorable for all four ground water types, becomes less favorable with increasing temperature consistent with observations regarding metamorphic N_2 [34], but is strongly favored for environments where H_2 concentrations are high. The rates for the Gas shift, Fischer–Tropsch and ammonia generation reactions are unknown and at the temperatures modeled in this report depend upon the catalyst available (e.g. metal oxides or sulfides).
12. The normally rapid and abiotic reduction of hematite by oxidation of HS^- to SO_4^{2-} (reaction (43) in Table 3 and Figure 2) is also not favored because of the high pH of these ground water types. The reaction becomes slightly less positive if goethite or amorphous $Fe(OH)_3$ is considered, but this reaction is extremely sensitive to the pH regardless of the choice of Fe(III) oxides and for this pH range the free energy is close to zero or positive.

The potential microbial power calculations revealed that the most exothermic reactions were not the most powerful reactions. The microbial redox reactions were ordered in Table 4 and Figure 3 to reflect their importance and this revealed the following:

1. Many of the H_2 oxidizing reactions, such as the reduction of S and SO_4^{2-} to HS^- (reactions (1) and (4) in Table 4 and Figure 3), the abiotic production of NH_3 (reaction (2) in Table 4 and Figure 3), methanogenesis and acetogenesis (reactions (7) and (12) in Table 4 and Figure 3) were the most powerful reactions despite the low free energy yields for some of these reactions. This directly corresponded to the high H_2 concentration as the power varied by four orders of magnitude from the dolomite water, 5×10^{-16} kJ cell^{-1} s^{-1}, to the highly saline water 5×10^{-12} J cell^{-1} s^{-1}, and with diffusivity as the power increases by a factor of two from 20 to 80 °C.
2. The anaerobic methane oxidation reaction (reaction (3) in Table 4 and Figure 3) is the third most powerful reaction despite its low free energy yield for highly saline water and its power increases by two orders of magnitude from the dolomite water, 5×10^{-14} kJ cell^{-1} s^{-1}, to the highly saline water, 5×10^{-12} kJ cell^{-1} s^{-1}. This is a reflection of the high concentrations and diffusivities of its reactants. As mentioned above, however, because the free energy of reaction (7) is negative, anaerobic methane oxidation via reverse methanogenesis cannot proceed.
3. Acetate oxidation coupled with the reduction of MnO_2 to Mn^{2+}, and reduction of S and SO_4^{2-} to HS^- (reactions (5), (10) and (13) in Table 4 and Figure 3) yields high potential power, 5×10^{-14} to 5×10^{-13} kJ cell^{-1} s^{-1}, because of their high free energies and the amount and diffusivity of acetate.
4. CO consuming reactions (reactions (6), (8), (9) and (11) in Table 4 and Figure 3), such as the Fischer–Tropsch reaction, are quite powerful reactions for the highly saline water. Their power values increase by three orders of magnitude from the dolomite water, 5×10^{-17} kJ cell^{-1} s^{-1}, to the highly saline water, 5×10^{-13} kJ cell^{-1} s^{-1}, and like the H_2 consuming reactions the value depends upon the CO concentration.
5. In the low salinity to highly saline ground water all of the nitrate-reducing reactions (reactions (14)–(22) and (28) in Table 4 and Figure 3) were nitrate limited and the annamox reaction nitrite limited (reaction (25) in Table 4 and Figure 3). In the dolomite water, however, the electron donor was limiting. Because these reactions yielded the greatest free energy, however, the power of nitrate reducing reactions ranged from 5×10^{-15} to 5×10^{-13} kJ cell^{-1} s^{-1}. These power values are comparable to those for the reduction of Fe(III) oxides (reactions (39) and (41) in Table 4 and Figure 3). The potential microbial power for Fe^{2+} oxidation by nitrate (reaction (28) in Table 4 and Figure 3) and reduction of hematite (reactions (30)

TABLE 4

POTENTIAL MICROBIAL POWER (5000 KJ CELL^{-1} S^{-1}) FOR FOUR TYPES OF GROUND WATER AT 20, 45 AND 80 °C

Microbial redox reactions	1. Do 20	2. Do 45	3. Do 80	4. LS 20	5. LS 45	6. LS 80	7. MS 20	8. MS 45	9. MS 80	10. Br 20	11. Br 45	12. Br 80
(1) $H_2 + S \rightarrow HS^- + H^+$	-1.8×10^{12}	-3.2×10^{12}	-5.8×10^{12}	-1.5×10^{13}	-2.5×10^{13}	-4.4×10^{13}	-2.8×10^{10}	**-4.0×10^{10}**	-6.9×10^{10}	-6.9×10^{10}	**-9.8×10^{10}**	-1.7×10^{09}
(2) $3H_2 + N_2 \rightarrow 2NH_3$	-6.8×10^{13}	-9.9×10^{13}	-1.3×10^{12}	-6.1×10^{14}	-7.9×10^{14}	-9.1×10^{14}	-1.9×10^{10}	-2.5×10^{10}	-3.8×10^{10}	-6.2×10^{10}	-8.0×10^{10}	-1.2×10^{09}
(3) $CH_4 + SO_4^{2-} \rightarrow H_2O + HCO_3^- + HS^-$	**-2.1×10^{11}**	-4.0×10^{11}	-7.3×10^{11}	-4.2×10^{11}	**-7.8×10^{11}**	-1.5×10^{10}	-1.3×10^{10}	**-1.9×10^{10}**	-3.7×10^{10}	-3.9×10^{10}	**-5.9×10^{10}**	-1.1×10^{09}
(4) $4H_2 + H^+ + SO_4^{2-} \rightarrow HS^- + 4H_2O$	-5.1×10^{13}	-7.5×10^{13}	-9.7×10^{13}	-4.3×10^{14}	-5.1×10^{14}		-1.7×10^{10}	**-2.2×10^{10}**	-3.3×10^{10}	-5.3×10^{10}	**-6.9×10^{10}**	-1.1×10^{09}
(5) Acetate $+ 4MnO_2 + 7H^+ \rightarrow 4Mn^{2+} + 4H_2O + 2HCO_3^-$	**-3.1×10^{11}**	-5.3×10^{11}	-8.7×10^{11}	-4.6×10^{11}	**-7.5×10^{11}**	-1.2×10^{10}	-9.4×10^{11}	**-1.3×10^{10}**	-2.0×10^{10}	-2.1×10^{10}	**-2.9×10^{10}**	-4.7×10^{10}
(6) $3H_2 + CO \rightarrow CH_4 + H_2O$	-7.1×10^{13}	-1.0×10^{12}	-1.4×10^{12}	-7.4×10^{14}	-9.9×10^{14}	-1.0×10^{13}	-1.7×10^{11}	-2.2×10^{11}	-3.2×10^{11}	-1.3×10^{10}	-1.7×10^{10}	-2.5×10^{10}
(7) $4H_2 + H^+ + HCO_3^- \rightarrow CH_4 + 3H_2O$	-3.6×10^{13}	-4.6×10^{13}	-4.1×10^{13}	-1.9×10^{14}			-1.2×10^{10}	**-1.4×10^{10}**	-1.8×10^{10}	-1.3×10^{10}	**-1.5×10^{10}**	-2.1×10^{10}
(8) $4CO + SO_4^{2-} + 4H_2O \rightarrow 4HCO_3^- + HS^- + 3H^+$	-2.2×10^{13}	-3.7×10^{13}	-6.1×10^{13}	-2.2×10^{13}	-3.8×10^{13}	-6.4×10^{13}	-9.2×10^{12}	-1.3×10^{11}	-2.2×10^{11}	-6.9×10^{11}	**-9.8×10^{11}**	-1.7×10^{10}
(9) $4CO + 5H_2O \rightarrow CH_4 + 3HCO_3^- + 3H^+$	-2.0×10^{13}	-3.2×10^{13}	-5.1×10^{13}	-1.9×10^{13}	-3.2×10^{13}	-5.4×10^{13}	-8.0×10^{12}	-1.1×10^{11}	-1.9×10^{11}	-5.9×10^{11}	-8.3×10^{11}	-1.4×10^{10}
(10) Acetate $+ 4S + 4H_2O \rightarrow 5H^+ + 2HCO_3^- + 4HS^-$	**-1.3×10^{11}**	-2.5×10^{11}	-4.8×10^{11}	-1.3×10^{11}	**-2.5×10^{11}**	-4.8×10^{11}	-2.2×10^{11}	**-3.5×10^{11}**	-7.0×10^{11}	-3.5×10^{11}	**-5.6×10^{11}**	-1.1×10^{10}
(11) CO + Hematite $+ 3H^+ \rightarrow 2Fe^{2+} + H_2O + HCO_3^-$	-2.4×10^{13}	-3.0×10^{13}	-2.6×10^{13}	-1.4×10^{13}	-1.4×10^{13}		-6.5×10^{12}	-6.4×10^{12}		-5.5×10^{11}	**-6.0×10^{11}**	-6.1×10^{11}
(12) $4H_2 + H^+ + 2HCO_3^- \rightarrow$ Acetate $+ 4H_2O$	-2.2×10^{13}	-2.2×10^{13}					-9.4×10^{11}	**-1.0×10^{10}**	-1.1×10^{10}	-4.8×10^{11}	**-5.5×10^{11}**	-6.1×10^{11}
(13) Acetate $+ SO_4^{2-} \rightarrow 2HCO_3^- + HS^-$	-2.4×10^{12}	-4.4×10^{12}	-8.3×10^{12}	-3.7×10^{12}	**-6.8×10^{12}**	-1.3×10^{11}	-9.1×10^{12}	-1.4×10^{11}	-2.6×10^{11}	-2.1×10^{11}	**-3.2×10^{11}**	-6.0×10^{11}
(14) $4CO + NO_3^- + 5H_2O \rightarrow 4HCO_3^- + NH_3 + 3H^+$	-7.1×10^{13}	-1.2×10^{12}	-2.0×10^{12}	-6.2×10^{13}	-1.1×10^{12}	-1.8×10^{12}	-2.4×10^{11}	**-3.4×10^{11}**	-5.7×10^{11}	-2.5×10^{11}	-3.5×10^{11}	-5.9×10^{11}
(15) $2.5CO + NO_3^- + 2H_2O \rightarrow 2.5HCO_3^- + 1.5H^+ + 0.5N_2$	-9.9×10^{13}	-1.6×10^{12}	-2.8×10^{12}	-8.5×10^{13}	-1.5×10^{12}	-2.5×10^{12}	-2.1×10^{11}	**-2.9×10^{11}**	-4.8×10^{11}	-2.1×10^{11}	-2.9×10^{11}	-4.9×10^{11}

(continued)

TABLE 4
CONTINUED

Microbial redox reactions	1. Do 20	2. Do 45	3. Do 80	4. LS 20	5. LS 45	6. LS 80	7. MS 20	8. MS 45	9. MS 80	10. Br 20	11. Br 45	12. Br 80
(16) $S_2O_3^{2-} + NO_3^- + 2H_2O \rightarrow 2SO_4^{2-} + H^+ + NH_3$	-2.2×10^{13}	-3.7×10^{13}	-6.1×10^{13}	-5.3×10^{12}	**-8.9×10^{12}**	-1.5×10^{11}	-2.1×10^{11}	**-2.9×10^{11}**	-4.8×10^{11}	-2.1×10^{11}	-2.9×10^{11}	-4.9×10^{11}
(17) $4H_2 + NO_3^- + H^+ \rightarrow NH_3 + 3H_2O$	**-3.1×10^{12}**	-5.2×10^{12}	-8.5×10^{12}	-3.9×10^{13}	-6.3×10^{13}	-1.0×10^{12}	-1.9×10^{11}	-2.6×10^{11}	-4.2×10^{11}	-2.0×10^{11}	-2.7×10^{11}	-4.4×10^{11}
(18) Acetate $+ NO_3^- + H_2O \rightarrow 2HCO_3^- + NH_3$	**-2.4×10^{11}**	-4.1×10^{11}	-7.0×10^{11}	-4.5×10^{12}	**-7.7×10^{12}**	-1.3×10^{11}	-1.4×10^{11}	-2.3×10^{11}	-3.9×10^{11}	-1.7×10^{11}	-2.4×10^{11}	-4.0×10^{11}
(19) $5H_2 + 2NO_3^- + 2H^+ \rightarrow N_2 + 6H_2O$	**-4.6×10^{12}**	-7.7×10^{12}	-1.3×10^{11}	-4.5×10^{12}	**-7.5×10^{12}**	-1.2×10^{11}	-1.3×10^{11}	-2.1×10^{11}	-3.4×10^{11}	-1.7×10^{11}	-2.4×10^{11}	-3.9×10^{11}
(20) $2S + 1.5NO_3^- + 3.5H_2O \rightarrow 2SO_4^{2-} + 2.5H^+ + 1.5NH_3$	**-5.7×10^{11}**	-9.6×10^{11}	-1.7×10^{10}	-4.2×10^{12}	**-7.3×10^{12}**	-1.2×10^{11}	-1.6×10^{11}	-2.2×10^{11}	-3.8×10^{11}	-1.6×10^{11}	-2.2×10^{11}	-3.7×10^{11}
(21) Acetate $+ 1.6NO_3^- + 0.6H^+ \rightarrow 2HCO_3^- + 0.8H_2O + 0.8N_2$	**-3.6×10^{11}**	-6.2×10^{11}	-1.1×10^{10}	-4.3×10^{12}	**-7.3×10^{12}**	-1.2×10^{11}	-1.3×10^{11}	-2.2×10^{11}	-3.7×10^{11}	-1.6×10^{11}	-2.2×10^{11}	-3.7×10^{11}
(22) $HS^- + NO_3^- + H_2O \rightarrow SO_4^{2-} + NH_3$	**-5.5×10^{11}**	-9.1×10^{11}	-1.6×10^{10}	-4.0×10^{12}	**-6.7×10^{12}**	-1.1×10^{11}	-1.5×10^{11}	-2.1×10^{11}	-3.4×10^{11}	-1.5×10^{11}	-2.1×10^{11}	-3.4×10^{11}
(23) $CO + 2H_2O \rightarrow HCO_3^- + H^+ + H_2$	-3.3×10^{14}	-5.9×10^{14}	-1.1×10^{13}	-4.2×10^{14}	-7.9×10^{14}	-1.5×10^{13}	-1.2×10^{12}	-1.9×10^{12}	-3.5×10^{12}	-9.0×10^{12}	-1.4×10^{11}	-2.5×10^{11}
(24) Acetate $+ H_2O \rightarrow CH_4 + HCO_3^-$	-1.1×10^{12}	-2.0×10^{12}	-3.7×10^{12}	-1.6×10^{12}	-2.8×10^{12}	-5.2×10^{12}	-3.2×10^{12}	-4.8×10^{12}	-9.1×10^{12}	-7.4×10^{12}	-1.1×10^{11}	-2.1×10^{11}
(25) $NO_2^- + H^+ + NH_3 \rightarrow 2H_2O + N_2$	-9.1×10^{13}	-1.5×10^{12}	-2.7×10^{12}	-3.8×10^{13}	-5.2×10^{13}	-1.1×10^{12}	-5.5×10^{12}	-7.6×10^{12}	-1.2×10^{11}	-8.3×10^{12}	-1.1×10^{11}	-1.9×10^{11}
(26) $S_2O_3^{2-} + 4H_2 \rightarrow 3H_2O + 2HS^-$	-4.9×10^{14}	-7.4×10^{14}	-9.2×10^{14}	-6.8×10^{14}	-9.9×10^{14}	-1.3×10^{13}	-8.4×10^{12}	-1.1×10^{11}	-1.8×10^{11}	-6.6×10^{12}	-9.0×10^{12}	-1.4×10^{11}
(27) 4Formate $+ H^+ + H_2O \rightarrow CH_4 + 3HCO_3^-$	**-5.9×10^{12}**	-9.1×10^{12}	-1.4×10^{11}	-1.8×10^{13}	-2.5×10^{13}	-2.8×10^{13}	-5.2×10^{13}	-6.0×10^{13}	-7.8×10^{13}	-3.7×10^{12}	-4.8×10^{12}	-7.1×10^{12}
(28) $5Fe^{2+} + NO_3^- + 12H_2O \rightarrow 5Fe(OH)_3 + 9H^+ + 0.5N_2$	**-3.2×10^{12}**	-5.7×10^{12}	-1.1×10^{11}	-5.9×10^{13}	-1.1×10^{12}	-2.0×10^{12}	-2.2×10^{12}	-3.3×10^{12}	-6.2×10^{12}	-2.1×10^{12}	-3.2×10^{12}	-5.9×10^{12}
(29) $S_2O_3^{2-} + H_2O \rightarrow SO_4^{2-} + H^+ + HS^-$				-6.6×10^{13}	-1.2×10^{12}	-2.5×10^{12}	-1.7×10^{12}	-2.7×10^{12}	-5.1×10^{12}	-1.0×10^{12}	-1.6×10^{12}	-3.0×10^{12}
(30) Acetate + 4Hematite $+ 15H^+ \rightarrow 8Fe^{2+} + 8H_2O + 2HCO_3^-$	-2.9×10^{12}											

(31) H_2 + Hematite + $4H^+ \rightarrow 2Fe^{2+} + 3H_2O$	-1.4×10^{12}									-6.2×10^{11}		
(32) Propionate + $3H_2O \rightarrow$ Acetate + $HCO_3^- + H^+ + 3H_2$					-1.4×10^{13}	-5.1×10^{13}						
(33) $2CO + O_2 + 2H_2O \rightarrow 2HCO_3^- + 2H^+$	-2.4×10^{61}	-4.3×10^{54}	-7.7×10^{46}	-2.1×10^{70}	-3.7×10^{64}	-6.6×10^{56}	-2.5×10^{74}	-3.6×10^{67}	-6.5×10^{59}	-2.3×10^{77}	-3.4×10^{70}	-6.1×10^{62}
(34) $2H_2 + O_2 \rightarrow 2H_2O$	-1.7×10^{61}	-3.1×10^{54}	-5.4×10^{46}	-1.0×10^{70}	-1.8×10^{64}	-3.1×10^{56}	-1.4×10^{74}	-2.1×10^{67}	-3.6×10^{59}	-1.3×10^{77}	-1.8×10^{70}	-3.1×10^{62}
(35) Acetate + $2O_2 \rightarrow 2HCO_3^- + H^+$	-1.5×10^{61}	-2.9×10^{54}	-5.5×10^{46}	-1.0×10^{70}	-2.0×10^{64}	-3.8×10^{56}	-9.6×10^{75}	-1.5×10^{67}	-3.0×10^{59}	-7.5×10^{78}	-1.2×10^{70}	-2.4×10^{62}
(36) $CH_4 + 2O_2 \rightarrow HCO_3^- + H^+ + H_2O$	-1.4×10^{61}	-2.6×10^{54}	-5.0×10^{46}	-9.1×10^{71}	-1.7×10^{64}	-3.4×10^{56}	-8.2×10^{75}	-1.3×10^{67}	-2.6×10^{59}	-5.9×10^{78}	-9.8×10^{71}	-2.0×10^{62}
(37) $S + 1.5O_2 + H_2O \rightarrow SO_4^{2-} + 2H^+$	-1.3×10^{61}	-2.6×10^{54}	-5.1×10^{46}	-9.0×10^{71}	-1.8×10^{64}	-3.5×10^{56}	-7.6×10^{75}	-1.3×10^{67}	-2.6×10^{59}	-4.5×10^{78}	-8.1×10^{71}	-1.8×10^{62}
(38) $S_2O_3^{2-} + 2O_2 + H_2O \rightarrow 2SO_4^{2-} + 2H^+$	-1.3×10^{61}	-2.5×10^{54}	-4.7×10^{46}	-9.2×10^{71}	-1.8×10^{64}	-3.5×10^{56}	-7.9×10^{75}	-1.3×10^{67}	-2.5×10^{59}	-4.7×10^{78}	-8.1×10^{71}	-1.7×10^{62}
(39) $HS^- + 2O_2 \rightarrow SO_4^{2-} + H^+$	-1.2×10^{61}	-2.3×10^{54}	-4.5×10^{46}	-7.4×10^{71}	-1.4×10^{64}	-2.8×10^{56}	-5.6×10^{75}	-9.3×10^{68}	-1.9×10^{59}	-3.0×10^{78}	-5.3×10^{71}	-1.1×10^{62}
(40) $2HS^- + 2O_2 \rightarrow S_2O_3^{2-} + H_2O$	-1.2×10^{61}	-2.2×10^{54}	-4.3×10^{46}	-5.6×10^{71}	-1.1×10^{64}	-2.2×10^{56}	-3.4×10^{75}	-5.8×10^{68}	-1.2×10^{59}	-1.2×10^{78}	-2.5×10^{71}	-6.0×10^{63}
(41) $2HS^- + O_2 + 2H^+ \rightarrow 2S + 2H_2O$	-9.2×10^{62}	-1.6×10^{54}	-2.9×10^{46}	-2.5×10^{71}	-3.8×10^{65}							
(42) HS^- + 4Hematite + $15H^+ \rightarrow SO_4^{2-} + 8Fe^{2+} + 8H_2O$												
(43) $4Fe^{2+} + O_2 + 10H_2O \rightarrow 4Fe(OH)_3 + 8H^+$			-3.7×10^{42}			-4.1×10^{52}			-2.0×10^{50}			

Microbial redox reactions have been ordered according to their power with the most powerful reactions for the 80 °C brine appearing first. The microbial reaction numbers and column heading numbers refer to Figure 3. The power is not reported for reactions for which the free energy was > -20 kJ mol^{-1}. Values in bold represent the top 10 values.

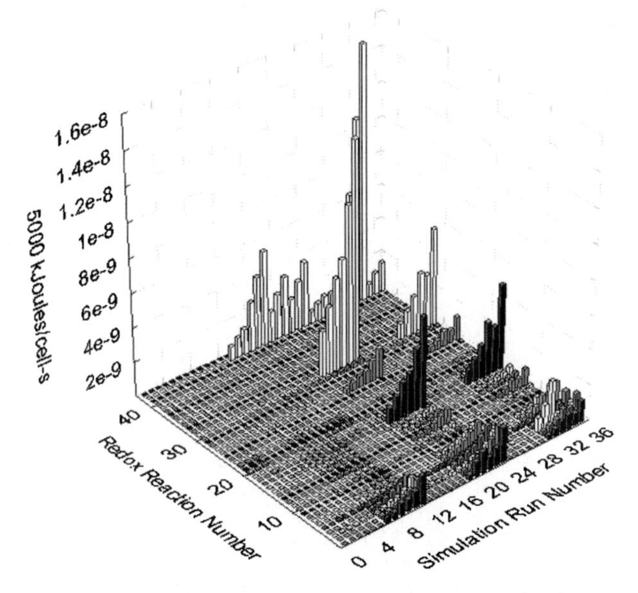

Figure 3: Free energy flux or "potential microbial power" (5000 kJ $cell^{-1}$ s^{-1}) for four different types of ground water and aquifers under ambient conditions at 25, 45 and 80 °C (Simulation Run Numbers 1–12), during simulated CO_2 injection (Simulation Run Numbers 13–24), and during post-injection equilibration with aquifer minerals (Simulation Run Numbers 24–36), based upon values in Tables 4, 6 and 8.

and (31) in Table 4 and Figure 3) are quite similar suggesting that low levels of nitrate could contribute to sustaining microbial Fe(III) reduction at higher temperatures. It is noteworthy that two Fe(III) reducing bacteria from the deep subsurface, *Bacillus infernus* [15] and *Thermus scotoductus* [14] were both capable of nitrate reduction as well.

6. The fermentative methanogenic reactions (reactions (24) and (27) in Table 4 and Figure 3), propionate reaction (reaction (32) in Table 4 and Figure 3) and thiosulfate disproportion (reaction (26) in Table 4 and Figure 3) yielded power levels ranging from 5×10^{-16} to 5×10^{-14} kJ $cell^{-1}$ s^{-1}.
7. All of the aerobic reactions (reactions (33)–(41) and (43) in Table 4 and Figure 3) were O_2 limited, which explains why their power levels are extremely low, $<5 \times 10^{-43}$ kJ $cell^{-1}$ s^{-1}, despite the high energy yield of aerobic reactions. In order for these reactions to be competitive with the above anaerobic reactions the O_2 concentrations need to be >0.1 μM.

If we select the 10 most powerful microbial redox reactions for each ground water type we come to the following conclusions:

1. For the dolomite ground water the total potential microbial power from the top 10 reactions was 10^{-12} kJ $cell^{-1}$ s^{-1}. The most powerful reactions are the oxidation of S and HS^- to SO_4^{2-} followed by nitrate reduction to NH_3. This appears consistent with the dominance of *Thiobacillus denitrificans* in the clone libraries from this aquifer [35]. Other potential metabolic reactions are the reduction of S to HS^-, Mn reduction and other nitrate reduction reactions. This suggests that the community would contain a diverse population of chemolithotrophs and heterotrophs and phylogenetically would probably be comprised of Proteobacteria. Because the oxidation of Fe by nitrate is among the top 10, the precipitation of $Fe(OH)_3$ is conceivable in which case microbial Fe reduction may also occur.

2. For the low salinity ground water the total potential microbial power from the top 10 reactions was 10^{-12} kJ cell^{-1} s^{-1}. The top 10 reactions were quite similar to those of the dolomite with one exception. The microbial reduction of SO_4^{2-} to HS^- was a more significant contributor to the total energy which is consistent with the appearance of SRBs in the 16S rDNA clone libraries (Kieft personal communication, 2004). It also means that the SO_4^{2-} generated by oxidation of S species to SO_4^{2-} with the reduction of nitrate could potentially fuel more sulfate reduction and form a sulfur cycle.
3. For the moderate salinity ground water the total potential microbial power from the top 10 reactions was 5×10^{-12} kJ cell^{-1} s^{-1}. The metabolic reactions are dominated by S and SO_4^{2-} reduction to HS^- and by reduction of CO_2 to methane and acetate. The change in microbial metabolic pathways is largely a reflection of the increasing H_2 concentrations. This appears consistent with the dominance of sulfate reducing members of the Firmicutes and the presence of methanogens in the clone libraries [8].
4. For the high salinity ground water the total potential microbial power from the top 10 reactions was 1.5×10^{-11} kJ cell^{-1} s^{-1}. The metabolic reactions remain dominated by S and SO_4^{2-} reduction to HS^- and by reduction of CO_2 to methane and acetate. SRBs appear to dominate the 16S rDNA clone libraries of the highly saline fracture water [36], but methanogens appear to be absent. One difference between the high salinity and moderate salinity ground water is that abiotic reactions appear competitive based upon their potential power. This appears to be consistent with isotopic data on hydrocarbons reported from these ground water types [37]. Another difference is that CO oxidation by SO_4^{2-} reduction to HS^- and by reduction of hematite appears to be competitive, but microorganisms capable of coupling these electron donors and acceptors have not been isolated to our knowledge. Finally, the highly saline water was CO_2 limited for CO_2 reducing reactions, an observation that bears some significance in terms of the injection of CO_2.
5. The microbial power for the more saline ground water types is greater than that of the dolomite and low salinity water and that power is concentrated into fewer reactions. The microbial power for the dolomite and low salinity water is more equally divided among the microbial redox reactions. This suggests that deeper, more saline ground water microbial communities are less diverse than the shallower, less saline ground water microbial communities, a trend which is borne out in the 16S rDNA clone libraries.

Injection of CO_2

The equilibration of the four ground water types with 200 bars of CO_2 decreased the pH to 2.7–3.3, and consequently increased the pe to 1–7, and dramatically increased dissolved CO_2 and HCO_3^- concentrations to 2.5–8 and 0.002–0.004 mol kg^{-1}, respectively. Solubilization of trace mineral phases affected the concentrations of trace metals and phosphate which obviously have potential impact upon microbial processes. For the purposes of this study we have focused on the first three effects, which had the following significant impact upon the acid and CO_2 producing microbial redox reactions:

1. The fermentation reactions of acetate to CH_4 and CO_2 (reaction (36) in Table 5) and propionate fermentation to acetate, CO_2 and H_2 (reaction (42) in Table 5) were no longer favorable for any of the ground water compositions. This would be a serious impediment to strictly aceticlastic methanogens, whereas the propionate reaction would be more dependent upon the PH_2.
2. The oxidation of reduced S compounds by O_2 (reactions (16)–(18), (21) and (25) in Table 5), which were marginally favorable in the highly saline ground water became endothermic with injection of the CO_2. Given that aerobic S oxidizers are not found in this ground water environment, this does not appear to be a significant perturbation.
3. Of the microbial reactions that were originally unfavorable prior to injection, the reduction of hematite to Fe^{2+} by oxidation of acetate (reaction (39) in Table 5) and H_2 (reaction (41) in Table 5) were far more exothermic due to the reduction in pH. The abiotic reduction of hematite by oxidation of HS^- (reaction (43) in Table 5) is also energetically favorable now.
4. The high CO_2 and HCO_3^- concentrations increased the free energy yield for CO_2 reducing methanogenic and acetogenic reaction (reactions (32) and (37) in Table 5). An increase in acetogenic activity may rescue the aceticlastic methanogens.
5. The aerobic oxidation of acetate (reaction (11) in Table 5) was less favorable because it is a proton and HCO_3^- producing reaction. Other acetate oxidation reactions, however, such as MnO_2 or nitrate reduction were more favorable.

TABLE 5
FREE ENERGY (KJOULE MOLE^{-1}) OF I REDOX REACTIONS FOR GROUND WATER EQUILIBRATED WITH 200 BARS OF CO_2.

Microbial redox reactions	13. Do 20	14. Do 45	15. Do 80	16. LS 20	17. LS 45	18. LS 80	19. MS 20	20. MS 45	21. MS 80	22. Br 20	23. Br 45	24. Br 80
(1) $5H_2 + 2NO_3^- + 2H^+ \rightarrow N_2 + 6H_2O$	−1246	−1122	−976	−1193	−1069	−924	−1305	−1182	−1036	−1321	−1197	−1053
(2) Acetate + $1.6NO_3^- + 0.6H^+ \rightarrow 2HCO_3^- + 0.8H_2O + 0.8N_2$	−907	−832	−745	−890	−815	−728	−897	−822	−735	−899	−824	−738
(3) $2S + 1.5NO_3^- + 3.5H_2O \rightarrow 2SO_4^{2-} + 2.5H^+ + 1.5NH_3$	−813	−747	−673	−799	−733	−655	−797	−733	−655	−797	−731	−653
(4) $4CO + NO_3^- + 5H_2O \rightarrow 4HCO_3^- + NH_3 + 3H^+$	−724	−653	−569	−710	−640	−553	−754	−683	−597	−777	−706	−621
(5) Acetate + $4MnO_2 + 7H^+ \rightarrow 4Mn^{2+} + 4H_2O + 2HCO_3^-$	−981	−895	−793	−1017	−932	−829	−966	−883	−781	−1012	−930	−837
(6) $2.5CO + NO_3^- + 2H_2O \rightarrow 2.5HCO_3^- + 1.5H^+ + 0.5N_2$	−634	−574	−503	−622	−562	−491	−650	−591	−519	−664	−604	−534
(7) $4H_2 + NO_3^- + H^+ \rightarrow NH_3 + 3H_2O$	−706	−631	−544	−670	−596	−507	−758	−683	−595	−772	−697	−610
(8) Acetate + $NO_3^- + H_2O \rightarrow 2HCO_3^- + NH_3$	−616	−566	−508	−606	−555	−496	−610	−560	−501	−614	−564	−506
(9) $S_2O_3^{2-} + NO_3^- + 2H_2O \rightarrow 2SO_4^{2-} + H^+ + NH_3$	−551	−501	−375	−564	−514	−456	−564	−515	−457	−560	−511	−453
(10) $HS^- + NO_3^- + H_2O \rightarrow SO_4^{2-} + NH_3$	−544	−496	−441	−538	−490	−432	−537	−489	−432	−537	−489	−432
(11) Acetate + $2O_2 \rightarrow 2HCO_3^- + H^+$	−344	−322	−438	−302	−282	−261	−101	−278	−261	−73	−71	−130
(12) Acetate + $4S + 4H_2O \rightarrow 5H^+ + 2HCO_3^- + 4HS^-$	−344	−322	−438	−302	−282	−261	−73	−278	−261	−73	−71	−130
(13) $5Fe^{2+} + NO_3^- + 12H_2O \rightarrow 5Fe(OH)_3 + 9H^+ + 0.5N_2$		−61	−73		−31	−44		−31	−42		−52	−58
(14) $NO_2^- + H^+ + NH_3 \rightarrow 2H_2O + N_2$	−355	−324	−286	−348	−317	−282	−356	−324	−289	−355	−324	−288
(15) $CH_4 + 2O_2 \rightarrow HCO_3^- + H^+ + H_2O$	−329	−306	−422	−293	−272	−250	−95	−271	−253	−66	−64	−121
(16) $S_2O_3^{2-} + 2O_2 + H_2O \rightarrow 2SO_4^{2-} + 2H^+$	−278	−257	−305	−260	−241	−221	−55	−233	−217	*−19*	*−18*	−77
(17) $HS^- + 2O_2 \rightarrow SO_4^{2-} + H^+$	−272	−252	−371	−234	−216	−197	−28	−207	−192	*4*	*4*	−56
(18) $2HS^- + 2O_2 \rightarrow S_2O_3^{2-} + H_2O$	−266	−247	−437	−209	−191	−173	−1	−180	−166	*27*	*26*	−35
(19) $2CO + O_2 + 2H_2O \rightarrow 2HCO_3^- + 2H^+$	−226	−204	−249	−203	−183	−159	−122	−200	−178	−118	−107	−123
(20) $4CO + SO_4^{2-} + 4H_2O \rightarrow 4HCO_3^- + HS^- + 3H^+$	−180	−157	−128	−172	−150	−121	−216	−194	−165	−240	−217	−189
(21) $S + 1.5O_2 + H_2O \rightarrow SO_4^{2-} + 2H^+$	−202	−191	−284	−172	−161	−151	*7*	−155	−148	*7*	*4*	−45
(22) $4CO + 5H_2O \rightarrow CH_4 + 3HCO_3^- + 3H^+$	−123	−103	−77	−114	−94	−68	−150	−130	−104	−170	−150	−124

(23) $2H_2 + O_2 \rightarrow 2H_2O$	−217	−193	−237	−183	−161	−136	−124	−201	−177	−115	−102	−117
(24) $S_2O_3^{2-} + 4H_2 \rightarrow 3H_2O + 2HS^-$	−167	−140	−37	−158	−131	−99	−247	−221	−189	−257	−230	−199
(25) $2HS^- + O_2 + 2H^+ \rightarrow 2S + 2H_2O$	−140	−123	−175	−125	−109	−92	−21	−104	−88	*−7*	*0*	−23
(26) $H_2 + S \rightarrow HS^- + H^+$	−108	−97	−119	−92	−81	−68	−58	−100	−89	−58	−51	−58
(27) $3H_2 + CO \rightarrow CH_4 + H_2O$	−109	−86	−58	−84	−61	−33	−153	−130	−102	−166	−143	−115
(28) $4H_2 + H^+ + SO_4^{2-} \rightarrow HS^- + 4H_2O$	−161	−135	−103	−132	−106	−75	−220	−194	−163	−234	−208	−178
(29) $3H_2 + N_2 \rightarrow 2NH_3$	−166	−141	−113	−148	−122	−90	−210	−185	−153	−222	−197	−167
(30) $4Formate + H^+ + H_2O \rightarrow CH_4 + 3HCO_3^-$	−102	−88	−72	−60	−47	−31	−65	−51	−36	−81	−68	−53
(31) $Acetate + SO_4^{2-} \rightarrow 2HCO_3^- + HS^-$	−72	−70	−67	−68	−66	−64	−73	−71	−69	−77	−75	−74
(32) $4H_2 + H^+ + HCO_3^- \rightarrow CH_4 + 3H_2O$	−104	−81	−52	−74	−50	−22	−154	−130	−102	−164	−141	−112
(33) $CO + Hematite + 3H^+ \rightarrow 2Fe^{2+} + H_2O + HCO_3^-$	−156	−133	−103	−162	−139	−110	−173	−151	−122	−170	−148	−121
(34) $CO + 2H_2O \rightarrow HCO_3^- + H^+ + H_2$	*−5*	*−6*	*−6*	*−10*	*−11*	*−12*	*1*	*0*	*−1*	*−1*	*−2*	*−3*
(35) $CH_4 + SO_4^{2-} \rightarrow H_2O + HCO_3^- + HS^-$	−57	−54	−51	−59	−56	−53	−67	−64	−61	−70	−68	−65
(36) $Acetate + H_2O \rightarrow CH_4 + HCO_3^-$	*−15*	*−15*	*−16*	*−9*	*−9*	*−10*	*−6*	*−7*	*−8*	*−7*	*−7*	*−8*
(37) $4H_2 + H^+ + 2HCO_3^- \rightarrow Acetate + 4H_2O$	−89	−65	−36	−65	−40	−11	−147	−123	−94	−157	−133	−104
(38) $S_2O_3^{2-} + H_2O \rightarrow SO_4^{2-} + H^+ + HS^-$	*−6*	*−5*	*66*	−26	−25	−24	−27	−26	−26	−23	−22	−21
(39) $Acetate + 4Hematite + 15H^+ \rightarrow 8Fe^{2+} + 8H_2O + 2HCO_3^-$	−515	−443	−352	−543	−473	−381	−549	−480	−392	−516	−449	−369
(40) $4Fe^{2+} + O_2 + 10H_2O \rightarrow 4Fe(OH)_3 + 8H^+$		*206*	*95*		*241*	*198*		*247*	*204*		*335*	*258*
(41) $H_2 + Hematite + 4H^+ \rightarrow 2Fe^{2+} + 3H_2O$	−151	−127	−97	−152	−128	−98	−174	−151	−121	−168	−146	−118
(42) $Propionate + 3H_2O \rightarrow Acetate + HCO_3^- + H^+ + 3H_2$	*60*	*42*	*20*	*39*	*21*	*−1*	*102*	*84*	*62*	*109*	*91*	*68*
(43) $HS^- + 4Hematite + 15H^+ \rightarrow SO_4^{2-} + 8Fe^{2+} + 8H_2O$	−443	−373	−285	−475	−407	−317	−476	−409	−323	−439	−374	−296
(44) $NH_3 + 1.5{*}O_2 \rightarrow NO_2^- + H^+ + H_2O$	*196*	*174*	*43*	*221*	*198*	*168*	*379*	*209*	*176*	*404*	*368*	*279*
(45) $4Mn^{2+} + NO_3^- + 5H_2O \rightarrow 4MnO_2 + 7H^+ + NH_3$	*364*	*330*	*284*	*411*	*377*	*333*	*398*	*322*	*280*	*398*	*367*	*331*
(46) $2NO_2^- + O_2 - > 2NO_3^-$	*153*	*140*	*54*	*166*	*152*	*134*	*260*	*147*	*128*	*274*	*250*	*193*
(47) $2Mn^{2+} + O_2 + 2H_2O \rightarrow 2MnO_2 + 4H^+$	*318*	*287*	*177*	*358*	*325*	*284*	*470*	*302*	*260*	*470*	*430*	*353*

The reactions are ordered from most negative to positive with respect to the free energy for the dolomite ground water at 20 °C. The microbial reaction numbers and column heading numbers refer to Figure 10. Values in italics are > -20 kJ mol^{-1} and therefore are not considered to be viable for microbial metabolism.

TABLE 6

POTENTIAL MICROBIAL POWER (5000 KJ CELL^{-1} S^{-1}) FOR FOUR TYPES OF GROUND WATER EQUILIBRATED WITH 200 BARS OF CO_2

Microbial Redox Reactions	13. Do 20	14. Do 45	15. Do 80	16. LS 20	17. LS 45	18. LS 80	19. MS 20	20. MS 45	21. MS 80	22. Br 20	23. Br 45	24. Br 80
(1) $H_2 + S \rightarrow HS^- + H^+$	-2.7×10^{-12}	-4.0×10^{-12}	-8.3×10^{-12}	-2.3×10^{-13}	-3.3×10^{-13}	-4.7×10^{-13}	-2.8×10^{-10}	**-8.3×10^{-10}**	-1.2×10^{-09}	-7.1×10^{-10}	**-1.1×10^{-09}**	-2.0×10^{-09}
(2) $3H_2 + N_2 \rightarrow 2NH_3$	-1.4×10^{-12}	-1.9×10^{-12}	-2.6×10^{-12}	-1.2×10^{-13}	-1.7×10^{-13}	-2.1×10^{-13}	-3.5×10^{-10}	**-5.1×10^{-10}**	-7.1×10^{-10}	-9.1×10^{-10}	**-1.4×10^{-09}**	-1.9×10^{-09}
(3) $CH_4 + SO_4^{2-} \rightarrow H_2O + HCO_3^- + HS^-$	**-4.7×10^{-11}**	-7.4×10^{-11}	-1.2×10^{-10}	-8.4×10^{-11}	**-1.3×10^{-10}**	-2.1×10^{-10}	-1.9×10^{-10}	**-3.1×10^{-10}**	-5.0×10^{-10}	-5.3×10^{-10}	**-8.7×10^{-10}**	-1.4×10^{-09}
(4) $4H_2 + H^+ + SO_4^{2-} \rightarrow HS^- + 4H_2O$	-1.0×10^{-12}	-1.4×10^{-12}	-1.8×10^{-12}	-8.2×10^{-14}	-1.1×10^{-13}	-1.3×10^{-13}	-2.7×10^{-10}	**-4.0×10^{-10}**	-5.7×10^{-10}	-7.2×10^{-10}	**-1.1×10^{-09}**	-1.6×10^{-09}
(5) Acetate + $4MnO_2 + 7H^+ \rightarrow 4Mn^{2+} + 4H_2O + 2HCO_3^-$	**-5.2×10^{-11}**	-8.0×10^{-11}	-1.2×10^{-10}	-6.8×10^{-11}	**-1.0×10^{-10}**	-1.6×10^{-10}	-1.3×10^{-10}	**-2.0×10^{-10}**	-2.9×10^{-10}	-3.4×10^{-10}	**-5.2×10^{-10}**	-7.9×10^{-10}
(6) $3H_2 + CO \rightarrow CH_4 + H_2O$	-9.0×10^{-13}	-1.2×10^{-12}	-1.4×10^{-12}	-6.9×10^{-14}	-8.4×10^{-14}	-7.7×10^{-14}	-2.5×10^{-10}	**-3.6×10^{-10}**	-4.8×10^{-10}	-6.8×10^{-10}	**-9.9×10^{-10}**	-1.3×10^{-09}
(7) $4H_2 + H^+ + HCO_3^- \rightarrow CH_4 + 3H_2O$	-6.4×10^{-13}	-8.4×10^{-13}	-9.1×10^{-13}	-4.5×10^{-14}	-5.2×10^{-14}	-3.8×10^{-14}	-1.9×10^{-10}	**-2.7×10^{-10}**	-3.5×10^{-10}	-5.1×10^{-10}	**-7.3×10^{-10}**	-9.8×10^{-10}
(8) $4CO + SO_4^{2-} + 4H_2O \rightarrow 4HCO_3^- + HS^- + 3H^+$	-2.0×10^{-13}	-2.9×10^{-13}	-4.0×10^{-13}	-1.4×10^{-13}	-2.1×10^{-13}	-2.9×10^{-13}	-6.0×10^{-12}	-9.1×10^{-12}	-1.3×10^{-11}	-4.7×10^{-11}	-7.1×10^{-11}	-1.0×10^{-10}
(9) $4CO + 5H_2O \rightarrow CH_4 + 3HCO_3^- + 3H^+$	-1.4×10^{-13}	-1.9×10^{-13}	-2.4×10^{-13}	-9.5×10^{-14}	-1.3×10^{-13}	-1.6×10^{-13}	-4.2×10^{-12}	-6.1×10^{-12}	-8.2×10^{-12}	-3.3×10^{-11}	-4.9×10^{-11}	-6.8×10^{-11}
(10) Acetate + $4S + 4H_2O \rightarrow 5H^+ + 2HCO_3^- + 4HS^-$	**-1.8×10^{-11}**	-2.9×10^{-11}	-6.6×10^{-11}	-2.0×10^{-11}	**-3.1×10^{-11}**	-4.9×10^{-11}	-9.7×10^{-12}	-6.2×10^{-11}	-9.8×10^{-11}	-2.4×10^{-11}	-4.0×10^{-11}	-1.2×10^{-10}
(11) CO + Hematite + $3H^+ \rightarrow 2Fe^{2+} + H_2O + HCO_3^-$	-6.9×10^{-13}	-9.9×10^{-13}	-1.3×10^{-12}	-5.4×10^{-13}	-7.8×10^{-13}	-1.0×10^{-12}	-1.9×10^{-11}	-2.8×10^{-11}	-3.8×10^{-11}	-1.3×10^{-10}	-1.9×10^{-10}	-2.7×10^{-10}
(12) $4H_2 + H^+ + 2HCO_3^- \rightarrow$ Acetate + $4H_2O$	-5.5×10^{-13}	-6.8×10^{-13}	-6.3×10^{-13}	-4.0×10^{-14}	-4.2×10^{-14}		-1.8×10^{-10}	**-2.6×10^{-10}**	-3.3×10^{-10}	-4.9×10^{-10}	**-6.9×10^{-10}**	-9.1×10^{-10}
(13) Acetate + $SO_4^{2-} \rightarrow 2HCO_3^- + HS^-$	-3.8×10^{-12}	-6.2×10^{-12}	-1.0×10^{-11}	-4.5×10^{-12}	-7.3×10^{-12}	-1.2×10^{-11}	-9.7×10^{-12}	-1.6×10^{-11}	-2.6×10^{-11}	-2.6×10^{-11}	-4.2×10^{-11}	-6.9×10^{-11}
(14) $4CO + NO_3^- + 5H_2O \rightarrow 4HCO_3^- + NH_3 + 3H^+$	-8.1×10^{-13}	-1.2×10^{-12}	-1.8×10^{-12}	-5.9×10^{-13}	-9.0×10^{-13}	-1.3×10^{-12}	-2.1×10^{-11}	-3.2×10^{-11}	-4.7×10^{-11}	-1.5×10^{-10}	-2.3×10^{-10}	-3.4×10^{-10}
(15) $2.5CO + NO_3^- + 2H_2O \rightarrow 2.5HCO_3^- + 1.5H^+ + 0.5N_2$	-1.1×10^{-12}	-1.7×10^{-12}	-2.5×10^{-12}	-8.3×10^{-13}	-1.3×10^{-12}	-1.9×10^{-12}	-2.9×10^{-11}	-4.4×10^{-11}	-6.5×10^{-11}	-2.1×10^{-10}	-3.2×10^{-10}	-4.7×10^{-10}
(16) $S_2O_3^{2-} + NO_3^- + 2H_2O \rightarrow 2SO_4^{2-} + H^+ + NH_3$	-2.4×10^{-13}	-3.7×10^{-13}	-4.7×10^{-13}	-5.3×10^{-12}	-8.1×10^{-12}	-1.2×10^{-11}	-2.8×10^{-11}	-4.3×10^{-11}	-6.3×10^{-11}	-1.9×10^{-11}	-2.9×10^{-11}	-4.3×10^{-11}
(17) $4H_2 + NO_3^- + H^+ \rightarrow NH_3 + 3H_2O$	-4.4×10^{-12}	-6.6×10^{-12}	-9.5×10^{-12}	-4.1×10^{-13}	-6.2×10^{-13}	-8.8×10^{-13}	-9.4×10^{-10}	**-1.4×10^{-09}**	-2.1×10^{-09}	-2.4×10^{-09}	**-3.6×10^{-09}**	-5.3×10^{-09}
(18) Acetate + $NO_3^- + H_2O \rightarrow 2HCO_3^- + NH_3$	**-3.3×10^{-11}**	-5.1×10^{-11}	-7.6×10^{-11}	-5.3×10^{-12}	-8.1×10^{-12}	-1.2×10^{-11}	-2.6×10^{-11}	-4.0×10^{-11}	-6.0×10^{-11}	-1.7×10^{-11}	-2.7×10^{-11}	-4.0×10^{-11}
(19) $5H_2 + 2NO_3^- + 2H^+ \rightarrow N_2 + 6H_2O$	-6.2×10^{-12}	-9.3×10^{-12}	-1.4×10^{-11}	-5.3×10^{-12}	-8.1×10^{-12}	-1.2×10^{-11}	-2.6×10^{-11}	-3.8×10^{-11}	-5.6×10^{-11}	-1.9×10^{-11}	-2.8×10^{-11}	-4.2×10^{-11}
(20) $2S + 1.5NO_3^- + 3.5H_2O \rightarrow 2SO_4^{2-} + 2.5H^+ + 1.5NH_3$	**-7.0×10^{-11}**	-1.1×10^{-10}	-1.6×10^{-10}	-4.6×10^{-12}	-7.0×10^{-12}	-1.1×10^{-11}	-2.3×10^{-11}	-3.5×10^{-11}	-5.3×10^{-11}	-1.5×10^{-11}	-2.3×10^{-11}	-3.5×10^{-11}
(21) Acetate + $1.6NO_3^- + 0.6H^+ \rightarrow 2HCO_3^- + 0.8H_2O + 0.8N_2$	**-4.8×10^{-11}**	-7.4×10^{-11}	-1.1×10^{-10}	-4.9×10^{-12}	-7.5×10^{-12}	-1.1×10^{-11}	-2.4×10^{-11}	-3.7×10^{-11}	-5.5×10^{-11}	-1.6×10^{-11}	-2.5×10^{-11}	-3.7×10^{-11}
(22) $HS^- + NO_3^- + H_2O \rightarrow SO_4^{2-} + NH_3$	**-7.0×10^{-11}**	-1.1×10^{-10}	-1.6×10^{-10}	-4.6×10^{-12}	-7.0×10^{-12}	-1.0×10^{-11}	-2.3×10^{-11}	-3.5×10^{-11}	-5.2×10^{-11}	-1.5×10^{-11}	-2.3×10^{-11}	-3.5×10^{-11}
(23) $CO + 2H_2O \rightarrow HCO_3^- + H^+ + H_2$												
(24) Acetate + $H_2O \rightarrow CH_4 + HCO_3^-$												
(25) $NO_2^- + H^+ + NH_3 \rightarrow 2H_2O + N_2$	-5.5×10^{-13}	-8.4×10^{-13}	-1.2×10^{-12}	-3.2×10^{-13}	-4.9×10^{-13}	-7.3×10^{-13}	-2.2×10^{-12}	-3.4×10^{-12}	-5.0×10^{-12}	-5.5×10^{-12}	-8.4×10^{-12}	-1.3×10^{-11}

(26) $S_2O_3^{2-} + 4H_2 \rightarrow 3H_2O + 2HS^-$	-7.4×10^{-14}	-1.0×10^{-13}	-4.6×10^{-14}	-9.7×10^{-14}	-1.4×10^{-13}	-1.7×10^{-13}	-3.1×10^{-10}	**-4.6×10^{-10}**	-6.6×10^{-10}	-7.9×10^{-10}	**-1.2×10^{-09}**	-1.7×10^{-09}
(27) 4Formate + $H^+ + H_2O \rightarrow CH_4 + 3HCO_3^-$	-8.4×10^{-12}	-1.2×10^{-11}	-1.7×10^{-11}	-2.5×10^{-13}	-3.2×10^{-13}	-3.6×10^{-13}	-5.3×10^{-13}	-7.1×10^{-13}	-8.3×10^{-13}	-3.3×10^{-12}	-4.7×10^{-12}	-6.1×10^{-12}
(28) $5Fe^{2+} + NO_3^- + 12H_2O \rightarrow 5Fe(OH)_3 + 9H^+ + 0.5N_2$		-1.1×10^{-12}	-2.3×10^{-12}		-4.6×10^{-13}	-1.1×10^{-12}		-6.9×10^{-13}	-1.6×10^{-12}		-3.5×10^{-12}	-3.8×10^{-12}
(29) $S_2O_3^{2-} + H_2O \rightarrow SO_4^{2-} + H^+ + HS^-$				-4.2×10^{-15}	-6.8×10^{-15}	-1.1×10^{-14}	-1.2×10^{-12}	-1.9×10^{-12}	-3.2×10^{-12}	-7.5×10^{-13}	-1.2×10^{-12}	-2.0×10^{-12}
(30) Acetate + 4Hematite + $15H^+ \rightarrow 8Fe^{2+} + 8H_2O + 2HCO_3^-$	**-2.7×10^{-11}**	-4.0×10^{-11}	-5.3×10^{-11}	-3.6×10^{-11}	**-5.3×10^{-11}**	-7.2×10^{-11}	-7.3×10^{-11}	-1.1×10^{-10}	-1.5×10^{-10}	-1.7×10^{-10}	-2.5×10^{-10}	-3.5×10^{-10}
(31) H_2 + Hematite + $4H^+ \rightarrow 2Fe^{2+} + 3H_2O$	**-1.3×10^{-11}**	-1.8×10^{-11}	-2.5×10^{-11}	-1.3×10^{-12}	-2.0×10^{-12}	-2.7×10^{-12}	-2.7×10^{-09}	**-4.0×10^{-09}**	-5.5×10^{-09}	-6.4×10^{-09}	**-9.3×10^{-09}**	-1.3×10^{-08}
(32) Propionate + $3H_2O \rightarrow$ Acetate + $HCO_3^- + H^+ + 3H_2$												
(33) $2CO + O_2 + 2H_2O \rightarrow 2HCO_3^- + 2H^+$	-2.6×10^{-56}	-4.0×10^{-52}	-8.1×10^{-36}	-2.3×10^{-59}	-1.1×10^{-54}	-5.2×10^{-49}	-1.4×10^{-74}	-3.9×10^{-55}	-5.8×10^{-51}	-1.4×10^{-77}	-2.1×10^{-70}	-4.0×10^{-59}
(34) $2H_2 + O_2 \rightarrow 2H_2O$	-2.5×10^{-56}	-3.7×10^{-52}	-7.7×10^{-36}	-2.1×10^{-59}	-9.4×10^{-55}	-4.4×10^{-49}	-1.4×10^{-74}	-3.9×10^{-55}	-5.8×10^{-51}	-1.3×10^{-77}	-2.0×10^{-70}	-3.8×10^{-59}
(35) Acetate + $2O_2 \rightarrow 2HCO_3^- + H^+$	-2.0×10^{-56}	-3.1×10^{-52}	-7.1×10^{-36}	-1.7×10^{-59}	-8.2×10^{-55}	-4.2×10^{-49}	-5.8×10^{-75}	-2.7×10^{-55}	-4.3×10^{-51}	-4.2×10^{-78}	-6.9×10^{-71}	-2.1×10^{-59}
(36) $CH_4 + 2O_2 \rightarrow HCO_3^- + H^+ + H_2O$	-1.9×10^{-56}	-3.0×10^{-52}	-6.9×10^{-36}	-1.7×10^{-59}	-7.9×10^{-55}	-4.1×10^{-49}	-5.5×10^{-75}	-2.6×10^{-55}	-4.1×10^{-51}	-3.8×10^{-78}	-6.2×10^{-71}	-2.0×10^{-59}
(37) $S + 1.5O_2 + H_2O \rightarrow SO_4^{2-} + 2H^+$	-1.6×10^{-62}			-1.3×10^{-59}	-6.3×10^{-55}	-3.3×10^{-49}		-2.0×10^{-55}	-3.2×10^{-51}			-3.4×10^{-60}
(38) $S_2O_3^{2-} + 2O_2 + H_2O \rightarrow 2SO_4^{2-} + 2H^+$	-1.6×10^{-56}	-2.5×10^{-52}	-5.0×10^{-36}	-1.5×10^{-59}	-7.0×10^{-55}	-3.6×10^{-49}	-3.2×10^{-75}	-2.3×10^{-55}	-3.5×10^{-51}	-1.1×10^{-78}	-1.7×10^{-71}	-1.3×10^{-59}
(39) $HS^- + 2O_2 \rightarrow SO_4^{2-} + H^+$	-1.6×10^{-56}	-2.4×10^{-52}	-6.0×10^{-36}	-1.4×10^{-59}	-6.3×10^{-55}	-3.2×10^{-49}	-1.6×10^{-75}	-2.0×10^{-55}	-3.1×10^{-51}			-9.1×10^{-60}
(40) $2HS^- + 2O_2 \rightarrow S_2O_3^{2-} + H_2O$	-1.5×10^{-56}	-2.4×10^{-52}	-7.1×10^{-36}	-1.2×10^{-59}	-5.6×10^{-55}	-2.8×10^{-49}		-1.7×10^{-55}	-2.7×10^{-51}			-5.6×10^{-60}
(41) $2HS^- + O_2 + 2H^+ \rightarrow 2S + 2H_2O$	-1.6×10^{-56}	-2.4×10^{-52}	-5.7×10^{-36}	-1.4×10^{-59}	-6.4×10^{-55}	-3.0×10^{-49}	-2.4×10^{-75}	-2.0×10^{-55}	-2.9×10^{-51}			-7.4×10^{-60}
(42) HS^- + 4Hematite + $15H^+ \rightarrow SO_4^{2-} + 8Fe^{2+} + 8H_2O$	**-8.2×10^{-10}**	-1.2×10^{-09}	-1.5×10^{-09}	-2.9×10^{-09}	**-4.2×10^{-09}**	-5.6×10^{-09}	-1.8×10^{-09}	**-2.7×10^{-09}**	-3.5×10^{-09}	-2.0×10^{-09}	**-2.9×10^{-09}**	-3.9×10^{-09}
(43) $4Fe^{2+} + O_2 + 10H_2O \rightarrow 4Fe(OH)_3 + 8H^+$												

Microbial redox reactions have been ordered according to their power with the most powerful reactions for the 80 °C brine appearing first. The microbial reaction numbers and column heading numbers refer to Figure 3. The power is not reported for reactions for which the free energy was > -20 kJ mol^{-1}. The values in bold represent the top 10 values.

The free energies of the nitrate reduction reactions were greater with the N_2 producing reactions being more favored than before. In terms of the potential microbial power values, the hematite reduction reaction by oxidation of HS^- (reaction (42) in Table 6) became the most powerful suggesting that this abiotic reaction will dominate in siliclastic aquifers where Fe(III) oxides and HS^- are present and will significantly ameliorate the low pH conditions. In the more saline aquifers where H_2 are high, the microbial reduction of Fe(III) oxides are equally important and will dominate if HS^- is limiting. These reactions will also raise the pH of the ground water and promote precipitation of the CO_2 as carbonate.

In terms of the available microbial power for the top 10 reactions, the CO_2 injection has increased power levels by a factor of 10. This is primarily the result of the reduction in pH. For many of the microorganisms, this pH range falls below their optimal growth regime so that the increased power may not be immediately available until the pH increases. Initially after CO_2 injection, therefore, an increase of the pH is anticipated due to abiotic redox reactions, such as redox reaction (42) in Table 6, or alteration of the aquifer minerals by the carbonic acid.

Dissolution of Aquifer Minerals

The low pH, CO_2 saturated, dolomitic ground water was reacted with dolomite and calcite to simulate a carbonate aquifer. The low pH, CO_2 saturated, low, moderate and highly saline ground water was reacted with albite and minor calcite to simulate a siliclastic aquifer [23]. The impacts on ground water chemistry were as follows:

1. For the dolomite system, the pH increased from 3.1 to 4.6, the pe decreased from 9.5 to 7.5 and the dissolved CO_2 decreased slightly from 7.99 to 7.91 M. As dolomite and calcite dissolved, chalcedony, kaolinite, hydroxyapatite, fluorite and various metal sulfide minerals precipitated until dolomite and subsequently calcite attained saturation. The reaction led to a net increase in porosity of 0.3%. The only significant difference in simulations at higher aquifer temperatures is that more carbonate precipitation occurred with no significant change in the porosity.
2. For the low salinity ground water, the pH increased from 2.9 to 7, the pe decreased from 5.7 to -2.8 and the dissolved CO_2 decreased from 7.99 to 0.2 M. In terms of pH and pe, these values are close to that of the initial ground water (Table 2). As albite and calcite dissolved, chalcedony, kaolinite and nahcolite, the Na bicarbonate mineral species, and minor sulfide mineral phases precipitated. The reaction led to a 20% reduction in porosity. For 80 °C, the reaction increased the pH to 6, decreased the pe to -1.9 and reduced the dissolved CO_2 to 1.3 M with no significant change in the porosity primarily because nahcolite did not precipitate.
3. For the moderate salinity ground water, the pH increased from 2.9 to 7, the pe decreased from 2 to -3.4 and the dissolved CO_2 decreased slightly from 7.99 to 0.2 M. In terms of pH and pe, these values are close to that of the initial ground water (Table 2). As albite and calcite dissolved, chalcedony, kaolinite, rhodochrosite and nahcolite and minor sulfide mineral phases precipitated. The reaction led to a 20% reduction in porosity. For 80 °C, the reaction increased the pH to 6, decreased the pe to -1.8 and reduced the dissolved CO_2 to 1.3 M with no significant change in the porosity primarily because nahcolite did not precipitate.
4. For the high salinity ground water, the pH increased from 2.7 to 6.9, the pe decreased from 2 to -3.3 and the dissolved CO_2 decreased slightly from 7.99 to 0.2 M. In terms of pH and pe, these values are close to that of the initial ground water (Table 2). As albite and calcite dissolved, chalcedony, kaolinite, rhodochrosite, dolomite, witherite and nahcolite and minor sulfide mineral phases precipitated. The reaction led to a 20% reduction in porosity. For 80 °C, the reaction increased the pH to 6, decreased the pe to -1.8 and reduced the dissolved CO_2 to 1.3 M with a 0.3% increase in the porosity primarily because nahcolite and the other carbonate minerals did not precipitate, although minor siderite did.

With the dissolution reactions ameliorating some of the effects of CO_2 injection, the only significant change in the microbial redox reactions were the following:

1. The fermentation reactions of acetate fermentation to CH_4 and CO_2 (reaction (36) in Table 7) and propionate fermentation to acetate, CO_2 and H_2 (reaction (42) in Table 7) still remain unfavorable for any of the ground water compositions. This would be a serious impediment to strictly aceticlastic

TABLE 7

FREE ENERGY (KJ MOL^{-1}) OF REDOX REACTIONS AFTER INTERACTION OF CO_2 SATURATED WATER WITH AQUIFER MINERALS

Microbial Redox Reactions	25. Do 20	26. Do 45	27. Do 80	28. LS 20	29. LS 45	30. LS 80	31. MS 20	32. MS 45	33. MS 80	34. Br 20	35. Br 45	36. Br 80
(1) $5H_2 + 2NO_3^- + 2H^+ \rightarrow N_2 + 6H_2O$	−1224	−1103	−961	−1140	−1023	−885	−1252	−1135	−997	−1268	−1151	−1013
(2) Acetate $+ 1.6NO_3^- + 0.6H^+ \rightarrow 2HCO_3^- + 0.8H_2O + 0.8N_2$	−888	−816	−732	−849	−773	−690	−856	−781	−697	−860	−783	−700
(3) $2S + 1.5NO_3^- + 3.5H_2O \rightarrow 2SO_4^{2-} + 2.5H^+ + 1.5NH_3$				−835	−765	−683	−833	−763	−680	−824	−754	−671
(4) $4CO + NO_3^- + 5H_2O \rightarrow 4HCO_3^- + NH_3 + 3H^+$	−705	−636	−555	−702	−614	−525	−744	−657	−567	−766	−679	−589
(5) Acetate $+ 4MnO_2 + 7H^+ \rightarrow 4Mn^{2+} + 4H_2O + 2HCO_3^-$	−907	−831	−741	−812	−764	−691	−797	−739	−662	−799	−742	−671
(6) $2.5CO + NO_3^- + 2H_2O \rightarrow 2.5HCO_3^- + 1.5H^+ + 0.5N_2$	−624	−565	−496	−622	−550	−476	−651	−579	−505	−664	−593	−518
(7) $4H_2 + NO_3^- + H^+ \rightarrow NH_3 + 3H_2O$	−686	−614	−530	−618	−552	−471	−705	−639	−557	−718	−651	−570
(8) Acetate $+ NO_3^- + H_2O \rightarrow 2HCO_3^- + NH_3$	−595	−547	−493	−556	−507	−453	−560	−511	−457	−563	−513	−460
(9) $S_2O_3^{2-} + NO_3^- + 2H_2O \rightarrow 2SO_4^{2-} + H^+ + NH_3$	−485	−446	−388	−567	−519	−460	−566	−518	−459	−558	−509	−451
(10) $HS^- + NO_3^- + H_2O \rightarrow SO_4^{2-} + NH_3$	−548	−499	−443	−540	−493	−436	−538	−491	−434	−534	−488	−430
(11) Acetate $+ 2O_2 \rightarrow 2HCO_3^- + H^+$	−483	−440	−415	−65	−64	−75	−36	−72	−83	−43	−39	−40
(12) Acetate $+ 4S + 4H_2O \rightarrow 5H^+ + 2HCO_3^- + 4HS^-$				−65	−64	−75	−28	−72	−83	−37	−32	−31
(13) $5Fe^{2+} + NO_3^- + 12H_2O \rightarrow 5Fe(OH)_3 + 9H^+ + 0.5N_2$	−132	−129	−127	−214	−186	−160	−168	−189	−163	−175	−188	−162
(14) $NO_2^- + H^+ + NH_3 \rightarrow 2H_2O + N_2$	−356	−324	−286	−349	−316	−280	−358	−325	−288	−359	−326	−289
(15) $CH_4 + 2O_2 \rightarrow HCO_3^- + H^+ + H_2O$	−470	−426	−400	−62	−63	−73	−36	−74	−83	−43	−40	−40
(16) $S_2O_3^{2-} + 2O_2 + H_2O \rightarrow 2SO_4^{2-} + 2H^+$	−373	−339	−310	−77	−76	−82	−43	−79	−85	−39	−35	−31
(17) $HS^- + 2O_2 \rightarrow SO_4^{2-} + H^+$	−436	−392	−365	−49	−50	−58	−15	−52	−59	*−15*	*−13*	*−11*
(18) $2HS^- + 2O_2 \rightarrow S_2O_3^{2-} + H_2O$	−499	−445	−420	−22	−25	−33	14	−26	−34	*8*	*8*	*10*
(19) $2CO + O_2 + 2H_2O \rightarrow 2HCO_3^- + 2H^+$	−296	−264	−238	−105	−86	−73	−110	−109	−97	−123	−102	−85

(continued)

TABLE 7
CONTINUED

Microbial Redox Reactions	25. Do 20	26. Do 45	27. Do 80	28. LS 20	29. LS 45	30. LS 80	31. MS 20	32. MS 45	33. MS 80	34. Br 20	35. Br 45	36. Br 80
(20) $4CO + SO_4^{2-} + 4H_2O \rightarrow 4HCO_3^- + HS^- + 3H^+$	−156	−137	−112	−161	−121	−89	−206	−166	−134	−232	−191	−159
(21) $S + 1.5O_2 + H_2O \rightarrow SO_4^{2-} + 2H^+$							*−18*			*−18*	*−16*	*−14*
(22) $4CO + 5H_2O \rightarrow CH_4 + 3HCO_3^- + 3H^+$	−123	−103	−77	−149	−108	−74	−184	−144	−110	−204	−164	−130
(23) $2H_2 + O_2 \rightarrow 2H_2O$	−287	−253	−226	−64	−55	−46	−91	−100	−92	−99	−88	−75
(24) $S_2O_3^{2-} + 4H_2 \rightarrow 3H_2O + 2HS^-$	−75	−61	−32	−105	−84	−60	−195	−174	−149	−207	−185	−160
(25) $2HS^- + O_2 + 2H^+ \rightarrow 2S + 2H_2O$	−210	−183	−163	0	0	0	*18*	*0*	*0*	*15*	*16*	*21*
(26) $H_2 + S \rightarrow HS^- + H^+$				−32	−27	−23	−43	−50	−46	−48	−42	−35
(27) $3H_2 + CO \rightarrow CH_4 + H_2O$	−109	−86	−58	−86	−61	−33	−155	−130	−102	−168	−144	−115
(28) $4H_2 + H^+ + SO_4^{2-} \rightarrow HS^- + 4H_2O$	−138	−114	−87	−78	−59	−35	−167	−148	−124	−183	−163	−140
(29) $3H_2 + N_2 \rightarrow 2NH_3$	−148	−125	−100	−96	−81	−57	−158	−142	−118	−167	−151	−127
(30) 4Formate + $H^+ + H_2O \rightarrow CH_4 + 3HCO_3^-$	−73	−63	−52	2	*19*	28	−2	*14*	*24*	*−21*	*−4*	*6*
(31) Acetate + $SO_4^{2-} \rightarrow 2HCO_3^- + HS^-$	−46	−48	−50	−15	−14	−17	−22	−20	−24	−28	−26	−30
(32) $4H_2 + H^+ + HCO_3^- \rightarrow CH_4 + 3H_2O$	−104	−81	−52	−65	−46	−20	−145	−126	−100	−156	−137	−111
(33) CO + Hematite + $3H^+ \rightarrow 2Fe^{2+} + H_2O + HCO_3^-$	−120	−102	−78	−87	−73	−58	−94	−83	−68	−101	−89	−73

(34) $CO + 2H_2O \rightarrow HCO_3^- + H^+ + H_2$	*−5*	*−6*	*−6*	*−21*	*−16*	*−14*	*−10*	*−5*	*−2*	*−12*	*−7*	*−5*
(35) $CH_4 + SO_4^{2-} \rightarrow H_2O + HCO_3^- + HS^-$	−33	−34	−35	−13	−13	−15	−22	−22	−24	−27	−27	−29
(36) Acetate + $H_2O \rightarrow CH_4 + HCO_3^-$	*−13*	*−14*	*−15*	−2	*−1*	−2	*0*	2	*1*	*−1*	*1*	*0*
(37) $4H_2 + H^+ + 2HCO_3^- \rightarrow$ Acetate + $4H_2O$	−91	−67	−37	−62	−45	−18	−145	−128	−100	−155	−138	−110
(38) $S_2O_3^{2-} + H_2O \rightarrow SO_4^{2-} + H^+ + HS^-$	*63*	*53*	*55*	−27	−26	−24	−28	−27	−26	−23	−21	−20
(39) Acetate + 4Hematite + $15H^+ 8Fe^{2+} + 8H_2O + 2HCO_3^-$	−371	−318	−252	−202	−185	−158	−193	−187	−160	−199	−190	−164
(40) $4Fe^{2+} + O_2 + 10H_2O \rightarrow 4Fe(OH)_3 + 8H^+$	*97*	*85*	*56*				*280*			*271*	*225*	*205*
(41) H_2 + Hematite + $4H^+ \rightarrow 2Fe^{2+} + 3H_2O$	−116	−96	−72	−66	−57	−44	−84	−79	−65	−89	−82	−68
(42) Propionate + $3H_2O \rightarrow$ Acetate + $HCO_3^- + H^+ + 3H_2$	*60*	*42*	*20*	*30*	*16*	*−3*	*93*	*80*	*60*	*100*	*86*	*66*
(43) HS^- + 4Hematite + $15H^+ \rightarrow SO_4^{2-} + 8Fe^{2+} + 8H_2O$	−325	−270	−202	−186	−171	−141	−171	−167	−137	−171	−164	−134
(44) $NH_3 + 1.5{*}O_2 \rightarrow NO_2^- + H^+ + H_2O$	*73*	*68*	*46*	*351*	*316*	*267*	*379*	*317*	*268*	*377*	*344*	*303*
(45) $4Mn^{2+} + NO_3^- + 5H_2O \rightarrow 4MnO_2 + 7H^+ + NH_3$	*313*	*284*	*248*	*256*	*256*	*238*	*237*	*229*	*205*	*237*	*229*	*212*
(46) $2NO_2^- + O_2 \rightarrow 2NO_3^-$	*78*	*77*	*64*	*281*	255	222	*288*	244	212	*284*	*260*	*233*
(47) $2Mn^{2+} + O_2 + 2H_2O \rightarrow 2MnO_2 + 4H^+$	*212*	*196*	*163*				*384*			*381*	*355*	*320*

The reactions are ordered from most negative to positive with respect to the free energy for the dolomite ground water at 20 °C. The microbial reaction numbers and column heading numbers refer to Figure 10. Values in italics are > -20 kJ mol^{-1} and therefore are not considered to be viable for microbial metabolism.

TABLE 8
POTENTIAL MICROBIAL POWER (5000 KJ CELL^{-1} S^{-1}) AFTER INTERACTION OF CO_2 SATURATED WATER WITH AQUIFER MINERALS

Microbial Redox Reactions	25. Do 20	26. Do 45	27. Do 80	28. LS 20	29. LS 45	30. LS 80	31. MS 20	32. MS 45	33. MS 80	34. Br 20	35. Br 45	36. Br 80
(1) $H_2 + S \rightarrow HS^- + H^+$				-1.6×10^{-13}	-2.3×10^{-13}	-3.2×10^{-13}	-2.1×10^{-10}	$\mathbf{-4.1\times10^{-10}}$	-6.4×10^{-10}	-8.3×10^{-10}	$\mathbf{-7.0\times10^{-10}}$	-1.2×10^{-09}
(2) $3H_2 + N_2 \rightarrow 2NH_3$	-1.2×10^{-12}	-1.7×10^{-12}	-2.3×10^{-12}	-1.6×10^{-13}	-2.2×10^{-13}	-2.6×10^{-13}	-2.6×10^{-10}	$\mathbf{-3.9\times10^{-10}}$	-5.5×10^{-10}	-9.6×10^{-10}	$\mathbf{-8.4\times10^{-10}}$	-1.5×10^{-09}
(3) $CH_4 + SO_4^{2-} \rightarrow H_2O + HCO_3^- + HS^-$	$\mathbf{-2.7\times10^{-11}}$	-4.7×10^{-11}	-8.1×10^{-11}				-8.3×10^{-10}	$\mathbf{-1.4\times10^{-09}}$	-2.6×10^{-09}	-5.2×10^{-10}	$\mathbf{-4.3\times10^{-10}}$	-7.9×10^{-10}
(4) $4H_2 + H^+ + SO_4^{2-} \rightarrow HS^- + 4H_2O$	-8.5×10^{-13}	-1.2×10^{-12}	-1.5×10^{-12}	-9.6×10^{-14}	-1.2×10^{-13}	-1.2×10^{-13}	-2.1×10^{-10}	$\mathbf{-3.1\times10^{-10}}$	-4.3×10^{-10}	-7.9×10^{-10}	$\mathbf{-6.8\times10^{-10}}$	-1.2×10^{-09}
(5) Acetate + $4MnO_2$ + $7H^+ \rightarrow 4Mn^{2+} + 4H_2O + 2HCO_3^-$	$\mathbf{-3.0\times10^{-11}}$	-4.6×10^{-11}	-7.0×10^{-11}	-5.4×10^{-11}	$\mathbf{-8.5\times10^{-11}}$	-1.3×10^{-10}	-1.1×10^{-10}	$\mathbf{-1.7\times10^{-10}}$	-2.5×10^{-10}	-1.6×10^{-10}	-2.5×10^{-10}	-3.8×10^{-10}
(6) $3H_2 + CO \rightarrow CH_4 + H_2O$	-9.0×10^{-13}	-1.2×10^{-12}	-1.4×10^{-12}	-1.4×10^{-13}	-1.7×10^{-13}	-1.6×10^{-13}	-2.5×10^{-10}	$\mathbf{-3.6\times10^{-10}}$	-4.8×10^{-10}	-9.7×10^{-10}	$\mathbf{-8.0\times10^{-10}}$	-1.3×10^{-09}
(7) $4H_2 + H^+ + HCO_3^- \rightarrow CH_4 + 3H_2O$	-6.4×10^{-13}	-8.4×10^{-13}	-9.1×10^{-13}	-8.0×10^{-14}	-9.5×10^{-14}	-6.9×10^{-14}	-1.8×10^{-10}	$\mathbf{-2.6\times10^{-10}}$	-3.5×10^{-10}	-6.7×10^{-10}	$\mathbf{-5.7\times10^{-10}}$	-9.7×10^{-10}
(8) $4CO + SO_4^{2-} + 4H_2O \rightarrow 4HCO_3^- + HS^- + 3H^+$	-1.7×10^{-13}	-2.6×10^{-13}	-3.5×10^{-13}	-1.8×10^{-13}	-2.3×10^{-13}	-2.8×10^{-13}	-1.1×10^{-11}	-1.6×10^{-11}	-2.1×10^{-11}	-5.8×10^{-11}	-7.2×10^{-11}	-1.0×10^{-10}
(9) $4CO + 5H_2O \rightarrow CH_4 + 3HCO_3^- + 3H^+$	-1.4×10^{-13}	-1.9×10^{-13}	-2.4×10^{-13}	-1.7×10^{-13}	-2.0×10^{-13}	-2.3×10^{-13}	-1.0×10^{-11}	-1.3×10^{-11}	-1.7×10^{-11}	-5.1×10^{-11}	-6.1×10^{-11}	-8.2×10^{-11}
(10) Acetate + 4S + $4H_2O \rightarrow 5H^+ + 2HCO_3^- + 4HS^-$				-4.3×10^{-12}	$\mathbf{-7.2\times10^{-12}}$	-1.4×10^{-11}	-3.7×10^{-12}	-1.6×10^{-11}	-3.1×10^{-11}	-7.3×10^{-12}	-1.1×10^{-11}	-1.8×10^{-11}
(11) CO + Hematite + $3H^+ \rightarrow 2Fe^{2+} + H_2O + HCO_3^-$	-5.4×10^{-13}	-7.6×10^{-13}	-9.9×10^{-13}	-3.9×10^{-13}	-5.5×10^{-13}	-7.2×10^{-13}	-2.1×10^{-11}	-3.1×10^{-11}	-4.3×10^{-11}	-1.0×10^{-10}	-1.3×10^{-10}	-1.8×10^{-10}
(12) $4H_2 + H^+ + 2HCO_3^- \rightarrow$ Acetate + $4H_2O$	-5.6×10^{-13}	-6.9×10^{-13}	-6.5×10^{-13}	-7.7×10^{-14}	-9.3×10^{-14}		-1.8×10^{-10}	$\mathbf{-2.7\times10^{-10}}$	-3.5×10^{-10}	-6.7×10^{-10}	$\mathbf{-5.7\times10^{-10}}$	-9.6×10^{-10}
(13) Acetate + $SO_4^{2-} \rightarrow 2HCO_3^- + HS^-$	-1.5×10^{-12}	-2.7×10^{-12}	-4.7×10^{-12}	-1.0×10^{-12}	-1.5×10^{-12}	-3.3×10^{-12}	-2.9×10^{-12}	-4.4×10^{-12}	-8.8×10^{-12}	-5.6×10^{-12}	-8.6×10^{-12}	-1.7×10^{-11}
(14) $4CO + NO_3^- + 5H_2O \rightarrow 4HCO_3^- + NH_3 + 3H^+$	-7.9×10^{-13}	-1.2×10^{-12}	-1.7×10^{-12}	-7.8×10^{-13}	-1.2×10^{-12}	-1.7×10^{-12}	-4.1×10^{-11}	-6.2×10^{-11}	-8.9×10^{-11}	-1.9×10^{-10}	-2.5×10^{-10}	-3.7×10^{-10}
(15) $2.5CO + NO_3^- + 2H_2O \rightarrow 2.5HCO_3^- + 1.5H^+ + 0.5N_2$	-1.1×10^{-12}	-1.7×10^{-12}	-2.5×10^{-12}	-1.1×10^{-12}	-1.7×10^{-12}	-2.4×10^{-12}	-5.8×10^{-11}	-8.7×10^{-11}	-1.3×10^{-10}	-2.7×10^{-10}	-3.6×10^{-10}	-5.2×10^{-10}
(16) $S_2O_3^{2-} + NO_3^- + 2H_2O \rightarrow 2SO_4^{2-} + H^+ + NH_3$	-2.7×10^{-13}	-4.1×10^{-13}	-6.0×10^{-13}	-7.1×10^{-12}	$\mathbf{-1.1\times10^{-11}}$	-1.5×10^{-11}	-2.8×10^{-11}	-4.2×10^{-11}	-6.1×10^{-11}	-1.9×10^{-11}	-2.8×10^{-11}	-4.2×10^{-11}
(17) $4H_2 + NO_3^- + H^+ \rightarrow NH_3 + 3H_2O$	-4.2×10^{-12}	-6.4×10^{-12}	-9.3×10^{-12}	-7.6×10^{-13}	-1.1×10^{-12}	-1.6×10^{-12}	-8.7×10^{-10}	$\mathbf{-1.3\times10^{-09}}$	-1.9×10^{-09}	-3.1×10^{-09}	$\mathbf{-2.7\times10^{-09}}$	-5.0×10^{-09}
(18) Acetate + $NO_3^- + H_2O \rightarrow 2HCO_3^- + NH_3$	$\mathbf{-2.0\times10^{-11}}$	-3.1×10^{-11}	-4.6×10^{-11}	-6.8×10^{-12}	$\mathbf{-1.0\times10^{-11}}$	-1.6×10^{-11}	-2.4×10^{-11}	-3.7×10^{-11}	-5.5×10^{-11}	-1.6×10^{-11}	-2.5×10^{-11}	-3.7×10^{-11}
(19) $5H_2 + 2NO_3^- + 2H^+ \rightarrow N_2 + 6H_2O$	$\mathbf{-6.1\times10^{-12}}$	-9.2×10^{-12}	-1.3×10^{-11}	-7.0×10^{-12}	$\mathbf{-1.1\times10^{-11}}$	-1.6×10^{-11}	-2.4×10^{-11}	-3.7×10^{-11}	-5.4×10^{-11}	-1.8×10^{-11}	-2.7×10^{-11}	-4.0×10^{-11}
(20) $2S + 1.5NO_3^- + 3.5H_2O \rightarrow 2SO_4^{2-} + 2.5H^+ + 1.5NH_3$				-6.4×10^{-12}	$\mathbf{-9.8\times10^{-12}}$	-1.5×10^{-11}	-2.4×10^{-11}	-3.7×10^{-11}	-5.5×10^{-11}	-1.6×10^{-11}	-2.4×10^{-11}	-3.6×10^{-11}
(21) Acetate + $1.6NO_3^-$ + $0.6H^+ \rightarrow 2HCO_3^- + 0.8H_2O + 0.8N_2$	$\mathbf{-3.0\times10^{-11}}$	-4.6×10^{-11}	-6.9×10^{-11}	-6.3×10^{-12}	$\mathbf{-9.8\times10^{-12}}$	-1.5×10^{-11}	-2.3×10^{-11}	-3.5×10^{-11}	-5.2×10^{-11}	-1.5×10^{-11}	-2.3×10^{-11}	-3.5×10^{-11}
(22) $HS^- + NO_3^- + H_2O \rightarrow SO_4^{2-} + NH_3$	$\mathbf{-7.0\times10^{-11}}$	-1.1×10^{-10}	-1.6×10^{-10}	-6.2×10^{-12}	$\mathbf{-9.5\times10^{-12}}$	-1.4×10^{-11}	-2.3×10^{-11}	-3.5×10^{-11}	-5.2×10^{-11}	-1.5×10^{-11}	-2.3×10^{-11}	-3.5×10^{-11}
(23) $CO + 2H_2O \rightarrow HCO_3^- + H^+ + H_2$			-1.9×10^{-14}									
(24) Acetate + $H_2O \rightarrow CH_4 + HCO_3^-$												
(25) $NO_2^- + H^+ + NH_3 \rightarrow 2H_2O + N_2$	-1.1×10^{-12}	-1.7×10^{-12}	-2.5×10^{-12}	-5.4×10^{-13}	-8.2×10^{-13}	-1.2×10^{-12}	-5.5×10^{-12}	-8.4×10^{-12}	-1.3×10^{-11}	-1.1×10^{-11}	-1.7×10^{-11}	-2.5×10^{-11}

Reaction												
(26) $S_2O_3^{2-} + 4H_2 \rightarrow 3H_2O + 2HS^-$	-4.1×10^{-14}	-5.6×10^{-14}	-5.0×10^{-14}	-1.3×10^{-13}	-1.8×10^{-13}	-2.1×10^{-13}	-2.4×10^{-10}	-3.6×10^{-10}	-5.2×10^{-10}	-8.9×10^{-10}	**-7.7×10^{-10}**	-1.4×10^{-09}
(27) $4\text{Formate} + H^+ + H_2O \rightarrow CH_4 + 3HCO_3^-$	-4.5×10^{-12}	-6.5×10^{-12}	-9.0×10^{-12}							8.8×10^{-14}		
(28) $5Fe^{2+} + NO_3^- + 12H_2O \rightarrow 5Fe(OH)_3 + 9H^+ + 0.5N_2$		-2.9×10^{-12}	-4.7×10^{-12}		-1.4×10^{-13}	-2.0×10^{-13}		-3.5×10^{-13}	-5.1×10^{-13}		-7.0×10^{-13}	
(29) $S_2O_3^{2-} + H_2O \rightarrow SO_4^{2-} + H^+ + HS^-$				-7.5×10^{-13}	**-1.2×10^{-12}**	-1.9×10^{-12}	-1.6×10^{-12}	-2.5×10^{-12}	-4.0×10^{-12}	-9.0×10^{-13}	-1.4×10^{-12}	-2.2×10^{-12}
(30) $\text{Acetate} + 4\text{Hematite} + 15H^+ \rightarrow 8Fe^{2+} + 8H_2O + 2HCO_3^-$	**-1.2×10^{-11}**	-1.8×10^{-11}	-2.4×10^{-11}	-1.3×10^{-11}	-2.1×10^{-11}	-3.0×10^{-11}	-2.6×10^{-11}	-4.2×10^{-11}	-6.0×10^{-11}	-4.0×10^{-11}	-6.4×10^{-11}	-9.2×10^{-11}
(31) $H_2 + \text{Hematite} + 4H^+ \rightarrow 2Fe^{2+} + 3H_2O$	**-9.2×10^{-12}**	-1.3×10^{-11}	-1.8×10^{-11}	-1.0×10^{-12}	**-1.5×10^{-12}**	-2.2×10^{-12}	-9.5×10^{-10}	-1.6×10^{-09}	-2.2×10^{-09}	-3.4×10^{-09}	**-3.2×10^{-09}**	-5.7×10^{-09}
(32) $\text{Propanoate} + 3H_2O \rightarrow \text{Acetate} + HCO_3^- + H^+ + 3H_2$												
(33) $2CO + O_2 + 2H_2O \rightarrow 2HCO_3^- + 2H^+$	-3.4×10^{-45}	-5.1×10^{-43}	-7.8×10^{-37}	-1.2×10^{-77}	-1.7×10^{-70}	-2.4×10^{-62}	-1.3×10^{-79}	-2.1×10^{-70}	-3.1×10^{-62}	-1.4×10^{-79}	-2.0×10^{-72}	-2.8×10^{-65}
(34) $2H_2 + O_2 \rightarrow 2H_2O$	-3.3×10^{-45}	-4.9×10^{-43}	-7.4×10^{-37}	-7.3×10^{-78}	-1.1×10^{-70}	-1.5×10^{-62}	-1.0×10^{-79}	-1.9×10^{-70}	-3.0×10^{-62}	-1.1×10^{-79}	-1.7×10^{-72}	-2.5×10^{-65}
(35) $\text{Acetate} + 2O_2 \rightarrow 2HCO_3^- + H^+$	-2.8×10^{-45}	-4.3×10^{-43}	-6.8×10^{-37}	-3.7×10^{-78}	-6.2×10^{-71}	-1.2×10^{-62}	-2.1×10^{-80}	-7.0×10^{-71}	-1.3×10^{-62}	-2.5×10^{-80}	-3.8×10^{-73}	-6.6×10^{-66}
(36) $CH_4 + 2O_2 \rightarrow HCO_3^- + H^+ + H_2O$	-2.7×10^{-45}	-4.1×10^{-43}	-6.5×10^{-37}	-3.6×10^{-78}	-6.1×10^{-71}	-1.2×10^{-62}	-2.1×10^{-80}	-7.2×10^{-71}	-1.4×10^{-62}	-2.5×10^{-80}	-3.9×10^{-73}	-6.5×10^{-66}
(37) $S + 1.5O_2 + H_2O \rightarrow SO_4^{2-} + 2H^+$												
(38) $S_2O_3^{2-} + 2O_2 + H_2O \rightarrow 2SO_4^{2-} + 2H^+$	-2.2×10^{-45}	-3.3×10^{-43}	-5.1×10^{-37}	-4.4×10^{-78}	-7.4×10^{-71}	-1.3×10^{-62}	-2.5×10^{-80}	-7.7×10^{-71}	-1.4×10^{-62}	-2.2×10^{-80}	-3.4×10^{-73}	-5.1×10^{-66}
(39) $HS^- + 2O_2 \rightarrow SO_4^{2-} + H^+$	-2.5×10^{-45}	-3.8×10^{-43}	-5.9×10^{-37}	-2.8×10^{-78}	-4.9×10^{-71}	-9.4×10^{-63}		-5.1×10^{-71}	-9.7×10^{-63}			
(40) $2HS^- + 2O_2 \rightarrow S_2O_3^{2-} + H_2O$	-2.9×10^{-45}	-4.3×10^{-43}	-6.8×10^{-37}	-1.3×10^{-78}	-2.4×10^{-71}	-5.4×10^{-63}		-2.5×10^{-71}	-5.5×10^{-63}			
(41) $2HS^- + O_2 + 2H^+ \rightarrow 2S + 2H_2O$	-2.4×10^{-45}	-3.5×10^{-43}	-5.3×10^{-37}									
(42) $HS^- + 4\text{Hematite} + 15H^+ \rightarrow SO_4^{2-} + 8Fe^{2+} + 8H_2O$	**-7.5×10^{-10}**	-1.1×10^{-09}	-1.3×10^{-09}	-1.3×10^{-09}	**-2.0×10^{-09}**	-2.8×10^{-09}	-6.6×10^{-09}	**-1.1×10^{-08}**	-1.5×10^{-08}	-7.9×10^{-10}	**-1.3×10^{-09}**	-1.8×10^{-09}
(43) $4Fe^{2+} + O_2 + 10H_2O \rightarrow 4Fe(OH)_3 + 8H^+$												

Microbial redox reactions have been ordered according to their power with the most powerful reactions for the 80 °C brine appearing first. The microbial reaction numbers and column heading numbers refer to Figure 3. The power is not reported for reactions for which the free energy was less negative than -20 kJ mol^{-1}. The values in bold represent the top 10 values.

methanogens, whereas the propionate reaction would be more dependent upon the PH_2. The oxidation of reduced S compounds by O_2 (reactions (16)–(18), (21) and (25) in Table 7), which were marginally favorable in the highly saline ground water remain endothermic after alteration of the siliclastic mineral assemblage. Given that aerobic S oxidizers are not found in this ground water environment, this does not appear to be a significant detriment. For the dolomite aquifer, the S oxidizing reactions remain exothermic despite the lower pH of the impacted system.

2. Of the microbial reactions that were originally unfavorable prior to injection, the reduction of hematite to Fe^{2+} by oxidation of acetate (reaction (39) in Table 7) and H_2 (reaction (41) in Table 7) are still exothermic due to the reduction in pH from 8 to 7. The abiotic reduction of hematite by oxidation of HS^- (reaction (43) in Table 7) is also energetically favorable.
3. The high CO_2 and HCO_3^- concentrations increased the free energy yield for CO_2 reducing methanogenic and acetogenic reaction (reactions (32) and (37) in Table 7) even after alteration of the aquifer mineral assemblage. An increase in acetogenic activity may rescue the aceticlastic methanogens. The extent to which these two reactions can be used to convert the CO_2 into methane and acetate depends upon whether an abiotically generated source of H_2 can be made available.

The most readily identified impact in Figure 3 on the potential microbial power is from reduction of hematite by HS^- oxidation (Table 8). The power levels were generally larger than in the original ground water systems and because of the reduction of one pH unit in the ground water, microbial Fe(III) reduction reactions were more significant. If sufficient electron donors are available for both biotic and abiotic reactions and sufficient Fe(III) bearing oxides are present in the aquifer (as is usually the case) then these reactions will restore the aquifer's pH to its initial value.

The dolomite aquifer was more severely impacted by the simulated CO_2 injection because the dissolution of the aquifer minerals failed to restore the pH to a range that is more commensurate with the pH ranges of some of the microorganisms. The most effective means of remediating this problem if it occurs in the real world is by the addition of H_2 to stimulate the CO_2 reducing methanogenesis and acetogenesis. If mafic igneous rocks are present that contain Fe bearing clinopyroxene, then the lower pH will automatically stimulate the release of H_2 by the oxidation of this ferrous iron to $Fe(OH)_3$ [38].

Another factor associated with the lower pH produced by CO_2 injection is that it facilitates proton pumping reactions across the cell membrane. Microorganisms need to maintain an internal pH that is 1–2 units less than the external pH in order for the proton pumps to generate ATP. For pH values approaching 8.5–9, this becomes problematic because high internal pH values affect the aqueous species of phosphate making it more difficult to synthesis ATP. The microorganism is then required to expend energy in ion transport across the membrane to correct for this problem. A more neutral pH of 6–7 alleviates this energy drain. The greater availability of energy will also facilitate the fixation of N_2 which would help support growth of the microbial population. The lower pH should also help solubilize phosphate for growth. In aquifers where organic acids are naturally more abundant and the pH typically lower, the impact of CO_2 injection should be less. For aquifers low in organic acids, CO_2 injection will lead to an increase in acetate if a sufficient source of H_2 is available. This in turn should lead to stimulation of overall microbial activity.

For long-term storage of CO_2 the activity of Fe(III) reducing microorganisms will increase the pH and, most likely, lead to the precipitation of various carbonates. Microbial biomass may become concentrated at the gas/water boundary where electron donor/acceptor fluxes will be highest. As readily available Fe(III) is depleted it can be introduced. If this is not feasible and sulfate is not a major constituent in the ground water, then methanogenic activity will begin to dominate and the proportion of CO_2 converted to CH_4 will depend upon the H_2 and acetate fluxes.

For rhizosphere and surface biosphere the most obvious impact would be due to a potential increase in crustal CH_4 flux and a decrease in H_2 flux. Since the fluxes of both gaseous species from fermentative communities in shallower, organic-rich aquitards are 10–100 times greater than the deep subsurface flux, this probably is not a showstopper.

CONCLUSIONS

Based upon the calculated potential microbial power for microbial redox reactions, the most readily identified impact of CO_2 injections on the subsurface microbial communities was the reduction of one pH unit for the ground water hosted in the siliclastic reservoir. The slightly lower pH is based upon the assumption, yet to be verified, that alteration of detrital feldspars to clay in equilibrium with calcite occurs on the time scale of the injection. The power levels for many of the microbial redox reactions were generally larger than in the original ground water systems but because of this reduction of one pH unit in the ground water, microbial Fe(III) reduction reactions were particularly more significant. If sufficient electron donors are available for both biotic and abiotic Fe(III) reducing reactions and sufficient Fe(III) bearing oxides are present in the aquifer (as is usually the case) then these reactions will restore the aquifer's pH to its initial, pre-injection value. CO_2 injection should cause a short term stimulation of Fe(III) reducing communities.

A dolomitic or carbonate aquifer may be more severely impacted by the simulated CO_2 injection because the dissolution of the carbonate failed to restore the pH to a range that is more commensurate with the pH ranges of some of the microorganisms. The most effective means of remediating this problem if it occurs in the real world is by the addition of H_2 to stimulate the CO_2 reducing methanogenesis and acetogenesis. If mafic igneous rocks host the groundwater and contain Fe bearing clinopyroxene, then the lower pH will automatically stimulate the release of H_2 by the oxidation of this ferrous iron to $Fe(OH)_3$ [38]. This in turn would lead to stimulation of methanogenic and acetogenic communities and a reduction of the injected CO_2. Fe(III) reducing microbial reactions may also be stimulated by the appearance of $Fe(OH)_3$ leading to Fe(III) reduction and an eventual increase in pH. The outcome of CO_2 injection in carbonate and mafic rock hosted aquifers is probably the least understood.

Another factor associated with the lower pH produced by CO_2 injection is that it facilitates proton pumping reactions across the cell membrane. Microorganisms need to maintain an internal pH that is 1–2 units less than the external pH in order for the proton pumps to generate ATP. For pH values approaching 8.5–9, this becomes problematic because high internal pH values affect the aqueous species of phosphate making it more difficult to synthesis ATP. The microorganism is then required to expend energy in ion transport across the membrane to correct for this problem. A more neutral pH of 6–7 alleviates this energy drain. The greater availability of energy will also facilitate the fixation of N_2 which would help support growth of the microbial population. The lower pH values should also help solubilize phosphate for growth. Overall CO_2 injection should increase the availability of N and P to microbial communities.

For shallow aquifers where organic acids are naturally more abundant and the pH lower than used in the simulations reported here, the impact of CO_2 injection should be less. For aquifers low in organic acids, CO_2 injection will lead to an increase in acetate through acetogenesis, if a sufficient source of H_2 is available. H_2 can be artificially provided through the introduction of zero valence Fe. This in turn should lead to stimulation of overall anaerobic microbial activity.

For long-term storage of CO_2 in siliclastic reservoirs the short-term enhancement of Fe(III) reducing microorganisms will increase the pH and most likely lead to the precipitation of various carbonates. Microbial biomass may become concentrated at the gas/water boundary where electron donor/acceptor fluxes will be highest. As readily available Fe(III) is depleted it can be introduced. If this is not feasible and sulfate is not a major constituent in the ground water, then methanogenic activity will begin to dominate and the proportion of CO_2 converted to CH_4 will depend upon the H_2 and acetate fluxes.

For rhizosphere and surface biosphere the most obvious impact would be due to a potential increase in crustal CH_4 flux for carbonate and mafic rock hosted aquifers and a decrease in H_2 flux in all cases. Since the fluxes of both gaseous species from fermentative communities in shallower, organic-rich aquitards are 10–100 times greater than the deep subsurface flux, this probably is not a showstopper. Nevertheless, this merits further investigation.

RECOMMENDATIONS

The next phase of modeling will simulate microbial reactions by using the potential microbial power value to select the relative rates among the different microbial redox reactions. Additional observation on the dissolved gas concentrations in an aquifer where CO_2 injection is occurring and comparing those measurements to a similar aquifer where CO_2 injection is not taking place would provide constraints for a model to takes into account the changes in dissolved gas concentrations and its impact on the microbial redox reactions.

In siliclastic aquifers where Fe and Al oxyhydroxides are present, surface protonation reactions may moderate pH changes. This will be included in the next phase of modeling. The next phase of modeling should be combined with kinetic expressions for mineral dissolution at ambient formation conditions to refine the rates of approach to equilibrium compared to the rates of gaseous CO_2 migration and heat advection.

The above analysis predicts changes in the gas and aqueous geochemistry and in the composition of the microbial community in response to CO_2 injection. These predictions could be readily tested by collection and geochemical and 16S rDNA analyses of formation fluids at a CO_2 injection site and control site. This would represent the first critical step in validation of the model's predictions. If the microbial factor turns out to be important, then these observations could also provide the foundation upon which experiments could be performed, initially in the lab, on configuration of the injection stream to enhance optimal microbial activity. Geochemical and 16S rDNA analyses of formation fluids from a CO_2 rich gas reservoir would supply critical observations pertinent to long-term residence of CO_2 and would expand our understanding of the deep subsurface carbon cycle.

ACKNOWLEDGEMENTS

This study was supported by a project subcontract from the US Department of Energy (DOE Contract No. DE-FC26-01NT41145) to BP Corporation North America Inc. I appreciate the efforts and suggestions by Dr Charles A. Christopher in recommending that I undertake such a study.

REFERENCES

1. K. Takai, D.P. Moser, M.F. DeFlaun, T.C. Onstott, J.K. Fredrickson, *Appl. Environ. Microbiol.* **67** (2001) 5750.
2. J. Lippmann, M. Stute, T. Torgersen, D.P. Moser, J. Hall, L.-H. Lihung, M. Borcsik, R.E.S. Bellamy, T.C. Onstott, *Geochim. Cosmochim. Acta* **67** (2003) 4597.
3. J.P. Amend, E.L. Shock, *FEMS Microbiol. Rev.* **25** (2001) 175.
4. T.C. Onstott, T.J. Phelps, T. Kieft, F.S. Colwell, D.L. Balkwill, J.K. Fredrickson, F.J. Brockman, in: J. Seckbach (Ed.), Enigmatic Microorganisms and Life in Extreme Environments, Kluwer, Dordrecht, 1998, p. 489.
5. T.C. Onstott, D.P. Moser, J.K. Fredrickson, F.J. Brockman, S.M. Pfiffner, T.J. Phelps, D.C. White, A. Peacock, D. Balkwill, R. Hoover, L.R. Krumholz, M. Borscik, T.L. Kieft, R.B. Wilson, *Environ. Microbiol.* **5** (2003) 1168.
6. R.J. Parkes, B.A. Cragg, S.J. Bale, J.M. Getliff, K. Goodman, P.A. Rochelle, J.C. Fry, A.J. Weightman, S.M. Harvey, *Nature* **371** (1994) 410.
7. K. Pedersen, J. Arlinger, S. Ekendahl, L. Hallbeck, *FEMS Microbiol. Ecol.* **19** (1996) 249.
8. L.-H., Lin, Radiolytic H_2 and Microbial Communities in the Witwaterstand Basin, PhD in Department of Geosciences, Princeton University, Princeton, 2003, p. 160.
9. B.J. Baker, D.P. Moser, B.J. MacGregor, S. Fishbain, M. Wagner, N.K. Fry, B. Jackson, N. Speolstra, S. Loos, K. Takai, B. Sherwood-Lollar, J.K. Fredrickson, D.L. Balkwill, T.C. Onstott, C.F. Wimpee, D.A. Stahl, *Environ. Microbiol.* **5** (2003) 267.
10. P. Reuter, R. Rubus, H. Wilkes, F. Aeckersberg, F.A. Ruiney, H.W. Jannasch, F. Widdel, *Nature* **372** (1994) 455.

11. J. Beeders, R.K. Nilsen, T. Thorstenson, T. Torsvik, *Appl. Environ. Microbiol.* **62** (1996) 3551.
12. G. Ravot, M. Majot, M.-L. Fardeau, B.K.C. Patel, P. Thomas, J.-L. Garcia, B. Ollivier, *Int. J. Syst. Evol. Microbiol.* **49** (1999) 1141.
13. K. Takai, D.P. Moser, T.C. Onstott, N. Speolstra, S.M. Pfiffner, A. Dohnalkova, J.K. Fredrickson, *Int. J. Syst. Bacteriol.* **51** (2001) 1245.
14. T.L. Kieft, J.K. Frredrickson, T.C. Onstott, Y.A. Gorby, H.M. Kostandarithes, T.J. Bailey, D.W. Kennedy, S.W. Li, A.E. Plymale, C.M. Spadoni, M.S. Gray, *Appl. Environ. Microbiol.* **65** (1999) 1214.
15. D.R. Boone, Y. Liu, Z. Zhao, D.L. Balkwill, G.R. Drake, T.O. Stevens, H.C. Aldrich, *Int. J. Syst. Bacteriol.* **45** (1995) 441.
16. Lovley, *Ann. Rev. Microbiol.* **47** (1993) 263.
17. C.R. Myers, J.M. Myers, *FEMS Microbiol. Lett.* **108** (1993) 15.
18. A.C. Greene, B.K.C. Patel, A.J. Sheehy, *Int. J. Syst. Evol. Microbiol.* **47** (1997) 505.
19. K.O. Stetter, R. Hubert, E. Blochl, M. Kurr, R.D. Eden, M. Fielder, H. Cash, I. Vance, *Nature* **365** (1993) 743.
20. R. Jakobsen, H.-J. Albrechtsen, M. Rasmussen, H. Bay, P.L. Bjerg, T.H. Christensen, *Environ. Sci. Tech.* **32** (1998) 2142.
21. N.L. Plummer, J.F. Busby, R.W. Lee, B.B. Hanshaw, *Water Resour. Res.* **26** (1994) 1981.
22. L.N. Plummer, *Water Resour. Res.* **13** (1977) 801.
23. H.C. Helgeson, A.M. Knox, C.E. Owens, E.L. Shock, *Geochim. Cosmochim. Acta* **57** (1993) 3295.
24. R.K. Thauer, K. Jungermann, K. Decker, *Bacteriol. Rev.* **41** (1977) 100.
25. B. Shink, *Microbiol. Mol. Biol. Rev.* **61** (1997) 262.
26. B.E. Jackson, M.J. McInerney, *Nature* **415** (2002) 454.
27. E.L. Cussler, Diffusion Mass Transfer in Fluid Systems, Cambridge University Press, Cambridge, 1984.
28. T.J. Phelps, E.M. Murphy, S.M. Pfiffner, D.C. White, *Microb. Ecol.* **28** (1994) 335.
29. T.L. Kieft, T.J. Phelps, in: P.S. Amy, D.L. Haldeman (Eds.), The Microbiology of the Terrestrial Deep Subsurface, CRC Lewis Publishers, Boca Raton, FL, 1997, p. 137.
30. J.G. Kuenen, M.S.M. Jetten, *ASM News* **69** (2001) 456.
31. B. Thamdrup, T. Dalsgaard, *Appl. Environ. Microbiol.* **68** (2002) 1312.
32. T. Dalsgaard, B. Thamdrup, *Appl. Environ. Microbiol.* **68** (2002) 3802.
33. T.M. Hoehler, M.J. Alperin, D.B. Alber, C.S. Martens, *Geochim. Cosmochim. Acta* **62** (1998) 1745.
34. D. Haendel, K. Muhle, H.-M. Nitzxche, G. Stiehl, U. Wand, *Geochim. Cosmochim. Acta* **50** (1986) 749.
35. M. Hochman, A Case for Biotic Speliogenesis in a Dolomite Aquifer in South Africa, B.A. Dissertation, in Department of Geosciences, Princeton University, Princeton, 2000, p. 132.
36. D.P. Moser, T.C. Onstott, J.K. Fredrickson, F.J. Brockman, D.L. Balkwill, G.R. Drake, S. Pfiffner, D.C. White, K. Takai, L.M. Pratt, J. Fong, B. Sherwood-Lollar, G. Slater, T.J. Phelps, N. Spoelstra, M. DeFlaun, G. Southam, A.T. Welty, B.J. Baker, J. Hoek, *Geomicrobiol. J.* **20** (2003) 1.
37. B. Sherwood-Lollar, S.K. Frape, S.M. Weise, P. Fritz, S.A. Macko, J.A. Welhan, *Geochim. Cosmochim. Acta* **57** (1993) 5087.
38. T.O. Stevens, J.P. McKinley, *Science* **270** (1995) 450.

Carbon Dioxide Capture for Storage in Deep Geologic Formations, Volume 2
D.C. Thomas and S.M. Benson (Eds.)

Chapter 31

FRAMEWORK METHODOLOGY FOR LONG-TERM ASSESSMENT OF THE FATE OF CO_2 IN THE WEYBURN FIELD

Mike Stenhouse[1], Wei Zhou[1], Dave Savage[2] and Steve Benbow[3]

[1]Monitor Scientific LLC, 3900 S. Wadsworth Blvd., Denver, CO 80235, USA
[2]Quintessa Limited, 24 Trevor Road, West Bridgford, Nottingham NG2 6FS, UK
[3]Quintessa Limited, Dalton House, New Town Road, Henley-on-Thames, Oxon RG9 1HG, UK

ABSTRACT

A key objective of the IEA Weyburn CO_2 Monitoring and Storage Project is to determine the long-term fate of CO_2 injected into the reservoir. Such a determination involves an evaluation of the potential for CO_2 to migrate away from the reservoir along both natural and artificial (wellbore) pathways to the environment, and relies on the technical input from a number of disciplines. These disciplines include geology and hydrogeology, geochemistry, geomechanics, reservoir modeling and wellbore technology. This paper describes the framework used for carrying out the long-term assessment, thus ensuring that work being carried out by other research workers is properly integrated into the CO_2 migration modeling. The discussion focuses on the various components of systems analysis, including features, events and processes and their incorporation into scenario development.

INTRODUCTION

Background

In July 2000, a 4-year research project to study geologic storage of CO_2 in the Weyburn oilfield was launched. A key objective of this multidisciplinary project is to determine the long-term fate of CO_2 injected into the reservoir. Such a determination involves an evaluation of the potential for CO_2 to migrate away from the reservoir along both natural and artificial (wellbore) pathways to the environment, and relies on the technical input from a number of disciplines. These disciplines include geology and hydrogeology, geochemistry, geomechanics, reservoir modeling and wellbore technology. The long-term assessment starts at the end of enhanced oil recovery (EOR) operations, the results of which are reported elsewhere [1]. Separate reservoir simulations, that were not a part of this study, were conducted to determine the conditions at the end of EOR operations.

CO_2 storage is still a developing field of research technology and so assessments associated with CO_2 storage are just beginning. In the particular case of Weyburn, long-term storage or storage of CO_2 would be an additional benefit of EOR. However, safety studies for the geological storage of CO_2 are unusual in that they need to consider the evolution of natural systems over timescales considerably in excess of those considered in typical engineering projects. Most environmental assessments address periods of tens or occasionally hundreds of years.

Opportunely, many of the advances made in the last 20 years in the field of safety assessments for the geological disposal of radioactive wastes can also be applied to CO_2 storage [2]. As for CO_2 storage, the final storage of nuclear waste requires an understanding of complex coupled physical–chemical–mechanical processes occurring over hundreds to tens of thousands of years. It is this field of work that provides the framework for the long-term assessment of the fate of CO_2 left in place in the Weyburn field at

---

Abbreviation: FEPs, features, events and processes.

the end of EOR operations. The reasons for this "transfer of technology" are three-fold:

- systems analysis provides a systematic framework for conducting safety assessments;
- systems analysis is used to identify features, events and processes (FEPs) over hundreds to thousands of years—the timescales of relevance in this project;
- the systems analysis approach is a useful method of documenting progress and why particular decisions were made.

SPECIFICS OF THE METHODOLOGY OF LONG-TERM ASSESSMENT

Components of Systems Analysis Approach

Systems analysis consists of several inter-related elements:

- definition of the "System" to be assessed;
- development of a list of FEPs which together describe the particular system being studied;
- differentiation between those FEPs which belong to the system itself and those which can be regarded as external to the system;
- identification of interactions between these FEPs;
- construction of scenarios;
- description of how the FEP–FEP interactions will be accommodated in the consequence analysis modeling to be undertaken for each scenario.

Each of these elements is discussed briefly below, providing examples relevant to the Weyburn Project, where appropriate. A more detailed account of these elements and the way in which they are combined in the systems analysis approach is described in Chapman et al. [3] and, more recently, in Stenhouse et al. [4].

Definition of the Weyburn System

One of the first steps in the methodology is to define what is meant by the "System" to be assessed. Figure 1 provides a schematic diagram of the basic components for the Weyburn System and their physical relationship; these components include:

- the *CO_2 storage reservoir*;
- the *geosphere*, which comprises a number of geological and hydrogeological units above and below the reservoir (not shown explicitly); and
- the surface or near-surface environment is also referred to as the *biosphere*.

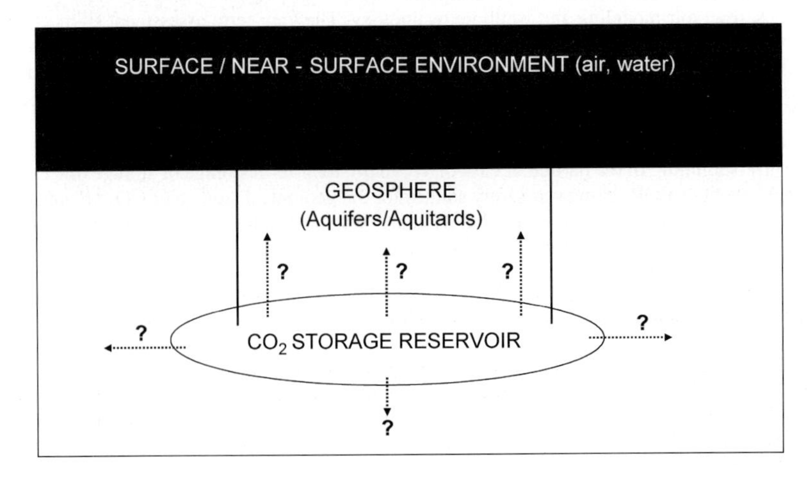

Figure 1: Schematic representation of the Weyburn CO_2 storage system.

The arrows shown in this schematic diagram are hypothetical representations of how CO_2 might migrate out of the storage reservoir. Two abandoned wells are also shown in Figure 1, representing wellbores as *potential* pathways for reservoir CO_2 to migrate to the surface or near-surface. Note that, although the geosphere is shown only as one uniform "compartment", the geosphere has been defined in much greater detail, so that the main features of the geosphere, principally those features that represent potential pathways or sinks for CO_2, may be incorporated in the migration modeling. Thus, Figure 2 shows the detailed layers of the System Model of the geosphere and biosphere, which comprise a series of aquitards and aquifers. The assessment area has been defined as covering an area 10 km beyond the outside of the EOR region (the perimeter is shown in red in Figure 2). Not included in this diagram are the numerous wells drilled through the area.

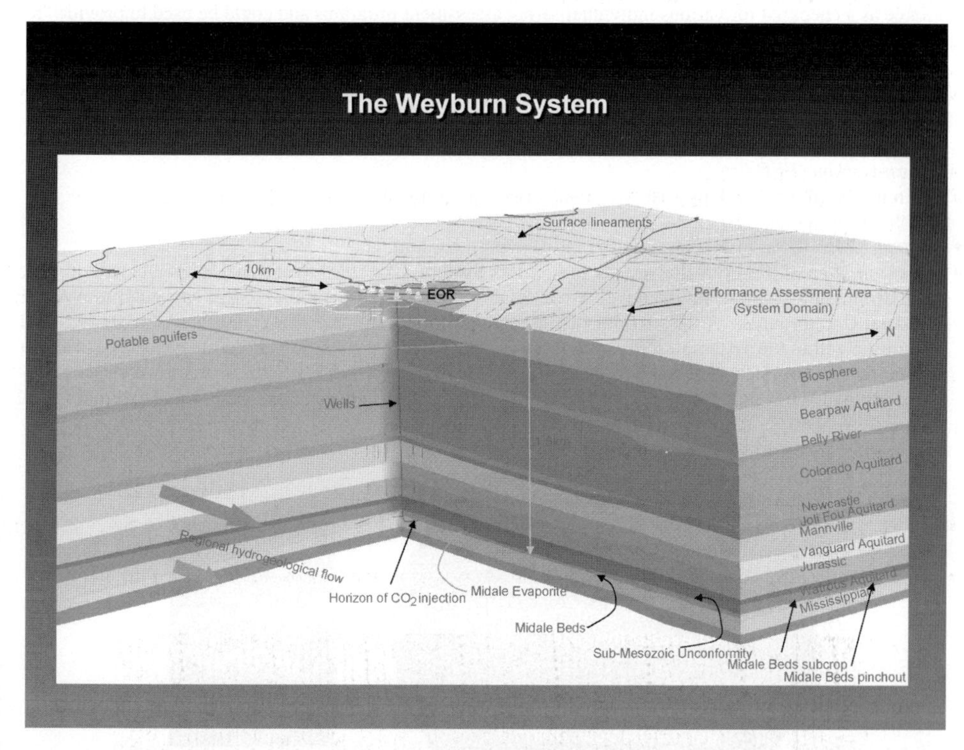

Figure 2: Weyburn System model—geosphere and biosphere (courtesy Steve Whittaker, Saskatchewan Industry and Resources). Note: The red perimeter defines the assessment area. Wells are not shown in this figure, for clarity.

FEPs

As stated above, FEPs is the acronym for *F*eatures, *E*vents or *P*rocesses, consisting of all factors that must be considered in describing/defining a system as well as assessing its performance.

- *Features* are typically specific components of the System being studied. For example, in the case of the geosphere, specific features would correspond to different geological and hydrogeological units, permeability and porosity of these units, and other important features such as faults and fractures. *Features* could also include inadequately sealed boreholes, and the quality (composition) of the injected carbon dioxide.
- *Events* are usually of short duration and can be of natural or human origin, such as seismic events, faulting, a well blow-out, or intrusion by people into the storage reservoir.

- *Processes* comprise the detailed individual scientific and engineering processes that govern the System. Examples are the variation of carbon dioxide's physical properties with pressure and temperature, multiphase flow of CO_2 and water, dissolution of CO_2 into the in situ reservoir fluids, and chemical reactions with reservoir and cap rocks. Examples of geochemical-type processes include the precipitation and dissolution of minerals.

FEP lists have been developed for safety assessment involving the final storage of nuclear waste in individual countries, not only by the national agency responsible for the waste management, but also by agencies responsible for overseeing or authorizing the process. Thus, Stenhouse et al. [5] compiled and categorized an FEP database consisting of FEPs from eight national and international FEP lists. Subsequently the Nuclear Energy Agency published an international FEP list database [6]. This list was available as a checklist for various individual safety assessment programs and could be used to provide "an aid to achieving and demonstrating comprehensiveness within an assessment".

Monitor Scientific developed a Weyburn-specific FEP List and Quintessa assembled independently a "generic" FEP database based on NEA's list but applicable for CO_2 disposal in general.

Weyburn working FEP list
FEPs in the Weyburn Working List were categorized in terms of:

- *System FEPs*: those FEPs that describe the Weyburn System, and
- *External FEPs*: those FEPs that are not part of this System. Examples of external FEPs are earthquakes, well drilling long into the future, development of new communities near the storage site and discovery of new mineral resources in the vicinity of the storage project. Such FEPs can affect CO_2 storage and migration within the system in some way, if they occur, thereby generating different *Scenarios*—ways in which the Weyburn System might evolve. For this reason, external FEPs are also known as scenario-generating FEPs. Figure 3 shows schematically the relationship between system FEPs and external FEPs.

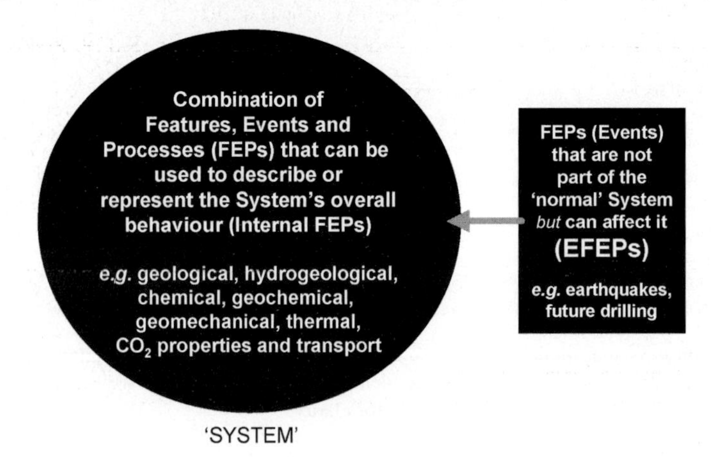

Figure 3: Relationship between system FEPs and external (scenario-generating) FEPs.

For convenience, the system FEPs were subdivided into a few arbitrary categories: geological, hydrogeological, chemical/geochemical, transport and miscellaneous. The resulting working list of FEPs for Weyburn was mapped to the generic FEP database (see below) to ensure that no relevant generic FEPs had been excluded from consideration in the Weyburn list.

The Weyburn-specific FEP list was also "mapped" to the FEPs generated at a workshop that was held in Rome, again to ensure that no relevant generic FEPs had been excluded from consideration in the Weyburn list.

The resultant, updated Weyburn-specific FEP list was reviewed at a Weyburn Workshop held in June 2002. Representatives of the Weyburn Project (Research Providers and the Management Committee) attended this Workshop, one of the objectives being to obtain a consensus on the working FEP list. The resultant working list of geosphere FEPs is reproduced here as Table 1.

TABLE 1
WORKING LIST OF WEYBURN SYSTEM GEOSPHERE FEPS

FEP title	FEP title
Geological units	*Chemical/geochemical*
A series of units representing aquitards and aquifers within the Weyburn System	Colloid formation and transport
Abandoned wells	Precipitation/dissolution of mineral (including surface processes)
Annular space (integrity/quality)	Dissolution/exsolution of CO_2
Corrosion of borehole metal	Gaseous contaminants
Expansion/collapse of corrosion products	Water chemistry
Degradation of borehole seal(s)	Purity of CO_2
Rock properties	*Properties and transport of CO_2 and other phases*
Mechanical properties of rock	Hydrodynamic flow
In situ stress distribution	Diffusion
Lithology and mineralogy	Dispersion
Lithification	Gas flow
Presence and nature of faults	Starting conditions (i.e. post-operational CO_2 distribution)
Presence and nature of fractures	Interfacial tension and wettability
Bounding seal system	Capillary pressure
Other geology	Bubble transport of CO_2
Natural seismicity	Transport of CO_2 (including multiphase flow)
Temperature/thermal field	*Other/miscellaneous*
Uplift and subsidence	Gas pressure (bulk gas)
Presence of unconformities	Pressure gradient
Desiccation of clay	Buoyancy
Hydrogeological properties	Coalescence of bubbles
Cross-formation flow	Release and transport of other fluids
Fluid characteristics of rock	Operational artifacts
Subsurface water flow	
Hydraulic pressure	
Hydrogeological properties of rock (basic)	
Brine displacement	
Mixing of water bodies	

Generic FEP database

The generic FEP database for the geological storage of CO_2 includes around 200 FEPs in a hierarchical structure, with FEPs grouped into categories such as "assessment basis", "external factors" and "boreholes" [7]. Each FEP has a text description and a discussion of its relevance to performance and safety. Key references in the published literature are included to enable retrieval of more detailed information for each FEP.

The database is available online and incorporates hyperlinks to other relevant sources of information (reports, websites, maps, photographs, videos, etc.). The database is searchable in a variety of ways and provides a centralized "knowledge base". Essentially, the list of FEPs defines the process system and represents all the factors that help define CO_2 behavior and migration.

The FEP database was expanded following an "FEP Workshop" held in Rome in January 2002 through the EC-funded Weyburn/Nascent projects clustering process. For example, a list of FEPs appropriate to generic CO_2 storage technologies was identified at this meeting.

FEP–FEP Interactions

The Weyburn FEPs discussed in the previous section do not exist in isolation, nor should they be treated as such. Rather, each of them may affect the system by influencing another FEP in some way, or by causing a more specific interaction on/with another FEP. For example, in the geosphere, the mineralogy of different rocks is one factor which will determine what rock–water interactions (geochemical interactions) occur; the basic chemistry (pH, major ions) of the groundwater is another.

Each of these interactions should be identified so that the total system can be described in a comprehensive way. Interactions between FEPs may be presented in a variety of ways, namely:

- a list identifying the interactions in terms of the initial and final FEPs;
- a diagram depicting individual FEPs as boxes, e.g. with interactions shown as arrows connecting two boxes; or
- an interaction matrix, whereby the FEPs are laid out in a two-dimensional matrix and interactions are represented by filled cells within this matrix.

Again, irrespective of the way in which these FEP interactions/influences are represented, the objective is to ensure that all possible/potential interactions are included. The mode of presentation is secondary, though important in providing some clear form of visual display that is as readily understood as possible. Such presentations are described by Stenhouse et al. [4].

Interactions between FEPs are often classified in terms of those which are highly important and those of low importance. Highly important is normally intended to mean that such interactions *must* be treated within the assessment, i.e. cannot be ignored. In contrast, to ignore FEP interactions of low importance should not affect the consequence analysis significantly. These classifications are rather arbitrary and depend on expert judgment but as long as each decision is documented, there is a sound basis for subsequent discussion and, where necessary, for revising a decision.

Figure 3 provides the interaction matrix for the Weyburn geosphere FEPs. The system FEPs of Table 1 appear vertically on the left-hand side and also horizontally along the top of the matrix. Any interaction between two FEPs is identified by a filled cell within the matrix.

Scenario Development

Even for a well-characterized CO_2 storage reservoir such as Weyburn, there are unavoidable uncertainties about the future state or evolution of the system. Such uncertainties arise from uncertainty about the importance (impact) or rate of various natural processes which will act on the system, the timing or frequency of certain natural phenomena (e.g. seismic events), and essentially unpredictable human activities in the future. In the assessment of the impacts of the final geological storage of nuclear wastes, uncertainty in future states has traditionally been handled by carrying out assessment calculations for a number of stylized conceptual descriptions of future state or evolution termed *scenarios*. Scenarios have become widely used in business and industry as planning and brainstorming tools and were first applied to the disposal of radioactive waste in the early 1980s by Sandia National Laboratory for the US Nuclear Regulatory Commission [8]. Regarding CO_2 storage, a scenario can be thought of as:

> a hypothetical sequence of processes and events, devised to illustrate a range of possible future behaviors and states of a carbon storage system, for the purposes of making or evaluating a safety case, or for considering the long-term fate of CO_2.

Scenarios form the basis for calculations of consequence analysis or risk. It is not necessary, or indeed possible in our view, to describe all possible scenarios. Thus, using the approach described by Chapman et al. [9], scenarios are viewed as *illustrative examples* of future behavior. There is no intent (indeed there is no possibility) to be either comprehensive or mutually exclusive, since there is no international consensus on applying probability theory to scenario analysis (see, e.g. NEA [10]). However, consideration of a set of scenarios should provide an adequately robust test of safety by addressing the most likely possible evolutions of the system together with less likely futures which exhibit features of possible concern [11].

Weyburn scenarios
A brainstorming session was held at EnCana (Weyburn Scenario Development Workshop, June 18, 2002) focusing on identifying scenario-generating events and characterizing them in terms of likelihood and severity of impact (consequence). The key output from this Workshop is the list of scenario-generating events provided in Table 2. A summary text description of the Base Scenario was also developed at this Workshop, and this is provided in Table 3.

TABLE 2
LIST OF SCENARIOS (SCENARIO-GENERATING EVENTS) IDENTIFIED FOR WEYBURN SYSTEM

Scenario-generating event	Scenario-generating event
Mining (salt dissolution and other resources)	Geothermally induced instability
Leaking wells (slow, fast—including self-propagating gas-pressure-driven fracture)	Igneous activity (causing change in thermal gradient)
Overpressuring of reservoir	Glaciation/unloading post-glaciation
Alternative techniques for resource recovery (CO_2 identified as resource)	Marine transgression
Tectonic activity (including seismic events)	Lack of quality control of injection
Fault movement/re-activation (covers undetected conductive feature)	Lack of records/knowledge
Influence of shallow trapping feature	Migration of CO_2 to other wells/ formations/surface
Accidental or intentional surface casing damage	CO_2 phase change, volumetric changes
Future drilling (above, to, through reservoir)	Displacement by other formation fluids
No wellbores (geosphere evaluation)	Unknown pyrite zone or similar (accelerated corrosion/degradation)
Extensive dissolution leading to subsidence	No surprises (no degradation of seals)
Open borehole (failure of top and bottom internal casing seals)	Favorable mineral/fluid chemistry (mineral fixation of CO_2)
Annular open borehole	Population changes above reservoir
Thermally induced fracture	Topographic changes
Additional CO_2 injection (>75-pattern)	Terrorist attack/sabotage
Blowdown (CO_2 recycle)	Change of supply of CO_2
Reversibility (CO_2 access)	Previously unobserved event
Exploration for oil/other resources	Gross exothermic reactions
Brines identified as resource	Meteorite impact
Other storage activities (concerning other fluids)	Political changes
Geothermal exploitation	

Note: Not all these scenarios will be addressed in the initial safety assessment; they are available, however, as the basis for future work.

Treatment of FEP–FEP Interactions: Modeling/Data Needs
The information contained in a FEP interaction matrix such as Figure 3 needs to be processed in order to show how each interaction will be dealt with during the assessment stage. The major ways in which these interactions translate to some form of action for the assessment are as follows:

- provision of data;
- one or more (robust) assumptions made;

TABLE 3
TEXT DESCRIPTION OF BASE CASE SCENARIO FOR WEYBURN CO_2 STORAGE SYSTEM

- The injected CO_2 starts off in the reservoir at the conclusion of commercial operations. (The CO_2 characteristics (pressure and phase distributions) at the end of EOR operations are predicted from reservoir simulations)
- Some CO_2 will exist as a supercritical fluid; some will be dissolved in oil and water phases; and some CO_2 may be mineralized. (The extent of mineralization is determined by geochemical modeling of conditions within the reservoir)
- The migration pathways are a combination of natural (geosphere) and manmade (abandoned wells). These two categories of migration pathways are treated independently, but eventually combined to represent the true long-term CO_2 storage conditions
- CO_2 can migrate from the reservoir by a number of different processes:
  - Pressure-driven flow
  - Density-driven flow
  - Diffusion

  Hydrodynamic flow (advection)
- CO_2 flux out of the reservoir is dependent upon the hydrogeological properties of rock in the Weyburn field and surrounding formations as well as the state of the wellbores (including annulus). The wellbore seals do not leak at time zero
- The Base Scenario takes into account all hydrogeological units above the reservoir and those units within the Mississippian below the reservoir. Note that the CO_2 may not reach many of these units
- CO_2–water–rock interactions can occur along the CO_2 fluid pathways. (Geochemical modeling is used to identify the chemical changes that occur and any resultant changes in hydrogeological properties caused by these chemical changes.) The timescale and pathways addressed by geochemical modeling are compatible with the corresponding predictions of CO_2 migration
- Long-term performance of abandoned wells:
  - Long term degradation of well seals (including annulus) and metal components will occur and will be governed by appropriate degradation rates consistent with the materials considered, e.g. corrosion rate of steel for casing metal
  - Such degradation may affect the CO_2 pathways and resultant flux; the impact of wellbore degradation will be reflected in modified transport properties of the wellbore (including annulus)
  - The responses of different formations to wellbore degradation or collapse are factored into the estimates of modified transport properties

- scoping calculations to provide bounding limits for one or more parameters; or
- detailed modeling.

For example, in the case of the influence of basic groundwater chemistry on precipitation/mineralization, geochemical modeling requiring solubility/thermodynamic data is needed. Similarly, as the result of an EFEP such as fault movement/activation, changes in the transport properties of the rock matrix (porosity, permeability) might be expected; in such a case, some bounding assumption may be made about the resultant increase in porosity/permeability.

In order to facilitate and document the process of identifying actions such as the examples discussed above, a spreadsheet was prepared outlining the assessment needs corresponding to the matrix shown in Figure 4. An extract from this spreadsheet is shown in Figure 5.

SUMMARY

The assessment of the long-term performance of geological systems for CO_2 storage safety is one of the most important issues for the feasibility of the widespread use of geologic storage. The systems analysis approach used for the long-term assessment of the fate of CO_2 in the Weyburn field is based on an understanding of the storage system constructed through an analysis of relevant FEPs the development

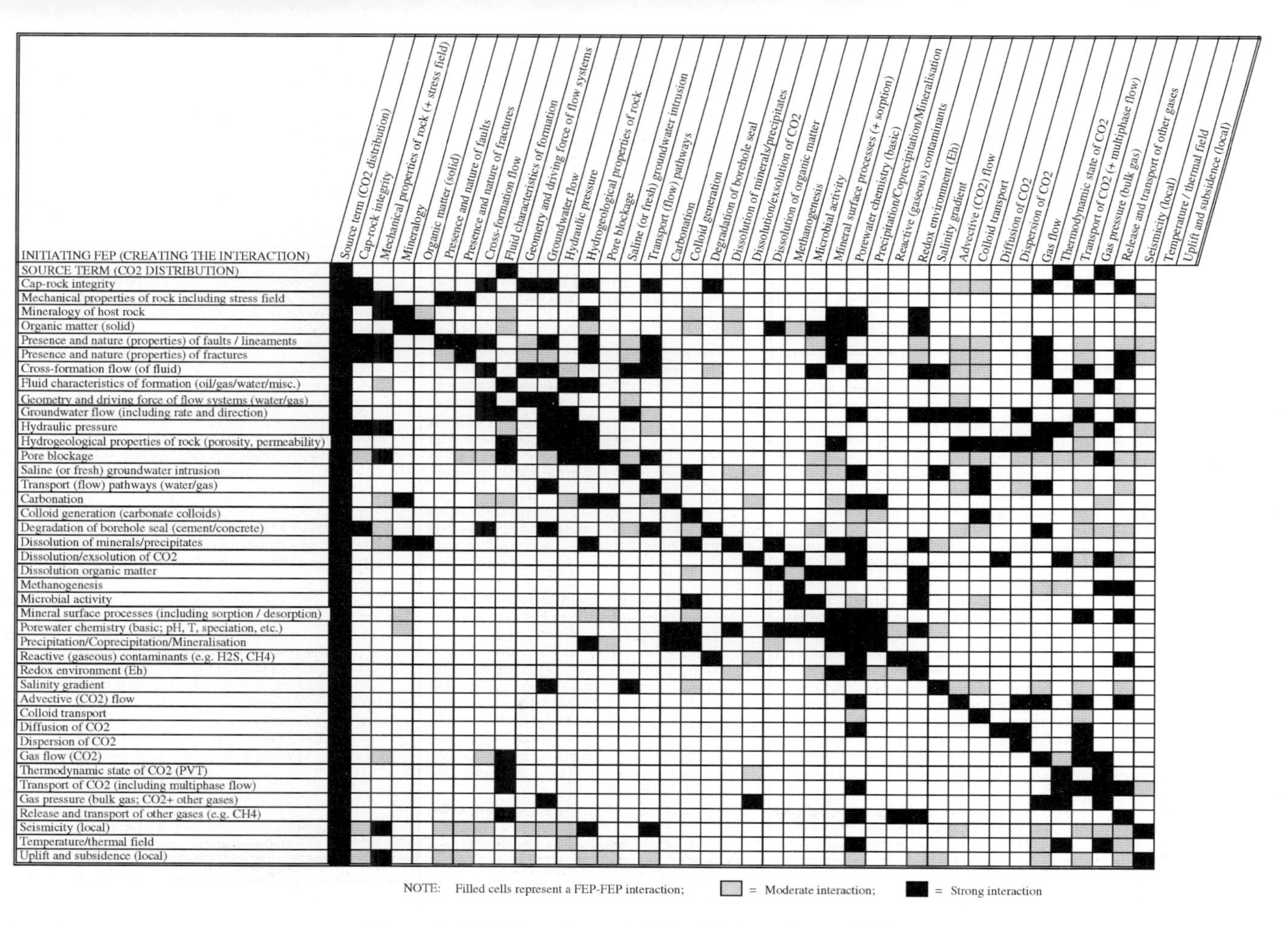

Figure 4: FEP interaction matrix for Weyburn System FEPs.

Initiating FEP	FEP Influenced / Affected	How Interaction is Treated
Source term (CO_2 distribution) *COMMENT: Basic input for long-term assessment*	Fluid characteristics of formation	Output from reservoir simulations of EOR period
	Thermodynamic state of CO_2	Output from reservoir simulations of EOR period
	Gas pressure (bulk gas)	Output from reservoir simulations of EOR period
Cap-rock integrity *COMMENT: For the Base Scenario, the cap-rock is assumed to be tight. Loss of integrity will enhance flow and transport of CO_2, both dissolved in groundwater and as a gas; migration not necessarily to the surface, but possibly to other groundwater bodies or formations, providing alternative pathways to the environment.*	Cross-formation flow	Incorporate additional formations as modelling results dictate
	Geometry and driving force of flow systems	Change hydraulic gradient as necessary
	Groundwater flow	Change hydrogeological parameters as necessary
	Hydrogeological properties of rock	Change hydrogeological parameters as necessary
	Transport (flow) pathways	Change hydrogeological parameters as necessary
	Degradation of borehole seal (cement / concrete)	Change transport properties with degradation
	Advective flow (CO_2)	Change hydrogeological parameters as necessary
	Colloid transport	Change hydrogeological parameters as necessary
	Gas flow (CO_2)	Change hydrogeological parameters as necessary
	Transport of CO_2 (including multiphase flow)	Change hydrogeological parameters as necessary
	Release and transport of other gases	Change hydrogeological parameters as necessary
Mechanical properties of rock (+ stress field) *COMMENT: Properties will have an impact on stability of formations, as will degradation of these properties*	Cap-rock integrity	Assume perturbation (increased stress) creates fractures in cap-rock
	Presence and nature (properties) of faults / lineaments	Assume perturbation (increased stress) generates fault
	Presence and nature (properties) of fractures	Assume perturbation (increased stress) creates fractures
	Seismicity (local)	Assume perturbation (increased stress) increases local seismicity
Mineralogy of host rock *COMMENT: Type of minerals (e.g. hard / soft) will affect stability of rocks, pore structure, chemical properties and the likelihood of colloids*	Mechanical properties of rock (+ stress field)	Qualitative assessment of stability of different formations
	Organic matter (solid)	Small interaction; ignore (reverse interaction important)
	Fluid characteristics of formation	Small interaction; ignore
	Hydrogeological properties of rock	Change hydrogeological parameters as mineralogy changes
	Colloid generation	Indirectly via porewater chemistry; consider carbonate colloids only
	Dissolution of minerals / precipitates	Chemical / geochemical modelling
	Mineral surface processes (including sorption / desorption)	Chemical / geochemical modelling; carbonate sorption / desorption
	Porewater chemistry (basic; pH; major ions etc.)	Chemical / geochemical modelling; rock-water interactions
	Redox environment	Chemical / geochemical modelling; rock-water interactions
Organic matter (solid) *COMMENT: Nature and quantity of solid organic components could affect rock-water interactions; effects are mainly chemical although mechanical stability expected to be a function of organic content.*	Mineralogy of host rock	Establish organic content of rocks
	Fluid characteristics of formation	Pore structure will change with changing organic content
	Hydrogeological properties of rock	Small interaction; ignore
	Colloid generation	Only carbonate colloids relevant; chemical modelling
	Dissolution of minerals / precipitates	Chemical / geochemical modelling; rock-water interactions
	Dissolution of organic matter	Chemical / geochemical modelling; include organics
	Methanogenesis	Chemical modelling to determine limits of occurrence
	Microbial activity	Assess importance of microbes in presence of solid organic matter
	Mineral surface processes (+ sorption / desorption)	Chemical / geochemical modelling; carbonate/CO_2 sorption
	Porewater chemistry (basic; pH; major ions etc.)	Chemical / geochemical modelling; include organic complexants
	Redox environment	Assess whether reducing / oxidising conditions exist

Figure 5: Extract of data modeling needs addressing FEP–FEP interactions.

of scenarios to represent the evolution of the system, and calculations of potential impacts using mathematical models to represent key processes. Over time, this methodology will be tested and if successful, confidence will build in our ability to accurately assess the health, safety and environmental risks of geologic storage projects.

ACKNOWLEDGEMENTS

We thank the Weyburn Project and the EC for funding the work carried out and reported in this chapter.

REFERENCES

1. W. Zhou, M.J. Stenhouse, S. Whittaker, D. Law, R. Chalaturnyk, W. Jazrawi, The IEA Weyburn Monitoring and Storage Project—modeling of the long-term migration of CO_2 from Weyburn, *Abstract #582 Accepted for GHGT7 Conference in Vancouver*, September 5–9, 2004, paper submitted for peer review, 2004.
2. D. Savage (Ed.), The Scientific and Regulatory Basis for the Geological Disposal of Radioactive Waste, Wiley, Chichester, UK, 1995, 437 pp.
3. N.A. Chapman, J. Andersson, P. Robinson, K. Skagius, C.-O. Wene, M. Wiborgh, S. Wingefors, Systems analysis, scenario construction and consequence analysis definition for SITE-94, SKI Technical Report No. 95:26, Swedish Nuclear Inspectorate, Stockholm, Sweden, 1995.
4. M.J. Stenhouse, P. Salter, W. Zhou, Monitor Scientific Report MSCI-2103 to CO_2 Capture Project, Surveillance, Monitoring and Validation Group, 2002.
5. M.J. Stenhouse, N.A. Chapman, T. Sumerling, Scenario development: FEP audit list preparation: methodology and presentation, SKI Technical Report No. 93:27, Swedish Nuclear Inspectorate, Stockholm, Sweden, 1993.
6. NEA, Disposal of radioactive waste: review of safety assessment methods, Report of Performance Assessment Advisory Group of the Radioactive Waste Management Committee, OECD Nuclear Energy Agency, Paris, France, 1991.
7. J.B. Riding, I. Czernichowski-Lauriol, S. Lombardi, F. Quattrocchi, C.A. Rochelle, D. Savage, N. Springer, The IEA Weyburn CO_2 monitoring and storage project—the European dimension, in: J. Gale, Y. Kaya (Ed.), *Sixth International Greenhouse Gas Control Conference*, Kyoto, Japan, Elsevier, Amsterdam, 2003, pp. 1629–1632.
8. R.M. Cranwell, R.W. Guzowski, J.E. Campbell, N.R. Ortiz, Risk methodology for geologic disposal of radioactive waste: scenario selection procedure, Sandia Report SAND80-1429, Sandia National Laboratory, Albuquerque, New Mexico, USA, 1982.
9. N.A. Chapman, J. Andersson, P. Robinson, K. Skagius, C.-O. Wene, M. Wiborgh, S. Wingefors, Devising scenarios for future repository evolution: a rigorous methodology, in: T. Murakami, R.C. Ewing (Eds.), *Scientific Basis for Nuclear Waste Management, Materials Research Symposium Series*, vol. 335, MRS, Pittsburgh, PA, 1995, pp. 495–502.
10. NEA, Features, Events and Processes (FEPs) for Geologic Disposal of Radioactive Waste, OECD Nuclear Energy Agency, Paris, France, 2000.
11. T. Sumerling, D.P. Hodgkinson, The treatment of uncertainty in future states for radioactive waste disposal, in: G. Apostolakis (Ed.), *Probabilistic Safety Assessment and Management*, vol. 1, Elsevier, New York, USA, 1991, pp. 641–648.

Carbon Dioxide Capture for Storage in Deep Geologic Formations, Volume 2
D.C. Thomas and S.M. Benson (Eds.)

Chapter 32

CO_2 STORAGE IN COALBEDS: RISK ASSESSMENT OF CO_2 AND METHANE LEAKAGE

Shaochang Wo[1], Jenn-Tai Liang[2] and Larry R. Myer[3]

[1]Institute for Enhanced Oil Recovery and Energy Research, University of Wyoming, 1000 E University Ave, Dept 4068, Laramie, Wyoming, 82071
[2]The University of Kansas, Lawrence, KS 66045, USA
[3]Lawrence Berkeley National Laboratory, Berkeley, CA 94720, USA

ABSTRACT

The practice of testing seal integrity is not routinely employed in coalbed methane projects. With injection of CO_2, changes in stress caused by potential high injection pressure and rate may open previously closed fractures and faults, thus generating new leakage pathways. The research presented in this chapter focuses on assessing potential leakage pathways and developing a probabilistic risk assessment methodology. A study was performed to evaluate geomechanical factors that need to be taken into account in assessing the risk of CO_2 leakage in CO_2 storage in coalbeds. The study revealed that geomechanical processes lead to risks of developing leakage paths for CO_2 at each step in the process of CO_2 storage in coalbeds. Risk of leakage is higher for old wells that are converted to injectors. Risks of leakage are much higher for open cavity completions than for cased well completions. The processes of depressurization during dewatering and methane production, followed by repressurization during CO_2 injection, lead to risks of leakage path formation by failure of the coal and slip on discontinuities in the coal and overburden. The most likely mechanism for leakage path formation is slip on pre-existing discontinuities that cut across the coal seam. A mathematical model for probabilistic risk assessment was developed. The model consists of six functional constituents: initiators, processes, failure modes, consequences (effects), indicators, and inference queries. Potential leakage pathways are usually coupled with identified failure modes. In assessing the risk of CO_2 storage in geological formations, inference rules can generally be categorized into seven different types. The inference logic of this model is based on set theory, which is superior to the traditional decision-tree based inference logic in terms of flexibility, generality, capability in dealing with uncertainties and handling large, complex problems, such as cascading phenomena. The model was designed to be implemented on a relational database.

INTRODUCTION

A recent report by Reeves [1] estimates that the total CO_2 storage potential in unmineable coalbeds in the US alone is about 90 gigatons, with the additional benefit of 152 trillion cubic feet of methane recovery. Methane production from coalbeds can be enhanced by injection of CO_2 to displace or N_2 to strip the methane from the coal and accelerate methane production at higher pressures (see Chapter 15). The mechanism by which CO_2 or N_2 can enhance the coalbed methane recovery process, and CO_2 is stored, is a complex mix of physical and chemical interactions that strive to achieve equilibrium simultaneously

---

Abbreviations: BP, British Petroleum; BLM, Bureau of Land Management; CCP, CO_2 Capture Project; CBM, Coalbed Methane; CRADA, Cooperative Research and Development Agreement; DOE, Department of Energy; ECBM, Enhanced Coalbed Methane Recovery; INEEL, Idaho National Engineering and Environmental Laboratory; JIP, Joint Industry Program; LBNL, Lawrence Berkeley National Laboratory; NETL, National Energy Technologies Laboratory.

in the sorbed state and the gaseous state. Coal has the capacity to hold considerably more CO_2 than either methane or nitrogen in the adsorbed state, in an approximate ratio of 4:2:1 for typical Fruitland coal in the San Juan basin [2–4]. This is because stronger forces of attraction exist between coal and CO_2 than between coal and methane or nitrogen. Two commercial demonstration projects of enhanced coalbed methane recovery (ECBM) by gas injection have been implemented at the Allison and Tiffany Units [2–6] in the San Juan basin.

Historically, methane seepage has been observed from the Pine River [7–9], South Texas Creek, Valencia Canyon, Soda Springs, and other areas [10–12] along the north and west Fruitland outcrops. Both of the Tiffany and Allison Units are located more than 15 miles away from any outcrop sites. It is very unlikely that injected CO_2 or N_2 could migrate to outcrops. However, simulation predicted that a large volume of methane and N_2/CO_2 breakthrough could occur if the N_2/CO_2 injection wells are placed too close to outcrops [13] (Chapter 15, this volume). Prior to any CO_2/N_2 being injected, methane leakage was observed in the CBM producing area. On July 23 1991, the Bureau of Land Management (BLM) issued a notice NTLMDO-91-1 in response to evidence of methane contamination in groundwater [10]. Since 1991, the BLM has aggressively implemented the terms and conditions of NTL MDO-91-1. The Colorado Oil and Gas Conservation Commission (COGCC) has also implemented and enforced similar requirements for gas wells on state and free lands. With the injection of CO_2 or N_2, the concern is that it could follow the methane leakage pathways to leak toward outcrops or the surface. In addition, repressuring coalbeds by CO_2 or N_2 injection will generate stresses and displacements in the coal seam and the adjacent overburden. The question is whether these stresses and displacements will generate new leakage pathways by failure of the rock or slip on pre-existing discontinuities such as fractures and faults.

In this study we evaluated the geomechanical factors which should be taken into account for assessing the risk of CO_2 leakage from coalbed storage projects. While conceptual and descriptive risk characterization is necessary and helpful in providing the baseline for quantitative risk assessments, decision makers need meaningful quantitative indicators, such as CO_2 leakage paths, leakage rate and volumes, CO_2 concentration at a leakage site, and remediation cost. In reality, quantifying site-specific risks is not easy. One must address uncertainties in almost all aspects of the project including site characterization, operations, and particularly in assessing the future evolution of the storage site. Probability-based risk assessment is considered as a meaningful and effective method for dealing with uncertainties. In this study, a mathematical model for probabilistic risk assessment was developed. Potential leakage pathways are assessed as failure modes. The model was designed to be implemented on a relational database.

NATURAL AND INJECTION-INDUCED LEAKAGE PATHWAYS

Coalbed reservoirs are self-contained petroleum systems, wherein the two critical petroleum system elements of source rock and reservoir rock are located together in a single geologic unit. Unlike conventional reservoirs, where gas or oil accumulated in a sedimentary porous rock below a low-permeability formation that acts as a seal, the majority of coalbed methane is adsorbed on the surface of the coal matrix and is not free to migrate until pressure is relieved by the withdrawal of water. For that reason, the seal integrity of coalbeds is generally not tested by the techniques that are used in conventional oil and gas reservoirs. In addition to naturally occurring microfractures (cleats), joints and faults may also be present in coalbeds, such as in the San Juan basin [14]. Joints and faults are larger scale fractures that typically cut across coalbeds and non-coal interbeds.

During the primary production in the San Juan Basin, methane seepage has increased at historic seepage sites. Inadequately cemented conventional gas wellbores and vertical microseepage are suspected of contributing to methane migration into surface soils and groundwater [10]. With the injection of CO_2 or N_2, the change in stress caused by high injection pressure and rate may open previously closed fractures and faults. To evaluate the geomechanical issues in CO_2 storage in coalbeds, it is necessary to review each step in the process of development of a CO_2 storage project and evaluate its geomechanical impact. A coalbed methane production/CO_2 storage project will be developed in four steps:

- drilling and completion of wells;
- formation dewatering and methane production;

- CO_2 injection with accompanying methane production; and
- possible CO_2 injection for storage only.

The approach taken in this study was to review each step, identify the geomechanical processes associated with it, and assess the risks that leakage would result from these processes.

Drilling and Completion Risks

Drilling issues

Wellbore instability is a geomechanical problem that can be encountered during drilling. Weak shale layers, weak coal layers, overpressure, and fault zones are common causes. Rock failure and displacements associated with wellbore instability generate potential leakage paths in the vicinity of the well. The risk of leakage will be minimized by cementing the casing. It is conventional practice to place cement behind production casing. Title 19 chapter 15 of the New Mexico Administrative Code states "cement shall be placed throughout all oil-and gas-bearing zones and shall extend upward a minimum of 500 ft above the uppermost perforation or, in the case of open-hole completion 500 ft above the production casing shoe". Alabama's regulations specific to coalbed methane operations have been used by other states as a model. Section 400-3 of the Rules and Regulations of the State Oil and Gas Board of Alabama states that the casing shall be cemented for 200 ft above the top of the uppermost coalbed which is to be completed, or for 200 ft above the production casing shoe in open hole completions. The production interval in cased hole completions need not be cemented.

When a coalbed methane project is converted to CO_2 storage, CO_2 will be injected under pressure. Wells used for injection in oil and gas formations are subject to additional regulations requiring periodic testing for leakage in the cased section. The type of testing which is required is set by individual states. In New Mexico, these tests can include the use of tracers to test for leakage in the annulus.

Injection of CO_2 also increases the risk of leakage in the annulus between casing and formation due to chemical dissolution of the cement. Experience in enhanced oil recovery has led to development of additives for cement used for CO_2 injectors. This experience should be applicable to coalbed methane CO_2 projects.

If old production wells or idle wells are used for CO_2 injection there is a risk that leakage paths may be present in the annular space between the casing and the rock due to deteriorated or missing cement. Casing bond logs and tracer tests can be used to evaluate the integrity of the cement in the annulus or the contact between casing and formation. If the integrity of the cement bond is inadequate, cement can be injected (squeezed) into the annulus. However, the process of seal formation in the annulus by cement squeeze behind casing is expensive and often only partially successful.

Because of the importance of the casing cement in minimizing the risk of CO_2 leakage, additional work should be directed toward development of recommendations for best practices. In particular, criteria for setting the height of the cement behind casing needs further study. Because of the substantial industry experience in water flooding and CO_2 enhanced oil recovery, a case history study of the performance of production casing cement would provide valuable data for a best practices study.

Conventional completions

A conventional completion for a coalbed methane project involves perforating or slotting the casing in the coal seam (Figure 1). Since the permeability of coal matrix is low, hydrofracturing is used to enhance permeability during dewatering and primary production. If the project is converted to CO_2 enhanced recovery and storage, pre-existing hydrofractures will enhance the injectivity of the CO_2. However, the risk of CO_2 leakage is also increased if hydrofractures extend into the overburden. Growth into the overburden can happen when the hydrofracture is initially created. In addition, since CO_2 is injected under pressure, fracture growth into the overburden could also occur during the enhanced recovery and storage phases of the project.

The potential for vertical extension of a hydraulic fracture is dependent upon several factors [15].

- *In situ stress state*. Higher horizontal stress in surrounding layers will impede vertical fracture growth, while lower horizontal stress tends to accelerate it. Higher pore pressure will enhance fracture growth.

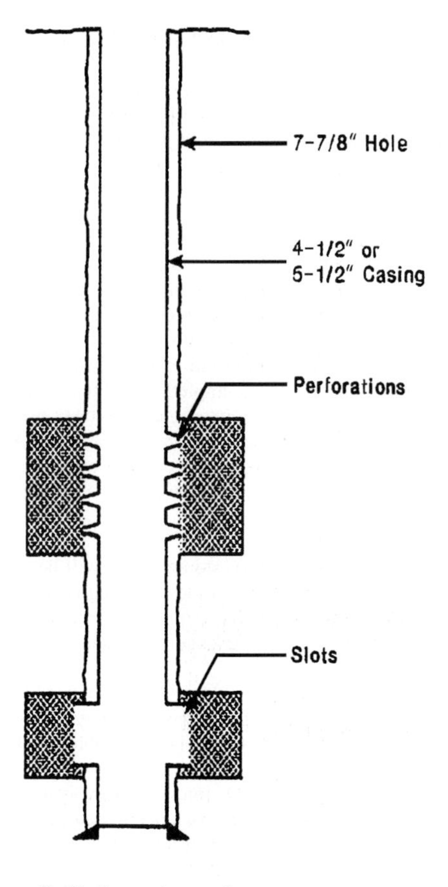

Figure 1: Schematic diagram of cased-hole completion for coalbed methane well [17].

On average, horizontal stress increases with depth but the lithology can affect in situ stress values. Pore pressures can also depart significantly from a "normal" hydrostatic gradient depending on the regional hydrologic setting as well as previous production and injection activities in the field.

- *Elastic moduli.* Vertical growth is impeded if the adjacent layer is stiffer than the coal seam. This is most likely to be the case if limestone or sandstone is the bounding strata. Siltstones and shale can vary widely in properties, but many are also stiffer than coals.
- *Toughness.* Higher fracture toughness will impede fracture growth. For large fractures, tensile strength is not a major factor [15]. The fracture toughness of coal is not well known. Atkinson and Meredith [16] compiled results of tests on four different coals. For Latrobe Valley Brown and Pittsburgh coal, values of "stress intensity resistance" ranged from 0.006 to 0.063 MPa $m^{1/2}$. However, for Queensland semi-anthracite and New South Wales black coal, values ranged from 0.13 to 0.44 MPa $m^{1/2}$. For comparison, values for sandstone, shale and limestone ranged from about 0.4 to 1.7 MPa $m^{1/2}$, with values for limestone generally being higher. This data indicates that some coals will have significantly lower

fracture toughness than typical bounding formations, and, therefore, there is a low risk of fracture growth out of interval.

- *Leakoff*. High fluid loss into the formation will retard growth of a fracture propagating into it.
- *Fluid flow*. Vertical fracture propagation will also be affected by the vertical component of fluid flow, which is affected by fracture opening and fluid properties. The effects of the fluid properties of CO_2 (particularly the non-wetting characteristics) on fracture propagation are a topic for further research.

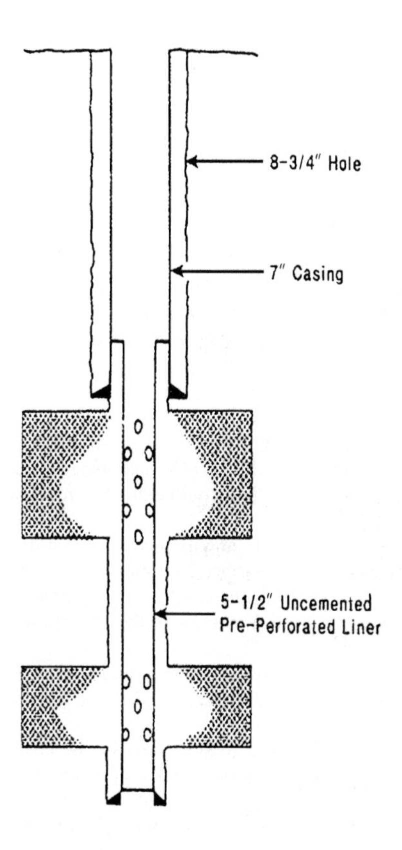

Figure 2: Schematic diagram of cavity completion for coalbed methane well [17].

Linear elastic fracture mechanics models have been developed to predict vertical fracture growth [18]. Ahmed et al. [19] developed expressions specifically for design in multiple zones. The approach is to first calculate the stress intensity factors for the top and bottom of the fracture. The stress intensity factor is a function of the height of the fracture the in situ horizontal effective stress, and the fluid pressure in the fracture. Fracture growth is predicted when the stress intensity factor exceeds a critical value given by the fracture toughness of the rock.

Risk of leakage will be reduced if the vertical extent of hydrofractures can be monitored. In cased wells measurement of fracture height, or detection of vertical propagation into bounding formations, is a challenging undertaking. Ahmed [18] and Anderson et al. [20] describe the use of radioactive tracers in conjunction with gamma ray logging. However, this technique only provides information in the near wellbore region.

In principle, seismic methods could be used to monitor the extension of a hydrofracture. Passive seismic techniques use seismic "events" generated by the fracturing process to locate the fracture. The fracture can also be imaged by a number of active seismic techniques. Though field experiments have been conducted, there is as yet no generally accepted seismic technique for determining fracture height. Nolte and Economides [21] describe a method for interpreting the downhole pressure decline during pumping to determine if a fracture has propagated into a bounding layer. The fracture extension may, however, not be vertical. Augmenting the pressure data analysis with other techniques such as passive or active seismic imaging may provide more information on the geometry of the propagating fracture.

Open cavity completions

A second type of completion for coalbed methane projects is the open hole cavity method (Figure 2). This technique was developed in the San Juan basin and is advantageous in areas where reservoir pressures are higher than normal. In such areas, casing is set above the coal seam and a cavity is generated by one of the two methods [22]. The first method is to drill through the coal seam underbalanced with water, air or foam. The excess formation pressure causes the coal to collapse into the wellbore. The coal is removed by displacing with drilling fluid and a perforated screen is set.

The second method uses pressure surges to collapse the coal. The well is shut in to build up pressure and then is abruptly released. Collapsed coal is then removed. This process can be repeated several times until the coal no longer collapses. Bland [22] reported that the effect could extend as much as 100 m into the coal seam.

Creation of a cavity can potentially cause failure and displacements in the overlying strata which provide pathways for CO_2, and increase the risk of leakage. Factors which influence the amount of disturbance in the overburden include the size and shape of the cavity, surge pressures, depth and in situ stress, layer thickness, rock strength and degree of natural fracturing in the overburden.

The process of pressure surging sets up high pore pressure gradients in the rock and corresponding flow lines as schematically illustrated in Figure 3a. Underbalanced drilling has the same effect though the pore pressure gradients would be lower. These pressure gradients cause fractures, joints, and cleats oriented perpendicular to the flow lines to open, leading to sloughing of the coal into the opening. The pressure gradients are also present in the overburden, so there is risk that this rock will also collapse into the cavity. The risk is highest for weak, thinly bedded, highly fractured shale. The risk is least for massively bedded sandstone and limestone.

The risk of overburden collapsing into the cavity increases as the cavity grows in width. As shown in Figure 3b of Chapter 33, removal of coal results in an unsupported span of layered overburden. As the span increases, so does the likelihood of finding fractures which define blocks. These blocks can be moved or removed by repeated surging. Since the interfaces between rock layers are weak, repeated surging would also tend to cause separation between layers producing more fluid pathways.

Creation of a cavity also results in a redistribution of the in situ stresses. This redistribution is very dependent upon the shape of the cavity as well as the relative magnitude of the vertical and horizontal far field stresses. The shape of the cavity formed by surging can be approximated by an ellipsoid with major axis equal to the thickness of the seam. The stress distribution around an elliptical (2D) cavity with major axis oriented parallel to the vertical far field stress is shown in Figure 4. It is seen that near the opening, in a direction along the minor axis the horizontal stress is less than the far field stress. Thus, the stress redistribution would be acting to further open fractures already opened by pressure surging. Similarly, along the major axis the vertical stress is less than the far field, increasing the risk that pressure surges would cause bedding plane partings.

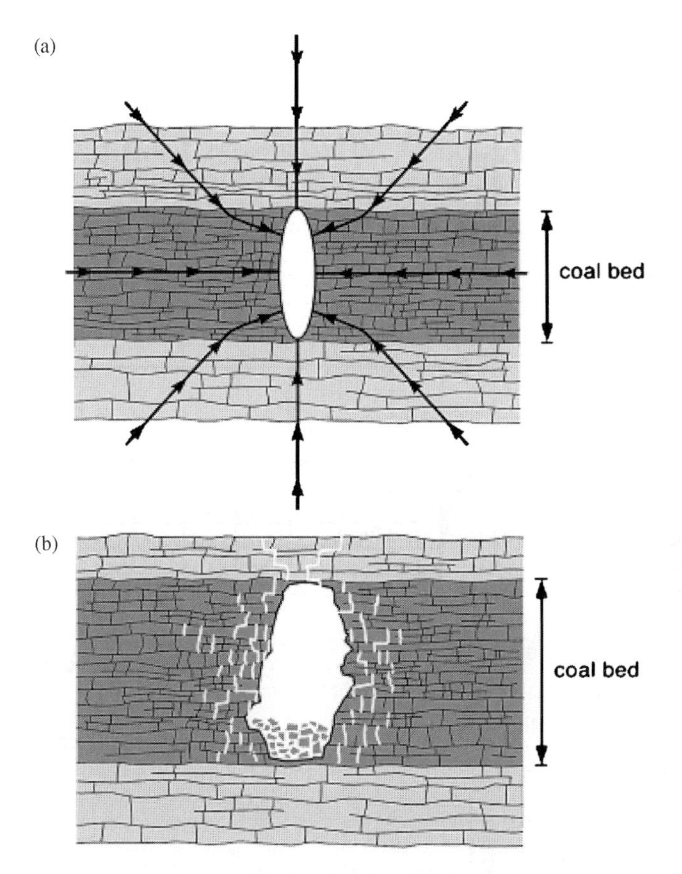

Figure 3: Schematic illustration of rock mass behavior associated with cavity completions in coalbeds. (a) Flow lines for water movement during surging. (b) Growth of cavity and fracturing in the coal and overburden.

Production and Repressurization Risks

The pore pressure reductions that occur during dewatering and methane production and pore pressure increase that occur during CO_2 injection, cause displacements in the reservoir and surrounding rock. A conservative assumption (to be discussed further) is that leakage will result if the rock fails or if slip occurs on pre-existing faults or discontinuities.

Failure and slip in a coal seam

A convenient way of assessing the potential for failure or slip is the Mohr diagram (Figure 5). A simple two-dimensional linear Mohr–Coulomb failure criterion is shown for illustration. The effective principal stress defined as total stress minus pore pressure is plotted on the horizontal axis and referred to as "normal stress". It is commonly assumed that an increase in pore pressure in the reservoir has an equal effect on both the components of principal stress, causing the Mohr circle to shift to the left, closer to failure, i.e. from I → II in Figure 5. This assumption has been employed in previous assessments of the potential for fault slip due to reservoir pressurization by CO_2 injection [25]. If pore pressures are reduced, it follows from this model that both the components of effective stress would be increased by the same amount, moving the Mohr circle away from failure.

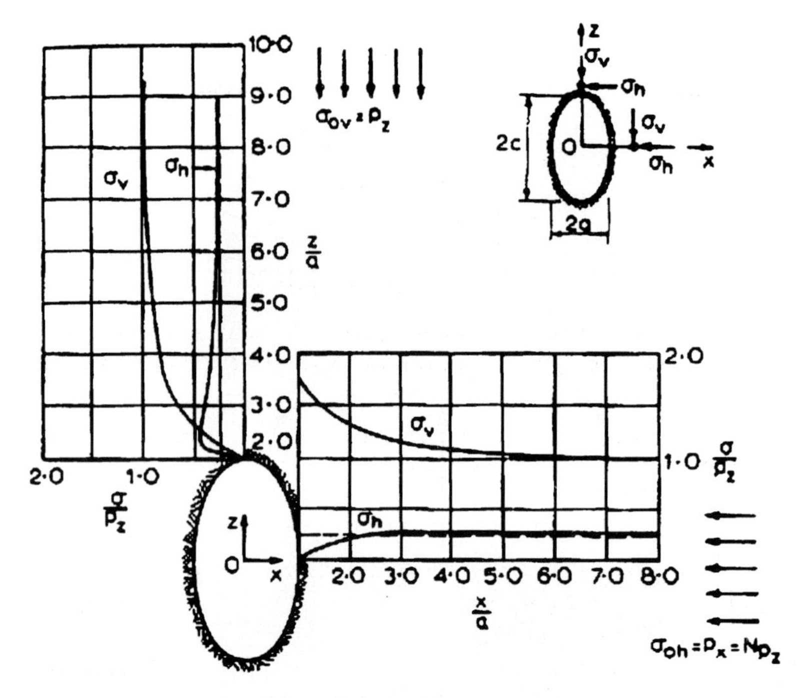

Figure 4: Stresses around an elliptical cavity $(a/c = 1/2)$ in homogeneous stress fields $(N = 0.25)$ [23,24].

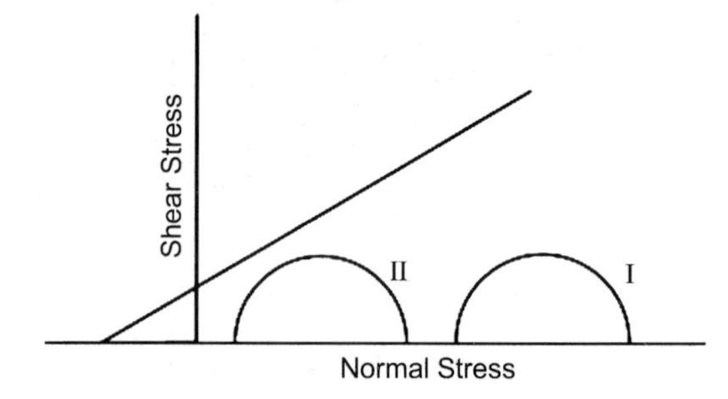

Figure 5: Mohr circles for initial (I) and final (II) stress state when it is assumed that a pore pressure increase affects both principal stresses equally.

Observations in a number of petroleum reservoirs [26,27] have shown that the reduction in pore pressure due to production causes a smaller change in horizontal stress than in vertical stress. The effect on the potential for failure is shown in Figure 6. Since pore pressures are decreasing, the Mohr circle moves to the right. However, since the change in horizontal effective stress is less than in the vertical effective stress, the circle actually gets closer to failure that is from I → III in Figure 6 of Chapter 33. Teufel et al. [28] showed that these effects were large enough to cause failure of the high porosity chalk in the North Sea Ekofisk reservoir. Streit and Hillis [29] further analyzed the effects on fault slip.

These relative changes in horizontal and vertical effective stresses are the result of the effects of far field (in situ) boundary conditions and poroelastic properties of the rock. Figure 7 shows that the rate of change

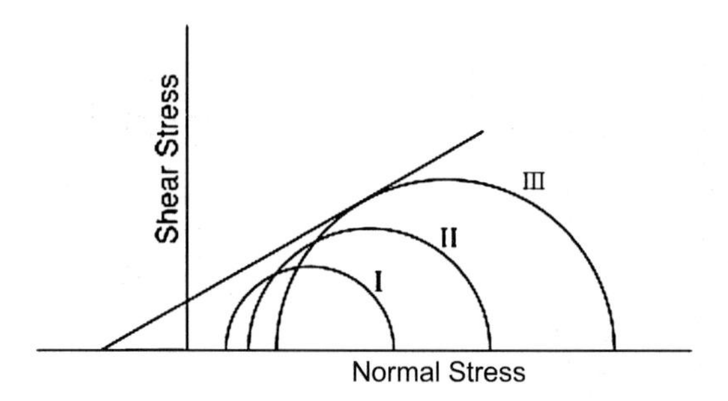

Figure 6: Mohr circles for initial (I), intermediate (II), and final (III) stress states for pore pressure reduction assuming that horizontal stresses are less affected than vertical stresses. Failure or slip occurs at III.

in horizontal stress with pore pressure, i.e. $\Delta\sigma_h/\Delta P$ where σ_h is horizontal stress and P is pore pressure, decreases as Poisson's ratio of the reservoir rock increases. Touloukian et al. [30] reported measured values of Poisson's ratio for coal of 0.2–0.4.

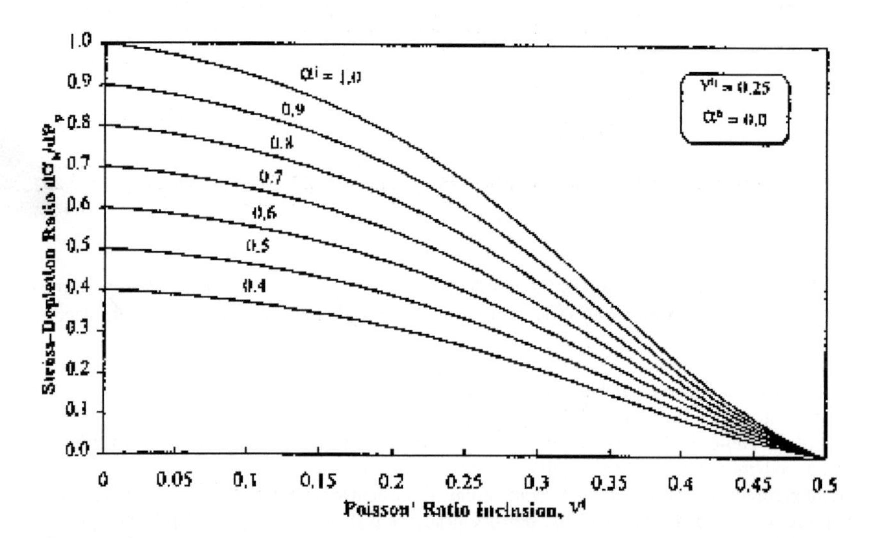

Figure 7: Effect of Poisson's ratio of the reservoir rock on rate of change in horizontal stress with pore pressure for a disc-shaped reservoir modeled as an inclusion (i) in a host (h) rock and various Biot coefficients [31].

The risk of failure or slip in the coal will depend on depth, in situ stress state, pressure drawdown, and coal strength and poroelastic properties. Conditions which result in large principal stress differences increase the risk of failure and slip. Tectonic activity will result in increased differential far field stresses. Large pore pressure drawdown will increase differential stress. Risk of failure increases for low strength coal. In situ stresses increase with depth, but the strength of rock increases with level of confinement. The risk of failure may or may not increase with depth depending on the amount of pore pressure drawdown and the magnitude of differences between components of in situ stress. The risk of slip on pre-existing discontinuities is increased for low cohesion and low frictional sliding resistance.

Injection of CO_2 for enhanced methane production and storage will increase pore pressures in the coal seam. In a poroelastic system, effective stress changes due to pore pressure drawdown are simply reversed by pore pressure increase due to injection. Thus, a Mohr circle which had moved closer to failure under drawdown would move farther from failure during injection until the original, pre-development pore pressures are obtained. Failure, however, is an inelastic process and, in general, results in a complex redistribution of stress in the system.

If pore pressures from CO_2 injection exceed pre-development levels, then there is a risk that slip will occur even though it had not occurred under drawdown conditions. This is conceptually illustrated in Figure 8, where the Mohr circle for pre-development stress state is labeled I. Dewatering and methane production moves the Mohr circle to the right (state II) under conditions in which the change in horizontal effective stress is less than the change in vertical effective stress. The maximum stress difference is not sufficient to cause failure or slip. Upon repressurization, assuming no inelastic effects, the Mohr circle returns to state I. If pressurization continues so that pore pressures rise above pre-development levels the Mohr circle moves to the left, resulting in the condition for failure or slip as indicated by state III in the figure. It has been assumed in this construction that the vertical effective stress changes more rapidly than the horizontal effective stress during pore pressure increase.

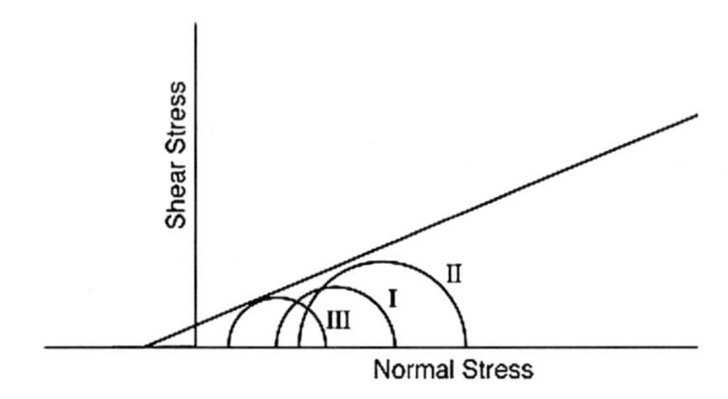

Figure 8: Mohr circles for initial (I), intermediate (II) and final (III) stress state when pore pressure first decreases (II) and then increases (III) with respect to initial conditions. Failure or slip occurs at III.

The approach outlined above can be used to make a preliminary assessment of the potential for slip on pre-existing discontinuities in the coal in the San Juan basin. Values of parameters used in the analysis are summarized in Table 1. A mean depth of 3200 ft and an initial reservoir pressure of 1500 psi before dewatering and methane production are assumed. The reservoir pressure is consistent with a normal hydrostatic gradient and observations in some areas of the San Juan basin. It is assumed that the maximum principal stress is vertical (S_V) and the density gradient is 1 psi per foot of depth. For purpose of this calculation the in situ stress, S_{hmin}/S_V, where S_{hmin} is the minimum horizontal stress, is assumed to be 0.7. The condition for slip on the discontinuity is given by a linear Mohr–Coulomb criteria with the conservative assumption that the cohesion is zero. A coefficient of friction, μ, of 0.6 is assumed. This value is frequently assumed in analyses of slip on faults in petroleum reservoirs [25,32]. It is also consistent with laboratory measurements of the strength of coal under confining pressures of several thousand psi [33].

The Mohr circle labeled by I in Figure 9 represents the initial stress conditions. It is assumed that pore pressures have equilibrated over a large area over time, so the initial major and minor principal effective stresses, σ_1 and σ_3, are given by subtracting 1500 psi from both S_V and S_{hmin}. It is then assumed that reservoir pressures are drawn down to 500 psi and there is a poroelastic effect in a finite-sized reservoir. From Figure 7, if the Poisson's ratio of the coal is 0.3, then $\Delta S_{hmin} = -0.53\Delta P$ (where P is reservoir pressure and " $-$ " refers to a decrease in P) and the Mohr circle moves to position labeled II. As seen in the figure, there is no slip. For a Poisson's ratio of 0.4, $\Delta S_{hmin} = -0.23\Delta P$ and the Mohr circle is given by II′ which is a more stable condition than that attained for Poisson's ratio of 0.3.

TABLE 1
SLIP ANALYSIS PARAMETER

Parameter	Value
Mean reservoir depth	3200 ft
Initial reservoir pressure	1500 psi
Post drawdown reservoir pressure	500 psi
Reservoir pressure after CO_2 injection	2000 psi
Poisson's ratio for coal	0.3, 0.4
Coefficient of friction for slip	0.6
In situ stress ratio (S_{hmin}/S_V)	0.7

Finally, it is assumed that CO_2 injection increases reservoir pressure to 2000 psi. Taking account of poroelastic effects and assuming a Poisson's ratio of 0.3 for the coal, the Mohr circle moves from II to III. For this case, there is still no slip on discontinuities. However, for Poisson's ratio of 0.4, $\Delta S_{hmin} = 0.23\Delta P$, and the Mohr circle moves from II′ to III′; intersecting the criterion for slip. During repressurization more stable conditions are attained if the Poisson's ratio of the reservoir material is low.

The dip of discontinuities upon which slip would occur can be determined from the intersection of the Mohr circle with the failure criteria. The equations for the two values of β corresponding to the points of intersection are [34]

$$2\beta_1 = \pi + \varphi - \sin^{-1}[(\sigma_m/\tau_m)\sin\varphi]$$

and

$$2\beta_2 = \varphi + \sin^{-1}[(\sigma_m/\tau_m)\sin\varphi]$$

where

$$\varphi = \tan^{-1}\mu$$

$$\sigma_m = \frac{1}{2}(\sigma_1 + \sigma_3)$$

$$\tau_m = \frac{1}{2}(\sigma_1 - \sigma_3)$$

For conditions represented by the circle III′ in Figure 9, slip would occur on discontinuities with dips between 50° and 70°.

Results of these analyses are very sensitive to the in situ stress state. The risk of slip is significantly reduced as $S_{hmin}/S_V \rightarrow 1$. If the stability analysis is repeated assuming $S_{hmin}/S_V = 1$, a common assumption in reservoir simulation, then no slip would be predicted for any of the reservoir pressure conditions. However, if $S_{hmin}/S_V = 0.6$, slip is predicted even under the assumed initial reservoir pressure of 1500 psi.

Failure and slip in the overburden

So far, the discussion has focused only on the risk of failure or slip within the coal seam. However, potential leakage paths require failure in slip in the bounding rock layers as well as in the coal seam. A possible, though least likely mechanism, is the propagation of a shear failure from the coal into the bounding rock. As discussed previously, fracture propagation into the bounding rock is impeded when the coal strength is less than the strength of the bounding rock.

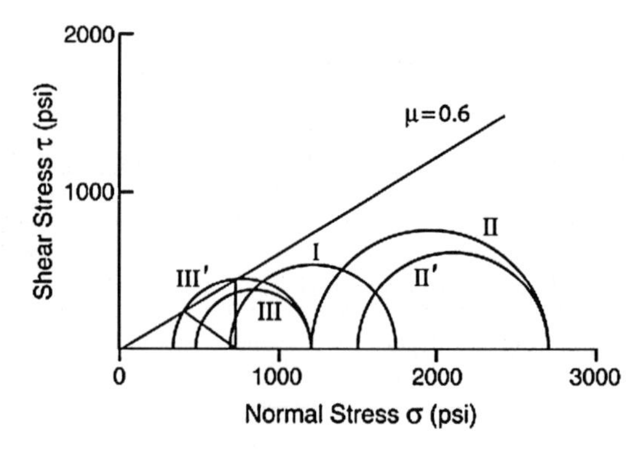

Figure 9: Mohr circles for slip on a discontinuity in a coal seam under conditions representative of the San Juan basin.

Volumetric changes in the reservoir have an important influence on displacements in the overburden. During production, there is a volumetric decrease in the reservoir due to pore pressure reduction. The amount of volumetric decrease is a function of the compressibility of the reservoir rock and its thickness. In coal there is an added component due to shrinkage from desorption of the methane. The volumetric decrease in the reservoir may cause subsidence of the overburden. On the flanks of the reservoir, bending of the overburden layers results in shear stresses which can cause failure or slip on pre-existing discontinuities. If the pore pressure distribution, and hence, volumetric deformation, in the reservoir is not uniform, shear displacements in the overburden will be introduced at places other than the flanks.

Repressurization of the reservoir may cause volumetric expansion and upward displacement, or heave, in the overburden. The effect on shear displacements is to reverse the sense of motion. Thus, shear displacement on a discontinuity can move in one direction during drawdown and reverse and move in the opposite direction during injection. An example of this is shown in Figure 10. The figure shows modeled

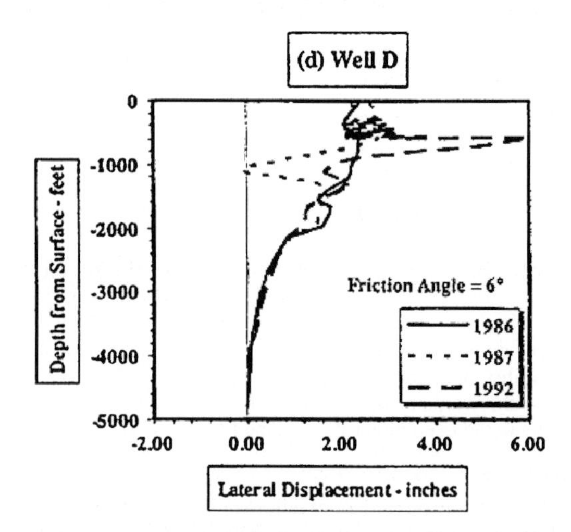

Figure 10: Numerical simulation of lateral displacement of a well in the South Belridge reservoir. Large lateral displacements at about 1000 ft depth occur due to slip on an interface with a friction angle of 6°. Lateral displacements reverse between the years of 1987 and 1992 [35].

well displacements due to shear on a weak zone in the overburden above the South Belridge oil reservoir. This reservoir has undergone pressure drawdown from production and then repressurization from aggressive water injection.

An example of the development of shear displacements near the interface between the reservoir and overburden when CO_2 is injected is shown in Figure 11. The figure shows results of a numerical simulation of injection of CO_2 from a single well into a brine-saturated layer. The shaded region in part b of the figure shows where shear stresses develop. The blue outline shows the extent of the CO_2 plume. The volumetric expansion of coal with CO_2 will have an additional component due to swelling associated with gas sorption. Experimental work indicates that CO_2 causes more volumetric changes than methane. This will further alter the distribution of volumetric expansion resulting from repressurization.

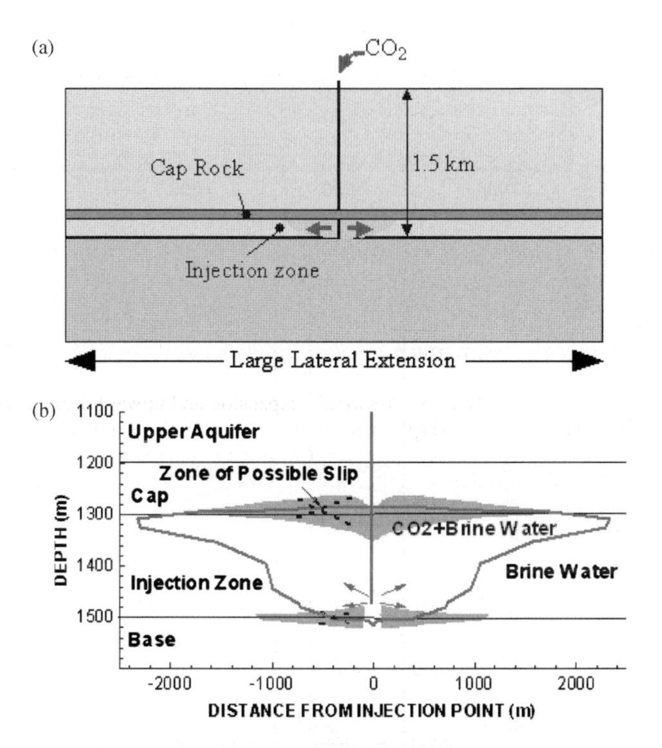

Figure 11: Results of numerical simulation of stresses and displacements due to injection of CO_2 into a brine saturated formation [36]. (a) The model. (b) Outline of plume and region where shear stresses could cause slip on discontinuities.

If a pre-existing discontinuity cuts across the coal seam, model results show that slip can occur in the overburden, outside of the region of pore pressure change. Figure 12a shows a model in which there is a pressurized region between two discontinuities ("faults") dipping at 45°. Calculations were carried out using the coupled hydrologic/geomechanical simulator TOUGH-FLAC [37]. The faults were represented by "slip lines" with a friction angle of 25°. Figure 12b shows the shear slip on the faults as a function of depth. Due to the symmetry of the problem, the sense of motion is in one direction on one fault and in the opposite direction on the other fault. It is seen that the magnitude of the slip is greatest within the region of pressure increase and tails off quickly outside the region.

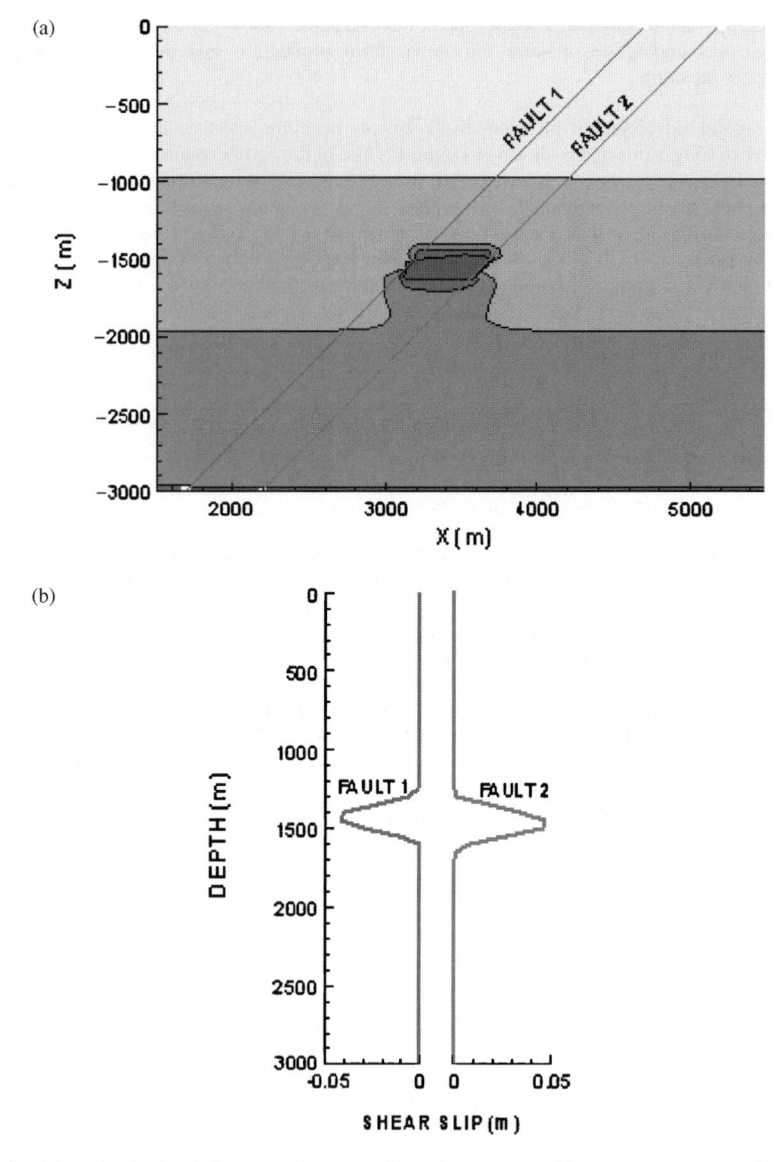

Figure 12: Numerical simulation of slip on discontinuities resulting from a pressurized region. (a) The model, showing a maximum pressure increase in the region of 2.6 times original pressure. (b) Shear slip on the faults.

Slip on pre-existing faults and other discontinuities which intersect the coal seam are viewed as a likely scenario for generation of possible leakage paths for CO_2. Numerical sensitivity studies should be performed to evaluate the effects the dip and frictional properties of faults for representative coal seam pressure changes. It is important to capture coal volumetric changes due to sorption and desorption as part of these models.

While slip on pre-existing discontinuities creates a potential leakage path, further analysis is required to evaluate whether or not fluid flow will occur in conjunction with the slip. The risk of leakage will be increased if the magnitude of the slip is on the order of bed thickness. Geologic studies of fault seals have shown that fault movement which brings sand layers into contact can lead to fluid flow across faults from higher to lower pressure sands.

The degree to which slip will increase the potential for flow along faults and discontinuities is much less well understood. Laboratory tests have shown that shearing a rock fracture in rock will increase its permeability as a result of dilatancy. Since fracture surfaces are rough, shear displacements can lead to an opening of the fracture and an increase in permeability. Less dilatancy would be expected for faults or discontinuities filled with clay gouge. The relationship between stress state, slip magnitude, fault and fracture surface geometry and changes in hydrologic properties of infilling materials is an area requiring substantial additional basic research.

Other Potential Failure Modes

The risk of methane emission is another environmental issue that must be considered during CO_2 enhanced coalbed methane production. Methane is the second most important greenhouse gas, responsible for about 15% of the greenhouse gas buildup in the atmosphere to date (Greengas.htm). Molecule for molecule, methane traps about 27 times more heat than CO_2.

Ideally, the majority of injected CO_2 will be trapped by adsorption onto the surface of coal matrices. However, CO_2 retention in a coalbed is largely dictated by how effectively the injected CO_2 contacts and interacts with the coalbed over the project lifetime. As observed in the Tiffany field, the early N_2 breakthrough and high N_2 cut indicated that the injected N_2 may only contact a small portion of the total available pay [13,42] (Chapter 15, this volume).

Coalbed water, whose salinity varies from fairly fresh to very saline, is a potential water resource for domestic, irrigation, industrial, mining, and thermoelectric use. In 1990, about 48 million gallons per day (MGD) of saline ground water was utilized in the United States as a source of public water supply, mostly for thermoelectric power. This represents a 28% increase since 1987 and a 178% increase since 1985. The potential future usage of coalbed water, therefore, must be taken into account in the selection of a coalbed CO_2 storage site.

Given the complex and unique nature of storing CO_2 in coalbeds, risks exist both during and after the injection of CO_2. Besides the risk scenarios common to other geological formations, storing CO_2 in coalbeds has five additional pitfalls that should be assessed carefully:

- insufficient CO_2–coal contact volume due to coalbed heterogeneity;
- injectivity loss due to coal swelling caused by CO_2 adsorption;
- CO_2 and methane leakage through pre-existing faults and discontinuities;
- CO_2 and methane seepage through outcrops; and
- CO_2 and methane desorption due to potential future coalbed water extraction.

Table 2 summarized the most likely failure modes that pertain to the operation of CO_2 storage in coalbeds. Along with the failure modes, their potential initiators and consequences are provided. The duration of a failure mode is indicated by short term (S), or long term (L), or both.

PROBABILISTIC RISK ASSESSMENT METHODOLOGY

Conceptually, a risk assessment methodology should include four major elements: hazard identification, event and failure quantification, predictive modeling, and risk characterization. The hazards of CO_2 exposure are well known and described in Benson et al. [39] and Chapter 27 of this volume. Similarly the hazards of methane releases are well known [7–12,38].

TABLE 2
SUMMARY OF FAILURE MODES PERTAINING TO CO_2 STORAGE IN COALBEDS

Failure modes	Event initiators	Consequences	Short/ long term
CO_2 pipeline failure	Corrosion, manufacturer's defects, earthquake, sabotage	Short-term release of concentrated CO_2 into atmosphere, human safety and health hazard	S
Compressor failure	Corrosion, improper maintenance, manufacturer's defects	Interruption of CO_2 injection	S
Well string failure (surface casing, intermediate casing, tubing, etc.)	Corrosion, manufacturer's defects	CO_2 migration out of zone, CO_2 migration into meteoric water, absolute open flow (AOF), human safety and health hazard	S&L
Cement failure	Corrosion, poor cement bond	CO_2 migration out of zone, CO_2 migration into meteoric water, AOF, human safety and health hazard	S&L
Seal failure	CO_2/H_2O/rock interactions, in situ stress by coal swelling, over pressurization	CO_2 migration out of zone, CO_2 migration into meteoric water, reduced sequestration capacity, diminished recovery, catastrophic CO_2 release into atmosphere, human safety and health hazard	S&L
Fracture extension within zone or into overburden	Injection above parting pressure, hydraulic fracturing, earthquake	Long-term CO_2 release into atmosphere, CO_2 migration into meteoric water, asset degradation, AOF, human safety and health hazard	S&L
Injectivity loss	Coal swelling caused by CO_2 adsorption	Lower-than-planned injection rate, asset degradation, early project termination	S
Insufficient storage capacity	Reservoir heterogeneity	Early project termination, asset degradation	S
Insufficient methane ecovery	Reservoir heterogeneity	Early project termination, asset degradation	S
Methane and CO_2 seepage through outcrops	Methane and CO_2 release paths leading to outcrops	Long-term methane and CO_2 release into atmosphere, human safety and health hazard	S&L
Methane and CO_2 seepage through out-of-area abandoned wells	Methane and CO_2 migration out of sequestration area, poor cement bond and wellbore integrity	Long-term methane and CO_2 release into atmosphere, methane and CO_2 migration into meteoric water, human safety and health hazard	S&L

(*continued*)

TABLE 2
CONTINUED

Failure modes	Event initiators	Consequences	Short/ long term
Seal penetration	Future oil and gas drilling activities into underlying reservoirs	Methane and CO_2 migration out of zone, methane and CO_2 migration into meteoric water, catastrophic methane and CO_2 release into atmosphere, human safety and health hazard	S&L
Annular cement failure in converting old wells to CO_2 injection wells	Deteriorated or missing cement in the annular space between the casing and the rock	CO_2 leakage into overlying formations from injection wellbore	S
Overlying strata displacement	Open cavity completion	Generating fractures and CO_2 leakage paths in overburden	S&L
Coal seam slip on pre-existing discontinuities	Tectonic activity, earthquake, formation pore pressure above pre-developed level due to CO_2 injection	Potential slip and methane and CO_2 leakage paths in the bounding rock layers, catastrophic methane and CO_2 release into atmosphere, human safety and health hazard	S&L
Hydrostatic pressure drop down in coal seam	Declined water table caused by coalbed water extraction, coalbed water leakage due to underlying strata displacements	Methane and CO_2 desorption from coal matrix, catastrophic methane and CO_2 seepage from outcrops and pre-existing leakage paths, human safety and health hazard	L

Identifying and quantifying potential failure modes (event and failure quantification) at a CO_2 storage site, during and after the injection operation, is an essential part of any risk assessment. In general, any potential breach of storage integrity and normal operation can be regarded as a potential failure mode. As illustrated in Figure 13, a failure can be caused by reservoir properties and natural events, but may also be caused by engineering failures. We use the following set of questions as the guideline in identifying potential failure modes:

- What can go wrong? What causes the failure?
- What is the likelihood of the failure happening?
- How much CO_2 (and methane) could be released?
- What are the consequences?
- What is the remediation cost if the failure is reparable?

Finding credible answers to these questions is often not easy. Reservoir simulation and predictive modeling will be required to estimate the quantity and rate of unintended CO_2 and methane release. In the final step, risk characterization, quantitative estimates of methane and CO_2 leakage will be compared to a set of criteria that define, for example, acceptable rates of leakage and CO_2 exposure.

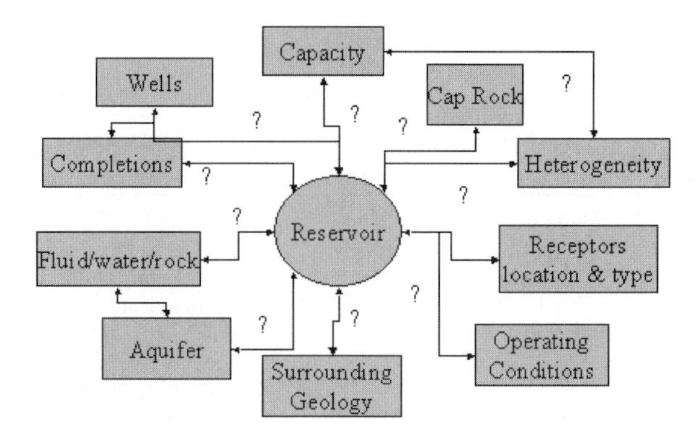

Figure 13: Examples of potential failure modes.

As discussed in Introduction, uncertainties are inherent in almost every aspect of the operation during and after the project lifetime. Consequently, a risk assessment process must be designed as a dynamic system capable of quickly redoing risk assessments when additional or updated data become available. In addition, failure modes can interact or cascade with one another. The consistency, transparency, and correctness in inference logic of such a risk assessment system must be validated. The traditional decision-tree approach, which is usually effective for simple problems, is inadequate in handling large, complex, and dynamic systems.

To make the risk assessment process rigorous and transparent, a mathematical model specifically designed for probabilistic risk assessment was developed. The guidelines for model development were

- generality and transparency;
- designed for implementation on a relational database;
- inference rules can be converted to and verified by set operations; and
- quantified indicators as model outputs.

The inference logic of this model is based on set theory, which is superior to the traditional decision-tree based inference logic, in terms of flexibility, generality, capability in dealing with uncertainties and handling large, complex problems, such as cascading phenomena. The model is also applicable for the risk assessment of CO_2 storage in other geological formations such as oil and gas reservoirs. For simplification, only CO_2 leakage is considered in the model configuration but methane leakage can be modeled similarly.

Mathematical Model

Model constituents

To create a rigorous inference system, various factors and terminologies from a real-world risk scenario must be abstracted to a limited set of functional constituents. In this model, six functional constituents have been identified. They are initiators, processes, failure modes, consequences (effects), indicators, and inference queries. In a database application, each constituent will be implemented as a database table. The six sets of constituents are symbolically defined by

(1) $I = \{i_1, i_2, i_3, \ldots\}$, Initiators.
(2) $P = \{p_1, p_2, p_3, \ldots\}$, Processes.
(3) $M = \{m_1, m_2, m_3, \ldots\}$, Failure Modes.
(4) $C = \{c_1, c_2, c_3, \ldots\}$, Consequences (effects).
(5) $D = \{d_1, d_2, d_3, \ldots\}$, Indicators.
(6) $Q = \{q_1, q_2, q_3, \ldots\}$, Inference queries.

The concept of a failure mode was already discussed and defined. Any cause leading to a failure mode is regarded as an initiator. Any effect, usually an adverse effect, is called a consequence. In cascading phenomena, a consequence of one failure mode can be the initiator of other failure modes. The fate and transport of CO_2 is represented by a set of processes. A process can represent a planned CO_2 path or an unintended CO_2 release path. To make the risk characterization transparent and meaningful, results from a risk assessment need to be organized and presented by meaningful indicators. Generally, indicators can be classified into two groups: descriptive indicators and performance indicators. Descriptive indicators provide mainly statistical information, such as averages, maxima, minima, and risk profiles while performance indicators compare different scenarios. An example of a performance indicator is the difference between the current CO_2 in-place (a specified scenario) and the maximum capacity (the base scenario). Other examples of indicators are activated initiators and their likelihood, affected processes (failure modes), consequences and associated severity scales, process tree, initiator–process–consequence diagram, consequence–process–initiator diagram, initiator–consequence diagram, overall risk index, sensitivity of initiators to the overall risk, and sensitivity of consequences to the overall risk. In some cases, additional information and criteria are required in the determination of certain initiators and consequences or in a decision-making process. These supplement information and criteria will be stored in the Inference Query table.

Inference rules

In the next step, relationships and connections between the constituents are converted to set operations, or so-called inference rules. Inference rules can be developed based on expert judgment, results of mathematical models or from statistical analysis of data from related experience. The quality of the information contained in the inference rules dictates the quality of the risk assessment. Over time, the quality of the information contained in the inference rules will improve if the experience from geologic storage projects is incorporated. Similarly, models are expected to improve as real-world data sets are used to calibrate and verify them.

For assessing underground CO_2 storage, inference rules can be categorized into seven different types.

(1) $P \Leftarrow \vec{F}_P(I)$, identify processes affected by each initiator.
(2) $M \Leftarrow \vec{F}_M(P)$, define failure modes associated with each process.
(3) $C \Leftarrow \vec{F}_C(P, M)$, identify consequences if a failure mode occurs.
(4) $I \Leftarrow \vec{F}_I(C)$, identify cascading effects.
(5) $D \Leftarrow \vec{F}_D(I, P, M, C)$, dynamically calculate and reevaluate indicators.
(6) $I \Leftarrow \vec{F}_I(Q)$, indirectly identify initiators.
(7) $C \Leftarrow \vec{F}_C(Q)$, indirectly identify consequences.

In a database application, inference rules of the same type will be implemented in one database table. Because processes are associated directly with CO_2 transport or release paths, the natural cascading flow path of CO_2 can be used in defining how processes are linked one to another. As a part of the process properties stored in the process table, the description of the connections to up-stream and down-stream processes are required information that will later be used in calculating the likelihood of cascaded CO_2 releases.

It is worth pointing out that, in general, decision-tree based rules can be converted to set operations as demonstrated in the following example (see Figure 14). A simple decision tree for identifying initiators consists of three Questions (criteria) and four possible Initiators (answers). As shown in Figure 14a relational table between the set of Questions and the set of Initiators is generated in which each column represents a possible decision route.

The relational table implies two sets of inference rules: (1) if a question is known, then it indicates possible initiators or (2) if an initiator occurs, then it gives the answers to the questions. In comparison, an inference rule can be represented by a route of a decision tree or equivalently by set operations as demonstrated in the following example.

Decision Tree: If Q_1 true and Q_2 false then I_2
Set Operation: $\{I_1, I_2\} \cap \{I_2\} = I_2$

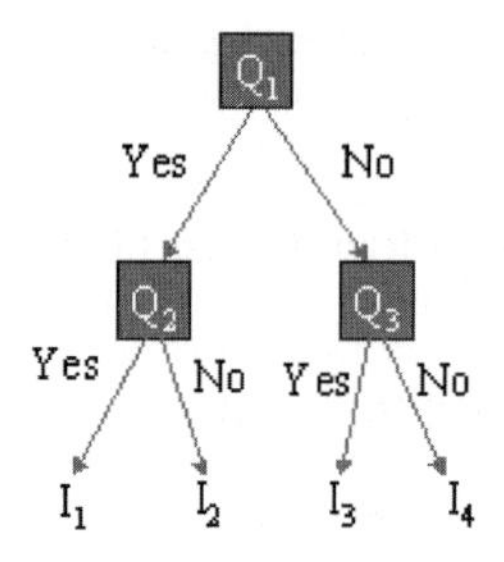

	I1	I2	I3	I4
Q1	Yes	Yes	No	No
Q2	Yes	No	NA	NA
Q3	NA	NA	Yes	No

Figure 14: The conversion between a simple decision tree and its relational table.

Severity scale of consequences

Quantifying the severity of a consequence is probably the most difficult task in risk assessment. We suggest using a numerical scaling system based on the set of criteria shown in Table 3. Ranking risks by standardized criterion such as these has already been used for other complex systems [40].

In addition to CO_2 release rate and remediate cost, other criteria in risk assessment may include adverse effect to human health, adverse effect to animals, potentiality in regulation breach, duration, cascading effect, undetectability, uncontrollability, and irreversibility. After the criteria are accepted, a numerical severity scale, for example between 0 and 1, will be defined according to the effective impact of a consequence based on expert knowledge, statistical data, and regulations. In practice, the average value of all severity scales can be used to indicate the overall severity scale of a consequence. The overall severity scale could also include a weighting factor that recognizes that not all factors are equally important (e.g. risk to human life may be weighted more strongly than undetectability).

$$\text{Average Severity Scale} = \frac{\sum_i S_i}{\text{Number of Criteria}}$$

TABLE 3
EXAMPLES OF CRITERIA FOR THE SEVERITY ASSESSMENT OF CONSEQUENCES

Criterion	**Severity scale (0–1)**
Adverse effect to human health	S_1
Adverse effect to animals	S_2
Potentiality of violating regulations	S_3
Duration	S_4
Cascading effect	S_5
Undetectability	S_6
Uncontrollability	S_7
Irreversibility	S_8

Likelihood of failures

For a given failure mode, M, we use *LIKELIHOOD(M)* to indicate the failure likelihood of M (*LIKELIHOOD(M)* ranges from 0 to 1). Let P be the process associated with the failure mode M. Without losing generality, we assume that the failure could be caused by each of n identified initiators, $\{I_1, I_2, \ldots, I_n\}$. The failure likelihood caused by initiator I_i alone is given by $LIKELIHOOD(I_i)$, $i = 1{-}n$.

The effective failure likelihood caused by the combined effects of the *n* initiators can be calculated from the following iterative procedure.

$$LIKELIHOOD(\{I_1, I_2, \ldots, I_i\})] = LIKELIHOOD(I_i) + LIKELIHOOD(\{I_1, I_2, \ldots, I_{i-1}\}) - LIKELIHOOD(I_i) * LIKELIHOOD(\{I_1, I_2, \ldots, I_{i-1}\}), i = 2\text{–}n.$$

We define the failure likelihood of *M* as

$$LIKELIHOOD(M) = LIKELIHOOD(P) * LIKELIHOOD(\{I_1, I_2, \ldots, I_n\})$$

where *LIKELIHOOD(P)* is the likelihood of CO_2 existence in process *P* and is defined by

$$LIKELIHOOD(P) = \begin{cases} 1, & \text{if } P \text{ is a planned } CO_2 \text{ path} \\ \prod_k LIKELIHOOD(M_k), & \text{otherwise} \end{cases}$$

In the above definition, M_k represents a preceding failure mode on the cascading CO_2 release path to the process *P* and *LIKELIHOOD(Mk)* is its failure likelihood.

Rate, cost, and effective severities
In probabilistic risk assessment, the failure likelihood of a failure mode is considered to be equally important as other factors in the evaluation of effective severities.

Let $Rate^*$ and $Cost^*$ be the estimated CO_2 release rate and remediation cost in the case where 100% failure occurs to the failure mode, *M*. After the failure likelihood of *M* is obtained, the effective CO_2 release rate can be evaluated by

$$Rate = Rate^* * LIKELIHOOD(M)$$

and the effective reparable cost of the failure mode can be estimated by

$$Cost = Cost^* * LIKELIHOOD(M)$$

For each of the identified consequences of *M*, its effective severity scale can then be evaluated by the geometric average of its severity scale and the failure likelihood of *M*,

$$\sqrt{(Severity\ Scale\ of\ Consequence) * LIKELIHOOD(M)}$$

Risk scenario simulator
A relational database, such as MS Access, is capable not only of managing large datasets but can also perform complex dataset operations. Because the inference rules of the model are represented as set operations, a database application of the model can entirely be coded by database language, which is referred to as the risk scenario simulator. A scenario simulation will consist of the following steps: (1) activating selected initiators, (2) identifying affected processes, (3) calculating the failure likelihood of each failure mode, (4) identifying their consequences, (5) estimating the effective CO_2 release rates, reparable costs, and the effective severity scales of consequences, and (6) repeating steps (1)–(5) if new initiators have been invoked by resulting consequences (cascading effects). Practically, once initiators are manually activated in step 1, the rest of steps and computational works can be performed by pre-stored procedures. A prototype application has been developed and will be discussed in the following section.

RESULTS AND DISCUSSION

Leakage Quantification
Identifying potential leakage pathways and estimating leakage flux are the two basic tasks for leakage evaluation. The severity of a leakage is directly related to the leakage rate. Structural geology and monitoring data at historic seep sites provide a useful indication of existing leakage paths and flux intensity. Reservoir modeling is an essential tool for quantitative predictions of CO_2 and methane transport in

sedimentary strata. The water–gas transport through the coal cleat system is normally described by Darcy's law for two-phase flow, which is applied by most current CBM simulators and is capable to predict CO_2 and methane seepage rates at outcrops. In contrast, modeling vertical seepage is much more difficult. Variations in overlying stratigraghic column and formation structures complicate the model settings. Unsaturated zones and fracture networks may cause further uncertainties in flow regimes. In fact, the majority of vertical seepage flux may largely be controlled by fracture networks [14,41–44].

Once gas seeps into fractures, the buoyancy force drives gas bubbles migrating upward to the surface. Brown [41] analyzed gas flow in fractures and proposed four mechanisms for gas migration in fractures. They are (1) continuous-phase gas migrating in fractures, (2) bubble ascent without wall or concentration effects (Stokes' law), (3) maximum velocity of isolated bubble ascent in fractures, and (4) steady ascent of bubbly water in a vertical fracture having infinitesimal bubble size and 18% gas concentration. Figure 15 shows the calculated gas migration velocities along a fracture for the four different mechanisms. By comparing reported seepage velocities (in the order of 100–10,000 m/yr), Brown [41] concluded that single-phase gas flow in fractures having half widths from 0.1 to 2 μm can be responsible for buoyancy-driven flow at rates equal to the range of reported seepage velocity (Figure 15).

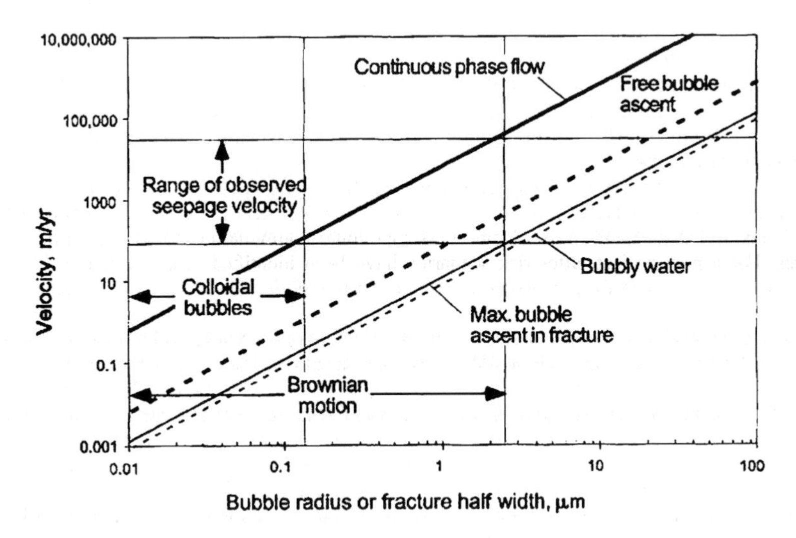

Figure 15: Comparison of calculated migration velocities for the different mechanisms and observed seepage velocities [41].

Leakage in wellbores can be detected by tracer tests, image and casing bond logs, and Bradenhead (casing) pressure tests. Bradenhead (casing) pressure monitoring is routinely required for gas wells in the San Juan Basin. A threshold pressure of 25 psig (2 psig in the critical areas) was established by the BLM in 1991 [10]. Therefore, the likelihood of gas leakage in the annular space can be directly evaluated according to the measured Bradenhead pressure, as shown in Figure 16.

In general, a failure mode's failure likelihood caused by an identified initiator can usually be represented as a function (cumulative probability distribution) of relevant parameters, such as simulated leakage flux, injection pressure, fracture density, and statistical data. Similarly, the severity scale of a consequence can be determined by its measurable parameters. In Figure 17, the severity scales to human safety and health are assessed by the CO_2 release rate and the distance to the source point.

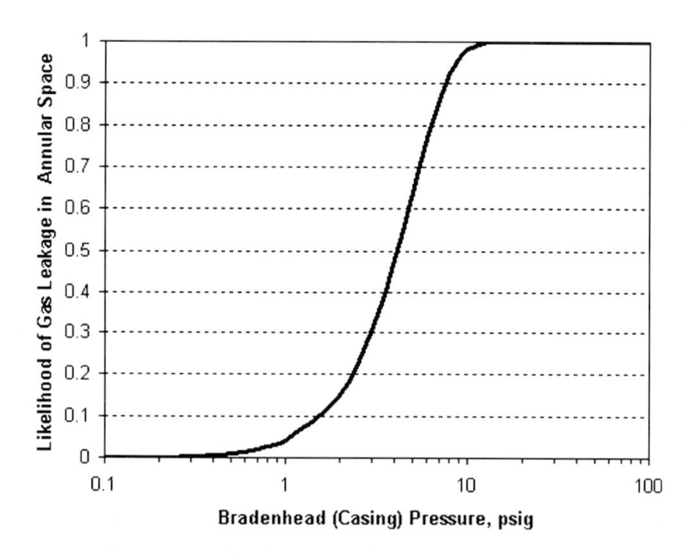

Figure 16: Bradenhead (casing) as an indicator of gas leakage in the annular space.

Software Tool Development

To demonstrate the applicability of the model, Microsoft Access was used as a platform to develop a prototype application. Complying with the major steps of the assessment procedure, the application consists of several modules that reside in the database as functional combinations of tables, forms, and stored procedures. For a new project, after risk scenarios have been identified, a user can use the main user interface and its pop-up interfaces to define processes, failure modes, initiators, and consequences.

Once the risk scenarios are defined and entered into the database, quantifying and tuning each failure mode is the main task performed via the Failure Mode form interface (see Figure 18). In the top-left corner of the form, there is a drop-down list for the selection of any defined failure modes. When a failure mode is selected, all the data and computations will be associated with the selected failure mode. By changing settings, for example, activating/deactivating initiators and consequences, changing likelihood values, changing maximum cost, and changing maximum CO_2 release rate, one can quickly perform risk quantification for different risk scenarios. The results are dynamically and visually presented by the severity matrix. After each failure mode has been properly tuned, one can simultaneously run all failure modes together to see the interaction and cascading effects between the failure modes.

Scenario simulation vs. Monte Carlo simulation

When the UPDATE button on the Failure Mode form is clicked, only a single scenario simulation will be performed. In addition, a built-in Monte Carlo simulation procedure is also provided. To perform a Monte Carlo simulation, the number of seeds (runs) has to be selected first from the drop-down list (Figure 18). By clicking the Monte Carlo simulation button, the Monte Carlo simulation will be performed and the results will be saved in the table of Monte Carlo-Failure Mode. The difference between a single scenario simulation and the Monte Carlo simulation is in how to select the failure likelihood of the initiators. In a single scenario simulation run, we use

$$I_i \text{ with } LIKELIHOOD(I_i), \quad i = 1\text{–}N$$

While for a series of Monte Carlo runs, we use

$$I_i \text{ with } LIKELIHOOD_{MC}(I_i), \quad i = 1\text{–}N,$$

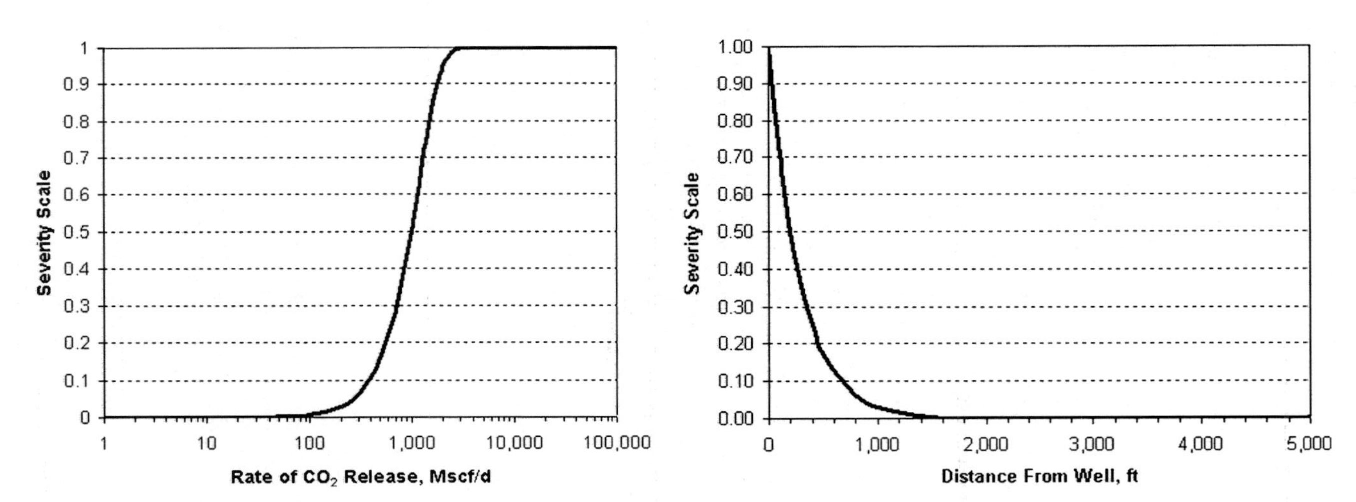

Figure 17: Rate effect (left) and distance effect (right) of point source CO_2 release into atmosphere on severity of human safety and health hazard.

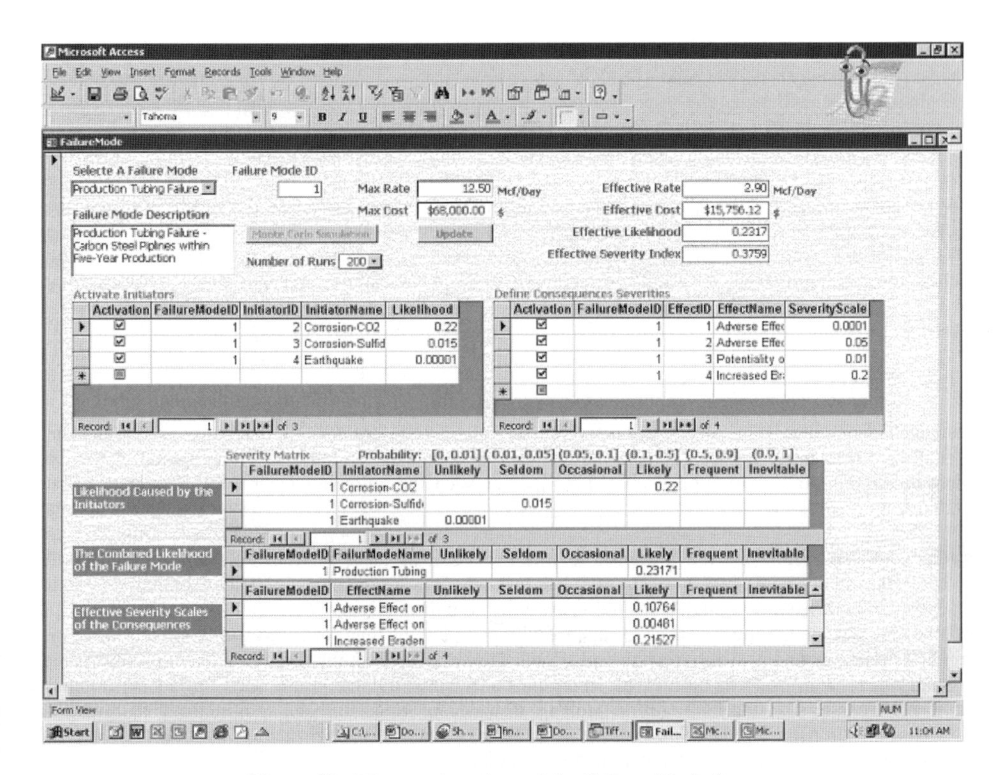

Figure 18: The user interface of the Failure Mode form.

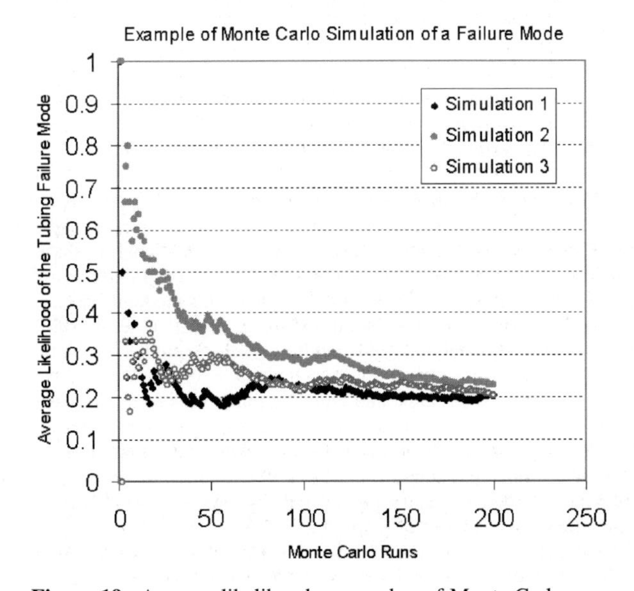

Figure 19: Average likelihood vs. number of Monte Carlo runs.

where

$$LIKELIHOOD_{MC}(I_i) = \begin{cases} 0 & \text{if } LIKELIHOOD(I_i) < Random() \leq 1 \\ 1 & \text{if } 0 \leq Random(\ ) \leq LIKELIHOOD(I_i) \end{cases}$$

To prevent over-sized tables, the Monte Carlo-Failure Mode table only keeps the results from the most recent run. Figure 19 shows the plot of the average likelihood versus the number of Monte Carlo runs where 200 seeds (runs) were used for all three Monte Carlo simulations. In this example of production tubing failure, three initiators were activated, Corrosion-CO_2, Corrosion-Sulfide, and Earthquake, with likelihood of 0.22, 0.015, and 0.00001, respectively for a 5-year production period. After 200 runs, the average likelihood from the Monte Carlo simulations all approach the combined likelihood of 0.23171.

CONCLUSIONS

Geomechanical processes lead to risks of developing leakage pathways for CO_2 and methane at each step in a coalbed methane project for methane production and eventual CO_2 storage. Though each of the risks identified in this study need to be evaluated for specific sites, the following general conclusions have been drawn from this review:

- Conventional techniques are available to minimize risk of leaks in new well construction though additional study should be devoted to establish best practices for the height of cement behind production casing; risk of leakage is higher for old wells converted to injectors.
- Risks of leakage are much higher for open cavity completions than for cased well completions.
- Coal properties and available technology should minimize the risk that hydrofractures, used as part of completion, will grow out of interval; techniques to monitor fracture height need further development.
- The processes of depressurization during dewatering and methane production, followed by repressurization during CO_2 injection, lead to risks of leakage path formation by failure of the coal and slip on discontinuities in the coal and overburden.
- The most likely mechanism for leakage path formation is slip on pre-existing discontinuities which cut across the coal seam. Sensitivity studies need to be performed to better evaluate this risk.
- Relationships between the amount of slip and the increase in flow (if any) along a discontinuity need to be developed.

The risk assessment methodology proposed in this study includes four major elements: hazard identification, event and failure quantification, predictive modeling, and risk characterization. The central part of the methodology is a mathematical model, wherein potential CO_2 and methane leakage pathways are defined by failure modes. The results from this work are summarized as follows:

- A mathematical model for probabilistic risk assessment was developed. The model consists of six functional constituents, initiators, processes, failure modes, consequences (effects), indicators, and inference queries. The model was designed to implement on a relational database.
- For assessing the risks of CO_2 storage in geological formations, inference rules can generally be categorized into seven different types. The inference logic of this model is based on set theory, which is superior to the traditional decision-tree based inference logic, in terms of flexibility, generality, capability in dealing with uncertainties and handling large, complex problems, such as cascading phenomena.
- The mathematical model provides a logic and computational basis for a risk-based scenario simulator.
- To demonstrate the applicability of the mathematical model, a prototype application was developed in Microsoft Access. The application consists of several modules that reside in the database as functional combinations of tables, forms, and stored procedures. An intuitive main user interface and its pop-up interfaces are created to facilitate the data input and risk assessment process. The application can perform both scenario simulations and Monte Carlo simulations.
- In addition to the risk scenarios common to other geological formations, storing CO_2 in coalbeds may face other pitfalls. The likely risks pertaining to CO_2 injection and storage in coalbeds include: insufficient CO_2–coal contact volume due to coalbed heterogeneity, injectivity loss due to coal swelling

caused by CO_2 adsorption, CO_2 and methane leakage through pre-existing faults and discontinuities, CO_2 and methane leakage through outcrops. In the long term, CO_2 and methane desorption caused by potential coalbed water extraction after the project lifetime is also a concern.

ACKNOWLEDGEMENTS

We are grateful to Daryl Erickson of BP America Inc. for providing helpful insights and the Tiffany field data. This work was supported in part by a Cooperative Research and Development Agreement (CRADA) between BP America Inc., as part of the CO_2 Capture Project (CCP) of the Joint Industry Program (JIP), and the US Department of Energy (DOE) through the National Energy Technologies Laboratory (NETL) under contract DE-AC07-99ID13727.

REFERENCES

1. S.R. Reeves, Enhanced CBM recovery, coalbed CO_2 storage assessed, *Oil Gas J.* **101** (27) (2003) 14.
2. D. Erickson, Overview of ECBM Commercial Demonstration Pilot at Tiffany Unit, Internal Report, BP America, 2002, October.
3. K.A.M. Gasem, R.L. Robinson Jr., S.R. Reeves, Adsorption of Pure Methane, Nitrogen, and Carbon Dioxide and Their Mixtures on San Juan Basin Coal, DOE Topical Report, 2002, May.
4. S.R. Reeves, Geologic Storage of CO2 in Deep, Unmineable Coalbeds: An Integrated Research and Commercial-Scale Field Demonstration Project, SPE 71749, Proceedings of the SPE Annual Technical Conference and Exhibition, New Orleans, September 30–October 3, 2001.
5. S.R. Reeves, C. Clarkson, D. Erickson, Selected Field for ECBM Recovery and CO2 Storage in Coal based on Experience Gained at the Allison and Tiffany Units, San Juan Basin, DOE Topical Report DE-FC26-00NT40924, September 30, 2002.
6. S. Reeves, A. Taillefert, L. Pekot, C. Clarkson, The Allison Unit CO_2—ECBM Pilot: A Reservoir Modeling Study, DOE Topical Report, 2003, February.
7. Amoco, Pine River Fruitland Coal Outcrop Investigation: Southern Rockies Business Unit, Amoco Production Company, Denver, CO, 1994.
8. Advanced Resources International Inc., Gas seepage in the Pine River Area, Colorado, prepared for the Geological and Reservoir Engineering Subcommittee of the Pine River Fruitland Coal Investigative Team, November 1994.
9. P. Oldaker, Monitoring data review, Pine River ranches, prepared for Colorado Oil and Gas Conservation Commission and Amoco Production Company, USA, 1999.
10. Bureau of Land Management (BLM), San Juan Field Office, Coalbed Methane Development in The Northern San Juan Basin of Colorado, December 1999.
11. Questa Engineering Corporation, The 3M coalbed methane reservoir model, Prepared for the Southern Ute Indian Tribe, Ignacio, Colorado, the Colorado Oil and Gas Conservation Commission, Denver, Colorado, and the Bureau of Land Management, Durango, Colorado, May 26, 2000.
12. L.L. Wray, Late Cretaceous fruitland formation geologic mapping, outcrop measured sections, and subsurface stratigraphic cross sections, Northern La Plata County, Colorado, Colorado Geological Survey Open File Report 00-18, Denver, Colorado, 2000.
13. S. Wo, J.T. Liang, Simulation assessment of N_2/CO_2 contact volume in coal and its impact on outcrop seepage in N_2/CO_2 injection for enhanced coalbed methane recovery, The 14th Improved Oil Recovery Symposium, Tulsa, Oklahoma, April 17–21, 2004.
14. C.R. Nelson, Geologic controls on effective cleat porosity variation in San Juan basin Fruitland Formation coalbed reservoirs, Tuscaloosa, Alabama, Proceedings, International Coalbed Methane Symposium, Paper, 108, 2001, pp. 11–19.
15. K. Ben-Naceur, Modeling of hydraulic fracture, *Reservoir Stimulation* (1989) 3-1–3-31.
16. B. Atkinson, P. Meredith, Experimental fracture mechanics data for rocks and minerals, *Fracture Mechanics of Rock* (1987) 477.
17. D. Murray, 1993, Coalbed methane reservoir evaluation and completion technology. Atlas of Major Rocky Mountain Gas Reservoirs, New Mexico Bureau of Mines and Mineral Resources, pp. 88–189, 1971.

18. U. Ahmed, Fracture-height predictions and post-treatment measurements, *Reservoir Stimulation* (1989) 10-1–10-3.
19. U. Ahmed, B.M. Newberry, D.E. Cannon, Hydraulic fracture treatment design of wells with multiple zones, SPE 13857, 1985.
20. J.A. Anderson, C.M. Pearson, A.S. Abou-Sayed, G.D. Myers, Determination of fracture height by spectral gamma log analysis. SPE 15439, Proceedings of 61th Annual Technical Conference, New Orleans, LA, 1986.
21. K. Nolte, M. Economides, Fracturing diagnosis using pressure analysis, *Reservoir Stimulation* (1989) 7-1–7-34.
22. D. Bland, Coalbed methane from the Fruitland Formation, San Juan Basin, New Mexico, North Mexico Geological Society 43rd Conference, 1992.
23. H. Poulos, E. Davis, Elastic Solutions for Soil and Rock Mechanics, Wiley, New York, 1974, p. 237.
24. K. Terzaghi, F.E. Richart, Stresses in rock about cavities, *Geotechnique* **3** (1952) 57–90.
25. C.M. Gibson-Poole, S.C. Lang, J.E. Streit, G.M. Kraishan, R.R. Hillis, Assessing a basin's potential for geological storage of carbon dioxide: an example from the Mesozoic of the Petrel Sub-basin, NW Australia. Presented at Proceedings of the Petroleum Exploration Society of Australia Symposium, Perth, Western Australia, 2002.
26. M.A. Addis, Reservoir depletion and its effect on wellbore stability evaluation, *Int. J. Rock Mech. Mining Sci.* **34** (3–4) (1997a) 423.
27. M.A. Addis, The stress-depletion response of reservoirs. SPE 38720, Proceedings of 72nd SPE Annual Technical conference and Exhibition, San Antonio, TX, 1997b.
28. L.W. Teufel, D.W. Rhett, H.E. Farrell, Effect of reservoir depletion and pore pressure drawdown on in situ stress and deformation in the Ekofisk field, North Sea, Proceedings of 32nd US Rock Mechanics Symposium, Norman, OK, 1991, pp. 63–72.
29. J.E. Streit, R.R. Hillis, Estimating fluid pressures that can induce reservoir failure during hydrocarbon depletion. Presented at SPE 78226. SPE/ISRM Rock Mechanics Conference, Irving, TX, 2002.
30. Y. Touloukian, W. Judd, R. Roy, Physical Properties of Rocks and Minerals, McGraw-Hill, New York, 1981, pp. 132–144.
31. M.A. Addis, X. Choi, J. Cumming, The influence of the reservoir stress-depletion response on the lifetime considerations of well completion design. SPE 47210, Proceedings of SPE/ISRM, Trodheim, Norway, 1998.
32. P. Peska, M.D. Zoback, Compressive and tensile failure of inclined well bores and determination of in situ stress and rock strength, *J. Geophys. Res.* **100** (1995) 12791–12811.
33. S.A.F. Murrell, The strength of coal in triaxial compression. Presented at Proceedings of Conference Mechanical Properties Non-Metallic Brittle Materials, Butterworths, London, England, 1958.
34. J.L. Jaeger, N.G.W. Cook, Fundamentals of Rock Mechanics, Chapman & Hall Ltd. and Science Paperback, 1971, p. 67.
35. L.B. Hilbert, J.T. Fredrick, M.S. Bruno, G.L. Deitrich, P.E. de Rouffignae, Two dimensional nonlinear finite element analysis of well damage due to reservoir compaction, well to well interactions and localization, *Proceedings of Second North American Rock Mechanics Symposium, Montreal, Balkema* **2** (1996) 1863–1870.
36. J. Rutqvist, C-F. Tsang, TOUGH-FLAC: A numerical simulator for analysis of coupled thermal hydrologic mechanical processes in fractured and porous geological media under multiphase flow conditions, TOUGH Symposium, Berkeley, CA, 2003.
37. J. Rutqvist, Y.-S. Wu, C-F. Tsang, G. Bodvarsson, A modeling approach for analysis of coupled multi-phase fluid flow, heat transfer, and deformation in fractured porous rock, *Int. J. Rock Mech. Mining Sci.* **39** (2002) 429–442.
38. R.W. Klusman, Evaluation of leakage potential from a carbon dioxide EOR/storage project, *Energy Conversion Manage.* **44** (2003) 1921–1940.
39. S.M. Benson, R. Hepple, J. Apps, C.-F. Tsang, Lessons Learned from Natural and Industrial Analogues for Storage of Carbon Dioxide in Deep Geological Formations, LBNL-51170, 2003.
40. Y.Y. Haimes, S. Kaplan, J.H. Lambert, Risk filtering, ranking, and management framework using hierarchical holographic modeling, *Risk Analysis* **22** (2) (2002) 000.
41. A. Brown, Evaluation of possible gas microseepage mechanisms, *AAPG Bulletin* **85** (11) (2000) 1775–1789.

42. K.T. Raterman, Assessing Reservoir Heterogeneity from a Single Well Injection Test, Internal Report, Amoco Production Company, Denver, CO, 1996.
43. C.M. Tremain, E. S, N.H. Laubach, I.I.I. Whitehead, Fracture (cleat) patterns in Upper Cretaceous Fruitland Formation coal seams, San Juan basin, in: W.B. Ayers Jr., W.R. Kaiser (Eds.), Coalbed Methane in the Upper Cretaceous Fruitland Formation, San Juan basin, New Mexico and Colorado, Bulletin, vol. 146, New Mexico Bureau of Mines and Mineral Resources, 1994, pp. 87–102.
44. J.C. Pashin, R.H. Groshong Jr., R.E. Carroll, Enhanced Coalbed Methane Recovery Through Storage of Carbon Dioxide: Potential for a Market-Based Environmental Solution in the Black Warrior Basin of Alabama, Tuscaloosa, AL, 2003.

Carbon Dioxide Capture for Storage in Deep Geologic Formations, Volume 2
D.C. Thomas and S.M. Benson (Eds.)

Chapter 33

RISK ASSESSMENT METHODOLOGY FOR CO_2 STORAGE: THE SCENARIO APPROACH

A.F.B. Wildenborg[1,*], A.L. Leijnse[1], E. Kreft[1], M.N. Nepveu[1], A.N.M. Obdam[1], B. Orlic[1], E.L. Wipfler[1], B. van der Grift[1], W. van Kesteren[2], I. Gaus[3], I. Czernichowski-Lauriol[3], P. Torfs[4] and R. Wójcik[4]

[1]Netherlands Institute of Applied Geoscience TNO-National Geologic Survey, P.O. Box 80015, 3508 TA, Utrecht, The Netherlands
[2]WL Delft Hydraulics, P.O. Box 177, 2600 MH Delft, The Netherlands
[3]Bureau de recherches géologiques et minières, BP 6009, 45060 Orléans Cedex 2, France
[4]Wageningen University and Research Centre, P.O. Box 9101, 6700 HB Wageningen, The Netherlands

ABSTRACT

The ambition of the R&D work presented here was to further develop the "scenario approach" as a methodology for the long-term safety assessment of underground CO_2 storage and to demonstrate its applicability in an example of safety assessment.

The developed methodology consists of three main parts: scenario analysis, model development and consequence analysis. The scenario analysis focuses on a comprehensive inventory of risk factors (Features, Events and Processes, FEPs) and subsequent selection of the most critical factors that will be grouped into discrete CO_2 leakage scenarios. Quantitative physico-mathematical models need to be developed to enable a quantitative safety assessment of the scenarios in the consequence analysis.

The developed method was successfully applied to two virtual settings in the southern part of the North Sea. In these examples, two leakage scenarios were considered, leakage up a fault and through a failed well. Modeling showed that CO_2 concentrations and fluxes in the biosphere were largest in the case of a leaking well, compared to the leaking fault. However, the duration of release of CO_2 to the biosphere was longer in case of the leaking fault. The assessed scenarios did not include any monitoring or mitigation measures and thus represent worst-case situations in this respect. The outcome of the assessment enables the development of a monitoring system and mitigation plan so that the safety risks can be adequately managed.

INTRODUCTION

The R&D work presented here was directed to the improvement of the HSE risk assessment methodology for storage of CO_2 in various geological media. The specific objectives of the study were:

- To develop a methodology and computational tools for HSE risk assessment of geological CO_2 storage in various geological media. The method and related tools must be applicable to site-specific

Abbreviations: BRGM, Bureau de Recherche Géologique et Minière (France); FEP, acronym for Feature, Event or Process; any factor that could potentially influence the future HSE performance of the CO_2 storage system; HSE, health, safety and environment; SA, safety assessment; SAMCARDS, acronym of the R&D-project presented here, the full title of which is Safety Assessment Methodology for Carbon Dioxide Storage; TNO, Netherlands Organization of Applied Scientific Research; TNO-NITG, Netherlands Institute of Applied Geoscience TNO.

**E-mail:* a.wildenborg@nitg.tno.nl; fax: +31 30 256 4605.

assessment of CO_2 storage in saline water bearing formations and gas fields, both in offshore and onshore settings.
- To demonstrate the method and tools by applying it two virtual storage sites in the southern North Sea region.

The research focused in particular on the potential *long-term* effects of subsurface CO_2 storage, i.e. the period after injection of CO_2. In the present work the *scenario approach* has been adopted, which was introduced earlier for and successfully applied to the long-term assessment of hazardous waste disposal [1] though CO_2 is not considered to be a hazardous waste. The full description of the methodology, the testing and its demonstration including input data can be found in the report compiled by TNO-NITG [9].

STUDY METHODOLOGY

The presented method for the assessment of long-term behavior of a CO_2 storage facility basically consists of three major phases, each of which can be divided in one or more sub-phases (see Figure 1).

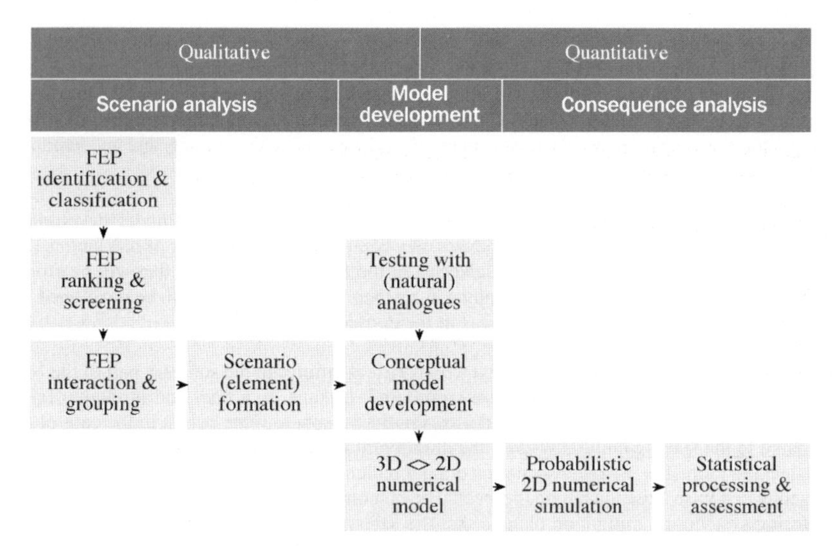

Figure 1: The scenario approach for safety assessment consists of three consecutive main phases, i.e. scenario analysis, model development and consequence analysis, each of which is divided into several sub-phases.

The core of the methodology is the systematic development of a limited number of scenarios that describe the possible future state or evolution of the storage site (scenario analysis). The basic elements for the development of the scenarios are features, events or processes (FEPs), a scenario consisting of an assemblage of interdependent FEPs. Once the scenarios have been defined, mathematical models are selected or developed that are able to quantify the consequences of these scenarios. Subsequently, the models are applied to quantify the consequences and assess the risks. A proper definition of the assessment basis is crucial for a successful execution of the safety assessment.

Assessment Basis

The constraints for the safety assessment are defined in the assessment basis (not presented in Figure 1). A well-focused definition of the assessment basis improves the quality of the work in all subsequent assessment phases, i.e. scenario formation, model development and consequence analysis. Its most crucial ingredients are:

1. *Assessment criteria*. Quantitative criteria that relate to acceptable levels of CO_2 exposure and acceptable consequences for health, safety and environment like the maximum acceptable CO_2 concentration or heavy metal concentration or the maximum individual lethality risk. These criteria can be defined in a safety or environmental regulation or in an industrial standard.
2. *Storage concept*. A clear description of the concept of underground CO_2 containment must be provided like the concept of structural trapping of CO_2, hydrodynamic trapping, dissolution trapping, mineral trapping or a combination of these. The specific requirements for the chosen storage concept must be elucidated and will vary depending on the storage concept that has been selected.
3. *Characteristics of the storage site*. A detailed description of the geological and geographical setting of the storage system including previous underground human activities in the area is very important to constrain the scope of the assessment. These concern the location, geological environment, lithology and past human underground activities. It is also important to have proper knowledge of the planned number of injection wells, the CO_2 injection rate over a certain time period and other design properties.
4. Additional items that can be included in the assessment basis are:
   - the times scale and spatial domain of the storage system;
   - type of assessment methodology to be used; and
   - any other requirement or constraint.

Scenario Analysis

A properly defined assessment basis establishes the starting-point for the scenario analysis. A scenario is a possible future state or evolution of the storage site that might lead to unintended leakage of CO_2 or to unintended (a) seismic movement of the earth's surface. Scenario analysis consists of two major phases, i.e. FEP analysis and scenario formation.

FEP database

The FEP database holds FEPs that may have a potential effect on the safety of the storage system. The current version of the database developed at TNO contains a total number of 665 FEPs that were extracted from various sources (see Table 1).

TABLE 1
NUMBERS OF FEATURES (F), EVENTS (E) AND PROCESSES (P)

F, E or P	Description	Number
F	All (static) factors and parameters describing the sequestration facility	239
E	Future occurrences, future changes to features (F) and future alteration of processes (P)	288
P	All surface and subsurface processes that describe the current and future physical, chemical and biological dynamical aspects of the sequestration facility	138
Total		665

The distinction between FEPs is made to support the scenario formation process subsequent to the FEP analysis. The status of features (F) is quite different from the status of events (E) and processes (P) in the database. Features are static input factors and/or parameters that characterize the state of the storage site. Features will be included in the reference and/or variant scenarios depending on the type of processes and events that will be incorporated. The reference scenario comprises events and processes (EPs) with a unit probability and represents the expected evolution of the storage system. Variant scenarios include—in addition to the EPs of the reference scenario—one or more EPs, the future occurrence of which is uncertain.

FEP analysis
The FEP database is used to support the FEP analysis process. It keeps track of all the steps and decisions that are made during the evaluation of individual FEPs. It is used to analyze interactions with other FEPs and supports the grouping process. The FEP grouping process is also supported by the visual analysis software "GRIN" [2]. This tool visualizes the interaction between FEPs as an influence diagram and provides options to present FEP groups. The analytical tools for the various stages of FEP analysis are provided either by the FEP database or by the visual analyzer "GRIN" (Figure 2).

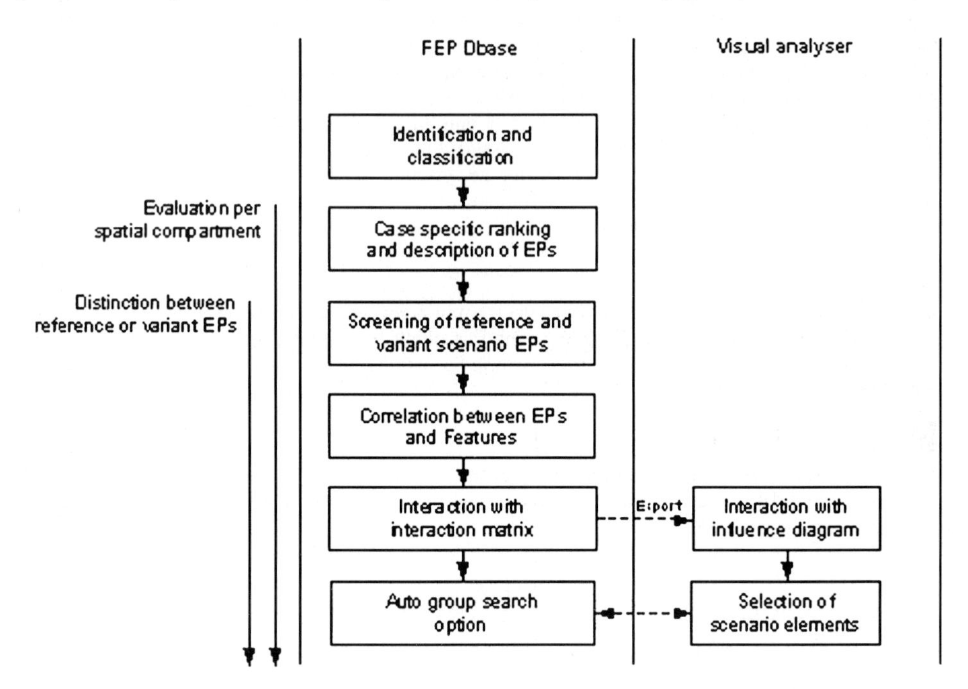

Figure 2: Scheme of the various phases in FEP analysis with special reference to the supporting software tools.

Identification and classification. All FEPs in the FEP database have a complete set of identification and classification attributes (Figure 3). The identification and classification of FEPs is performed in a qualitative generic way, independent of the storage site or assessment basis.

Ranking. The generic identification and classification attributes are used as the starting-point for assigning case-specific descriptions during the ranking phase. This and next phases of the scenario analysis are performed by experts or expert groups, for example in workshops. In the ranking phase, it is allowed to add case-specific information to the database. The expert evaluator has the option to split generic FEPs into several more detailed, case-specific FEPs, to which he can assign different semi-quantitative probability and impact levels.

During the ranking process, a distinction is made between features as static factors, on one hand, and EPs as dynamic factors on the other hand. Only the EPs will be ranked. The EPs represent potential future changes and dynamical aspects of the storage facility that may lead to unintended leakage of CO_2 or to unintended

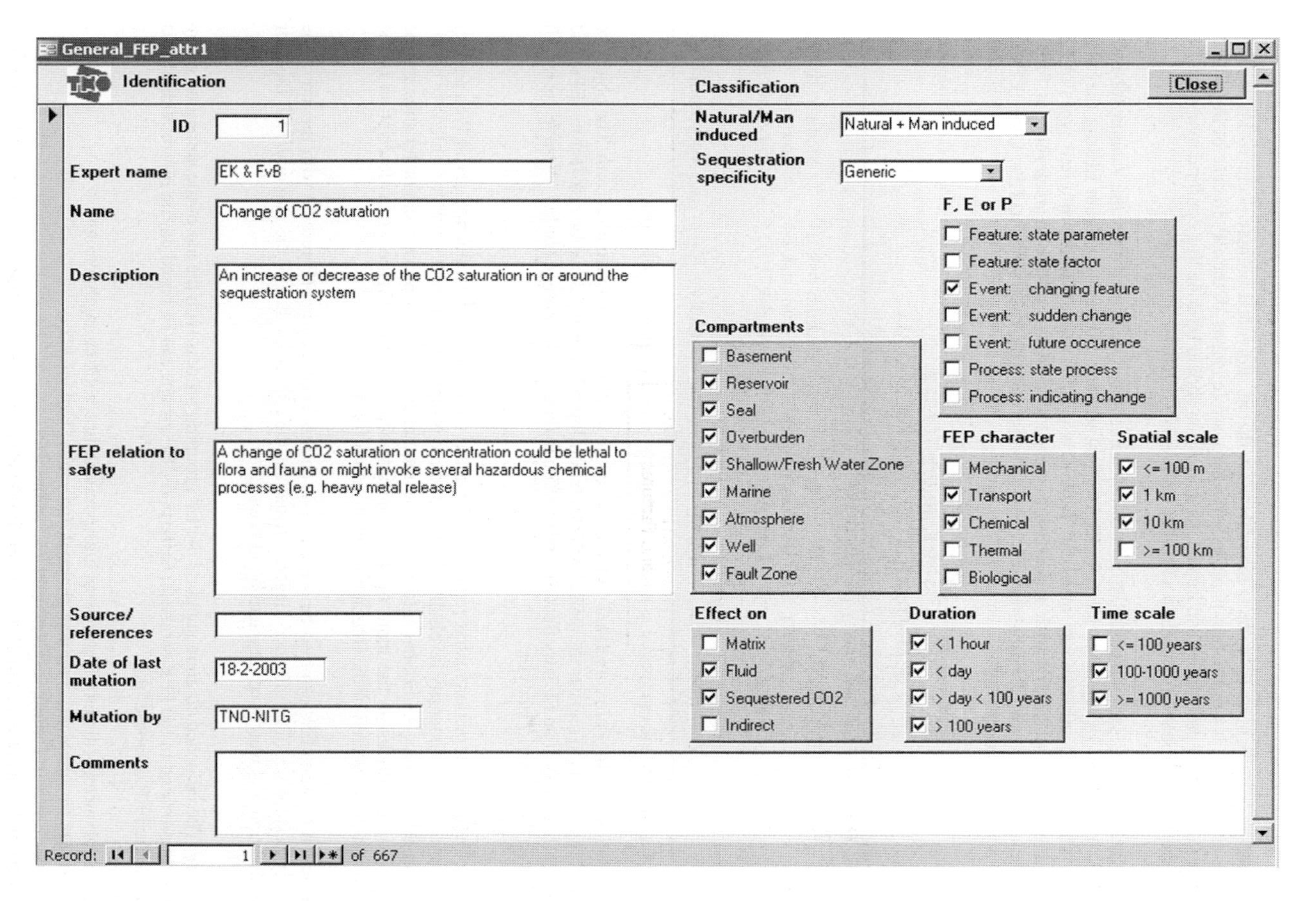

Figure 3: Example of one of the input screens in the CO_2 FEP database showing stored generic information; identification attributes are shown to the left and classification attributes to the right.

seismic movement of the earth's surface. The most important attributes that are determined in the ranking phase are (1) the semi-quantitative probability that an EP will occur, (2) potential impact if the EP occurs, and (3) the relevance for assessment. These three attributes are assessed based on expert opinion.

The estimated probability of an EP reflects the probability that an individual EP may occur within the time frame of the assessment. No distinction is made between possible causes of the EP. In case of uncertainty with respect to the actual probability and impact, its estimation should be done in a conservative way. This means that the actual probability and impact might be overestimated.

Screening. Based on the semi-quantitative probability and impact, resulting from the ranking phase, a distinction is made between reference scenario EPs, variant scenario EPs and irrelevant EPs for the safety assessment. The semi-quantitative risk matrix in Table 2 is used to categorize the different types of EPs during the screening process. EPs with a low risk or very low risk are considered irrelevant for further analysis. Remaining EPs with a probability of very likely are the reference scenario EPs. Other EPs are categorized as variant scenario EP.

TABLE 2
SEMI-QUANTITATIVE RISK MATRIX

		Potential impact		
Significant	High risk	High risk	Medium risk (l)	Medium risk (s)
Marginal	Medium risk (l)	Medium risk (l)	Medium risk (s)	Low risk
Negligible	Low risk	Low risk	Low risk	Very low risk
Likelihood				
	Very likely	Likely	Unlikely	Very unlikely

Medium risk FEPs are sub-categorized as either (l) large or (s) small medium risk categories.

F-EP correlation. Features are correlated with the EPs in the F-EP screening evaluation form. The objective of the screening form is to register the cause–effect relationship between the dynamic risk factors (EPs) and static factors (features). If an EP has effect on one or more features of the storage facility, these features will be included in the scenario analysis.

Interaction. The FEP interaction matrix represents the relative intensity of the influence of an EP on another EP (see Figure 4). Three intensity levels are identified: three is high intensity and one is low intensity of this cause–effect relationship. Additional information on mutual features and process characteristics can be retrieved by double clicks on the input fields of the interaction matrix (Figure 4). A description of the interaction can be registered in the interaction information form.

The interaction between EPs can also be presented as an influence diagram (Figure 5) with the aid of the visual analysis software. The influence diagram visualizes the risk magnitude of the individual EP together with the direction and weight of interaction between EPs.

EP grouping. The influence diagram supports the EP grouping process with the aid of the automatic group search option provided in the FEP database. Criteria for EP groups can be based on the information that is available in the FEP database like

- common parameters (distinct features such as permeability, rock strength, etc.),
- process types (mechanical, chemical, thermal, hydraulic, biological),
- effect type (on matrix, fluid, sequestered CO_2, indirect),
- time scale of EP occurrence (in 100 years, in 1000 years, or in 10,000 years),
- duration scale of EP while occurring (hours, days, centuries and longer), and
- spatial scale (1 m, 1 km, 10 km, basin scale).

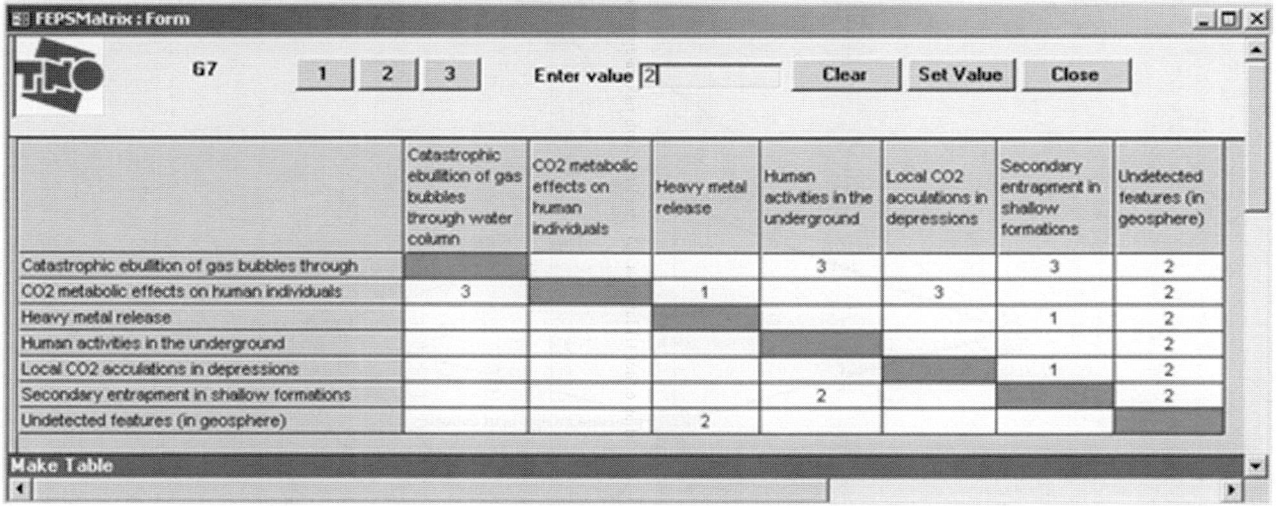

FEPSMatrix : Form

G7 | 1 | 2 | 3 | Enter value 2 | Clear | Set Value | Close

	Catastrophic ebullition of gas bubbles through water column	CO2 metabolic effects on human individuals	Heavy metal release	Human activities in the underground	Local CO2 acculations in depressions	Secondary entrapment in shallow formations	Undetected features (in geosphere)
Catastrophic ebullition of gas bubbles through				3		3	2
CO2 metabolic effects on human individuals	3		1		3		2
Heavy metal release						1	2
Human activities in the underground							2
Local CO2 acculations in depressions						1	2
Secondary entrapment in shallow formations				2			2
Undetected features (in geosphere)			2				

Make Table

Figure 4: Example of an interaction matrix in the CO_2 FEP database.

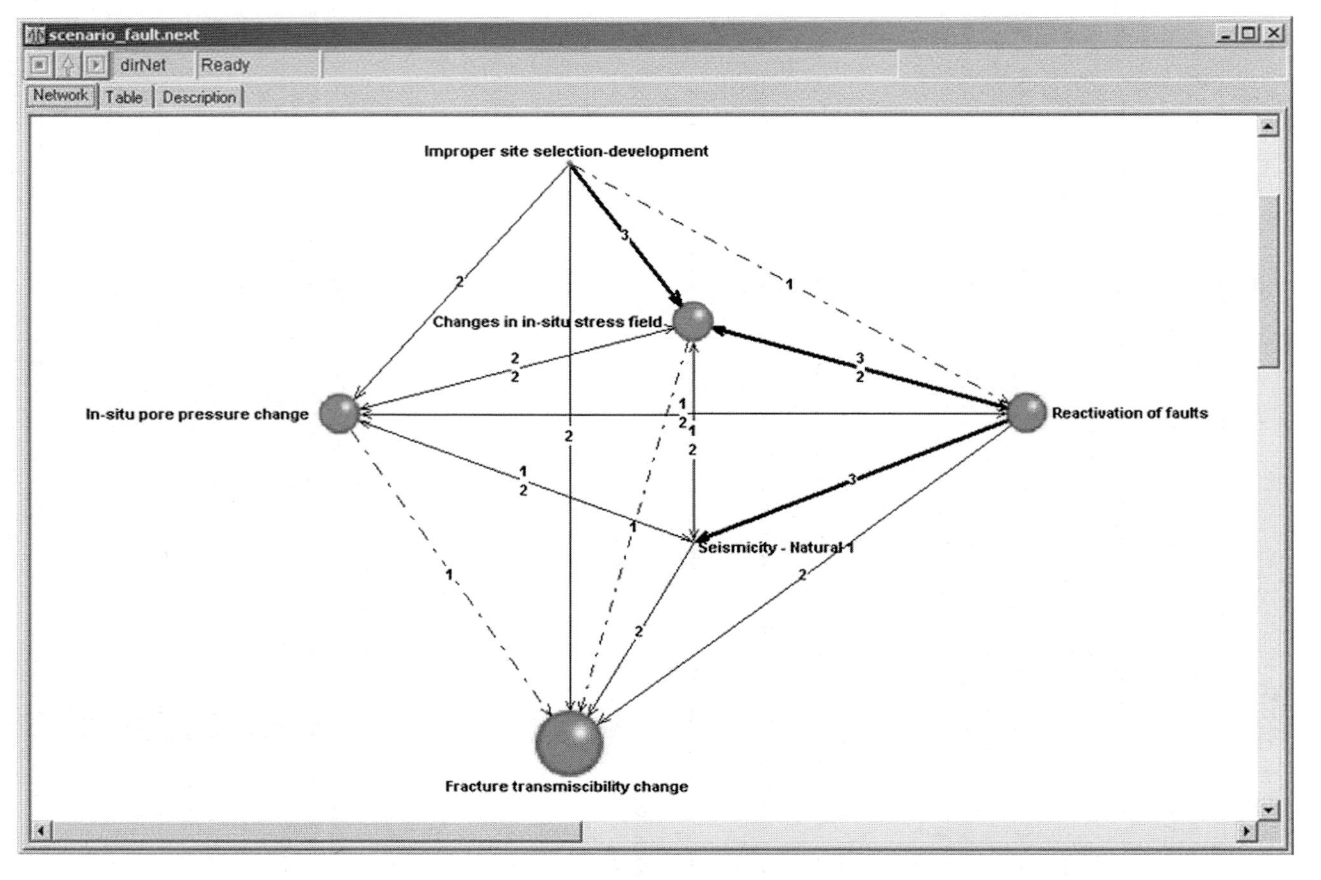

Figure 5: Example of an influence diagram.

Scenario elements. The way EP groups are assigned to specific compartments, depends on the type of compartment. An EP group is a combination of interrelated EPs that affect

1. the integrity of the containment zone consisting of the reservoir, seal, fault and well/engineering compartments,
2. the migration of CO_2 from the zone from the containment zone to the biosphere
3. the biosphere (see also Section on "Assessment basis"), which consists of the shallow/fresh water compartment, marine compartment and atmospheric compartment.

Per compartment one or more different EP groups (or scenario elements) can be defined depending on the spatial and temporal (in-)consistency of the individual EPs. Huge groups might be split in subgroups for mere practical reasons. Scenario elements are presented as tables and influence diagrams.

The core of a scenario element is formed by the EP or a group of EPs that directly affect the integrity of the containment zone, the migration of CO_2 in the overburden to the biosphere or health, safety and environment in the biosphere. Secondly, the EP or group of EPs that initiate or drive the EPs mentioned above should be identified. The grouping must be such that the resulting combination of EPs is consistent in time and space.

Scenario formation
Scenarios are formed through the logical combination of the scenario elements resulting in a complete description of a potential future state or evolution of the storage facility for every scenario. Temporal and spatial consistency of the assembled scenario elements must be checked. No specific software tools have been developed to support the scenario formation process itself. In this study, the scenarios are presented in a tree diagram.

The individual EPs in the conceptual models of individual scenarios are either represented as

- a parameter,
- a process or equation representing a physical law,
- a boundary condition, and
- other, e.g. a conservatively determined constant.

Or not represented.

The transfer of individual FEPs in a scenario element to the conceptual model representing the scenario element is discussed with the aid of tables. An example is given in Table 3.

Model Development
Scenarios are the starting-points for the development of conceptual physical/chemical models, on the basis of which mathematical models are constructed or selected from existing software libraries. A complete analysis of each of the scenarios requires simulations with the individual models for the different compartments that play a role in the transport of CO_2 from the geosphere to the biosphere.

In general, the inputs that are required for such models are inherently uncertain. Consequently, CO_2 fluxes and concentrations predicted with these models are uncertain. Quantification of this uncertainty requires Monte Carlo type simulations with the mathematical models. If these simulations are carried out with the complicated models for the different compartments, the computer resources required for such an analysis will be tremendous. Therefore, simplification of the models for the different compartments is then necessary, introducing more uncertainty in the final results. There are a number of ways these simplified models can be obtained:

- *Reducing the dimensionality of the problem.*A typical example of reduction of the dimensionality is the description of the transport of CO_2 in a radial symmetric system (2D) rather than in a fully 3D mode. This can, for example, be done for an injection well that starts leaking because of degradation of the well cementing and casing. However, one cannot give general rules when this simplification can be adopted.

TABLE 3
EXAMPLE OF A TABLE SHOWING THE INDIVIDUAL REPRESENTATION OF FEPS

A/G	Scenario element	FEPs included	Parameter change	Physical law/ equation	Boundary condition	Not represented	Other
A, G	Reservoir/ Seal-Ref-	Changed fluid chemistry	×				
		Alkalinity change	×				
		Chemical equilibrium reactions		×			
		Kinetics of chemical reactions		×			
		In situ pore pressure change	×				
		Stress change	×				
		P and T phase behavior of the CO_2-reservoir system		×			
		Water mediated transport of contaminants				×	

In this example the representation of reference scenario FEPs is listed.

In each specific case (site), transport paths for the CO_2 have to be studied with a full-scale model before adopting this approach.

- *Lumping of the effect of certain processes*. In case the dimensionality of the problem cannot be reduced, a different approach needs to be taken. That is, e.g. the case with the leaking fault scenario. The structure of the fault in relation to the injection well and the shape of the reservoir do not allow for a 2D radial symmetric description of the CO_2 transport. Since the vertical resistance to flow is one of the important properties of the system, lumping the horizontal layers in the model to a smaller number of layers, making sure the vertical resistance is the same, can be considered. Also, the lateral extend of the model might be reduced, thus constructing a simplified model that is still 3D, but requires much less grid blocks.
- *Neglecting the effect of certain processes*. Some of the physical processes taking place in the system are either highly non-linear, or pose numerical constrictions on the solution. A typical example of such processes is the dissolution of CO_2 in the water phase. One could consider neglecting this process, as was done in the reservoir/seal/overburden model.

For risk assessment, simplifications can only be accepted if they do not lead to underestimation of the CO_2 fluxes and/or concentrations. For some of the simplifications mentioned above the effect is obvious. If, for example, we neglect the dissolution of CO_2 in the reservoir and overburden, we are certain that the transport of CO_2 in this compartment will be overestimated. For other simplifications, this is not obvious, like the reduction of the number of vertical layers and the lateral extend in the case of the fault leakage scenario. In all cases, however, the simplified models should be calibrated on the basis of the results obtained with comprehensive models.

The Safety Assessment (SA) model quantifies the risks of the individual HSE scenarios. It is based on the results of the underlying simplified compartment models and Monte Carlo simulation with these models. Monte Carlo simulation is necessary to quantify the effect of the parameter uncertainty. The difficult part in this Monte Carlo simulation is to determine the probability distributions of the relevant physical parameters. In most cases these will be generated by expert judgment. Sensitivity analysis with the simplified models will be carried out to determine the number of required Monte Carlo simulations obtaining a good estimate of the probability distributions of the relevant outputs. The safety model comprises representations of all relevant components: the stored CO_2, the reservoir, the seal, the overburden, the soil and the atmosphere. It handles both the uncertainty in the input parameters and the uncertainty generated by the simplification of the detailed compartment models. Limited detailed modeling of individual processes is, however, still necessary to prove that the processes incorporated in the safety model have a sound physical basis. Basically, the safety model will generate probability distributions of CO_2 fluxes on the basis of limited input. It is based on interpolation using Parzen density functions [3].

Figure 6 gives an overview of the different model concepts that are being used in the construction of the safety assessment model and the interrelation between the concepts.

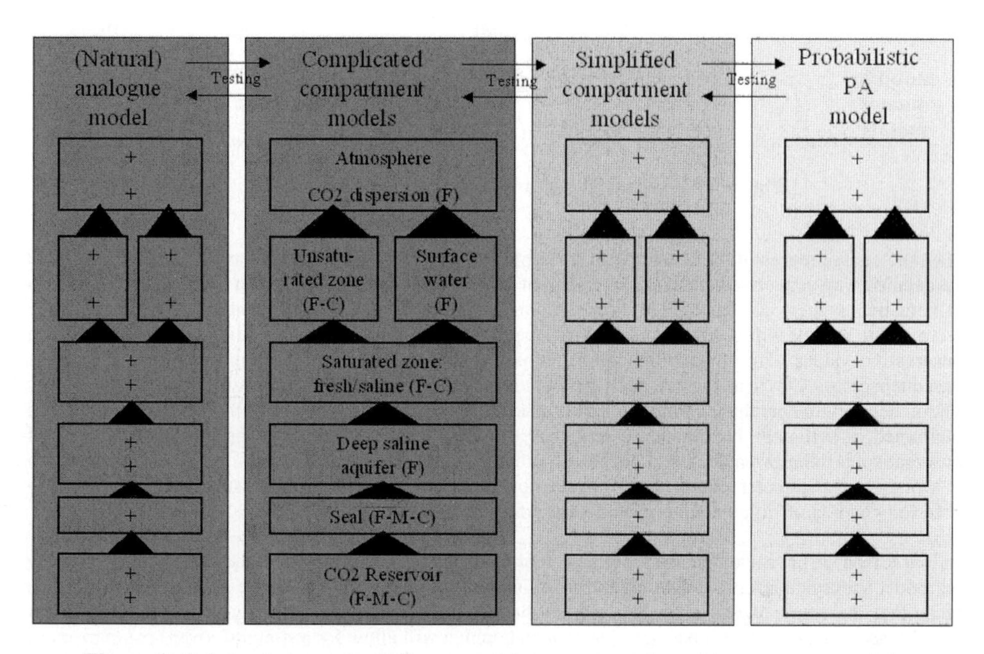

Figure 6: Relation between the different models to construct the safety assessment model.

Testing and model validation is a very important and extensive activity in the model development. Test cases should relate to a specific storage option and a specific regional geological setting. Case histories for short-term assessment (<100 years) can be found in already existing CO_2 storage projects (SACS or Weyburn). For the long-term assessment, natural analogues of CO_2 storage may be useful.

Consequence Analysis

Before the consequences of leakage can be assessed, one has to define the basis on which risks to human health and to the environment should be assessed. Figure 7 gives an overview of the assessment variables in different compartments. In almost all compartments, the concentration of CO_2 in the gas phase is an important assessment variable, because this determines among other things the risk to living creatures. Also the concentration of CO_2 in the water phase can play an important role. For example, in the shallow subsurface the CO_2 concentration in the water phase has an effect on the possible mobility of heavy metals, which might threaten the drinking water supply in populated areas.

The analysis of the consequences of the scenarios can be performed in two modes:

- Deterministic
- Probabilistic

Fresh groundwater (LBNL & TNO)	Soil (LBNL & TNO)	Basement constructions (LBNL)	Earth's surface (TNO)	Atmosphere (LBNL)	Sub-seafloor (WL)	Marine hydrosphere (WL)
[CO_{2gas}]	[CO_{2gas}]	[CO_{2gas}]	Seismicity index	[CO_{2gas}]	[CO_{2gas}]	[CO_{2gas}]
[$CO_{2dissolved}$]	[$CO_{2dissolved}$]		Subsidence		[$CO_{2dissolved}$]	[$CO_{2dissolved}$]
Ph	Ph		Uplift		Ph	Ph
Mobility heavy metals						

Figure 7: Assessment variables in the different compartments.

When the consequences of the scenarios are deterministically analyzed the relevant processes are modeled in a detailed way using fixed and time-independent parameter values. The selection of these fixed values is problematic as they can change in time or simply are unknown. Consequently, the results of the deterministic models will be highly uncertain. An approach often followed then is the selection of so-called conservative values. This means that realistic parameter values are selected in such a way that the consequences for CO_2 leakage are over-estimated. There, however, is not always a simple monotonic relation between the parameter values and the consequences. Even if conservative parameter values can be determined, it will still be difficult to compare the results for different scenarios as the amount of conservatism is unknown and most probable are different for the different scenarios. Furthermore, selection of "worst case" parameter values for all physical parameters in the system will result in a highly unlikely scenario, which is of little value to use as a basis for comparison.

For that reason we have adopted a probabilistic approach, in which the problem of the parameter uncertainty and model simplification is handled by a probabilistic interpolation technique (see previous section). This means that the result of this approach will consist of probability density functions for the assessment variables such as CO_2 fluxes, concentrations and pH, which will allow for a straightforward comparison of results for different scenarios.

RESULTS AND DISCUSSION

The example of a safety assessment is presented to illustrate the applicability of the developed methodology. The outcome of this assessment example should certainly not be considered as a formal site-specific safety assessment and thus should not be used for the actual assessment of the storage option or a site in the considered part of the North Sea region.

Assessment Basis

Assessment criteria

At the time of writing no formally accepted safety criteria for underground CO_2 storage are available. Benson et al. [4] did a literature review on safety and environmental aspects of underground CO_2 storage. Regulation in the US prescribes maximum limits of CO_2 concentration for various exposure times ranging from 5000 to 50,000 ppm. In the current assessment a concentration criterion of 10,000 ppm has been used. The criterion of maximum concentration of heavy metals in groundwater has been set according to regulations in the Netherlands.

Storage concept

In the current assessment the classical structural trap concept was adopted consisting of a CO_2 reservoir, a top seal and side seals. The majority of the stored CO_2 is assumed to be present as free gas.

Setting of the storage site
A domal trap structure typical for the UK sector in the southern North Sea [5] was considered. The reservoir consists of Bunter sandstone and the seal is predominantly rock salt. Exploration wells may be present that transect the reservoir. This geological setting was placed in two different geographical settings, a marine and a continental environment. The marine environment is typical of shallow waters in the southern North Sea. The continental setting is typical of a lowland area in the south-western Netherlands.

Timescale and spatial domain of the storage system
Potential risks within the next 10,000 years after termination of CO_2 injection will be assessed.

Scenarios and Related FEPs
A FEP analysis was performed with the aid of the FEP database and influence diagram software. FEPs were screened and assigned to either the reference scenario or to the variant scenarios. FEP groups were identified for the individual spatial compartments, resulting in one or more scenario elements per compartment. The elements have been logically combined in discrete scenarios. The objective is to identify the most critical scenarios.

Reference scenario
This scenario includes all EPs that are very likely to occur and might affect seal integrity and migration of CO_2 to the biosphere. The reference scenario EPs are included in both the reference scenario itself and all variant scenarios. The EPs assigned to the reference scenario are given in Table 4.

The following reference scenario EPs have not been evaluated in the conceptual models: soil mechanical behavior of CO_2 in the onshore shallow subsurface compartment, platform legs penetrating the overburden in the offshore shallow subsurface compartment and phase behavior of CO_2 in the atmospheric compartment. For purely practical reasons the atmospheric compartment has not been incorporated in the current safety assessment.

Variant scenarios
Next to the reference scenario, the scenario analysis has resulted in the identification of variant scenarios, the occurrence of which is uncertain. One or more scenario elements have been defined for each compartment or group of compartments and subsequently, scenarios have been constructed through assemblage of the scenario elements.

Reservoir/seal, fault and well compartments. About 40 EPs have been identified in the reservoir/seal compartment, the fault compartment and the well compartment, that could potentially affect the seal integrity. This number of EPs has been considered far too large for individual assessment and had to be split in subgroups of EPs.

The guiding principle here is to

1. identify those EPs that directly affect seal integrity and,
2. identify other EPs that initiate or force the EP that is directly affecting seal integrity.

The combination of these two types of EPs forms the central part of a variant scenario. Other interrelated EPs that are not directly affecting seal integrity or initiating the deterioration of the seal integrity will be included in the scenario element as well.

Two examples of scenario elements that represent potential reduction of seal integrity have been selected for further quantitative analysis:

- *Well scenario element.* Degradation of cement and casing might lead to unintended leakage of CO_2 to the biosphere.
- *Fault scenario element.* An undetected fault might lead to unintended leakage of CO_2 to the biosphere.

TABLE 4
REFERENCE SCENARIO EPS AND THEIR REPRESENTATION IN THE SAMCARDS CONCEPTUAL MODELS

On/ Off	Scenario element	FEPs included	Parameter change	Physical law/ equation	Boundary condition	Not represented	Other
On, Off	Reservoir/ Seal-Ref-	Flow and fate of CO_2 over multiple phases		×			
		Changed fluid chemistry	×				
		Alkalinity change	×				
		Chemical equilibrium reactions		×			
		Kinetics of chemical reactions		×			
		In situ pore pressure change	×				
		Stress change	×				
		P and T phase behavior of the CO_2-reservoir system		×			
On, Off	Fault	None					
On, Off	Well/ engineering	None					
On, Off	Overburden-Ref-	Flow and fate of CO_2 over multiple phases		×			
		Phase behavior of CO_2		×			
On	Shallow subsurface-Ref-	Flow and fate of CO_2 over multiple phases		×			
		Phase behavior of CO_2		×			
		Soil mechanical behavior				×	
		Platform legs penetrating the overburden				×	
Off	Marine compartment-Ref-	Flow and fate of CO_2 over multiple phases		×			
		Phase behavior of CO_2		×			
		Soil mechanical behavior	×				
		Wind induced transport in water column		×			
		Tidal driven transport		×			
On, Off	Atmosphere-Ref-	Phase behavior of CO_2				×	

In the first column: On, onshore case; Off, offshore case.

Overburden and biosphere compartments. In the overburden compartment about 10 EPs have been identified that could affect the migration of CO_2 to the biosphere. Except for the fault and well scenario elements that transect the overburden, no variant scenario elements additional to the reference scenario element of the overburden have been selected for further quantitative analysis. The variant scenario EPs in the overburden have incorporated in the fault leakage scenario.

Four EPs have been identified for the shallow subsurface and marine compartments, of which secondary entrapment of CO_2 is considered to be most relevant. A variant scenario element with this particular EP has been constructed for the marine compartment.

The following variant EPs have not been evaluated in the conceptual models: undetected features, future man induced EPs (e.g. drilling, interference other projects), improved cap rock integrity, meteorite impact and local CO_2 accumulations in depressions. As explained earlier, the atmospheric compartment has not been included in the present safety assessment.

The summary of the identified scenario elements is given in Table 5. A scenario tree based on various possible future states of the individual model compartments is given in Figure 8.

TABLE 5
DESCRIPTION OF SCENARIO ELEMENTS

Compartment	**Reference scenario element**	**Variant scenario element**
Atmosphere	Regular atmospheric transport of CO_2 as a gas phase. *Not represented as scenario element*	Depression element: potential of accumulation of CO_2 in depressions under stable atmospheric conditions. *Not represented as scenario element*
Shallow subsurface	Multi-phase transport of CO_2 in a layered aquifer/aquitard system, slow process	Human intrusion element: unforeseen and sudden release of CO_2 from secondary entrapped CO_2 accumulations triggered by human activities in combination with neglect. *Not represented as scenario element*
Marine	Multi-phase transport of CO_2 in a layered aquifer/aquitard system below the seabed. Depending on the CO_2 flux, the majority of CO_2 will dissolve in the water column	Local ebullition of gas bubbles from secondary entrapped CO_2 accumulations as a result of natural processes or triggered by human activities in combination with neglect
Overburden	Transport of CO_2 in a layered aquifer/aquitard system, slow process	*See also leaking fault scenario element*
Fault	*No element*	Leaking fault scenario element: transmissibility increase as a result of natural and man-induced events and processes followed by transport of CO_2 from the reservoir into the overburden along the fault plane
Well	*No element*	Leaking well scenario element: release of CO_2 from the reservoir into the overburden along the well trajectory as a result of chemical processes (e.g. metallic corrosion and cement degradation) around the well bore
Reservoir/seal	Transmissibility increase of the seal as a result of interacting chemical and mechanical processes	*See also leaking fault and leaking well scenario elements*

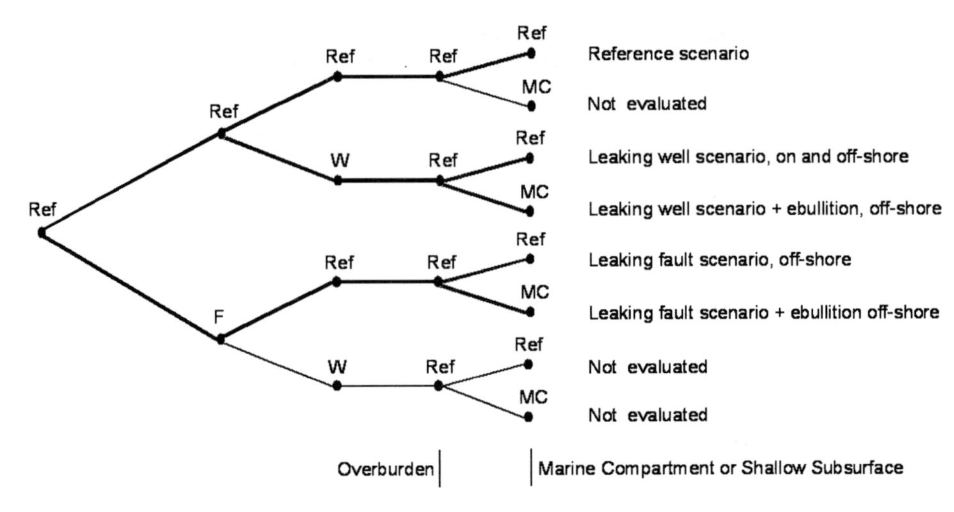

Figure 8: Scenario tree diagram resulting after combination of scenario elements.

HSE Consequences

Three different scenarios have been considered for the analysis of the consequences of CO_2 storage in the deep subsurface:

- The reference scenario, where the natural barrier is assumed to be intact.
- The leaking well scenario, where it is assumed that a conventional well completion degrades as a consequence of contact with high concentrations of CO_2. This will result in largely increased permeability around the well, thus creating a potential pathway for the CO_2 to the biosphere.
- The leaking fault scenario, where a fault in the vicinity of the injection well acts as a potential natural pathway for CO_2 to the atmosphere.

Reference scenario

For the Monte Carlo simulations in the reference scenario all relevant parameters are assumed to have uncertainty associated with them. These parameters are: the shale vertical permeability, the porosity, the reservoir sand horizontal permeability, the salinity and the seal vertical permeability. Table 6 shows the mean values and the probability distributions associated with these input parameters.

TABLE 6
PROBABILITY DISTRIBUTION OF SYSTEM PARAMETERS FOR THE REFERENCE SCENARIO

Parameter	Units	Type of distribution	Low	Mean	High
Salinity	kg/m^3	Triangular	8.5	10.5	12.5
Seal vert permeability	mD	Lognormal	−0.5 In-unit	0.0001	+0.5 In-unit
Shale vert permeability	mD	Lognormal	−0.5 In-unit	0.01	+0.5 In-unit
Sand horn permeability	mD	Lognormal	−0.5 In-unit	100.0	+0.5 In-unit
Porosity average	Proporation	Triangular	0.12	0.17	0.20

The Monte Carlo simulation of the reference case results in a total containment of CO_2 within the reservoir and seal layers for all parameter realizations. No CO_2 migration is detected directly above the seal. A typical result is depicted in Figure 9, which shows that CO_2 partly migrated and partly remains in the reservoir.

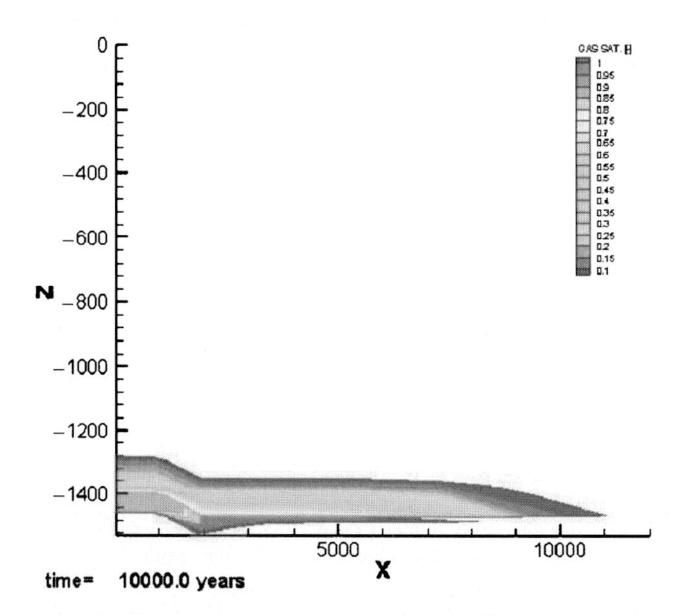

Figure 9: Areal distribution of CO_2 in reservoir and seal 10,000 years after start of injection of 1 Mt CO_2 per year for 100 years in the reference scenario.

Since no CO_2 appears above the seal in the reference scenario, no further probabilistic treatment of the biosphere compartments is necessary.

Well leakage scenario

For the Monte Carlo simulations for the leaking well scenario the relevant parameters that have uncertainty associated with them are: the shale vertical permeability, the porosity, the reservoir sand horizontal permeability, the salinity and the well zone permeability. For all parameters the same distribution as in the reference scenario were taken. The maximum well cement permeability is assumed to be log normally distributed with a mean of 10,000 mD [6] and a standard deviation of 0.5 (on a ln scale). Figure 10 shows a typical CO_2 distribution at 10,000 years after injection in the well leakage scenario.

The fluxes at 300 m below the surface/seafloor, as calculated by the reservoir/seal/overburden model (SIMED; [7]) are characterized by a limited number of characteristic values. For the 1000 Monte Carlo simulations carried out, the statistics of these parameters can be determined. Each of the results of the simulations with the reservoir/seal/overburden model has been used as input for both the marine compartment model (DELFT3D; [8]) and the continental shallow subsurface model (performed by LBNL). Stochastic analyses of the results of these models have been carried out for both environments individually.

Marine environment. Making a probability function from the 4D data allows us to compute marginal distributions and a distribution for the sum of the “build-up” time and the “decay” time of CO_2 in seawater. In the well leakage scenario the added CO_2 concentration in the water is of order of a few times 10^{-5} the normal concentration of HCO_3^- in water, the value of which is considered to be negligible (Figure 11, left).

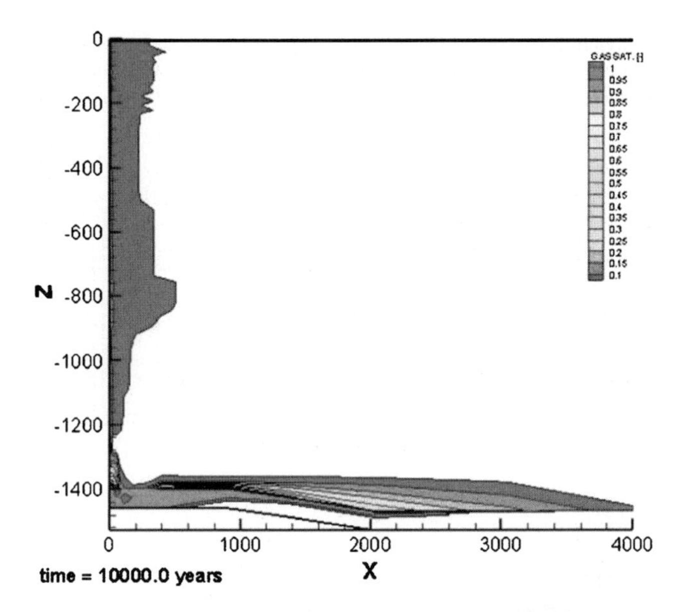

Figure 10: The CO_2 distribution pattern 10,000 years after injection for the well leakage scenario.

Transport process in seawater effectively dilutes CO_2 passing through the sea bottom. The surface area influenced by the surplus CO_2 is substantial at a lower limit of 1 g/m^3 extra CO_2.

An interesting special case of the well leakage scenario occurs when upward migrating CO_2 is trapped secondarily below shallow clay layers, below which CO_2 is able to accumulate. The CO_2 release happens only after pressure build-up. Cracks or channels do form when the clay is made to yield to the pressure, and substantial amounts of CO_2 are released in about a week's time. Part of this bulk release now gets into the atmosphere as well; it is more than the seawater can absorb.

The 2D probability density distribution (CO_2 concentration in water, kg of CO_2 released to the atmosphere) has been constructed in order to generate the marginal distribution of the CO_2 release into the atmosphere (Figure 11, right). An order of magnitude of typical CO_2 releases of 10^8 kg must be expected if this scenario occurred.

Continental environment. In the leaking well scenario has been investigated what would happen in the unsaturated zone, 1 m below ground level. As is seen from Figure 12 (left), the molar fraction of CO_2 exhibits a bi-modular distribution. So far, we have not been able to come up with an explanation for this bimodality. Figure 12 (right) pertains to the mass fraction of CO_2 dissolved in ground water at 40 m below ground level.

At both levels, the CO_2 content just below surface and CO_2 concentrations in water at 40 m, lateral spreading is quite small. The time scales at which the enhanced CO_2 concentrations are noticeable are up to a few thousand years.

Fault leakage scenario

For the Monte Carlo simulations for the leaking fault scenario the relevant parameters that have uncertainty associated with them are: the shale vertical permeability, the porosity, the reservoir sand horizontal permeability, the salinity, the distance from the well to the fault and the fault vertical permeability. For all parameters the same distribution as in the reference scenario were taken. The distance from the well to the fault is assumed to be normally distributed with a mean of 2000 m and a standard deviation of 50 m. The fault vertical permeability is assumed to be log normally distributed with

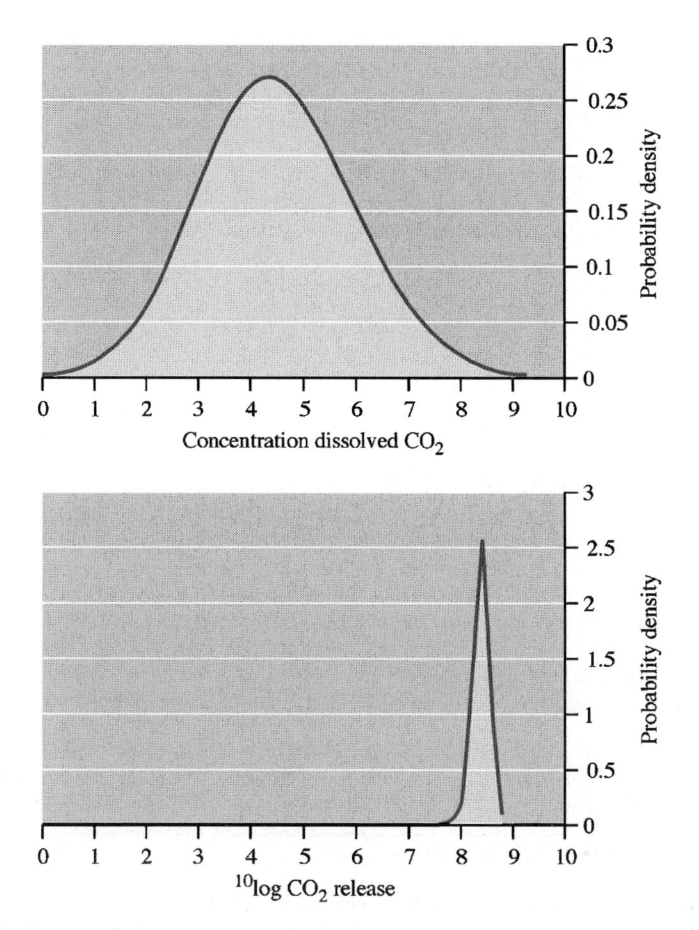

Figure 11: Marginal distributions for the well leakage scenario in a marine setting: CO_2 concentration in water in units of standard CO_2 concentration in seawater $\times 10^{-5}$ due to gradual CO_2 release from the seafloor (top) and $^{10}\log CO_2$ release (kg) to the atmosphere due to episodic release of CO_2 from the sea bottom in a week's time (bottom).

a mean, which is dependent on the surrounding lithology (1 mD in case of rock salt, 10 mD in case of shale/claystone and 25 mD in case of chalk), and a standard deviation of 2.3 on a ln scale. This standard deviation corresponds with a factor 10 variation in permeability. Figure 13 shows a typical spatial distribution of CO_2 after 10,000 years for one of the simulations carried out with the reservoir/seal/overburden model.

Each of the results of the simulations with the reservoir/seal/overburden model has been used as input for the marine compartment model. A stochastic analysis of the results of this model has been carried out. The CO_2 concentration increase in seawater due to the fault leakage scenario is typically an order of magnitude less than for the well leakage scenario, and is considered to be negligible.

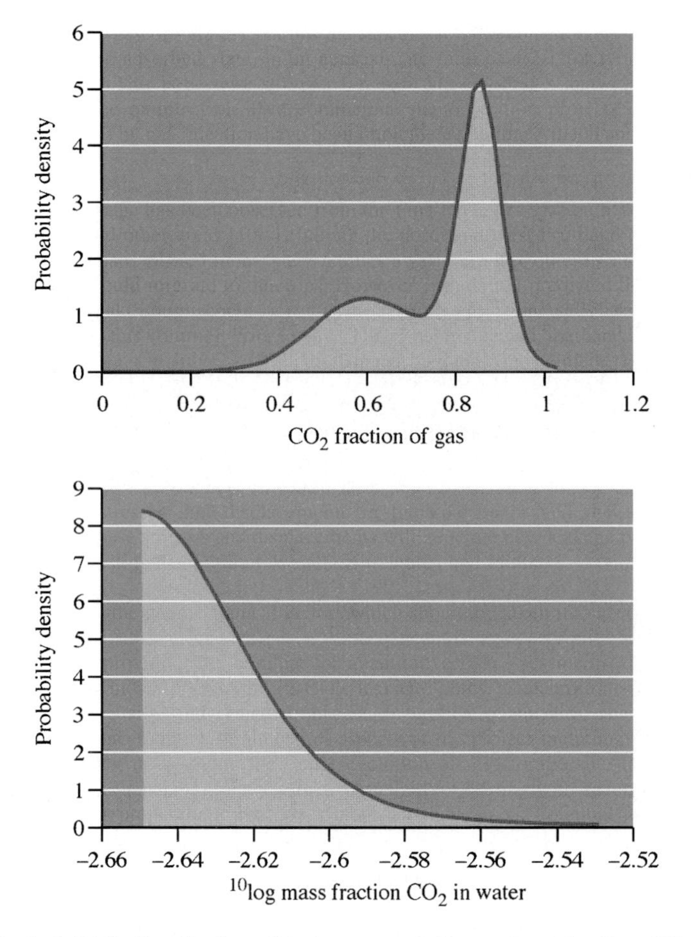

Figure 12: Marginal distributions for the well leakage scenario in a continental setting: CO_2 molar fraction in gas phase in unsaturated zone at 1 m below surface (top) and ^{10}log mass fraction CO_2 dissolved in water at 40 m below surface (bottom).

Secondary entrapment of CO_2 below shallow clay layers may also happen in the fault leakage scenario. Again here, the CO_2 releases to the atmosphere are somewhat less than in the well leakage scenario. Note that not the flux but the resulting CO_2 concentration is determining the impact in the atmosphere.

Discussion

Geologic storage of CO_2 must be secure for hundreds to several thousands of years. Little data is available on this timescale to support safety assessment. Furthermore, monitoring of the storage site will probably not take place indefinitely and the storage site will not only be subject to internal engineering factors of the storage system but also to external natural and human-induced factors.

These factors create special requirements for the methodology and tools for the long-term safety assessment of underground CO_2 storage. To start with, the method should be comprehensive and include all factors that could potentially affect the long-term safety of the site. These factors are both of storage engineering and of external human and natural origin. The method should account for the many

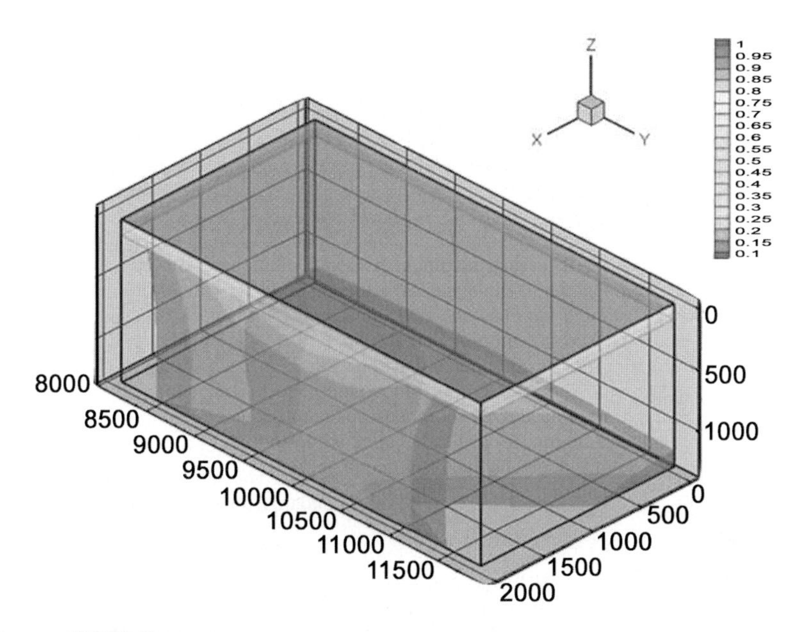

Figure 13: The CO_2 distribution pattern 10,000 years after injection for the fault leakage scenario.

uncertainties that are inherent to long-term assessment. The method should be based on sound physico-mathematical insight and should preferably not use "black-box" models. As case studies of long-term underground CO_2 storage are not yet available, natural or industrial analogues should be found that enable testing of the methods.

The objective of the prsesent research was to develop a methodology and related tools that can be used for the long-term safety assessment of underground CO_2 storage and to demonstrate its technical applicability by applying it in a safety assessment of a virtual storage site. As discussed, the overall methodology for long-term safety assessment of underground CO_2 storage is available and can be readily applied. The three individual basic components of the method, i.e. scenario analysis, model development and consequence analysis, were developed and applied to a realistic example.

Definition of the assessment basis forms the very crucial initial step in the safety assessment. It is extremely important to put substantial effort in this first step, because it contributes significantly to the success of the assessment. The assessment basis relies heavily on the results of the site characterization, information on the design of the storage facility, a clear understanding of the storage concept and knowledge of the HSE criteria that will be applied. A good assessment basis enables the definition of the containment zone and the biosphere in the domain of the storage facility and the assessment criteria that should be applied. Furthermore, it provides decision rules for the screening of the safety factors or FEPs.

Scenario analysis leads to the definition of possible future states or evolution of the storage facility that could lead to unintended leakage of CO_2 (or to unintended (a) seismic movement of the earth's surface). It consists of two steps: FEP analysis and scenario formation. A FEP database and several supporting tools for the FEP analysis were developed and tested during two workshops. The workshops did not allow testing the full cycle of scenario analysis, though the outcome was promising. It was noted that the FEP database contained ambiguous descriptions of FEPs that caused problems in ranking the FEPs. Also the decision rules for screening of the FEPs should be more clearly stated and more effort should be put in the scientific rationale for assessing individual FEPs.

Detailed process models have been constructed for the different spatial compartments. Simplifications of these models have been constructed on the basis of results obtained with the full-scale detailed models. These simplified models gave results in terms of CO_2 fluxes comparable to the fluxes obtained with the detailed models. Carrying out Monte Carlo simulations with the simplified compartment models did not pose major problems. Statistical analysis of the results of the Monte Carlo simulation has been carried out with the stochastic safety model. This turned out to be a fast and easy tool to determine probabilities of occurrence of CO_2 concentrations exceeding standard values.

The application of the methodology in the consequence analysis is promising. In addition to the reference scenario, two leakage scenarios, i.e. the well leakage and the fault leakage scenario, were defined and quantitatively assessed. The results for the reference scenario showed that seepage of CO_2 to the biosphere would not occur within the period simulated (10,000 years). This was true for all (1000) parameter realizations considered. Statistical analysis was therefore not necessary. For both the well leakage and the fault leakage scenarios, CO_2 concentrations and fluxes showed a large variation as a result of parameter uncertainty in the compartment models. The safety model could easily evaluate probabilities of CO_2 concentrations exceeding certain standards, or CO_2 fluxes exceeding prescribed values.

The gradual release of CO_2 in the well leakage scenario has negligible effects on the marine environment. Sudden release of CO_2 from shallow secondary accumulations just below the sea-bottom results in migration of CO_2 to the atmosphere via the seawater. The impact of atmospheric CO_2 release on safety was not analyzed. Gradual release from a well leads to significant increase of the CO_2 concentration in the unsaturated zone within a limited area. The leaking fault scenario is less hazardous than the leaking well scenario.

Risk management of well leakage can be improved through proper design and implementation of new wells and plugs, and through proper assessment and remediation of old wells combined with a dedicated monitoring system and remediation plan. Risk management of fault leakage should primarily focus on proper site selection, site characterization and testing that is combined with the development of a dedicated monitoring system and a remediation action plan. The effects of monitoring and remediation on lowering the safety risks were, however, not part of the current study.

CONCLUSION

A workable method and supporting tools for the long-term safety assessment of underground CO_2 storage has become available that has been applied successfully to both a virtual onshore site and a virtual offshore site in the southern North Sea region. The method is the amalgamation of qualitative (scenario analysis) and quantitative risk assessment (model development and consequence analysis), which can be applied in all phases of the life cycle of a CO_2 storage facility.

RECOMMENDATION

Safety assessment is a crucial part of the risk management of future underground storage operations (Figure 14), the further development of which should be the prime focus in future research on CO_2 storage. The outcome of the assessment defines the scope for site selection and characterization, design of the facility, testing of the facility, injection operations, abandonment of the site and the period after abandonment. The development of monitoring and remediation plans is directed by the results of the safety assessment.

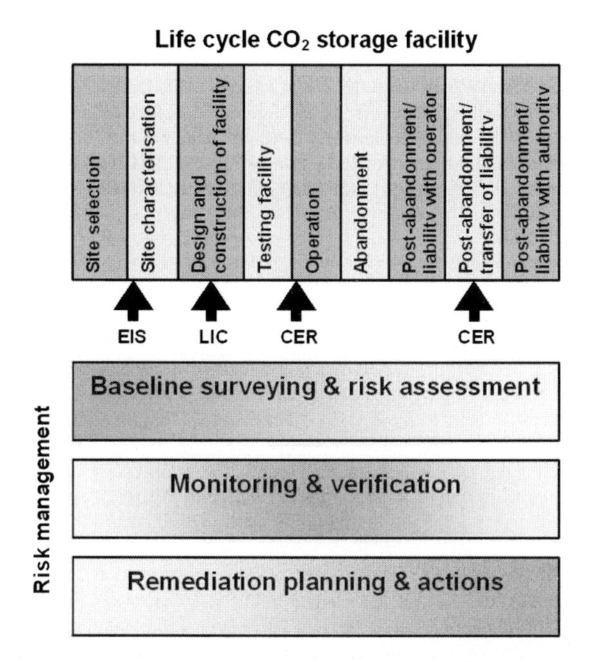

Figure 14: Risk management is the integration of risk assessment, monitoring and remediation through the whole life cycle of a storage facility. Important regulatory milestones for risk management are the Environmental Impact Statement (EIS), Licensing, Certification for (tradable) Carbon Credits and Certification for transfer of liability from operator to competent legal authority.

ACKNOWLEDGEMENTS

We greatly acknowledge the financial support of the CCP consortium and TNO and their support to publish results of the SAMCARDS project in this volume. The constructive remarks of the reviewers are greatly appreciated and have lead to significant improvement of this chapter.

REFERENCES

1. D.P. Hodgkinson, T.J. Sumerling, A review of approaches to scenario analysis for repository safety assessment, OECD Nuclear Energy Agency, Proceedings of the Paris Symposium on Safety Assessment of Radioactive Waste Repositories, 9–13 October 1989, 1990, pp. 333–350.
2. V. Pechenkin, GRIN: GRaphical INterface, Saratov State Technical University, Russia, 1998, http://www.geocities.com/pechv_ru/.
3. R. Wójcik, P. Torfs, Manual for the Use of the PARDENS Tool, Wageningen University and Research Centre, Wageningen, 2003.
4. S.M. Benson, R. Hepple, J. Apps, C.F. Tsang, M. Marcelo Lippmann, Lessons learned from natural and industrial analogues for storage of carbon dioxide in deep geological formations, Report prepared by LBNL, Earth Sciences Division for CCP-project, No. LBNL-51170, 2002, 227pp.
5. M. Brook, K. Shaw, C. Vincent, S. Holloway, 2003. The potential for storing carbon dioxide in the rocks beneath the UK southern North Sea, in: Gale, J. and Kaya, Y (eds.), GHGT-6, Kyoto, 1–4 October 2002, pp. 333–338.
6. E. Lindeberg, 2003, SINTEF, personal communication.

7. M. Stevenson, V. Pinczewski, SIMED—multicomponent coalbed gas simulator, User's Manual Version 1.21. APCRC Restricted Report No. 022, May 1995, University of New South Whales, Australia, 1995.
8. Vatvani, 1999, Delft3D-FLOW version 3.05, Users Manual, WL|Delft Hydraulics.
9. A.F.B. Wildenborg, A.L. Leijnse, E. Kreft, M.N. Nepveu, A.N.M. Obdam, E.L. Wipfler, B. Van der Grift, C. Hofstee, W. Van Kesteren, I. Gaus, I. Czernichowski-Lauriol, P. Torfs, R. Wójcik, B. Orlic, 2003, Safety assessment methodology for CO_2 storage (SAMCARDS). CCP Report Reference 2.1.1, DOE Award Number: DE-FC26-01NT41145, 149 p. and 18 enclosures.

Carbon Dioxide Capture for Storage in Deep Geologic Formations, Volume 2
D.C. Thomas and S.M. Benson (Eds.)

Chapter 34

KEY FINDINGS, TECHNOLOGY GAPS AND THE PATH FORWARD

Scott Imbus[1] and Charles Christopher[2]

[1]Chevron Texaco Energy Technology Company, Bellaire, TX, USA
[2]CO_2 Management, Innovation, Improved Recovery, BP Americas Inc., Houston TX, USA

Options for large-scale geological storage of CO_2 emissions have proceeded from concept development and capacity inventories in the 1990s to systematic site characterization and strategies for injection, long-term monitoring and risk assessment in recent years. To date, the only purpose-built CO_2 storage facility is the 1 million tonne/year Sleipner–Utsira project in the Norwegian North Sea. Although the project is deemed successful, it is doubtful that numerous projects of the scale or considerably larger such projects will be permitted without extensive technical due diligence.

In the constellation of industry, academic and government programs addressing geological CO_2 storage, the role assumed by the CCP Storage Monitoring and Verification (SMV) program over 2000–2004 is unique. The risk-based approach adopted entailed identifying technical gaps and addressing them by leveraging the existing natural and industrial analog knowledge base and developing new R&D avenues. Whereas some projects were based on a specific asset or storage venue type, the applications developed are universally applicable. The present chapter outlines the key findings of the SMV program and identifies needs for further R&D needed to support pilots, demonstration and commercial projects.

The SMV program was comprised of some 30 projects organized along four technical areas.

1. "Integrity"—assessing the competence of natural and engineered systems to retain CO_2 over extended periods
2. "Optimization"—strategies for improving the efficiency and economics of CO_2 transportation and storage
3. "Monitoring"—identification of techniques suitable for tracking CO_2 movement within (performance) and outside (leakage or seepage) the injection target
4. "Risk Assessment"—development of concepts, protocols and methodologies to quantify probability and impact of CO_2 leakage from storage sites.

INTEGRITY

The SMV integrity studies included characterization of naturally charged CO_2 systems, a survey of the natural gas storage industry, evaluations of reservoir and cap rock property responses to CO_2 injection and the stability of well materials in the presence of carbonated water. Key findings are given below.

- The suitability of natural systems to retain CO_2 over extended periods of time is predictable using 3D structural and stratigraphic models combined with fluid migration history analysis.
- The basin to reservoir scale geohydrologic model and simulation of CO_2 storage in the Forties Field serves as a prototype for systematic assessment of prospective geological storage sites.
- A survey of the natural gas storage industry, comprising >600 facilities operated over 90 years in North America and Europe, documents few gas migration incidents and an excellent HSE record. Site selection, operation and leak detection, intervention and remediation techniques used by the gas storage are applicable to CO_2 storage.
- Assessments of rock response to CO_2 injection through core flood experiments, geomechanical models and geochemical/geomechanical simulations identify and begin to quantify risks for failure through fracturing and fault reactivation.

- Experiments demonstrate that well materials currently in use are subject to rapid degradation through carbonic acid attack, particularly in the case of flowing water.

The importance of integrated geological characterization of prospective CO_2 storage sites from the systems scale to the reservoir scale is highlighted by the SMV integrity studies. There is a particular need to obtain reservoir and cap rock samples for geomechanical and geochemical testing under CO_2-flooded reservoir conditions and matching of observed behavior using simulations. Further work with natural gas storage and EOR analogs will likely reveal additional details of geologic features and operational parameters necessary for appropriate selection and safe operation of CO_2 storage facilities. Well integrity issues are clearly becoming more of a concern than geological integrity issues. Development of new, resistant materials and sealants and modification of existing construction and completion protocols are essential. Novel technologies for rehabilitation of old wells and leakage detection and intervention are essential needs for CO_2 storage facility development in depleted oil and gas fields.

OPTIMIZATION

The SMV optimization studies sought to leverage industry experience of gas injection, identify operational parameters that ensure rapid and secure CO_2 immobilization and realize cost reduction opportunities in CO_2 capture, transportation and injection. Key findings include the following.

- A survey of the CO_2 EOR experience, centered on the Permian Basin for ~3 decades, shows that performance issues are mostly attributable to inadequate reservoir characterization. Leakage and other untoward incidents have not been reported (although monitoring for CO_2 leakage is not in widespread use). The development of acid gas (CO_2 and H_2S form gas processing) injection programs in North America and elsewhere demonstrates that more hazardous gases can be safely injected and stored given appropriate pre-injection characterization, well construction design and testing, controlled injection testing and long-term monitoring.
- Injection of CO_2 into depleted gas fields is promising as infrastructure is in place and gas containment is proven. Experiments and models demonstrate that CO_2 compatibility with remnant hydrocarbon gases is predictable and that, given the high compressibility of CO_2, storage capacity may approach five times that of the original hydrocarbon charge.
- Injection of CO_2 into saline formations comprises the largest volume opportunity for CO_2 storage although compared to oil and gas reservoirs, reservoir data and infrastructure are often lacking and economic offsets are unavailable. Nevertheless, well-planned saline formation CO_2 injection projects could minimize costs and maximize storage through efficient well placement and operating parameters. Two independent reservoir simulations that variously incorporated multiple water–CO_2 interaction phenomena (e.g. buoyant flow, solubility trapping, pore space capillary trapping and mineralization) show that injection at the base of the aquifer slows the progress of CO_2 migration to the top of the reservoir and contact with abandoned wells. A considerable proportion of the CO_2 is immobilized in the decadal timeframe and the vast majority in the millennial timeframe. Immobilization of CO_2 via mineralization is probably minor and effective over the 10,000 + year timeframe.
- The success of CO_2 injection into coal beds for the purpose of enhanced coal bed methane recovery (ECBM) and permanent CO_2 storage relies on appropriate coal characterization, production history (primary production and N_2 injection) and facility installation and operation.
- Opportunities to reduce CO_2 capture cost by injecting less than pure CO_2 streams (~5% SO_x, NO_x) into aquifers or CO_2 EOR fields are unlikely to damage clastic reservoir or substantially affect oil production. The effects of such contaminants, particularly in the presence of water, on surface equipment (pipelines, compressors, etc.), however, is a concern.
- Costs associated with long distance pipeline transportation of CO_2 from the capture point to storage reservoir may determine the economic and technical feasibility of a CO_2 storage facility. New experimental and theoretical data on water solubility in CO_2 and predicted corrosion/erosion rates demonstrate that existing specifications for expensive alloy steels currently in place may be relaxed given some circumstances, particularly in offshore, northern water environments.

Early opportunities for geological CO_2 storage, particularly in regimes without carbon taxes or restrictions, will focus on settings with enhanced resource recovery potential. Existing enhanced recovery projects, particularly EOR are promising, but need more study related to storage security in more diverse reservoir/cap rock types (e.g. clastic as opposed to carbonates prevalent in the Permian Basin). Separation of CO_2 and other impurities during gas processing to meet pipeline specifications and its subsequent injection into saline aquifers would add a relatively small incremental cost. Credits might be obtained for associated SO_x and NO_x disposal. Accurate reservoir characterization and predictions of CO_2 behavior in the subsurface will, along with establishing best practices for facility operations and abandonment, ease the regulatory approval of CO_2 storage projects and ensure good performance and long-term safety. The poor geographic match between industrial CO_2 sources and suitable geologic sinks in many areas of the World will require new thinking on transportation systems. Adapting existing operation practices to extend the use of conventional materials such as carbon steel in pipelines and identifying alternative transportation schemes (e.g. shipping) will determine the technical and economic viability of many CO_2 capture and storage schemes.

MONITORING

The SMV monitoring program evaluated a broad range of existing and novel technologies that might be used to improve the cost effectiveness and safety of geological CO_2 storage. These technologies ranged from remote detection of injected CO_2 effects on the surface to direct detection near the surface to alternatives for subsurface imaging of CO_2 movement. Key findings are given below.

- Existing monitoring techniques vary widely in resolution and cost. Successful application depends on site-specific geologic and geographic features and required resolution level over time.
- The satellite-based InSAR technique may have the resolution necessary to detect ground movement from CO_2 injection if topographic effects are minimized. Remote geobotanical acquisition produces detailed surface images but relies on indirect effects of CO_2 on plant life or soils that, unless extreme, must be surveyed over an extended period of time.
- Near-ground direct CO_2 laser spectroscopy detection techniques are already in commercial use. Their ability to detect CO_2 depends on the rate, magnitude and type (diffuse or point) of seepage and local topography and atmospheric conditions.
- Conventional time lapse (4D) seismic techniques have a proven ability to image CO_2 movement in the subsurface but are expensive, logistically difficult over the long term and in some areas restricted due to environmental impacts. Non-seismic geophysical methods may have the resolution to detect subsurface CO_2 movement inexpensively over long periods without little impact on the surface.
- Addition of natural and synthetic tracers to injected CO_2 could be used to monitor the movement of injected CO_2 within target reservoirs. This would allow for detection of leaks from well bores and faults and in predicting gas break through in time to adjust operating parameters. The Mabee Field case study identified an isotope of Xe as the ideal noble gas tracer based on distinctiveness from natural reservoir and atmospheric noble gases and cost/availability.

An ideal monitoring system for a given CO_2 storage project would include the necessary resolution based on local subsurface and surface features, cost effectiveness and robust and stable operation with minimal environmental impact. Meeting these criteria would probably require some redundancy (subsurface imaging, tracers and surface collection and detection) with reliance of different techniques over short and long terms. Updating and calibration of reservoir simulations to match monitoring results will be necessary to verify CO_2 storage for carbon credits and ultimately facility abandonment. Development of inexpensive, instrumented monitoring wells and dual use wells (injection and post-injection monitoring) may be a cost effective, long-term solution to reservoir surveillance.

RISK ASSESSMENT

Risk assessment methods have long been applied to familiar hazards. The SMV risk assessment program includes a HSE perspective on handling and storage of CO_2 and other industrial materials, simulations showing the behavior of CO_2 in the vadose zone and atmosphere, strategies for early detection, intervention and remediation of CO_2 leakage and the development of two comprehensive methodologies tailored to geologic CO_2 storage. Key findings include the following.

- An initial survey of natural and industrial analogs to CO_2 handling, storage and HSE/regulatory implications has become a much-cited contribution to geologic CO_2 storage and provided guidance to the selection and execution of several of the subsequent SMV projects.
- The comprehensive risk assessment methodology developed by TNO included features, events and processes (FEP) development and application over a multi-compartment model. Testing of the model predicted no leakage over the 10,000 year timeframe (the consequence analysis was therefore not performed). Further testing is recommended.
- The INEL probabilistic methodology, in addition to its capability of predicting the likelihood and consequences of CO_2 leakage over multiple compartments, allows testing of well placement options and operation parameters for safe and effective CO_2 storage in coal beds.
- The concentration of CO_2 within the vadose zone and topographic lows with eventual atmospheric dispersion was simulated for specific sites. This simulation approach, in addition to its capability to identify site-specific risks of CO_2 concentration near the surface, provides an instructive visualization tool for regulators and the public.
- The impact of CO_2 injection on subsurface ecology showed that, depending on lithology and water chemistry, some classes of organisms will be favored at the expense of others. Whereas local extinction of useful organisms may not be an issue, possible operational parameters may be affected via microbial gas generation and porosity and permeability changes.
- Pre-injection assessment of potential leakage pathways and their impact on economic and HSE interests comprise the basis for early leakage detection, intervention and remediation planning.

The credibility of the "holistic", risk-based approach to CO_2 storage encompassing the SMV integrity, optimization, monitoring and risk assessment studies is a key contribution to the science and technology of geological CO_2 storage. Logical steps in progressing risk assessment for CO_2 storage include standardization of FEPs, benchmarking of independently developed methodologies and quantifying and bracketing risks relative to familiar hazards. The development of technologies that prevent or allow response to leakage will facilitate project approval, safeguard economic and HSE interests and ensure verification and favorable liability release terms.

THE PATH FORWARD

Progress in advancing the technology and acceptance of geological CO_2 storage has accelerated in recent years to the point that several pilot/demonstration and a few commercial projects are underway or planned for the near future. By 2010, geologic storage is expected on the 10 million tonne/year scale. To reach the 1 billion tonne/year scale required to achieve mid-century stabilization targets, key technical issues related to storage capacity and security need to be resolved, and integrated evaluation protocols developed. Initiation of large-scale storage will be facilitated by the example of successful projects and creative approaches to source–sink matching and infrastructure development.

Key technical R&D needs include:

- integrated site evaluation protocols including accurate 3D structural/stratigraphic models and fluid flow paths/history that can be used for multi-compartment risk assessment;
- integration of experimental data and simulations to predict physicochemical rock response to perturbations from CO_2 injection and document the types and rates of CO_2 immobilization mechanisms;
- development of well technologies including resistant materials and construction/completion procedures, leakage intervention strategies and old well remediation;
- detailed leveraging of EOR and natural gas storage site characterization, operation and intervention/remediation protocols, optimization of oil production versus storage maximization;
- systematization of near and long-term monitoring and verification technology resolution with guidelines for site-specific suitability based on FEPs; validation of long-term CO_2 immobilization models and development of criteria for safe facility abandonment and liability release;
- benchmarking of CO_2 storage methodologies and quantification of storage risk relative to familiar hazards and those associated with climate change;
- economic tradeoffs, process integration and infrastructure considerations for CO_2 capture, transportation and storage.

There is good reason to be optimistic that geological CO_2 storage can substantially reduce atmospheric emissions in the next 10–50 years. Compared to geological storage, ocean storage presents serious environmental risks, mineral storage is slow and terrestrial storage is inefficient and probably temporary. Given the present and anticipated scale of anthropogenic CO_2 emissions, however, a portfolio approach to carbon mitigation that also includes conservation, renewables and nuclear energy will be required. The evolution of a hydrogen economy, the ultimate approach to carbon mitigation, will nevertheless require fossil fuel use and subsequent CO_2 storage. To make large-scale geologic CO_2 storage a reality, technical developments such as those outlined above need to be applied to moderate regulations and ensure public acceptance. Collaboration of industry, governments, academic institutions and environmental NGOs has begun in earnest and should continue to expand.

AUTHOR INDEX

VOLUME 1 and 2

SUBJECT INDEX

VOLUMES 1 and 2